MW01625322

EXPLORING

Zoology

Second Edition

A Laboratory Guide

David G. Smith

Michael P. Schenk
University of Mississippi Medical Center

MORTON PUBLISHING
925 W. Kenyon Avenue, Unit 12
Englewood, CO 80110

www.morton-pub.com

Book Team

Publisher	Douglas N. Morton
President	David M. Ferguson
Acquisitions Editor	Marta R. Martins
Project Manager	Melanie Stafford
Associate Project Manager	Rayna S. Bailey
Production Manager	Joanne Saliger
Production Assistant / Interior Design and Composition	Will Kelley
Illustrations	Michael P. Schenk; Imagineering Media Services, Inc.

Printed in the United States of America

10 9 8 7 6 5 4 3 2 1

ISBN-13: 978-161731-156-7

Library of Congress Control Number: 2013944121

Preface

Exploring Zoology: A Laboratory Guide, Second Edition, is designed to provide a comprehensive, hands-on introduction to the field of zoology. A basic understanding of, and appreciation for, the diversity of animals is fundamental in establishing a secure foundation for more advanced courses in biology. This manual provides a diverse series of observational and investigative exercises, delving into the anatomy, behavior, physiology, and ecology of the major invertebrate and vertebrate lineages, and thereby giving a broad survey of the animal kingdom in an evolutionary context. We emphasize the use of a mixture of living and preserved specimens. This approach provides students with the opportunity to observe natural behaviors and processes in living organisms and to use the more practical means of prepared specimens to study anatomical structures.

The manual follows a logical progression, beginning with animal cells and tissues, then advancing phylogenetically from protists through chordates. The content and breadth of the material is geared to the university level and designed to be easily adapted for use with any introductory zoology text and with various course plans. We have intentionally included more material than reasonably can be covered in a typical one-semester general zoology course, providing instructors the flexibility to select the chapters and exercises that best meet their unique approaches, schedules, and available resources.

Major changes to the second edition include:

- Updated taxonomy for many groups, including significant revisions to protist classification.
- Coverage of many new animal groups including: ctenophores, rotifers, gastropods, trilobites, and jawless fishes, and significantly enhanced coverage of birds.
- New chapter covering taxonomy and systematics.
- Increased student engagement—additional questions throughout chapters require students to describe, interpret, or draw what they see.
- Many chapters contain "Laboratory Practical Challenge" questions to replicate the method of assessment and type of questions students may be asked on lab practicals.
- Seventy-five new and improved illustrations and photos.
- Geologic time scale included in Appendix.

We have taken great care to provide information in an engaging, user-friendly way that students and instructors alike will appreciate. As before, many features of this laboratory guide facilitate access to the information presented:

- Chapters begin with objectives to focus students' attention on essential material.
- Accompanying each exercise are materials lists of specimens and other supplies needed.
- Procedural steps are numbered chronologically and set off from the main text.
- Important terms are boldfaced to highlight their significance and facilitate review.
- Full-color illustrations and photos are used extensively to clarify procedures and to identify organisms and their anatomy.
- Tables are used throughout to conveniently summarize key information.
- Dissection diagrams provide visual clarification to ensure proper specimen dissection.
- "Check Your Progress" questions following major sections within chapters allow students to gain confidence in their understanding of the material before proceeding.
- Each chapter ends with a series of "Questions for Review" to reinforce key concepts and objectives.
- A comprehensive glossary containing definitions of key terms is provided for quick reference.

Ideally, students should read Chapter 1, "Fundamental Laboratory Skills," before coming to the first laboratory period to lay the necessary groundwork for microscopy, the metric system, preparing wet mounts, and the like. Subsequent chapters build on these skills as students progress to microscopic investigations of cells and tissues and advance sequentially through the major animal groups.

We hope that *Exploring Zoology: A Laboratory Guide*, Second Edition, will provide instructors and students with an illuminating, hands-on view into the fascinating world of animals. More than ever in this edition, we have tried to take students on an instructive journey through the animal world that is both stimulating and rewarding, and we welcome your comments and suggestions for improving future editions of this book.

—David Smith and Michael Schenk

Acknowledgments

Laboratory manuals such as this become the product of such an enormous collective effort that it is difficult for authors to thank all of the people who have had a hand in bringing a book like this to print. In this era of instantaneous electronic correspondence, many collaborative relationships are forged by conference call, fax, and e-mail, and blossom into successful partnerships without contributors ever meeting in person. The fact that many of these individuals may personally go unknown in no way diminishes our gratitude for their instrumental efforts on this project.

We are indebted to many individuals who participated in the compilation and review of this material, without whose assistance, diligence, and dedication we certainly would not have been able to complete this book. As always, we would like to thank everyone on our book team at Morton Publishing for the opportunity to continue to produce educational materials for the next generation of students and for supporting us through this entire endeavor: Doug Morton, David Ferguson, Marta Martins, Rayna Bailey, and Melanie Stafford. Joanne Saliger and Will Kelley applied their masterfully creative talents to produce a spectacular design and cover for the book. Gretchen Bernard, Lee Boyd, Nancy Butler, Quent Lupton, and Michelle Zurawski extensively reviewed the first edition and provided in-depth and insightful comments to improve the presentation of the material and ensure the accuracy of the second edition.

We would also like to thank Carolina Biological Supply Company and especially Tim Atkinson for their generosity in donating the preserved specimens that were dissected. Byron Adams and John Crawley supplied us with many of the beautifully detailed photographs used throughout this book. Charles Runyan and Robert Gray provided photographic advice and assisted in color balancing of images, Bill Armstrong provided photographic assistance with some invertebrate specimens, and Kyle Cunningham provided computer graphic assistance in laying out the laboratory exercise images. Jon Glase, Mel Zimmerman, and Jerry Waldvogel graciously permitted the use of their ideas on orientation behaviors, and Charles Drewes was overly generous with his time and provided wonderfully creative ideas for investigations into the behavior and physiology of sponges and blackworms. Finally, we want to thank Nikon Instruments, Inc. for providing images of their microscopes.

About the Authors

David Smith received his BS in biological sciences from Clemson University in 1992 and his MS in zoology from the University of Oklahoma in 1995, where he taught zoology labs for six years. He accepted a faculty appointment at the University of Richmond as the director of Biology Laboratories and taught laboratory courses there for another four years. David coauthored his first zoology laboratory manual in 1996, launching a writing career that has spanned more than a decade and has produced six major titles to date. He has since moved out of the classroom to concentrate his efforts on authoring educational materials that are designed to improve the quality of the laboratory experience for zoology students.

Michael Schenk received his undergraduate degree from the University of Minnesota and his BSMI and MSMI from the Medical College of Georgia in 1973/1974. He is a certified medical illustrator. He has received numerous awards for his illustrations and photography and was inducted as a Fellow in the Association of Medical Illustrators in 2002. Mike has held illustrator positions at the University of Virginia; the University of Minnesota College of Veterinary Medicine, where he started a department; the University of Texas Health Science Center Dallas, where he was an assistant professor in medical illustration; and currently is director of Biomedical Illustration Services at the University of Mississippi Medical Center in Jackson. Over the course of his career he has created thousands of illustrations in all scientific disciplines, and his work has been published by every major publisher.

Exploring Zoology: A Laboratory Guide, Second Edition, represents the fourth major title on which David and Michael have collaborated with Morton Publishing.

Contents

Laboratory Safety

Safety guidelines in biology laboratories are often taken with a tongue-in-cheek approach, but should be taken seriously. Many of you casually think, "What could happen to me in zoo lab?" Although you certainly don't often run the risk of catching yourself on fire or blowing up something (these risks are more akin to a chemistry lab), there are plenty of avenues for accidents of all sorts in the zoology lab. Integral to learning about zoology is accepting the responsibility that comes with doing zoology. In your pursuit of knowledge about the animal kingdom, you will immerse yourself in observations, dissections, and experiments, each with their own specific protocols, techniques, and inherent perils. Don't become complacent about the purported sanctuary of the zoology lab. Accidents can happen—and will happen if you let down your guard.

A common complaint is that too many rules take the fun out of lab. Common sense dictates that some degree of "looseness" must be sacrificed to gain the necessary level of safety that will ensure a positive lab experience. The laboratory is no place for a carefree, haphazard attitude. In the proper perspective, however, these basic guidelines will keep you safer and happier and will teach you the appropriate protocols that allow zoologists to study their organisms safely and effectively.

The following list of basic safety rules for the laboratory is offered as a guide to make your laboratory experiences safe and enjoyable. It is by no means a complete list but, rather, a starting point upon which the instructor can build a tailored list to suit your specific laboratory. Remember—your best defense against accidents in the lab and your greatest asset in dealing with situations when they arise is common sense. If that fails you, alert your instructor!

Basic Laboratory Safety Guidelines

- Never eat, drink, or smoke in the laboratory to prevent intake of chemicals or pathogens.
- Keep your hands away from your mouth, eyes, and nose as much as possible during lab.
- Wear close-toed shoes that adequately protect your feet.
- Wear protective gloves and/or goggles when handling or dissecting preserved specimens.
- Keep track of the materials on your workbench, and keep your workbench uncluttered.
- Place any disposable or broken glass in their properly designated containers.
- Place disposable scalpel blades, razor blades, syringe needles, and other sharp metal objects in their properly designated containers.
- Replace dull or broken scalpel blades (have your instructor demonstrate this process). Accidents occur more frequently with dull scalpels because more force is needed to cut with them, increasing the chance of slipping.

- Know the locations of the first-aid kit and eyewash fountain in your laboratory and know how to use them. Your instructor should discuss these safety items during the first laboratory period. If they are not discussed, ask about them.
- Report any electrical anomalies to your instructor immediately (for example, frayed electrical cords, bare wires, broken plugs, foreign objects in sockets, faulty switches).
- Alert your instructor in the event of an accident—no matter how harmless it seems! There may be unseen dangers of which you are not aware.
- Report any contact with human blood to your instructor immediately.
- Clean your lab bench and other work surfaces at the end of each lab period.
- Wash your hands carefully with soap and warm water, and rinse them thoroughly before leaving the laboratory.

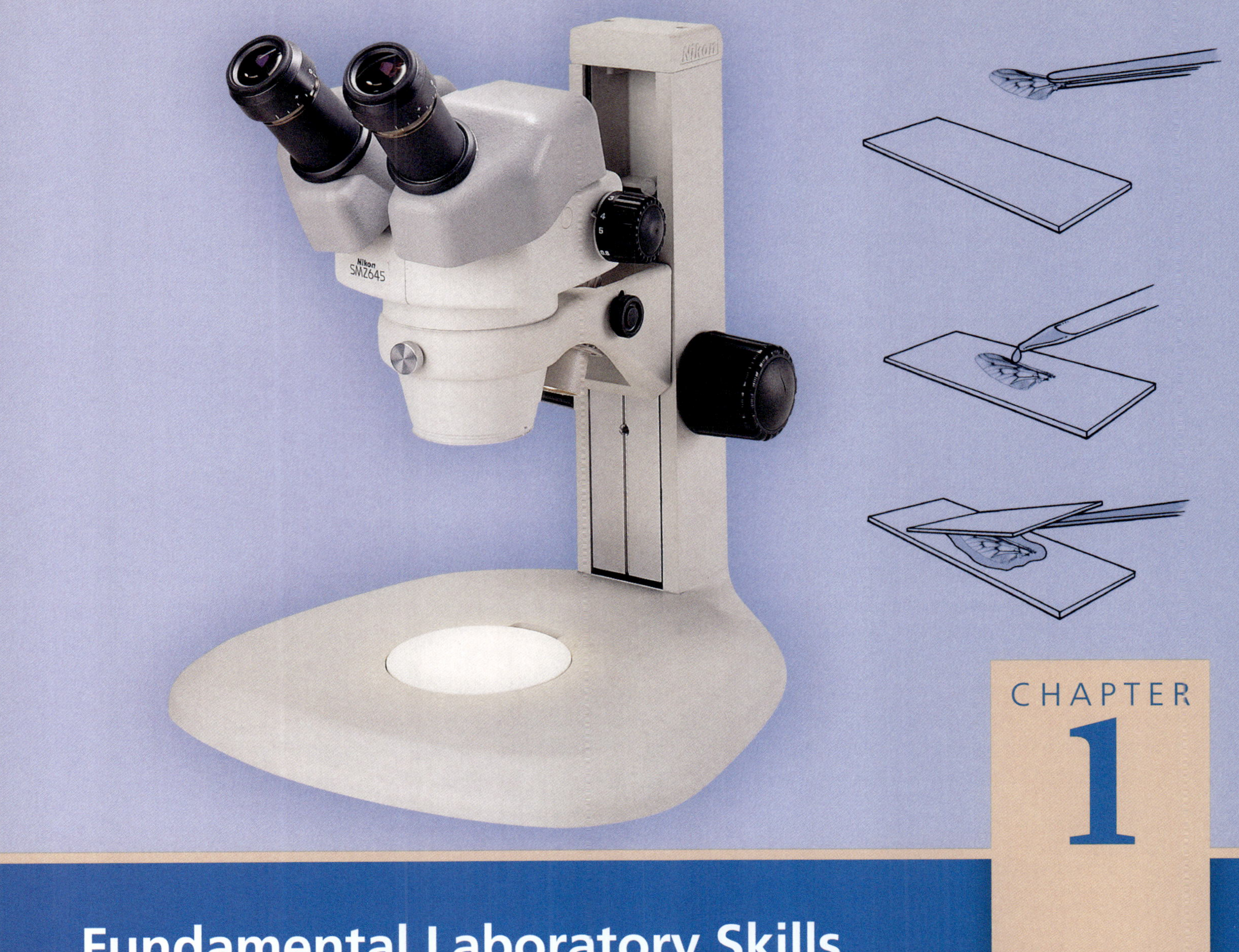

CHAPTER 1

Fundamental Laboratory Skills

After completing the exercises in this chapter, you should be able to:

1. Understand metric weights and measurements.
2. Identify the parts of a compound microscope and understand how to use it.
3. Make a wet mount of a specimen on a microscope slide.
4. Identify the parts of a stereoscopic microscope and understand how to use it.
5. Discuss basic dissection techniques.
6. Interpret references to body symmetry, body planes, and body regions of animals.

Exercise 1-1 Metric Weights and Measurements

Materials Needed

- ❑ metersticks
- ❑ metric rulers
- ❑ staples
- ❑ graduated cylinders (50 ml, 100 ml, 1,000 ml)
- ❑ coffee cups
- ❑ gallon jugs
- ❑ paper clips
- ❑ electronic or triple beam balances

Scientists throughout the world use a standardized system of weights and measurements, the metric system. This system has been adopted by virtually every major country around the world with the exception of the United States. As a result, U.S. students often are not as familiar with the relationships of metric units. Yet, the metric system has crept into our society in a few areas (2-liter soda bottles, 35 mm cameras, 9 mm handguns, etc.). In general, all scientific measurements that you make should be in metric units, but you occasionally may obtain measurements in English units or be given them from another source, so a conversion table is provided in Table 1-1 for your reference. Because you probably have had some prior training with the metric system, the attention to metrics will be brief here, serving as a review of the basic concepts of the metric system.

In the metric system the basic unit of length is the meter (m), the basic unit of volume is the liter (L), the basic unit of mass is the gram (g), and the basic unit of temperature is the degree Celsius (°C). The metric system is conveniently based on units of 10, which simplifies conversions from one metric unit to another. Simply moving a decimal point to the right or the left is usually all that is needed to convert from one metric unit to another.

Units of area are obtained by squaring the respective metric unit of length (for example, 25 cm^2). Units of volume are obtained either by cubing the respective metric unit of length (for example, 1 cm^3) or by measuring the displacement of the item in a fluid volume (for example, 1 ml).

The metric system was designed around the basic physical properties of water—one of the most abundant compounds on our planet. At sea level, under standard atmospheric pressure (one atmosphere), water boils at 100°C and freezes at 0°C. One gram (1 g) of water at 4°C occupies one cubic millimeter (1 mm^3,) or one milliliter (1 ml,) of volume. Fluid measurements in cubic centimeters (cm^3) are commonly abbreviated with the designation "cc."

Procedure

1 Obtain a meterstick or metric ruler from your instructor and measure the following items:

Length:

Pencil = ________ cm

Staple = ________ mm

Your height = ________ m

Area:

Student ID card = ________ cm^2

Table surface = ________ m^2

2 Obtain graduated cylinders of several sizes from your instructor, and measure the volume of fluid that the following items can hold:

Coffee cup = ________ ml

Gallon jug = ________ L

Note

When reading the volume of fluid in a graduated cylinder, position your eyes level with the water line in the cylinder, and record your measurement by aligning with the bottom of the meniscus (the curved surface of fluid caused by surface tension between the fluid and the walls of the cylinder).

3 If an electronic balance or a triple beam balance is available, record the mass of the following objects:

Paper clip = ________ mg

Quarter = ________ g

Coffee cup (empty) = ________ kg

Pencil = ________ g

4 Using Table 1-1 as a guide, convert the following measurements:

187 cm = ________ m	37.9 fl. oz. = ________ ml	32.3 lb. = ________ kg
763 mm = ________ in.	4.7 L = ________ gallons	37°C = ________ °F
42 yd. = ________ m	4,845 g = ________ kg	−15°F = ________ °C
6.2 mi. = ________ km	0.32 kg = ________ g	100°F = ________ °C

Table 1-1 Conversion Table of Metric Units and Their English Equivalents

Units of Length	Units of Volume	Units of Mass	Units of Temperature
1 m = 39.4 in. = 1.1 yd. = 100 cm = 1,000 mm = 0.001 km	1 liter = 1,000 ml = 1,000 cm^3 = 2.1 pints = 0.26 gal. = 35 fl. oz.	1 g = mass of 1 cm^3 of water at 4°C 1 g = 0.035 oz. = 0.001 kg	°C = 5/9 (°F − 32) °F = (9/5)°C + 32 Ex: 37°C = 98.6°F (body temp) 0°C = 32°F (H_2O f.p.) 100°C = 212°F (H_2O b.p.)
1 cm = 0.394 in. = 10 mm	1 ml = 0.035 fl. oz. = 1 cm^3 (1 cc)	1 kg = 1,000 g = 2.2 lb.	
1 mm = 0.0394 in.	1 gal. = 3.85 L	1 oz. = 28.35 g	
1 nm = 10^{-9} m = 10^{-6} mm 1 yd. = 0.91 m	1 fl. oz. = 28.6 ml 1 qt. = 0.943 L	1 lb. = 0.45 kg	
1 ft. = 30.5 cm			
1 mi. = 1.61 km			
1 in. = 2.54 cm			

Exercise 1-2 The Compound Microscope

Materials Needed

- ❑ compound microscope
- ❑ slide of letter "e"
- ❑ slide of crossed threads
- ❑ slide of ruler section (or clear plastic ruler)

The compound microscope is a tool that you will use repeatedly in the laboratory to reveal tiny structures and details that cannot be seen with the unaided eye. Unfortunately, many students do not refine their microscope skills to their maximum potential and, thus, consistently miss a substantial portion of the material presented in laboratory exercises that utilize microscopes. Dozens of hours in front of the microscope will be necessary to polish your technique to the point where everything you look at is in sharp contrast and clearly focused. Whether you have or have not used a microscope before, you should read through this section carefully and perform the accompanying exercises. There is something for everyone to learn here. Your first goal is to familiarize yourself with the mechanical parts of the compound microscope.

Procedure

1 Obtain a compound microscope from the cabinet. *Be sure to carry it with both hands* while transporting it to your laboratory bench. Microscopes are expensive, precision instruments that may be damaged easily by dropping or excessive jarring.

2 Label the diagram of the compound microscope in Figure 1.1 with the terms for the major parts. Because numerous brands and models of microscopes are available, some features on your microscope may differ slightly from the one illustrated in the figure. Your instructor will point out any differences between your microscope and the one illustrated.

ocular lenses—lenses nearest the eye through which you look

objective lenses—lenses of different magnification that work in conjunction with ocular lenses to magnify the image, located just above the stage

body—housing that keeps ocular and objective lenses in proper alignment

nosepiece—revolving housing that supports objective lenses

arm—supports microscope body, stage, and adjustment knobs

coarse-focus adjustment knob—moves stage up or down to focus image

fine-focus adjustment knob—permits precise focusing

stage—supports slides

stage clips (may be absent)—hold slide in steady, stationary position

stage adjustment knobs—move stage to center slide under objective lens

condenser—lens mounted beneath stage that focuses the light beam on the specimen

iris diaphragm—mounted beneath stage near condenser; regulates amount of light illuminating specimen

condenser adjustment knob—moves condenser lens up or down to focus light (not visible on diagram)

illuminator—source of light

base—supports microscope unit

light intensity adjustment dial—rheostat (dimmer switch) that permits further adjustment of light intensity

power switch—turns microscope light on or off

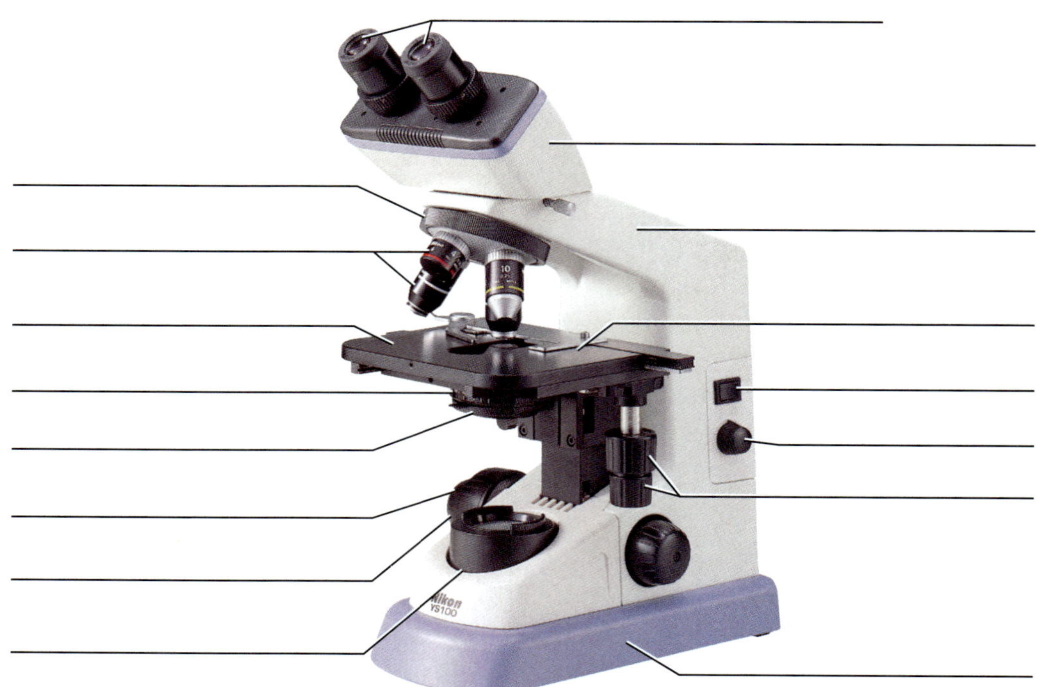

1.1 Label the parts of the compound binocular microscope.

Photo courtesy of Nikon Instruments Inc. (http://www.nikonimagelibrary.com), Melville, NY, USA.

Care and Use of the Microscope

After familiarizing yourself with the basic parts of the microscope, you are ready for some hands-on use.

Procedure

1 Clean the ocular and objective lenses by gently wiping them with dry lens paper.

2 Unwrap the cord, plug in the microscope, and turn on the power switch. If you do not see any light coming from the illuminator, check to ensure that the light intensity slide near the power switch (if present) is not set to the lowest setting. If that fails, alert your instructor, because you may have a burned-out bulb.

3 Rotate the nosepiece until the low-power objective lens "clicks" firmly into place directly over the stage. On most microscopes the lowest-power lens is either a 4× or a 10× lens. The number will be stamped on the side of the lens. On some microscopes, the lowest-power lens is referred to as the "scanning lens."

4 Turn the coarse-focus adjustment knob to completely raise the stage until there are only a few millimeters between the stage and the objective lens. On some microscopes, adjustment of the coarse-focus knob moves the entire microscope body rather than the stage.

5 Place a slide of the letter "e" on the stage with the "e" facing upright and centered with the "e" directly over the hole in the stage and surrounded by a circle of light.

6 Adjust the iris diaphragm to the midway position, allowing a moderate amount of light to penetrate the slide. You may open or close the iris diaphragm later as needed to fine-tune the clarity of the image. As a rule of thumb, the higher the magnification you use, the more light you will need to see an image clearly.

7 Look through the ocular lenses at the letter "e" and adjust the distance between the ocular lenses to match the distance between your pupils. Resist the temptation to close one eye when looking through the microscope! This is not an effective alternative to using both eyes and will eliminate the advantages of binocular 3-D vision, as well as contribute to eyestrain and even headaches.

8 While looking through the ocular lenses at the "e," rotate the coarse-focus adjustment knob until the image of the "e" comes into focus. If you do not see the "e" the slide may be off center. Be sure that the slide is aligned directly beneath the objective lens and held firmly by the stage clips (if present).

9 Remember—always focus by *increasing* the distance between the stage and the objective lens. Never focus in the other direction (decreasing the distance between the stage and the objective lens). This avoids the financial expense to you of broken slides or damaged objective lenses!

10 Use the fine-focus adjustment knob to finely focus the image in view.

11 Check to see if one of your ocular lenses has an independent focus adjustment. Usually one ocular will have a series of "tick marks" on the side with a "0" and "+" and "−" signs. When the "0" is aligned with the indicator mark on the side of the ocular lens, both ocular lenses have equivalent focal distances. If you wear contact lenses to correct your vision or you have 20/20 vision, you will want the indicator mark on "0" so both oculars have the same focal plane. If you wear eyeglasses, it is recommended that you remove them when using the microscope, because this optical adjustment will allow you to correct for differences between your eyes.

12 Rotate the condenser adjustment knob to increase contrast and sharpen the image. The condenser is an often overlooked but extremely valuable component of the microscope that should not be neglected. You will find its use absolutely necessary to obtain sharp images with certain slides.

13 Adjust the position of the slide on the stage so the image of the "e" is centered perfectly in your field of view. Only the central portion of the entire field of view will be visible as you increase magnification.

14 Carefully rotate the next objective lens (10× or 40×) until it clicks firmly into place *while viewing the slide from the side* to ensure that the lens does not contact the slide. ***Do not lower the stage to accommodate the next lens.*** If the slide is too thick to allow the next lens to swing into position, ***do not*** use that lens!

15 From this point on, use only the fine-focus adjustment knob to focus the image.

16 Because microscope lenses are adjusted at the factory to be **parfocal,** the plane of focus and center of field of view should be nearly identical between different objective lenses. As a result, only minor adjustments should be necessary to bring the object into sharp focus and to center it in your field of view.

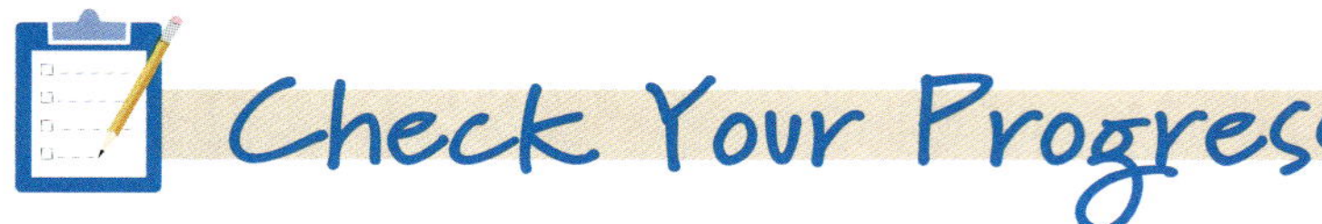

1.1 Describe the orientation of the letter "e" as it appears through the ocular lenses.

1.2 As you move the slide toward the right of the stage, to which direction does the image of the "e" move when viewed through the microscope?

1.3 As you move the slide away from you on the stage, to which direction does the image move?

Using High-Power Lenses

When using high-power objective lenses (40× and 100×), keep in mind that the **working distance** between the objective lens and the slide is only a few millimeters at most. Therefore, the fine-focus adjustment is sufficient to bring images into clear view. *Never use the coarse-focus adjustment with high-power lenses.*

Procedure

1 If your microscope is equipped with a 40× lens, once the image is in focus under medium power, rotate the 40× objective into place while viewing the slide from the side to check for adequate clearance. If the objective does not rotate into place without clearing the slide, do not force it. Ask your lab instructor to help you.

2 If you cannot locate the object through the ocular lenses and you find that the 40× objective is more than a centimeter away from the slide, you have passed the focal plane of the lens. Therefore, start from the beginning again with the lowest-power objective and progressively work your way back to the 40× lens. Be sure not to lower the stage (or raise the objectives) once you have focused the image and wish to change to the next objective.

3 Your microscope may be equipped with a 100× lens. This is a special type of lens known as an *oil immersion lens*. The working distance is so small between this lens and the slide, and the magnification is so great, that it is necessary to place a drop of immersion oil between the coverslip of the slide and the 100× lens to obtain a clear image. At such high magnification the distortion of light rays caused by light passing through glass, then air, then into glass again is perceptible and causes a blurred image. Immersion oil has the same optical density

as glass and, thus, stray light rays are not lost or distorted. ***Do not attempt to use the 100× lens without assistance from your instructor.*** 1

Magnification

Together, the ocular and objective lenses constitute the magnifying system of your microscope. The initial magnification of the objective lens provides an image with good detail but is too small for easy examination. The ocular lenses supply secondary magnification of the initial image so the details are clear enough for normal viewing. The resulting image is a magnification of a magnified image, and because the properties of magnification are multiplicative, you can easily calculate the total magnification of a specimen by multiplying together the independent magnification values of each lens. Remember that the magnification of objectives and oculars is stamped on them.

In Table 1-2, record the values for the total magnification of each of your microscope lens pairs.

Table 1-2 Compound Microscope Lens Magnification

Objective Lens in Place	Objective Lens Magnification		Ocular Lens Magnification		Total Magnification
Low power		×		=	
Medium power		×		=	
High power		×		=	

Field of View

The circular field that you see when looking through the microscope is described as the **field of view**. The diameter of the field of view changes with different magnifications. While still observing the slide of the letter "e," rotate the low-, medium-, and high-power objective lenses into place and compare the amount or proportion of the entire letter that is visible under each one.

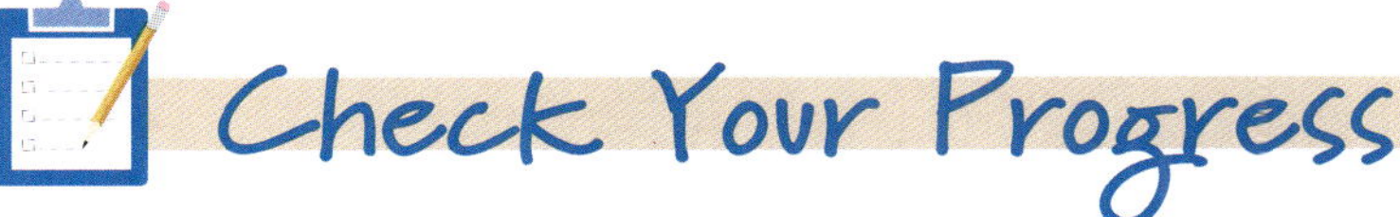

2.1 Under which objective lens is the field of view largest?

2.2 Under which objective lens is the field of view smallest?

2.3 If you did not know what you were looking at already, could you still determine if it was an "e" using high power alone? If so, how?

2.4 Which lens (low-, medium-, or high-power) gives you the largest working distance?

Knowing that a higher-power lens has a smaller field of view is informative but is not precise enough to allow you to calculate the relative size of objects you are viewing. To do this, you must precisely know the diameter of each field of view for your microscope. An easy way to determine the size of the field of view of each objective lens is to place a clear plastic ruler (or prepared slide of a section of a ruler) on the stage, much like you would a slide, and view it through the ocular lenses.

Procedure

1. Obtain a slide of a section of a ruler or a small, clear plastic ruler and place it on the stage.
2. Using low power, view the ruler through the ocular lenses and estimate the size of the field of view by measuring the diameter in millimeters. You should estimate this value to the nearest 0.1 mm.
3. Repeat this procedure with the medium-power lens in place.
4. Because the ruler (or slide) is too thick to observe using the higher-power lenses, ***do not attempt to measure the field of view of the 40× or 100× lenses.*** Your instructor can provide the values for the field of view diameters of those lenses if present on your microscope.

Check Your Progress

3.1 The diameter of the field of view of your low-power lens is ________ mm.

3.2 The diameter of the field of view of your medium-power lens is ________ mm.

3.3 Convert these values to micrometers (µm). Remember, there are 1,000 µm in 1 mm.

a. Low-power lens (4×) = ________ µm

b. Medium-power lens (10×) = ________ µm

c. High-power lens (40×) = ________ µm (*Obtain this value from your instructor.*)

d. Oil immersion lens (100×) = ________ µm (*Obtain this value from your instructor.*)

3.4 As a rule, as magnification increases, diameter of field of view ____________.

Depth of Field

The thickness of an image that is acceptably in focus is referred to as the **depth of field** of a lens. Put another way, it's the measure of the distance in front of and beyond the exact focal point of the lens that remains reasonably in focus. Depth of field varies with the magnification of the objective lens used. You can establish the differences in depth of field of your objective lenses by viewing a slide containing overlapping objects.

Procedure

1. Obtain a slide with a few strands of overlapping colored threads.
2. First view this slide on low power.

3 Concentrate on a section where the overlapping of the threads can be seen. Are all three colored threads reasonably in focus using low power?

4 Now switch to medium power and try to determine which thread is on top of the other two. As you focus through the image, some threads will be in focus and others will be blurred.

5 Now switch to high power and repeat this procedure. Can all three threads be in focus at the same time with the high-power lens?

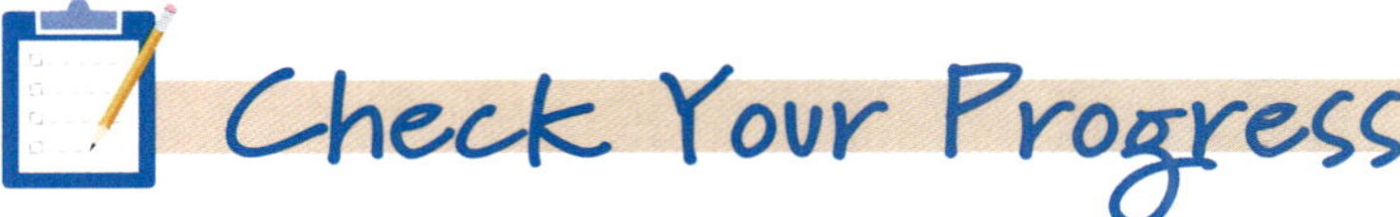

Check Your Progress

4.1 Which lens has the greatest depth of field?

4.2 As a rule, as magnification increases, depth of field ______________.

Exercise 1-3 Making a Wet Mount

Materials Needed

- glass slides
- coverslips
- pond water culture
- compound microscope
- plastic droppers

A common way to view living organisms or tissues with the microscope is to make a **wet mount** of the specimen. This technique allows you to observe movements and properties of living specimens that are impossible to view with prepared slides. Figure 1.2 depicts the proper technique for preparing a wet mount of a biological specimen for viewing with the compound microscope. In this exercise, you will examine a drop of pond water using your microscope. It is very important to practice proper microscope technique to reinforce the lessons that you have learned thus far.

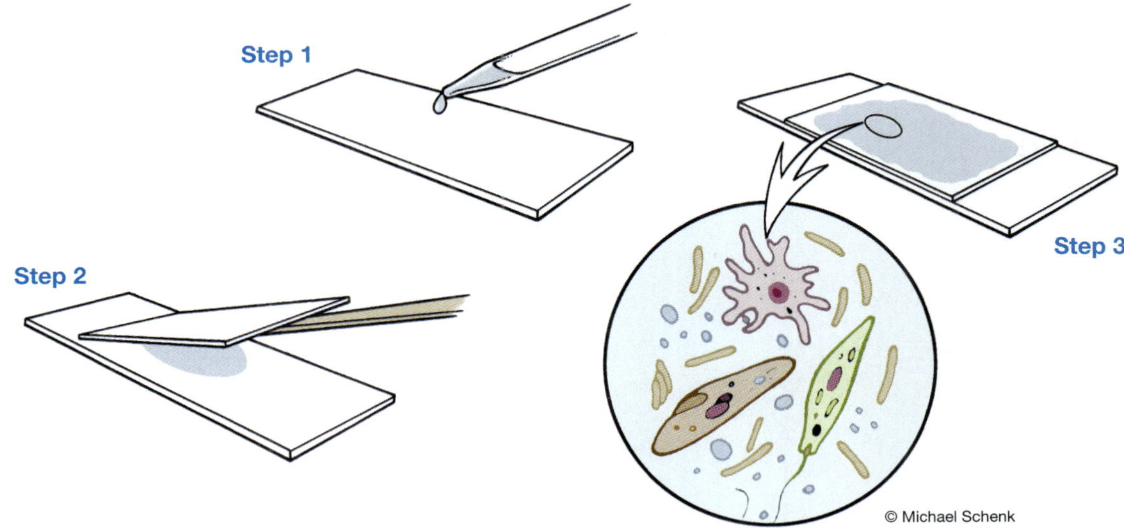

1.2 Procedure for preparing a wet mount of a biological specimen.

Procedure

1 Place a drop of pond water culture in the center of a clean glass slide.

2 Carefully add a coverslip by placing one edge along the drop and gently lowering it onto the slide.

3 Press *gently* on the coverslip to remove any tiny air bubbles that may have been trapped in the process.

4 View this slide using low power.

5 Experiment with different intensities of light, and condenser and iris diaphragm settings, to maximize the clarity and contrast of the specimens in view.

6 Switch to higher magnifications and readjust the condenser and iris diaphragm settings.

7 After answering the following questions, ask your instructor where to dispose of wet-mount slides.

Check Your Progress

5.1 Which level of magnification requires the most illumination for the best clarity and contrast?

5.2 Why is it imperative that you place a coverslip over the drop of fluid when making a wet mount?

5.3 If you have a specimen that is transparent or thin, would you need to open the iris diaphragm (to allow in more light) or close the iris diaphragm (to allow in less light)?

5.4 Examine your slide, and sketch at least five of the organisms that you see in the space provided. You may see a mixture of plant, animal, and protozoan material on your slide.

Exercise 1-4 The Stereoscopic or Dissecting Microscope

Materials Needed

- ❏ stereoscopic microscope
- ❏ prepared slides of various organisms
- ❏ prepared specimens of various organisms, feathers, fish scales

A stereoscopic microscope (often called a dissecting microscope) is another type of microscope that you will use frequently in the laboratory. Stereoscopic microscopes differ in several important ways from compound microscopes. Knowing the differences between these two types of microscopes, and the advantages and disadvantages of each, allows you to determine which type of microscope to use under different circumstances to reveal the most detail and provide the best visual experience.

Stereoscopic microscopes have a much larger working distance than compound microscopes and are designed for viewing whole specimens that are too large, too thick, or too opaque for study with a conventional compound microscope. This large working distance permits easy manipulation and dissection while you are simultaneously viewing the specimen. The large working distance also allows for illumination of the specimen from above (reflected light), as well as from below (transmitted light). Generally, reflected light reveals surface features on the specimen better than does transmitted light. Transmitted light usually is preferable for viewing internal structures on extremely thin or transparent specimens. You always should experiment with the direction of light for each situation and specimen to determine which method provides the clearest view.

Stereoscopic microscopes are always binocular. Each ocular lens views the specimen at a slightly different angle through the objective lenses, providing a three-dimensional view of the specimen with a large depth of field. This is in contrast to the compound microscope, which produces a flat, two-dimensional image with a comparatively narrow depth of field.

Also, unlike the compound microscope, the internal arrangement of prisms and lenses within the stereoscopic microscope produces an image that is not inverted or backward. Thus, moving an object in a given direction on the stage also moves the image in the same direction. This property allows for more intuitive manipulation and dissection of objects on the stage.

These advantages of the stereoscopic microscope are offset by lower resolution and lower magnification than a compound microscope. Most stereoscopic microscopes are limited to a magnification range of 4× to 50×. Although this limits the amount of detail you can see, it does provide for a larger field of view. Many models of stereoscopic microscopes have an "infinite zoom" feature not found on compound microscopes, which allows for a smooth progression from lower to higher power without losing sight of the specimen while switching lenses. This feature can be extremely useful for zeroing in on specific structures for study.

Procedure

1. Obtain a stereoscopic microscope from the cabinet. ***Be sure to carry it with both hands*** while transporting it to your laboratory bench.
2. Carefully clean any exposed lenses by gently wiping them with dry lens paper, as you did with your compound microscope.
3. Use Figure 1.3 to familiarize yourself with the parts of your microscope.
4. Examine several of the specimens provided by your instructor until you are comfortable using this type of microscope. Remember to experiment with both reflected and transmitted light with each specimen to determine which method reveals the greatest detail and best overall image.

1

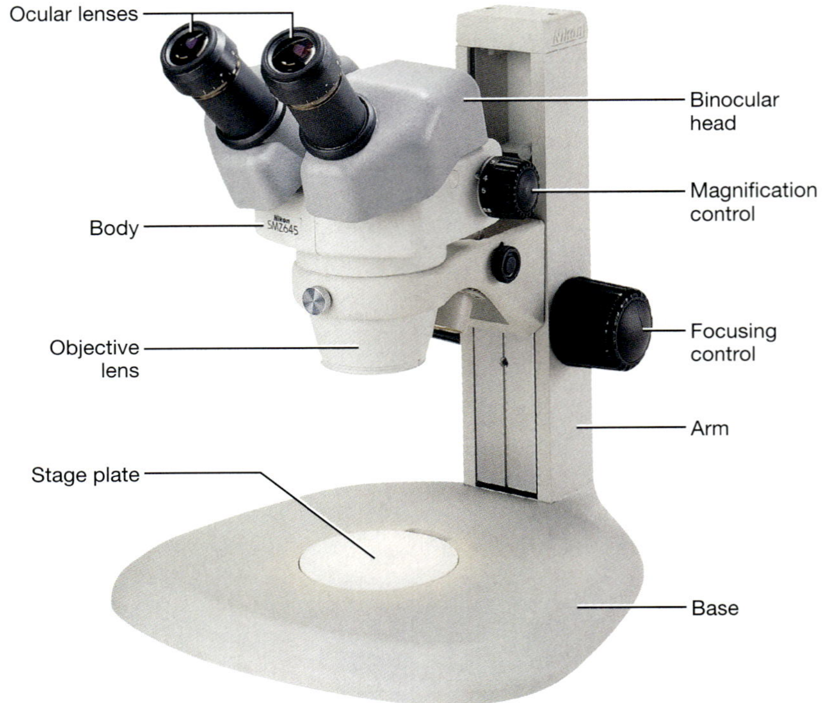

1.3 Stereoscopic microscope (also called a dissecting microscope).
Photo courtesy of Nikon Instruments Inc. (http://www.nikonimagelibrary.com), Melville, NY, USA.

Check Your Progress

6.1 What is the magnification range of your stereoscopic microscope?

____

6.2 How does the image through a stereoscopic microscope move when the specimen is moved to the right or left? Up or down?

____

____

6.3 For thin, transparent specimens, which method of illumination is generally better—transmitted or reflected light? Which method of illumination is better for larger, thicker specimens?

____

____

6.4 Use the words *lower/smaller* or *higher/larger* to correctly complete the following sentence: When compared to most compound microscopes, stereoscopic microscopes have a ____________ working distance, ____________ depth of field, ____________ field of view, ____________ magnification, and ____________ resolution.

Exercise 1-5 Basic Dissection Techniques

Because animal dissections will constitute a major component of this laboratory guide, you will have to hone your dissection skills early to be successful in this laboratory. A brief review of basic dissection techniques and suggestions will help build your proficiency, ensuring that you obtain the maximum benefit from your studies of the specimens that are presented in this book.

1. Always practice safe hygiene when dissecting. Wear appropriate protective clothing, gloves, and eyewear, and ***never*** place your hands near your mouth or eyes while handling preserved specimens. If fumes from your specimen irritate your eyes, ask your instructor about the availability of goggles.
2. If, at any time, you feel light-headed or woozy, alert your lab partner and instructor immediately. If you need to be excused to get some fresh air, have your lab partner or instructor accompany you for safety.
3. Read all instructions ***carefully*** before making any incisions. Be sure you understand the direction and depth of the cuts to be made, because many important structures may be damaged by careless or imprecise cutting.
4. Use scissors, a teasing needle, and a blunt dissecting probe whenever possible. Despite the popularity of scalpels, they often do more harm than good and should not be relied upon as your primary dissection tool. Remember—the purpose of dissection is to reveal organs and structures in their natural, intact state for observation, without cutting or destroying them.
5. Resist the temptation to stick scalpels or teasing needles into the rubber or wax bottom of your dissecting pan. This unnecessarily dulls instruments. Sharp tools are essential to performing clean, precise dissections.
6. When instructed to "expose" or "view" an organ, remove all of the membranous tissues that typically cover the organ (fat, fascia, etc.) and separate the "target" organ from neighboring structures. Your goal should be to expose the organ or structure as completely as possible without damaging it.
7. When working in pairs, an effective strategy is for one of you to read aloud the directions from the book while the other performs the dissection. These roles should be traded from section to section to give both of you a chance to participate.
8. Refer to illustrations and photographs frequently, but focus primarily on the specimen. Remember—pictures are intended to help you in your dissections but do not substitute for the study of real specimens.

Check Your Progress

7.1 True or false? A scalpel should be your primary dissection instrument. ____________

7.2 True or false? It's permissible to have a small snack or drink during the lab if your dissection specimens are preserved in nontoxic solutions. ____________

7.3 What should you do if you begin to feel light-headed during a dissection?

a. Take a few deep breaths and hope the wooziness goes away.
b. Alert your lab partner and instructor immediately.
c. Lie down on the floor and close your eyes.
d. Switch roles and have your partner perform the dissection, while you read the instructions.

Exercise 1-6 Body Symmetry, Body Planes, and Body Regions

Materials Needed

- ❑ preserved sponges, sea anemones, sea urchins, sand dollars, insects, and vertebrates (or pictures of representative animals)
- ❑ Play-Doh® or modeling clay (2 colors)
- ❑ kitchen knife or scalpel

Body Symmetry

The arrangement of body parts in animals differs, and these differences can be described best in relation to certain reference planes, or axes. The three major categories of body symmetry in animals are: asymmetry, radial symmetry, and bilateral symmetry (Fig. 1.4).

Asymmetry—lack of symmetry; irregular arrangement of body parts with no plane of symmetry to divide them into similar halves

Radial symmetry—arrangement of body parts around a central axis; any plane passing through the central axis divides the body into two similar halves

Bilateral symmetry—division of body parts into similar halves (mirror images) by a single plane of symmetry

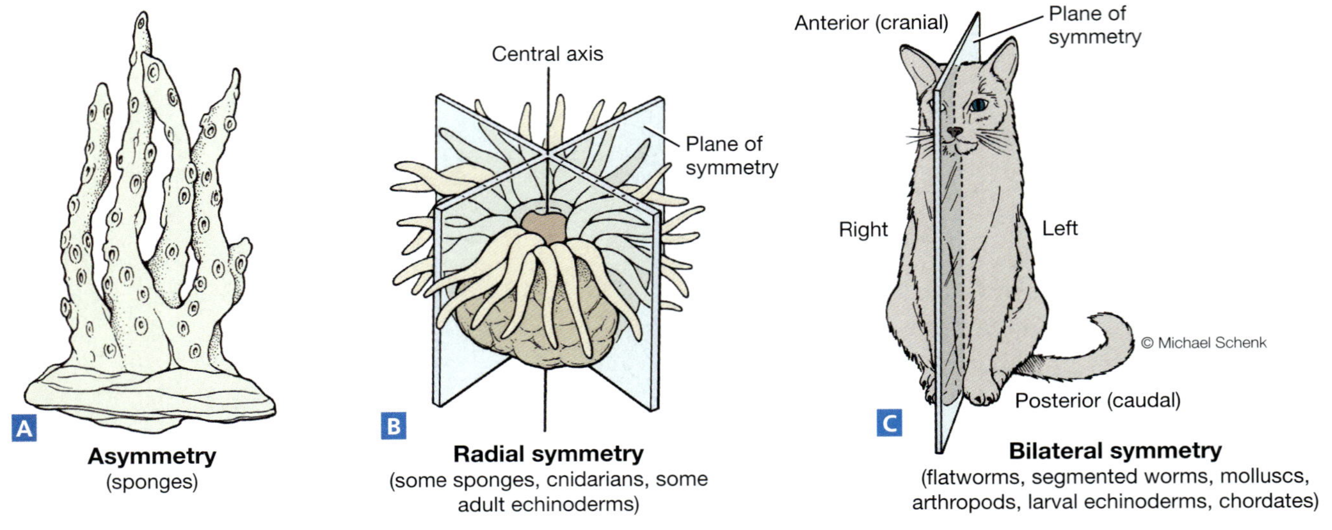

1.4 Patterns of body symmetry in animals: (**A**) asymmetry, (**B**) radial symmetry, and (**C**) bilateral symmetry.

Check Your Progress

8.1 Your instructor has set out a number of different organisms. Categorize each according to its body symmetry.

Asymmetrical:

Radially symmetrical:

Bilaterally symmetrical:

8.2 Which type of symmetry do you think is most prevalent among animals?

8.3 Radial symmetry is most common among sedentary organisms or organisms that drift passively with water currents. Speculate about the adaptive value of radial symmetry for these organisms.

Body Planes and Regions

Anatomical references to body planes and regions often differ between radially symmetrical and bilaterally symmetrical animals, as well as between **quadrupedal** animals and bipedal animals (such as humans). For example, the ventral surface of a dog is equivalent to the anterior surface of a human and the oral surface of a jellyfish. Many of the animals you will encounter in this manual are bilaterally symmetrical and some are quadrupedal; therefore, the following terms will be used to refer to the regions of the body and the orientation of the organs and structures you will identify.

A section perpendicular to the long **axis** of the body separating the animal into anterior and posterior portions is called a **transverse plane** (Fig. 1.5). The terms *anterior* and *posterior* refer to the head and tail regions, respectively. When referring to quadrupedal animals, the terms *cranial* and *caudal* may be appropriately substituted for anterior and posterior. A longitudinal section separating the animal into right and left sides is called a **sagittal plane.** The sagittal plane running down the midline of the animal has a special name, the *median plane*. Structures that are closer to the **median plane** are referred to as **medial.** Structures farther from the median plane are referred to as **lateral. Dorsal** refers to the back or upper surface of the body, and **ventral** refers to the front or lower side of the body. A longitudinal section dividing the animal into dorsal and ventral parts is called a **frontal plane. Proximal** refers to a point of reference nearer the median plane or point of attachment on the body than another structure (for example, when your arm is

1 extended, your elbow is proximal to your hand). **Distal** refers to a point of reference farther from the body's median plane or point of attachment than another structure (for example, when your arm is extended, your elbow is distal to your shoulder). **Rostral** refers to a point closer to the tip of the nose. If you still feel a little confused about the terminology associated with animal body planes and regions, try the simple exercise that follows.

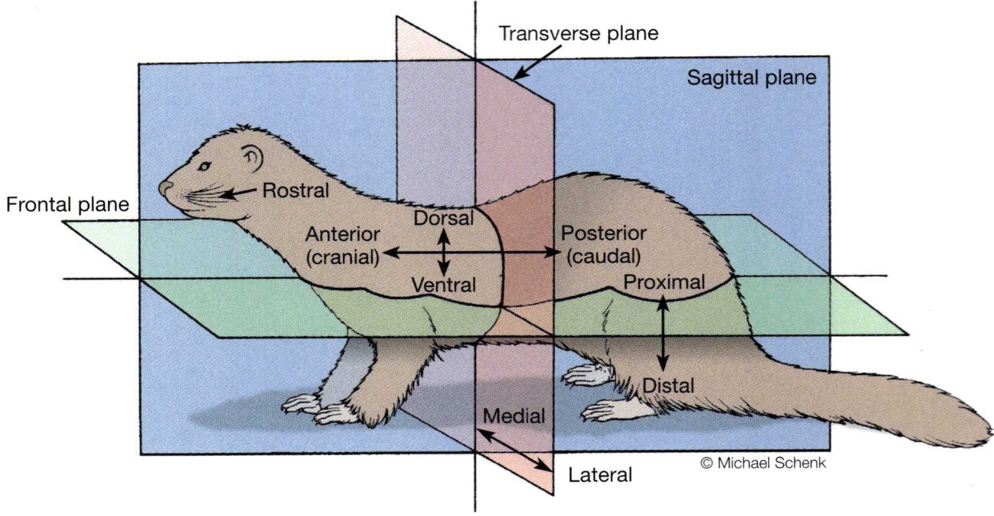

1.5 Body planes and body regions in animals.

Procedure

1. Obtain two different colors of Play-Doh® or modeling clay.
2. Construct a stout, wormlike organism with a distinct head, using one color to represent the gut and a different color for the outer body layer.
3. Using a kitchen knife or scalpel, make a cut passing through the *frontal plane* of your imaginary organism, separating it into *dorsal* and *ventral* halves. (In humans these would be referred to as posterior and anterior halves, respectively.)
4. Make another cut passing through the *transverse plane* of the ventral (or anterior) half. This separates the ventral half of the organism into *cranial* and *caudal* sections.
5. Finally, make a longitudinal cut through the *sagittal plane* of the remaining cranial section, separating the "head" into left and right halves through the median plane of the organism.
6. In the space provided below, sketch your organism's head after all of the above cuts have been made.

Name ______________________ Section ______________

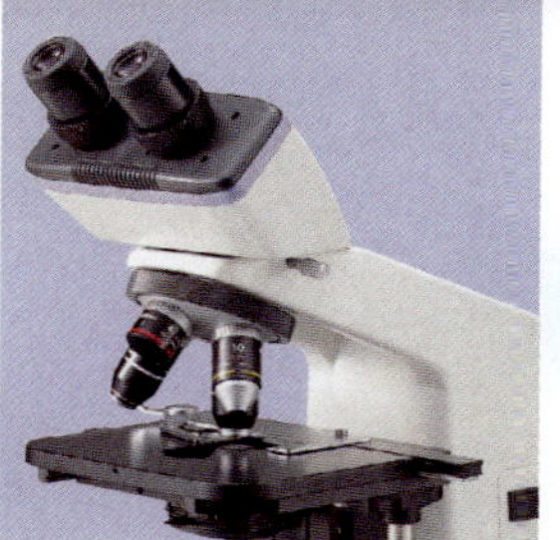

1

1 Questions for Review

1 How many grams are there in 6.23 kg? __________ g

2 How much does 1 cm^3 of water weigh? __________ g

3 Which distance is greater, 45.6 km or 24 miles? __________

4 If you have a microscope with a 5× ocular lens and a 40× objective lens, what is the total magnification of the image?

____

5 Describe what is meant by the term *working distance*.

____

____

6 List advantages and disadvantages of using a stereoscopic microscope compared to a compound microscope.

Advantages:

____

____

Disadvantages:

____

____

7 Describe the change in each of the following as magnification decreases:

a. field of view __________ (*increases* or *decreases*)

b. depth of field __________ (*increases* or *decreases*)

c. working distance __________ (*increases* or *decreases*)

d. light intensity requirement __________ (*increases* or *decreases*)

8 Describe the procedure for making a basic wet mount of a biological specimen.

9 List the three types of body symmetry, and for each type list an animal that has that kind of symmetry.

Symmetry Type	*Animal*
1.	
2.	
3.	

10 Place the following terms in their correct locations on Figure 1.6.

anterior (cranial)	medial	sagittal plane
posterior (caudal)	lateral	proximal
dorsal	transverse plane	distal
ventral	frontal plane	

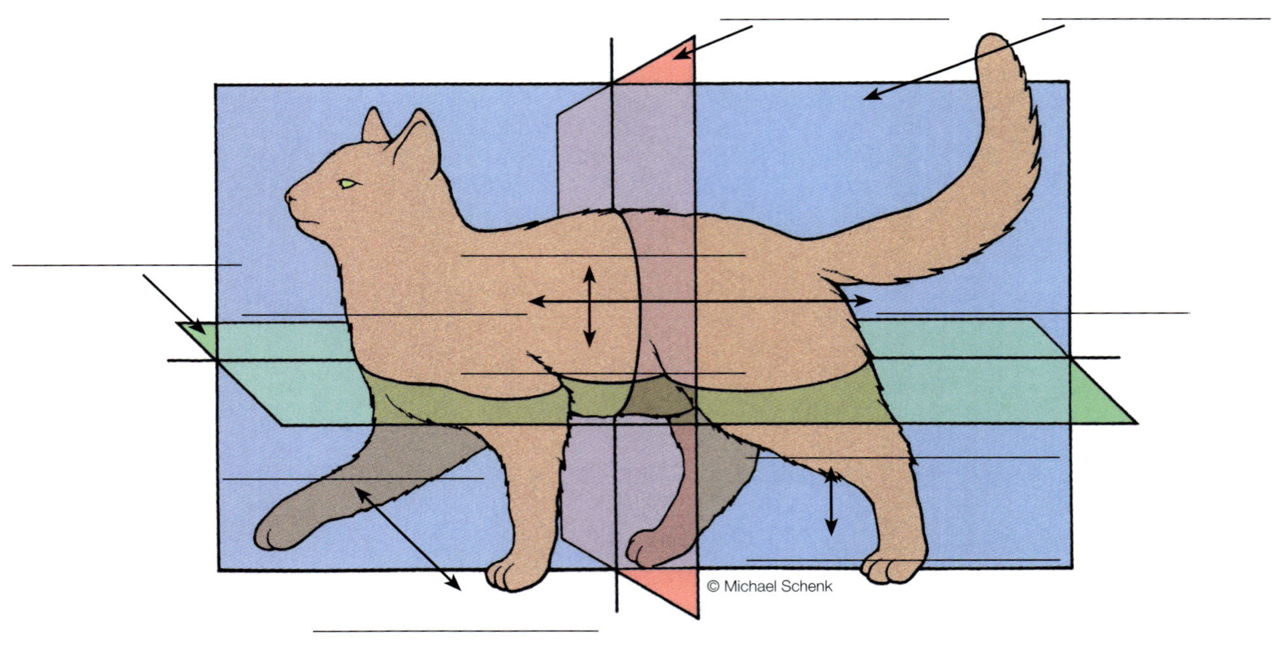

1.6 Label the diagram of body planes and body regions.

CHAPTER

2

Animal Cells and Tissues

After completing the exercises in this chapter, you should be able to:

1 List the three unifying concepts of the modern cell theory.

2 Recognize the principal organelles of animal cells visible with the compound microscope.

3 Discuss representative examples, specific functions, locations in the body, and major histological features of the four general categories of adult vertebrate tissues.

4 Distinguish between the different subcategories of epithelial tissue and describe the functions of each type

5 Distinguish between the different subcategories of connective tissue and describe the functions of each type.

6 Describe the histological structure of bone and explain the role of Haversian canals, lacunae, canaliculi, and lamellae.

7 Describe the composition of human blood and identify erythrocytes, leucocytes, and platelets in microscopic preparations.

8 Distinguish between skeletal, cardiac, and smooth muscle in microscopic preparations.

9 Describe the structure of a neuron and give the function of each main part.

Understanding Cells: The Cell Theory

2

Before tackling the complexity of an entire living organism, it is useful to gain an understanding of the basic structural and functional units of animals—cells. The **cell** is the fundamental biological unit in living organisms and thus represents a tremendous milestone in the long evolutionary history of life on this planet. The present structure of the basic cell evolved some 3.5 billion years ago, and since that time few changes have occurred *within* cells. Once a stable, functional, and efficient design for a basic cell had arisen, most evolutionary change thereafter occurred at higher levels, such as the tissue level, organ level, and organismal level, rather than at the subcellular and biochemical levels. Still, however, these "macroscopic" variations are all ultimately driven by subcellular alterations in the DNA of organisms, which produce the vast diversity in organismal life seen on the multicellular level.

Because of their small size, cells were not first identified until the development of crude microscopes in the 17th century. Robert Hooke first described cells in 1665 by observing thin slices of cork and noticing the tiny, empty compartments, which he called *cellulae* (Latin for "small rooms"). Living cells eventually were observed a few years later by a Dutch naturalist, Antonie van Leeuwenhoek. Yet, for nearly 200 years after these pioneering observations, biologists did not understand the significance of cells to life.

The modern cell theory has its beginnings in the 19th century, and generally is credited to the advances of three German biologists: Matthias Schleiden (in 1838), Theodor Schwann (in 1839), and Rudolf Virchow (in 1855). Their work gave rise to the modern **cell theory**, the unifying concept that states:

1. All organisms are composed of cells,
2. cells are the basic living units of organization and function in all organisms, and
3. all cells arise only by the division of previously existing cells.

An important concept in cell biology is that regardless of what kind of cell one is observing, certain functions and structures are indistinguishable. Replication of cells is the same in an amoeba or a flatworm or a human. The mitochondrion from an earthworm is indistinguishable in its shape and function from the mitochondrion of a bird or a human. Similarities in cellular structure between such widely disparate groups of organisms illustrate how evolution has been conservative at the biochemical and cellular levels. Even though cells have the same basic components and work under the same principles, cell exteriors have become markedly diverse in appearance, reflecting specialization to myriad different tasks and functions. To gain an appreciation for the variety of specializations that cells have undergone, you will examine several representative animal cells and tissues in the following exercises.

The three most easily identifiable parts of the animal cell are the central **nucleus**, the peripheral **cytoplasm**, and the darker, surrounding **plasma membrane**. The nucleus is enclosed by the thin **nuclear membrane**, which regulates the passage of substances into and out of the nucleus. Within the nucleus are darkly stained clumps of **chromatin** and a small, spherical structure called the **nucleolus** that manufactures ribosomes and exports them to the cytoplasm, where they play a role in protein synthesis. Rarely will other cellular structures be visible within animal cells with the compound microscope.

Exercise 2-1 Animal Cells

Materials Needed

- ❑ compound microscope
- ❑ slide of unfertilized sea star eggs
- ❑ glass slides
- ❑ coverslips
- ❑ toothpicks
- ❑ methylene blue stain
- ❑ distilled H_2O (or isotonic saline solution)

Procedure

1 Obtain a prepared slide of sea star eggs.

2 Use your low-power (or scanning) objective lens to locate a single, undivided spherical cell similar to the one shown in Figure 2.1.

3 After you have found such a cell, examine it using medium, then high power.

After you feel comfortable identifying the basic parts of a cell on a prepared slide, you are ready to look at cells from your own body.

Procedure

1 Obtain a clean microscope slide and place a small drop of distilled water (or isotonic saline solution) on the slide.

2 Use a clean toothpick to gently scrape the inside of your cheek.

3 Carefully swirl the scrapings containing cheek epithelial cells in the drop of liquid on the slide. Identify the nucleus, nucleolus, nuclear membrane, plasma membrane, and cytoplasm.

4 Add one small drop of methylene blue stain to the solution.

5 Carefully lower a coverslip over the drop and gently press down on the coverslip to flatten the cells and remove any air bubbles.

6 Use the low-power (or scanning) objective lens to locate a single flat cell or small group of cells.

7 Next, view the cells using medium and finally high power.

8 Remember to adjust the condenser and iris diaphragm as needed on your microscope to increase contrast and obtain the sharpest possible image.

9 Identify the same structures in these cells as you did in the sea star cell.

10 Sketch a few of these cells in the space provided, and label all visible parts.

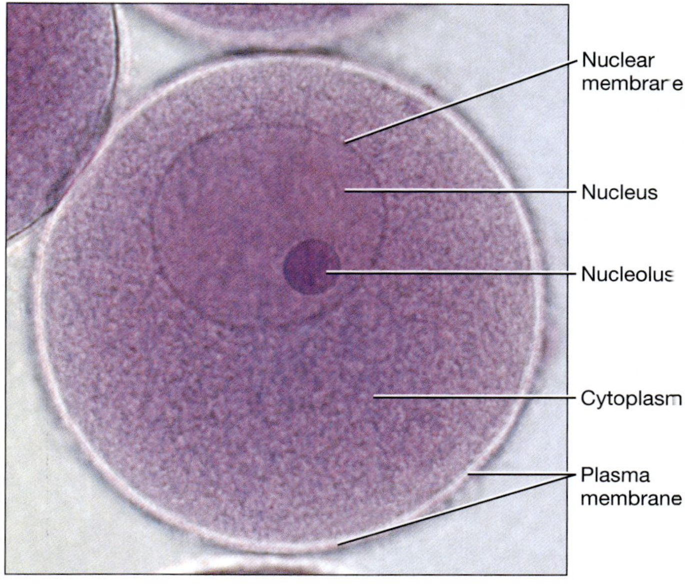

2.1 Generalized animal cell depicting structures visible with the compound microscope.

Exercise 2-2 Vertebrate Tissues

Materials Needed

- ❑ compound microscope
- ❑ simple squamous epithelium slide
- ❑ simple cuboidal epithelium slide
- ❑ simple columnar epithelium slide
- ❑ stratified squamous epithelium slide
- ❑ hyaline cartilage slide
- ❑ elastic cartilage slide
- ❑ ground bone slide
- ❑ adipose tissue slide
- ❑ loose connective tissue slide
- ❑ regular dense connective tissue slide
- ❑ mammalian blood smear slide
- ❑ skeletal muscle slide
- ❑ smooth muscle slide
- ❑ cardiac muscle slide
- ❑ neuron slide

All animals are multicellular organisms, and in all but the simplest animals, their great numbers of cells are arranged into discrete functional units called tissues. Sponges are a notable exception to this rule, because they lack true tissue-level organization. A **tissue** is defined as a group of similar cells that work together to perform a common function for an animal. True tissues have complex physiological connections between neighboring cells that allow these cells to communicate and exchange materials with each other.

Early in embryonic development, the cells of the growing embryo differentiate into distinct germ layers. As the embryo grows, these germ layers further differentiate into specialized tissue layers that form the organs and organ systems of the body. Biologists classify adult animal tissues into four principal types, based on their structure and function (Fig. 2.2).

1. **Epithelial tissues**—cover external surfaces for protection or line the internal surfaces of body cavities and vessels. They function as barriers, and in secretion and absorption.
2. **Connective tissues**—bind, support, and protect body parts and systems.
3. **Muscle tissues**—permit movement of the animal through its environment and/or movement of substances through the animal.
4. **Nervous tissues**—initiate and transmit electrical nerve impulses to and from the body parts and store information in the form of biochemical compounds. They serve in communication and control of the animal.

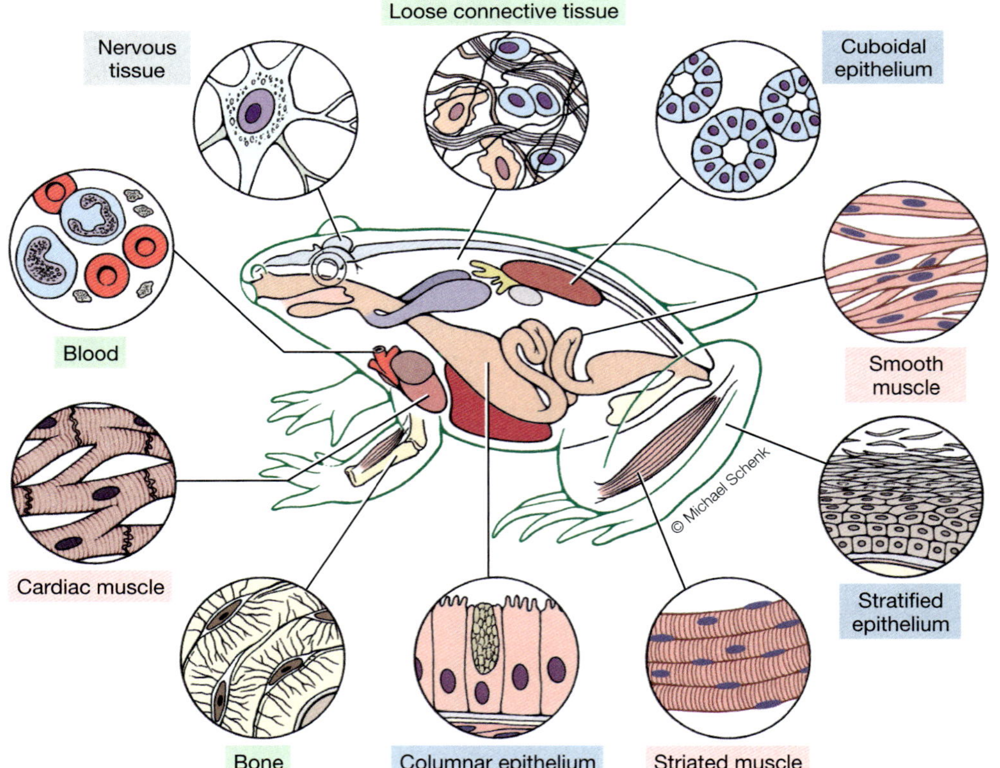

2.2 General vertebrate tissue types and their locations in the body: epithelial (blue), muscle (red), connective (green), and nervous (gray).

Epithelial Tissues

Epithelial tissues cover external surfaces for protection or line the internal surfaces of body cavities and vessels. They are typically arranged into tightly packed layers of cells with little or no intercellular space. They are further categorized based on the shapes of the cells and the number of layers of cells that constitute the tissue.

Simple Epithelium

Simple epithelial tissues consist of a single layer of cells and are classified based on their shapes. Squamous epithelium is composed of flattened, irregularly shaped cells. These cells typically have a two-dimensional appearance in the microscope. When viewed from the side, they often are difficult to distinguish; a thin band of cytoplasm with a small bulge where the nucleus appears is usually all that is identifiable.

Simple squamous epithelium would be represented by a single layer of flattened cells (Fig. 2.3A). The alveoli of lungs and the inner walls of arteries are examples of simple squamous epithelium. Cuboidal epithelium and columnar epithelium contain cells that are thicker and fuller and have the three-dimensional appearances that their names suggest. Thus, **simple cuboidal epithelium** would be represented by a single layer of box-shaped cells (Fig. 2.3B), and **simple columnar epithelium** would contain a single layer of elongated, rectangular cells (Fig. 2.3C). Simple cuboidal epithelium is found in the tubules of the mammalian kidney, for example, and simple columnar cells are prevalent in the inner lining of the stomach and intestines in mammals. Epithelial tissues typically exist in simple layers when absorption or diffusion across the tissues is necessary.

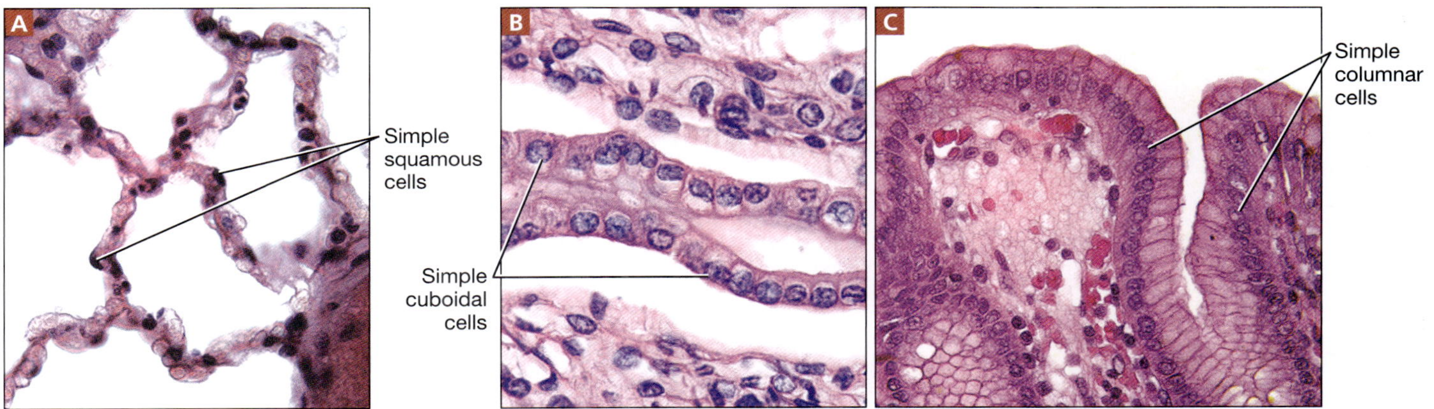

2.3 Epithelial tissues: (**A**) simple squamous, (**B**) simple cuboidal, and (**C**) simple columnar.

Procedure

1. Examine slides of simple squamous, simple cuboidal, and simple columnar epithelium. Note the distinctive shapes of the cell types that make up each tissue.
2. Sketch and label representative examples of each of these three types of tissue in the space below.

2

Stratified Epithelium

Stratified epithelial tissue derives its name from the layered arrangement of the cells in these tissues (Fig. 2.4). In many cases, these tissues are composed of more than one type of cell (for example, several layers of squamous cells followed by several layers of cuboidal cells). Epithelial tissues typically exist in stratified layers to serve as barriers against foreign substances and injury. For example, the skin consists of an outer layer of **stratified squamous epithelium** to protect against impact, abrasion, radiation, desiccation, and infection.

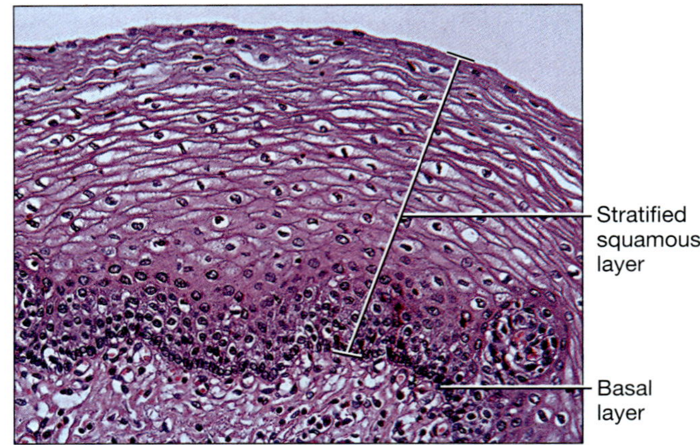

2.4 Stratified squamous epithelium.

Procedure

1 Examine a prepared slide of stratified squamous epithelium.

2 Sketch and label a representative sample of the tissue below.

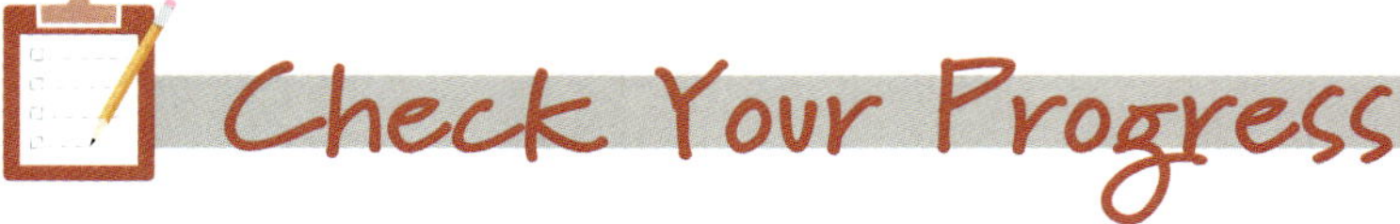

Check Your Progress

1.1 To which subgroup would you classify the cheek cells that you prepared earlier? Explain why.

Connective Tissues

Tissues that bind organs together, hold organs in place, support body structures, and store nutrients are grouped into the general category of connective tissues. This category contains a wide variety of specialized tissue types, including tendons, cartilage, fat, blood, and bone. At first glance, it may seem unlikely that these different tissues would be grouped together; yet all connective tissues share a common structural feature: They all contain cells that are widely separated by an extracellular **matrix** secreted by the living cells. In bone, this extracellular matrix contains crystals that make the bone hard; in blood, the extracellular matrix is plasma; and in cartilage, the extracellular matrix is composed of a gelatinous glycoprotein.

Cartilage

A common connective tissue in vertebrates is **cartilage**. Cartilage is composed of widely spaced cells within a gelatinous, glycoprotein matrix that provides firm, but flexible support. Embedded within this matrix are hollow chambers, called

lacunae, that contain the **chondrocytes**, or cartilage-producing cells. Although there is no direct blood supply to chondrocytes in lacunae, they are able to remain alive by exchanging oxygen, nutrients, carbon dioxide, and waste with surrounding blood vessels by diffusion through the gelatinous matrix.

One specific type of cartilage is **hyaline cartilage**, found between bones, where it cushions the surfaces of joints. Its intercellular matrix is composed primarily of chondrin with thin collagen fibers to provide support and suppleness (Fig. 2.5A). **Elastic cartilage** contains fine collagen fibers and numerous elastic fibers that provide greater elasticity to this cartilage (Fig. 2.5B). Elastic cartilage is much more flexible than hyaline cartilage and can be found in the ear, nose, and voice box of humans.

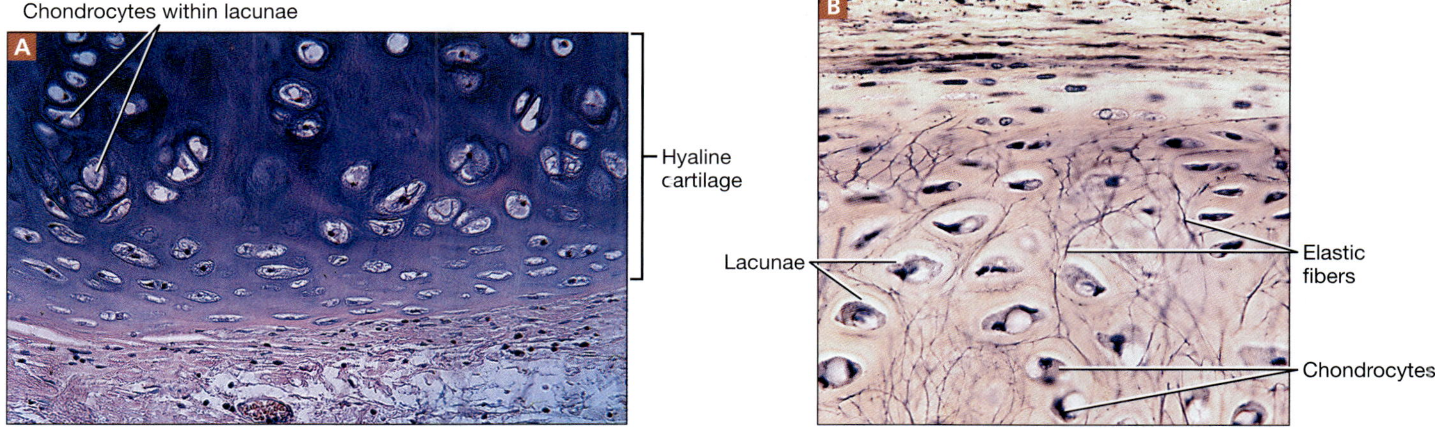

2.5 Cartilage (**A**) hyaline, and (**B**) elastic.

Procedure

1 Examine slides of hyaline and elastic cartilage. Notice the differences in appearance between the two types of cartilage.

2 Sketch and label representative examples of hyaline and elastic cartilage in the space provided.

Bone

Bone is one of the most specialized structural connective tissues, and the strongest of them. In addition to providing structural support for the body, bone is responsible for storing calcium that can be withdrawn by the body as blood calcium levels drop, and for producing red blood cells in the bone marrow. As bone grows, the **osteocytes** (bone-producing cells) secrete a hard, calcified matrix that forms thin, concentric layers called **lamellae**, giving bone its characteristic appearance (Fig. 2.6). These lamellae form layered rings around tiny, narrow pathways called **Haversian canals**. As in cartilage, the cells that secrete the matrix—in this case, osteocytes—are housed individually in lacunae. In living bone, the Haversian canals align themselves parallel to the long axis of the bone and contain blood vessels and nerves that exchange nutrients and wastes and communicate with the osteocytes.

Diffusion of gases and nutrients through the hard, calcified matrix is not possible in bone, as in the more permeable gelatinous matrix of cartilage. Instead, nutrients are transported to the osteocytes through tiny, fingerlike channels in the lamellae called **canaliculi** that form a miniature canal system linking neighboring osteocytes for communication and nutrient transfer (Fig. 2.7).

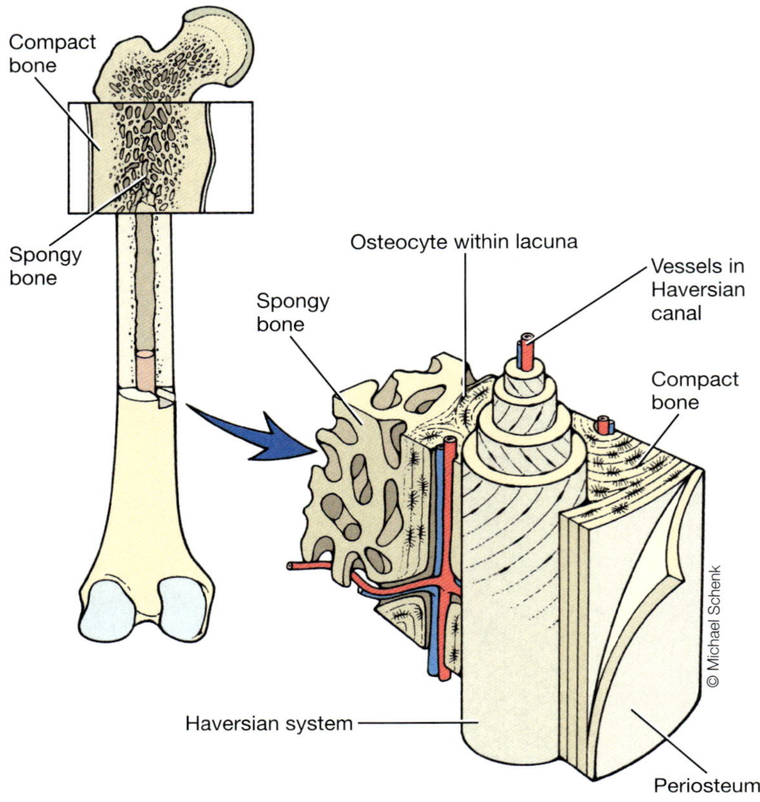

2.6 Internal structure of bone and arrangement of osteons.

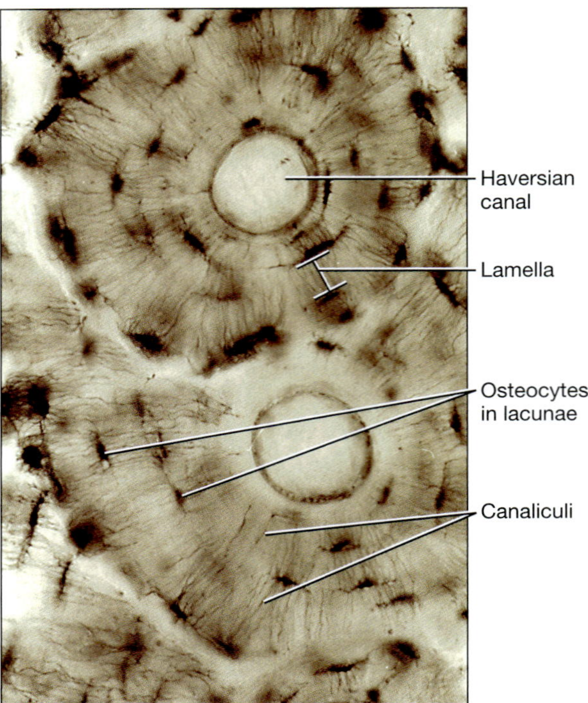

2.7 Transverse sections of two osteons in bone tissue.

Procedure

1 Examine a slide of ground bone.

2 Sketch and label a small section of bone depicting at least one Haversian canal and several lamellae. Also label several lacunae and canaliculi.

Adipose Tissue

Adipose tissue is a type of connective tissue that stores or sequesters food for the body in the form of fat droplets. Each cell in adipose tissue contains a large, oil-filled **vacuole**, giving adipose cells the appearance of being empty spaces (Fig. 2.8). Adipose cells often become so full of fat that the cytoplasm, nucleus, and other organelles are pressed tightly against the outer margin of the cell membrane. Often, nuclei are barely visible as small bulges along the margins of the cells. When the body needs the energy stored in fat, adipose cells secrete their fatty acids into the bloodstream where they are carried to and metabolized by muscles and other target organs. In addition to storing energy reserves for the body, adipose tissue has the effects of insulating the body, and of providing cushioning and support for organs such as the kidneys and mammary glands.

As an adult, you have a fixed number of fat cells in your body. Depending on your diet, metabolism, and exercise regimen, these cells either enlarge as they store more fat or shrink as their energy stores are metabolized. Try as you might, you never can get rid of your fat cells through diet and exercise; you can only shrink them and compact them into thin layers by emptying them of their reserves.

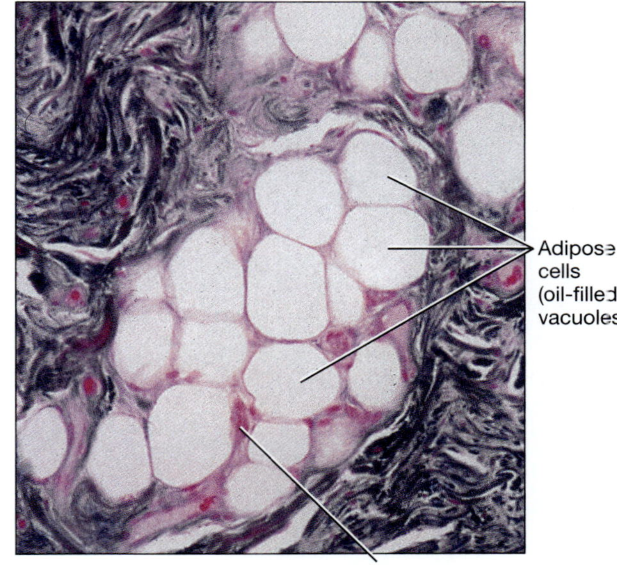

2.8 Adipose connective tissue. Each clear sphere is a single adipose cell containing a large, oil-filled vacuole and a nucleus pressed against the side of the cell.

Loose Connective Tissue

Another connective tissue found in all vertebrates is **loose connective tissue.** Its name derives from its appearance—loosely scattered cells surrounded by a clear, gelatinous matrix (Fig. 2.9). Similar to the matrix found in cartilage, the loose connective tissue matrix contains thin elastic fibers and thicker nonelastic fibers composed of **collagen.** The cells that secrete collagen and other fibrous proteins are known as **fibroblasts.** Incidentally, collagen is the most abundant protein in the animal kingdom—found in animal bodies wherever both flexibility and resistance to stretching are required. But it is not developed as elaborately in invertebrate tissues. Loose connective tissue in vertebrates is primarily responsible for holding other organ tissues together and in place and can be found beneath the skin and between many organs in the body.

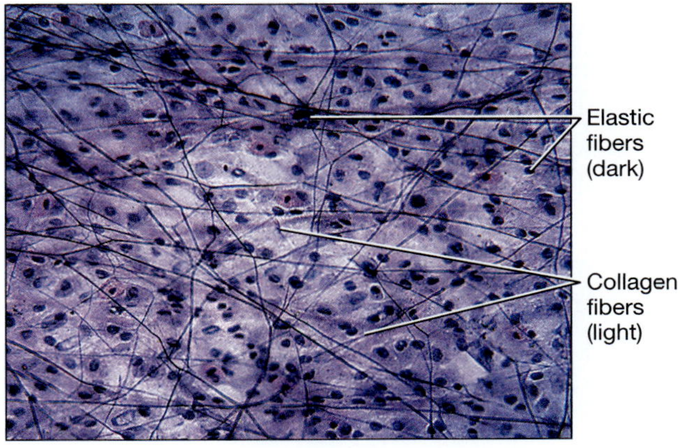

2.9 Loose connective tissue depicting elastic and collagen fibers.

Dense Connective Tissue

Dense connective tissue contains tightly packed collagen fibers, making it stronger than loose connective tissue. Dense connective tissue is further classified based on the arrangement of collagen fibers into *regular* and *irregular* types. The collagen fibers of regular dense connective tissue are extremely long, densely packed, and parallel, like the strands of wire in a cable, forming structures that are extremely resistant to stress (Fig. 2.10). **Tendons**, the cords that connect muscles to bones, and **ligaments**, the attachments that connect bones to one another, both consist of regular dense connective tissue.

In contrast, the collagen fibers of irregular dense connective tissue lack a parallel arrangement. Instead, these fibers have many different orientation patterns, often being arranged in bundles distributed in all directions throughout the tissue, as in the dermis of the skin. This type of connective tissue produces the tough outer coverings of fascia that help hold together organs such as kidneys, muscles, and nerves.

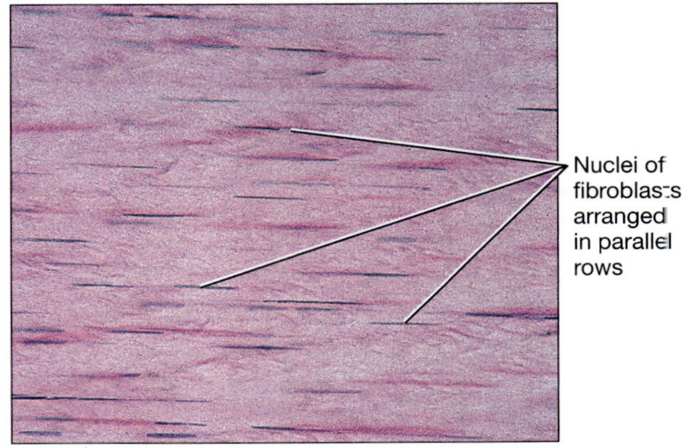

2.10 Dense regular connective tissue. Note parallel arrangement of fibroblast nuclei.

Procedure

1 Examine slides of adipose tissue, loose connective tissue, and regular dense connective tissue.

2 Sketch a representative portion of each tissue below, and label any visible fibers and cell types.

Blood

Despite its fluid nature, **blood** is classified as a type of connective tissue. Its cells and the fluid matrix, **plasma**, in which they are suspended course through blood vessels transporting oxygen, carbon dioxide, nutrients, electrolytes, hormones, metabolic wastes, and practically any other substance that cells use or produce. Plasma contains many proteins, including fibrinogen, produced by the liver, which helps blood to clot; albumin, also produced by the liver, which exerts an osmotic force needed for fluid balance; and antibodies produced by lymphocytes and needed for immunity.

The plasma of many invertebrates also contains free-floating respiratory pigments such as **hemoglobin** and **chlorocruorin** (found in annelids), and **hemocyanin** (found in molluscs and arthropods). The respiratory pigment in vertebrates, hemoglobin, is not found floating freely in the plasma but, instead, is packaged within the red blood cells.

Mammalian red blood cells, **erythrocytes**, appear as tiny, light pink, biconcave discs (Fig. 2.11). They are the most numerous type of cell in blood, with 4–5 billion erythrocytes per milliliter of blood in adult humans, and they contain hemoglobin to reversibly bind and transport oxygen and carbon dioxide. Their characteristic biconcave shape provides a higher surface-to-volume ratio, increasing the diffusion rates of oxygen and carbon dioxide into and out of the cells.

Leukocytes
Erythrocytes

2.11 Electron micrograph of blood cells in the lumen of a blood vessel.

In birds, amphibians, and most invertebrates, mature red blood cells are round or oval and contain large nuclei, allowing them to replicate through mitosis. In mammals, however, as erythrocytes mature, their nuclei are ejected from the cells, giving them a limited life span of around 120 days. In humans, more than 2.4 million "worn out" red blood cells are removed from the blood every second by phagocytic cells in the liver and spleen. To balance this loss, an equal number of new erythrocytes must be produced every second by the bone marrow and released into the bloodstream.

White blood cells make up less than 1% of the cells in human blood (4–10 million per milliliter), yet play a vital role in defending the body against disease. White blood cells, or **leukocytes**, are generally larger than erythrocytes and contain distinct, purplish nuclei. Each of the several different kinds of leukocytes plays a specific role in defending the body against invading microorganisms and other foreign substances. The relative sizes of their cells and unique shapes of their nuclei are helpful characteristics for identifying the different types of white blood cells.

Neutrophils (Fig. 2.12A) are the principal phagocytic cells in blood and are specialized for seeking out and ingesting foreign bacterial cells and dead host cells. In addition to being the most abundant type of white blood cell, they are generally about twice the size of a red blood cell and have nuclei with two to five distinct lobes. Much less common, **basophils** (Fig. 2.12B) are also about twice the size of red blood cells but typically have unlobed nuclei and a more granular appearance when stained. They are instrumental in the inflammatory response of allergic reactions and help to prevent blood from inappropriately clotting within blood vessels.

Eosinophils (Fig. 2.12C) also play a role in the allergic response and help defend against parasites. They, too, are about twice the size of red blood cells and their nuclei usually possess two lobes. **Lymphocytes** (Fig. 2.12D) are slightly larger than red blood cells and have spherical nuclei that almost completely fill the interior of the cell, leaving little visible cytoplasm. Lymphocytes produce antibodies that recognize and destroy foreign cells. **Monocytes** (Fig. 2.12E), the largest of the white blood cells, spend a brief time (one–three days) developing in the blood before exiting circulation and completing their development in the tissues. There they greatly enlarge and become macrophages—giant scavenger cells that voraciously engulf bacteria, dead cells, and other cellular debris. Table 2-1 provides a summary of the different types of white blood cells and their distinguishing characteristics.

Extremely small **platelets** also should be visible scattered among the other blood elements. Typically there are 150 to 400 million platelets per milliliter of human blood. Platelets, also called *thrombocytes*, are not whole cells at all but, rather, fragments of cytoplasm from a type of cell found in the bone marrow called a megakaryocyte. Platelets play an important role in the clotting process by accumulating at the injured site of a broken blood vessel and forming a plug by sticking to each other and to surrounding tissues.

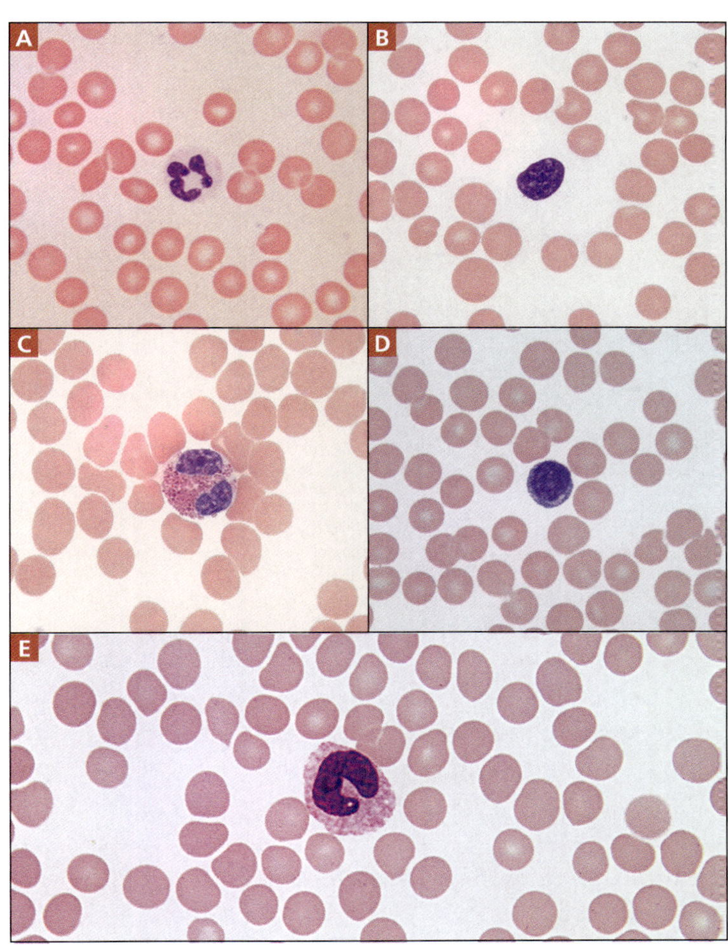

2.12 Fluid connective tissue (blood) depicting erythrocytes, tiny platelets, and the five types of leukocytes: (**A**) neutrophil, (**B**) basophil, (**C**) eosinophil, (**D**) lymphocyte, and (**E**) monocyte.

Table 2-1 Comparison of Types of Human White Blood Cells

Cell Type	Diameter (μm)	% Abundance	Nucleus	Function
Neutrophil	12–15	55–65	2–5 lobes	Bacterial phagocytosis
Basophil	12–15	0.5–1	Unlobed	Inflammatory & allergic response
Eosinophil	12–15	1–3	2 lobes	Allergic response & parasite defense
Lymphocyte	7–18	25–33	Large, spherical	Produce antibodies for immune response
Monocyte	12–20	3–7	Horseshoe-shaped	Phagocytosis

Procedure

2

1 Examine a prepared slide of blood.

2 Sketch a small region of the slide of blood in the space provided. Label the following: erythrocytes, leukocytes, and platelets.

3 Carefully scan the entire slide to identify and sketch a neutrophil, eosinophil, basophil, monocyte, and lymphocyte.

4 Examine three different fields of view and count the number of leukocytes visible in each field. Record your observations below.

Check Your Progress

2.1 Which type of connective tissue provides the most rigid support?

____

2.2 Which type of connective tissue stores lipids?

____

2.3 Tendons and ligaments are composed of ______________________________.

2.4 Are nuclei present in mammalian red blood cells?

____

2.5 Which general type of blood cell is the most numerous (erythrocyte, leukocyte, or platelet)? the least numerous?

____

2.6 List the five categories or types of leukocyte and their functions.

1. ______________________________
2. ______________________________
3. ______________________________
4. ______________________________
5. ______________________________

2.7 Which type of leukocyte is the most numerous? the least numerous?

2

2.8 What general structural features do all connective tissues share in common?

Muscle Tissues

The characteristic feature of muscle tissue is its ability to contract and thus create movement. These movements often propel an animal through its environment, but just as often propel substances through the animal's body. The interaction of **actin** and **myosin** filaments, which occur in abundance and in uniform orientation in muscle cells, is responsible for the contractility of muscle tissues. Three types of muscle tissue occur in vertebrates: smooth muscle, skeletal muscle, and cardiac muscle. Similar to the case with adipose tissue, your body contains a finite number of muscle cells. As you exercise muscles, the cells that make up those muscles grow in response to the stress. What is perceived as a muscle getting bigger is actually the enlargement of existing cells rather than the addition of new muscle cells.

Smooth Muscle

The simplest type of muscle tissue is **smooth muscle.** Lacking striations and generally confined to regions of the body under autonomic nervous control, smooth muscle fibers are long and spindle-shaped and contain single nuclei (Fig. 2.13A). Smooth muscle tissue can be found in the bladder, the uterus, the stomach, and in blood vessels. Contractions of smooth muscle are characteristically slow and rhythmic.

Procedure

1 Examine a slide of smooth muscle and sketch and label several smooth muscle cells in the space provided.

2 From what region of the body was your sample of smooth muscle taken?

Skeletal Muscle

Skeletal muscle is composed of long, unbranched muscle fibers that are actually composites of many individual muscle cells, giving these fibers their multinucleated appearance (Fig. 2.13B). As skeletal muscles develop during embryonic growth, groups of individual muscle cells fuse and their nuclei get pushed toward the outer margins of the forming muscle fiber. This bundling of cells enhances the strength and speed with which these fibers are able to contract. Skeletal muscle fibers make up the muscles attached to our skeletons and are under voluntary control. Skeletal muscle has a characteristic striated appearance, caused by the precise alignment of actin and myosin filaments along the **sarcomeres** of each **myofibril**.

2

Procedure

1 Examine a slide of skeletal muscle.

2 Sketch several skeletal muscle cells and label the nuclei, muscle fibers, and striations on your drawing.

Cardiac Muscle

A second type of striated muscle is found in the walls of the heart—**cardiac muscle.** Unlike skeletal muscle, cardiac muscle is not under voluntary control and its nuclei are not located on the periphery of the cells. Its steady, rhythmic contractions are controlled by a confluence of **ganglia** embedded in the muscle of the heart itself. Cardiac muscle also is composed of bands of muscle fibers that branch and reunite with one another to form a continuous network of muscle tissue (Fig. 2.13C).

The cells of these highly branched fibers are partially connected to one another by intercalated discs—bands between individual cells that appear as particularly dark, bold striations in stained slides of cardiac muscle. **Intercalated discs** are gap junctions that allow communication between the cells of the muscle fibers and permit cardiac muscle to depolarize quickly and contract as a unit, much like skeletal muscle, though unlike the slow, wavelike contractions of other involuntary muscle.

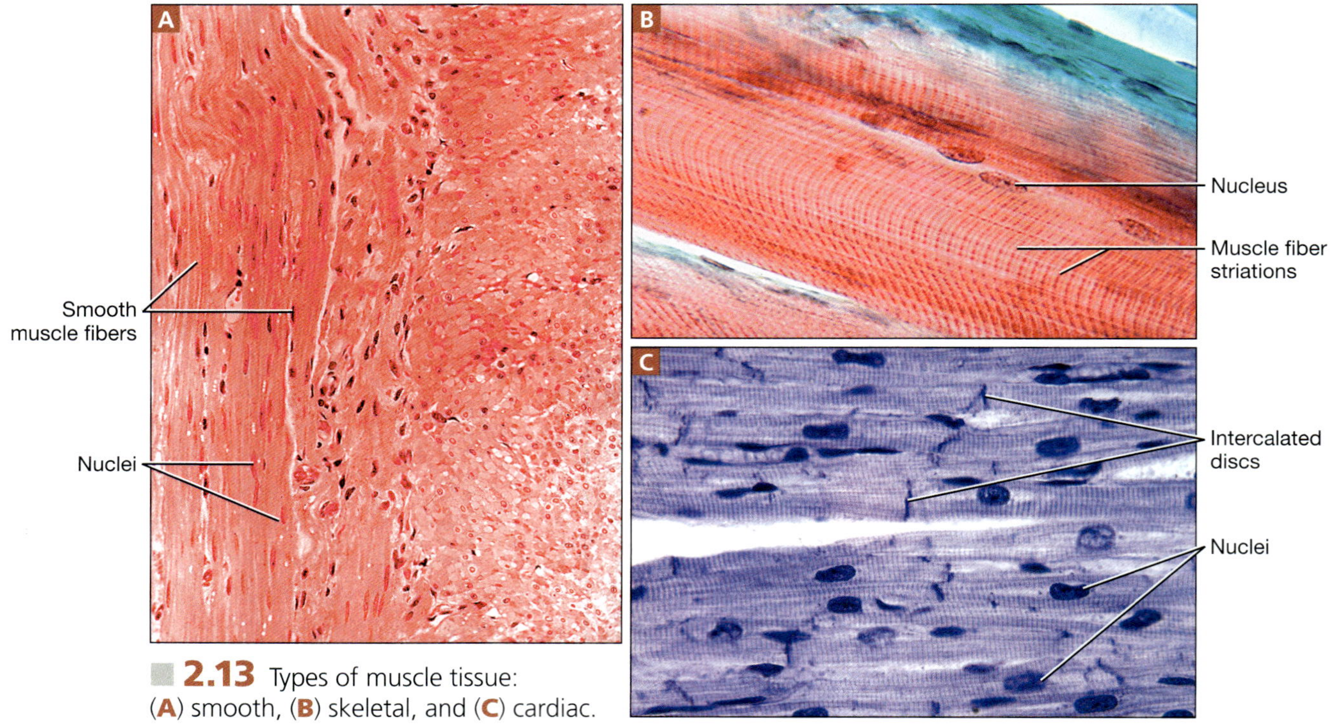

2.13 Types of muscle tissue: (**A**) smooth, (**B**) skeletal, and (**C**) cardiac.

Procedure

1 Examine a slide of cardiac muscle.

2 Sketch a representative section of cardiac muscle and label the nuclei, intercalated discs, and striations on your drawing.

Nervous Tissue

Nervous tissue is specialized for reception of stimuli and conduction of impulses from one region to another, and it consists of two basic kinds of cells: neurons and supporting cells called glial cells. **Neurons** are the functional units of the nervous system. A typical neuron is made up of (1) a **cell body** containing the nucleus and other organelles, (2) a long **axon** that transmits electrical impulses away from the cell body, and (3) short extensions called **dendrites** that typically receive electrical impulses from neighboring neurons or sensory receptors and transmit them to the cell body (Fig. 2.14).

Glial cells assist in propagating nerve impulses and provide a nutritive role for neurons (Fig. 2.15). Neuron cell bodies are located only in the brain and spinal cord; thus, many axons must be of considerable length (some up to 1 meter long in humans!) to reach from the spinal cord to the extremities of the body. In many areas, scores of axons are bundled together in cable-like nerve fibers to traverse the great distances required to reach throughout the body. The axons of these nerve cells have sheaths coated with a lipid called **myelin**, giving them special electrical properties that greatly enhance the transmission speed of nerve impulses through these fibers.

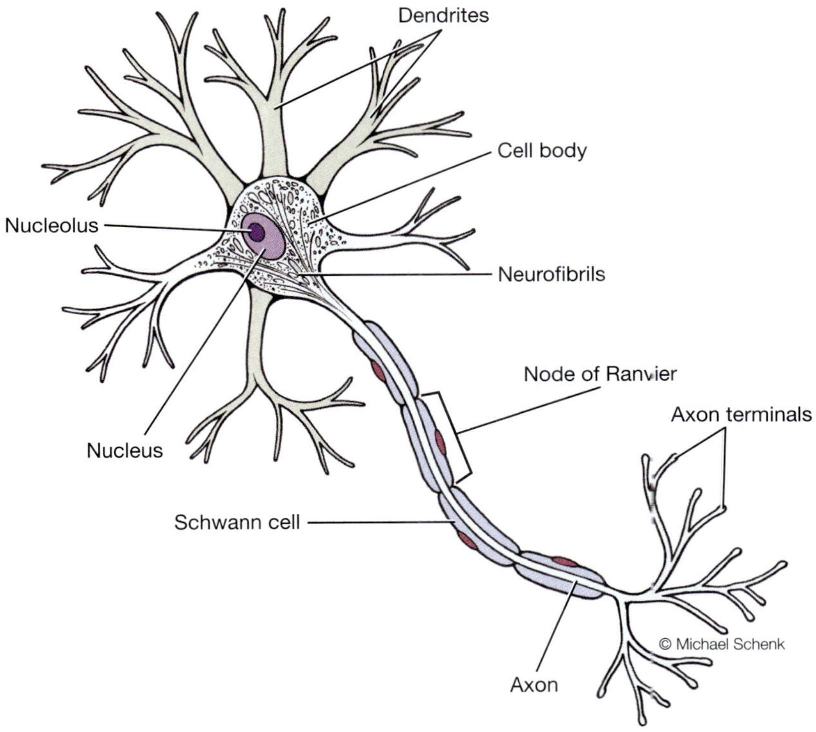

2.14 Structure of a generalized neuron.

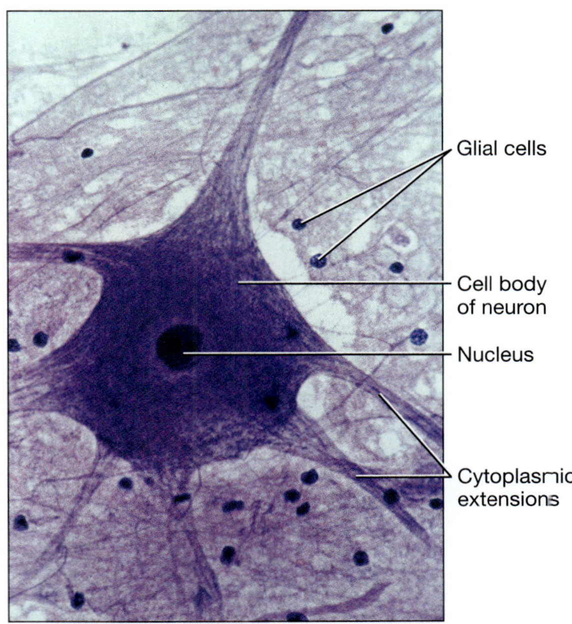

2.15 Nervous tissue: neuron with glial cells.

Procedure

1 Examine a slide of a neuron smear. Locate a single neuron in which the cell body, axon, and dendrites are visible (Fig. 2.15).

2 Sketch this neuron and its supporting glial cells below. Label the cell body, axon, dendrites, and glial cells.

When you have finished examining of all the vertebrate tissue slides, fill in Table 2-2 with the name of each type of tissue studied, a location in the body where each type of tissue may be found, and the function of each type of tissue.

Table 2-2 Comparison of Vertebrate Tissues

After viewing the slides of vertebrate tissues, fill in the appropriate information in the table below.

Tissue Studied	Location in Body	Function

Name ______________________ Section ______________

2 Questions for Review

1 State the three premises of the modern cell theory.

1. ______________________

2. ______________________

3. ______________________

2 Give an example of an epithelial tissue designed to "keep things out" of the body.

____

3 Give an example of an epithelial tissue designed to serve as a surface for diffusion of substances.

____

4 A common myth is that if you stop exercising your muscle will turn into fat. Explain why this myth is not true.

____

5 When your body metabolizes fat for energy, it gets 9 Cal./g, as opposed to only 4 Cal./g from protein (muscle). Knowing this, explain why your body uses adipose tissue rather than muscle tissue as the "default" energy store.

____

6 Unlike mammalian red blood cells, the red blood cells of amphibians contain large nuclei. Speculate about the consequences on the relative oxygen-carrying capacity of amphibian red blood cells and the effects this could have on the potential metabolic rates of both animal groups.

____

7 Match the type of blood cell with its function. (Some letters will be used more than once.)

______ erythrocyte	**a.**	Instrumental in clotting response
______ neutrophil	**b.**	Involved in inflammatory and immune response
______ lymphocyte	**c.**	Phagocytosis of foreign bacteria and dead cells
______ platelet	**d.**	Transports O_2, CO_2, and nutrients
______ basophil		
______ monocyte		

8 Describe two similarities and two differences between bone and cartilage.

____

____

____

____

9 Contrast the three types of vertebrate muscle tissue.

____

____

____

____

____

10 Identify each of the specific tissue types indicated below. (Be as specific as you can.)

______________ ______________ ______________ ______________

____

CHAPTER 3

Reproduction, Regeneration, and Cell Division

After completing the exercises in this chapter, you should be able to:

1 Describe the cell cycle and explain the principal events of its four main stages.

2 Understand the principal events that occur in prophase, metaphase, anaphase, and telophase of mitosis.

3 Identify mitotic stages of division in a microscopic preparation of animal cells.

4 Describe major differences between mitosis and meiosis.

5 Understand the basic processes of spermatogenesis and oogenesis in mammals.

6 Understand the differences between haploid and diploid cells and their significance in reproduction.

In Chapter 2 you learned that groups of cells work together as tissues to perform specific functions for animals. The process by which animals develop from a single fertilized egg cell into complex, multicellular organisms with dozens of different tissues specialized for vastly different roles is the topic of this and the next chapter. Because all cells are formed by the division of preexisting cells, cells divide to enable organisms to grow, repair damaged parts, and reproduce. In this chapter you will study two mechanisms of cell division—mitosis and meiosis. You also will examine the role that cell division plays in the processes of growth, regeneration, and reproduction. In the next chapter you will examine the continuing process of embryonic development in two organisms with different developmental pathways.

Growth and Regeneration

All animals depend on cell division for their growth and repair processes. As cells reach a certain size, they must either stop growing or divide. As you grow, heal wounds, and regenerate worn out or damaged tissues, new cells are added to your body through a process of cell division and replication known as **mitosis.**

Each cell has a precise set of genetic information built into its chromosomes. For a cell to replicate successfully, it must have the ability to leave a functional, duplicate copy of its genetic material and an adequate supply of organelles in each daughter cell to allow those cells to function normally and properly interpret the genetic code they have been given. Most cells in your body have the ability to replicate mitotically, although a few cells, such as red blood cells, muscle cells, and brain cells, are exceptions to this rule and lose their ability to replicate as they mature and become highly specialized for their specific roles.

In many simple organisms a form of reproduction known as **asexual reproduction** may occur through mitotic division. A single parent cell or organism either splits, buds, or fragments to produce two or more offspring (Fig. 3.1). Because these cells are the result of mitotic division, each new offspring is genetically identical to the parent cell or organism. Asexual reproduction is an adaptation for animals that cannot move about and search for mates, especially when population densities are low and mates are not readily available. This can be a fast and efficient way to reproduce, but it severely limits genetic diversity because all offspring are clones of the parent. Among species that reproduce asexually through mitosis, the only way that genetic variation can occur is by mutation, a permanent change in the genes.

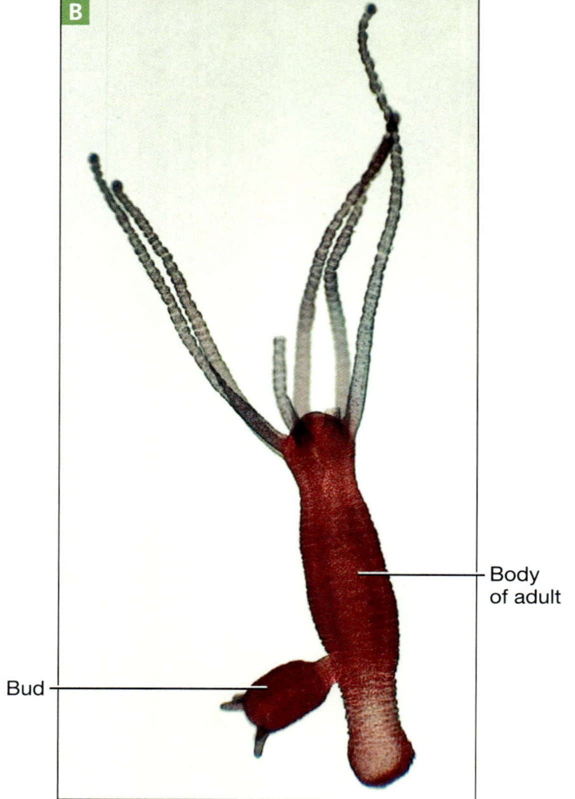

3.1 Examples of asexual reproduction: (**A**) *Paramecium* splitting in half to produce two separate cells, (**B**) part of the *Hydra's* body growing outward to form a bud that later separates and develops into a new individual.

The Cell Cycle

The events that encompass the entire life cycle of a cell from one division to the next are called, collectively, the **cell cycle.** The actual events of the cell cycle, including mitosis, are not discrete events but, rather, occur in a continuous sequence. We artificially separate the cell cycle into discrete stages and assign them labels merely for the convenience of discussing and organizing this complex process (Fig. 3.2). Mitosis occurs during only a small portion of this cycle. The remainder of the cell cycle, **interphase**, is devoted to growth of the cell, synthesis of genetic material and other organelles, and taking care of the cellular duties for which that particular cell is designed. For example, some epithelial cells in the lining of the intestine absorb nutrients; others secrete saliva in the mouth, or filter blood in the kidney tubules.

Interphase traditionally is broken down into three separate subphases: G_1 (gap) phase, S (synthesis) phase, and G_2 phase. The G_1 phase is a period of rapid growth, active protein synthesis, and formation of new cell organelles. Toward the end of the G_1 phase increased activity of enzymes is required for DNA synthesis, making it possible for the cell to enter the S phase. The S phase is the period during which DNA and other molecules making up the chromosomes are synthesized. DNA replication also occurs during this phase. As the cell enters the second gap phase, G_2 phase, the proteins necessary for cell division, such as actin and tubulin, are synthesized along with other materials necessary to form the mitotic spindles. These are the final steps in the cell's preparation for division.

Mitosis follows the G_2 phase and involves the actual division of chromosomal material into two distinct nuclei. Mitosis is generally further subdivided into four phases that collectively take up only a relatively short span of the entire cell cycle—usually less than 10%. Nuclear division is followed by **cytokinesis**, division of the cytoplasm, in which the cell membrane pinches apart, forming two physically separate cells.

3

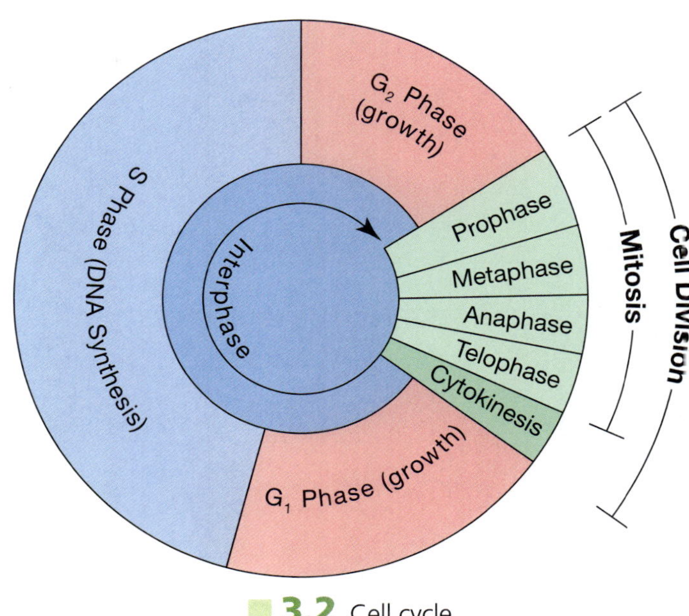

3.2 Cell cycle.

Exercise 3-1 Stages and Events of Mitosis

Materials Needed

- ❑ slide of whitefish blastula cells
- ❑ compound microscope

In this exercise you will examine a prepared slide of whitefish cells at the blastula stage. This stage is an embryonic phase during which the cells are dividing rapidly. When this embryo was preserved and sliced into thin sections, most of its cells were undergoing mitosis. Thus, the slide you will examine contains an unusually high number of cells arrested in mitotic division, making it ideal for studying this process.

Procedure

1 Obtain a prepared slide of a whitefish blastula.

2 Use your low-power (or scanning) objective lens to locate a single section (or disc) of cells and examine this section carefully using high magnification (400×).

3 Use Figure 3.3 to help you locate several cells in interphase and each of the stages of mitosis described on page 41.

Note

You may see small purple dots in and around many of the cells. These dots are spheres of oil that serve as food packets for the developing embryonic cells. Do not confuse these spheres with chromosomes or other nuclear material.

4 If you do not see a nucleus or stained chromosomal material in the cell, do not attempt to categorize that cell. Often when the blastula is cut into extremely thin sections, the slice of cell represented on the slide does not contain the nucleus—remember a cell is spherical, and you are only viewing a thin slice of that sphere.

3.3 Cell division in blastomeres of the whitefish: (**A**) early prophase, (**B**) late prophase, (**C**) metaphase, (**D**) anaphase, (**E**) telophase, and (**F**) cytokinesis.

Interphase

This stage, the longest of the cell cycle, is characterized by cells with distinct nuclear membranes enclosing lightly stained chromatin and prominent nucleoli. At this point the DNA is loosely coiled into a thin spaghetti-like mass within the nucleus, often too dispersed to visualize. DNA replication occurs during this phase. Centrioles also replicate but remain invisible.

Prophase

The DNA coils tightly into chromosomes that are actually pairs of **sister chromatids** replicated during interphase. Sister chromatids are joined together at the **centromere,** a point of constriction along the chromosome (Fig. 3.4). These chromosomes often appear as dark, rod-shaped structures. The nuclear membrane disintegrates and the nucleolus disappears. The two pairs of **centrioles** move to opposite poles of the cell and form a proteinaceous matrix of microtubules between them that will become the **spindle fibers.** When the centriole pairs reach opposite ends of the cell, they anchor themselves to the cell membrane with an array of microtubule filaments, known as **asters,** radiating outward from the centrioles.

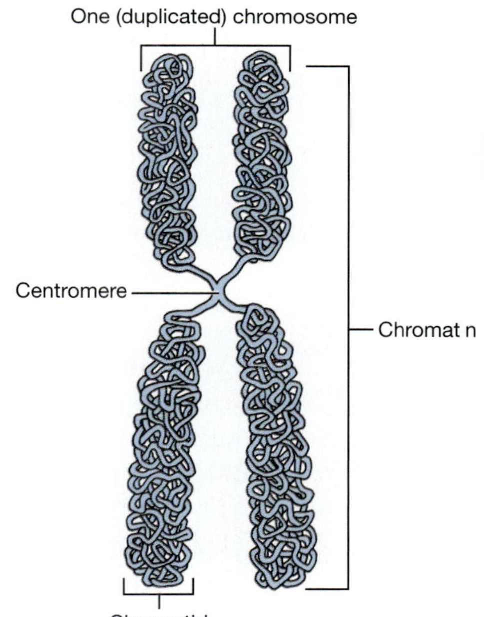

3.4 Tightly coiled chromosome as it appears in prophase. At this stage, each chromosome consists of two identical sister chromatids attached at the centrally located and constricted centromere.

Metaphase

The chromosome pairs align along the equatorial plane of the cell, and each pair of sister chromatids attaches to a spindle fiber at its **kinetochore** (a site of attachment at the chromosome's centromere). Spindle fibers and asters surrounding the centriole pairs typically are visible during this stage of mitosis.

Anaphase

The centromeres divide, separating sister chromatids from one another. As contractions of the microtubules shorten the spindle fibers, these single-stranded chromosomes are pulled toward opposite ends of the cell.

Telophase

Single-stranded chromosomes complete their migration toward opposite poles of the cell and the events of prophase are reversed. The chromosomes slowly disperse into their original, uncoiled state, the microtubule networks begin to collapse, the spindle fibers disintegrate, nucleoli reappear, and a nuclear membrane forms around each of the two bodies of chromosomal material.

Cytokinesis

This stage begins as a distinct **cleavage furrow** develops along the equatorial plane of the dividing cells and begins to pinch inward. Eventually the cleavage furrow will completely divide the cytoplasm into two distinct daughter cells, each containing a complete set of chromosomes identical to those of the parent cell.

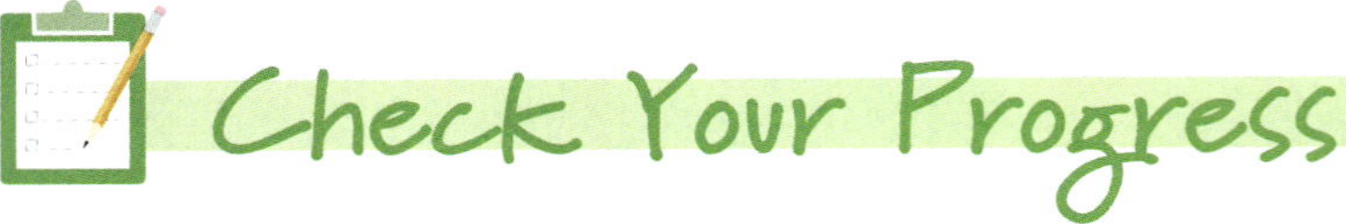

3

1.1 If a cell has 20 chromosomes during G_1 of interphase, how many chromosomes would be present during prophase?

1.2 How many chromosomes would each of the two new cells have that resulted from the mitotic division of this cell?

1.3 The different stages of mitosis are not of equal duration; some stages are relatively long while others are quite brief. Because the frequency of cells in each stage is proportional to the relative duration of that stage, you can estimate the relative length of a stage by counting the number of cells arrested in that particular stage. Based on the number of cells you observed in each stage of mitosis (excluding interphase), which stage do you think takes the longest time to complete? (You may want to review the slide again and count a random sample of 100 cells to answer this question.)

1.4 Which stage takes the shortest time to complete?

Exercise 3-2 Stages and Events of Meiosis

Not every cell in our bodies divides into two identical replicas of itself. Earlier we touched on the mechanism of asexual reproduction. As we know, however, most animals reproduce through **sexual reproduction**—a complex process involving the production of specialized sperm and egg cells and their fusion to form a **zygote**. For such a process to work, a type of cell division that enabled cells to combine their genetic material without doubling the number of chromosomes had to be possible. Otherwise, each successive generation of offspring would have twice the number of chromosomes in its cells as the previous generation! Biologically, this is an intolerable situation for all except a few unique species. Some species have evolved tolerances to tetraploidy (four sets of chromosomes) and other unusual chromosome combinations, but these instances are rare.

Clearly, then, the mechanism of cell division that produces egg and sperm cells is fundamentally different from the mitotic division responsible for growth and maintenance of our bodies. In fact, **meiosis** is a specialized type of cell division that produces **haploid gametes**, cells that contain half the number of chromosomes as the somatic cells. Later, after fertilization occurs, the genetic complement of the zygote is restored to the diploid condition.

The events of meiosis are similar to the events of mitosis, with four important differences.

1. Meiosis involves two successive nuclear and cytoplasmic divisions, generally producing four daughter cells.
2. Despite two successive nuclear divisions, the DNA and other chromosomal components are duplicated only once, during interphase preceding the first meiotic division.
3. Each of the four cells produced by meiosis contains the haploid chromosome number of the parent cell.
4. During meiosis, the genetic information from homologous chromosomes is shuffled, so each resulting haploid cell has a potentially unique combination of genes.

Meiosis generally consists of two successive nuclear divisions known as first and second meiosis, or meiosis I and meiosis II (Fig. 3.5). In prophase I of meiosis, the homologous pairs of double-stranded chromosomes actually join together in a process called **synapsis**, forming tetrads. During metaphase I of meiosis, the tetrads align along the equator of the cell and **crossing over** occurs—a process in which small sections of chromosome are exchanged between neighboring homologous chromosomes, shuffling some of the genetic information between chromosome pairs (Fig. 3.6).

Unlike anaphase in mitosis, no division of centromeres occurs during anaphase I of meiosis. Instead, the centromeres of homologous chromosomes simply move apart and the two double-stranded chromosomes move to opposite poles. The cell divides, creating two daughter cells that *do not* undergo replication of the genetic material. Then during meiosis II, the double-stranded chromosomes in each daughter cell divide again, this time separating into single-stranded chromosomes that move to opposite poles, producing four haploid daughter cells, each with half the number of single-stranded chromosomes as the original cell.

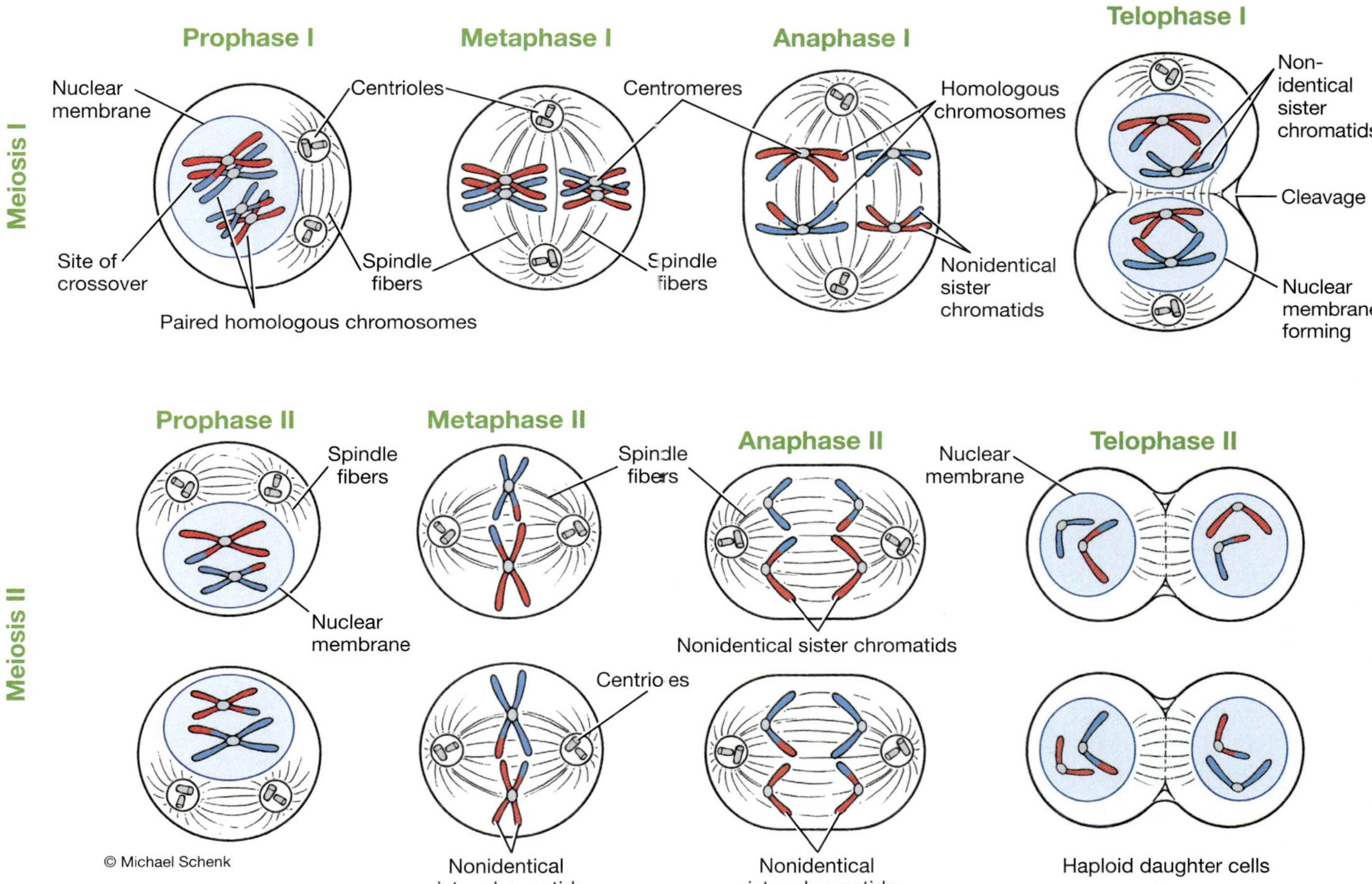

3.5 Stages of meiosis. *Parent cell* with two pairs of homologous chromosomes. **Prophase I**: DNA has replicated during preceding interphase. Cell now consists of four double-stranded chromosomes that pair up (synapsis) to form tetrads and begin to exchange genetic material (crossing over). Nuclear membrane dissolves and spindle fibers begin to form. **Metaphase I**: Centromeres migrate to opposite poles of the cell. Tetrads migrate along spindle fibers and align at the equator of the cell while crossing over continues. **Anaphase I**: Homologous chromosome pairs separate and move toward poles. **Telophase I**: Homologous chromosome pairs reach poles of cell. New nuclear membranes temporarily appear and parent cell begins to separate into two daughter cells. **Prophase II**: Following a brief interphase with no DNA replication, nuclear membranes dissolve, centrioles and spindle fibers reappear. **Metaphase II**: Each double-stranded chromosome migrates along spindle fibers to equator of cell. **Anaphase II**: Centromeres divide and single-stranded chromosomes (sister chromatids) move toward opposite poles. **Telophase II**: Nuclear membranes re-form, centrioles and spindle fibers disappear, and cytokinesis results in four haploid cells.

3

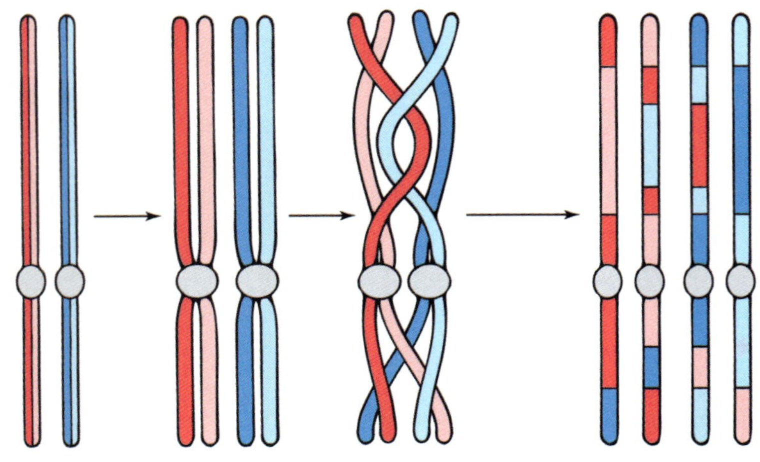

3.6 Crossing over during meiosis results in genetic recombination of paternal and maternal genes when exchanges occur between non-sister chromatids. This causes a reshuffling of the genetic material that each haploid cell receives, promoting genetic variability among offspring. Exchanges between sister chromatids are without genetic significance because sister chromatids are identical.

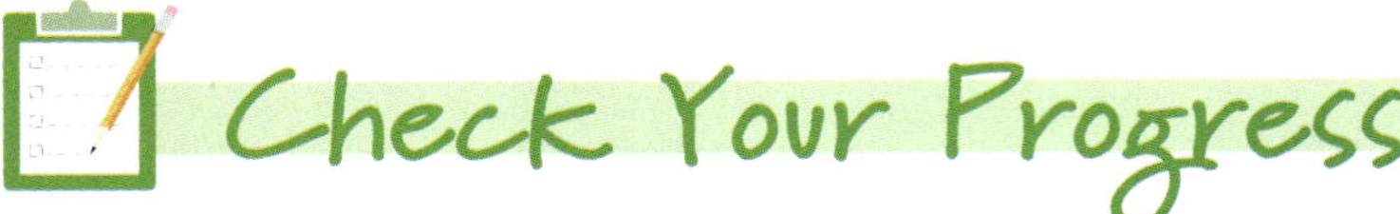

Check Your Progress

2.1 How does meiosis create genetic diversity for natural selection to operate?

2.2 What are the major differences between mitosis and meiosis?

1.

2.

3.

4.

2.3 Which type of crossing over is more significant genetically—crossing over between sister chromatids or crossing over between non-sister chromatids? Why?

2.4 Do tetrads form during mitosis?

Exercise 3-3 Gametogenesis

Materials Needed

- ❑ testis slide
- ❑ ovary slide
- ❑ compound microscope

Because sexual reproduction is characterized by the fusion of two haploid sex cells to form a diploid zygote, it follows that meiosis must be the method of cell division responsible for producing these haploid gametes. The process of gamete formation is known as gametogenesis. Male gametogenesis is called **spermatogenesis**, and female gametogenesis is referred to as **oogenesis.** Although the process of gametogenesis follows the same basic pathway in males and females, there are significant differences (Fig. 3.7). For instance, spermatogenesis is ongoing in men, whereas oogenesis is cyclic in women. To gain a clear understanding of these processes, we will examine gametogenesis in the two sexes separately, using mammalian tissues as our model.

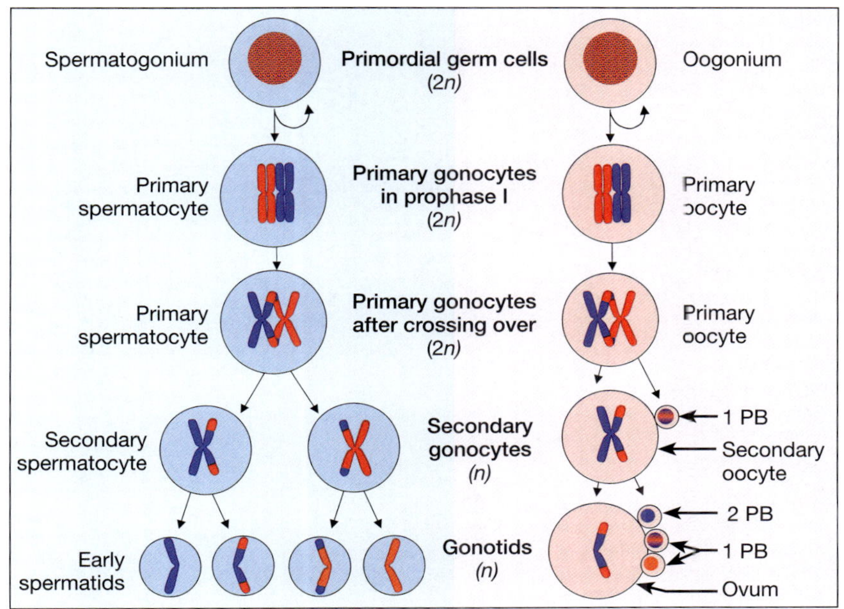

3.7 Comparison of spermatogenesis (blue) and oogenesis (pink). Notice that primordial germ cells (PGCs) divide by mitosis to form more PGCs and primary gonocytes (spermatocytes or oocytes). The primary gonocytes undergo meiosis to form four haploid cells. One spermatogonium yields four mature sperm, whereas one mature ovum and three nonviable polar bodies (PB) are generated from a single oogonium.

Spermatogenesis

Spermatogenesis, the meiotic production of sperm cells, occurs within the tiny, coiled seminiferous tubules of the testes. The process begins when diploid **spermatogonia** (singular = spermatogonium) divide mitotically to produce either more diploid spermatogonia (cells that later can divide to produce sperm or more spermatogonia) or diploid **primary spermatocytes** (Fig. 3.7). The primary spermatocytes undergo meiosis I to become haploid **secondary spermatocytes.** The second meiotic division transforms each secondary spermatocyte into two haploid **spermatids.** Spermatids differ from one another genetically because random segregation and assortment of the chromosomes during metaphase I of meiosis, combined with the process of crossing over, shuffles the parent genomes. Thus, any given spermatid usually contains genetic material from both the individual's mother and father—the particular combination being purely a matter of chance.

In the final phase of maturation the spermatids differentiate into **spermatozoa,** mature sperm cells, and are stored in the coiled tubules of each epididymis that cups around a testis (Fig. 3.8). From an original diploid spermatogonium, four haploid spermatozoa are produced. In human males, spermatogonia retain their ability to divide mitotically throughout life and continue to produce sperm precursors as well as more spermatogonia. Thus, sperm production occurs continuously from the time of sexual maturation until death, taking between 65 to74 days for mature sperm to complete the developmental transformation from spermatogonia. Around 12 million mature sperm are produced every day through this process and remain viable in the epididymis for only a few days before being reabsorbed by the body.

A single spermatozoan is only two-thousandths of an inch long, and a single ejaculation can contain 200–500 million of them. In the biological sense, sperm are cheap—they require little energy investment to produce and are in a virtually limitless supply throughout life. This represents a marked difference from the human egg as we will see in the next section.

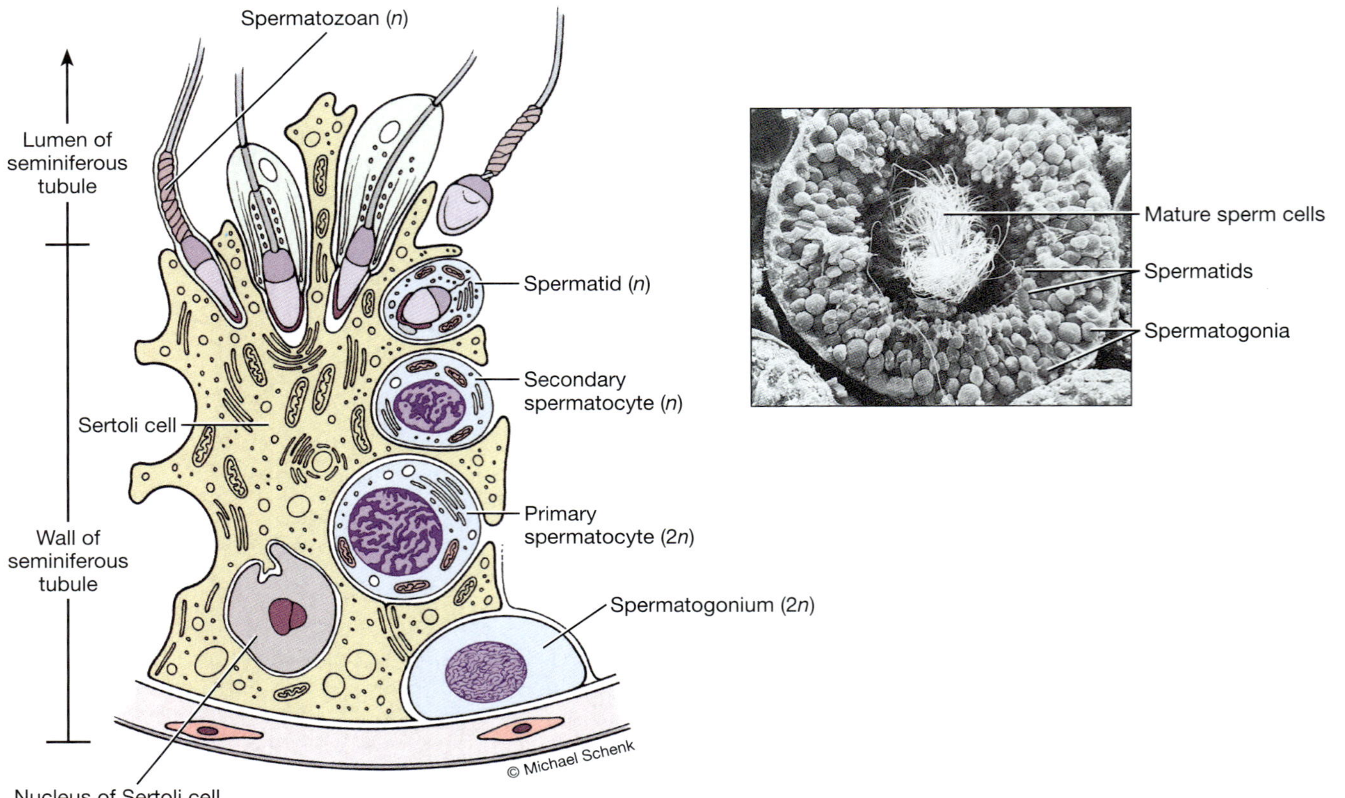

3.8 Section through mammalian seminiferous tubule showing spermatogenesis, with accompanying electron micrograph of a single seminiferous tubule in cross section.

Procedure

1 Observe a slide of a testis using high magnification, and locate several sections through the seminiferous tubules (Fig. 3.8).

2 Are primary spermatocytes and spermatogonia located near the periphery of the tubules or near the middle of the tubules?

3 Are maturing sperm located near the periphery of the tubules or near the middle of the tubules? Why do you suppose this is the case?

Oogenesis

The meiotic production of eggs (or ova) is known as oogenesis. Within the ovaries, diploid **oogonia** (singular = oogonium) divide mitotically to produce diploid **primary oocytes** (Fig. 3.9). These primary oocytes immediately enter prophase I of meiosis, at which point their development stops. Primary oocytes remain arrested in prophase I of meiosis until the appropriate hormonal cues cause them to resume division. In humans some of these primary oocytes will remain suspended in prophase I of meiosis for up to 50 years! Only after the primary oocyte receives appropriate hormonal signals from the body will it progress through the latter stages of meiosis I.

Contrary to what happens in spermatogenesis, four mature eggs are not produced from one primary oocyte. The first nuclear division during meiosis I occurs near the margin of the cell rather than at the cell's equator, and grossly unequal portions of cytoplasm are allocated to each of the resulting cells. The daughter cell receiving the bulk of this cytoplasm becomes a haploid **secondary oocyte**, and the one that receives virtually no cytoplasm forms the first **polar body** (Fig. 3.7). Quite often, this is the stage at which fertilization of the oocyte by the sperm occurs. In many organisms the second phase of meiosis will not be initiated unless fertilization does occur.

The second meiotic division, which occurs in the secondary oocyte, also involves an asymmetric allocation of cytoplasmic resources and results in a large haploid **ootid** (which later will become the ovum) and a second, nonfunctional polar body. Occasionally the first polar body also undergoes meiosis II and divides into two more nonfunctional polar bodies. Eventually the polar bodies disintegrate, leaving only a large ovum well stocked with enough nutrients and reserves to make it through the first few divisions after fertilization.

In human females the proliferation of oogonia is limited to a short period of time before birth. Thus, by the time a female is born, she contains all of the primary follicles she will ever have available to her. This finite number of egg precursors numbers around 400,000 at birth, but only a few hundred of these cells will ever mature into ova and be released into the oviducts. Couple this with the fact that each egg is an extremely large cell packed full of cytoplasm containing nutrients, mitochondria, and other organelles necessary for the early stages of embryonic development, and you can see that unlike sperm, eggs are not cheap (biologically speaking)! They exist in a limited supply and require large energetic investments to produce and release.

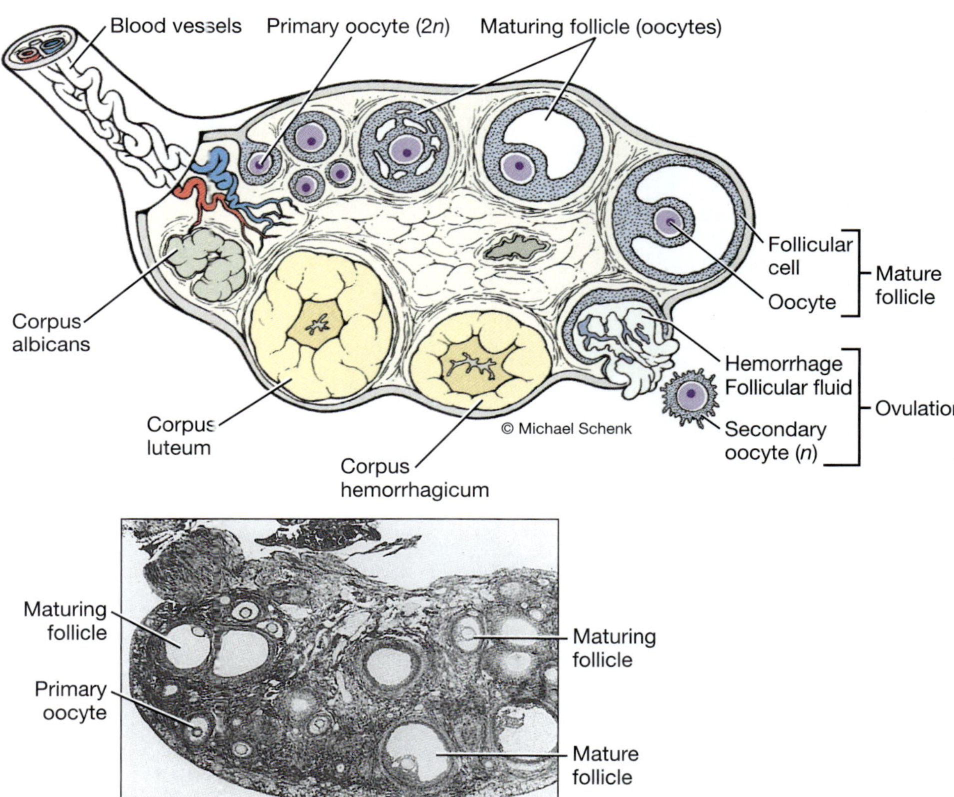

3.9 Illustration of mammalian ovary showing follicle development, ovulation, and formation of the corpus luteum.

Photomicrograph of mammalian ovary courtesy of Turtox, Inc.

Procedure

1 Observe a slide of an ovary, using low- or medium-power magnification.

2 Locate ova in different stages of development.

3 Notice that as ova mature and the follicle surrounding them enlarges, they migrate outward, toward the periphery of the ovary (contrary to the pathway of maturing sperm cells). At the point of ovulation, the ovum will erupt from the ovary and be swept into an oviduct by currents created by cilia lining each oviduct.

Check Your Progress

3.1 If a spermatogonium has 20 chromosomes, how many chromosomes will be present in a mature sperm cell from that spermatogonium?

3.2 How many sperm cells will one spermatogonium produce?

3.3 If an oogonium has 20 chromosomes, how many chromosomes will be present in an ovum from that oogonium?

3.4 How many ova will one oogonium produce?

3.5 Describe how the production of nonfunctional polar bodies is adaptive for the resulting ovum.

Name ______________________________ Section ______________

3 Questions for Review

1 In your study of mitosis, why was the blastula used?

---

2 Interphase sometimes has been referred to as a "resting" phase. Explain why this term *inaccurately* describes interphase.

---

3 The correct sequence of stages in mitosis is:

a. prophase, telophase, anaphase, interphase, metaphase.
b. interphase, prophase, metaphase, anaphase, telophase.
c. interphase, metaphase, prophase, telophase, anaphase.
d. prophase, telophase, metaphase, anaphase, interphase.
e. metaphase, prophase, interphase, anaphase, telophase.

4 A diploid nucleus at early mitotic prophase has ______ set(s) of double-stranded chromosomes; a diploid nucleus at mitotic telophase has ______ set(s) of single-stranded chromosomes.

a. 1; 1
b. 1; 2
c. 2; 1
d. 2; 2

5 During the S period of interphase:

a. cell growth takes place.
b. chromosomes divide.
c. replication of DNA occurs.
d. both a and c.
e. a, b, and c are all correct.

6 Match the stages of mitosis given below to their appropriate images.

a. prophase **b.** telophase **c.** anaphase **d.** metaphase

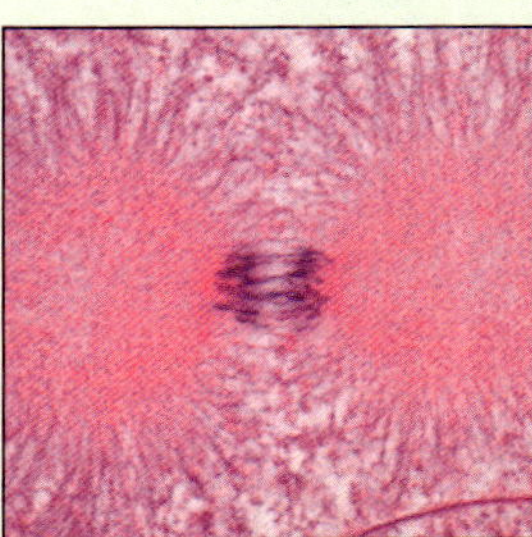
______

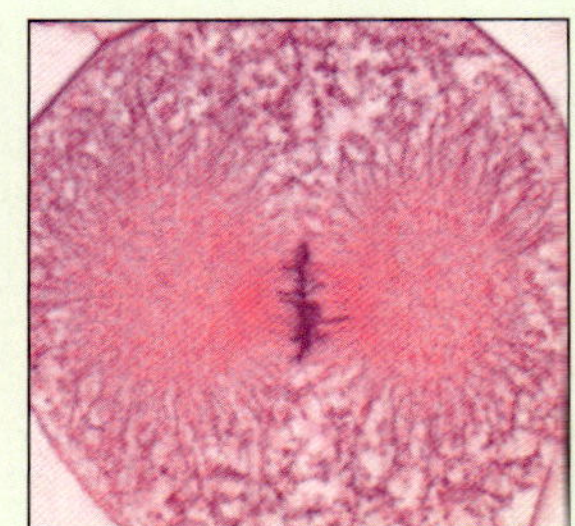
______

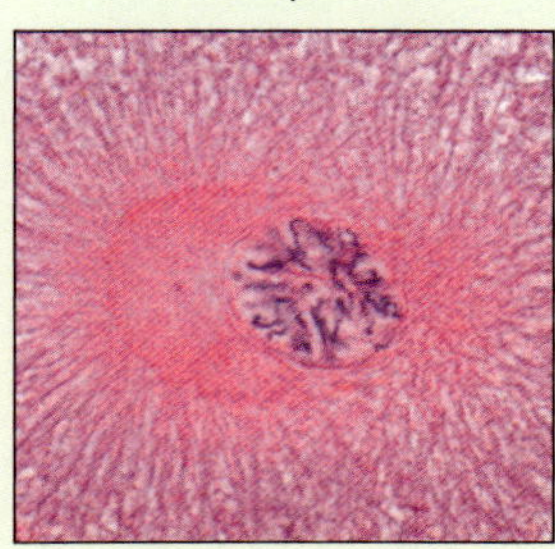
______

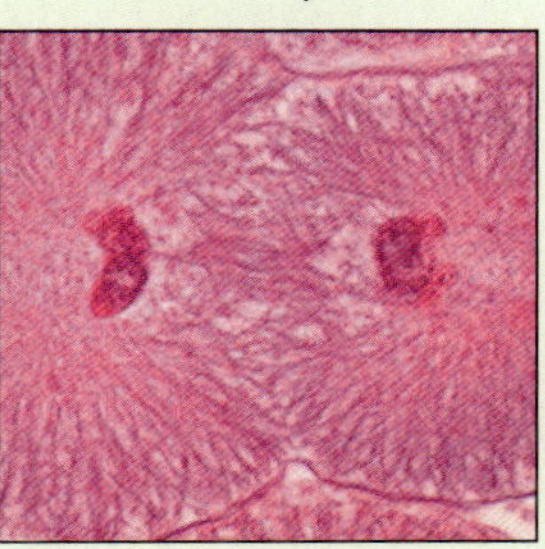
______

3

7 What would happen if a cell underwent mitosis but not cytokinesis?

8 Where in your body would you most likely find cells in active mitosis? Where would you find cells undergoing meiosis?

9 Why is the process of crossing over important to the survival and evolution of a sexually reproducing species?

10 Define spermatogenesis and oogenesis.

11 Which of the following statements concerning gametogenesis is *false*?

a. Gametogenesis results in the production of haploid cells.
b. Spermatogenesis results in the production of four spermatozoa.
c. Oogenesis results in the production of four ova.
d. Both mitosis and meiosis are involved in gametogenesis.

12 How old is a mature sperm cell in a 38-year-old man? How old is an ovulated oocyte of a 38-year-old woman? What reproductive consequences might this have?

CHAPTER 4

Animal Development

After completing the exercises in this chapter, you should be able to:

1 Name and describe the four main types of yolk distribution found in animal eggs.

2 Describe the basic patterns of cleavage observed in animal embryos.

3 Discuss the mechanisms involved in fertilization.

4 Understand the general sequence of embryonic development in animals.

5 Identify the three embryonic germ layers from microscopic preparations and discuss their fates.

6 Identify and describe the major stages of embryonic development in sea stars.

7 Identify and describe the major stages of embryonic development in frogs.

8 Compare and contrast embryonic development in sea star, sea urchin, and frog embryos.

During the process of sexual reproduction, sperm and egg fuse to form a new cell. Typically the nuclear material from the haploid sperm is donated to and combines with the haploid nuclear material of the egg to restore the diploid condition in the fertilized egg, or zygote. From this point, the single cell develops into a magnificently complex, multicellular organism capable of carrying out the daily functions required for life on its own.

Three major processes govern the basic developmental sequence of animals:

1. **cell division**, which creates new cells through mitosis;
2. **cell migration**, which causes groups of cells to organize and move, to create an animal's body shape; and
3. **cell differentiation**, which causes cells to develop different morphological features for specialization toward specific tasks.

We will examine each of these three processes in detail using two different animal models, an echinoderm (sea star) and a vertebrate (frog). First we must discuss some of the basic concepts that apply to all developmental pathways.

Fertilization

In human males, up to 200–500 million **sperm** are released in a single ejaculation. Not all sperm come out alive or swim straight, but if only half do, this usually is enough for successful fertilization. Sperm swim in and are nourished and protected by the more alkaline **semen**, which partially counteracts the natural acidity of the vagina. The sperm must swim vigorously to survive this first leg of their journey, because their maximum life span in the vagina is only a few hours. Although only one sperm cell will fertilize an egg, one cell cannot make it to the egg on its own. By swimming together, at a painstaking rate of about 8 inches an hour, the sperm slowly move the semen toward the egg. They swim from the vagina, through the cervix, to the uterus, and finally to the fallopian tubes—their direction of travel governed by powerful chemical hormones released by the egg. Sperm that swim slowly or are not strong enough to make it to the fallopian tubes perish along the way, and because only one fallopian tube contains the egg, many sperm swim futilely up the wrong tube.

Once inside the fallopian tubes, the conditions for survival become more favorable, and the remaining sperm cells' life span is increased to about five days. This means that sperm can be in the fallopian tube for up to five days in advance of ovulation and still fertilize the egg. By the end of their journey, only about 200 sperm will finally reach the egg.

When the remaining sperm cells reach the egg, the act of **fertilization** involves several complex steps that occur in a precise sequence to ensure that only one sperm and egg fuse to form the zygote. **Polyspermy**, the entrance of more than one sperm, would be detrimental to normal development, and animals have universal mechanisms to prevent this occurrence. Almost instantaneously, as a sperm cell bonds to the surface of the egg, receptor proteins on the plasma membrane of the egg recognize the sperm cell (if it is of the same species as the egg) and erect a **fertilization cone** to draw the head of the sperm inside the egg's plasma membrane. Within seconds, the electrical potential of the egg's plasma membrane changes polarity to block any other sperm from penetrating. This mechanism is called the **fast block** to polyspermy.

Shortly thereafter, cortical granules just beneath the egg's plasma membrane release their contents into the membrane, causing a **fertilization membrane** to form around the entire surface of the egg, creating a more permanent boundary to polyspermy. Fusion of the nuclear material of the sperm and egg follows and the fertilized egg, now referred to as a **zygote**, becomes activated to undergo cleavage.

Influence of Yolk on Cleavage Patterns

All fertilized eggs contain an embryo and a food reserve, called yolk, to nourish the developing embryo. Despite its obvious benefit, yolk actually serves as an impediment to the dividing cells of the embryo; therefore, the manner in which a fertilized egg divides depends greatly on the amount and distribution of yolk within the egg. In general, the **cleavage furrows** of cell division can more easily penetrate regions of the egg with very little, evenly distributed yolk. Thus, in eggs with large amounts of heavily concentrated yolk, cell division often is confined to the small portion of the egg containing the embryo.

Because of this influence of yolk, eggs are classified into four main types based on the amount and distribution of their yolk (Fig. 4.1).

1. **Isolecithal eggs** have very little yolk evenly distributed throughout the egg. Humans and sea stars both possess isolecithal eggs. Eggs of this type generally undergo **holoblastic cleavage**, meaning that the

cleavage furrow extends completely through the egg and completely separates daughter cells.

2. **Mesolecithal eggs** have a moderate amount of yolk concentrated at one end of the egg, referred to as the vegetal hemisphere, or pole. Amphibians possess mesolecithal eggs. Even though they have more yolk than isolecithal eggs, mesolecithal eggs generally undergo holoblastic cleavage, though not as symmetrically.
3. **Telolecithal eggs** have a large amount of yolk, which occupies most of the cell volume. The embryo is confined to a small disc at the opposite pole of the cell, the animal pole. Birds and reptiles possess telolecithal eggs. Because of the large, heavily concentrated yolk, cleavage is only partial, stopping at the border between the embryo and the yolk, and daughter cells remain connected to each other. This type of cleavage is referred to as **meroblastic cleavage,** or incomplete cleavage.
4. **Centrolecithal eggs** also have a large amount of yolk, but the embryo forms a thin layer around the outside of the large, central yolk mass. Most arthropods have centrolecithal eggs.

The pattern of cleavage is also dictated by the plane of division within the developing embryo (Fig. 4.1). When cleavage planes occur perpendicular or parallel to each other, it is referred to as **radial cleavage.** When cleavage planes occur at oblique angles, it is referred to as **spiral cleavage.** When cleavage planes are restricted to a dome of cytoplasm at one end of the yolk-laden cell, it is referred to as **discoidal cleavage.** Because of the unique distribution of yolk in centrolecithal eggs, cleavage occurs only in the cell's periphery. This pattern is known as **superficial cleavage.**

In the exercises that follow you will examine the embryonic development of two different animals, a sea star and a frog. Echinoderms and vertebrates are thought to have a common, recent evolutionary ancestor because they share a similar developmental pathway, and because complex developmental pathways tend to be highly conserved during the course of evolution. The sea star and frog are both **deuterostomes,** meaning that the mouth develops from the second embryonic opening. Their developmental processes do vary somewhat, however, which makes these two specimens useful models for studying particular aspects of animal development.

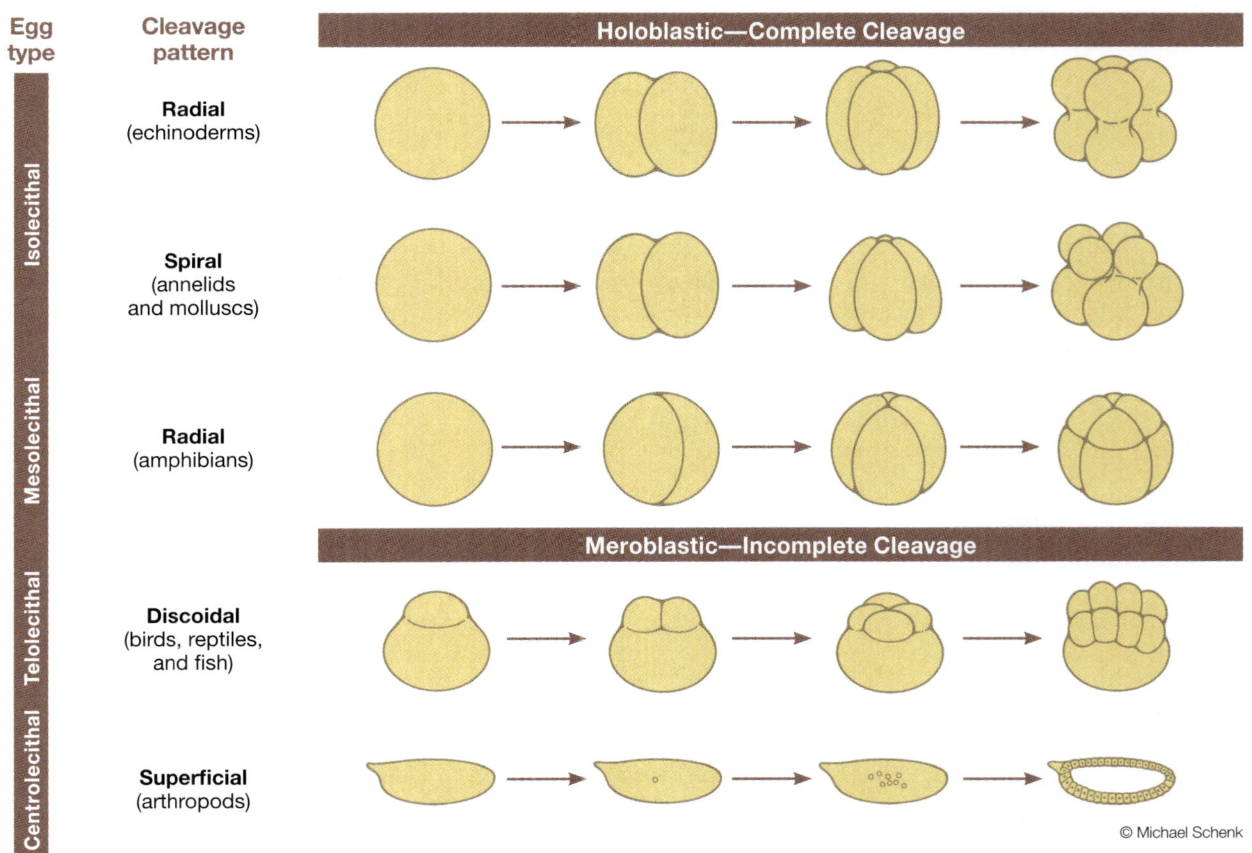

4.1 Cleavage patterns in animals depend greatly on the amount and distribution of yolk within the unfertilized egg.

Exercise 4-1 Sea Star Embryonic Development

Materials Needed

- ❑ prepared slide of sea star developmental stages
- ❑ compound microscope

Sea stars of the genus *Asterias* are familiar marine invertebrates we commonly refer to as "starfish" (though they are not fish), and are members of the phylum Echinodermata (Chapter 14).

4

Procedure

1 Obtain a prepared slide of *Asterias* containing a mixture of sea star developmental stages. As you scan the slide using low power, you will notice that your slide contains dozens of embryos in different stages of development.

2 Identify each of the stages discussed below using medium or high magnification.

Unfertilized Egg

In this stage, the egg appears as a large, nearly spherical cell with a distinct nucleus and nucleolus (Fig. 4.2A). Tiny granules are visible within the cytoplasm of the cell. Notice how these granules are uniformly distributed throughout the cytoplasm. These are particles of yolk for the developing embryo. Remember from our previous discussion of yolk distribution and cleavage patterns that sea stars possess isolecithal eggs.

Fertilized Egg

After the sperm penetrates the egg, a thin fertilization membrane is erected by the egg, ensphering it to prevent multiple fertilizations by the many nearby sperm (polyspermy). A distinct nucleus is not visible while the nuclear material from sperm and egg fuse to restore the zygote to the diploid chromosome number (Fig. 4.2B).

Two-Cell through Eight-Cell Stages

Shortly after fertilization, the zygote begins to divide through mitosis. Notice that the pattern of **cleavage** (cell division) completely separates the cells into roughly equal-sized, but distinct, cells (Fig. 4.2C–E). This pattern of development, known as *radial holoblastic cleavage*, is common in isolecithal eggs with a small amount of evenly distributed yolk. Notice also that the fertilization membrane remains intact through these early cleavages.

At this early stage of development, it is still possible for nearby sperm to penetrate these cells consequently upsetting the genetic balance of the zygote. The persistence of the fertilization membrane prevents this from occurring. Finally, notice that as these cells divide, they do not get bigger. The existing cytoplasm is partitioned equally among the new cells with very little growth during these early stages of embryonic development.

Morula (16–32 Cell Stages)

At the **morula** stage the embryo is represented by a solid mass of tiny cells, called **blastomeres**, that continue to divide at a furious pace. Still, the overall size of the embryo has not increased; rather, the size of the individual blastomeres has decreased to accommodate the fixed amount of cytoplasm being divided among cells. The fertilization membrane is still usually visible throughout this stage (Fig. 4.2F–G).

Blastula (64+ Cell Stages)

The **blastula** stage is characterized by a migration of cells toward the periphery of the embryonic sphere, creating a hollow cavity deep within the spheroid of cells (Fig. 4.2H). During this stage the blastula breaks free from the envelopment of the fertilization membrane and is able to swim about freely using ciliary movements of blastomeres on the embryo's surface. The hollow, fluid-filled cavity within the blastula is referred to as the **blastocoel** (Fig. 4.3A).

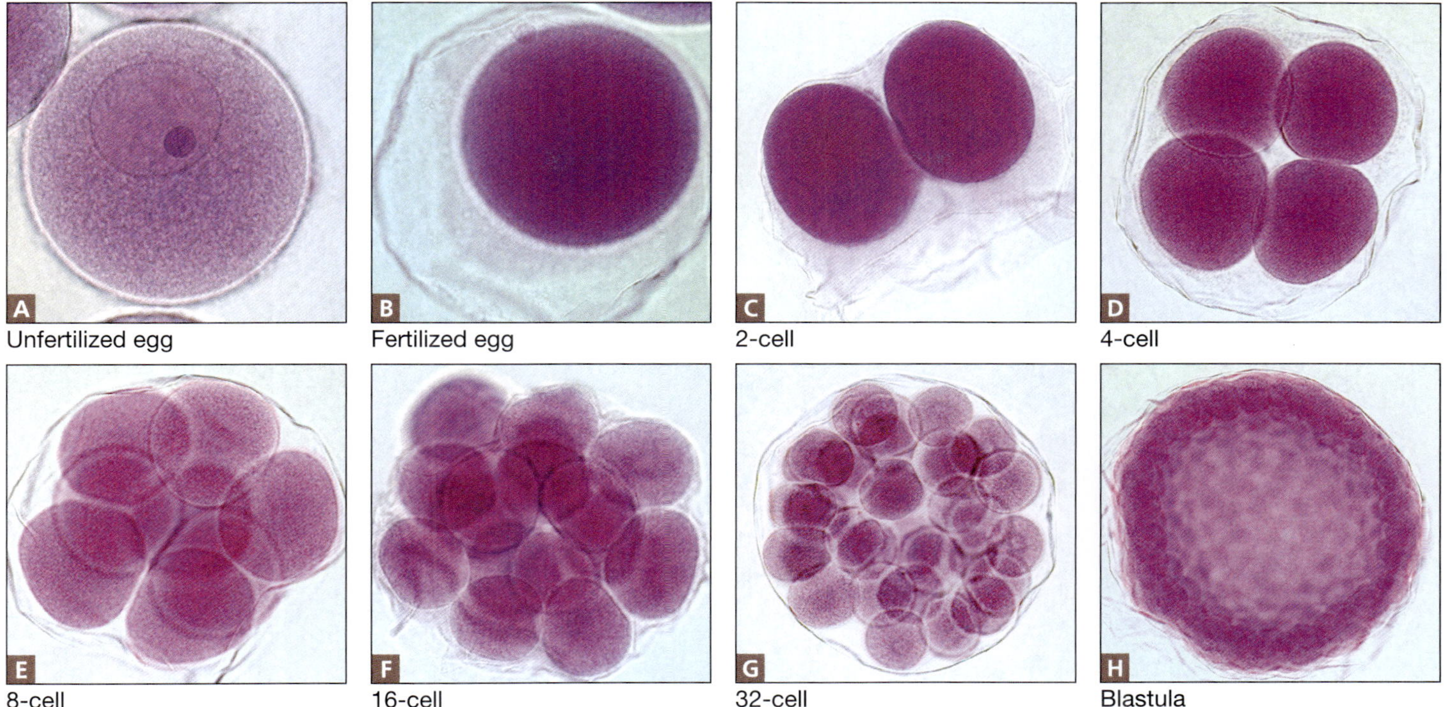

4.2 Photo plate of early embryonic development in the sea star: (**A**) unfertilized egg, (**B**) fertilized egg, (**C**) 2-cell stage, (**D**) 4-cell stage, (**E**) 8-cell stage, (**F**) 16-cell stage, (**G**) 32-cell stage, (**H**) blastula.

Early Gastrula

Several hours after the formation of the blastula, a small depression begins to appear at one end of the embryo, and cells begin to organize, move, and start to create the geometry of the animal's body (Fig. 4.3B). Remember that cell migration is one of the three major factors governing embryonic development. This invagination and migration of cells toward the center of the blastocoel marks the onset of **gastrulation.**

As gastrulation proceeds, more and more cells stream inward, deepening the invagination. The opening to the outside of the embryo, marking the site of the inward migration of cells, is called the **blastopore** (Fig. 4.3C). In sea stars and other deuterostomes this opening will become the anus. The hollow tube that is created by the arrangement of invaginated cells is called the **gastrocoel**, or **archenteron** (meaning primitive gut).

Late Gastrula

By now, the individual cells are so small as to be indistinguishable from each other under normal magnification. At most, they appear as a flecking on the surface of the embryo. The archenteron has folded inward to the point of nearly connecting with the opposite end of the gastrula, and the shape of the gastrula has become markedly elongated (Fig. 4.3D). No longer reminiscent of a spherical ball of cells, the gastrula continues to develop toward the larval stage. During late gastrulation **cell differentiation** begins, and the primary tissue layers of the embryo start to form. The outer layer begins to develop into **ectoderm**, while the inner layer differentiates into **endoderm.** The third embryonic layer, **mesoderm**, will develop later between these two existing layers from cells that disassociate from the endodermal layer (Fig. 4.4).

Bipinnaria Larva

By the larval stage, **morphogenesis** (the development of body shape) and **organogenesis** (the differentiation of organ tissues) have begun. The larval sea star has a complete digestive tract, including an anus and a functional mouth lined with cilia for sweeping in organic particles (Fig. 4.3E–G). Soon after this stage, the young sea star will undergo another developmental transformation as it grows arms and assumes the familiar shape of a starfish (Fig. 4.3H).

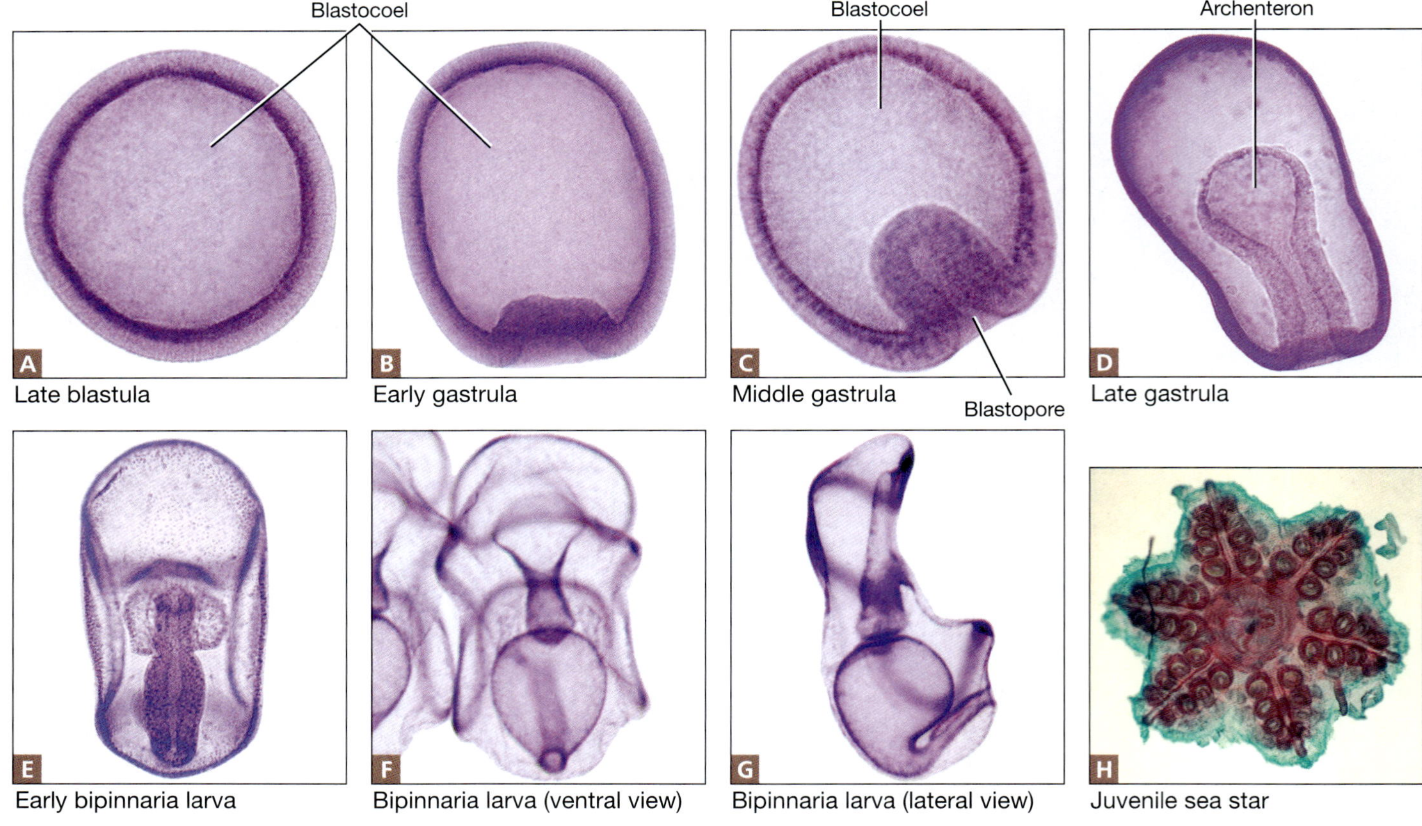

4.3 Photo plate of gastrulation in the sea star: (**A**) late blastula, (**B**) early gastrula, (**C**) middle gastrula, (**D**) late gastrula, (**E**) early bipinnaria larva, (**F**) bipinnaria larva, ventral view, (**G**) bipinnaria larva, lateral view, (**H**) juvenile sea star.

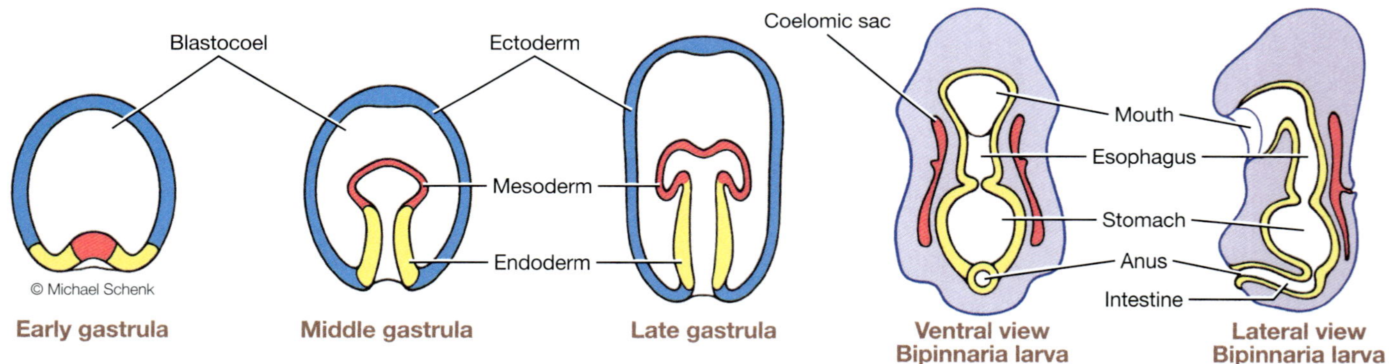

4.4 Gastrulation in the sea star: ectoderm is blue, mesoderm is red, and endoderm is yellow.

Exercise 4-2 Frog Embryonic Development

Materials Needed

- ❏ prepared slide series of amphibian developmental stages
- ❏ compound microscope
- ❏ dissecting microscope

Frogs belong to the vertebrate class Amphibia within the phylum Chordata (Chapter 19). Females of many frog species produce relatively large eggs (around 2 mm in diameter), and fertilization and embryonic development occur externally, making the frog another useful model for studying developmental pathways.

4

Procedure

1 Obtain a prepared slide series of frog embryonic development.

2 Using either the compound or dissecting microscope, depending on the thickness of the slide, identify each of the stages discussed below.

3 Additional models of these developmental stages or plastic-embedded specimens may be available for study.

Unfertilized Egg

Frog eggs are mesolecithal, meaning that they contain a moderate amount of yolk. This yolk is segregated toward one pole of the egg known as the **vegetal pole**, which appears lighter in coloration. The darker pole, or **animal pole**, represents the portion of the egg where the embryo will develop (Fig. 4.5). Frog eggs are also enclosed in a gelatinous covering to protect them from bacteria, fungi, UV radiation, desiccation, and to deter predation. The entire course of development takes place with the embryo fully encased in this coating of jelly. Not until the embryo reaches the tadpole stage does it break free of its protective barrier and swim away.

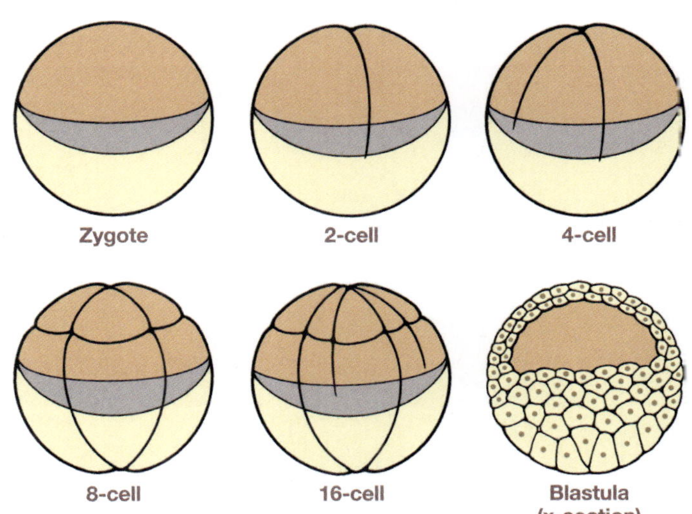

4.5 Cleavage stages of the frog depicting the brownish animal pole; the yellow, yolk-laden vegetal pole; and the gray crescent.

Fertilized Egg

When fertilization occurs, the fertilization membrane lifts away from the surface of the egg and the egg rotates so that the heavier, yolk-filled vegetal pole is downward. Later, a **gray crescent** appears along the margin of the animal-vegetal axis on the opposite side of the egg from the entry point of the sperm (Fig. 4.6A–B). In many cases this feat of counter-alignment involves rotation of these pigmented zones, since the sperm may penetrate the egg at any place along its surface.

Early Cleavage

Cleavage in the frog embryo is radial and holoblastic, similar to that in the sea star. The first two cleavage planes are perpendicular to one another, producing four cells of roughly equal size (Fig. 4.6C–D). The third cleavage occurs parallel to, but slightly above, the equator that separates the animal and vegetal poles (Fig. 4.6E). From this point forward, the division of cells will follow this unequal pattern with cells in the animal pole dividing at a faster rate than cells in the vegetal pole, which are laden with yolk.

Late Cleavage

By the later stages of cleavage, the appearance of the two poles differs markedly. The animal pole is composed of considerably smaller, more numerous cells while the vegetal pole contains relatively few, large cells containing yolk granules (Fig. 4.6F–H).

Blastula

4

With the formation of the blastocoel, the blastula stage is achieved. Notice that the blastocoel in the frog embryo is not centrally located as it is in the sea star. It is offset toward the animal pole of the embryo because of the presence of the large yolk cells of the vegetal pole (Fig. 4.6I–J). Also remember that the blastula is not much larger than the fertilized egg because the cells become smaller with each successive division as their cytoplasm is partitioned into more and more cells.

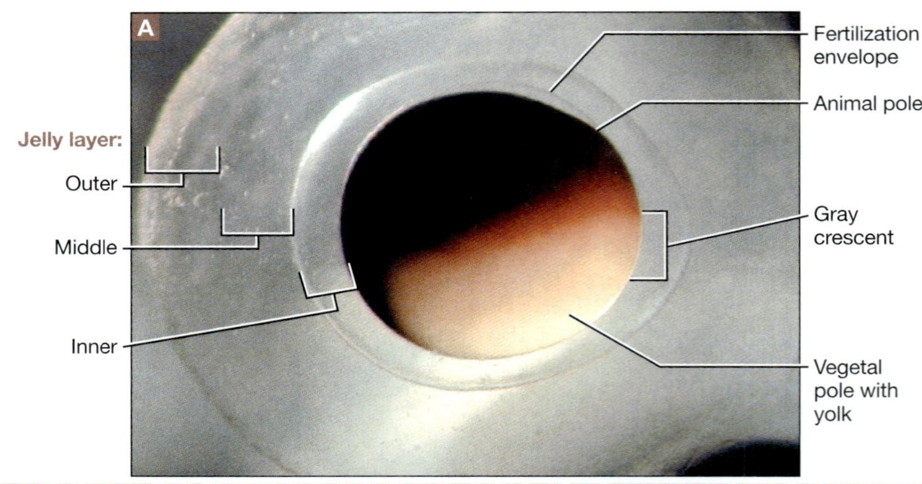

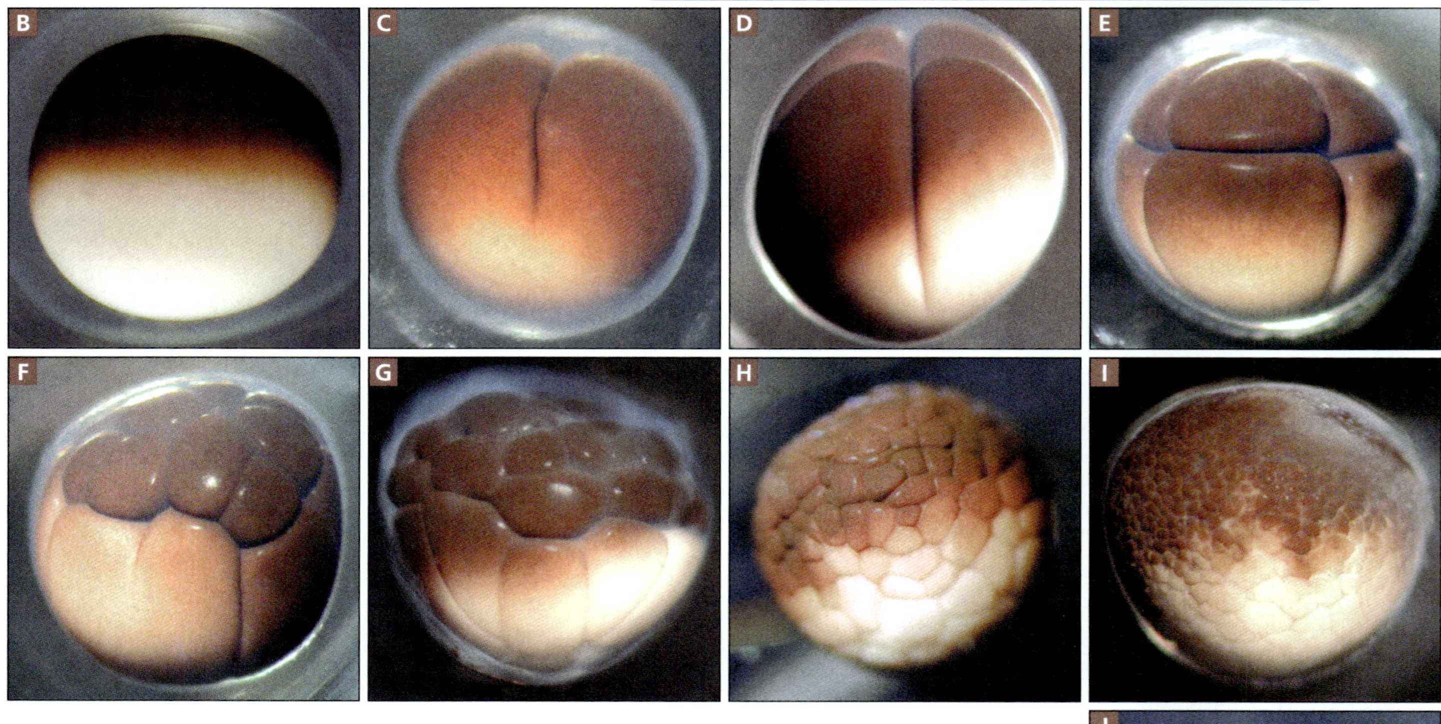

4.6 Cleavage in the leopard frog, *Rana pipiens*: (**A** and **B**) fertilized egg, (**C**) 2-cell stage, (**D**) 4-cell stage, (**E**) 8-cell stage, (**F**) 16-cell stage, (**G**) 32-cell stage, (**H**) 64-cell stage, (**I**) early blastula, and (**J**) late blastula.

Gastrula

As in the sea star, the onset of gastrulation is characterized by a migration of surface cells inward toward the blastocoel. In the frog, this forms a crescent-shaped line along the surface of the blastula known as the **dorsal lip**, which marks the opening of the blastopore (Fig. 4.7A). As the surface cells migrate inward, a slight depression forms on the surface of the blastula. This depression gradually enlarges to form the archenteron (gastrocoel) within the gastrula (Fig. 4.7B). Because vegetal pole cells are so large, blastomeres do not invaginate as they do in the sea star; rather, they grow down over the larger yolk cells, enveloping them as development progresses.

By the late gastrula stage the vegetal pole cells have become almost completely enveloped by migrating surface cells from the animal pole and only a small, circular **yolk plug** remains visible at the opening to the blastopore (Fig. 4.7C). By this stage, the three primary **germ layers** have begun to differentiate. The outer surface of the gastrula is covered by ectoderm, the archenteron (gastrocoel) is lined with endoderm, and the thin layer of cells sandwiched between the ectoderm and endoderm constitutes mesoderm (Fig. 4.8).

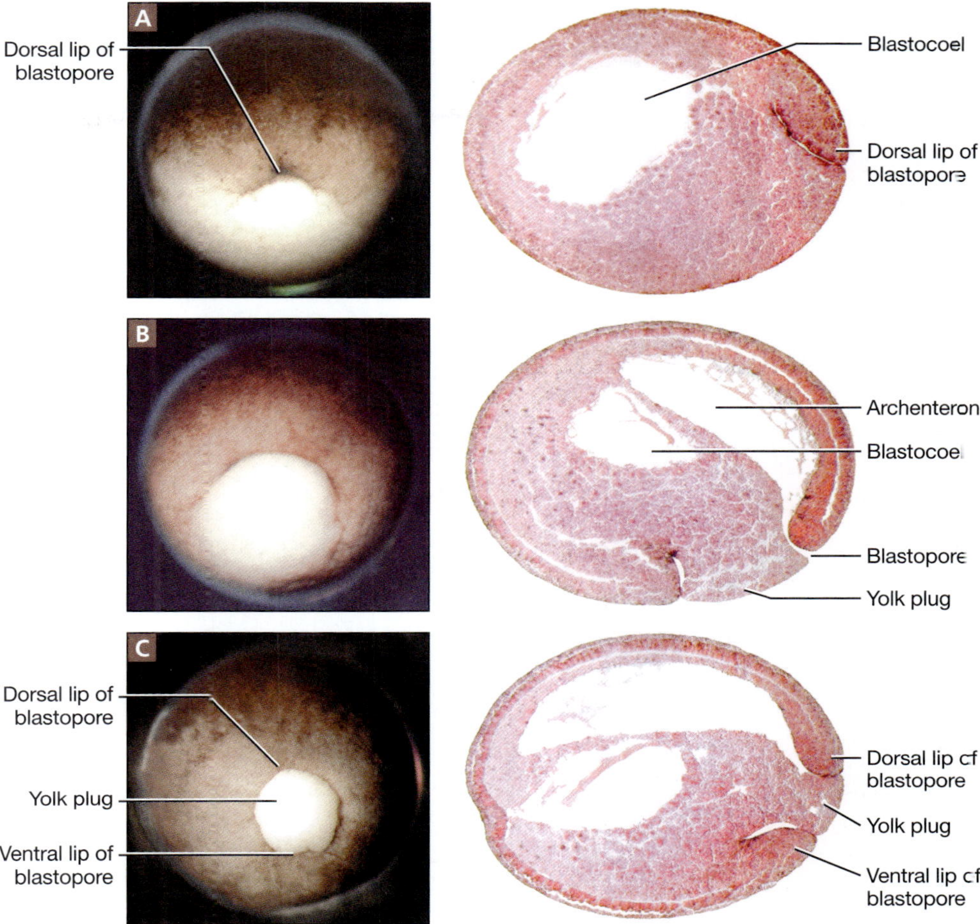

4.7 Gastrulation in the leopard frog, *Rana pipiens*: (**A**, **B**, and **C**) external views and sagittal sections show inward migration of cells along the dorsal lip of the blastopore, and increasing development of the archenteron, followed by yolk plug formation.

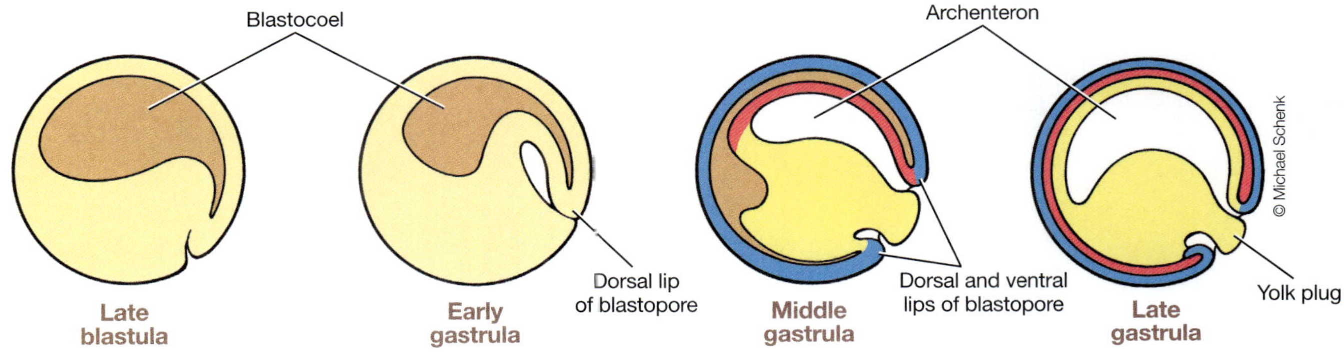

4.8 Gastrulation in the frog: ectoderm is blue, mesoderm is red, and endoderm is yellow.

Neurula

Several hours after gastrulation, the first visible elements of the nervous system begin to appear as a result of ectodermal cells along the mid-dorsal region of the embryo thickening to form two enlarged ridges on the surface, the **neural folds**, which border a depression, the **neural groove** (Fig. 4.9). This marks the onset of **neurulation**. Eventually the neural folds will meet and fuse together, forming the enclosed **neural tube** that will develop into the brain and spinal cord. Beneath the developing neural tube a long, cylindrical section of mesodermal cells is differentiating to form the **notochord**, which later will develop into vertebrae (Fig. 4.10).

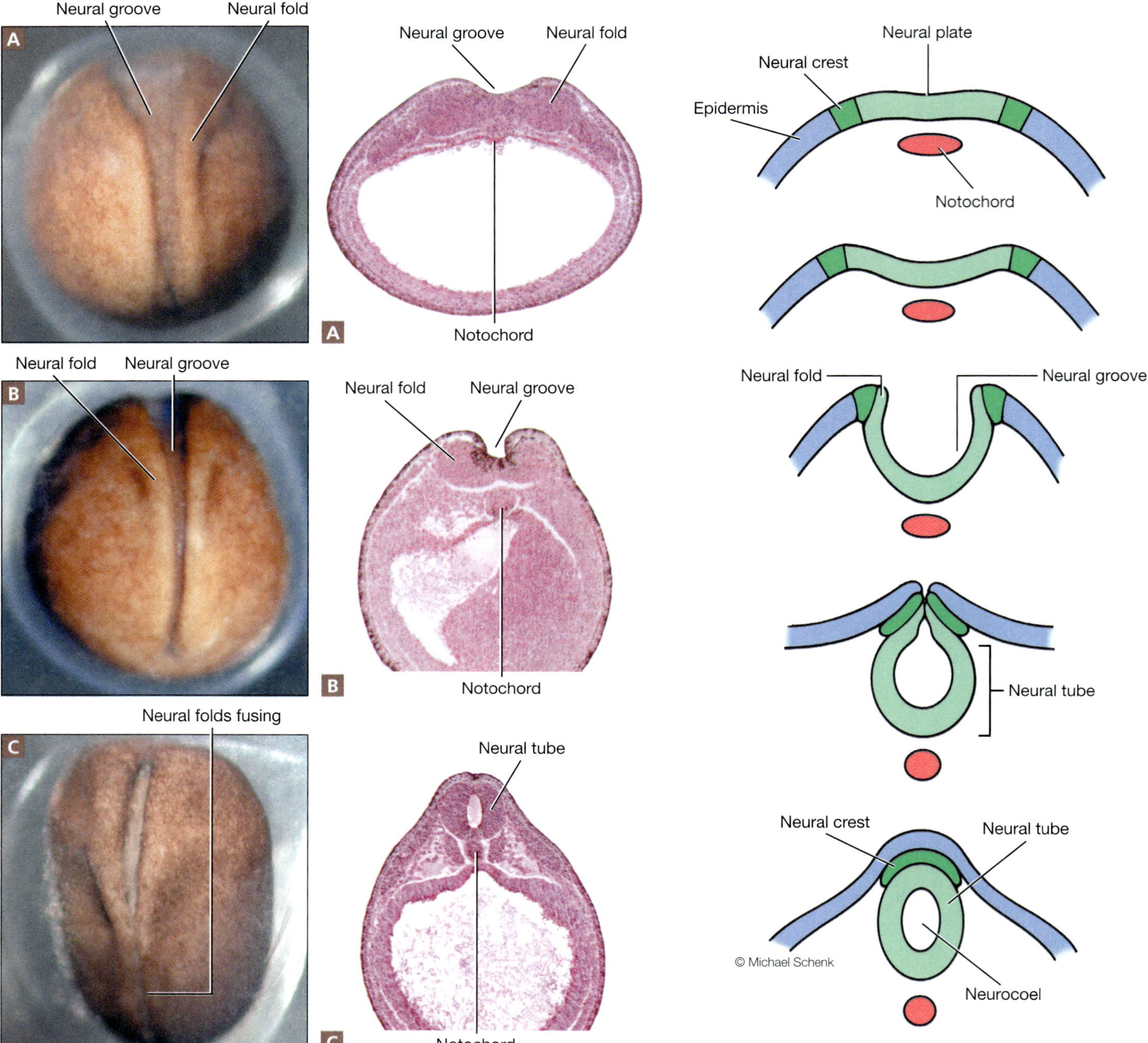

4.9 Neurulation in the leopard frog, *Rana pipiens*: (**A**, **B**, and **C**) external views and transverse sections depict formation and eventual closure of the neural groove along the dorsal surface of the embryo, and further differentiation of internal germ layers.

4.10 Development of spinal cord in the leopard frog, *Rana pipiens*.

Larva

Several days after fertilization the embryo begins to elongate and takes on the more familiar tadpole shape (morphogenesis). The posterior end develops a pronounced **tail**, and **gills** and a **mouth** develop at the anterior end. What is left of the **yolk mass** remains as a slight bulge in the abdomen of the developing tadpole (Fig. 4.11). Digestive, respiratory, muscular, and sensory systems begin to develop (organogenesis) as cell differentiation continues, and the tadpole begins to wiggle within its jelly coat. Once the tadpole emerges from its gelatinous enclosure, the yolk reserves will be depleted and it must find food for itself for the next several weeks (or months, depending on the species) until it has stored enough energy to metamorphose into a young frog.

4

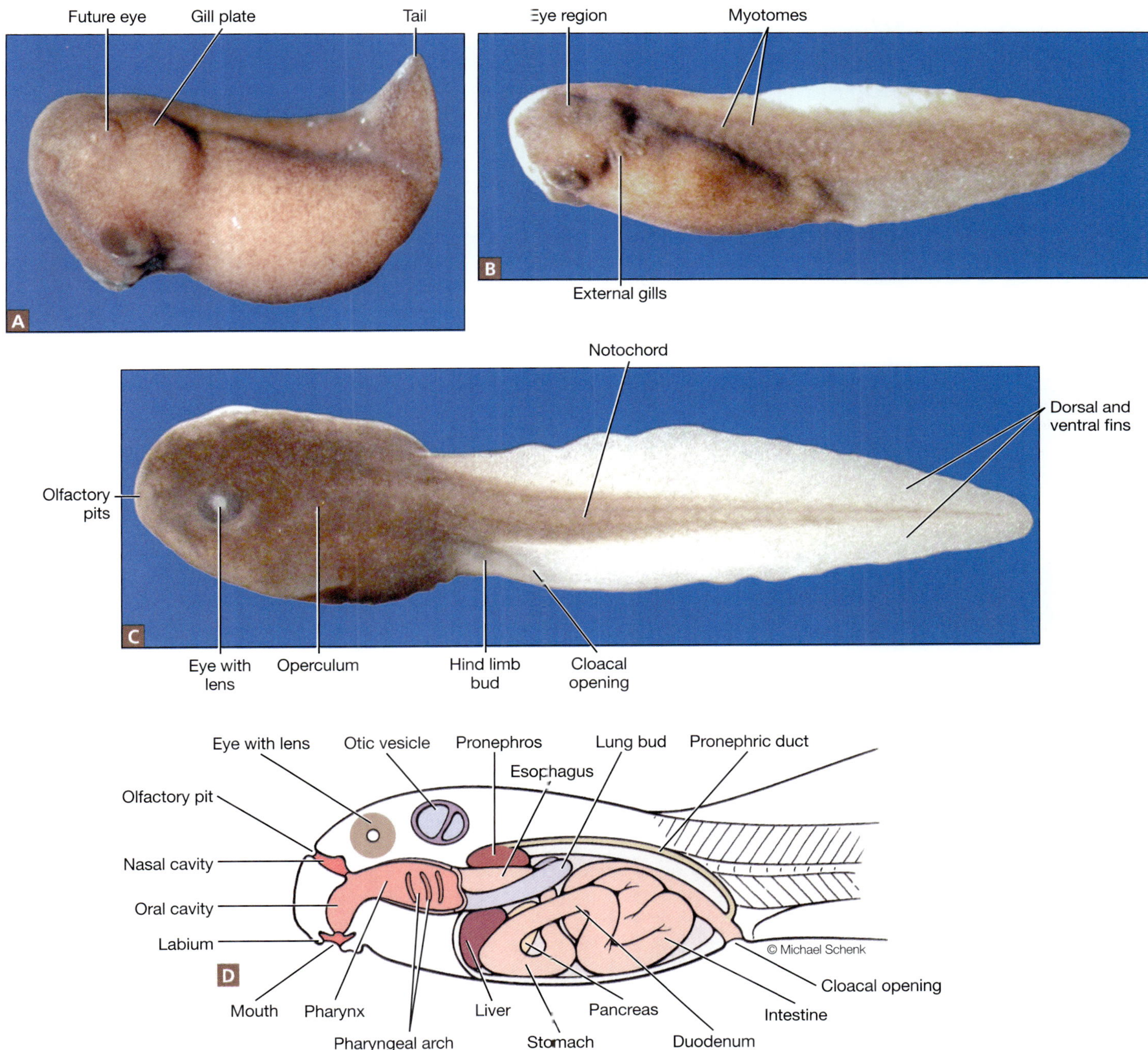

4.11 Organogenesis in the leopard frog, *Rana pipiens*: (**A**) At 4 mm, the frog embryo begins to contract muscles and wiggle around in its jelly coat. (**B**) By 7–8 mm, the major sensory organs have developed and the mouth can open. (**C**) Tadpole at the 11 mm stage is fully developed and can swim and feed. (**D**) All major organs have developed and are functional by this stage.

Check Your Progress

4

1.1 When cleavage planes occur perpendicular or parallel to each other, cleavage is characterized as ________. When cleavage planes are restricted to a dome of cytoplasm at one end of the cell, cleavage is characterized as ________.

1.2 Which is larger, a sea star zygote or a frog zygote?

1.3 How does the pattern of cleavage differ in sea stars and frogs?

1.4 How does the formation of the blastocoel differ in sea stars and frogs?

1.5 How does the process of gastrulation differ in sea stars and frogs?

Exercise 4-3 In Vitro Fertilization with Sea Urchins

Materials Needed

- ❑ live sea urchins
- ❑ 35 ml petri dishes (sterile)
- ❑ 50 ml and 100 ml beakers (sterile)
- ❑ sterile sea water
- ❑ 4% potassium chloride solution (0.5 M)
- ❑ 5 ml syringes with ½-inch, 25-gauge needles
- ❑ sterile pipettes
- ❑ toothpicks
- ❑ glass depression slides
- ❑ coverslips
- ❑ compound microscope

Much like sea stars, sea urchins have been extremely useful organisms for embryologists in studying the process and mechanisms of development. Earlier you viewed prepared slides of developing sea stars and frogs. In this exercise you actually will perform in vitro fertilization (that is, outside the body), using sea urchin eggs and sperm to create zygotes whose development you will track for the next several days.

Fertilization is a complex set of steps involving: (1) contact between sperm and egg, (2) entry of the sperm into the egg, (3) fusion of the egg and sperm pronuclei, which restores the diploid condition and results in the formation of a zygote, and (4) activation of developmental mechanisms that initiate successive divisions of the new cell (by means of mitosis) to form the multicellular embryo.

This process occurs naturally with great ease and miraculously with clocklike precision millions of times a day throughout the animal world. Unfortunately, conducting this exercise in the unnatural conditions of the laboratory requires painstaking attention to detail and protocol. To achieve successful fertilizations that can be monitored for several days, you will have to maintain an aseptic environment that substitutes for the natural conditions in which these animals would normally develop.

Precautions for Successful Development

1. *Cleanliness.* Use only dishes, pipettes, beakers, seawater, etc., that are sterile (either new or recently autoclaved). Do not cross-contaminate containers by using a pipette for more than one container. Label your containers and pipettes to avoid confusion!
2. *Avoid overinsemination.* Under laboratory conditions, very few sperm are necessary for successful fertilization. Use a *dry* toothpick to introduce sperm. If the egg suspension becomes noticeably milky after adding sperm, you have added too many.
3. *Avoid overcrowding.* Eggs spawned under natural conditions never have to face the overcrowded conditions of a beaker or petri dish. Minimize this by placing very few eggs in a single culture. Anything more than a thin layer on the bottom is too much.
4. *Keep temperatures constant.* Remember—marine animals are accustomed to environments with stable temperatures. Pay strict attention to the temperature requirements for the species you use! Moreover, slides heat up quickly under the intense light on the microscope stage, and culture solutions evaporate quickly, so do not expect to follow development for a considerable length of time on a single slide. You will have to create "stock" solutions of embryos from which you can pipette out small aliquots for successive viewing.

4

Obtaining Gametes

Note

Check with your instructor to determine whether you will perform the initial injections or your instructor will perform these steps for you.

Procedure

1 Place a live sea urchin *oral side up* in a shallow dish of seawater.

2 Inject approximately 2 ml of 4% potassium chloride (KCl) into the *oral* side of the urchin in the soft, fleshy tissue adjacent to the mouth. This chemical induces the release of gametes through the genital pores on the aboral side of the animal. It may take a few minutes for gametes to start "flowing."

3 Place the sea urchin *oral side up* over a dry petri dish until it begins to shed gametes through the pores on its aboral side into the dish (Fig. 4.12). Occasionally a second injection of KCl is necessary to induce gamete release.

4 If an urchin has not begun to shed gametes after 10 minutes, inject another 1–2 ml of KCl.

After injecting the urchins, the first task is to identify the gametes. The sex of the sea urchin can be determined by the color, texture, and size of the gametes. The color will vary depending on the species, so be sure you know what species you are using! In general, eggs are much larger and will precipitate to the bottom of a beaker quickly, whereas the tiny sperm cells will form a homogenous suspension in water.

5 *Male urchins*: Leave oral side up over a *dry*, sterile petri dish and allow the sperm to collect on the bottom of the dish (Fig. 4.12). Cover this mixture when done, to maintain sterility. Sperm may be dispensed later using *dry* toothpicks.

6 *Female urchins*: Suspend them over the opening of a beaker (oral side up) so their genital pores are in contact with the sterile seawater in the beaker (Fig. 4.12). This will permit the eggs to drift to the bottom.

7 For an added measure of security after the eggs accumulate on the bottom, you may carefully decant the seawater, leaving the eggs in the bottom of the beaker, and replace the discarded water with new, sterile seawater. This step is particularly important if you wish to save the eggs.

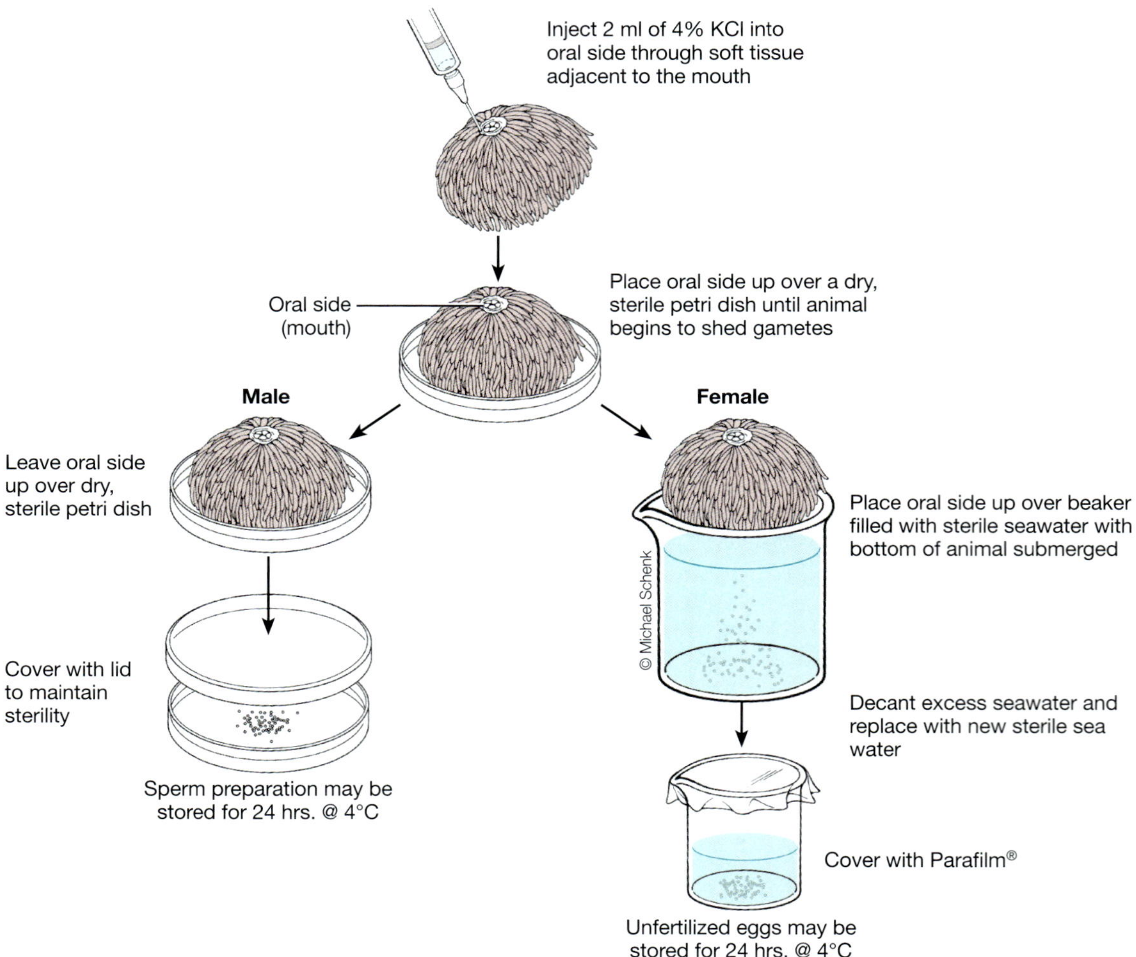

4.12 Experimental setup and procedure for obtaining gametes from live sea urchins.

Fertilization

Procedure

1 Use a sterile pipette to transfer a few eggs from the beaker to a clean glass depression slide. Holding the pipette vertically will allow the eggs to settle to the tip of the pipette to facilitate depositing on the slide (Fig. 4.13).

2 Fertilize the eggs by dipping the end of a *dry* toothpick into a petri dish containing sperm and swirling it gently through the seawater on the slide (Fig. 4.13).

3 Cover with a coverslip.

4 Scan the slide quickly using low power to locate eggs with sperm swimming nearby.

5 Observe carefully and try to find eggs in the process of being fertilized.

6 If fertilization has not occurred within 5 minutes, try again, using fresh sperm and eggs.

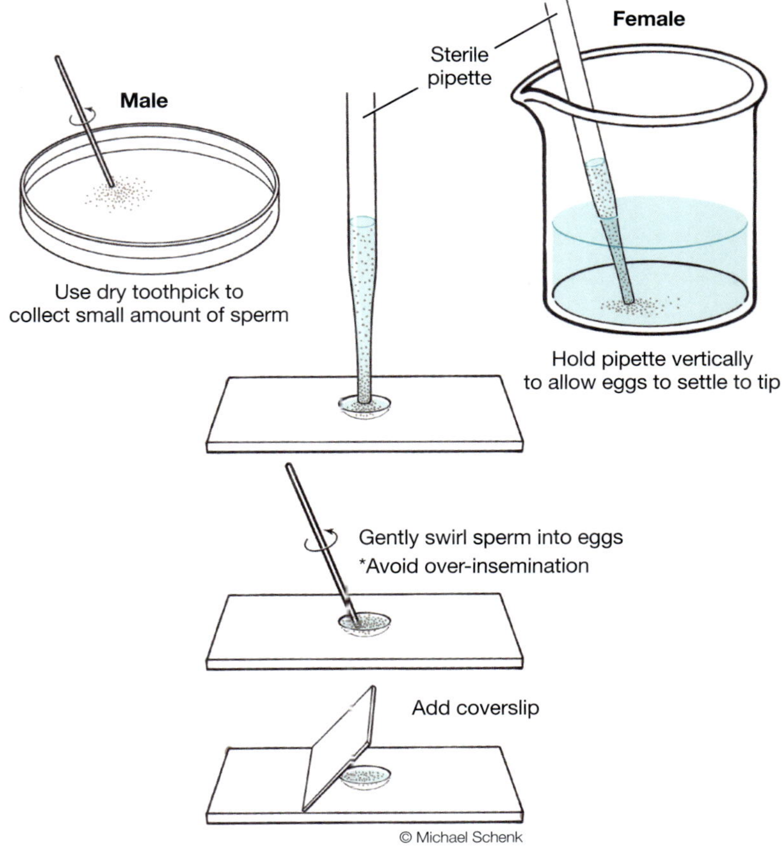

4.13 Experimental procedure for in vitro fertilization using gametes from sea urchins.

The actual fertilization event may occur too rapidly to observe. But, soon after a sperm cell successfully penetrates the egg, you will see a thin fertilization membrane begin to lift away from the outer margin of the egg at the site where the sperm entered. The fertilization membrane will gradually envelop the entire egg in a protective barrier, preventing other sperm from entering. Likewise, you can recognize eggs that have been fertilized already by the presence of a fertilization membrane. You also may see many unsuccessful sperm attached to the sticky fertilization membrane trying vainly to deposit their genetic payload in the egg.

Your instructor will make a stock culture of fertilized eggs from which samples can be taken and viewed throughout the remainder of the laboratory period to monitor subsequent stages of development. Table 4-1 provides developmental schedules for a number of commonly used sea urchin and sand dollar species.

4

Table 4-1 Sea Urchin and Sand Dollar Developmental Timetable

Times are approximate and temperature-dependent; development will progress more slowly at lower temperatures

	Strongylocentrotus droebachiensis	*Strongylocentrotus purpuratus*	*Echinarachnius parma*	*Arbacia punctulata*	
Stage	**4°C**	**10°C**	**15°C**	**20°C**	**23°C**
Fertilization	0	0	0	0	0
2-Cell Stage	5 hrs.	3.5 hrs.	1.5 hrs.	1 hr.	50 min.
4-Cell Stage	8 hrs.	5 hrs.	3 hrs.	1.8 hrs.	78 min.
8-Cell Stage	10.5 hrs.	6 hrs.	4 hrs.	2.4 hrs.	1.7 hrs.
16-Cell Stage	14 hrs.	8.5 hrs.	5.5 hrs.	3 hrs.	2.3 hrs.
Morula	18 hrs.	16 hrs.	6 hrs.	4–5 hrs.	3–4 hrs.
Blastula	50 hrs.	28 hrs.	7–8 hrs.	6–10 hrs.	5–6 hrs.
Gastrula	3–4 days	50–57 hrs.	25–30 hrs.	15–20 hrs.	12–15 hrs.
Pluteus Larva	11 days	5 days	2–3 days	48 hrs.	20–24 hrs.
Metamorphosis	7–22 weeks	9–12 weeks	n/a	n/a	5–16 weeks

Cleavage

The rate of cleavage in the embryo will depend on the temperature and pH of the surrounding medium. In sea urchins, cleavage is almost identical to that of sea stars until the fourth cleavage. At this point, sea urchin eggs begin to divide unevenly.

Procedure

1 Thirty minutes after fertilization, examine another wet mount from the original container of fertilized eggs. Can you detect any changes in the cytoplasm of the egg?

2 If a sample of fertilized eggs from 3–5 hours ago is available, try to locate embryos in their fourth cleavage (16-cell stage). Can you see the large and small cells?

3 At regular intervals throughout the remainder of the lab period, make wet mounts to determine whether subsequent cleavages have taken place. If you are careful (and lucky) you actually may see a cleavage event in progress. Be certain to keep accurate records of both time intervals and appearance of the embryo.

4 Your instructor may have cultures of fertilized eggs in various stages of development from previous labs. Make wet mounts from each of these cultures, carefully noting the time of fertilization indicated on the container and calculating the age (in days and hours) of the culture at the time you examine it. What other developmental stages do you recognize?

5 Record the time (from fertilization) when each of these stages was first seen. Combine your results with data from other members of the class. Using these data, you should be able to devise a developmental schedule for the species of sea urchin with which you have been working under the specific conditions in your lab.

Name ______________________ Section ______________

4 Questions for Review

4

1 Describe the four types of animal eggs based on the amount and distribution of their yolk.

1. ______________________
2. ______________________
3. ______________________
4. ______________________

2 How does the amount and distribution of yolk affect cleavage?

____

3 What is meant by the term *deuterostome*?

____

4 Define the following terms:

blastocoel ______________________

archenteron ______________________

blastomere ______________________

blastula ______________________

morula ______________________

gastrula ______________________

dorsal lip ______________________

neural fold ______________________

4

5 Label the three embryonic germ layers on the diagrams of the sea star gastrula and frog gastrula depicted below.

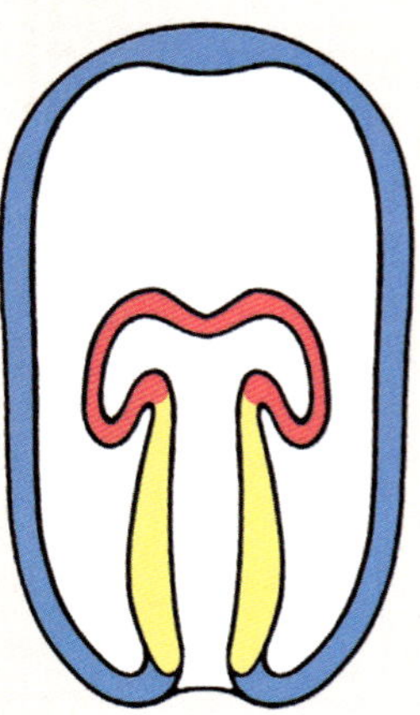

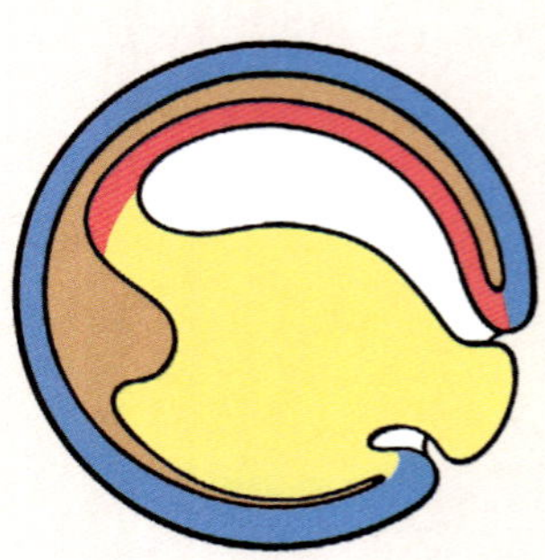

6 Does the sea star embryo grow appreciably during early embryonic development? Explain.

7 Are all of the organ systems fully developed in free-swimming larvae? Which organ systems are likely to develop first?

8 How does temperature affect embryonic development rates in sea urchins?

9 Would you expect this temperature-related pattern to be true in frogs as well? Explain.

Name ______________________ Section ______________

4 Laboratory Practical Challenge

The following evaluation is intended to replicate the method of assessment and type of questions you will be asked on lab practicals. If you have difficulty with these questions, you may benefit from additional review of the specimens provided in lab.

1 Identify/categorize the type of egg based on its distribution of yolk.

____

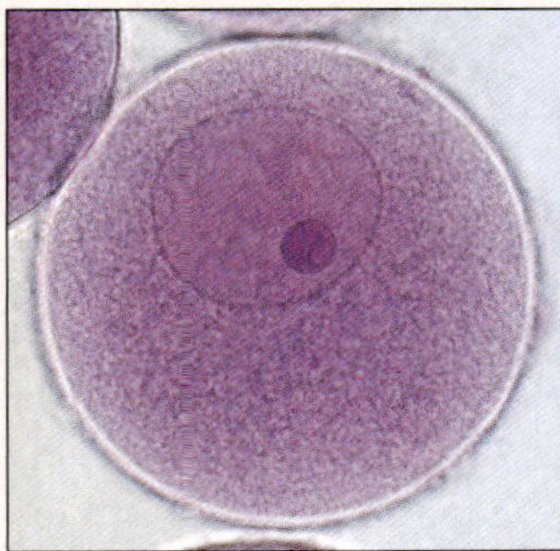

2 Label the following structures on the sea star embryo: blastopore, blastocoel, archenteron.

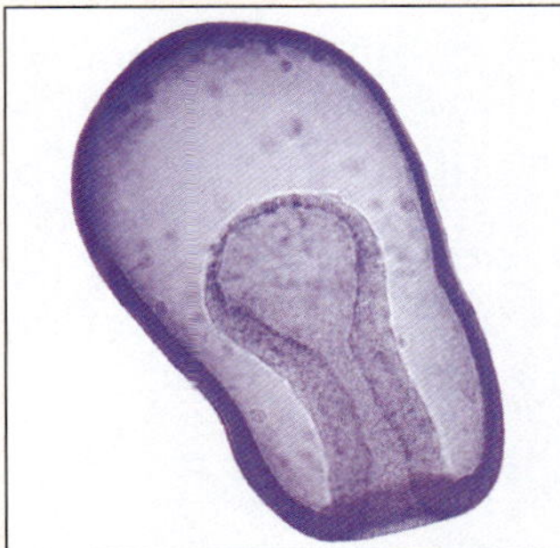

3 Identify/categorize the type of egg based on its distribution of yolk AND label the following regions: vegetal pole, animal pole.

____

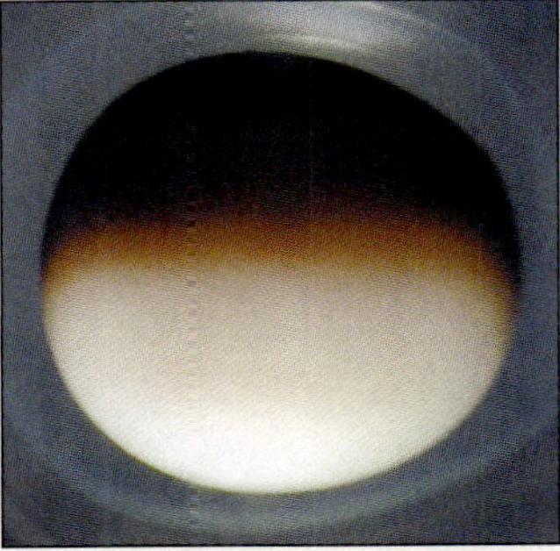

4 Identify the name of the developmental stage depicted.

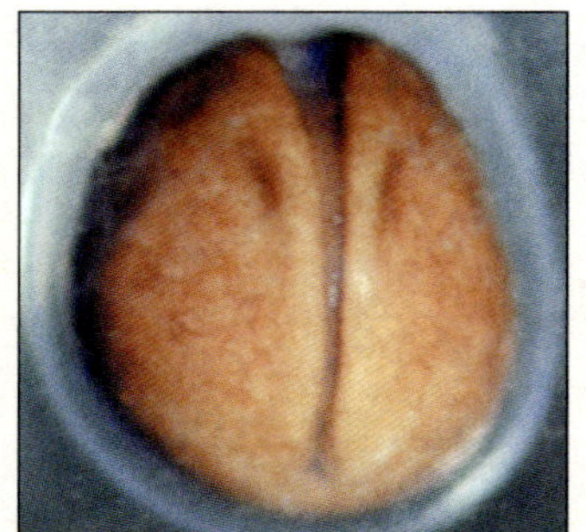

CHAPTER 5

Taxonomy and Systematics

After completing the exercises in this chapter, you should be able to:

1 Understand the modern taxonomic hierarchy and binomial nomenclature.

2 Understand how phylogenies depict relationships between taxa and represent evolutionary histories.

3 Recognize monophyletic, paraphyletic, and polyphyletic groupings.

4 Identify sister taxa and common ancestral traits of taxa.

5 List sources of information that modern systematists use to build phylogenies.

6 Construct phylogenies from morphological data.

5

Scientists face a monumental challenge in systematically identifying, naming, describing, categorizing, and studying millions of different species of living organisms (likely tens of millions if you include microbes!). More than 1.6 million species of plants and animals already have been described by biologists, and hundreds of new species are discovered every year. Scientists called taxonomists, who specialize in identifying and cataloging new species, utilize set principles and rules for naming and classifying organisms, whereas scientists in the field of systematics are concerned with the diversity and relatedness of organisms. Systematists, as they are called, study the evolutionary history of groups of organisms and establish hypotheses about who is more closely related to whom. The exercises that follow will introduce the basic principles of these two subdisciplines, classical taxonomy and systematics, to provide a foundation for learning about the diversity of animals that are covered throughout the remainder of this laboratory guide.

Exercise 5-1 Introduction to Classical Taxonomy

Naming Earth's Living Things

The subdiscipline of biology that deals with naming and classifying living organisms is called **taxonomy**. This system has its roots in the mid-18th century when the Swedish scientist Carl Linnaeus set out on a seemingly daunting mission to name and catalog every known living organism—at the time, only about 12,000 species of plants and animals were known to science. He realized the confusion that can result from having several common names for the same creature. Take, for example, the cougar found throughout the Americas, from Canada to South America (Fig. 5.1). In addition to being called a cougar, it has more than 40 other common names in English, including puma, mountain lion, shadow cat, panther, and catamount, not to mention numerous common names in other languages. But, this species of cat has only one scientific name, *Puma concolor*. When scientists in Mexico, Canada, or the United States (or any other country, for that matter) refer to *Puma concolor*, everyone knows exactly which animal is the subject of discussion.

5.1 Like many animals, the familiar cougar has many different common names but only one valid scientific name, *Puma concolor*. A standardized naming system that is universally accepted among scientists serves to minimize the potential confusion brought about by multiple names and language barriers.

Latin is used as the conventional standard for scientific names because Linnaeus established this system at a time when Latin was still widely used among scientists in the Western world. The two-part format for a scientific name is referred to as **binomial nomenclature**. The first word of the name is the **genus** (plural = genera) to which a species belongs. The second word designates a particular species of the genus and is called the **species epithet**. Notice that the first letter of the genus is always capitalized and the entire binomial is always italicized (or underlined if you handwrite the name). The practical importance of combining the genus name and species name in the binomial is that, if we know two organisms are part of the same genus, then we immediately know that they will share many similarities, but not as many as will members of the same species. For example the gray wolf, *Canis lupus*, is a member of the genus *Canis*, along with foxes, coyotes, jackals, and all domestic dogs. So right away we know that all of these organisms share many physical and behavioral characteristics in common due to their inclusion within the same genus. Furthermore, we know to expect at least a few significant differences between the various species within this genus.

Hierarchical Classification

In addition to naming species, Linnaeus also grouped them into a hierarchy of increasingly inclusive categories. There are eight basic categories in use in the modern taxonomic system: species, genus, family, order, class, phylum, kingdom, and domain. Each of these categories is more inclusive than the category that precedes it. Figure 5.2 depicts how this taxonomic system is applied to the colorful lazuli bunting. This particular species of songbird common in the western United States belongs to the genus *Passerina* with six other species of buntings. The genus *Passerina* is a small part of the family Cardinalidae that contains 10 other genera. This family, in turn, is part of the order Passeriformes, the largest order of birds with a total of 87 families of "perching birds." As we continue up the taxonomic hierarchy, each category (or **taxon**) is more inclusive than the one beneath it, all the way to the highest categories, the kingdom Animalia, which includes all animals, and the domain Eukarya, which includes all organisms with membrane-enclosed cell organelles (protists, plants, fungi, and animals). Note that in the Linnaean system, taxa above the genus level are not italicized, though they are capitalized.

So what is the basis for placing organisms into particular categories such as families and orders? Linnaeus and his contemporaries named and classified thousands of organisms long before evolution became widely accepted as a central concept of biology. Without an understanding of the principles of evolution and modification by descent to guide them, these naturalists described features of organisms and grouped them according to the similarities that seemed most important to them—most often physical or anatomical traits. Today, taxonomists attempt to classify organisms in accordance with how closely they are related, with the goal that our biological classification systems reflect the evolutionary relationships among organisms.

Common name

Lazuli Bunting

Taxonomic classification

Domain:	Eukarya
Kingdom:	Animalia
Phylum:	Chordata
Class:	Aves
Order:	Passeriformes
Family:	Cardinalidae
Genus:	*Passerina*
Species:	*amoena*

Binomial name

Passerina amoena

5.2 Classification of the lazuli bunting based on the taxonomic hierarchy developed by Linnaeus.

Let's return to our example of the gray wolf, *Canis lupus*. It turns out that modern domestic dogs, *Canis familiaris*, are closely related to gray wolves, and are, in fact, descendants of wolves. The differences that exist between them today are the product of a mere 15,000 years or so of selective breeding and domestication by humans. This close evolutionary relationship is likewise reflected in the classification of these animals by their inclusion within the same genus, *Canis*.

Domestic dogs are also related to domestic cats, *Felis catus*, if we look far back enough in time. In this case, we must go back a lot further than 15,000 years; we must go back perhaps 60 million years! So there is a big difference in how closely related dogs and wolves are as opposed to dogs and cats. Once again, we see that this more distant evolutionary relationship is reflected in our classification of these animals. Dogs and cats are not only in different genera, but occupy different families: Canidae (dogs) and Felidae (cats). They both are included within the same order, Carnivora, along with a host of other carnivores such as lions, bears, seals, weasels, hyenas, and raccoons. Thus, in an ideal taxonomic system, the further up the taxonomic hierarchy you must go to find a common category shared by two organisms, the longer it has been since they shared a common ancestor, generally speaking. This generalization obviously has limitations and numerous exceptions, but it serves as a guiding principle for naming and categorizing living organisms into a framework that reflects the relationships among those organisms (Fig. 5.3).

Blue-webbed flying tree frog,
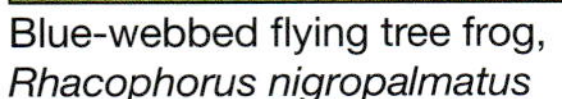
Rhacophorus nigropalmatus

Canyon tree frog, *Hyla arenicolor*

American bullfrog, *Rana catesbeiana*

Poison-dart frog, *Dendrobates sp.*

African clawed frog, *Xenopus laevis*

Woodhouse's toad, *Bufo woodhousii*

5.3 All frogs and toads share many superficial similarities, yet taxonomists face the challenge of distinguishing different species, genera, families, etc., and must identify the key characteristics for proper identification and classification. The frogs and toads depicted in this figure are all members of the order Anura, but each species belongs to a different genus and family.

Check Your Progress

1.1 Name an animal with more than one common name, and list its alternate names.

1.2 Write the following scientific binomial in the proper format: species = capensis, genus = delphinus.

1.3 Find the common name and scientific name of your state bird, tree, and flower, and list them below.

1.4 Which one of the following taxonomic levels would be the most inclusive and thus contain the most diverse assemblage of organisms: genus, order, family, or class?

Exercise 5-2 Interpreting and Constructing Phylogenies

Linking Classification and Systematics

Although a well-devised system of taxonomic classification can give general ideas about the evolutionary relationships among groups of organisms, it is the field of **systematics** that tries to ascertain the precise *degrees* of relatedness that exist between entire lineages of organisms. Systematists attempt to establish the truth about which groups gave rise to other groups, who is more closely related to whom, and which traits are more ancestral and which are more recently derived. They create hypotheses about these evolutionary relationships that are depicted in **phylogenetic trees** (also called phylogenies or cladograms) that show the order in which lineages split over the course of time.

A phylogenetic tree is a diagram with a series of dichotomous, or two-way, branch points in which each branch point (or **node**) represents the divergence of two evolutionary lineages from a common ancestor (Fig. 5.4). A particular tree may portray the evolution of all life, of major evolutionary lineages such as all vertebrates, or of only a small group of organisms, such as a genus of caterpillars. Understanding a phylogeny is a lot like reading a family tree. The base of the tree represents the ancestral lineage, and the tips of the branches represent the descendants of that ancestor. As you move from the base to the tips, you are moving forward in time. Groups, or taxa, that share an immediate common ancestor (node) are referred to as **sister taxa** and represent each other's closest relatives. Which node in Figure 5.4 represents the common ancestor to taxa B through E? Which taxon would be considered taxon A's closest relative?

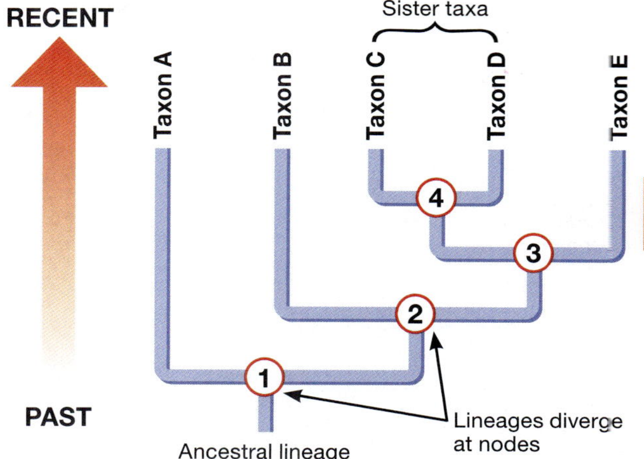

5.4 Phylogenies trace patterns of shared ancestry between lineages of organisms as you move up the tree from the ancestral lineage at the bottom to more recent taxa at the top. Nodes represent the branching points where evolutionary lineages split. Node **1**, for example, represents the common ancestor of taxa A through E.

Most systematists today believe that taxonomic groups should be monophyletic. A **monophyletic** group contains all the descendants of a particular ancestor and no other organisms (Fig. 5.5A). It may help to remember that a monophyletic group (*mono* means one) can be removed from a phylogenetic tree with a single "cut" to one branch of the tree. A group that contains some, but not all, of the descendants of a particular ancestor is said to be **paraphyletic** (Fig. 5.5B). A group consisting of members that do not share the same common ancestor is referred to as **polyphyletic** (Fig. 5.5C). Polyphyletic groups have at least two separate evolutionary origins, often requiring independent evolutionary acquisition of similar features (such as the wings in bats, birds, and some insects).

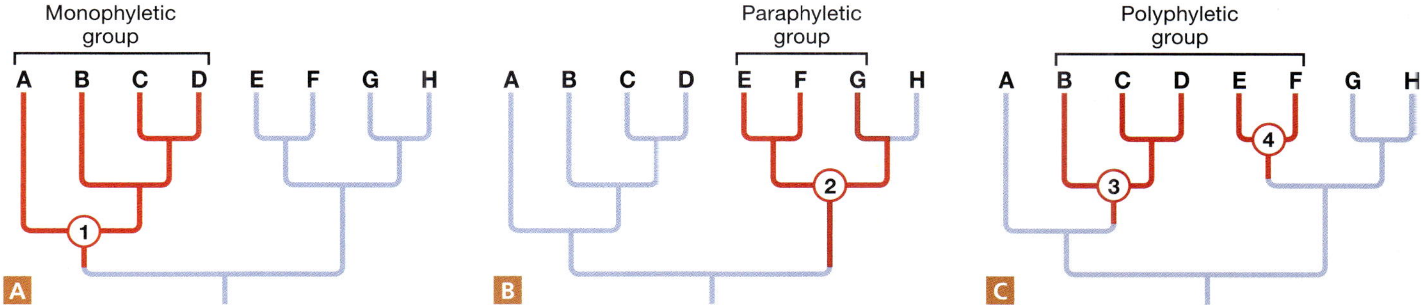

5.5 Comparison of monophyletic, paraphyletic, and polyphyletic groups: (**A**) monophyletic group that consists of a common ancestor **1** and all of its descendants; (**B**) paraphyletic group that consists of an ancestral group **2** and some, but not all, of its descendants; (**C**) polyphyletic group in which B, C, and D share common ancestor **3**, but E and F have a different common ancestor **4**.

How Phylogenies Are Reconstructed

To reconstruct phylogenies, systematists analyze evolutionary changes in the traits (or characters) of related organisms. **Characters** are heritable features that can be used to study variation within and among species, such as anatomical characteristics, genetic sequences, behaviors, or even ecological factors. The first step is to determine which character state or variant was present in the common ancestor of the entire group; this is known as the **ancestral character state.** A trait that differs from its ancestral form is called a **derived character state.** For example, more than 54,000 species of animals on Earth possess a vertebral column (better known as a backbone). The ancestor to all fishes, amphibians, reptiles, and mammals had such a structure, making the vertebral column an ancestral character for *all* vertebrates. However, only *some* vertebrate lineages evolved four limbs (and became known as tetrapods). Thus, the presence of four limbs is a derived character of *some* vertebrates.

5

Figure 5.6 depicts the results of the continuation of this stepwise process of evermore selective grouping to reveal that only some tetrapods developed mammary glands (mammals), only some mammals became carnivores, and only some carnivores became aquatic. The presumption in reconstructing phylogenies is that shared derived characters are evidence of common ancestry, and the more derived character states two organisms share, the more recently they will have shared a common ancestor compared to other organisms in the phylogeny. Based on the phylogeny depicted in Figure 5.6 which mammal shares the most recent common ancestor with bears (or to put it another way, what is the sister taxon to bears)?

A phylogenetic classification system ignores superficial similarities among organisms. Like fish, walruses swim in the water, but walruses share far more derived characteristics with bears than they do with fish—enough to indicate to scientists that walruses have returned to the water in relatively recent times, following a period in which their ancestors lived on land. Notice that we did not say that walruses evolved *from* bears. We should not assume that a taxon on a phylogenetic tree evolved from the taxon next to it. We can only infer that the lineage leading to bears and the lineage leading to walruses both split off from a common ancestor at some point in the past.

To say that certain animals under consideration have derived character states, we must compare them to an outgroup—in this case the lancelet. An **outgroup** is a species from an evolutionary lineage that is known to have diverged before the lineage that includes the species we are studying. Lancelets suit this purpose in our example because they are believed to be more distantly related to vertebrates than vertebrates are to each other. For all six characters listed in Figure 5.6, "absence" is the ancestral character state because this is the condition found in the outgroup. Thus, the derived traits in this example are those that have been acquired by other members of the lineage since they separated from the ancestral line that leads to lancelets.

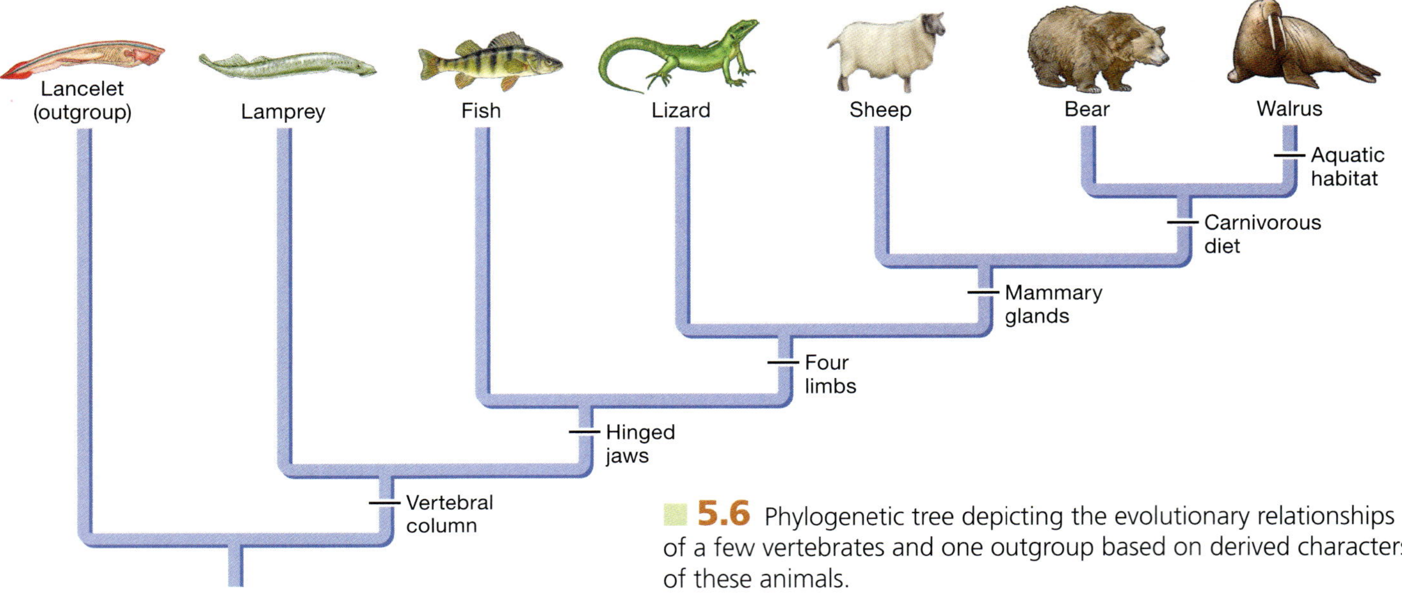

5.6 Phylogenetic tree depicting the evolutionary relationships of a few vertebrates and one outgroup based on derived characters of these animals.

Table 5-1 Shared Derived Character States

A zero (0) indicates a character state is absent; a one (1) indicates that it is present.

Characters	Taxa					
	Lancelet	Lamprey	Fish	Lizard	Sheep	Bear
Vertebral column	0	1	1	1	1	1
Hinged jaws	0	0	1	1	1	1
Four limbs	0	0	0	1	1	1
Mammary glands	0	0	0	0	1	1
Carnivorous diet	0	0	0	0	0	1

5

Reconstructing phylogenies often comes down to counting the number of derived character states present in each taxon. The results are generally depicted in a character table as shown in Table 5-1. The taxon with no derived character states, the lancelet, is the outgroup. The lamprey has only one derived character state, a vertebral column, that happens to be shared with all other taxa—a good indicator that this trait evolved next in the lineage. Thus, we place lampreys next to lancelets on our phylogeny and note that vertebral columns evolved sometime *after* the split from the common ancestor of lancelets and lampreys. Fish possess two derived character states, a vertebral column and hinged jaws, and thus would be the next descendant in the lineage of organisms. Again, we would place the character state of "hinged jaws" after the split from the common ancestor of lampreys and fish. And so we continue, until we have placed all of the taxa and all of their derived character states on our phylogeny.

Because a phylogenetic tree is a proposition about evolutionary relationships, we must use characters that are reliable indicators of common ancestry. Therefore, we must identify **homologous** characters—shared traits in different groups that are similar because they were inherited from a common ancestor. This is in contrast to **analogous** characters, which are similar structures that have separate evolutionary origins. An example of a homologous character can be found in the forelimb bones of tetrapods. Frogs, crocodiles, birds, bats, and humans all have the same basic bones in their forelimbs—bones that they each inherited from the common ancestor of all tetrapods (Fig. 5.7).

We also can see an example of analogous structures in this example. When we examine bird wings and bat wings closely, we see key differences. A bat's wing consists of a flap of skin stretched between the body and the bones of the fingers and arm. A bird's wing, however, consists of feathers extending along the arm. These structural dissimilarities suggest that bird wings and bat wings were not inherited from a common ancestor with wings. Bird and bat wings are therefore *analogous*—they have separate evolutionary origins, but are superficially similar because they evolved to serve the same function (Fig. 5.8). So we can say that the *bones* of bird and bat forelimbs are homologous structures because forelimbs were inherited from a common ancestor with forelimbs, but bird and bat *wings* are analogous structures because they were not inherited from

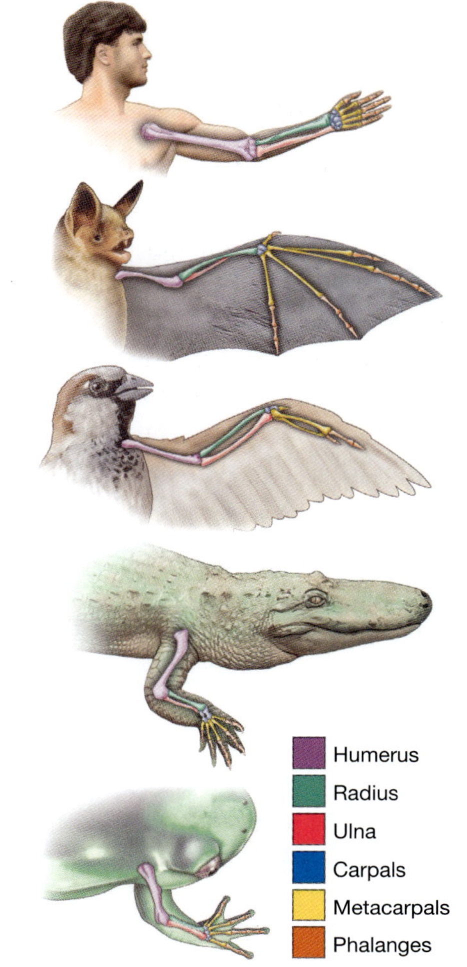

5.7 Forelimb bones of tetrapods are homologous structures inherited from a common ancestor. Bones shown in the same color are homologous.

a common ancestor with wings. Would you consider the wings of a moth and the wings of a bird analogous or homologous? Why? What about the heart of a chicken and the heart of a fish? Why?

5

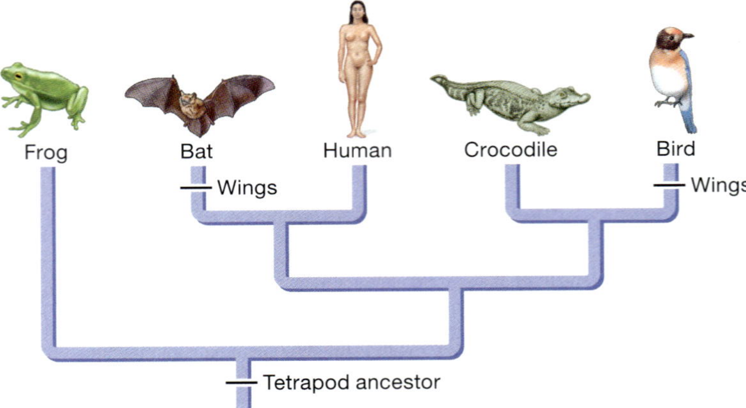

5.8 Although the wing bones of bats and birds are homologous, having been inherited from a common tetrapod ancestor, the wings themselves are not homologous because they evolved independently in bats and in birds from the forelimbs of a nonflying tetrapod ancestor.

Sources of Phylogenetic Information

You may be wondering what sources of information systematists use in charting the evolutionary histories of organisms and constructing phylogenies. How do they decide on the groupings of organisms? Rather than being limited only to the observable physical characteristics of species, as were scientists during Linnaeus's time, systematists today have a much wider and more sophisticated range of resources to draw upon to construct phylogenies that more accurately reflect patterns of ancestry.

1. **Radiometric dating**—a technique for determining the age of objects by measuring the decay of radioactive elements within them.
2. **Fossils**—comparing characteristics of different fossils of known ages to determine which species predate others.
3. **Comparative morphology**—comparing the extent to which groups of organisms share similar (homologous) anatomical features to determine their degree of relatedness.
4. **Comparative embryology**—examining the developmental patterns of organisms to make judgments about lines of descent.
5. **Biochemical and molecular analysis**—comparing DNA, RNA, protein sequences, enzymes, and metabolic pathways among organisms to make judgments about lines of descent.

An important point to remember is that phylogenies are *hypotheses* put forth by scientists as the most plausible explanation of the evolutionary relationships among organisms based on the best evidence available at that time. As additional research is conducted and new evidence is uncovered, scientists often propose alternate hypotheses, and our phylogenies shift to reflect these changes in the body of scientific knowledge. Sometimes this means changing just a branch or two; other times it means reconstructing an entire phylogenetic tree. Often it means that disagreements arise among scientists about which phylogeny is more accurate. Sometimes it means that information you read in one book disagrees with that in another book. This apparent lack of finality is actually by design, it's built right into the system of science and illustrates a fundamental ideology—that any conclusion or "fact" in science is subject to modification based solely on the best evidence available.

Building Simple Phylogenies

Now it's time to put this knowledge of systematics to use to construct your own phylogenetic trees. As you build your trees keep the following rules in mind:

1. All taxa are placed on the endpoints of the phylogeny, never at nodes.

2. Each node must have a list of character states common to all taxa above the node (unless the character is later modified).
3. All character states appear on the phylogeny only once (unless the character state was derived separately by more than one group).

Procedure

1 Use the data in Table 5-2 to complete the phylogeny depicted in Figure 5.9.

2 Place each animal correctly at the ends of the branches and label the branching points (nodes) with the appropriate character states. **HINT:** A node may have more than one character that defines that branch.

3 The taxa across the top of Table 5-2 are not arranged in order, so you will have to determine the correct order of placement on your phylogeny. Remember to start your phylogeny with the taxon that has the fewest derived character states (the outgroup) and work your way up to the taxon with the most derived character states.

4 You may assume that the absence of a trait represents the ancestral character state.

Table 5-2 Shared Derived Character States of Sponges

Use the data to construct a phylogenetic tree representing the relationships of these four taxa. A zero (0) indicates a character state is absent; a one (1) indicates that it is present.

Characters	Choanoflagellates (outgroup)	Hexactinellida	Demospongiae	Calcarea
Spicules with 6 rays	0	1	0	0
Spicules composed of silica	0	1	1	0
Internal system of canals	0	1	1	1
Pores	0	1	1	1
Flagellated collar cells	1	1	1	1

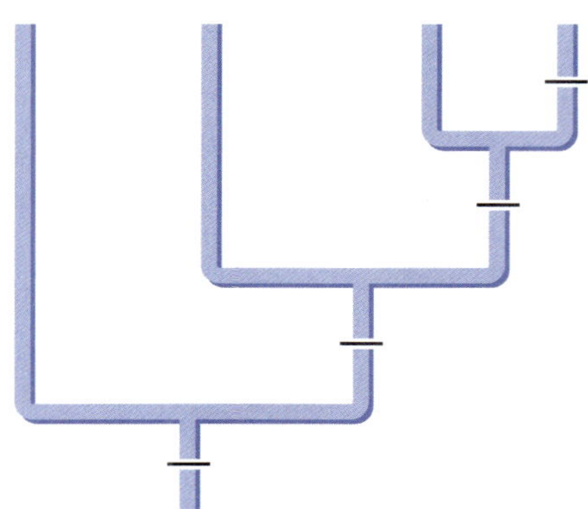

5.9 Complete the phylogeny using the data presented in Table 5-2.

5 Next, use the data in Table 5-3 to complete the phylogeny depicted in Figure 5.10.

6 Place each animal correctly at the ends of the branches and label the branching points (nodes) with the appropriate character states.

7 Once again, the taxa across the top of Table 5-3 are arranged in random order, so you will have to determine the correct order of placement on your phylogeny.

8 In this example, we have added another element to the characters listed in the data table. Instead of just determining the presence or absence of a trait, we have included some characters with more complex character states. You may assume that the character states of the outgroup are ancestral character states and any that differ from the outgroup are considered to be derived.

9 As before, remember to start your phylogeny with the taxon that has the fewest derived character states and work your way up to the taxon with the most derived character states.

Table 5-3 Shared Derived Character States of Echinoderms

Use the data to construct a phylogenetic tree representing the relationships of these six taxa.

Characters	Homalozoa (outgroup)	Brittle Stars	Sea Urchins	Sea Stars	Sea Lilies	Sea Cucumbers
Water vascular system	Present	Present	Present	Present	Present	Present
Number of ambulacral grooves	3	5	5	5	5	5
Ambulacral grooves extend from oral to aboral pole	No	No	Yes	No	No	Yes
Ambulacral grooves closed	No	Yes	Yes	No	No	Yes
Madreporite	Oral	Aboral	Aboral	Aboral	Oral	Internal (near oral)
Endoskeleton	Present	Present	Present	Present	Present	Highly reduced

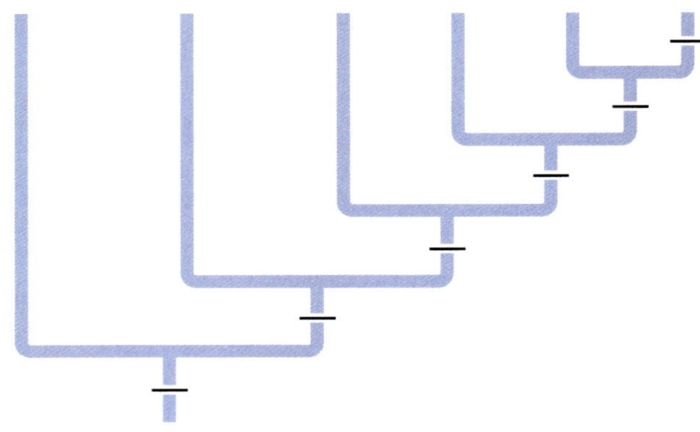

5.10 Complete the phylogeny using the data presented in Table 5-3.

10 In this final example, you will have to fill in the data table of character states and draw the entire phylogeny on your own! If you understand the basic principles of creating phylogenies that we have covered thus far, this shouldn't be too difficult.

11 Think about the familiar animals listed in Table 5-4 and establish which character states are present and which are absent in each taxon. Use a zero (0) to indicate absence of a trait, and a one (1) to indicate the presence of a trait.

12 Next, use the data in Table 5-4 to create a phylogeny for these animals in the space below. As before, you may assume that the absence of a trait represents the ancestral character state.

13 The taxa across the top of Table 5-4 are again arranged in random order, so you will have to determine the correct order of placement on your phylogeny. Remember to start your phylogeny with the taxon that has the fewest derived character states and work your way up to the taxon (or taxa) with the most derived character states.

Table 5-4 Shared Derived Character States of Vertebrates

Fill in the table below using a zero (0) if the character state is absent and a one (1) if it is present.

Characters	Salamander	Frog	Lamprey	Fish	Shark	Bird	Mouse	Gorilla
Jaws								
Bony skeleton								
Lungs								
Amniotic egg								
Fur								

14 We have added one more twist to this collection of animals. You will see that multiple taxa in this example have the same number of shared derived character states, meaning that they will have a slightly different branching relationship than in the previous examples. This additional factor will test your ability to construct an accurate phylogeny based on the data presented!

15 When in doubt about a particular grouping, follow the principle of **parsimony**. In other words, choose the simplest arrangement of branches capable of explaining the data. Even though similar structures may evolve independently in separate lineages through convergent evolution (as we saw earlier with wings), this is assumed to be a rare event. Most major morphological features are assumed to have evolved or to have been lost only rarely. Therefore, unless your evidence suggests otherwise, choose a branching pathway that minimizes the number of times a trait is postulated to have arisen (or been lost).

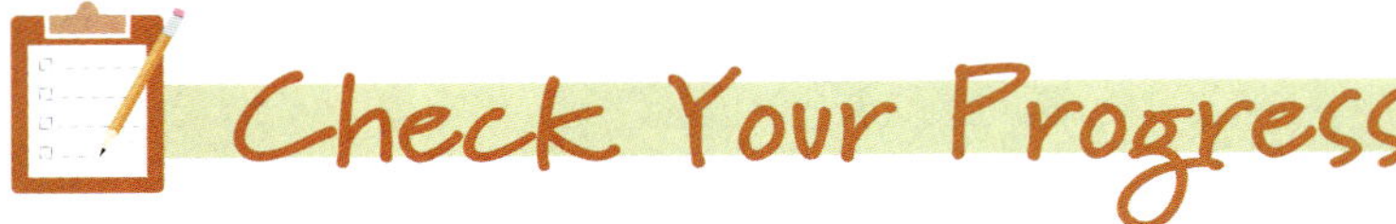

2.1 Identify the sister taxa in the phylogeny you created from Table 5-4.

2.2 Which of the taxon below is most closely related to a mouse?

a. shark

b. fish

c. frog

2.3 Which taxon should be designated as the outgroup?

2.4 List all traits present in the direct common ancestor of frogs.

Using Molecular Evidence to Determine Evolutionary Relationships

In the examples thus far we have focused primarily on physical characteristics of organisms to construct phylogenies, primarily because they are readily observable and are easily pictured in our mind's eye when thinking of these animals. But as we know, scientists have newer, more sophisticated methods to supplement the physical data gathered by comparative morphology. There has been a revolution over the past few decades in the field of molecular biology as many new techniques are being discovered and refined to allow scientists to make increasingly more accurate judgments about the evolutionary relationships among organisms. In this final section we will review one example of how genetic data are being used to construct phylogenies.

As you know, every living thing on Earth uses the same basic genetic code in strands of DNA contained within cells and organized into functional units we call genes. We also know that genes mutate over time, and these mutations result in changes in the base pairs (A, T, C, and G) along the DNA's double helix. It turns out that some genes have a constant rate of mutation, meaning that the changes to these base pairs occur in an almost "clocklike" fashion. In fact, when scientists speak of "molecular clocks," this is exactly the sort of phenomenon they are discussing. If we can isolate genes with "molecular clock" properties and compare their genetic sequences among different organisms, we can determine the degree of relatedness between the organisms. The greater the number of base differences we see in similar genes (that is, the greater the number of mutations), the longer we can assume it has been since those two species split from a common ancestor.

A gene known as the *cytochrome c oxidase* gene exists in a wide range of organisms, including insects, fishes, birds, humans, and even yeast. When we compare the number of base differences in this gene in humans with the same gene in pigs, we see there are 13 base differences. But when we compare the cytochrome c oxidase genes in humans and moths, there are 36 base differences. Figure 5.11 displays the base differences of the cytochrome c oxidase gene in several other familiar organisms that have been studied. These data suggest something that most of us would say we already know—that humans and pigs are more closely related than humans are to moths, and much more closely related than humans are to yeast! However, this type of comparison is even more useful when comparing species that we aren't as sure of their evolutionary relationships. It also illustrates how we strengthen our hypotheses about the evolutionary relationships among organisms when systematists reconstruct a phylogeny based on one line of evidence that agrees with phylogenies created by other lines of evidence. In this case, what we learned from comparing genetic differences in the cytochrome c oxidase gene matches precisely with prior evidence about the evolutionary relationships among these organisms that has been collected by radiometric dating, the fossil record, comparative embryology, and comparative morphology.

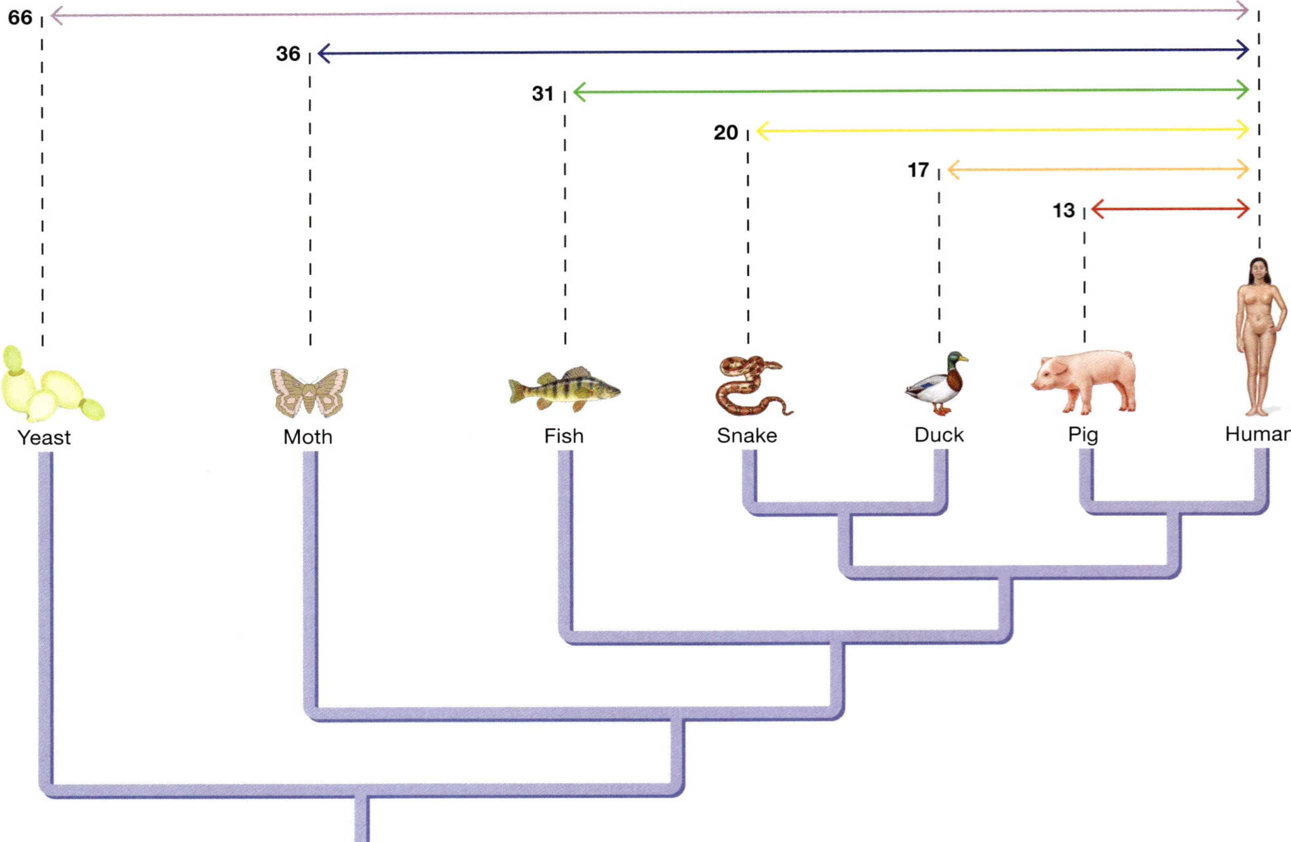

5.11 Number of base differences in the cytochrome c oxidase gene among a diverse assemblage of organisms. Because the mutation rate of this gene is fairly constant over time, scientists can conclude that the larger the number of base differences in the genes between organisms, the longer it has been since those species shared a common ancestor.

Name ______________________ Section ____________

5 Questions for Review

1 Use the data in Table 5-5 to complete the phylogeny depicted in Figure 5.12, and then use the completed phylogeny to answer questions 2 through 8.

Table 5-5 Shared Derived Character States of Aquatic Vertebrates

Use the data to construct a phylogenetic tree representing the relationships of these seven taxa. A zero (0) indicates a character state is absent; a one (1) indicates that it is present.

Characters	Lancelets (outgroup)	Coelacanths	Hagfishes	Ray-finned Fishes	Lampreys	Sharks	Lungfishes
Lobed fins	0	1	0	0	0	0	1
Ability to breathe air	0	0	0	0	0	0	1
Vertebrae	0	1	0	1	1	1	1
Jaws	0	1	0	1	0	1	1
Dorsal nerve cord	1	1	1	1	1	1	1
Paired appendages	0	1	0	1	0	1	1
Lung or swim bladder	0	1	0	1	0	0	1
Paired sense organs	0	1	1	1	1	1	1
Tripartite brain	0	1	1	1	1	1	1

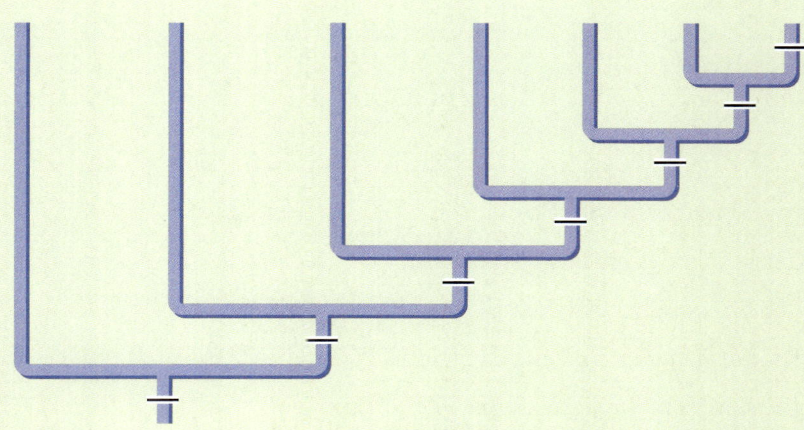

5.12 Complete the phylogeny using the data presented in Table 5-5.

2 List three homologous structures shared by sharks and ray-finned fishes.

3 List all traits present in the direct common ancestor of lampreys.

5

4 What is the sister taxon to lungfishes?

5 Are the aquatic vertebrates depicted in Figure 5.12 monophyletic, paraphyletic, or polyphyletic? How do you know?

6 Which taxon represents the outgroup in Figure 5.12?

7 If a gene common to all of the aquatic vertebrates depicted in Figure 5.12 was sequenced, which taxon would you predict to have the most base pair differences (most mutations) in the DNA of this gene compared to lungfishes? Why?

8 Which taxon would you predict to have the fewest base pair differences in the DNA of this gene compared to lungfishes? Why?

9 Put the following taxonomic categories in order from most inclusive to least inclusive: genus, species, family, kingdom, class, phylum, order, domain.

10 Which of the following statements about binomial nomenclature is *false*?

a. Both genus and species names are always italicized.

b. The name following the genus is referred to as the species epithet.

c. Both genus and species names are always capitalized.

d. The genus category is more inclusive than the species category.

CHAPTER 6

Animal-like Protists

After completing the exercises in this chapter, you should be able to:

1 Characterize the features common to all protists.

2 Identify the basic anatomy of representatives from the major taxa within each subgroup of animal-like protists.

3 Recognize and distinguish members of each subgroup from one another.

4 Discuss general physiological processes in animal-like protists, such as feeding, reproduction, locomotion, and osmoregulation.

6

In the last decade, advances in genetic, molecular, and biochemical research have almost completely revised our understanding of the evolutionary relationships among unicellular, eukaryotic organisms and have caused significant modifications in the classification of the former kingdom Protista. The formal terms *protist* and *protozoa* have been largely abandoned, and major revisions to the "roots" of the evolutionary tree of life have brought about entirely new groupings of these organisms, with many lineages of protists now being recognized as kingdoms in their own right. The term *protist* is still used by biologists, but only in an informal sense to refer to eukaryotes that are not plants, animals, or fungi.

Protists occur nearly everywhere on Earth where there is available water. They live in oceans, streams, lakes, puddles, damp soils, moist bark, underneath rocks, and even in the body fluids of plants and animals as symbionts or **parasites**. Despite the enormous diversity that exists within this assemblage, most protists are **unicellular**, and all are **eukaryotic**, meaning their cells contain membrane-bound nuclei and other membrane-bound organelles. Because protists must perform all the essential biological functions of life within the limits of a single cell, these cells are generally much more complex than individual cells of a multicellular organism. Yet unlike multicellular organisms, protists do not have separate organs, but instead have many kinds of specialized subcellular organelles used for support, locomotion, reproduction, defense, osmoregulation, nutrient acquisition, and even sensing their environment.

Protists are more nutritionally diverse than other eukaryotes. Some are plantlike in nature—sedentary, **autotrophic** organisms that make their own food through photosynthetic processes. Others are more animal-like—**motile, heterotrophic** organisms that ingest large food particles and digest them intracellularly. Others still are fungus-like in nature and absorb small organic compounds requiring no further breakdown. Several are even able to combine photosynthesis and heterotrophic nutrition. Some protists store their reserves of glucose as **starches** (as do plants), whereas others employ **lipids** or lipid by-products (as do animals).

Reproductive methods and life cycles also are highly varied among protists. All protists can undergo **asexual reproduction** by mitosis (generally by simple fission), yet many are able to undergo **sexual reproduction** through a combination of **meiosis** and nuclear exchange. Many protists also can form **cysts** that allow them to lie dormant for long periods of time to escape harsh environmental conditions that they would not survive in an active metabolic state.

Evolutionary History

Current theories on what gave rise to the enormous diversity of protists center around two key events. Changing environmental conditions on Earth some 1.2–2 billion years ago prompted a progression of adaptive responses in ancient prokaryotes that led to the eventual rise of eukaryotes. One key event in the evolution of eukaryotic protists appears to have been the invagination of outer cell membranes that allowed the cytoplasm and cell interior to become partitioned, paving the way for specialized regions capable of division of labor within the cell. Another key event was the incorporation of certain symbiotic prokaryotes with specific properties that benefited the host cells. Evidence abounds for this process, known as **endosymbiosis**, whereby unicellular organisms engulf other cells, which become endosymbionts and ultimately organelles in the host cell. It is now widely accepted among zoologists that organelles such as mitochondria and plastids in protists (and in plants and animals, for that matter) came about through this process.

It is important to recognize that the classification of protists continues to change, and any attempt to draw accurate phylogenetic depictions is severely limited by our current understanding of the specific relationships among the members of these broad groups. Protists are most certainly **polyphyletic**, meaning all modern protists do not share the most recent common ancestor. Some protist groups are, in fact, more closely related to plants, fungi, or animals than they are to other protists. Presently, the most widely accepted hypothesis depicts five monophyletic supergroups of eukaryotes, with bacteria and archaea as offshoots (Fig. 6.1). Notice the prominent, multicellular organisms that we know best (plants, animals, and fungi) are merely tips of just a few branches on this tree.

As a zoology course, our ultimate focus is the study of *animal* life, thus a complete survey of all protists would be neither possible nor prudent. Instead, this chapter serves to acquaint you with some of the more significant groups of "animal-like" protists, classified within three of these supergroups: Excavata, Chromalveolata, and Unikonta.

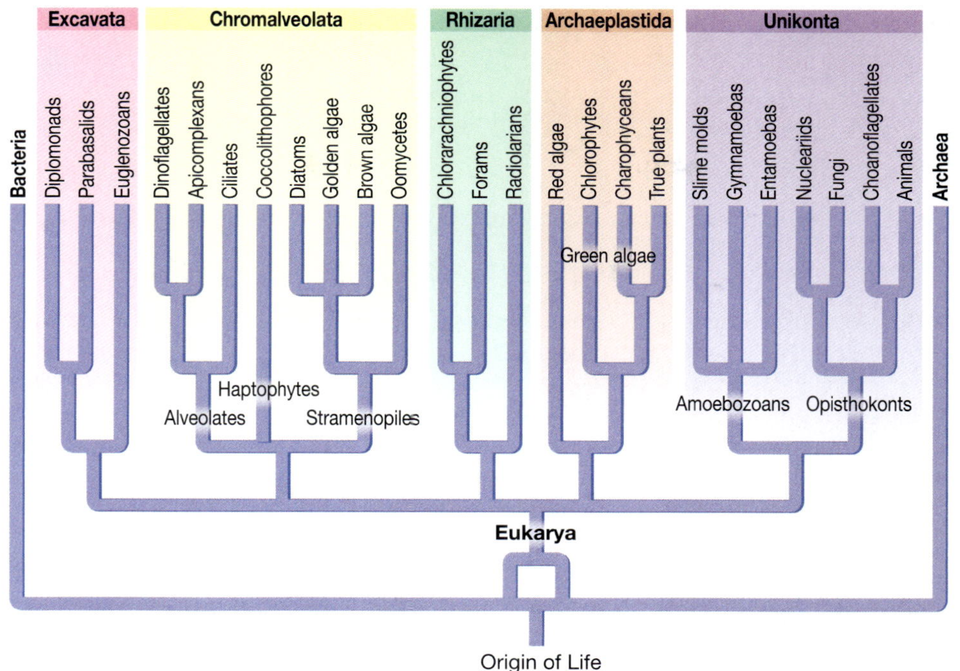

6.1 Phylogenetic hypothesis for the relationships among major eukaryotic groups (shown by the five shaded supergroups). Notice that the large, multicellular organisms that we know best (plants, animals, and fungi) are merely tips of just a few branches on the tree.

Exercise 6-1 Phylum Euglenozoa (Euglenoids, Kinetoplastids)

Materials Needed

- ❑ slide of *Euglena* w.m.
- ❑ live culture of *Euglena*
- ❑ slide of *Trypanosoma* w.m.
- ❑ blank slides
- ❑ coverslips
- ❑ plastic droppers
- ❑ methylcellulose, Protoslo™, or Detain™
- ❑ compound microscope

Though there are many protists with flagella, two subphyla of flagellates, the **euglenoids** and **kinetoplastids**, make up the phylum Euglenozoa, within the supergroup Excavata. The Excavata derive their group name from the "excavated" feeding groove that runs along one side of the cytoskeleton of the cell body. Euglenozoans are unicellular, motile flagellates that reproduce asexually by binary fission and possess flexible cell membranes. Many species are heterotrophic, some are parasitic, and others are photosynthetic.

Euglenoids (ex. *Euglena*)

Procedure

1 Obtain a prepared slide of a whole mount (w.m.) of *Euglena* and examine it using the medium- or high-power lens of your microscope.

2 *Euglena*, a complex unicellular organism with many visible organelles, typifies this subgroup.

The characteristic feature of euglenoids is an anterior pocket that bears one or two **flagella** extending from the reservoir in the *anterior* end of the organism (Fig. 6.2). Near the base of the primary flagellum is a pigmented **eyespot** that serves as a photoreceptor, providing chemical information to the cell about the intensity of light in its environment. A large, central **nucleus** and many large **chloroplasts** should also be evident, indicating that *Euglena* can photosynthesize. In addition to autotrophic means of food production, *Euglena* can, when kept in dark environments, resort to heterotrophic methods of food acquisition—absorbing organic molecules from its environment. Although plantlike in their ability to photosynthesize, euglenoids are more animal-like in their method of food storage, storing glucose in the form of lipids rather than as starch.

3 Using a plastic dropper, place a drop of culture medium from a container of live *Euglena* on a clean microscope slide.

4 Add a coverslip and observe under medium or high power.

5 If the organisms are moving too quickly to observe easily, make another slide, this time adding a drop of methylcellulose, Protoslo™, or Detain™ to slow their movement.

6 Notice the fluid, corkscrew-like motion that *Euglena* exhibits. In which direction does the primary flagellum move the organism?

The world of swimming flagellates contains two types of flagella. One type, the whiplash flagellum, pushes the organism through the medium. A familiar example of this is a human sperm cell. The other type is the tinsel flagellum, which pulls the organism through its environment. Often a large, circular, **contractile vacuole** will be evident on your specimen near the anterior pocket. Its function is to pump out the excess water that continuously diffuses into the organism as a consequence of the osmotic gradient faced by organisms living in a hypotonic (freshwater) environment.

7 On a live *Euglena* identify as many of the other organelles depicted in Figure 6.2 as possible.

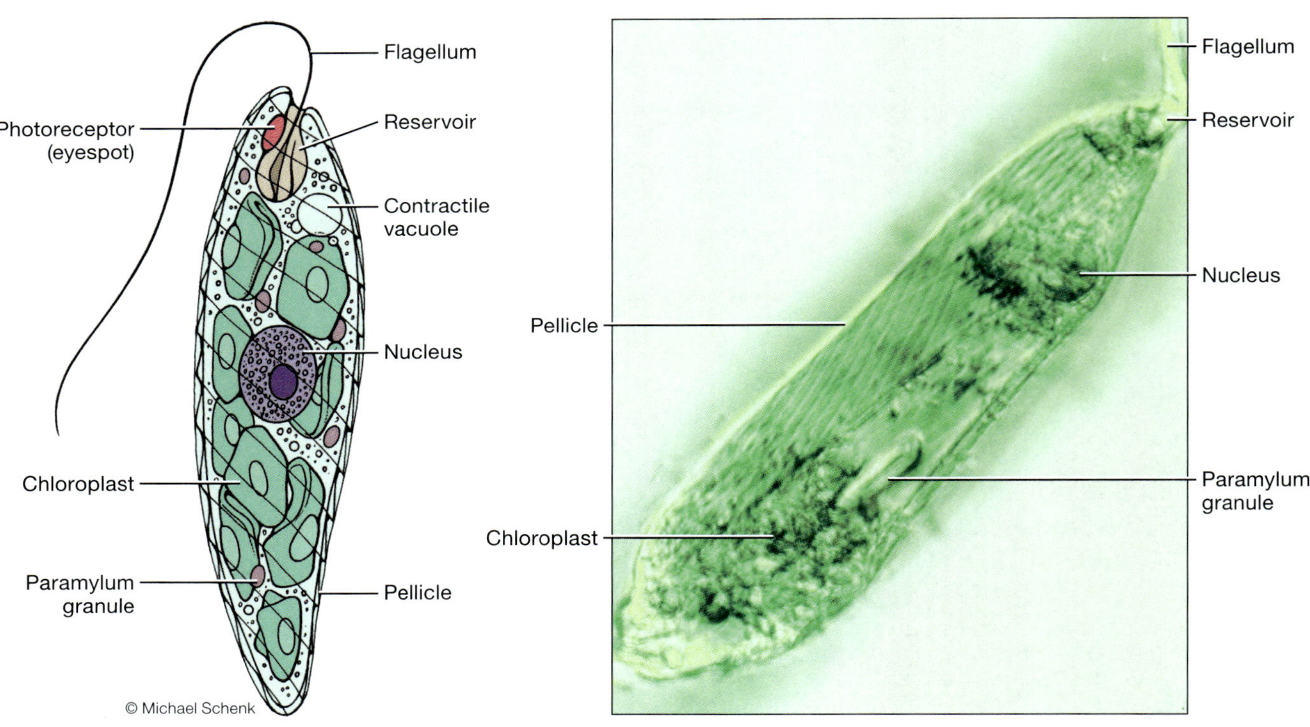

6.2 *Euglena*, a typical euglenozoan.

1.1 Which type of flagellum, whiplash or tinsel, does *Euglena* possess?

1.2 What color is the photoreceptive eyespot of *Euglena*?

1.3 How might such an eyespot be advantageous for an autotrophic organism?

1.4 In what ways might *Euglena* be considered plantlike? In what ways is it animal-like?

Kinetoplastids (ex. *Trypanosoma*)

Kinetoplastids, another subphylum of Euglenozoans, are unicellular, parasitic flagellates characterized by a single, large mitochondrion containing a **kinetoplast**—a unique organelle that houses extra-nuclear DNA. Many kinetoplastids, such as *Trypanosoma*, are human pathogens. *Trypanosoma brucei* causes African sleeping sickness, a human disease spread by the bite of the tsetse fly. One reason that this disease is still so widespread today in parts of Africa is that the molecular composition of these pathogens changes rapidly, preventing immunity from developing in hosts.

Procedure

1. Obtain a whole mount (w.m.) slide of *Trypanosoma* and examine it using high power.
2. Notice that a single field of view contains many trypanosomes, found among the red blood cells of their host (Fig. 6.3). Trypanosomes reproduce in the blood but do not harm the blood cells. Instead, they absorb nutrients from the host's blood, multiply in the bloodstream, and produce toxins that ultimately reach and affect the nervous system.

The trypanosome body is elongated and slightly twisted, bearing a distinct nucleus near the center (Fig. 6.3). A single flagellum runs closely along a fold of the plasma membrane, forming the **undulating membrane**, and extends beyond one end of the body. The kinetoplast, the characteristic feature of this group, is located at the end of the body opposite the flagellum, but is usually too small to be seen with a compound microscope.

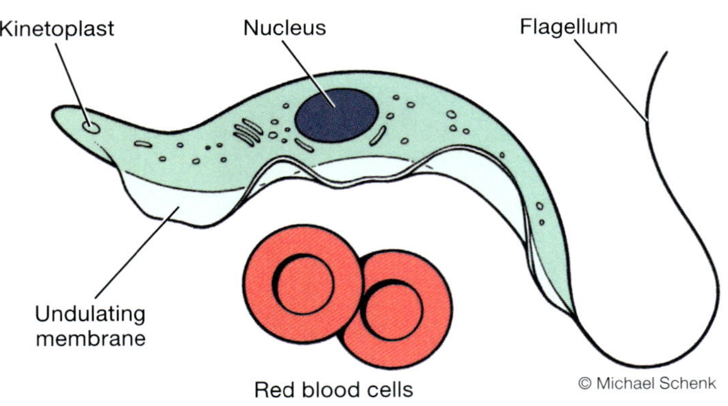

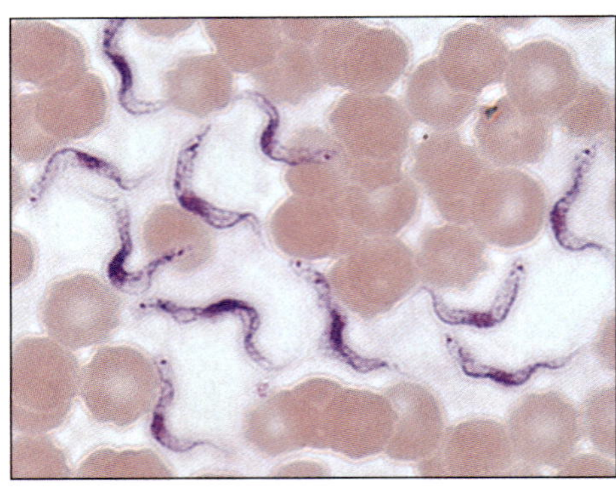

6.3 Trypanosome parasites in blood (*Trypanosoma brucei*)—the causative agent of African sleeping sickness in humans. Nuclei, flagella, and kinetoplasts are visible on the specimens depicted in the micrograph.

Check Your Progress

2.1 Describe the length and width of a trypanosome compared to a red blood cell.

2.2 Are trypanosomes larger or smaller than *Euglena*?

2.3 Do trypanosomes possess chloroplasts? How do they obtain nutrients?

2.4 How are kinetoplastids similar to euglenoids?

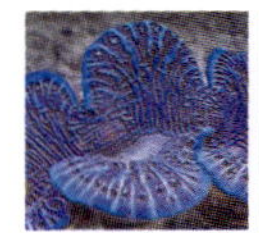

Exercise 6-2 Alveolata (Dinoflagellates, Apicomplexans, Ciliates)

Another monophyletic clade of protists (a superphylum within the Chromalveolata) called Alveolata includes three traditional phyla: dinoflagellates, apicomplexans, and ciliates. All members of this group of unicellular protists have small cavities under their cell surfaces called **alveoli.** In the ciliates, the alveoli produce pellicles; in the parasitic apicomplexans, the alveoli serve a structural function; in the dinoflagellates, the alveoi produce thecal plates.

Materials Needed

- ❑ slide of dinoflagellates w.m.
- ❑ live culture of *Peridinium* or *Ceratium*
- ❑ slide of *Plasmodium* w.m.
- ❑ slide of *Paramecium* w.m.
- ❑ slide of *Paramecium* undergoing fission
- ❑ slide of *Paramecium* undergoing conjugation
- ❑ live culture of *Paramecium*
- ❑ live culture of *Vorticella*
- ❑ methylcellulose, Protoslo™, or Detain™
- ❑ yeast mixture dyed with Congo red
- ❑ blank slides
- ❑ coverslips
- ❑ plastic droppers
- ❑ dissecting needle
- ❑ compound microscope

6

Dinoflagellates (Phylum Dinoflagellata)

Approximately 1,000 species of dinoflagellates have been described, and they are among the most important primary photosynthetic producers of organic matter in the oceans. Most species are free-living components of marine plankton, but some species are important endosymbionts of other marine organisms such as cnidarians and molluscs (Fig. 6.4). Dinoflagellates are mostly marine, unicellular autotrophs whose flagella arise from two **perpendicular grooves** formed along internal **cellulose plates.** One flagellum lies in a groove around the equator of the cell, and the other extends from the end of the cell in the other groove. Their primary photosynthetic pigment is **chlorophyll C.** Reserves of glucose may be stored as starch or lipids.

When nutrients are abundant, certain marine dinoflagellates undergo large population "blooms" in such enormous numbers that they color the ocean waters a deep reddish hue for miles. The toxins and anoxic conditions produced by these **red tides** often poison massive numbers of fish and make shellfish in these areas unfit for human consumption.

6.4 Blue coloration of this giant clam, *Tridacna*, is caused by symbiotic dinoflagellates that live on the clam's skin.

Procedure

1. Obtain a prepared slide of dinoflagellates and examine with medium or high power.
2. Look for the two perpendicular grooves and the rigid cellulose plates. It may even be possible to make out flagella on well-prepared specimens (Figs. 6.5 and 6.6).
3. Next make a wet mount of a living dinoflagellate (either *Peridinium* or *Ceratium*) and examine it under medium or high power.
4. If the organisms are moving too quickly to observe easily, make another slide, this time adding a drop of methylcellulose, Protoslo™, or Detain™ to slow their movement.

   Do dinoflagellates push or pull themselves through the water with their flagella? How does this compare to euglenoid movement?

Describe the number and arrangement of grooves in the cellulose plates of your specimens.

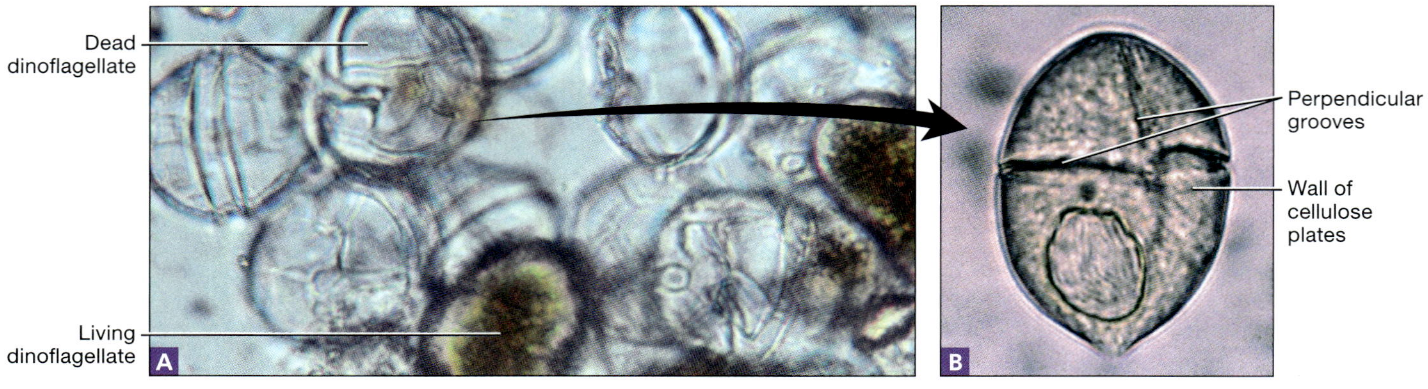

6.5 (**A**) Dinoflagellate, *Peridinium*. (**B**) As in many dinoflagellates, the cell wall of *Peridinium* is composed of overlapping cellulose plates.

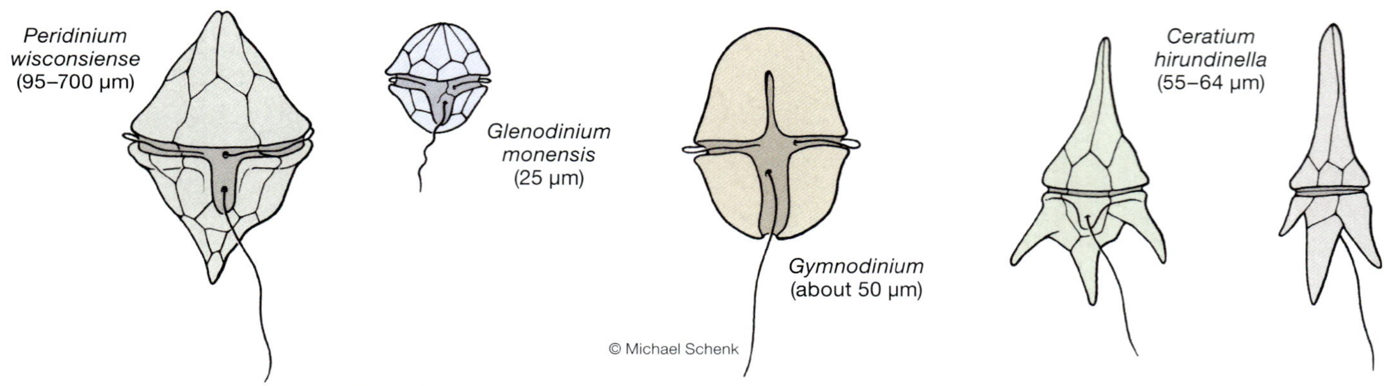

6.6 Common dinoflagellates of the phylum Dinoflagellata.

Check Your Progress

3.1 What ecological importance do dinoflagellates have?

3.2 What ecological problems do they occasionally cause?

3.3 Which type of flagellum, whiplash or tinsel, do dinoflagellates possess?

Apicomplexans (Phylum Apicomplexa)

Members of the phylum Apicomplexa are unicellular parasites characterized by an apical complex of organelles that they use to penetrate the host cell. Apicomplexans (or sporozoans, as they are commonly called) infect nearly every major group of animals, from simple invertebrates to humans. So far, about 3,900 different species have been identified—most infecting a different host! Though apicomplexans are all nonmotile, they are highly specialized for their parasitic lifestyle. They rely on their hosts for both nutrients and dispersal and often have complex life cycles that involve several host species.

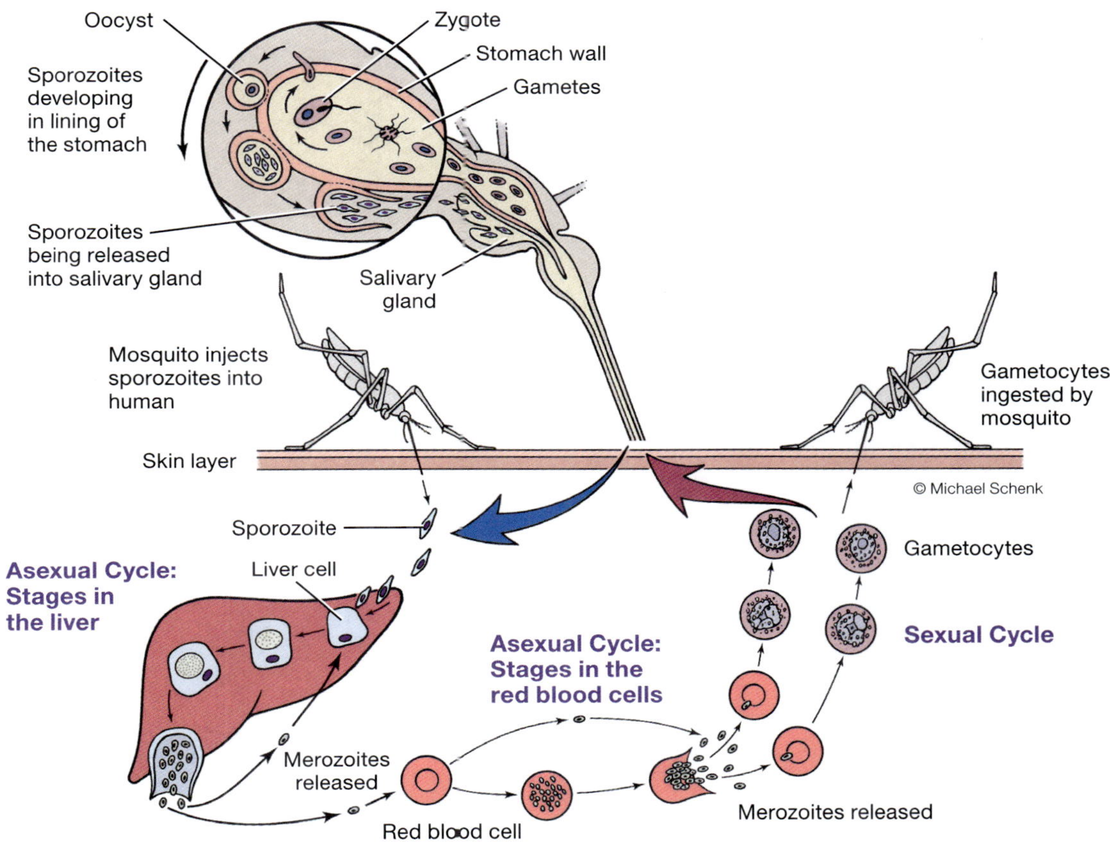

6.7 Life cycle of *Plasmodium*, the sporozoan parasite responsible for malaria in humans. More than 1 billion people per year suffer from malaria, and around 2 million people die every year as a result of this illness.

Procedure

1 Study the life cycle of *Plasmodium* in Figure 6.7. *Plasmodium* is a blood parasite carried by a secondary host, the female *Anopheles* mosquito, which transmits the parasite to humans through its saliva. The mosquito serves as a **vector** for the parasite, harboring the parasite and transmitting it while remaining unaffected by the parasite. But humans suffer gravely from the ensuing fever, chills, shaking, and delirious fits that typify the unfortunate victims of malaria—a result of the toxins released by these parasites into the host's bloodstream.

2 Obtain a prepared slide of a blood smear containing *Plasmodium* and view it using the high-power lens of your microscope. Organisms should be visible on your slide *inside* each red blood cell as tiny black rings with small dots (Fig. 6.8). Do not be confused by the larger white blood cells with oddly shaped, purple nuclei; these are normal constituents of blood. *Plasmodium* infects only the red blood cells. Notice, too, that unlike *Trypanosoma*, *Plasmodium* actually enters red blood cells. Depending upon the stage of infection depicted on your slide, you may observe one or more stages in the life cycle of *Plasmodium*.

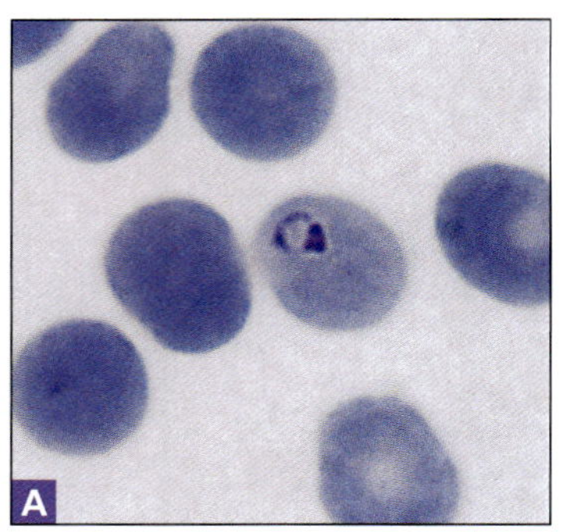

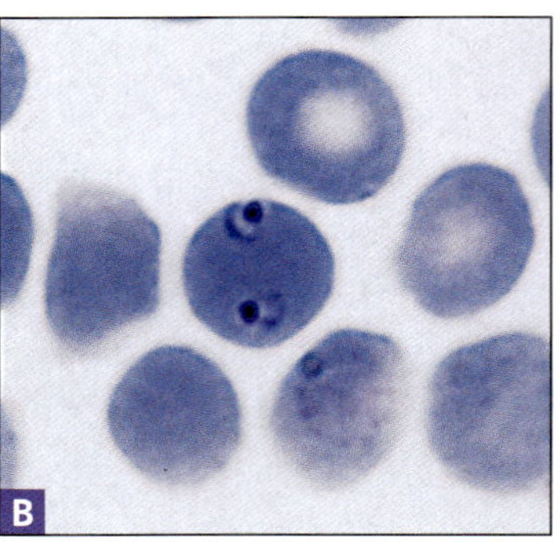

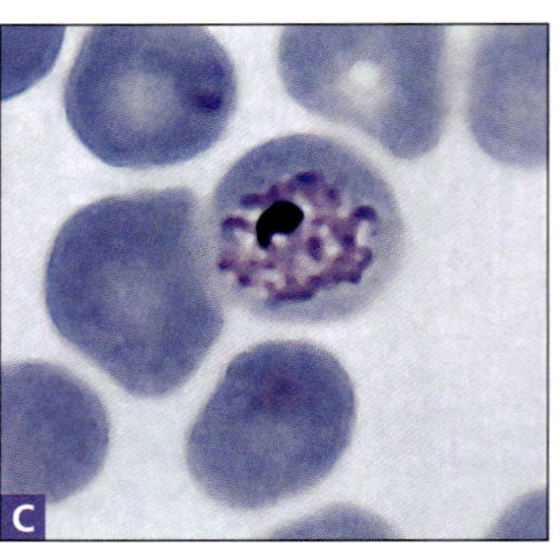

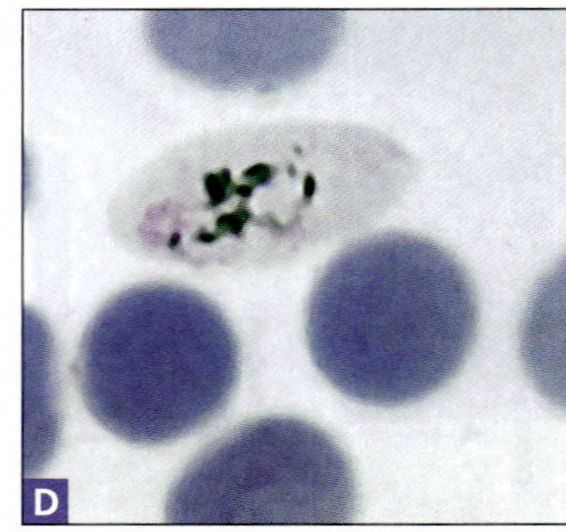

6.8 Several stages in the life cycle of the sporozoan *Plasmodium falciparum*, a causative agent of malaria in humans. (**A**) ring stage within a red blood cell, (**B**) single red blood cell with a double infection, (**C**) developing schizont, and (**D**) a gametocyte.

Check Your Progress

4.1 Do apicomplexans (such as *Plasmodium*) possess organelles for locomotion?

4.2 Does every red blood cell contain a parasite? Estimate the percentage of red blood cells that are infected.

4.3 Can you detect different stages of infection on the same slide? If so, how many?

Ciliates (Phylum Ciliophora)

Paramecium

More than 8,000 described species make up the subgroup Ciliophora. They are **unicellular** heterotrophs that possess **cilia** for locomotion and feeding, and have two types of nuclei—**micronuclei** and **macronuclei**. Most live as solitary organisms in freshwater, but their specific arrangements of cilia allow ciliates to be specialized for different lifestyles. Some are completely covered by cilia (*Paramecium*, for example), whereas in other species the cilia are clustered into a few rows along the body, such as *Vorticella*. Using their micronuclei, ciliates undergo an elaborate process of sexual gene shuffling called **conjugation**—a process we will examine more closely in *Paramecium*.

Procedure

1 Obtain a prepared slide of a whole mount of *Paramecium* and examine under medium power.

2 Use Figure 6.9 to assist you in identifying the major organelles of this unicellular protist:

pellicle—stiff outer covering that maintains basic cellular shape

cilia—hairlike projections used for locomotion and feeding

macronucleus—organelle containing many copies of a few genes; primarily controls metabolic processes of cell

micronucleus—typical eukaryotic nucleus containing entire genome; essential for genetic recombination

oral groove—lateral depression into which food is swept by ciliary currents

cytopharynx—tubular invagination lined with cilia where food enters and food vacuoles form

food vacuole—small, spherical organelle containing enzymes to digest food

contractile vacuole—one or more spherical organelles that pump out water to maintain the internal osmotic balance of the cell

anal spot (cytoproct)—site where indigestible matter is expelled from the cell

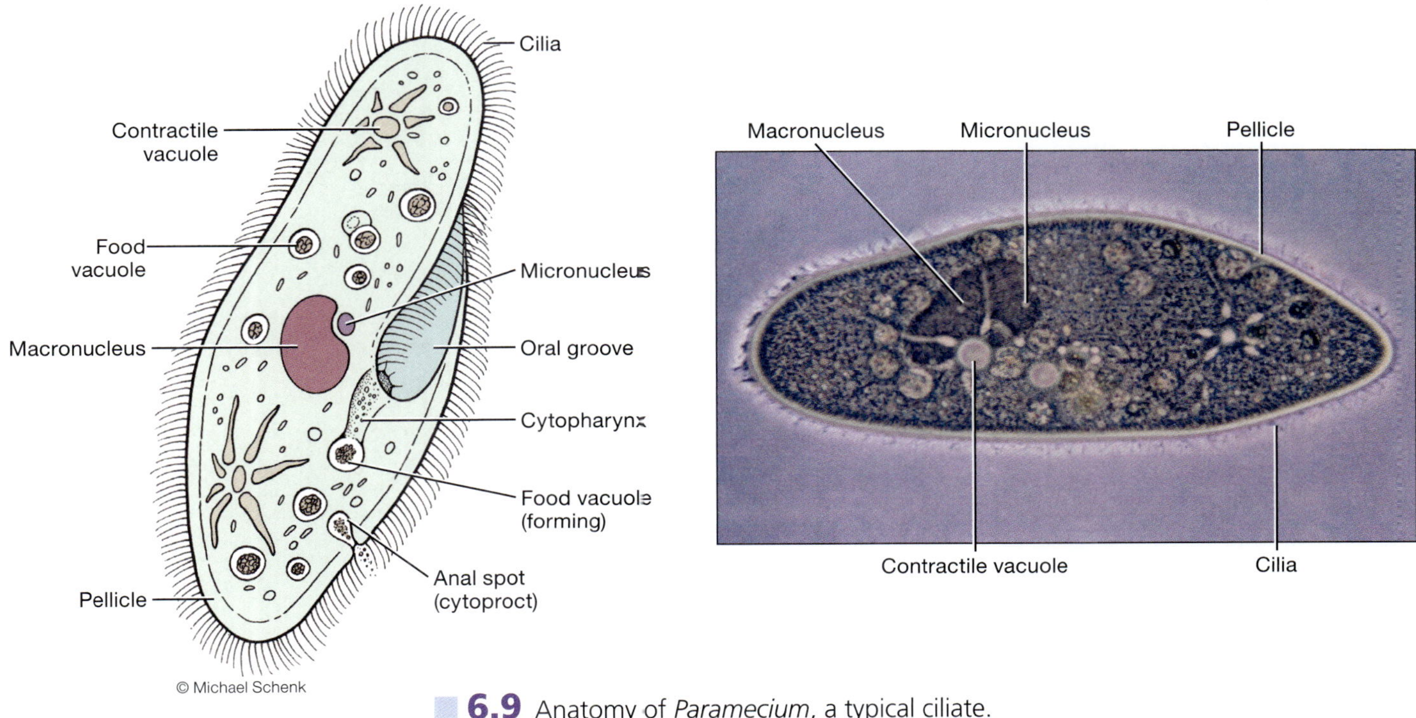

6.9 Anatomy of *Paramecium*, a typical ciliate.

3 Next obtain a prepared slide of *Paramecium* undergoing **binary fission** and examine under medium power. During this asexual process the micronucleus divides by mitosis, while the macronucleus and remainder of the original cell simply split in half to produce two genetically identical daughter cells. Notice the plane of division of the cell during fission (Fig. 6.10).

4 Now obtain a prepared slide of conjugating *Paramecium* and examine under medium power. Your slide may show pairs of cells at several different stages of this process.

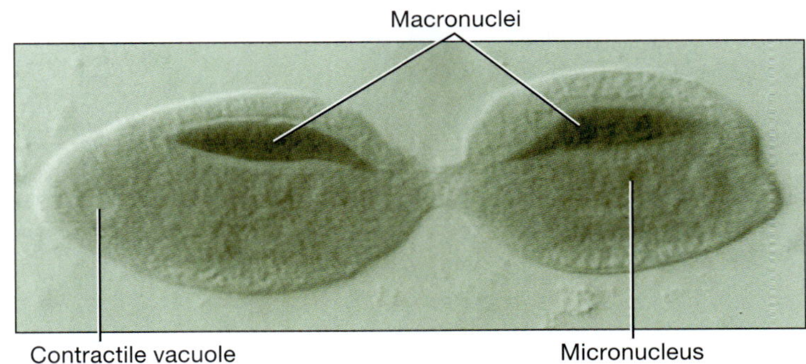

6.10 *Paramecium* undergoing binary fission—a method of asexual reproduction.

During conjugation, two individuals align longitudinally and attach temporarily (Fig. 6.11). The micronuclei in each cell undergo meiosis, creating four haploid micronuclei per parent micronucleus (ciliates often have more than one micronucleus). Three of these micronuclei disintegrate, leaving a single remaining haploid micronucleus in each cell. Each remaining micronucleus divides again mitotically, producing a haploid "male" and "female" pronucleus. The two cells then simultaneously exchange their "male" haploid pronuclei.

Next the remaining "female" and acquired "male" pronuclei in each cell fuse, forming a new, genetically "revised," diploid micronucleus in each cell. Finally, macronuclei re-form, the two cells separate, and they then usually undergo two more consecutive mitotic cell divisions, producing four daughter cells per parent cell.

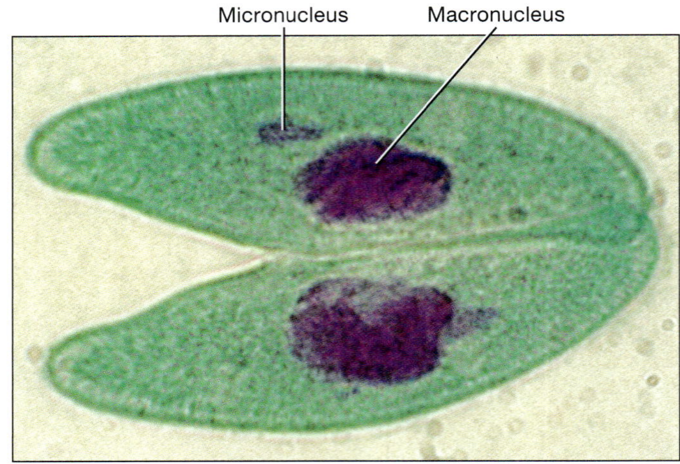

6.11 Two *Paramecium* undergoing conjugation—a form of sexual reproduction.

5 Next, make a wet mount from a culture of living *Paramecium*. Add a drop of methylcellulose, Protoslo™, or Detain™ to the slide, place a coverslip on top, and examine under medium power.

6 Identify as many of the structures depicted in Figure 6.9 as possible. If you watch carefully for a minute or two, you should be able to see the contractile vacuoles in action as they fill with water and contract to pump water out of the cell.

7 To observe the feeding mechanism in *Paramecium*, add to your slide a drop of yeast mixture dyed with Congo red. You should be able to observe the small yeast particles being swept into the oral groove and along the cytopharynx. Pink food vacuoles containing the stained yeast will form at the base of the cytopharynx as the organism begins to digest its meal.

Incidentally, Congo red is a pH indicator that is red at pH levels greater than 5.2 and blue at pH levels less than 3.0. If you watch carefully, you may be able to observe a color change in the food vacuole as the acidity levels increase during intracellular digestion. Often you will see the food vacuole change from red to purple as the pH drops below 5.2. Occasionally you will see the food vacuole change from purple to blue, indicating a highly acidic pH level below 3.0.

Check Your Progress

5.1 Give the function of each organelle listed below:

a. oral groove ______

b. micronucleus ______

c. macronucleus ______

d. contractile vacuole ______

e. food vacuole ______

f. cilia ______

5.2 Do the cilia of *Paramecium* beat in unison or in small groups?

______

5.3 Describe the process of feeding in *Paramecium*.

______

______

5.4 Is the plane of division during fission along the longitudinal or the transverse axis?

5.5 Why is conjugation considered a form of sexual reproduction?

5.6 What evolutionary benefit does conjugation provide that fission does not?

6

Vorticella

Next you will examine *Vorticella*, a sessile, stalked ciliate commonly found in stagnant bodies of freshwater. *Vorticella* typically remains attached to aquatic vegetation by a **contractile stalk** and possesses a funnel-shaped cell body with a ring of cilia around the larger **peristome** (open end). A contractile body permits *Vorticella* to push its cell body farther away from both the substrate and neighboring individuals to compete for food. The cilia beat rapidly to create currents that pull food particles into the peristome, where they are channeled to the **buccal cavity** for digestion.

Procedure

1 Prepare a wet mount from a culture of living *Vorticella*, add a coverslip, and examine under medium or high power (Fig. 6.12).

2 Closing the iris diaphragm and adjusting the condenser on your microscope to increase image contrast usually improves observation of these small, transparent organisms. Do the cilia surrounding the peristome of *Vorticella* beat in random fashion or in coordinated waves?

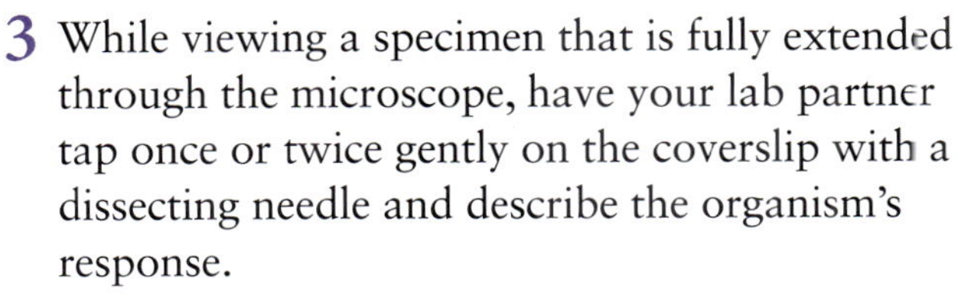

3 While viewing a specimen that is fully extended through the microscope, have your lab partner tap once or twice gently on the coverslip with a dissecting needle and describe the organism's response.

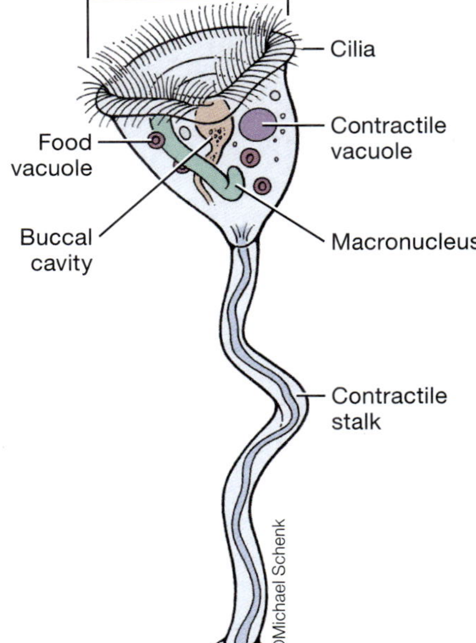

6.12 *Vorticella*, another common freshwater ciliate.

Exercise 6-3 Unikonta (Amoebas, Choanoflagellates, Animals, Fungi)

Materials Needed

- ❑ slide of *Amoeba* w.m.
- ❑ live culture of *Amoeba*
- ❑ carmine crystals in small bottle
- ❑ dissecting needle
- ❑ small squares of paper towel
- ❑ dropper bottle of distilled water
- ❑ blank slides
- ❑ coverslips
- ❑ plastic droppers
- ❑ compound microscope

The supergroup Unikonta has been one of the most surprising lineages to spring out of the wealth of molecular, biochemical, and genetic data that scientists have been gathering on protist groups over the past decade. This group of eukaryotes includes many specialized protists (such as amoebozoans and choanoflagellates), as well as all fungi and all animals! Animals, fungi, and a few protist groups are collectively grouped together as Opisthokonts, while many of the amoeboid protists are grouped together as Amoebozoans (Fig. 6.1). As strange as these relationships may seem, the validity of these assemblages is now strongly supported by numerous molecular studies.

Amoebas

Amoebas are unicellular heterotrophs that lack **cell walls**. Despite their simplistic appearance, amoebas are highly specialized protists, tailor-made for their preferred habitats. Their methodical creeping motion and manner of engulfing food particles equip them for life at the bottom of lakes and ponds amidst rich sources of sedentary organic particles. They have flexible plasma membranes that form **pseudopodia**—cytoplasmic extensions used for feeding and locomotion. Amoebas lack the ability to reproduce sexually, and instead, reproduce by binary fission. The majority of species are free-living, though some are parasitic, such as *Entamoeba histolytica*, the causative agent of amoebic dysentery in humans—a disease commonly spread through contaminated sources of drinking water or food.

Entamoeba histolytica is one of many species of protists that can form a cyst by secreting a proteinaceous cell wall around itself and entering a state of dormancy. In the cyst stage, amoebas can survive prolonged periods of dryness and can even be dispersed by wind. If the cysts are ingested by a host, the cyst is activated and the amoeba crawls out and burrows into the host's intestinal wall, causing inflammation of the intestinal lining, severe cramps, and diarrhea, sometimes resulting in death by dehydration!

Procedure

1 Obtain a prepared slide of amoebas and scan the slide using the low-power objective to locate several stained amoebas.

Although every specimen on your slide is likely captured in a different shape, many common features should be evident. A large, darkly stained, oval nucleus and numerous smaller, spherical **food vacuoles** should be visible (Fig. 6.13). Occasionally, spherical contractile vacuoles will be present on prepared specimens. Pseudopodia also should be evident along the borders of the **plasma membrane**.

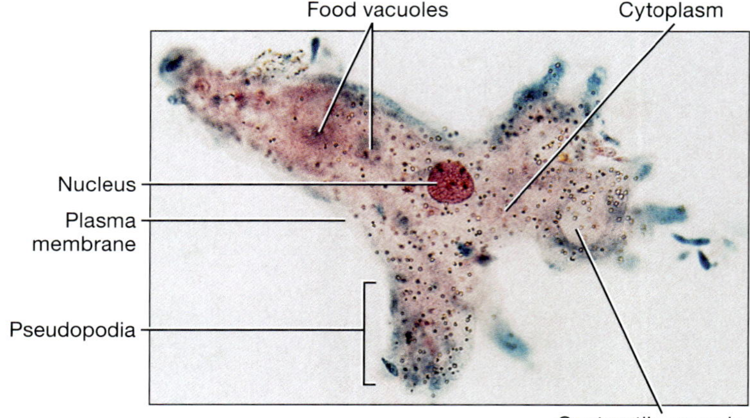

6.13 Anatomy of *Amoeba proteus*, a representative amoeba.

2 Now that you have a feel for what amoebas look like, obtain a drop of culture medium containing live amoebas and place on a slide with a coverslip.

3 If you are preparing the wet mount yourself, carefully draw a small volume of culture from the bottom of the container using a plastic dropper without stirring, bubbling, or otherwise disturbing the culture medium. Though often difficult to find, amoebas typically crawl along the bottom of culture dishes and congregate near food sources or debris in the dish.

4 Scan your slide using low power to locate live amoebas. Unlike the stained specimens on the prepared slide you viewed earlier, they will not be colored. Instead, they will be gray and semitransparent with a granular appearance.

5 Once you have located an amoeba, move to medium power, close the iris diaphragm, and adjust the condenser to increase the contrast (live amoebas are nearly transparent!). You may see very little outward movement initially, but upon closer examination under medium- or high-power magnification, **cytoplasmic streaming** should be visible.

6 Identify the following structures on your live amoeba: nucleus, food vacuoles, contractile vacuole, pseudopodia.

7 If you locate a clear, spherical contractile vacuole, watch it closely for a minute or two. You may see it contract as it pumps water out of the cell for the purpose of maintaining the proper internal ion concentrations inside the amoeboid cell.

Amoebas engulf their food by a process called **phagocytosis** in which extensions of the plasma membrane (pseudopodia) elongate and surround the food item. The portion of plasma membrane enclosing the food particle pinches off and forms a food vacuole inside the amoeba, where digestive enzymes begin breaking down the food.

8 To demonstrate this process, place a drop of distilled water against one edge of the coverslip nearest the amoeba.

9 Use the tip of a dissecting needle to pick up a few carmine crystals and deposit them into the water droplet.

10 Hold a small square of paper towel against the other side of the coverslip, *opposite* the water droplet containing the carmine, and draw the suspension beneath the coverslip.

11 Observe the amoeba and you may be lucky enough to catch it in the act of feeding! If so, describe the feeding process as it unfolds.

Check Your Progress

6.1 What is phagocytosis? What function does it serve?

6.2 Relative to other unicellular protists, how would you characterize the size of amoebas?

6.3 Why would excess water tend to accumulate in amoebas?

6.4 Do amoebas have a rigid cell wall? How can you tell?

6.5 Do amoebas have a "permanent" anterior end like many other motile protists? Explain.

Choanoflagellates

6

One lineage of protists in the supergroup Unikonta occupies a very special position in the evolutionary tree of life—the choanoflagellates. According to recent molecular studies of ribosomal RNA, this unique assemblage of single-celled eukaryotes represents the sister group to animals (Fig. 6.1). Choanoflagellates are sessile, aquatic organisms (either solitary or colonial) with each cell bearing a single flagellum surrounded by a thin collar (Fig. 6.14). The rhythmic beating of each flagellum creates a current that draws water into the collar where tiny organic particles (typically bacteria) are collected by microvilli lining the collar.

Of particular significance to biologists are the similarities between choanoflagellates and the choanocytes, or collar cells, of sponges. Many biologists have speculated that early sponges arose from assemblages of choanoflagellates and eventually gave rise to all other animal groups. When the molecular sequences of a variety of protist groups were compared to animal sequences, the data showed clearly that one group of protists, the choanoflagellates, was the most likely ancestor of sponges. Were ancient choanoflagellates the ancestors of all animal life? Perhaps. There are many sound scientific arguments in favor of this hypothesis and other good cases to be made against this position. The majority of research at this point in time seems to support the hypothesis that a little more than 600 million years ago, the first sponges evolved from colonial choanoflagellates similar to ones living today, and the evolutionary lineage that would eventually give rise to all other animals was off to a start.

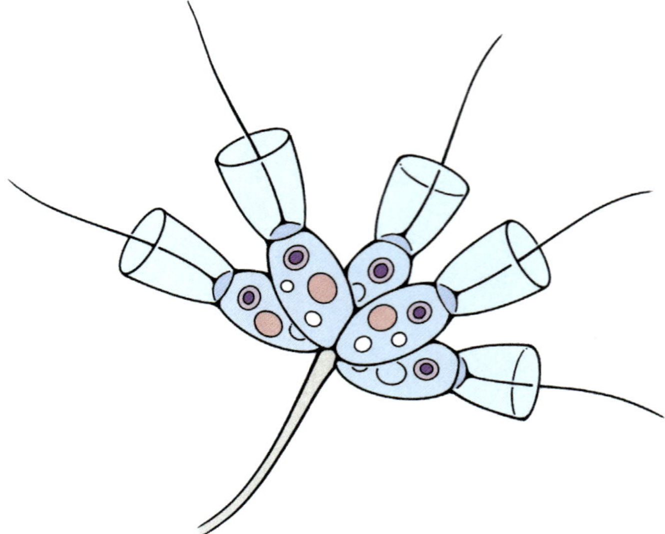

6.14 Choanoflagellates living today are thought to resemble the ancient eukaryotes that formed the choanocytes of the first sponges.

Name ____________________ Section ____________

6 Questions for Review

1 Name three general methods of locomotion exhibited by protists and examples of organisms that utilize each method.

Method of Locomotion *Example*

1. ____________________

2. ____________________

3. ____________________

2 What structures or features do all protists have in common?

____

____

____

3 Match the group with its correct description or characteristic.

	Group		Description
______	kinetoplastid	**a.**	unicellular heterotroph with two types of nuclei
______	euglenoid	**b.**	flagella arise from grooves in cellulose plates
______	ciliate	**c.**	nonmotile parasite with penetrating apical region
______	dinoflagellate	**d.**	flagella arise from anterior pocket
______	apicomplexan	**e.**	unicellular parasite with one large mitochondrion
______	choanoflagellate	**f.**	unicellular heterotroph that moves using pseudopodia
______	amoeba	**g.**	sessile protist with single flagellum surrounded by collar

4 Give two examples of parasitic protists and the disease they each cause.

1. ____________________

2. ____________________

5 Contrast the processes of binary fission and conjugation in ciliates.

____

____

____

____

6 Using *Paramecium* as an example, explain how a protist eats, respires, digests food, reproduces, and maintains an internal water balance (osmoregulation), all within its single-celled body.

7 According to current theory, which group of protists is most closely related to animals, and may have given rise to early sponges? What similarities does this group of protists share with modern sponges?

Name ______________________ Section ______________

6 Laboratory Practical Challenge

The following evaluation is intended to replicate the method of assessment and type of questions you will be asked on lab practicals. If you have difficulty with these questions, you may benefit from additional review of the specimens provided in lab.

1 Identify the phylum (or supergroup) to which this organism belongs AND identify the *function* of the structure indicated by line A.

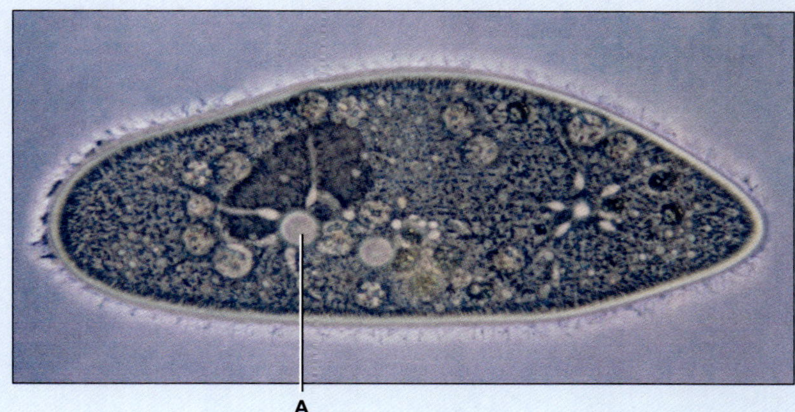

2 Identify the phylum (or supergroup) to which this organism belongs AND identify its mode of nutrition acquisition.

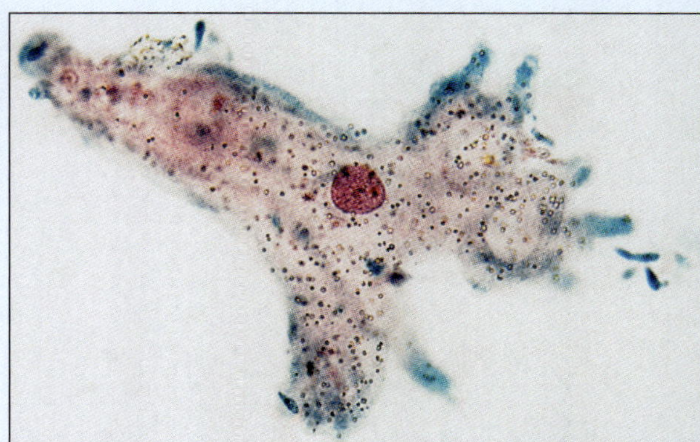

6

3 Identify the phylum (or subphylum) to which this organism belongs AND identify the name of the structure(s) it uses for locomotion.

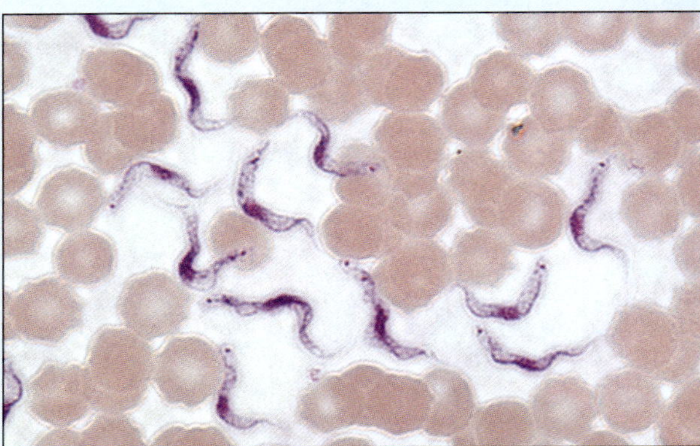

4 Identify the phylum (or supergroup) to which this organism belongs AND identify its mode of nutrition acquisition.

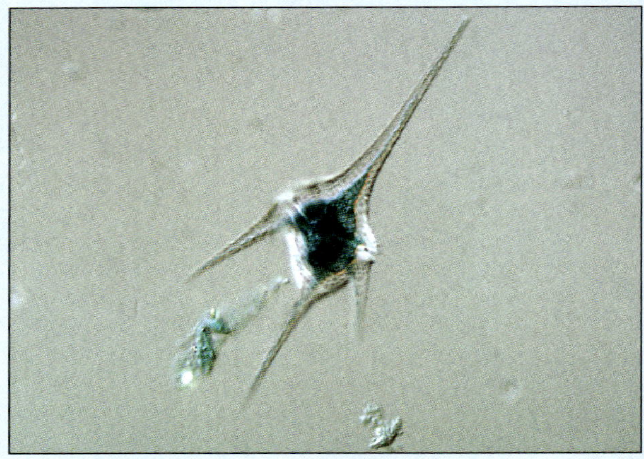

Sponges

After completing the exercises in this chapter, you should be able to:

1 List the distinguishing characteristics of the phylum Porifera.

2 Understand the evolutionary position of sponges on the phylogeny of major animal phyla.

3 Recognize and describe the basic anatomy and cellular organization of sponges.

4 Recognize the skeletal elements of sponges and be able to describe their composition.

5 Describe the pattern of water flow through the sponge and its significance in feeding, respiration, waste elimination, and reproduction.

6 Understand the phenomenon of reaggregation in sponges.

7

Sponges (phylum Porifera) are entirely aquatic, mostly marine animals, with only a small percentage (around 150 species) found in freshwater lakes and streams. They range in size from a few millimeters across to more than a meter in diameter and vary tremendously in color and shape (Fig. 7.1). Some species display low, encrusting body forms, and others are taller and more upright. They may have a hard, coarse, almost abrasive texture, or be soft and flexible.

Despite this diversity, all sponges share many anatomical similarities that tie in to their common lifestyle.

1. All sponges are sedentary animals that have no body symmetry and lack tissues, organs, or organ systems. As a result, all major biological functions occur at the cellular level.
2. Their simple bodies consist of just four primary cell types arranged around a system of pores and canals.
3. All sponges feed by filtration—their collar cells create water currents that pull in organic particles.
4. Sponges have an inner meshwork of microscopic **spicules**, or collagen fibers, that serves as an internal skeleton.

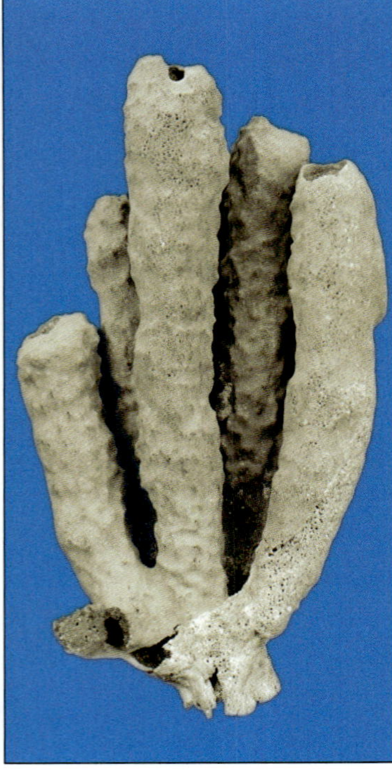

7.1 Sponges, members of the phylum Porifera.

Sponge Body Types

Sponges are categorized into three principal body types based on the degree of convolution or folding of their body walls (Fig. 7.2).

1. An **ascon** sponge has no folding to its body wall and is simply a hollow tube just a few cells thick with a prominent osculum at the top. Ascon sponges are generally quite small, due to the constraints of having such a small surface-to-volume ratio.
2. Sponges with a simple infolding to their body walls are known as **sycon** sponges. The increased surface-to-volume ratio inherent in this design allows sycon sponges to be larger.
3. The vast majority of sponges living today are organized in an even more complex way—with body folds that are themselves folded, resulting in a series of chambers connected by canals. This body plan is known as a **leucon** body type. The familiar bath sponges are the most prominent example of the leucon body plan.

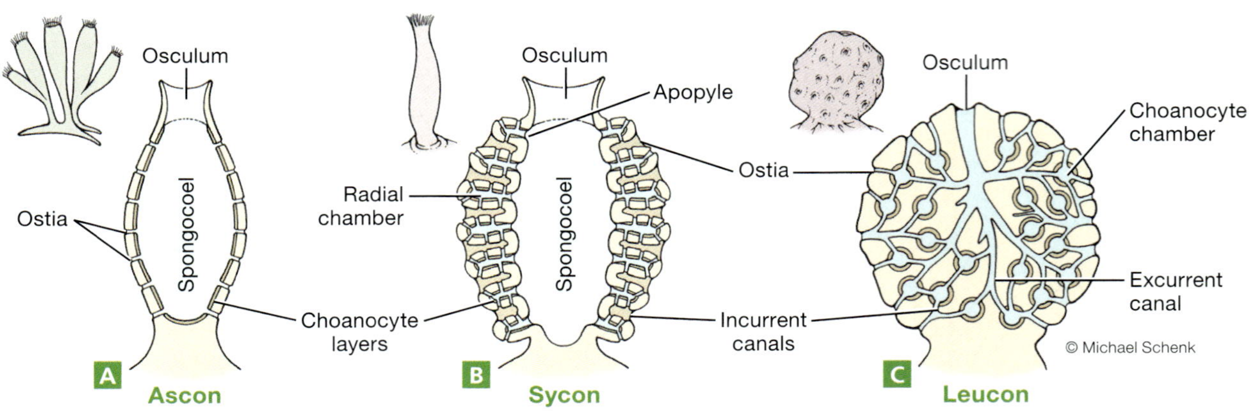

7.2 Sponge body types: (**A**) ascon, (**B**) sycon, and (**C**) leucon.

More than 5,600 species of sponges have been described, with the vast majority belonging to only one class, Demospongiae. The sponges used for bathing, painting, and other common commercial uses are all members of this class. Within phylum Porifera, three classes of sponges are commonly recognized and are primarily distinguished by the type of skeleton and the arrangement of spicules (Table 7-1).

Table 7-1 Classes of Phylum Porifera

Class and Representative Animals	Characteristics
Calcarea (≈ 400 sp.) *Scypha*, *Leucosolenia*	Marine sponges that possess small, needle-shaped, calcium carbonate spicules with three or four rays; may have ascon, sycon, or leucon body type
Hexactinellida (≈ 500 sp.) *Euplectella* (Venus flower basket)	Marine sponges with six-rayed spicules composed of silica; either sycon or leucon body type
Demospongiae (≈ 4,700 sp.) *Spongilla*, bath sponges, commercial sponges	Often brilliantly colored, mostly marine sponges with skeletons composed of spongin (a flexible, structural protein), spicules, or both; leucon body type; some freshwater species inhabit streams, ponds, and lakes

Evolutionary History

Sponges represent the simplest and most primitive animals living today and, thus, are important in considering the evolution of animals. The earliest sponge fossils date back 600 million years, making sponges the oldest known animals. These sponges were very simple in body form, similar to the ascon sponges of today. Natural selection favored body plans with increased surface area, giving rise over time to increasingly complex sponge body plans. Sponges' intermediate level of cellular organization hints at the evolutionary ties between unicellular protists and higher multicellular animals.

The **choanocytes**, or collar cells, of sponges are nearly identical to one type of colonial protist living today, known as choanoflagellates (Fig. 7.3). In fact, many biologists have speculated that early sponges arose from assemblages of choanoflagellates and eventually gave rise to all other animal groups—the one-ancestor theory. Based on certain aspects of cellular structure and chemistry, other biologists have favored the multiple-origins theory suggesting that all animals arose from two or three different single-cell ancestors.

Because fossil evidence from that long ago is sparse and incomplete, biologists turned to molecular and biochemical research to fill in the gaps and provide answers to these remaining questions. Recent molecular sequences of genes for ribosomal RNA and for several other ancient molecules fundamental to all cells show that an ancient spongelike ancestor was the first animal, and that all other animal groups descended from this ancestor (Fig. 7.4).

When the molecular sequences of a variety of protist groups also were compared to animal sequences, the data

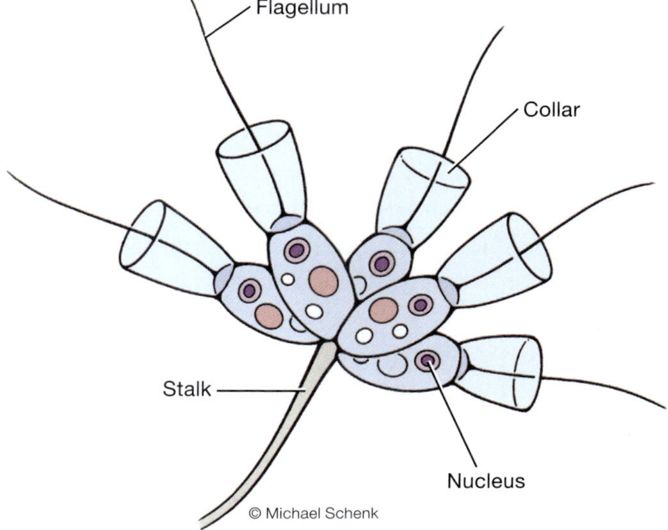

7.3 Choanoflagellates living today are thought to resemble the ancient organisms that formed the choanocytes of the first sponges.

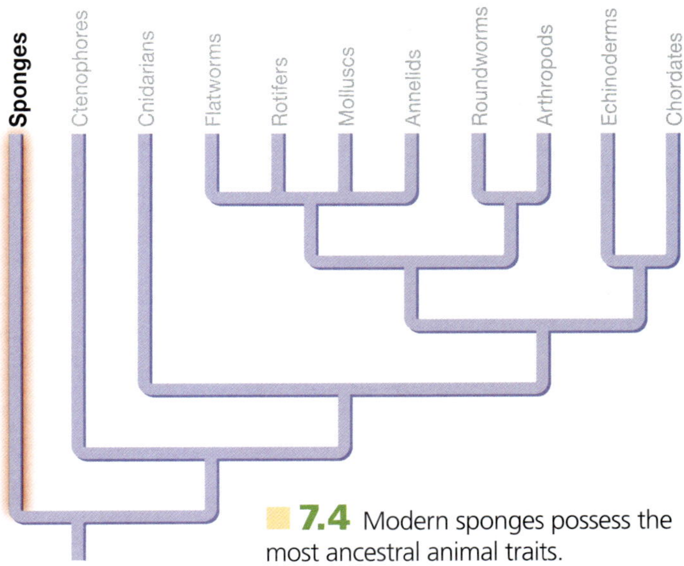

7.4 Modern sponges possess the most ancestral animal traits.

showed clearly that one group of protists, the choanoflagellates, was the most likely ancestor of sponges. This makes choanoflagellates our nearest living protist relative and provides evidence that strongly supports the one-ancestor theory, with ancient sponge ancestors positioned correctly at the base of the animal tree of life. Figure A.1 in the Appendix provides a more complete phylogeny depicting the evolutionary relationships among the major animal phyla covered in this laboratory guide.

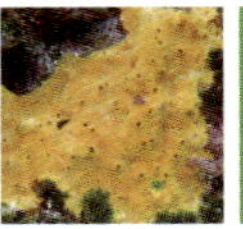

Exercise 7-1 Sponge Anatomy

Materials Needed

- ❑ slides of *Grantia* (or *Scypha*) c.s. and l.s.
- ❑ compound microscope

Procedure

7

1 Obtain a prepared slide of a longitudinal section (l.s.) through *Grantia, Scypha*, or another sycon sponge. As you view the slide, try to visualize the path of water flowing from the outside of the sponge toward the inside.

Water flow is *crucial* to understanding the anatomy of sponges and the reasons for their cellular organization. Because sponges do not move, every aspect of their existence—feeding, respiration, waste elimination, reproduction, even dispersal—hinges on water moving through them.

The more complex the sponge body type (that is, the more folds the body wall has), the more efficient the sponge is at extracting oxygen and food particles from the water. Ascon sponges have only slight folding of their canal systems and relatively large spongocoels, meaning that very little of the water that flows through their bodies comes in contact with their cells. This means that they miss a lot of the food and oxygen that passes through them and this becomes a determining factor in their small body sizes. Leucon sponges, on the other hand, have complex, intricately folded canal systems, allowing more of the passing water to come into contact with a larger proportion of their cells. This results in more food and oxygen being extracted from a given amount of water and permits these sponges to attain larger body sizes—up to 1 meter in height and diameter in some species.

The sponge you are viewing (*Grantia* or *Scypha*) is a sycon sponge, with an intermediate body type and simple folding of its body walls (Figs. 7.5 and 7.6). Water enters the sponge through pores on the body surface called **ostia** (singular = ostium). Ostia channel water down the **incurrent canals** to a larger number of tiny pores scattered along the folds of the incurrent canals. These tiny pores are actually openings in elongated, doughnut-shaped cells called **porocytes** (Fig. 7.7). The porocytes channel water into flagellated chambers known as **radial canals**. Specialized cells called choanocytes (believed to have evolved from choanoflagellates) lining the interior surface of the radial canals trap small food particles with their flagella and engulf them through phagocytosis. The beating of the choanocytes' flagella actually creates the water currents that flow through the sponge body. The choanocytes do not digest the captured food particles themselves, but instead pass them to another type of cell known as the amoebocyte. **Amoebocytes** are mobile cells that reside in the gelatinous matrix between the choanocytes and the **pinacocytes**, the cells making up the outer layer of the sponge. Amoebocytes carry food to other cells within the sponge body. Amoebocytes have another rather amazing function—they can undergo developmental changes to transform into any other cell type that may be required for the sponge to function properly.

As oxygen-rich water is passed through the sponge body, simple diffusion of oxygen into the individual sponge cells occurs, while carbon dioxide diffuses out of the sponge cells into the surrounding water. Likewise, metabolic wastes produced by cellular metabolism also diffuse out of individual cells into the surrounding water. Remember—all major physiological functions in sponges occur at the cellular level. Water, now devoid of most of its dissolved oxygen and food particles, passes out of each radial canal though a larger opening called the **apopyle** into the **spongocoel** where it collects. Finally the water is pushed out of the large **osculum**, which is usually located at the top of the sponge.

2 Now examine a prepared slide of a cross section (c.s.) through this same species of sponge (Fig. 7.6). Identify all of the regions and structures discussed previously, keeping in mind that the orientation of this section is perpendicular to the longitudinal section viewed earlier.

3 Use Figures 7.6 and 7.7 to assist you in identifying the structures and regions on the cross section.

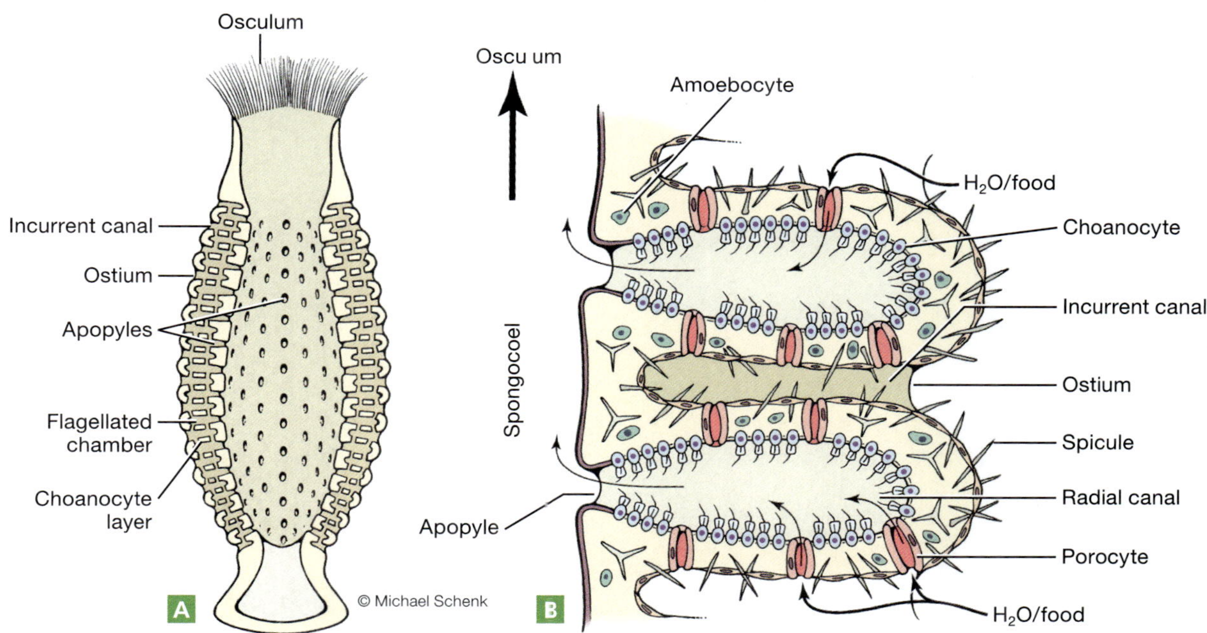

7.5 Longitudinal sections through (**A**) a sycon sponge, and (**B**) close-up depicting organization of the canals and chambers comprising the body wall of a simple sponge.

7.6 Cross section through (**A**) a simple sponge, and (**B**) close-up depicting arrangement of the different cell types.

7.7 Cell types and skeletal structure of the basic sponge body wall.

Reproduction in Sponges

Reproduction in sponges can take several different forms. Sponges have strong powers of regeneration, probably because of the interchangeability of their cells and the absence of specialized tissues. Wounded sponges can regenerate tissue and skeleton to replace the damaged part. These same abilities enable some species to reproduce asexually by fragmentation or budding. The fragments break off, drift away, and then settle to start new colonies.

The formation of gemmules is another form of asexual reproduction seen in sponges. This is common for freshwater sponges but rare in marine species. The **gemmule** is a densely matted, hardened ball containing amoebocytes surrounded by a coating of collagen and spicules. In response to adverse environmental conditions, the sponge body may disintegrate, leaving gemmules behind. These protected balls can survive through the winter and then form a new adult sponge in the spring.

Occasionally, sponges reproduce sexually. When environmental conditions are favorable, certain choanocyte cells in the sponge lose their collars and flagella and divide meiotically to become haploid sperm that are released into the water. These sperm cells enter a different sponge body, where they are captured and transferred to amoeboid cells that travel through the sponge body to an embedded haploid egg. The fertilized egg develops into a simple, ciliated larva that eventually breaks free and swims for a few hours before it settles to the bottom and gradually matures into a new sponge.

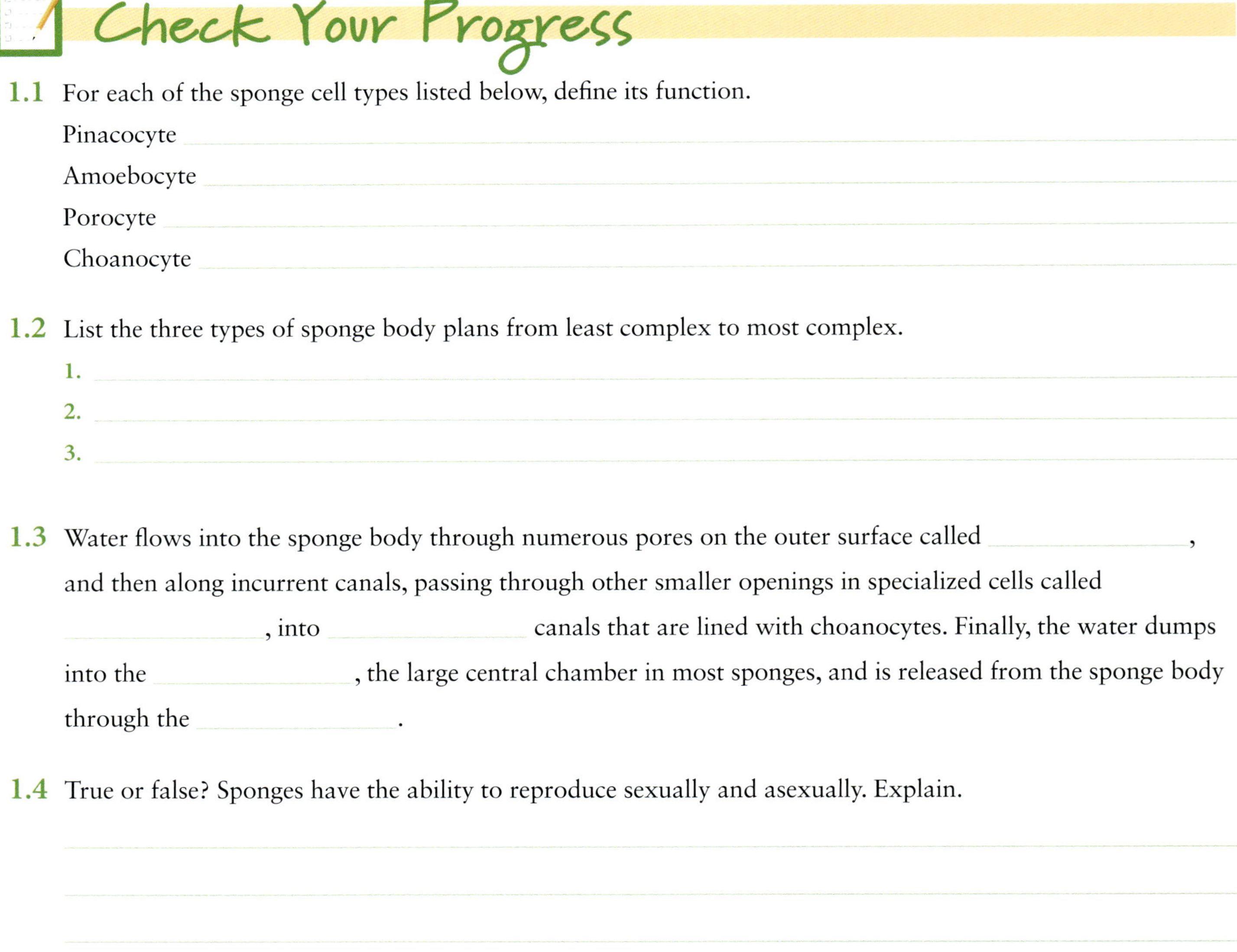

Check Your Progress

1.1 For each of the sponge cell types listed below, define its function.

Pinacocyte ____________________

Amoebocyte ____________________

Porocyte ____________________

Choanocyte ____________________

1.2 List the three types of sponge body plans from least complex to most complex.

1. ____________________
2. ____________________
3. ____________________

1.3 Water flows into the sponge body through numerous pores on the outer surface called ____________, and then along incurrent canals, passing through other smaller openings in specialized cells called ____________, into ____________ canals that are lined with choanocytes. Finally, the water dumps into the ____________, the large central chamber in most sponges, and is released from the sponge body through the ____________.

1.4 True or false? Sponges have the ability to reproduce sexually and asexually. Explain.

____

____

____

Exercise 7-2 Observation of Spicules: The Sponge Skeleton

Materials Needed

- ❑ preserved or living sponge
- ❑ dropper bottle of bleach solution
- ❑ dropper bottle of water
- ❑ glass depression slides
- ❑ coverslips
- ❑ compound microscope

One of the characteristics zoologists use to classify sponges into taxonomic groups is spicule structure. Spicules (and spongin) are the skeletal elements of sponges and are secreted by **sclerocytes**. In many sponges this skeleton is composed of hard, crystalline spicules formed from either calcium carbonate or silicon (Fig. 7.8). Other sponges, such as the exotic bath sponges that formerly were popular consumer items, secrete a proteinaceous material called **spongin** that is more flexible. Spicules come in many shapes and sizes and may have a haphazard arrangement within the sponge or be intricately woven to form dazzling geometric patterns (Fig. 7.9).

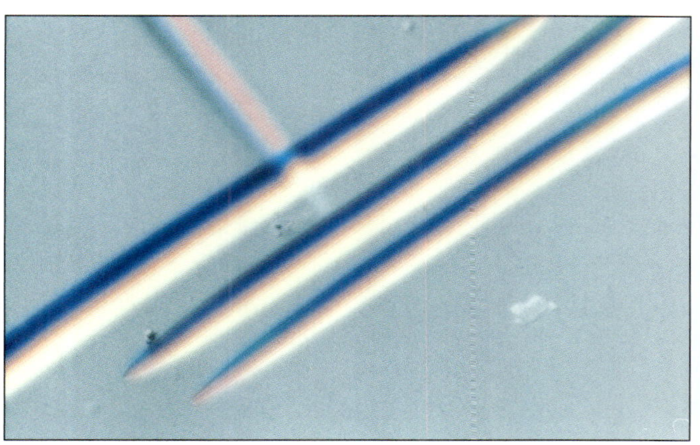

7.8 Spicules of freshwater sponges.

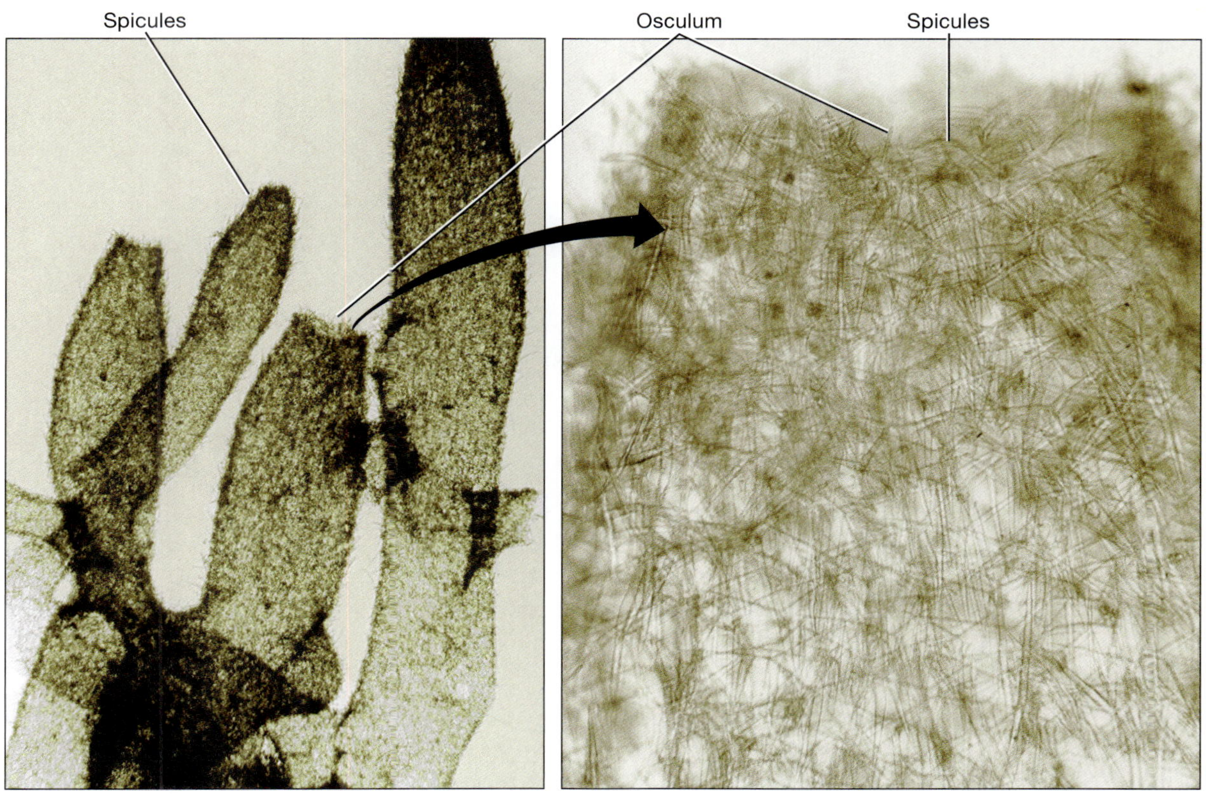

7.9 Micrographs of *Leucosolenia*, an ascon sponge, showing the arrangement of its spicules.

Procedure

1 After obtaining a small piece of living or preserved sponge (*Grantia* will work fine for this exercise), place a drop of water in the depression well of the slide (Fig. 7.10).

2 Use a second microscope slide to gently crush a small piece of sponge *on the edge of the depression slide*.

3 Scrape this crushed preparation into the water.

4 Add 1–2 drops of bleach to this mixture to dissolve the cells and proteinaceous material, thereby exposing the spicules.

5 Cover with a coverslip.

6 Examine this preparation under low power, then higher magnification, to view the spicules, and then sketch several of the spicules in the space below.

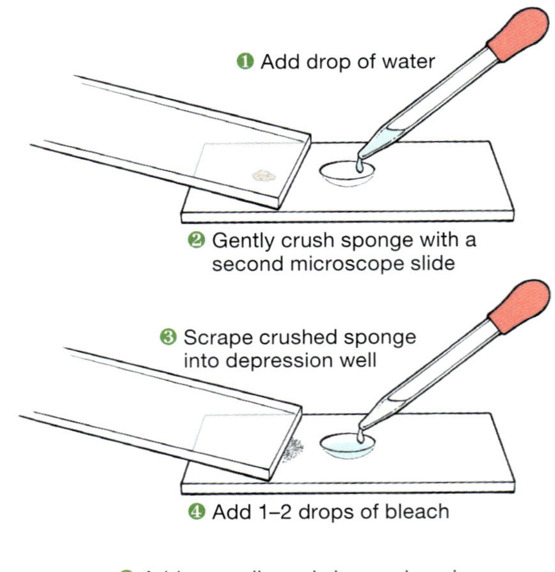

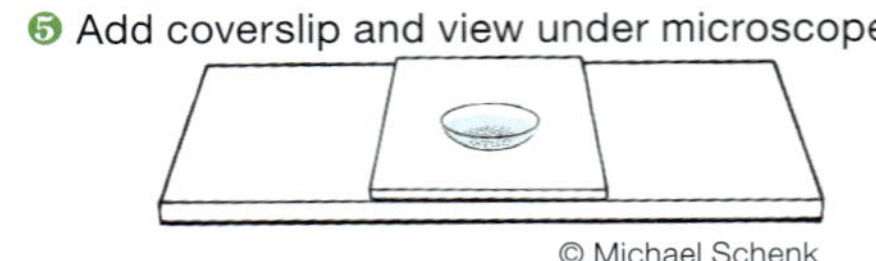

7.10 Procedure for isolating sponge spicules for microscopic examination.

7

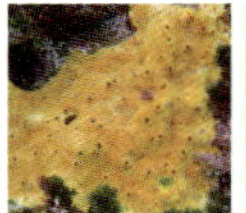

Exercise 7-3 Observing Water Flow in Freshwater Sponges

Materials Needed

- ❑ small, live freshwater sponges
- ❑ clear petri dish or finger bowl
- ❑ pond water
- ❑ powdered algae (*Spirulina*)
- ❑ small metal spatula
- ❑ dissecting microscope

Despite typically being much smaller and less abundant than their marine cousins, freshwater sponges also play an important ecological role in their ecosystems. Their apparent lack of behavior is deceiving, because these simple creatures are constantly pulling water through their porous bodies to extract suspended food particles and oxygen. While this is a necessary process for sponges to feed and breathe, it has the secondary effect of creating small microcurrents in the water surrounding sponges, which keeps the water from becoming stagnant and contributes to the natural filtration process of lakes and ponds. These microcurrents play a vital role in freshwater ecosystems, because sponges keep moving "new" water into their immediate vicinity and continue pushing away "old," oxygen-depleted water.

But how could a sponge the size of a postage stamp have any dramatic effect on the massive volume of water surrounding it? In this next exercise you will employ a simple method for visualizing the volume of water that a sponge can filter through its body in a given time frame. You may be surprised at the staggering volume of water that such a small sponge can move over the course of a day!

Procedure

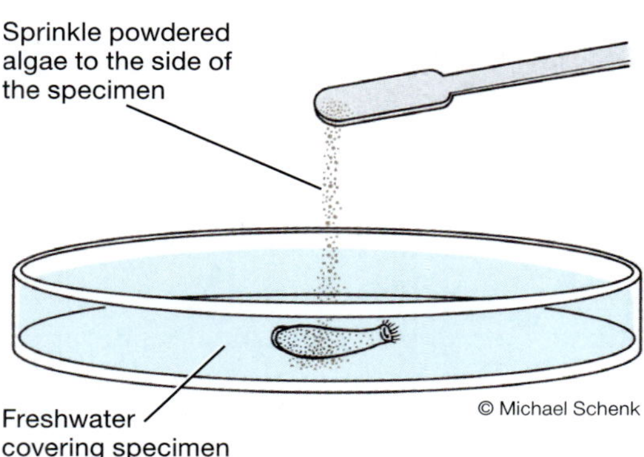

7.11 Experimental procedure for observing water flow in living freshwater sponges.

1. Obtain a small, living, freshwater sponge and submerge it in a clear petri dish or finger bowl filled with pond water. Ideally, your sponge should have one distinct osculum, preferably accentuated by a thin, transparent, chimneylike tube projecting from the osculum.
2. Observe the sponge with a dissecting microscope to locate the osculum.
3. Rotate the sponge so the osculum is pointing sideways in the dish (perpendicular to the plane of view of the microscope).
4. Adjust the level of water in the dish so the sponge is submerged approximately 1 cm below the water surface.
5. Using a small spatula, lightly dust the surface of the water directly over the sponge with a tiny amount of powdered algae (*Spirulina*) (Fig. 7.11).
6. Gently stir the water to suspend the algal particles in the solution. The algal particles will vary in size, and some of the larger particles will be visible as they "shoot" out of the osculum.
7. Adjust the focal plane of the dissecting microscope so the algal particles expelled from the osculum are in focus, and continue to observe this phenomenon to get an estimate of how quickly particles are cycled through the sponge.

Check Your Progress

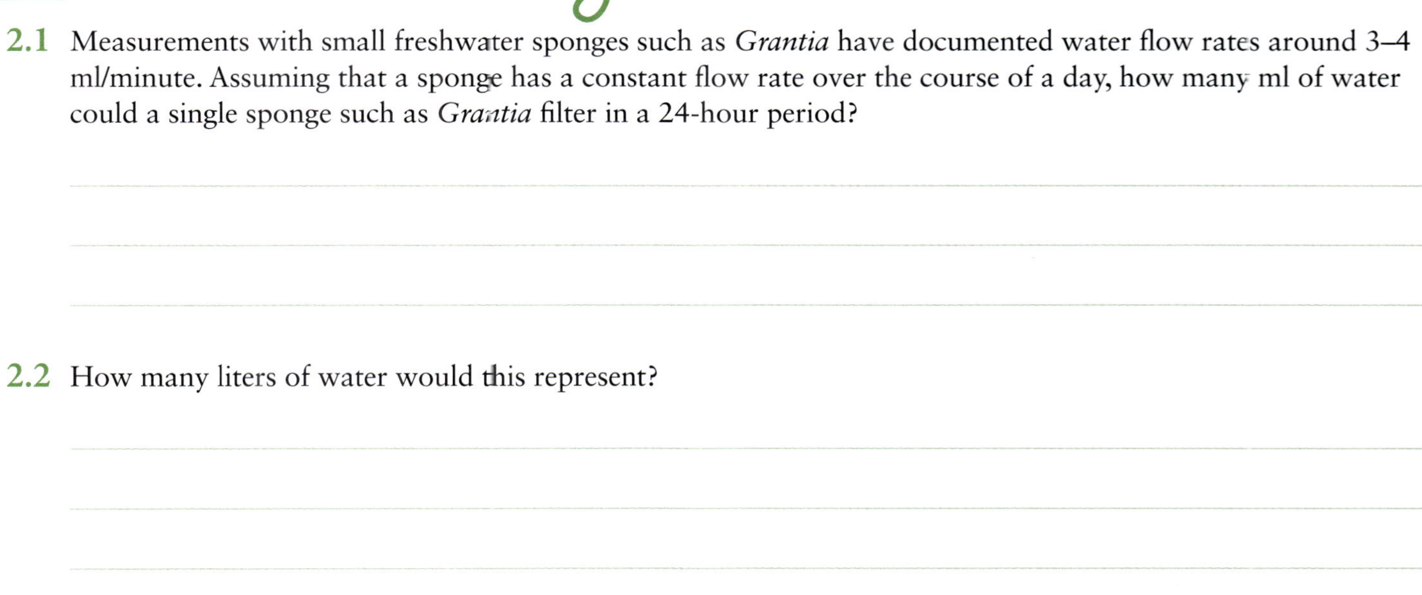

2.1 Measurements with small freshwater sponges such as *Grantia* have documented water flow rates around 3–4 ml/minute. Assuming that a sponge has a constant flow rate over the course of a day, how many ml of water could a single sponge such as *Grantia* filter in a 24-hour period?

2.2 How many liters of water would this represent?

Exercise 7-4 Cellular Reaggregation in Sponges

Materials Needed

- ❑ living marine sponges
- ❑ Ca^{2+}/Mg^{2+} free seawater
- ❑ normal seawater
- ❑ silk blotting cloth or cheesecloth
- ❑ plastic droppers
- ❑ blank slides
- ❑ coverslips
- ❑ 50 ml beakers
- ❑ compound microscope

One of the most fascinating properties of sponges that illustrates the loose physiological connections between their cells is their remarkable ability to reaggregate. If the cellular connections that hold the sponge cells together are broken, the sponge will fall apart into individual cells. These individual cells, however, can survive for some time and can reassemble spontaneously to form a functional sponge again. This feat cannot be accomplished by "higher" organisms with true tissues; the physiological connections between their cells are too permanent to permit disruption—even for a brief time.

In this experiment you will observe the reaggregation phenomenon and will test the specificity of this process. Two species of sponges (probably *Microciona prolifera* and *Haliclona occulata*) have been soaking overnight in Ca^{2+}/Mg^{2+} free seawater. Calcium and magnesium are necessary elements for intercellular adhesion; therefore, deprivation of these ions facilitates dissociation of the cells.

7

Procedure

1 Prepare individual suspensions of dissociated cells of these two species by pressing a small piece of each sponge through a piece of silk blotting cloth (or cheesecloth) directly into a beaker containing 25 ml of normal seawater (Fig. 7.12).

2 Since *Microciona prolifera* is bright red and *Haliclona occulata* is purple, you will be able to tell the two suspensions of dissociated cells apart. One will appear pink and the other blue.

3 Using a plastic dropper, remove a sample from one of the beakers.

4 Place a drop on a slide, cover it, and examine it with a microscope. Can you distinguish the individual cells?

5 Remove 1.0 ml from each beaker and mix the two samples together in a third small beaker containing 25 ml of normal seawater.

6 After 30, 60, and 90 minutes, remove small volumes of the suspension and make a wet mount. Look for thin protoplasmic extensions, called **filopodia,** put out by the small clumps of aggregating cells. Can you determine any differences in the size of aggregates? Are the aggregates formed from a combination of the red and purple cells or is there a segregation of cell types?

7 Record your observations in Table 7-2, and sketch a few cell clumps on the next page.

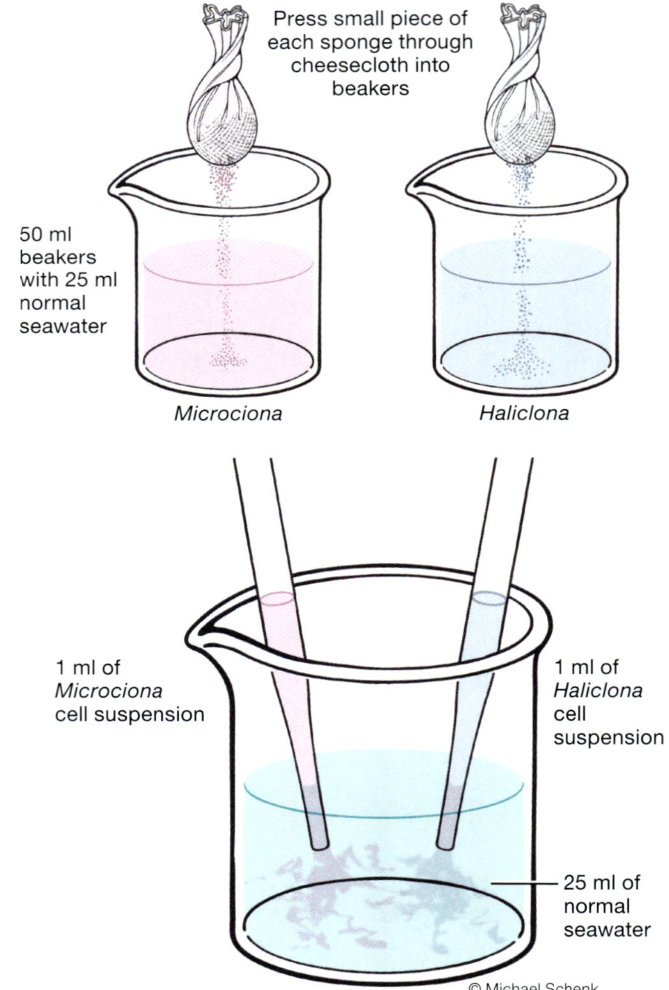

7.12 Experimental setup for examining reaggregation in living sponges.

Table 7-2 Reaggregation of Sponges in Normal Seawater

	Time		
	30 min.	**60 min.**	**90 min.**
Observations of sponge reaggregation			
Drawings of aggregates			

7

Notes

7

Name ______________________________ Section ______________

7 Questions for Review

1 Complete Summary Table A-1 in the Appendix, filling in the characteristics of sponges in the appropriate row.

2 What is the advantage of a folded or convoluted body wall in sponges?

3 Why are sponges placed in a taxonomic group beside the other animals and considered an evolutionary dead end by many zoologists?

4 What cell type in sponges is responsible for producing the water current through the sponge? What other important function do these cells serve?

5 What cell type in sponges is responsible for distributing nutrients to other cells? What other important roles do these cells perform?

6 Briefly discuss how sponges rely on water moving through their bodies for feeding, respiration, waste elimination, reproduction, and dispersal.

7 Discuss the ecological benefits that the presence of sponges may have on aquatic ecosystems.

8 Why were living sponges placed in Ca^{2+}/Mg^{2+} free seawater overnight? Hypothesize about the outcome of that exercise if this step had been omitted.

Name ______________________________ Section ______________

7 Laboratory Practical Challenge

The following evaluation is intended to replicate the method of assessment and type of questions you will be asked on lab practicals. If you have difficulty with these questions, you may benefit from additional review of the specimens provided in lab.

1 On Figure A, label the following structures: osculum, ostium, spongocoel.

2 On Figure B, label the following structures: choanocyte, spicule, porocyte.

3 On Figure B, draw arrows to represent the flow of water through the sponge.

4 On Figure B, what is the function of the structure indicated by line A?

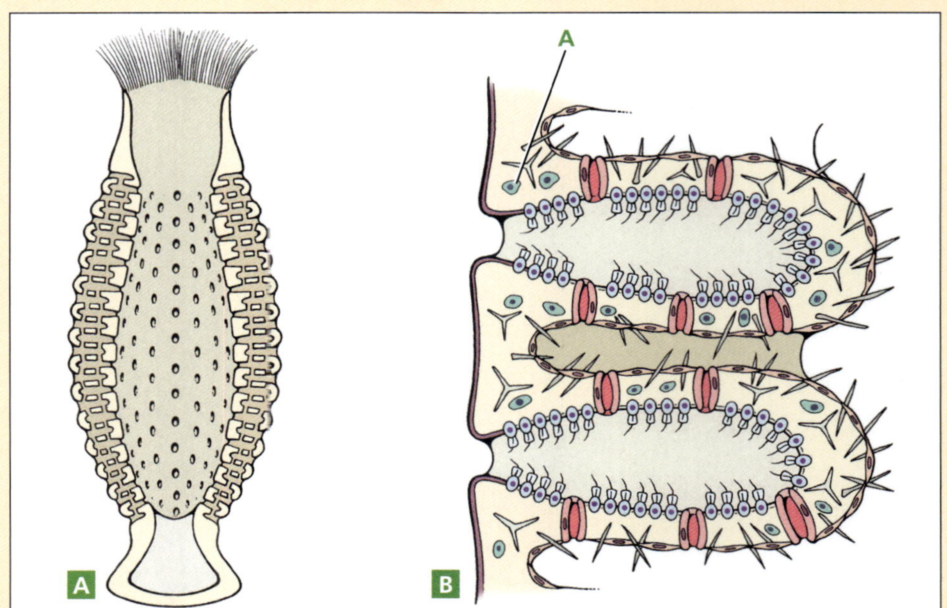

7

7

5 Identify the sponge body type depicted below.

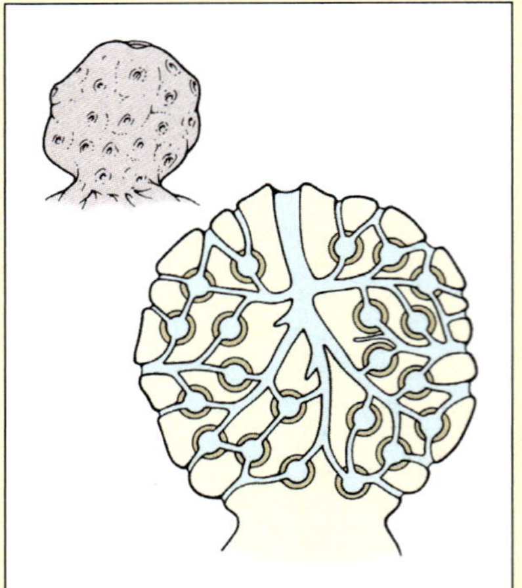

CHAPTER 8

Ctenophores and Cnidarians

After completing the exercises in this chapter, you should be able to:

1 List the distinguishing characteristics of the radiate phyla: Ctenophora and Cnidaria.

2 Identify specimens from three major classes of cnidarians.

3 Understand the evolutionary position of ctenophores and cnidarians on the phylogeny of major animal phyla.

4 Understand the difference between the polyp and the medusa body forms.

5 Understand the meaning of diploblastic tissue organization.

6 Identify the basic anatomy of representatives from three major classes of cnidarians.

7 Describe the life cycle of *Obelia*.

8 Categorize and describe the proximate mechanisms of feeding behavior in cnidarians.

Cnidarians and ctenophores are the simplest living animals with true tissues and thus form the base of the evolutionary group Eumetozoa—the "true animals." They are nearly all marine, with only a few known freshwater species of cnidarians (Fig. 8.1). The phylum name Cnidaria is derived from the unique stinging cells, called **cnidocytes**, present on the body and tentacles of the members of this group. Ctenophores are commonly called comb jellies (or comb jellyfish) due to the presence of eight ciliated bands superficially resembling combs that encircle the body. The phylum name Ctenophora actually means "comb bearer."

8.1 Underwater reefs often contain a diverse assemblage of cnidarians, including a variety of anemone and coral species.

Members of these two phyla have no blood vessels or specialized organs. Like sponges and protists, ctenophores and cnidarians depend on simple diffusion for gas exchange across their tissues and for elimination of metabolic wastes. Body walls in these organisms are thin enough that no cell is too far from either the outer epidermal or the inner gastrovascular surface for effective diffusion.

To simplify our discussion of these two groups, we will focus first on the phylum Cnidaria and later in the chapter take a closer look at ctenophores. Cnidarians have **radial symmetry** and lack any definite concentration of nervous tissue or sensory organs. Instead, they possess a nerve net that connects the uniformly distributed sensory cells and motor neurons to provide for a limited number of coordinated movements and behaviors. They are nearly all carnivorous but do not actively pursue their prey. Instead, they must rely on passing prey to make contact with their tentacles for capture.

Unlike sponges, the body wall of cnidarians is arranged in two discrete tissue layers, known as a **diploblastic** arrangement, consisting of an outer **epidermis** and an inner **gastrodermis**. Sandwiched between these two cellular layers is an inert, gelatinous layer called **mesoglea**. The gastrodermis lines the central digestive cavity, the **gastrovascular cavity**, which has only a single opening—the mouth—that is used both to ingest food and expel undigested waste.

Cnidarians have two basic body forms—polyps and medusae (Fig. 8.2). The **polyp** form generally is represented by a cylindrical organism that remains attached to the substrate by a short stalk and has a mouth that faces away from the substrate (usually upward). The **medusa** is a more circular, umbrella-shaped form with a mouth on the underside. Medusae generally are free-floating with tentacles that hang downward, surrounding the mouth. In either form, body shape is maintained to

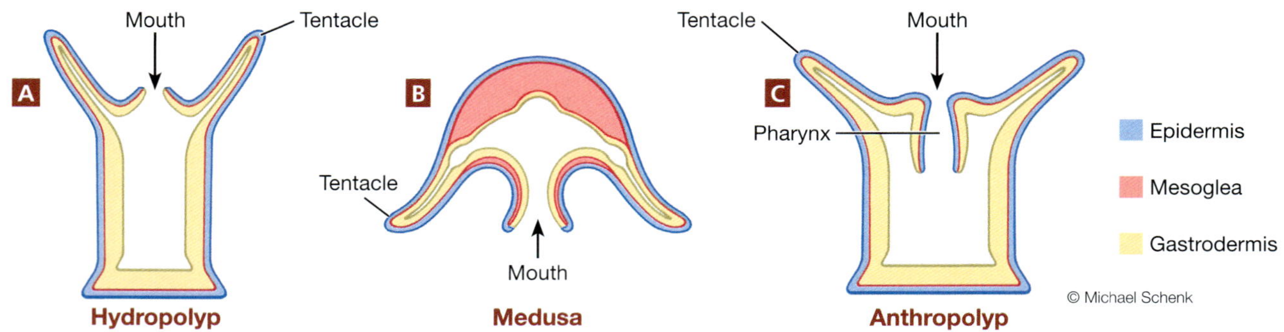

8.2 Basic body forms of the phylum Cnidaria: (**A**) polyp form as seen in hydrozoans, (**B**) medusa form as seen in jellyfish, and (**C**) polyp form as seen in sea anemones.

some degree by the viscosity of the gelatinous mesoglea and by the regulation of fluid in the gastrovascular cavity, an example of a **hydrostatic skeleton**.

More than 9,000 species of cnidarians have been described, with the vast majority belonging to only two classes, Hydrozoa and Anthozoa. Four of the five classes of cnidarians are listed in Table 8-1. The classes are distinguished primarily by the dominance of the polyp or medusa stage in the life cycle. In hydrozoans the polyp is the predominant form in the life cycle, and in scyphozoans and cubozoans the medusa form predominates. Anthozoans exist only as polyps; they have completely lost the medusa stage.

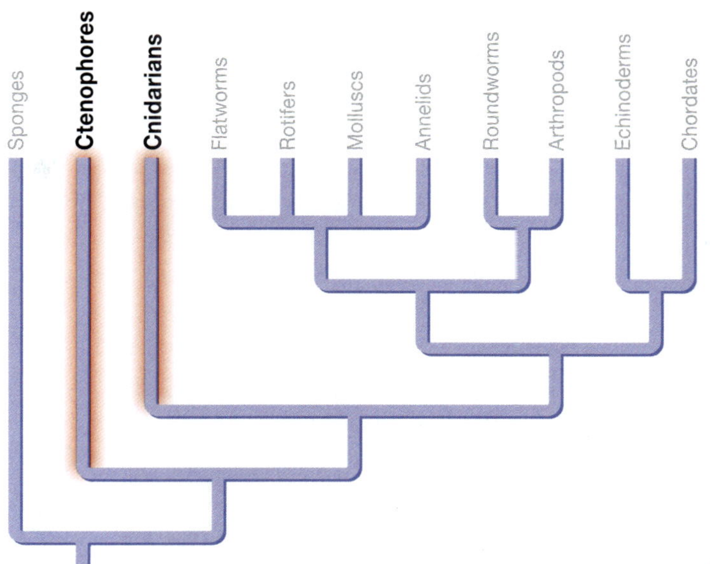

8.3 Radiate animals (ctenophores and cnidarians) exhibit a level of complexity above that of sponges but below that of the other animals.

Evolutionary History

Ctenophores and cnidarians occupy positions directly after the sponges on the evolutionary tree (Fig. 8.3). We see a few evolutionary trends from these two groups to later phyla, such as the emergence of discrete tissue layers, limited mobility, and the precursors to organ systems (digestive tissues, reproductive tissues, nervous tissues, and sensory cells). A major evolutionary advancement seen in ctenophores and cnidarians, compared to their predecessors, is their ability to digest food extracellularly within the gastrovascular cavity. Digestive enzymes are released from gastrodermal cells and partially break down the food, followed by phagocytosis and intracellular digestion. This capability allows these two groups of animals to digest much larger prey than sponges or protists can digest, allowing the cnidarians and ctenophores to expand their ecological niches.

Keep in mind that in discussions of any animal group, you must be careful not to view the currently living members of these "less complex" phyla as ancestral or primitive. Modern species are the product of hundreds of millions of years of evolutionary changes and, thus, may be very different from their ancestral forms. Because soft-bodied animals such as amoebas, sponges, ctenophores, and cnidarians do not fossilize well, it is difficult to know exactly what the first unicellular or multicellular organisms looked like. For a more complete phylogeny depicting the evolutionary relationships among the major animal phyla covered throughout this laboratory guide, review Figure A.1 in the Appendix.

Table 8-1 Major Classes of Phylum Cnidaria

Class and Representative Animals	Characteristics
Hydrozoa (≈ 2,700 sp.) *Hydra*, *Obelia*, Portuguese man-of-war	Alternation of polyp and medusa stages in most species, medusa has velum, mostly marine with a few freshwater species, colonial or solitary
Scyphozoa (≈ 200 sp.) Jellyfish (*Aurelia*)	Solitary, free-floating medusa most prominent form, medusa without velum, mostly marine, mesogleal layer generally thick
Cubozoa (≈ 10 sp.) Box jellyfish (*Chironex*)	Solitary, free-swimming, medusa most prominent form, medusa has velarium (similar to hydrozoan velum), small body (generally 2–3 cm) with square-shaped bell, tropical or subtropical
Anthozoa (≈ 6,200 sp.) Sea anemones (*Metridium*), corals, sea fans	No medusa stage—polyp stage predominates life cycle, entirely marine, may be colonial or solitary, gastrovascular cavity divided by septa into chambers to increase digestive surface area

Exercise 8-1 Hydrozoans (Class Hydrozoa)

Materials Needed

- ❑ slides of *Hydra* w.m., c.s., and l.s.
- ❑ slide of *Obelia* polyp w.m.
- ❑ slide of *Obelia* medusa w.m.
- ❑ compound microscope

Most cnidarians in the class Hydrozoa undergo a regular alternation of generations between body forms that utilize **asexual reproduction** in the polyp form and **sexual reproduction** in the medusa form. In general the polyp is the predominant body form, and in many groups polyps are assembled into colonies, as in the case of *Obelia*. Members of the genus *Hydra* are unusual in that they do not produce medusae. Rather, *Hydra* exist as single, mobile polyps that reproduce either sexually (through production of sperm and eggs) or asexually (through budding). In either scenario a new polyp is the result.

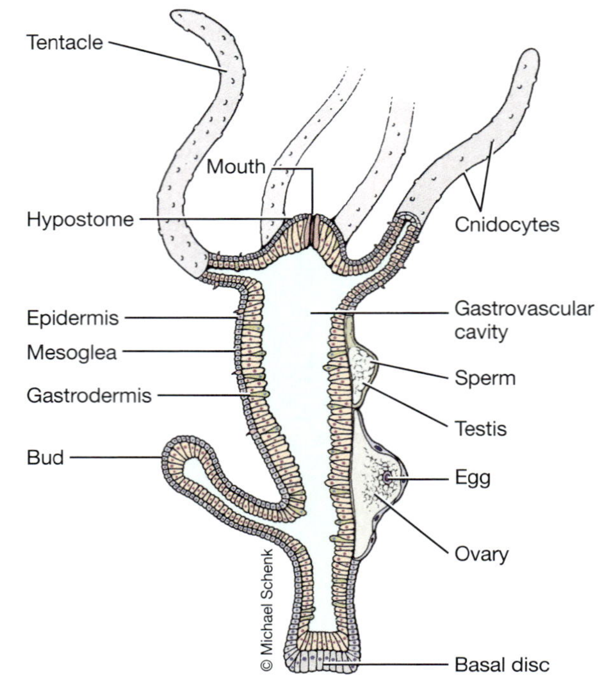

8.4 Longitudinal section through *Hydra* illustrating the two tissue layers characteristic of all cnidarians.

8 Solitary Hydrozoans

Procedure

1 Obtain a prepared slide of a whole mount (w.m.) of *Hydra littoralis*. These small hydrozoans exist as individual polyps in shallow, freshwater ponds and streams. Though diminutive in size, they are formidable predators of small aquatic invertebrates as they sit motionless among the submerged rocks, twigs, and vegetation with their tentacles outstretched, waiting for prey to pass too closely.

2 Using a microscope, examine the w.m. slide and locate the following structures: **tentacles, hypostome, basal disc, bud** (may not be present on every specimen), **gonads** (ovaries or testes). Anatomical structures of *Hydra* are represented in Figures 8.4 and 8.5 and defined in Table 8-2. How many tentacles are present on your specimen?

____

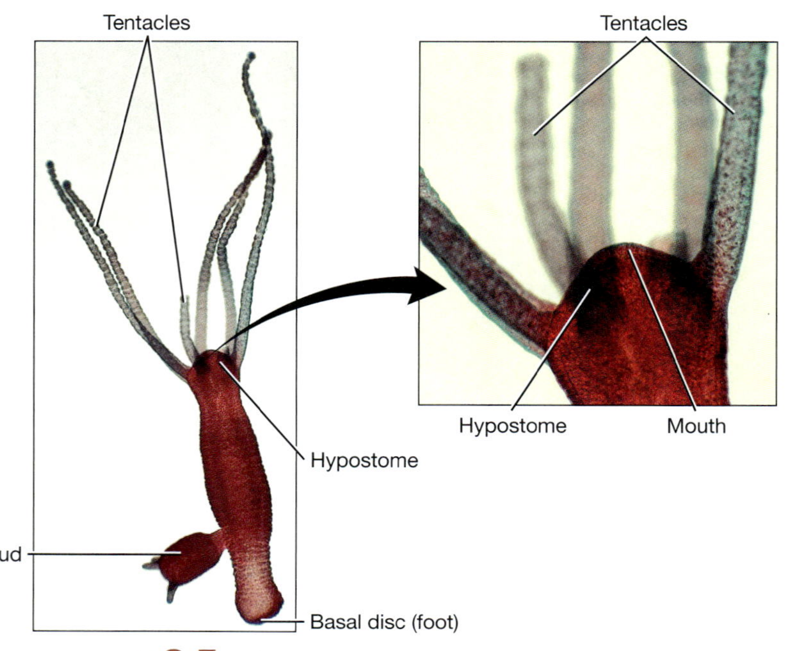

8.5 Solitary hydrozoan, *Hydra*, budding.

3 If available, examine a longitudinal section (l.s.) through *Hydra*. In addition to the structures visible on the whole mount, locate the following: mouth, gastrovascular cavity, epidermis, gastrodermis, and mesoglea.

4 Finally, examine a cross section (c.s.) through *Hydra* and locate the following structures: cnidocytes, gastrodermis, epidermis, mesoglea, and gastrovascular cavity, using Figure 8.6 as a guide. Which tissue layer is thicker on your specimen, the epidermis or gastrodermis? How large is the gastrovascular cavity in comparison to the tissue layers?

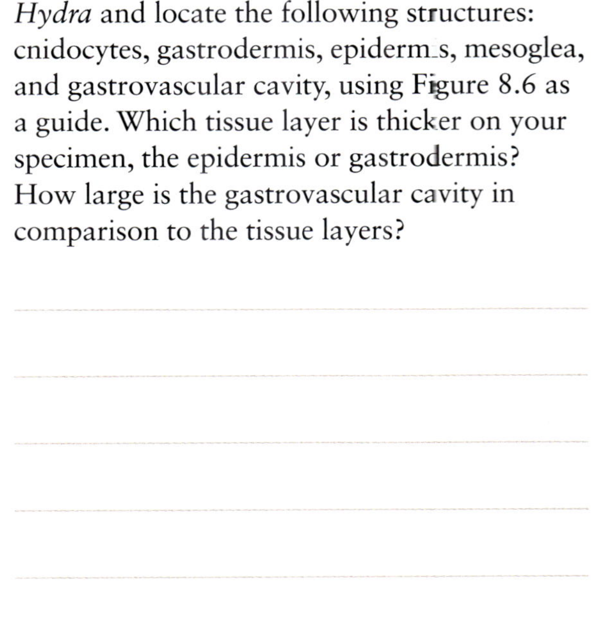

Table 8-2 Anatomy of *Hydra*

Structure	Function
Tentacles	Defense and prey capture
Mouth	Ingestion of food and elimination of indigestible particles (egestion)
Hypostome	Elevated mound of tissue that expands or contracts to regulate size of mouth opening
Cnidocytes	Specialized stinging cells located in the epidermal layer of the tentacles and body wall
Gastrovascular cavity	Chamber within which extracellular digestion of prey occurs; only opening is through the mouth
Bud	Product of asexual reproduction; will fall off when mature and become a self-sufficient organism
Gonads (testes and ovaries)	Organs for sexual reproduction; *Hydra* are dioecious, meaning that an organism has either testes or ovaries (male or female), but not both
Basal disc	Specialized region for attachment to the substrate
Epidermis	Outer tissue layer; specialized for protection
Gastrodermis	Inner tissue layer; specialized for digestion
Mesoglea	Inert, acellular, jellylike substance that aids in supporting the body

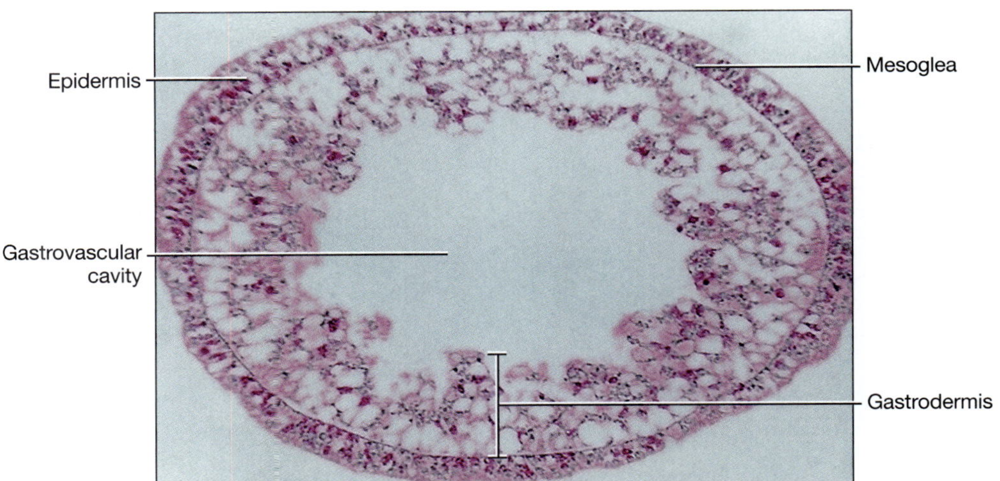

8.6 Cross section through the body of *Hydra*.

Reproduction in Solitary Hydrozoans

Hydrozoans may be either **monoecious** (having both testes and ovaries) or **dioecious** (having separate male and female sexes), depending on the species. They typically reproduce asexually by **budding** when environmental conditions are optimal (Fig. 8.5). In this scenario a part of the body wall grows outward, develops tentacles and a mouth, and eventually detaches from the parent. When environmental conditions become unfavorable (for example, when food and dissolved oxygen becomes scarce), they rely on sexual reproduction. Females develop an **ovary** that produces a single egg, and males form **testes** that produce sperm (Fig. 8.7). After fertilization, the zygote leaves the parent and remains within a protective shell for some time before developing into a new individual.

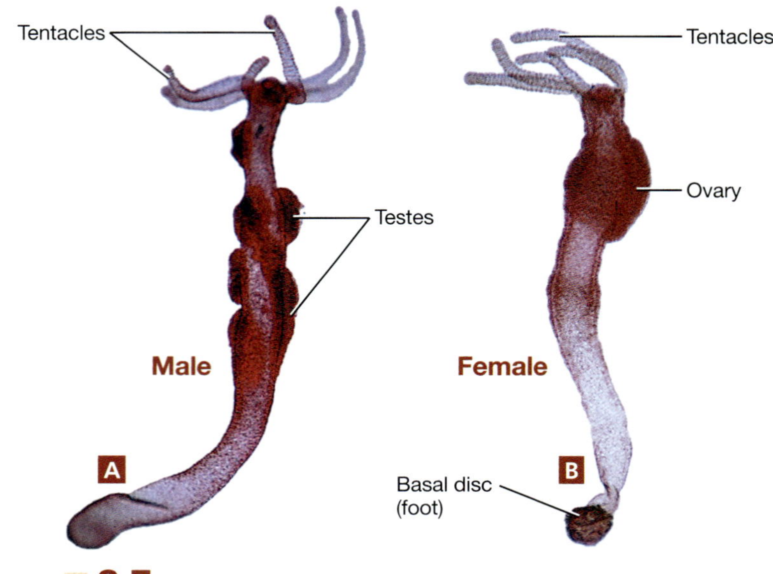

8.7 *Hydra*: (**A**) male with testes, (**B**) female with ovary.

The Cnidocyte: A Closer Look

Cnidocytes are clearly one of the defining characteristics of cnidarians, much like collar cells and spicules define the sponges. Cnidarians use these tiny cells both for capturing prey and also as a deterrent to predators. Hundreds or thousands of cnidocytes may line the tentacles and outer surface of the cnidarian body. Each cnidocyte contains a stinging capsule within it, known as a **nematocyst**. When the small trigger on the cell is stimulated, a rapid increase in osmotic pressure within the cnidocyte causes the nematocyst to discharge violently, exploding out of the cell and releasing a long, hollow, threadlike filament, driving its barbs and poison into the flesh of its victim (Fig. 8.8).

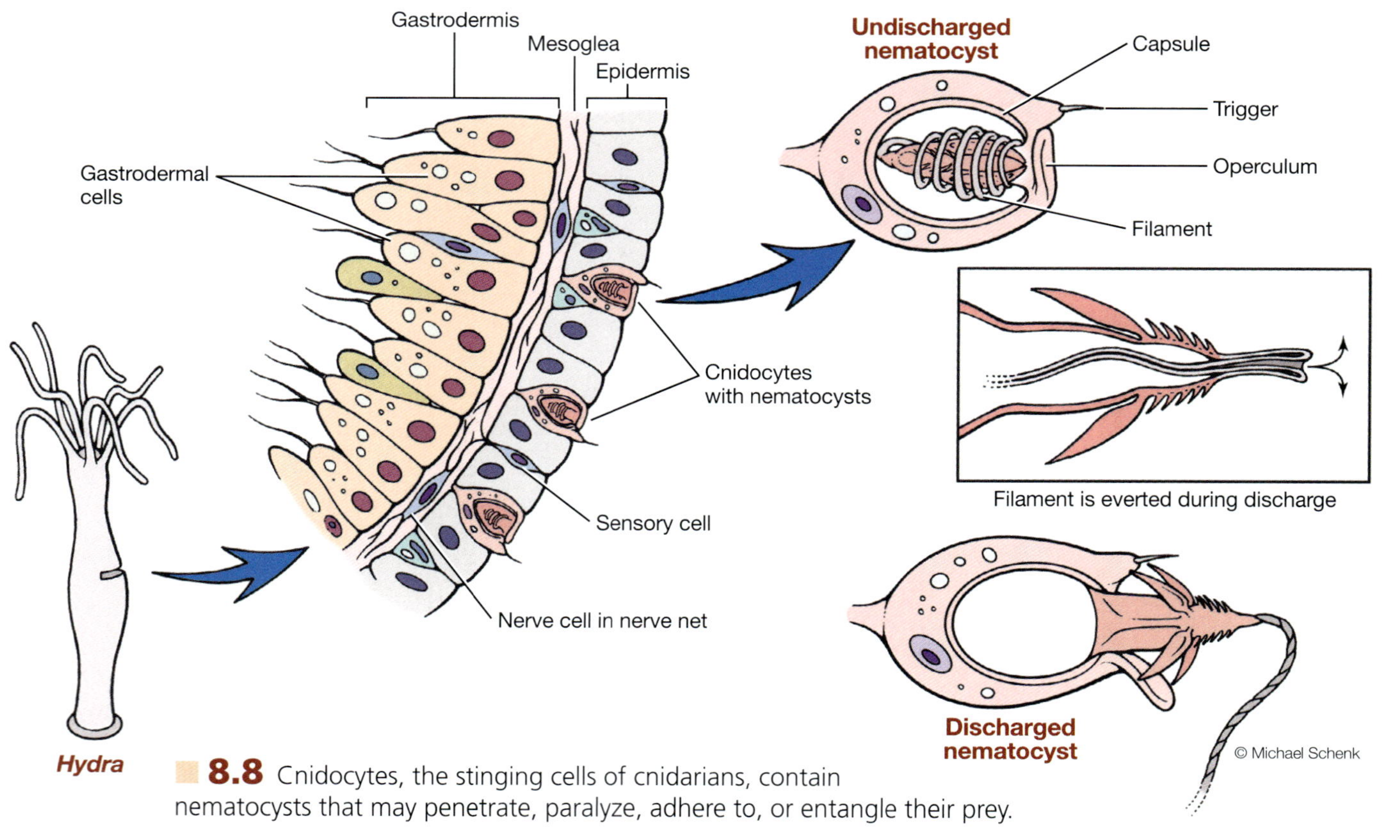

8.8 Cnidocytes, the stinging cells of cnidarians, contain nematocysts that may penetrate, paralyze, adhere to, or entangle their prey.

Some nematocysts are tipped with filaments that are sticky; some are long and coil around their prey; others bear barbs or spines and inject paralytic toxins into their prey. In many species they are capable of penetrating the hard shells of crustaceans, and even human skin, sometimes producing a painful wound or, in extreme cases, death.

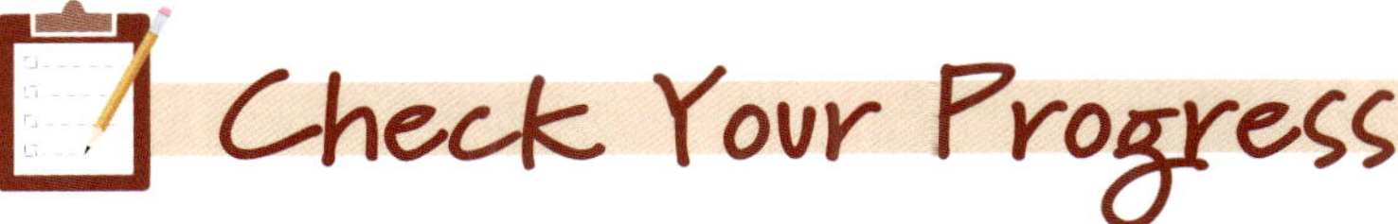

Check Your Progress

1.1 Draw a generalized sketch of *Hydra* and label the following structures: mouth, hypostome, tentacles, basal disc, bud, gonads.

1.2 In which tissue layer are the cnidocytes located? Why?

1.3 Briefly explain the mechanism of nematocyst discharge.

1.4 Does *Hydra* reproduce sexually? Asexually? Give evidence to support your answers.

1.5 What type of digestion is *Hydra* capable of: intracellular, extracellular, or both?

Colonial Hydrozoans

Next we will examine a more typical member of the class Hydrozoa. Members of the genus *Obelia* are colonial hydrozoans connected by branches of a common gastrovascular cavity—**coenosarc**—making them all part of a larger functioning body. This cooperative venture allows certain polyps to develop into highly specialized feeding polyps, and others to lose the ability to feed altogether in exchange for an enhanced ability to reproduce.

The life cycle of *Obelia* illustrates an alternation of the sexual (medusa) and asexual (polyp) stages commonly seen in hydrozoans (Fig. 8.9). The asexual stage consists of a colony composed of both feeding polyps and reproductive polyps. The medusa stage of most hydrozoans typically is a short-lived stage devoted primarily to reproduction. Free-swimming male and female medusae bud off from the reproductive polyps and release haploid sperm and eggs that fuse (externally) to form diploid zygotes. Each zygote develops into a ciliated, swimming **planula larva** that settles to the bottom of the ocean floor, attaches itself to the substrate, and begins to create a new generation of polyps through asexual budding.

8.9 Life cycle of *Obelia* showing asexual reproduction in the polyp generation and sexual reproduction in the medusa generation.

Procedure

1 Obtain a whole-mount (w.m.) slide of a colony of *Obelia* polyps. Examine it using low power, and identify the following structures depicted in Figure 8.10 and defined in Table 8-3: **hydranth** (feeding polyp), tentacles, hypostome, **gonangium** (reproductive polyp), **medusa buds**, coenosarc, and perisarc. In the space below, record the number of medusa buds you count on one gonangium.

2 Now examine a w.m. slide of an *Obelia* medusa using medium power. Locate the following structures that are defined in Table 8-3: tentacles, manubrium, and mouth. (Fig. 8.11).

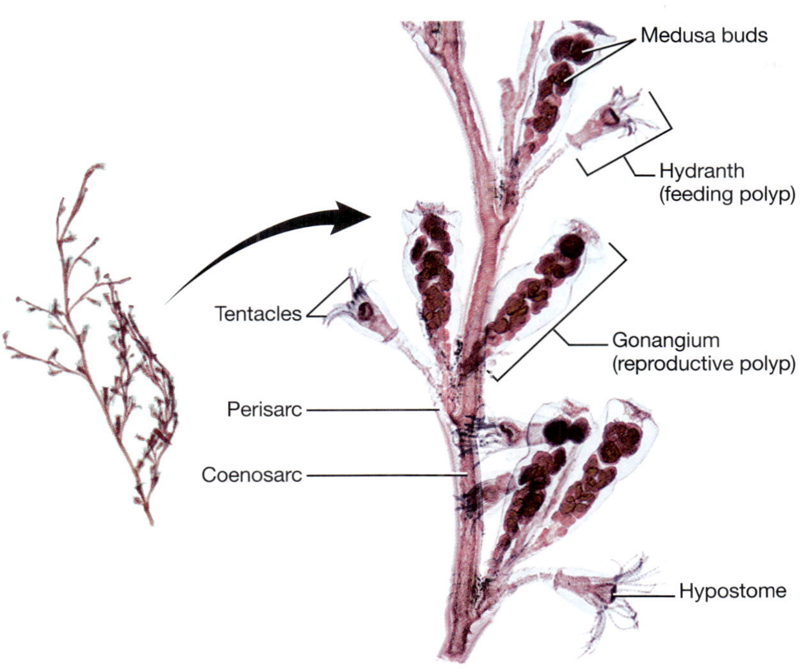

8.10 Colonial hydrozoan, *Obelia*.

Tentacles
Radial canal
Manubrium

8.11 Medusa of the colonial hydrozoan, *Obelia*.

8

Table 8-3 Anatomy of *Obelia*

Body Form	Structure	Function
Polyp	**Hydranth** (feeding polyp)	Polyp specialized for food acquisition
	Tentacles	Defense and prey capture
	Hypostome	Elevated mound of tissue that expands or contracts to regulate size of mouth opening
	Gonangium (reproductive polyp)	Polyp specialized for reproduction
	Medusa buds	Product of asexual reproduction; medusae will be released from the gonangium when mature and will produce either sperm or eggs, which fuse with the respective gamete forming a zygote that will develop into a new polyp
	Coenosarc	Common chamber within which extracellular digestion occurs; nutrients are distributed throughout organism
	Perisarc	Translucent outer covering of organism; serves protective function
Medusa	**Tentacles**	Defense and prey capture
	Manubrium	Stalk of fleshy tissue that supports the mouth
	Mouth	Ingestion of food and egestion of indigestible particles
	Gonads	Organs for sexual reproduction; either testes or ovaries

Check Your Progress

2.1 List several ways in which *Obelia* differs from *Hydra*.

Exercise 8-2 True Jellyfish (Class Scyphozoa)

Materials Needed

- ❑ preserved scyphozoans
- ❑ dissecting microscope

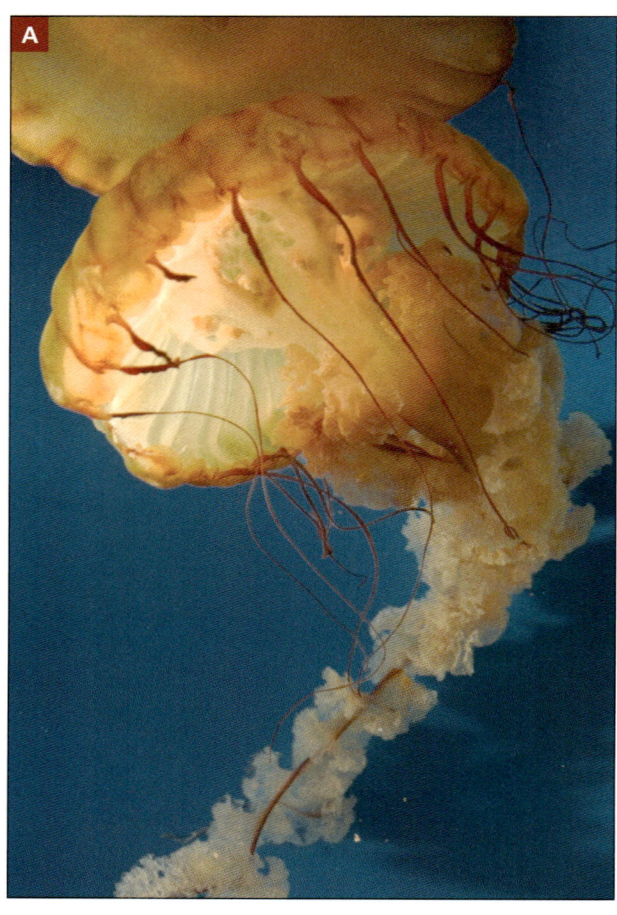

8.12 Representatives of the class Scyphozoa: (**A**) sea nettle, *Chrysaora fuscescens*, from the Pacific coast of North America, and (**B**) red-striped jellyfish, *Chrysaora melanaster*, from the Bering Sea. Individuals of the latter species often reach 30 cm in diameter and have tentacles that may stretch to 6 meters.

Cnidarians in the class Scyphozoa are called jellyfish because of their thick layer of gelatinous mesoglea (Fig. 8.12). In the open oceans, where they thrive, most species range in size from 2–30 cm in diameter. The largest scyphozoans, belonging to the genus *Cyanea*, may exceed 2 meters in diameter and have tentacles more than 30 meters long! In scyphozoans, the medusa stage dominates the life cycle, with the polyp stage relegated to an inconspicuous, short-lived larval form that matures quickly into a polyp that buds off young medusae.

Aurelia, the moon jellyfish, is a common, widely distributed genus within this class that typifies the anatomy of scyphozoans. You will not have to dissect your study specimen, because their transparent bodies readily show most of their anatomy.

Procedure

1 Obtain a preserved specimen of *Aurelia* (or similar scyphozoan medusa).

2 Carefully examine it and identify the following structures using Figure 8.13 (anatomy defined in Table 8-4): mouth, **oral arms**, **marginal tentacles**, gonad, **gastric pouch**, **radial canals**, and **circular canal.**

8.13 Anatomy of the scyphozoan jellyfish, *Aurelia*.

Table 8-4 Anatomy of *Aurelia*

Structure	Function
Marginal tentacles	Provide sensory information and used in defense and locomotion
Mouth	Ingestion of food and elimination of indigestible particles (egestion)
Oral arms	Defense and prey capture
Gonads (testes and ovaries)	Organs for sexual reproduction; gametes are released into the gastric pouches and exit the body through the mouth
Gastric pouch	One of four divisions of the gastrovascular cavity for digestion of food
Radial canals	Extensions of the gastric pouches that radiate outward from the pouches and distribute nutrients throughout body
Circular canal	Circular extension of the gastric pouches that distributes nutrients to outer rim of jellyfish

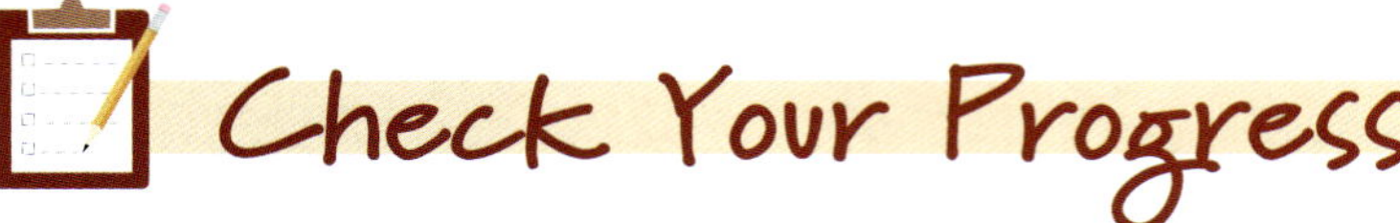

3.1 Sketch your specimen and label the following structures: oral arms, marginal tentacles, gonads.

3.2 Compare and contrast the scyphozoan medusa with the medusa bud of *Obelia*.

Exercise 8-3 Sea Anemones and Corals (Class Anthozoa)

Materials Needed

- ❑ preserved sea anemones
- ❑ latex gloves
- ❑ dissecting microscope

Anthozoans are represented by sessile organisms in the polyp stage, existing as solitary individuals (sea anemones) or as true colonies of dozens to thousands of individuals (corals) (Fig. 8.14). Their life cycle does not include a free-swimming medusa stage. Of the cnidarians, the anthozoans are the most numerous and most specialized, and their anatomy reflects this highly specialized life.

You will examine a member of the genus *Metridium*, a common North Atlantic sea anemone. Most of the internal anatomy can be viewed in specimens that have been bisected longitudinally, so further dissection is probably not necessary.

Two examples of anemone, disc anemones and a yellow anemone.

A group of anemones, *Anthopleura*, in a tide pool with other tidal organisms.

Polyps of glove xenia, *Xenia umbellata*.

Mushroom coral, *Rhodactis*.

8.14 Representatives of the class Anthozoa.

8

Procedure

1. Obtain a preserved specimen of *Metridium* that has been bisected longitudinally.
2. Examine this specimen and locate the following structures depicted in Figures 8.15–8.17 (anatomy defined in Table 8-5): tentacles, **oral disc**, mouth, **ostium**, **pharynx**, retractor muscles, gonad, gastrovascular cavity, primary septum, secondary septum, **pedal disc**, and **acontia**.
3. Notice that the large gastrovascular cavity of your specimen is partitioned by primary septa, inward extensions of the body wall. Openings in these septa, called **ostia**, allow water to circulate throughout the body. Smaller, **secondary septa** (or incomplete septa) further subdivide the main chambers. What do you suppose is the function or purpose of these folds of the gastrovascular cavity wall (the septa)?
4. Count the number of primary septa on your specimen and record that value below. Compare your finding with other lab groups. Do all of the sea anemone specimens in your lab have the same number of primary septa?

8

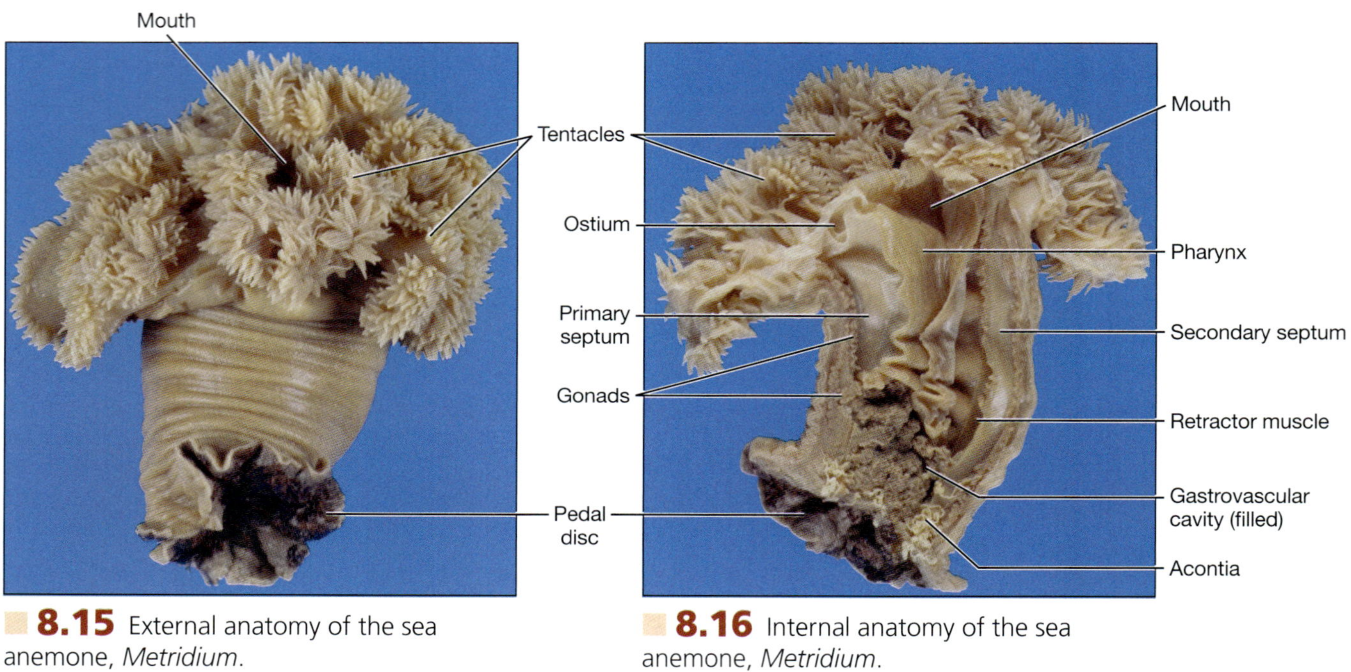

8.15 External anatomy of the sea anemone, *Metridium*.

8.16 Internal anatomy of the sea anemone, *Metridium*.

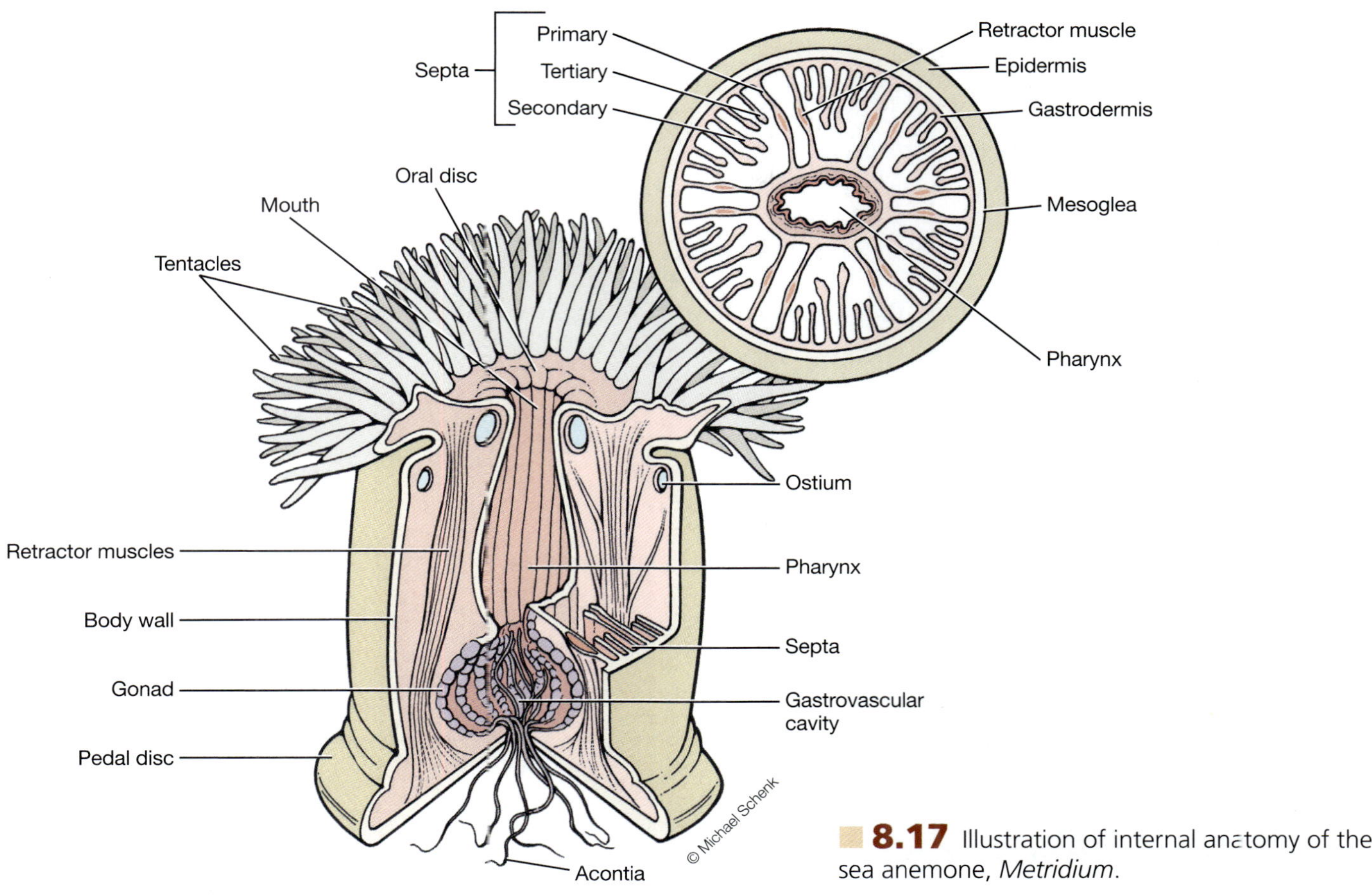

8.17 Illustration of internal anatomy of the sea anemone, *Metridium*.

8

Table 8-5 Anatomy of *Metridium*

Structure	Function
Tentacles	Defense and prey capture
Mouth	Ingestion of food and elimination of indigestible particles (egestion)
Oral disc	Raised portion of the mouth (equivalent to the hypostome of hydrozoans)
Gonads (testes and ovaries)	Organs for sexual reproduction; gametes are released into the gastric pouches and exit the body through the mouth
Ostium	Pore that allows circulation of fluids between adjacent body sections
Pharynx	Muscular portion of the gastrovascular cavity for pulling prey inward and expelling indigestible particles
Retractor muscles	Expand and contract the body
Gastrovascular cavity	Specialized chamber for extracellular digestion of prey
Primary septum	One of six thin, vertical walls that divides body into sections, providing support and increased surface area for digestion
Secondary septum	One of several thin, incomplete vertical walls that further subdivides body, providing support and increased surface area for digestion
Pedal disc	Tough, fleshy base that attaches organism to rocks or sandy ocean floor; most anemones can use their pedal discs to move slowly, gliding along the ocean floor
Acontia	Contain cnidocytes and may be extended through the mouth to aid in subduing live prey or to provide additional defense against predators

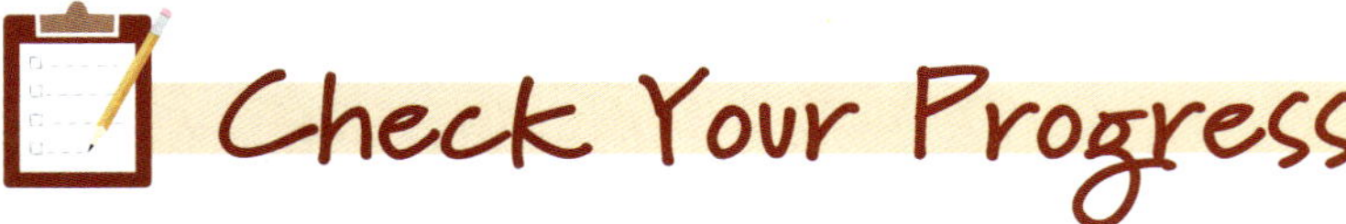

4.1 Match the structure with its correct function.

_____ pedal disc	**a.** Thin, vertical wall that divides body and provides increased surface area for digestion
_____ acontia	**b.** Extensions around mouth for defense and capture of prey
_____ tentacles	**c.** Threadlike filaments that may be extended through the mouth to aid in capture of prey and defense
_____ primary septum	**d.** Pore that permits circulation of fluid between chambers
_____ ostium	**e.** Tough, fleshy base for attachment

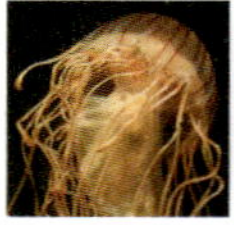

Exercise 8-4 Comb Jellies (Phylum Ctenophora)

Materials Needed

- ❑ preserved comb jellyfish (*Pleurobrachia* or other species)
- ❑ latex gloves
- ❑ dissecting microscope
- ❑ small finger bowl

The phylum Ctenophora contains approximately 100 described species of small, marine invertebrates with transparent, bioluminescent, gelatinous bodies commonly referred to as "comb jellies," "sea walnuts," or simply ctenophores. Their phylum name is derived from the unique distinguishing characteristic that all members of the phylum possess—eight rows of evenly spaced ciliated bands (comb plates) that encircle the body longitudinally (Fig. 8.18).

Ctenophores propel themselves through the water by synchronized, wavelike beating of cilia on these comb plates. Members of most species possess either a pair of tentacles or oral lobes that are used to collect food; however, ctenophores lack the stinging cnidocytes characteristic of cnidarians. Instead these carnivorous predators capture prey by discharging a sticky adhesive from specialized cells that are activated by physical contact, draw their victims into their mouths, and swallow them whole. Amazingly, some species of ctenophores that feed on small cnidarians can collect undischarged cnidocytes from their prey and incorporate these cells into their own epidermal tissue as a defense mechanism!

Like cnidarians, ctenophores have a decentralized nerve net without a brain, and the body wall is composed of two layers: an outer epidermis and inner gastrodermis, with a gelatinous layer of collenchyme (similar to mesoglea) sandwiched between the two tissue layers. Unlike cnidarians, ctenophores have a complete digestive system, consisting of a mouth, pharynx, branching gastrovascular canals, and an anal pore through which undigested material is expelled from the body. Most ctenophores are monoecious (hermaphroditic), and sperm ducts and oviducts transport sperm and eggs, respectively, from the testes and ovaries to the surrounding seawater where fertilization occurs externally.

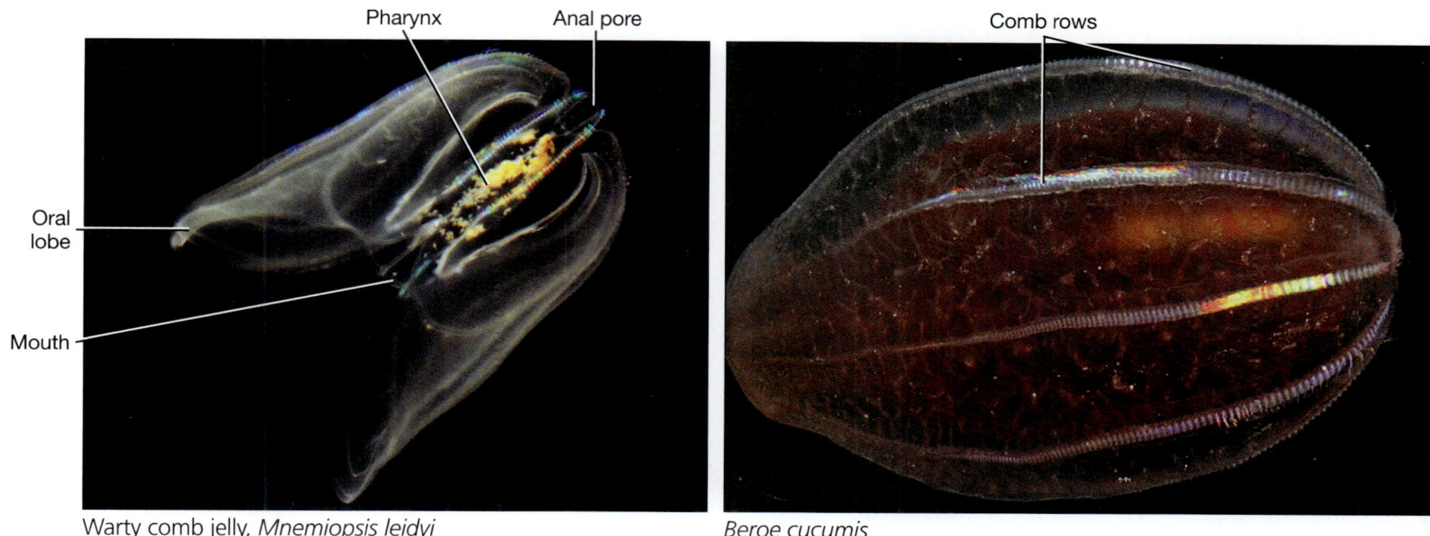

Warty comb jelly, *Mnemiopsis leidyi*

Beroe cucumis

8.18 Comb jellies (phylum Ctenophora) are thought to be a sister taxon to cnidarians; they have transparent bodies with comblike bands of cilia and bioluminescent properties.

8

Procedure

1 Obtain a preserved specimen of *Pleurobrachia* or other species of comb jelly and place it in a small finger bowl.

2 Examine this specimen with the dissecting microscope and locate the following structures depicted in Figure 8.18: oral lobes or tentacles (if present), mouth, comb rows, pharynx, and anal pore. In the space provided below, sketch your specimen and label the structures listed above.

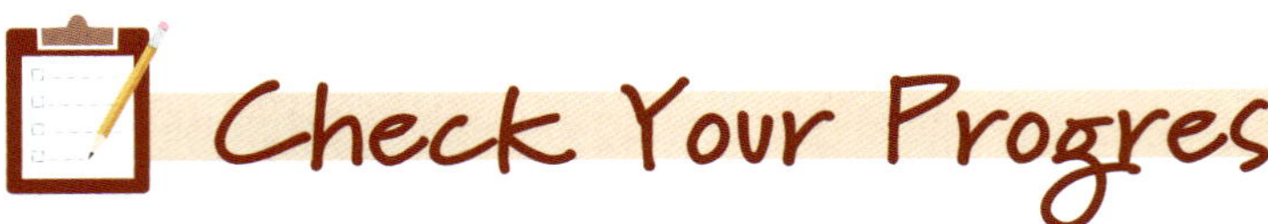

5.1 List two similarities between cnidarians and ctenophores.

1. ______________________________

2. ______________________________

5.2 What characteristics of ctenophores cause zoologists to classify them in a separate phylum from cnidarians?

____

____

Exercise 8-5 Investigating Cnidarian Feeding Behavior

Materials Needed

- ❑ live *Hydra littoralis*
- ❑ live blackworms (*Lumbriculus variegatus*)
- ❑ dropper bottle of tyrosine solution (0.9 g/L)
- ❑ dropper bottle of glutathione solution (1.53 g/L)
- ❑ dropper bottle of distilled water
- ❑ five Syracuse dishes
- ❑ 50 ml beaker
- ❑ plastic droppers
- ❑ stopwatches
- ❑ filter paper
- ❑ razor blades
- ❑ fine-tipped forceps
- ❑ dissecting microscope

Behavior is simply an animal's response to stimuli in its environment, and the basic physiological unit of behavior is the motor system. At its simplest level the motor system consists of sensory structures, neurons, and muscles (or glands) whose actions are coordinated to bring about certain behaviors.

When zoologists study behaviors, they distinguish between two basic causes of a behavior. The **proximate cause** of a behavior is the series of immediate physiological events that led up to the specific behavior. The **ultimate cause** of a behavior is the resultant selective advantages that have promoted that behavior to remain in the animal's repertoire of possible responses and to evolve over time.

Think of proximate causes as explaining *how* the behavior occurs and ultimate causes as explaining *why* the behavior exists. A brief example should remove any confusion. Some species of night-flying moths have ears located behind their wings that can detect sounds only in the ultrasonic range (20 KHz–40 KHz). This is well outside the range of sounds that the moths themselves can produce, making these moths deaf to their own sounds! It is, however, within the exact range that night-flying bats emit their sonar pulses. Thus, when these moths detect pulses of sound in the ultrasonic range (emitted by bats), they instantaneously initiate a complex series of aerial avoidance maneuvers involving flips, spins, and dives that allow them to avoid becoming a late-night snack for the bat.

What is the *proximate cause* of this behavior, or *how* does it occur? Sensory receptors in the moth's ears are stimulated by the sound vibrations, causing action potentials in the moth's auditory neurons to be generated and propagated through the nervous system to motor neurons in the moth's flight muscles, bringing about the subsequent array of aerial gymnastics. Does the moth have to think about what it's doing? Of course not; this is a hardwired behavior. It occurs whether a bat is actually making the sound or a sophisticated machine in the laboratory is producing the high-frequency pulses. What, then, is the *ultimate cause* of this behavior, or *why* does it occur? Night-flying bats do produce sounds in these frequencies while hunting, so moths that (1) had some capacity for detecting these sounds, and (2) had reflexive avoidance behaviors in response to these sounds enjoyed greater survival than their less fortunate relatives and consequently left more offspring of their own that had the genetic predisposition for effective ears and effective moves.

The cnidarian neuromuscular system is far simpler than that of a moth—or a bat, for that matter—but it represents an ideal model for analyzing the general principles of simple behaviors. When an object brushes against one of its tentacles, the *Hydra* reacts by "stinging" the object with its nematocysts. But the *Hydra* attempts to eat only *some* of the objects it stings—edible, organic objects, but not sand particles or inorganic debris. So how can an animal without a brain possibly know what is edible and what is not?

Prior research has demonstrated that *Hydra* have the capacity to sense the presence of certain chemicals in their environment through receptor proteins on the surface of specialized sensory neurons in the epidermis. Perhaps *Hydra* can "recognize" chemicals that only living organisms would release when punctured by the nematocysts. Such a proximate cue would be useful to prevent *Hydra* from wasting energy and potentially endangering itself by attempting to consume every piece of debris that bumped against its tentacles.

If we pursue this line of thinking, we might ask what types of chemicals would be good candidates for identifying an object as living? Amino acids are among the simplest chemical compounds common to all living creatures. Could amino acids act as a proximate cue for the feeding response in *Hydra*, providing them with an effective way to differentiate between living, organic prey and nonconsumable "trash" that just happened to bump into their tentacles?

We will test this question using the two amino acids glutathione and tyrosine—two amino acids present in the body fluids of small aquatic worms and arthropods. Glutathione is a tripeptide: Glu-Cys-Gly; tyrosine is a single amino acid. Our process will employ the scientific method to determine if the amino acids glutathione and tyrosine can serve as proximate triggers of the *Hydra* feeding response.

Before proceeding, your instructor should discuss any additional background regarding this experiment, and you should devise an appropriate scientific question and testable hypothesis for the experiment. You will set up five dishes, each containing one or two *Hydra*. Each dish will receive one of the following substances: (1) distilled water, (2) sections of blackworm, (3) glutathione, (4) tyrosine, or (5) glutathione + tyrosine. You will categorize any behaviors seen and record the degree and duration of each response.

Figure 8.19 depicts the array of feeding and resting postures and behaviors that you may observe in your specimen during the course of a feeding event. Use these diagrams to categorize the behaviors seen during the experiment. To quantify these behaviors, you should record in Table 8-6 (1) the intensity (on a scale of 1–10), (2) the frequency, and (3) the duration of each reaction. Before beginning the experimental protocol, complete the following section using the steps of the scientific method to approach this investigation systematically.

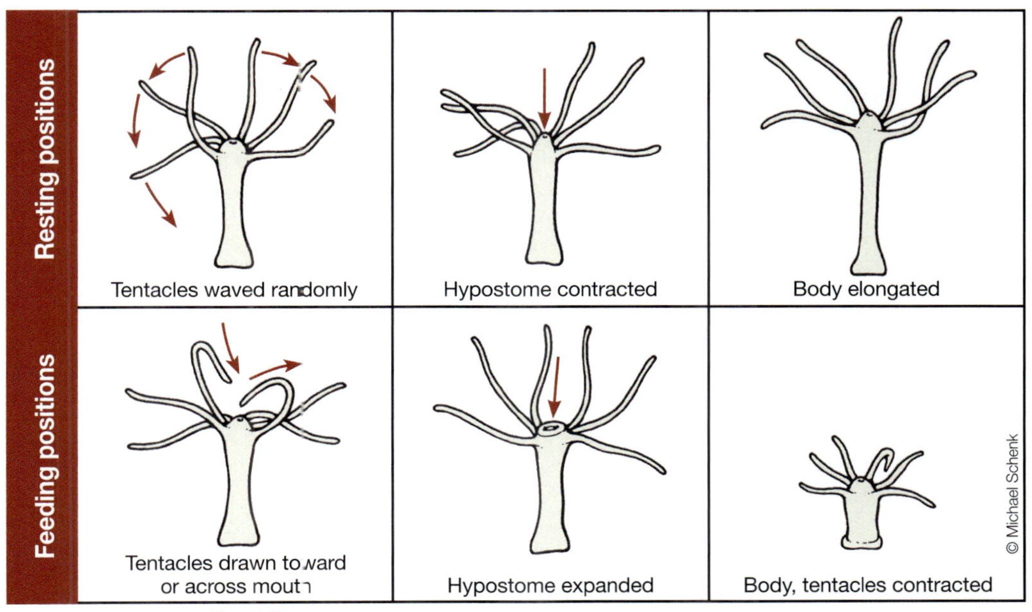

8.19 Typical feeding and resting postures seen in *Hydra*.

Using the Scientific Method

8

1 What is the general question that this experiment addresses?

2 Record the hypothesis you wish to test.

3 Identify the independent variable in this experiment.

> **Tip**
>
> The *independent variable* is the element that you change or manipulate and the *dependent variable* is the element that you measure to determine how it is affected by the changes you made to the independent variable.

4 Identify the dependent variable(s) in this experiment.

5 What are the levels of treatment (that is, levels of the independent variable)?

6 What will serve as the control treatments? (**HINT:** there is a positive control and a negative control.)

> **Tip**
>
> A *positive control* shows that the experiment is capable of producing a positive result when a phenomenon is expected, whereas a *negative control* is designed to demonstrate a lack of response or negative result when there should be no effect.

7 Identify several controlled variables.

Experimental Protocol

1 Work in groups of four for this experiment. Obtain five living *Hydra* from your instructor for your group.

2 Use a plastic dropper to transfer *Hydra* individually to a small amount of pond water in five Syracuse dishes, and place each dish under a separate dissecting microscope.

3 Adjust the mirror and lighting on each dissecting microscope to achieve optimum illumination of the specimen. These animals are likely to contract into a ball shape as a result of being transferred to a new environment. Therefore, you should transfer the animals to your experimental setup (that is, dish and microscope) and then turn off the microscope light. Leave the *Hydra* undisturbed for 4–5 minutes.

4 While the *Hydra* are "resting," use this time to prepare small sections of the blackworms for feeding. Because adult blackworms are a bit too large for our *Hydra*, you will have to cut them into smaller pieces and hand-feed them to the *Hydra* using extremely fine-tipped forceps.

5 Place a blackworm on wet filter paper on the stage of a dissecting microscope and slice it into sections that are 3–4 segments long (1 mm long) using a sharp razor blade (Fig. 8.20). The wet paper inhibits the worm's undulating movements, making slicing easier.

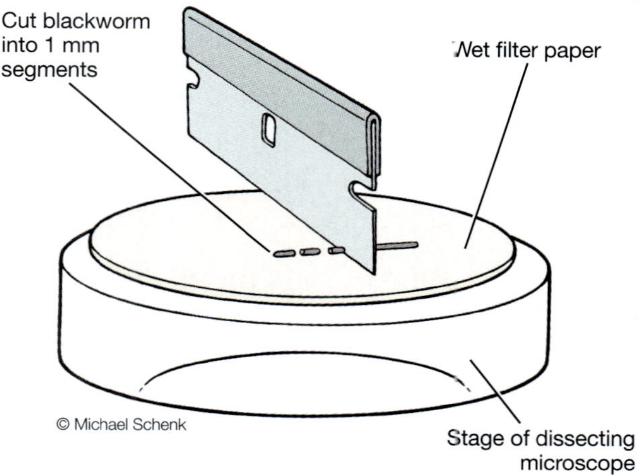

8.20 Procedure for slicing blackworm segments.

6 After 4–5 minutes, return to the dissecting microscope and observe one *Hydra* with subdued light. Critical to the success of this experiment, each *Hydra* has to be fully "relaxed," with its body and tentacles outstretched, before you proceed with subsequent steps. If any of your *Hydra* are not in this position, give them more time to rest and choose another specimen in the meantime. Remember—bumping them or jostling their dish usually will cause them to contract, so be careful with your movements around each *Hydra*.

7 One *Hydra* per group will serve as a control treatment for the experiment (Treatment 1). Using the dropper bottle, place 3 drops of distilled water into the dish *near* (but not directly on) the animal. What, if any, is the response? How long does the response to the disturbance last?

8 Record your observations in Table 8-6.

9 Repeat the stimulus and note how long it takes the *Hydra* to recover.

10 Use fine forceps to carefully position a small piece of the blackworm next to one of the outstretched tentacles of another *Hydra* (Treatment 2) and *very carefully and very gently* touch the worm segment to the tentacle. What is the immediate reaction? Does it differ from the reaction to distilled water? If the distilled water control animal (Treatment 1) has recovered sufficiently, you may repeat this step with this animal.

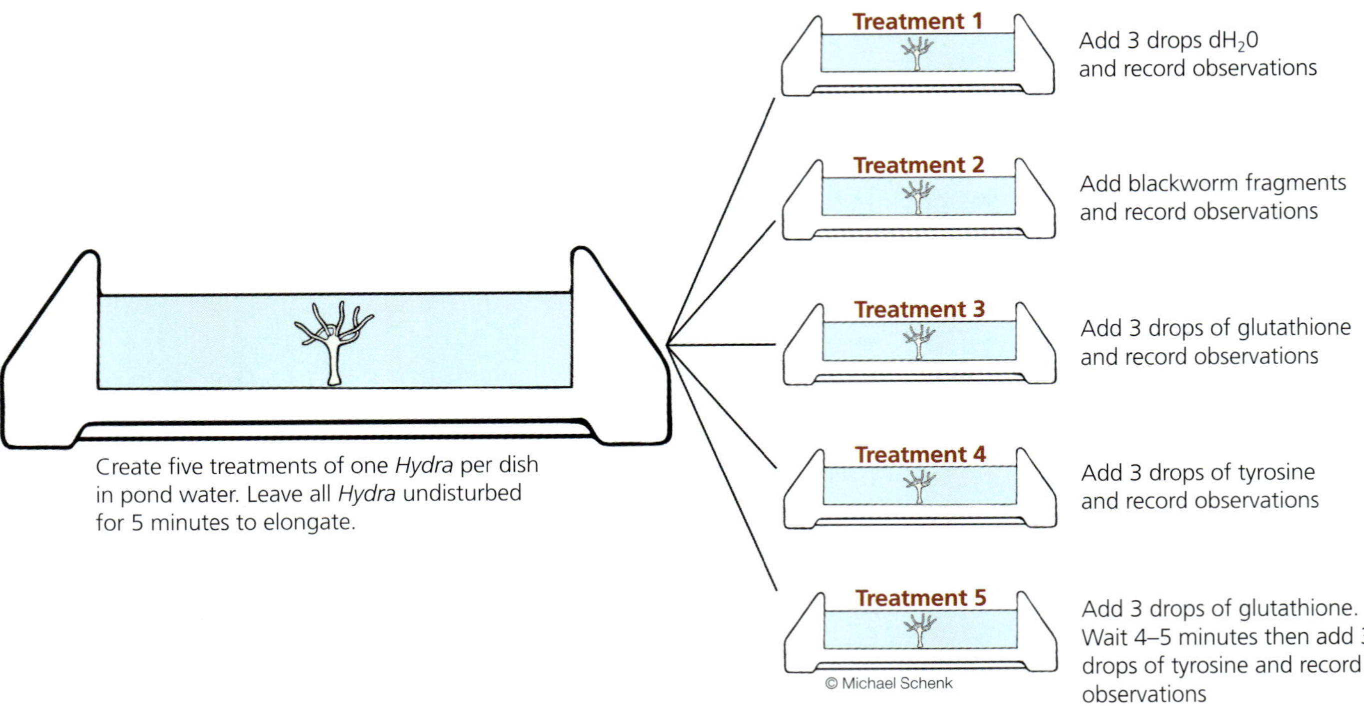

8.21 Experimental setup for examining feeding behaviors in *Hydra*.

8

11 Watch carefully to observe the natural feeding behaviors of *Hydra*. These behaviors will serve as the "standard" to which you will compare the *Hydra's* response to the different chemicals (Fig. 8.21).

12 Record your observations in detail in Table 8-6.

13 To one of the other *Hydra* in a separate dish, add 3 drops of the amino acid glutathione next to the animal (Treatment 3). Observe the reaction carefully.

14 Record your observations in Table 8-6.

15 To another *Hydra* in another dish, add 3 drops of the amino acid tyrosine next to the animal (Treatment 4). Observe the reaction carefully.

16 Record your observations in Table 8-6.

17 To the remaining *Hydra* in the last dish, add 3 drops of glutathione next to the animal, wait 4–5 minutes, and then add 3 drops of tyrosine (Treatment 5). Observe the reaction carefully.

18 Record your observations in Table 8-6.

Table 8-6 Effects of Distilled Water, Amino Acids, and Live Prey on Feeding Responses in *Hydra*

*For each treatment, you should record (**1**) the intensity (on a scale of 1–10), (**2**) the frequency, and (**3**) the duration of each response.*

Potential Behaviors	Treatments				
	#1 Distilled Water	#2 Blackworm Fragments	#3 Glutathione	#4 Tyrosine	#5 Glutathione + Tyrosine
Tentacles waved randomly	Intensity Frequency Duration	Intensity Frequency Duration	Intensity Frequency Duration	Intensity Frequency Duration	Intensity Frequency Duration
Tentacles drawn toward or across mouth	Intensity Frequency Duration	Intensity Frequency Duration	Intensity Frequency Duration	Intensity Frequency Duration	Intensity Frequency Duration
Hypostome expansion (mouth open)	Intensity Frequency Duration	Intensity Frequency Duration	Intensity Frequency Duration	Intensity Frequency Duration	Intensity Frequency Duration
Hypostome contraction (mouth closed)	Intensity Frequency Duration	Intensity Frequency Duration	Intensity Frequency Duration	Intensity Frequency Duration	Intensity Frequency Duration
Body elongation	Intensity Frequency Duration	Intensity Frequency Duration	Intensity Frequency Duration	Intensity Frequency Duration	Intensity Frequency Duration
Body contraction	Intensity Frequency Duration	Intensity Frequency Duration	Intensity Frequency Duration	Intensity Frequency Duration	Intensity Frequency Duration

Check Your Progress

6.1 Is the feeding behavior an all-or-nothing response, or can you distinguish discrete phases of the behavior?

6.2 Does *Hydra* respond to chemical cues in its environment?

6.3 What conclusions can you draw about the kinds of receptors involved in the induction of feeding behavior and their distribution in the body?

8

6.4 Are amino acids most likely the proximate cause of the feeding response seen in *Hydra*?

6.5 Describe the complete cycle of feeding responses of *Hydra* in terms of physical and chemical stimulation.

Name ______________________________ Section ______________

8 Questions for Review

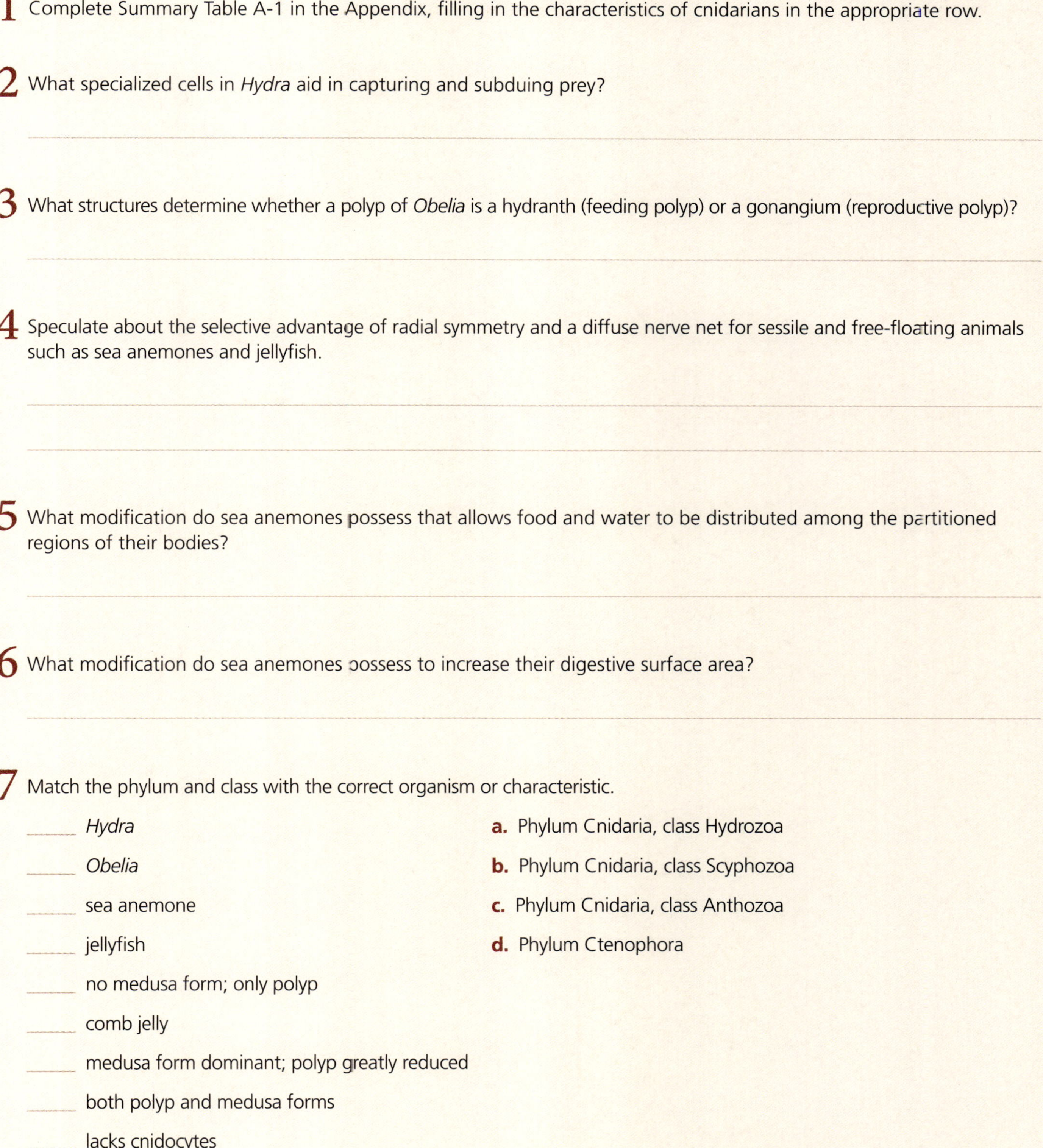

1 Complete Summary Table A-1 in the Appendix, filling in the characteristics of cnidarians in the appropriate row.

2 What specialized cells in *Hydra* aid in capturing and subduing prey?

____

3 What structures determine whether a polyp of *Obelia* is a hydranth (feeding polyp) or a gonangium (reproductive polyp)?

____

4 Speculate about the selective advantage of radial symmetry and a diffuse nerve net for sessile and free-floating animals such as sea anemones and jellyfish.

____

____

5 What modification do sea anemones possess that allows food and water to be distributed among the partitioned regions of their bodies?

____

6 What modification do sea anemones possess to increase their digestive surface area?

____

7 Match the phylum and class with the correct organism or characteristic.

_____ *Hydra*

_____ *Obelia*

_____ sea anemone

_____ jellyfish

_____ no medusa form; only polyp

_____ comb jelly

_____ medusa form dominant; polyp greatly reduced

_____ both polyp and medusa forms

_____ lacks cnidocytes

a. Phylum Cnidaria, class Hydrozoa

b. Phylum Cnidaria, class Scyphozoa

c. Phylum Cnidaria, class Anthozoa

d. Phylum Ctenophora

8 Explain the difference between the proximate cause of a behavior and the ultimate cause.

9 As you walk around your backyard on a damp night with a flashlight pointed at the ground, you notice earthworms quickly recoiling into their burrows as you walk near them. Speculate about the proximate and ultimate causes of this common behavior.

Name ______________________ Section ______________

8 Laboratory Practical Challenge

The following evaluation is intended to replicate the method of assessment and type of questions you will be asked on lab practicals. If you have difficulty with these questions, you may benefit from additional review of the specimens provided in lab.

1 Identify the phylum and class of the specimen depicted below.

____

2 What is the function of the structure indicated by line A?

____

3 Identify the structure indicated by line B.

____

4 How many tissue layers does this organism have?

____

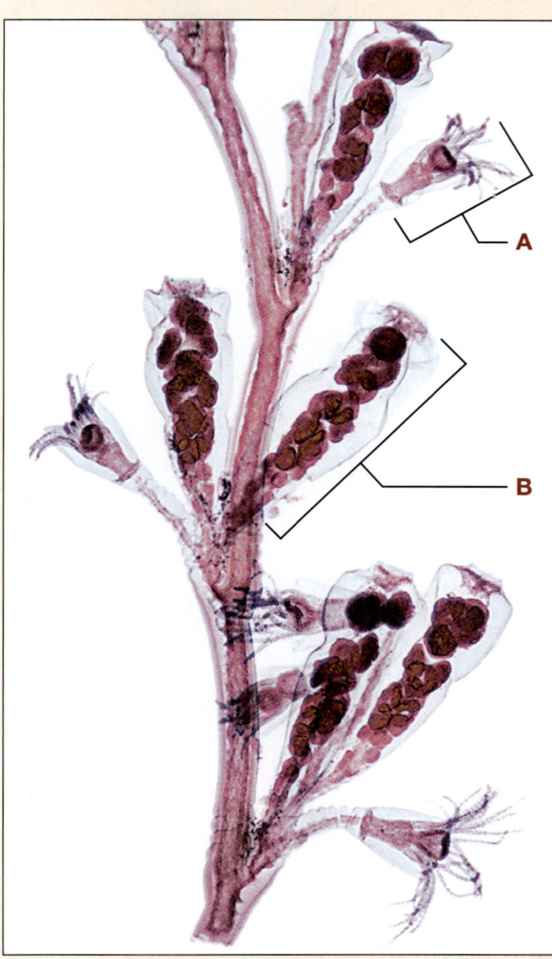

5 On the figure below, label the following structures: primary septum, secondary septum, ostium, tentacles, and pedal disc.

6 What is the function of the structure indicated by line C?

7 Identify the space indicated by line D.

8

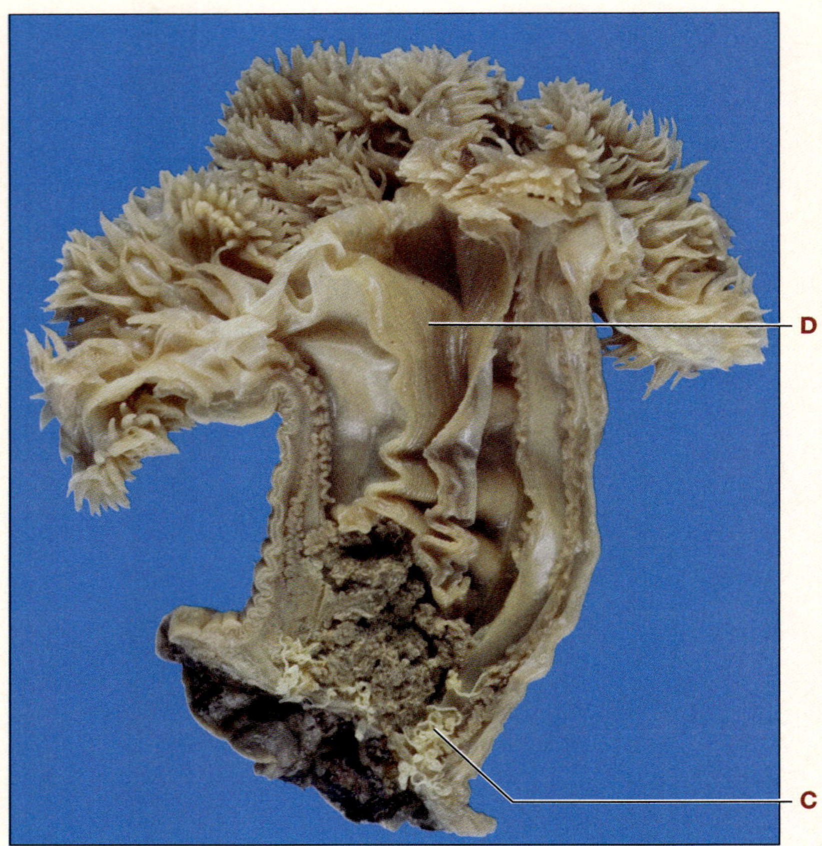

CHAPTER 9

Flatworms

After completing the exercises in this chapter, you should be able to:

1. List the distinguishing characteristics of the phylum Platyhelminthes.
2. Identify specimens from the three major classes of flatworms.
3. Understand the evolutionary position of flatworms on the phylogeny of major animal phyla.
4. Understand the concept of cephalization and its evolutionary significance.
5. Describe the three types of body cavity plans that characterize triploblastic animals.
6. Identify the basic anatomy of representatives from the three major classes of this phylum.
7. Discuss the evolutionary modifications for parasitic or free-living lifestyles that members of each class possess.
8. Describe the life cycles of a two-host and three-host parasite.
9. Understand the basic behavioral patterns of planaria and the degree to which these behaviors can be modified through learning.
10. Understand the concept of regeneration and the degree of flexibility present in flatworms for regenerating lost body parts.

An estimated 20,000 species of flatworms belong to the phylum Platyhelminthes, ranging from a few millimeters in length to more than 20 meters long! Despite this wide range in *length*, flatworms are never more than a few millimeters thick. Like sponges, ctenophores, and cnidarians, they lack circulatory systems for transporting oxygen and nutrients throughout their bodies. Thus, all cells must be close to the surface of the animal for effective diffusion and transport of nutrients. Flatworms are soft-bodied, usually **monoecious**, and often parasitic, living within the bodies of their hosts. Like cnidarians, flatworms have a **gastrovascular cavity** with a single opening, but it is more highly branched and extends throughout more of the body.

In addition to the **epidermis** and **gastrodermis**, flatworms possess a third tissue layer sandwiched between the two, called **mesoderm.** During embryonic development this third tissue layer differentiates into muscles, making flatworms the first highly motile group of animals that we have seen. Another advancement seen in flatworms is an excretory system consisting of a network of tubules with blind pouches called flame cells. **Flame cells** function primarily in fluid balance within the body, with **excretion** being a secondary role. The majority of metabolic waste is eliminated from the body by diffusion through the body wall.

Flatworms have **bilateral symmetry** and a simple, ladderlike nervous system—two longitudinal nerve cords connected by a series of transverse nerves. Free-living species typically possess a concentration of nervous tissue and sensory structures at the anterior end of the body—a condition known as **cephalization.**

Despite sharing these characteristics common to all flatworms, members of each class occupy radically different ecological niches, and natural selection has molded them in different ways to produce specific adaptations to these unique lifestyles. The three major classes of flatworms covered in this chapter include the free-living Turbellaria (planarians), Trematoda (parasitic flukes), and Cestoda (parasitic tapeworms) (Table 9-1).

Animal Body Plans

As you may recall, cnidarians and ctenophores are diploblastic, and sponges lack true tissues altogether. Animals, such as flatworms, containing three layers of tissue are known as **triploblastic.** In fact, most animals with which we are familiar are triploblastic. Within the triploblastic animal species, different developmental strategies have evolved for packaging the tissues and compartmentalizing the body space. As a result, three different body plans have arisen among triploblastic animals: **acoelomate, pseudocoelomate** and **eucoelomate** (Fig. 9.1). All flatworms are acoelomate, meaning that they have no coelom. A **coelom** is a cavity lying within the mesoderm. In flatworms the mesoderm is solid—filled with muscle fibers and other loose tissues. The gut (or digestive cavity) does not count as a body cavity when determining the type of coelom that an animal possesses.

Table 9-1 Major Classes of Phylum Platyhelminthes

Class and Representative Animals	Characteristics
Turbellaria (≈ 3,000 sp.) Planarians	Free-living flatworms with a ventral mouth
Trematoda (≈ 10,000 sp.) Flukes	Internal parasites with leaf-shaped body and anterior mouth
Cestoda (≈ 3,500 sp.) Tapeworms	Internal parasites with long, ribbonlike body composed of proglottids and no mouth

Body Plans of Triploblastic Animals

Acoelomate: animal whose central space is filled with tissue (mesoderm). No true body cavity exists. ***Example***: flatworms.

Pseudocoelomate: animal with a central body cavity that lies between the gastrodermis and the mesoderm. ***Example***: roundworms (nematodes), rotifers.

Eucoelomate: animal with a central body cavity that lies within mesoderm. ***Examples***: earthworms, molluscs, arthropods, echinoderms, chordates.

Evolutionary History

The first animals to evolve cephalized, bilaterally symmetrical bodies were probably most similar to the small, free-living turbellarian flatworms alive today. Their diminutive, muscular bodies would have been well suited to root around the sand on the ocean floor, ingesting small organic particles. Bilateral symmetry and cephalization are closely related to the animal's tendency when swimming or crawling to keep the same end of the body forward and the same surface downward toward the substrate. Organisms that moved consistently with the same part of the body forward benefited from strong selective pressure over successive generations for sensory organs and nervous systems to become more concentrated at the forward end, where environmental conditions are first being met.

It is generally agreed that the parasitic lifestyle of tapeworms and flukes represents a derived characteristic, yet biologists are not so quick to agree on the exact relationship of the acoelomate body plan to other animal phyla. The classical view is that modern organisms exhibiting acoelomate, pseudocoelomate, and eucoelomate body plans evolved simultaneously and independently from an ancestral triploblastic acoelomate. Conversely, the currently accepted hypothesis places flatworms together with five other invertebrate phyla in the Lophotrochozoa, an assemblage that is not based on the type of body cavity present, but instead derives its name from two other features of its members (Fig. 9.2). Some lophotrochozoans develop a structure called a lophophore, a horseshoe-shaped crown of ciliated tentacles used in feeding. Individuals in other phyla (such as molluscs and annelids) go through a distinctive larval stage called the trochophore. According to this view, the acoelomate body plan may be a secondarily derived characteristic, in which case the ancestors of modern flatworms would have possessed body cavities (true coeloms or pseudocoeloms) that eventually were eliminated during the course of flatworm evolution. For a more complete phylogeny depicting the evolutionary relationships among the major animal phyla covered throughout this laboratory guide, review Figure A.1 in the Appendix.

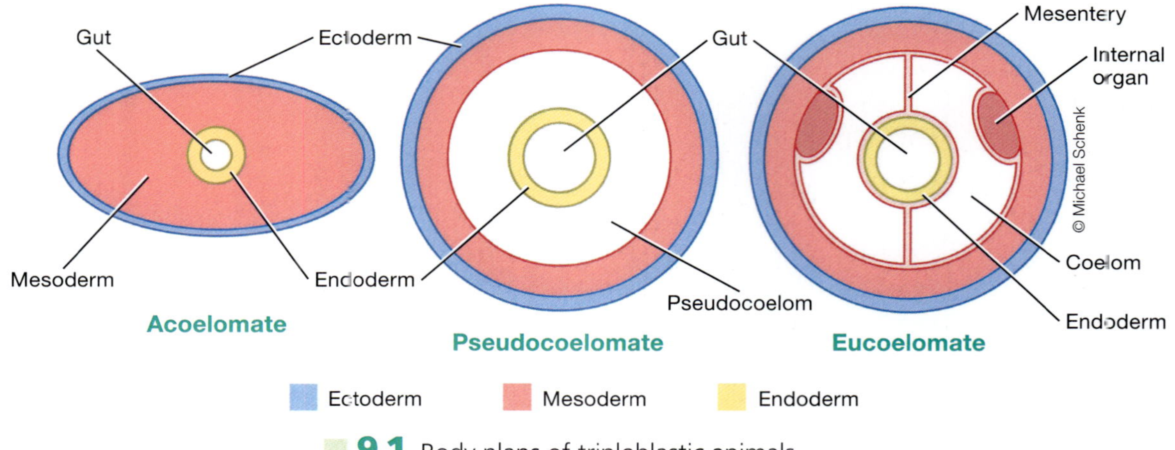

9.1 Body plans of triploblastic animals.

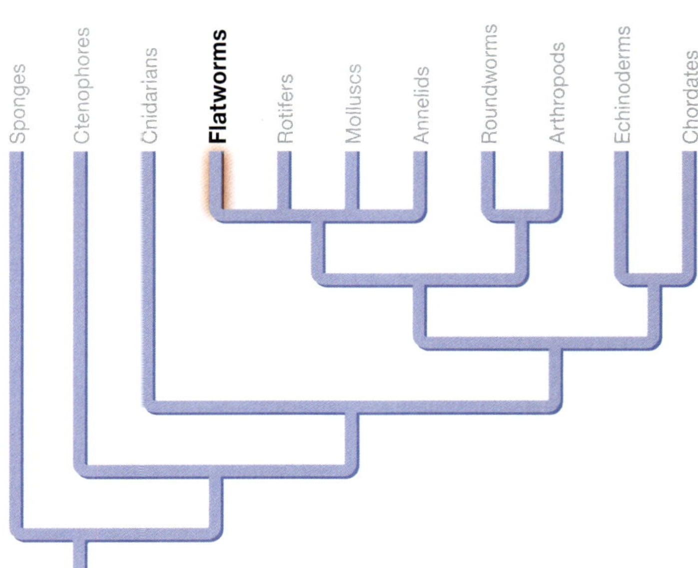

9.2 Modern flatworms are grouped with a collection of other invertebrate phyla into the Lophotrochozoan assemblage based on shared characteristics that indicate close evolutionary relationships.

Exercise 9-1 Planaria (Class Turbellaria)

Materials Needed

- ❑ slides of planaria w.m. and c.s.
- ❑ compound microscope

9

Members of the class Turbellaria (commonly called planaria) are free-living flatworms that typically inhabit freshwater streams and ponds, oceans, and moist terrestrial environments, where they may be found secreted under submerged rocks, foliage, and moist debris.

Planaria are primarily carnivorous, feeding on small crustaceans, protists, roundworms, and insects. Mucous secretions from specialized gland cells allow them to entangle small prey. The planaria projects its muscular pharynx outward through its ventral mouth and uses it like a flexible straw to suck in its food. Often, the planaria wraps its body around its prey to further ensnare it. When not in use, the pharynx is enclosed within the pharyngeal cavity or sheath. The mechanism for extending and retracting the pharynx is diagrammed in Figure 9.3.

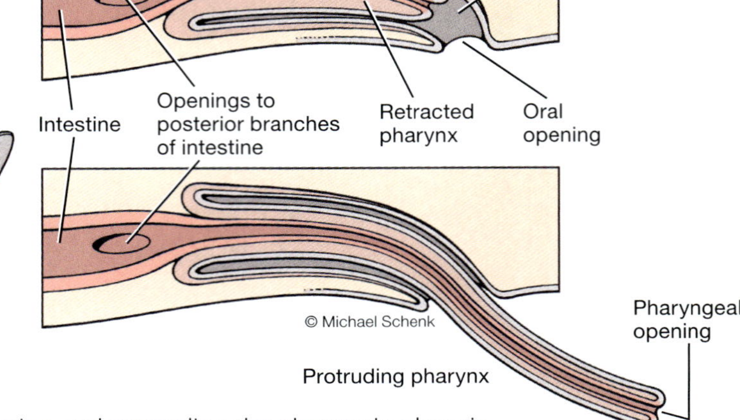

9.3 Mechanism for retracting and protruding the pharynx in planaria.

Planaria possess a simple excretory system (not visible on prepared slides) consisting of small structures called flame cells that are situated at the terminal ends of a branching network of tubules. These tubules collect metabolic wastes from nearby tissues through diffusion, and the flame cells create a current that pushes these wastes along the tubule system and out of the animal through tiny pores distributed along the body surface (Fig. 9.4).

The primary function of flame cells is in **osmoregulation** (water balance), with **excretion** being a secondary role. Nonetheless, the development of an excretory system designed to collect, concentrate, and eliminate metabolic wastes was a major evolutionary step that was one of the prerequisites for animals to evolve larger body sizes.

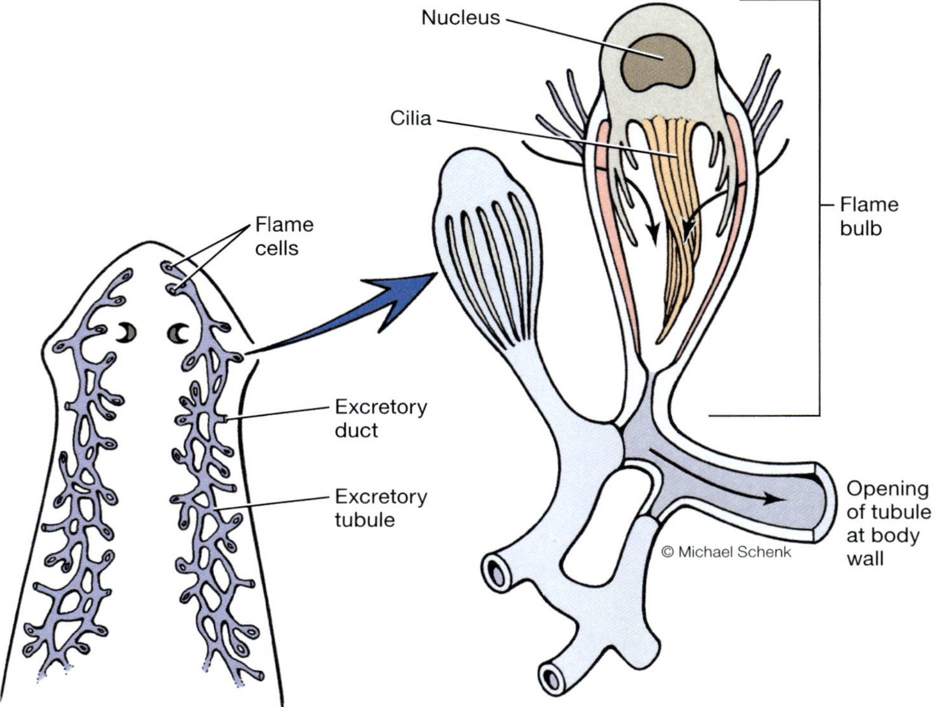

9.4 Flame cells and the excretory system in planaria. Arrows depict fluids filtering through membranous folds in the flame cell.

Procedure

1 Obtain a whole mount (w.m.) slide of the common planarian, *Dugesia*, and view it using low power.

2 Locate the anterior end of the body, where you will see a conglomeration of sensory structures, including eyespots and auricles (Fig. 9.5). The **eyespots** of a planaria sense light (only shadows and direct light, not images), and the **auricles** are chemoreceptors that detect dissolved chemicals in the water. Together, these sensory structures provide information about the outside world to the planaria's cerebral ganglia ("brain") and ladderlike nervous system, not visible on the slide (Fig. 9.6).

3 A long, lightly stained tube, the **pharynx**, should be visible near the middle of the planaria. This structure is a muscular extension of the gastrovascular cavity that is extended from its sheath in the body for feeding.

4 Notice the darkly stained, highly branched gastrovascular cavity that spreads throughout the entire body. Planaria lack a circulatory system to deliver nutrients to their individual cells, so their gastrovascular cavity must reach throughout the body to minimize the distance for nutrients to diffuse from the gastrovascular cavity to the neighboring tissues. Extracellular digestion takes place in the gastrovascular cavity, and nutrients are absorbed directly by individual cells. Undigested food is eliminated through the pharynx.

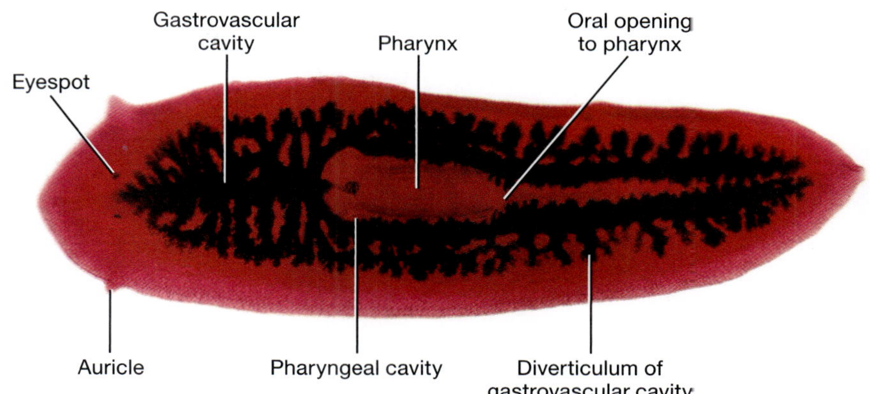

9.5 Major anatomical structures visible in stained whole mount of planaria.

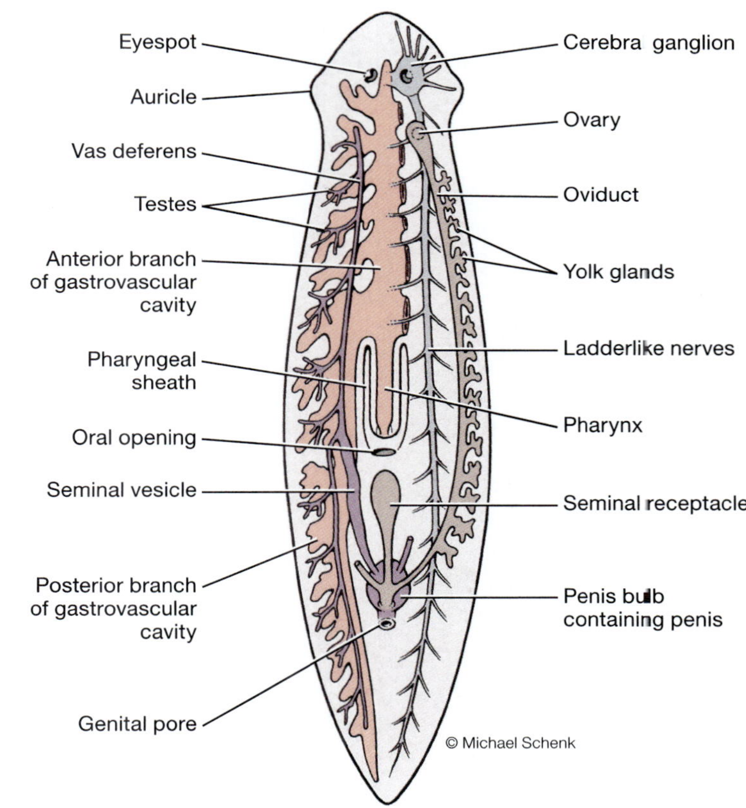

9.6 Major organ systems and structures present in planaria.

5 Obtain a cross-section (c.s.) slide through a planaria and view it using medium power. If you have several cross sections on your slide, first choose one through the middle region of the planaria that captures the tubular pharynx. Sections closer to the posterior end of the planaria will show the same tissue layers, but without the pharynx.

6 Use Figure 9.7 to identify the following tissue layers and structures: epidermis, mesoderm, gastrodermis, pharynx, pharyngeal cavity, and gastrovascular cavity.

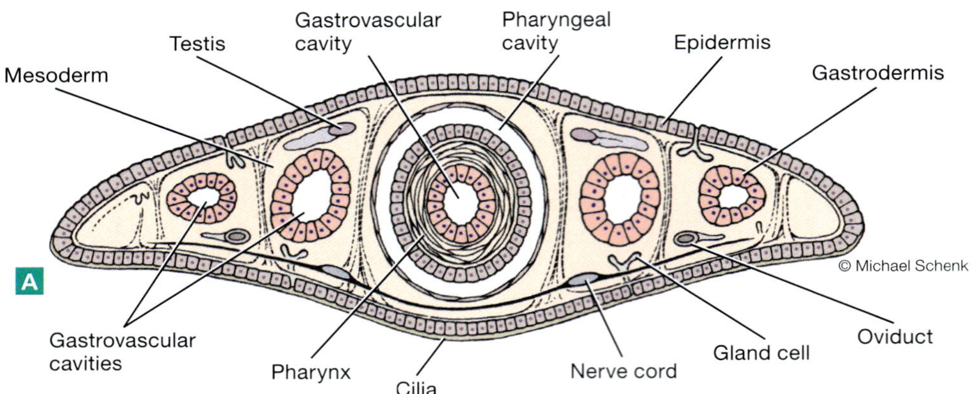

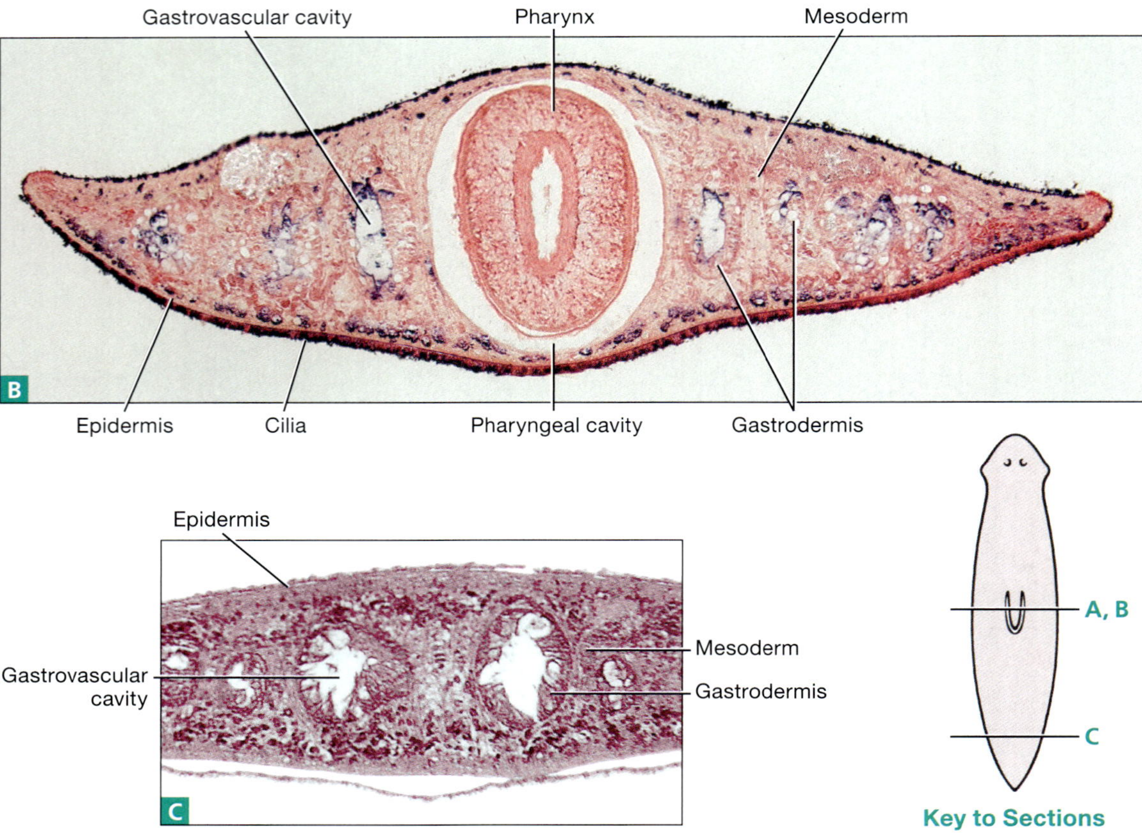

9.7 Cross sections through (**A** and **B**) pharyngeal region and (**C**) posterior region of planaria.

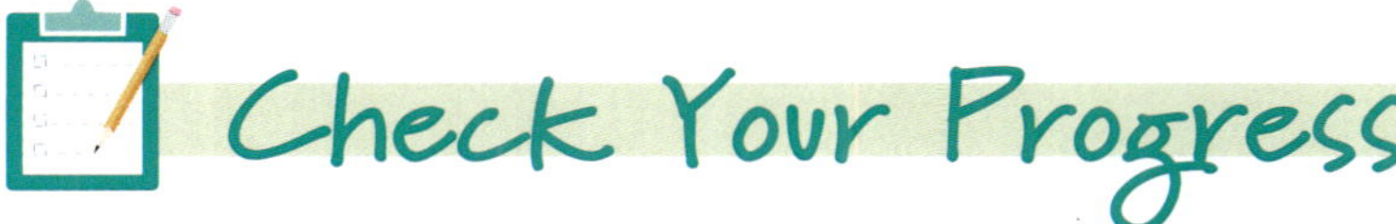

Check Your Progress

1.1 Define the following terms:

auricle

pharynx

acoelomate

monoecious

1.2 List several features of planaria that suggest adaptation to a free-living lifestyle.

1.3 Why do you suppose the pharynx is centrally located on the body, as opposed to being located at the anterior end of the animal?

9

1.4 Why is the gastrovascular cavity so extensively branched?

1.5 Flatworms were the first organisms to evolve specialized cells for excretion (flame cells). What specific purpose does an excretory system serve?

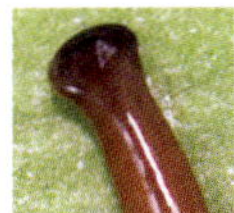

Exercise 9-2 Flukes (Class Trematoda)

Materials Needed

- ❑ w.m. slide of *Fasciola* or *Clonorchis*
- ❑ compound microscope

Flukes are parasitic flatworms belonging to the class Trematoda. They usually attach by means of suckers to their host's internal organs, such as intestines, lungs, blood vessels, bladder, or liver, where they feed by robbing the host of its body fluids and nutrients. Some species have complex life cycles, often involving several host species. Flukes do not obtain the incredible sizes for which some species of tapeworms are notorious; the largest flukes measure only about 7 centimeters in length.

As parasites, flukes display several specific adaptations for this lifestyle. **Suckers** are present for attachment to the host's inner body wall or organs, and flukes are protected by a thin, outer **cuticle** that prevents the digestive enzymes of the host from dissolving them. Another major adaptation of flukes (and other parasites such as tapeworms) is their prolific production of eggs. Flukes are essentially reproductive factories, continuously churning out thousands of eggs a day, often for many years. Even with such huge numbers, only a tiny percentage of eggs are fertilized, make it out of the host, hatch into free-swimming larvae, find another suitable host, mature, leave that host's body, find another host, and reproduce to complete the life cycle. Most flukes are monoecious and are capable of self-fertilization—another adaptation to the often solitary lifestyle of an **endoparasite** (a parasite that lives inside its host).

Procedure

1 Obtain a whole mount (w.m.) slide of a trematode (either *Fasciola* or *Clonorchis*) and view it using low power. Members of these two genera are liver flukes.

2 First locate the oral sucker and ventral sucker—the two primary structures for attachment to the host's tissue.

3 The digestive system includes the oral opening, muscular pharynx, a short **esophagus**, and two branches of the gastrovascular cavity. Because flukes have an incomplete digestive system, like all flatworms, the intestinal branches are basically a simple gastrovascular cavity.

4 The male reproductive organs are the highly branched **testes**, **vas deferens**, and **genital pore**.

5 The female reproductive organs include an **ovary**, a highly coiled **uterus**, a seminal receptacle, paired yolk glands, yolk ducts, and a genital pore. Why do you suppose so much of the body cavity in flukes is devoted to reproductive structures?

6 The osmoregulatory structures are **excretory canals**, a common bladder, and a posterior excretory pore.

7 Identify these structures on your w.m. slide using Figure 9.8 and Table 9-2 as references.

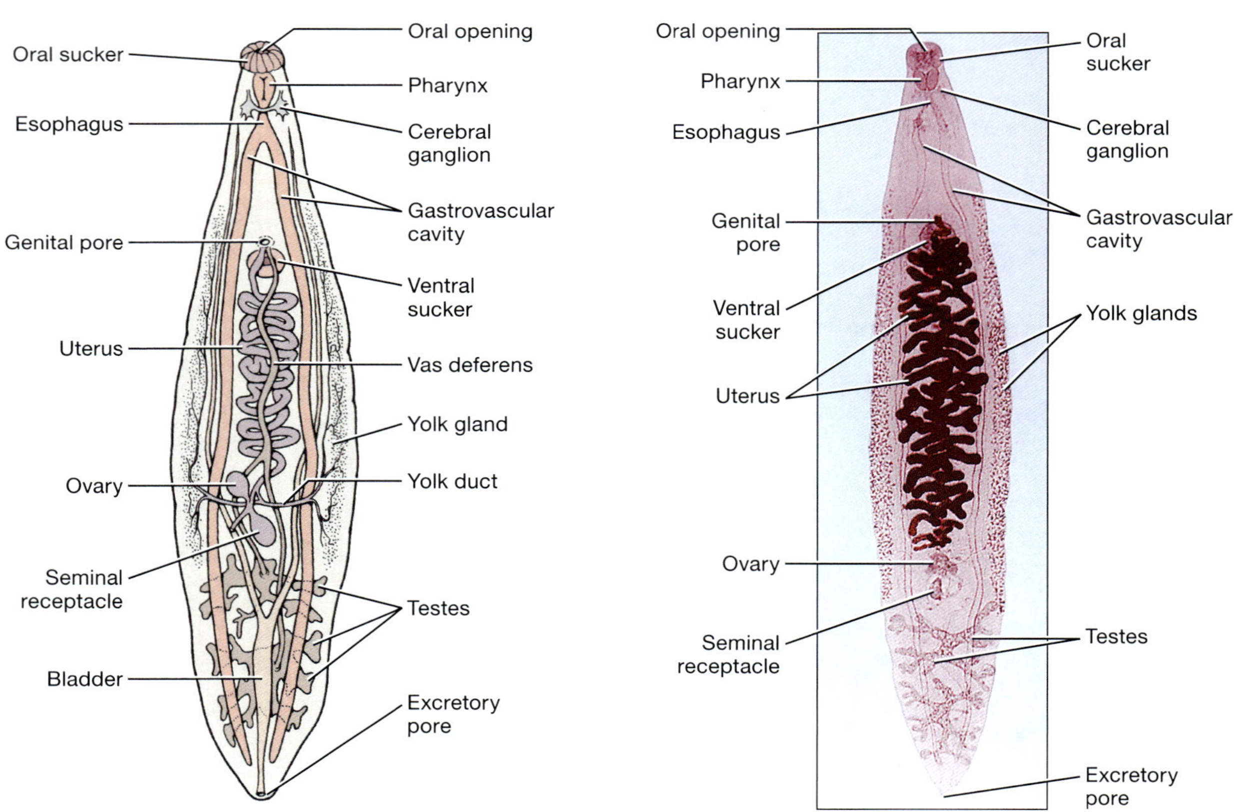

9.8 Anatomy of a common liver fluke, *Clonorchis sinensis*.

Table 9-2 Anatomy of a Liver Fluke

Structure	Function
Oral sucker	Specialized for attachment to host; used in feeding
Ventral sucker	Secondary point of attachment to host
Pharynx	Muscular tube for pumping in blood and body fluids from host
Gastrovascular cavity	Forked tube for digestion and distribution of nutrients throughout body (reduced in flukes)
Testes	Produce sperm
Vas deferens	Stores sperm and transports sperm to genital pore
Genital pore	Receives sperm and moves it to the seminal receptacle
Ovary	Produces eggs
Seminal receptacle	Receives sperm and stores it (often for the life of the fluke) eventually moving it to the uterus where eggs are fertilized
Yolk gland and yolk duct	Eggs are combined with yolk and shelled before passing to the uterus
Uterus	Shelled, fertilized eggs are stored here until release through the genital pore
Excretory canals and bladder	Collect metabolic waste from the flame cells
Excretory pore	Releases metabolic waste products out of the body

Sheep Liver Fluke

The sheep liver fluke, *Fasciola hepatica*, a common parasite of sheep and cattle, demonstrates the complex life cycle of a two-host parasite (Fig. 9.9). Adult liver flukes live in the liver and bile ducts of their hosts, where they feed on blood and produce as many as 20,000 to 50,000 eggs per day for several years of their lives! The eggs eventually enter the intestinal tract with the bile and are expelled with the animal's feces. If the feces drop onto wet pasture, the eggs hatch, releasing free-living **miracidia** larvae. Miracidia larvae swim until they find their intermediate host snail species. Once a snail is located, this larval stage invades its intermediate host, where it passes through several additional stages and multiplies asexually. A single miracidium larva can produce up to 4,000 **cercaria** larvae.

These free-swimming larvae leave the snail and attach to grasses or other aquatic vegetation, where they are ingested by their primary host, the sheep, as it grazes along wet, marshy pastures or streams. Once inside the sheep's digestive system, they encyst in the gut and complete their final stage of development, transforming into the adult liver fluke to begin the cycle again.

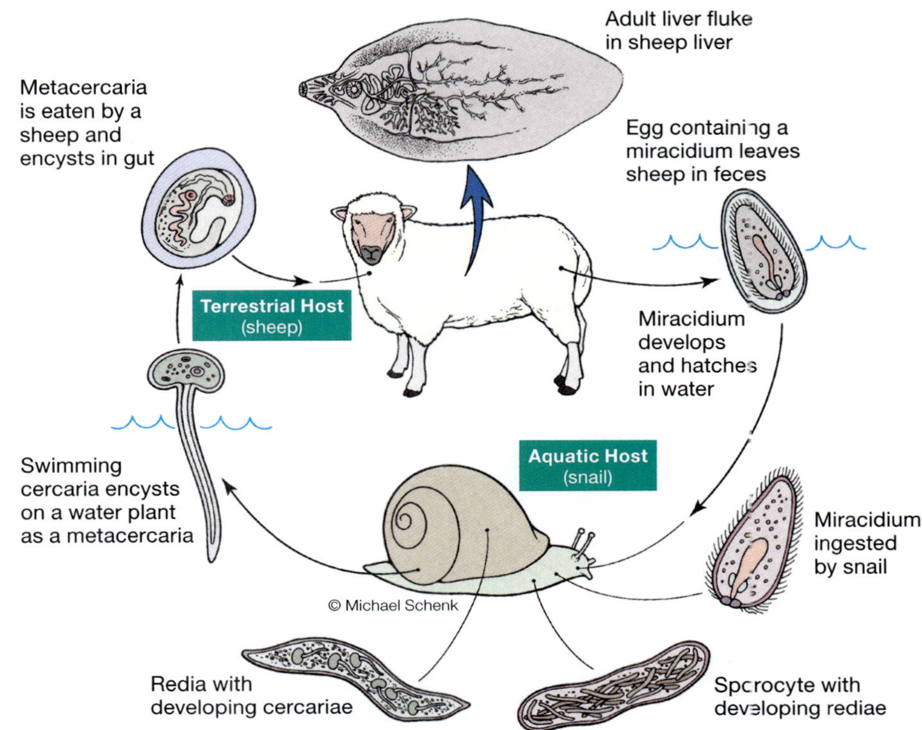

9.9 Two-host life cycle of the sheep liver fluke, *Fasciola hepatica*.

Human Liver Fluke

The human liver fluke, *Clonorchis sinensis*, is a common human parasite throughout much of eastern Asia. Its life cycle includes three parasitic stages: a primary host (humans) and two intermediate hosts (snail and fish). Humans acquire the parasite from eating raw or undercooked fish infected with the encysted **metacercaria** of this organism. Once inside the human digestive tract, the outer protective wall of the cyst is digested, releasing the metacercaria larva (Fig. 9.10). The larva migrates through the bile duct to the liver, where it develops into an adult liver fluke and begins to reproduce. Fertilized eggs are released into the host's bile duct, travel into the intestine, and are passed out of the host's body with the feces. The eggs are eaten by snails living in streams and rivers that contain concentrations of human feces from the people in these areas using the waterways as toilets.

Inside the snail, the eggs hatch and the first intermediate larval stage, the miracidium, develops. Miracidia pass through several more stages (sporocyst, redia, and cercaria), dividing asexually to produce thousands of additional larvae. The free-swimming cercaria larvae leave the snail and swim until they find the second intermediate host, a specific species of fish. The cercaria larvae burrow through the fish's skin and embed themselves in the muscle tissue, where they transform into the dormant metacercaria cyst and await consumption by humans to complete the life cycle.

9

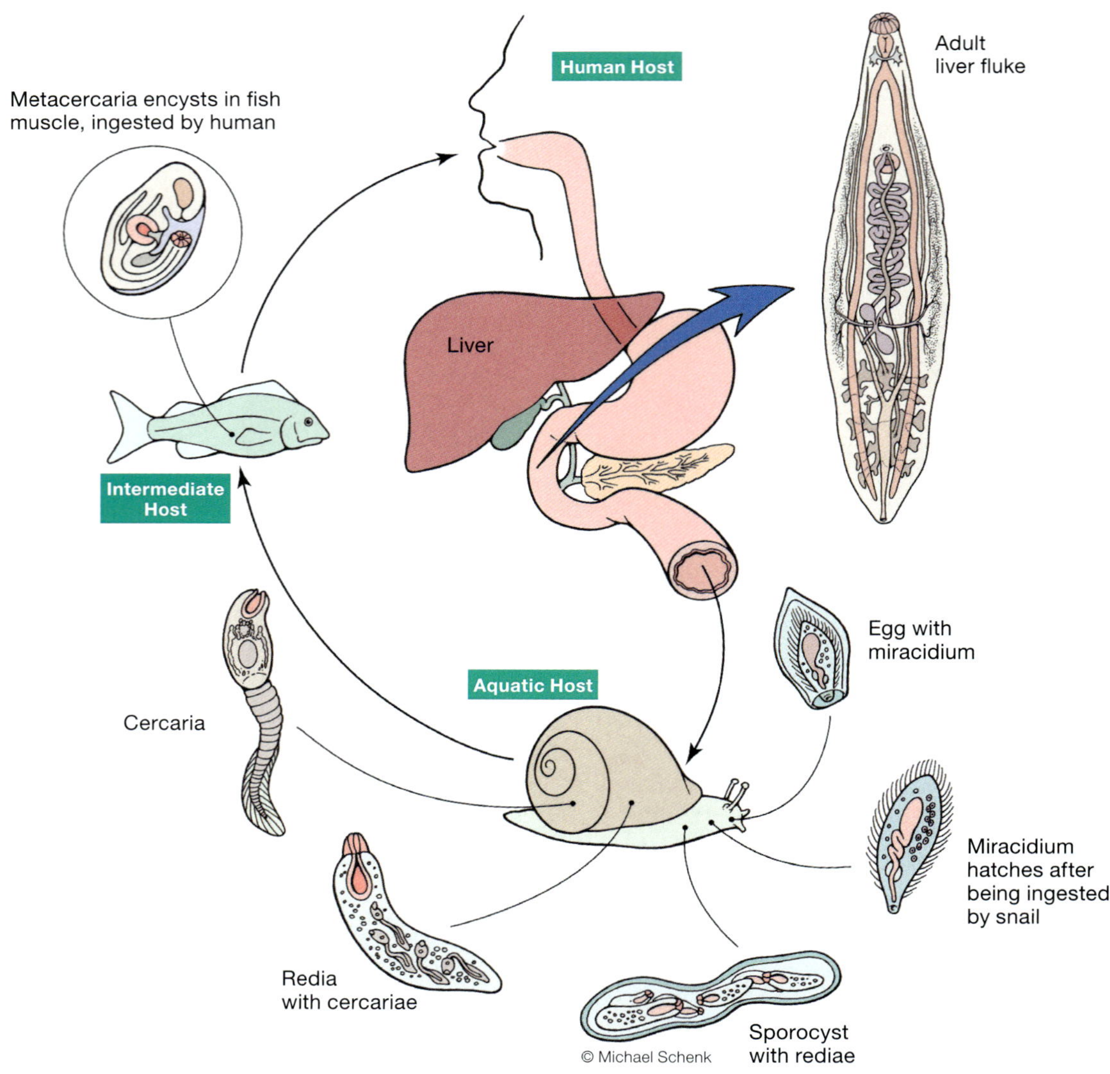

9.10 Three-host life cycle of the human liver fluke, *Clonorchis sinensis*.

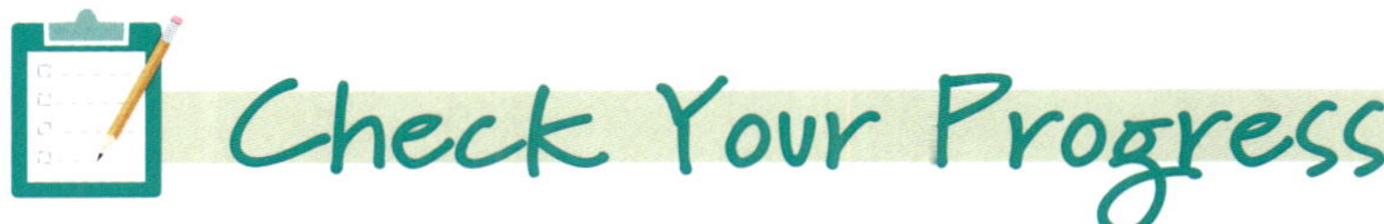

Check Your Progress

2.1 List several anatomical and behavioral features of trematodes that suggest adaptation to a parasitic lifestyle.

Anatomical Features *Behavioral Features*

2.2 Which organs occupy the majority of the body space of the fluke? Explain.

2.3 Is the digestive system of flukes complete or incomplete?

9

2.4 What are the two intermediate larval stages in the life cycle of flukes?

2.5 What defensive provision do flukes possess for protection against the digestive enzymes of their hosts?

Exercise 9-3 Tapeworms (Class Cestoda)

Materials Needed

- ❑ slide of *Taenia* proglottids
- ❑ compound microscope

Members of the class Cestoda, commonly referred to as tapeworms, represent the most specialized group of parasitic flatworms. They have many remarkable adaptations that allow them to be successful endoparasites (parasites that live within the body of their host). Tapeworms have no mouth or gastrovascular cavity; instead nutrients are absorbed directly through the epidermis. They have a thin but tough protective coating on their bodies, known as the cuticle, that prevents them from being digested. The **scolex**, or head, is modified for attachment to the host and usually possesses an array of **hooks** and suckers that embed the anterior end of the tapeworm deep inside the intestinal wall of the host (Fig. 9.11). Remember—tapeworm suckers are for attachment only, not feeding.

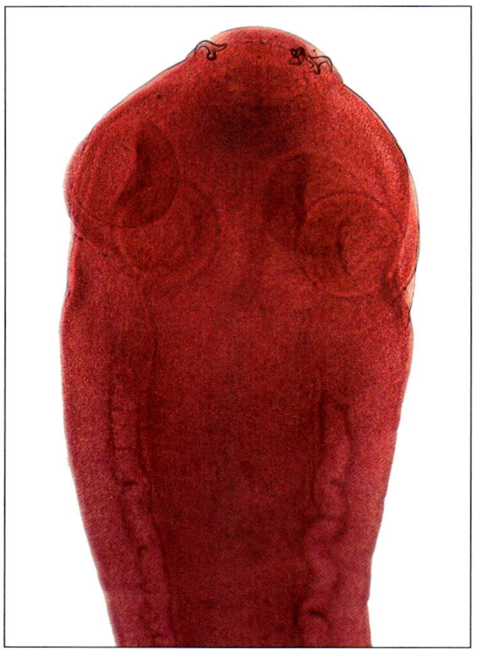

Taenia pisiformis

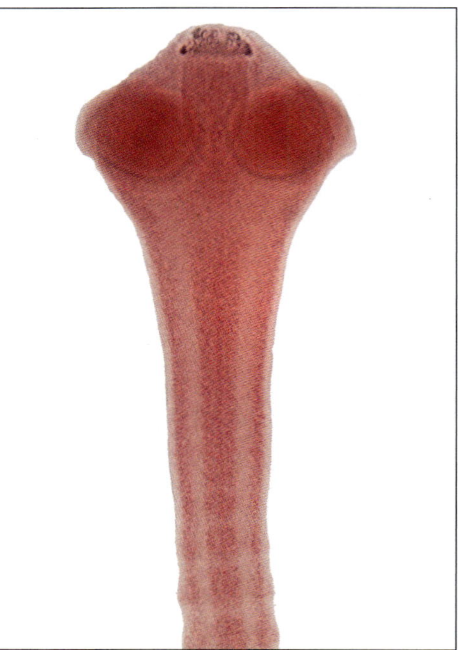

Dipylidium caninum

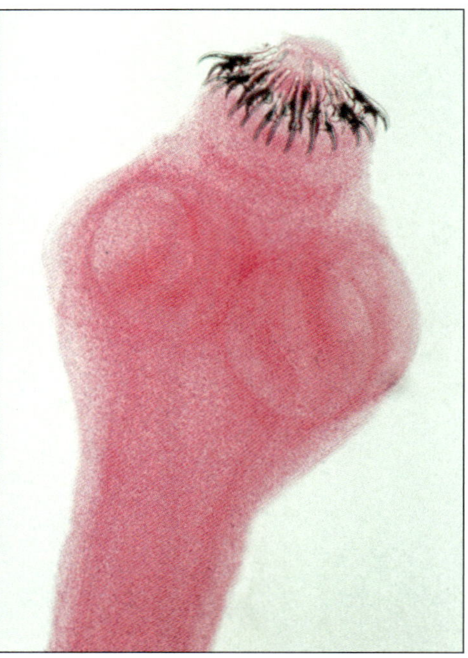

Taenia solium

9.11 Scolexes of tapeworms range in size and shape and may possess an assortment of hooks and/or suckers for attaching to the bodies of their hosts.

The tapeworm body is a series of continuously growing segments, called **proglottids**, that remain attached to one another as they develop. New proglottids form behind the scolex. As proglottids mature, they enlarge and are shifted toward the rear of the tapeworm. In some species this chain of proglottids may reach lengths over 20 meters!

Although self-fertilization is possible, cross-fertilization between two separate tapeworms is the more common mechanism of reproduction. The testes usually mature earlier than the ovaries, so a single proglottid cannot fertilize its own eggs, but the sperm from one proglottid may fertilize another proglottid on that same tapeworm. Sperm from one proglottid must swim through the genital pore of a more posterior proglottid (containing mature eggs) to fertilize the eggs. **Gravid** proglottids at the terminal end of the body containing fertilized eggs drop off the tapeworm, and the eggs are passed out of the host with feces. Although each proglottid represents a functionally separate reproductive unit, all of the proglottids are connected by a common series of excretory canals and longitudinal nerves.

Procedure

1. Obtain a slide of stained proglottids from the tapeworm *Taenia pisiformis*, the dog tapeworm. Examine the slide using low power.
2. On the anterior section of the tapeworm identify the following structures depicted in Figure 9.12 and defined in Table 9-3: scolex, hooks, suckers, neck, and immature proglottids.
3. Identify the excretory canals running longitudinally along the outer margins of the body.
4. Locate a mature proglottid on the slide. Identify the following structures: uterus, ovary, vagina, testes, vas deferens, and genital pore. *Taenia pisiformis* has only one genital pore per proglottid, but this is not the case with all tapeworm species. In fact, the tapeworm *Dipylidium caninum* derives its common name, the double-pored tapeworm, from its two opposing genital pores on each proglottid (Fig. 9.13).
5. Finally, locate a gravid proglottid and identify the well-developed uterus that is packed full of fertilized eggs. Higher magnification will reveal the astounding number of eggs that just one proglottid can contain, up to 100,000 in many cases! A single tapeworm can produce hundreds of millions of eggs in one year.

Hooks
Sucker
Neck

Sucker
Hooks
Scolex

Immature proglottid

Uterus
Transverse excretory canal
Longitudinal excretory canal
Vas deferens
Genital pore
Testes
Cirrus
Nerve cord
Vagina
Ovary
Oviduct
Yolk gland
Yolk duct

Mature proglottid

Genital pore
Uterus with eggs
Gravid proglottid

© Michael Schenk

9.12 Representative body sections and anatomy of a common parasitic tapeworm, *Taenia pisiformis*.

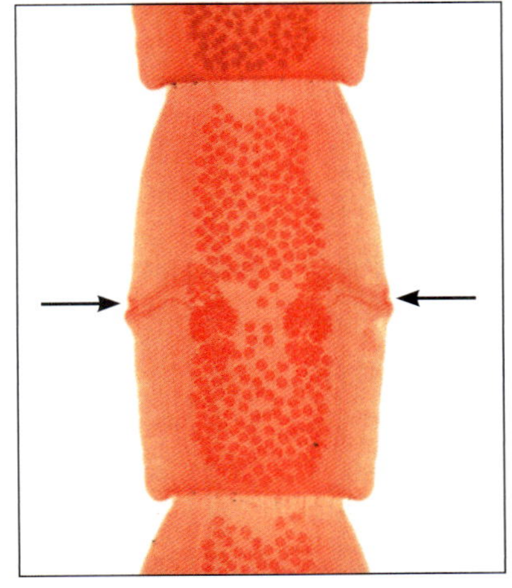

9.13 This species of tapeworm, *Dipylidium caninum*, has two genital pores per proglottid (indicated by arrows), giving it the common name "double-pored tapeworm."

Table 9-3 Anatomy of a Tapeworm (*Taenia pisiformis*)

Structure	Function
Scolex	Anterior end of tapeworm; lacks sensory structures but possesses modifications for attachment to intestinal wall of host
Hooks and suckers	Modified structures on scolex for attachment to host
Neck	Constricted portion signifying posterior end of scolex; marks the site of origin of immature proglottids
Immature proglottid	Newly produced segment of the tapeworm that has undeveloped reproductive organs
Excretory canals	Longitudinal channels running along the outer margins of the body that deliver metabolic waste products out of the tapeworm
Mature proglottid	Tapeworm segment that has functional reproductive organs
Yolk gland and yolk duct	Eggs are combined with yolk and shelled before passing to the uterus
Uterus	Shelled, fertilized eggs are stored here until proglottid drops off of tapeworm body
Ovary	Produces eggs
Vagina	Point of entry into female reproductive tract through which sperm travel to reach the eggs; joins with the genital pore
Testes	Produce sperm
Vas deferens	Canal through which sperm pass as they exit the proglottid through the genital pore
Genital pore	External opening common to the male and female reproductive tracts of the tapeworm; sperm exit and enter proglottids through this opening
Gravid proglottid	Tapeworm segment containing ripe fertilized eggs; this segment is ready to drop off the tapeworm body and be eliminated from the host in feces

Check Your Progress

3.1 Describe the changes that occur in tapeworm proglottids as they mature.

3.2 Why is there an entire reproductive system in each proglottid?

3.3 Tapeworms have been shaped by evolution for one purpose—reproduction. Describe how their body plan is set up to maximize their efficiency for obtaining nutrition and translating this energy into reproductive output.

Exercise 9-4 Investigating Flatworm Behaviors

Chemosensory and Feeding Behaviors

Materials Needed for All Procedures

- ❑ live planaria (*Dugesia tigrina*)
- ❑ pond water (or springwater)
- ❑ clear petri dishes
- ❑ plastic droppers

Procedure

1 Place a few live flatworms in a petri dish containing pond water, and use the dissecting microscope to observe the paths of the animals. Are the paths straight or meandering? Explain why.

Additional materials needed:

- ❑ small pieces of raw beef or chicken liver
- ❑ hard-boiled egg yolks
- ❑ toothpicks
- ❑ dissecting microscopes

2 Next, using a toothpick, place a very small piece of hard-boiled egg yolk at the far side of the petri dish away from the flatworms.

3 Observe their responses for the next several minutes. Does the path of the flatworms change after adding the egg yolk?

4 If not, try gently swirling the water or moving the egg yolk closer to the animals.

5 What sensory structures do the flatworms use to detect the presence of the egg yolk?

6 Now obtain a new dish with a few more flatworms. Place a small piece of raw liver or beef in this dish and observe the response. Is it similar to the response seen with the egg yolk? Does the response occur more quickly with one "food" source than the other?

Photosensory Behaviors

The eyespots of planaria are simple photoreceptors. Although they cannot form true images as your eyes can, they can detect shadows and distinguish between light and dark.

Direct illumination

Procedure

1 Place a planaria in a clear petri dish containing a small amount of pond water and let it crawl around under subdued light.

2 Using a strong, pinpointed light source, shine the beam directly at the anterior end of the worm (Fig. 9.14). What response does the animal exhibit? Why?

Additional materials needed:

- ❑ black-and-white petri dishes
- ❑ light stands
- ❑ light source with pinpoint beam
- ❑ light source with broad beam
- ❑ black paper
- ❑ white paper

3 Next shine the light on the animal from the side. How does it react? Why?

4 Repeat these tests several times until you have sufficient data to be able to generalize about the nature of planaria's reaction to direct illumination. What conclusions can you make about their response to bright light?

9.14 Procedure for examining responses of planaria to direct illumination.

Background pattern

Procedure

1 Place three or four flatworms in pond water in a petri dish that has been painted to render half of its surface black and half of its surface white. If the "bright" side of the dish has not been painted white, you may place a piece of white paper under the dish to further brighten its surface.

2 Partially cover the dish with black paper so the black surface of the dish is shaded from the overhead light, and illuminate the white portion of the dish with an overhead light source with a broad beam (Fig. 9.15).

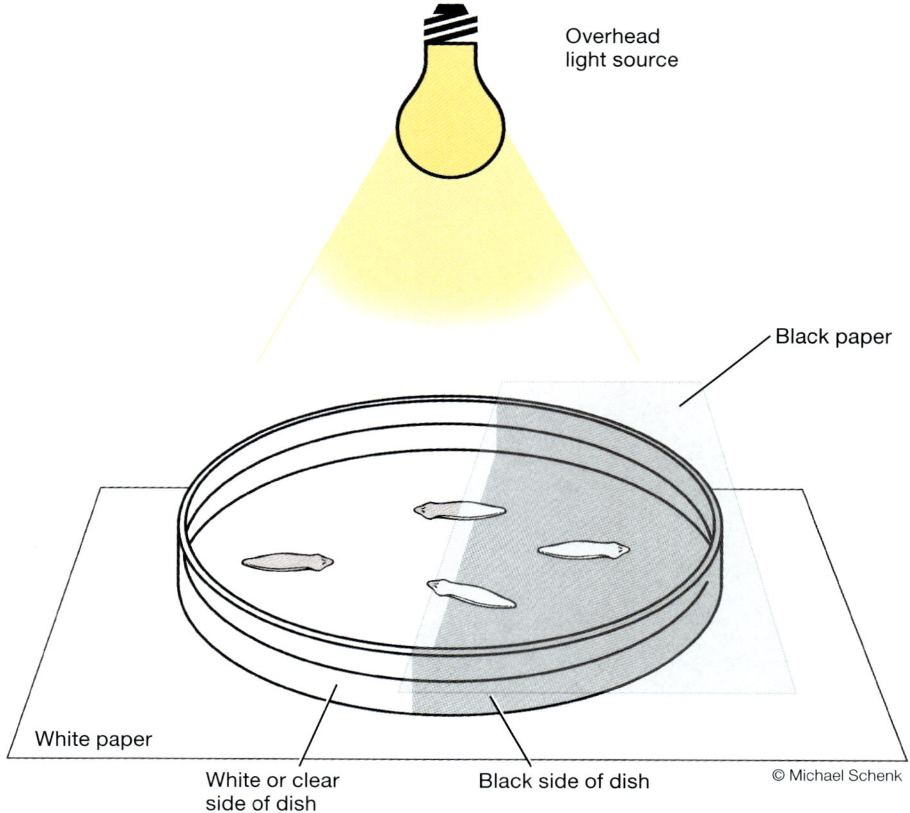

9.15 Experimental setup for examining responses of planaria to light and dark backgrounds.

3 If headless flatworms are available, place two or three headless flatworms in a separate black-and-white petri dish. Make sure that each dish is illuminated evenly from above.

4 From your data on direct illumination, where would you predict that the animals will aggregate in such a dish, and why?

5 Would you expect the headless flatworms to respond differently? Explain.

6 Let the dishes stand for about 20 minutes, and proceed to the next exercise during this time.

7 After 20 minutes, check the dishes to see where the intact and headless flatworms are found. How do you explain these results?

Tactile Responses

Procedure

Additional materials needed:
- ❑ dissecting needles
- ❑ fine-tipped forceps
- ❑ covered glass petri dishes containing crushed ice
- ❑ sterile razor blades
- ❑ dissecting microscope

1 Obtain a live planaria in a petri dish containing pond water. Touch the worm very gently with the sharp tip of a teasing needle, but *do not puncture the body.*

2 What is the reaction of the flatworm to this tactile stimulus?

3 Are the responses uniform throughout the body? If not, what portions of the body are most sensitive?

4 Next, cut a planaria into anterior and posterior halves. Before cutting, you must slow the planaria by cooling it. Use a plastic dropper to place the animal on a clean, covered, glass petri dish filled with crushed ice, and position the dish on the stage of the dissecting microscope.

5 Remember that the light of the microscope will produce heat that will counteract the cooling effect of the chilled dish, so work quickly but carefully.

6 While observing through your dissecting microscope, make the cut using a clean razor blade (Fig. 9.16). Fine-tipped forceps or dissecting needles may be helpful to hold the animal stationary while cutting.

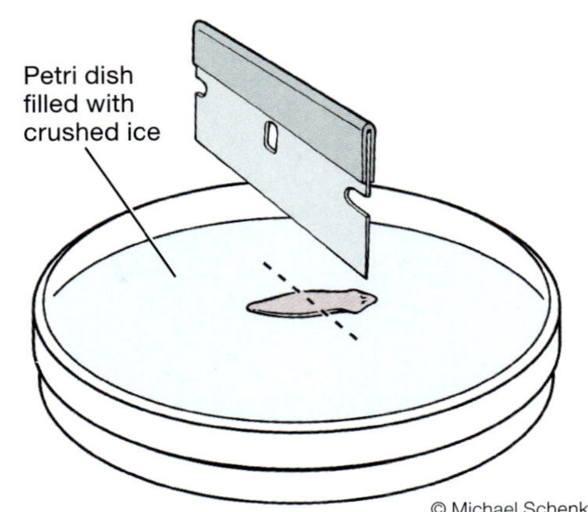

9.16 Procedure for cutting planaria into anterior and posterior halves.

7 After the planaria has recovered from being cut and has started to move normally again, touch the two halves with the dissecting needle (as in Step 1). Is there any difference in the responses compared to those of whole flatworms?

8 Speculate about the distribution of tactile receptors on the body of planaria.

A Simple Model for Learning

Procedure

1 Obtain a planaria and transfer it to a small petri dish containing pond water.

2 Expose the flatworm to a brief flash of pinpointed light. Is there any response?

Additional materials needed:

- ❑ covered glass petri dishes containing crushed ice
- ❑ sterile razor blades
- ❑ blunt probe
- ❑ light source with narrow, pinpoint beam

3 Now repeat the stimulus and follow it with a gentle poke from your probe, but *do not puncture the animal*. What is the response?

4 Give another brief flash of light, wait 2 seconds, and poke the flatworm again.

5 Repeat these coupled-stimulus events 100 more times. (This may seem like a lot, but these animals aren't that intelligent!).

6 After 100 coupled-stimulus events, administer the flash of light *without* the poke. What is the response? Does this demonstrate learning?

7 Continue to give the first stimulus (the light) without poking the flatworm. How many flashes of light are required for the flatworm to stop reacting to the uncoupled stimulus?

8 Train two more flatworms, using 100 coupled-stimulus events.

9 After 100 events, cut one worm into anterior and posterior halves (as described in steps 4–6 in the previous exercise, "Tactile Responses") and test each half to the first stimulus—the light.

10 Do both halves respond to the first stimulus? Explain.

11 Put the two halves in two different petri dishes and carefully label them.

12 Place the other trained flatworm in a third dish and label it. Your instructor will show you where to store these animals. Next week you will test these animals to see (1) if the control animal has retained its memory and (2) whether both halves of the regenerated animal retain the memory or whether long-term memory is restricted to the end of the bisected planaria that had the original brain.

Technically, the type of learning that flatworms display is a simple form of conditioning, much like the type of learning demonstrated in the classic study published in 1903 by Ivan Pavlov, who discovered that dogs became conditioned to salivate to auditory stimuli—a ringing bell. The dogs received a food reward each time they heard the bell, and over time began to associate the bell with food. Although this is not an incredibly sophisticated form of behavior, the fact that flatworms can be conditioned to "anticipate" adverse stimuli demonstrates that even their simple nervous system is capable of integrating coupled stimuli and drawing associations between them, especially if these stimuli occur close together in time. Thus, the basic, ladderlike nervous system that we see in modern flatworms probably represents something similar to the evolutionary precursor that led to the more complex sensory systems of other animals.

9

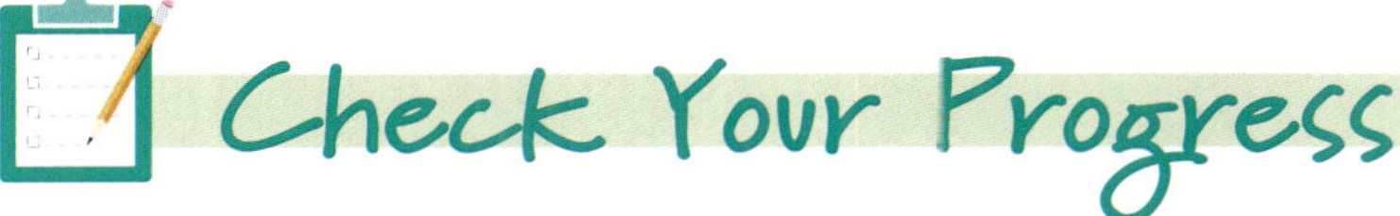

4.1 Are planaria capable of learning?

4.2 Contrast this type of learning with that demonstrated by mammals.

Exercise 9-5 Investigating Regeneration in Flatworms

Materials Needed

- ❑ covered glass petri dishes containing crushed ice
- ❑ glass slides
- ❑ sterile razor blades (or microknives)
- ❑ dissecting needles (or microneedles)
- ❑ snap-cap vials or Syracuse dishes
- ❑ pond water (or springwater)
- ❑ plastic droppers
- ❑ fine-tipped forceps
- ❑ dissecting microscope

Regeneration is the biological process of regrowing and reshaping tissues into exact replicas of missing body parts. Free-living planaria offer an ideal system for investigating the process of regeneration. You may think that cutting a flatworm into pieces is cruel, but this seemingly crude practice actually is a natural part of flatworm reproduction. They reproduce asexually through a process known as fission by adhering tightly to the substrate and constricting bands of muscular tissue to literally pinch themselves into two pieces, each of which will regenerate the missing body parts and become a fully functional flatworm again. Reproduction in this manner produces genetically identical individuals that are true clones.

The first stage in the regenerative process is wound closure and wound healing. Immediately after you make your cuts, you may observe this process, using the dissecting microscope. One to two days after the initial wounds have closed and have begun to heal, a series of complex events occurs in which cells with hypermitotic capabilities migrate to the tip of the damaged area and begin the secondary process of regrowth and reformation of the missing tissues. This process may take anywhere from two to three weeks to reach completion, depending on the temperature of the environment and the size of the piece to be regenerated.

Procedure

1 Before you begin, decide on the type of cut you intend to make. You may wish to cut the animal into equal-sized anterior and posterior halves or left and right halves. Or you may wish to separate just the head from the rest of the body. Alternatively, you may cut your planaria into several pieces transversely. Refer to Figure 9.17 for some examples of suggested cuts.

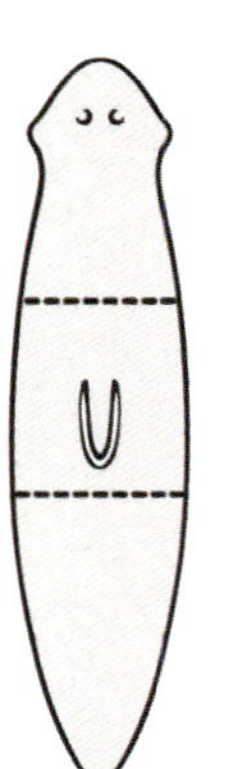
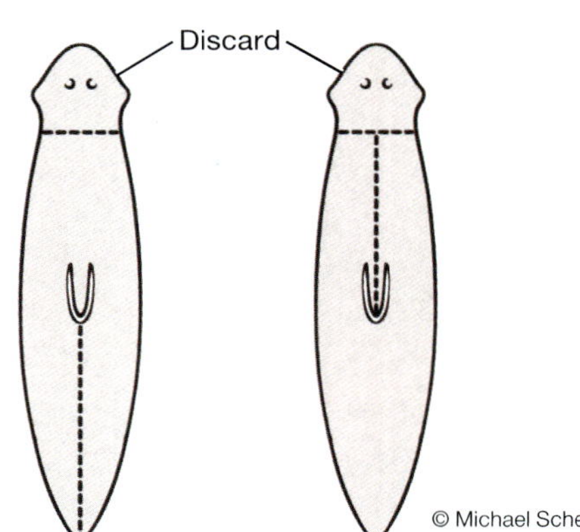

9.17 Suggested cuts to demonstrate regeneration in planaria.

2 Before cutting, you must slow the planaria by cooling it. Use a plastic dropper to place the animal on a clean, covered, glass petri dish filled with crushed ice, and position the dish on the stage of the dissecting microscope.

3 Remember that the light of the microscope will produce heat that will counteract the cooling effect of the chilled dish, so work quickly but carefully.

4 Use a microknife (or razor blade) to make the prescribed cut(s) through your specimen (Fig. 9.18). You may find it helpful to have a teasing needle in the other hand to steady the animal during cutting.

5 When the animal is stretched out, quickly make your cut. If you find that the actual cut you made does not match your initial plans, don't worry. Simply amend the drawing to depict the actual cut you made.

6 Transfer the pieces to a drop of springwater on a clean glass slide (not on ice) and continue to observe them under the dissecting microscope.

7 Carefully observe the cut edge of one of the pieces. You should notice muscular contractions in the body wall of the planaria that will slowly close the wound during the next 10 minutes. Until the wound is completely closed, some body fluids may be seen leaking from the wound. Once the wound is closed, adjacent epithelial tissue begins to actively spread over the wound to begin the healing process. This stage usually takes another 10 minutes to complete.

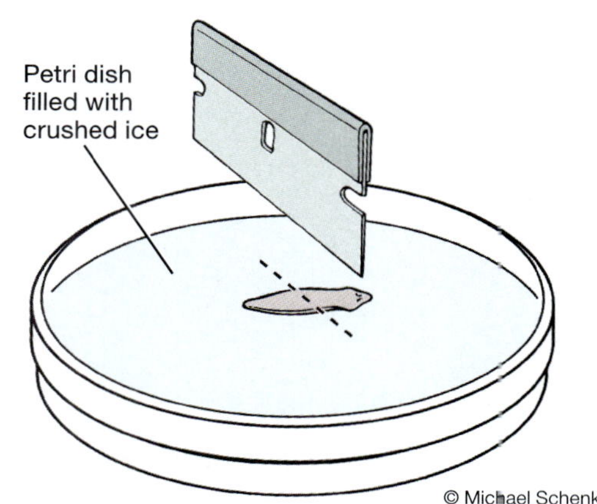

9.18 Procedure for cutting planaria on a chilled petri dish.

8 Don't let your specimen dry out on the slide during this time period.

9 After you have observed wound closure and the initial stages of wound healing, you are ready to transfer your pieces to separate containers for storage. Place each piece in a separate container partially filled with springwater, and cover the dishes to reduce evaporation. Keep them in a cool place with subdued light. Be sure to label the containers with your group name and section number and keep a record of which piece was placed in each dish.

10 You will be monitoring your regenerating pieces over the next several weeks. Your instructor may give you specific instructions for maintaining these specimens. It is imperative that the water in the dishes be changed every few days and that you *do not feed* your specimens while they are regenerating. They will not eat during this period, and any food you provide will simply foul the water with bacteria that could cause infection in your specimens and, potentially, death.

11 At the beginning of each of the next several laboratory periods, examine your specimens under the dissecting microscope and check the progress that each piece has made. In the space below, record the date, draw a picture of each piece, and give a summary of the changes seen.

Date:	Diagram of Planaria:	Description of Progress:

9

Name ______________________ Section ______________

9 Questions for Review

1 Complete Summary Table A-1 in the Appendix, filling in the characteristics of flatworms in the appropriate row.

2 A true coelom is completely lined with:

a. epidermis

b. cilia

c. gastrodermis

d. mesoderm

e. flagella

3 Contrast the nervous system seen in planaria (*Dugesia*) with that seen in *Hydra*.

4 Distinguish between the processes of egestion (or defecation) and excretion, using the flatworm as a model for both processes.

5 Define cephalization and discuss its significance.

6 What is the evolutionary advantage for bilaterally symmetrical, motile animals such as flatworms to have a concentration of nervous tissue and sensory organs located at their anterior end?

9

9

7 Match the class with the correct organism or characteristic.

______	*Fasciola hepatica*	**a.** Class Turbellaria
______	*Dugesia*	**b.** Class Cestoda
______	parasitic flukes	**c.** Class Trematoda
______	tapeworms	
______	free-living, carnivorous, aquatic flatworms	
______	oral and ventral suckers present; oral sucker used for feeding	
______	body consists of segment-like proglottids	

8 The cuticles of flukes and tapeworms are highly resistant to enzyme action. Why is this feature an important evolutionary modification for endoparasites?

____

____

9 Why does the tapeworm lack a mouth, well-developed sensory structures, and a digestive system?

____

____

10 Briefly describe the life cycle of the sheep liver fluke, *Fasciola hepatica*.

____

____

____

____

11 Based on your observations of planaria behaviors, predict how a planaria may react to each of the following stimuli, and indicate which sensory structure would be responsible for the response.

a. A bright, sunny day

____

b. A dead earthworm at the bottom of the lake

____

c. A small minnow nibbling at the flatworm's tail

____

Name ______________________________ Section ______________

9 Laboratory Practical Challenge

The following evaluation is intended to replicate the method of assessment and type of questions you will be asked on lab practicals. If you have difficulty with these questions, you may benefit from additional review of the specimens provided in lab.

1 Identify the phylum and class of the specimen depicted below.

____

2 What is the function of the structure indicated by line A?

____

3 Identify the structure indicated by line B.

____

4 To what system does the structure indicated by line C belong?

____

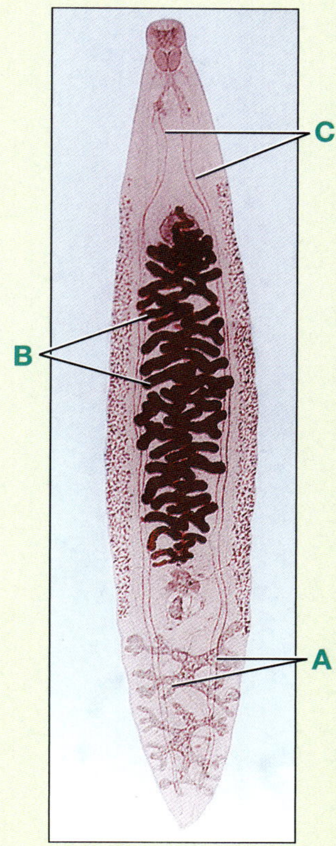

5 On the figure below, label the following tissue layers: epidermis, mesoderm, gastrodermis.

6 What is the function of the structure indicated by line D?

7 Identify the space indicated by line E.

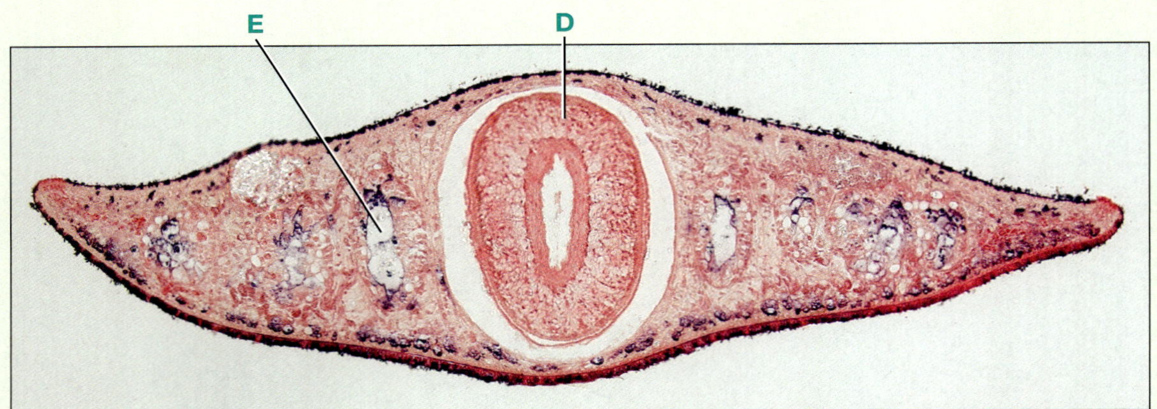

9

CHAPTER 10

Roundworms and Rotifers

After completing the exercises in this chapter, you should be able to:

1 List the distinguishing characteristics of the phylum Nematoda and phylum Rotifera.

2 Discuss several examples of parasitic nematodes.

3 Understand the evolutionary position of roundworms and rotifers on the phylogeny of major animal phyla.

4 Explain the evolutionary advantages of a complete digestive tract and body cavity.

5 Recognize the basic external and internal anatomy of a common roundworm and rotifer.

6 Differentiate male and female roundworms.

7 Characterize locomotion in roundworms and describe the effects of temperature on the metabolic rate of ectothermic organisms.

The phylum Nematoda is composed of a mixture of free-living and parasitic roundworms. Approximately 25,000 species have been described in this vast phylum, but biologists estimate that there may be 4–50 times that number of species in existence today. Whereas flatworms show tremendous diversity in body form, roundworms tend to display more similarities than differences among species. They all possess unsegmented, **bilaterally symmetrical**, tapered, tubular bodies devoid of appendages and covered by a thin, tough **cuticle** secreted by the epidermis. The cuticle is permeable only to water, gases, and some ions, and thus serves as a protective coating, especially in parasitic forms. In nematodes the sexes are usually separate (**dioecious**), with females generally being larger than males. They have a layer of longitudinal muscles beneath the epidermis but lack circular muscles, giving them a characteristic twitching motion as they thrash from side to side. This single layer of muscles is separated from the **complete digestive tract** by a body cavity known as a **pseudocoelom.**

Remember from the previous discussion of body cavities that a true coelom is lined on both sides by mesoderm. In roundworms the pseudocoelom lies between a layer of mesoderm (muscle tissue) and a layer of gastrodermis (intestine). Nematodes have an excretory system but lack circulatory and respiratory systems, because the fluid of the pseudocoelom performs the majority of gas and nutrient transport through the body.

Although nematodes usually are only a few millimeters thick, they range in length from a few millimeters to several meters. One parasitic species that inhabits the **placenta** of female sperm whales may attain a length of 9 meters! The infectious potential of parasitic nematodes is equally staggering. For instance, *Onchocerca*, a nematode that migrates through the body to its victim's eyes, occurs in 38 countries worldwide, including areas of Africa, South America, and the Middle East, and is a major cause of blindness. More than 123 million people live in endemic areas, and an estimated 18 million people are infected currently. Of those infected, an estimated 270,000 are blind and an additional 500,000 are estimated to have severe visual impairment. Infection rates in hyperendemic villages often reach 100%, with 50% of the villagers age 40 or older rendered blind.

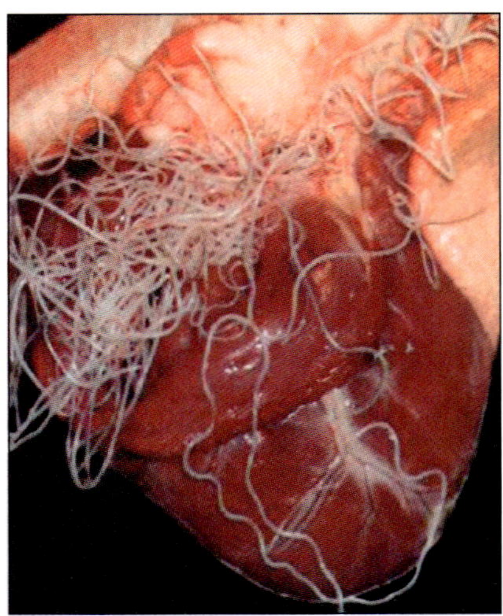

10.1 Dog heart infested with heartworms, *Dirofilaria immitis.*

The heartworm, *Dirofilaria immitis*, is a common parasitic roundworm that infects dogs, cats, and other mammals, including humans, and spreads from host to host through the bite of mosquitoes. The parasite is called "heartworm" because the adult resides primarily in the pulmonary artery of its host, where it can live for many years, producing thousands of live young each day, and resulting in potentially fatal health problems for the host (Fig. 10.1).

The pinworm, *Enterobius vermicularis*, is another parasitic nematode species, affecting some 500 million persons, particularly children, in temperate regions of the globe. These small roundworms (2–13 mm) live in the rectum of their victims and can produce severe abdominal pain, nausea, diarrhea, and itching around the anal region. While an infected person sleeps, female pinworms leave the rectum through the anus and deposit eggs on the surrounding skin. Children scratch their itchy, rectal area, contaminating their fingers, and often transfer the eggs to clothing, bedding, toys, or other surfaces where the eggs can survive for up to two weeks and usually are spread by direct **ingestion**. In the United States, pinworms are the most common roundworm infection. Young school-age children have the highest infection rates, with nearly 50% of children in some groups being infected.

One of the most serious diseases caused by nematodes in the United States is trichinosis. *Trichinella spiralis* lives in the intestines of pig hosts. Pregnant female roundworms burrow into the intestinal wall of their host and produce around 1,500 live young. These young roundworms enter lymph channels and migrate to muscle tissue, where they form highly resistant cysts that may remain alive and dormant for several years (Fig. 10.2). Humans become infected by eating undercooked pork containing these cysts.

Not only are their sheer numbers and infection rates bewildering, but roundworms' ability to reproduce in large numbers is equally impressive. A single female *Ascaris* releases some 200,000 fertilized eggs per day (that's 73 million per year!). Females of the guinea worm, *Dracunculus medinensis*, live just beneath the skin in their human hosts. When the skin comes into contact with water (as when the host bathes), the nematode protrudes its posterior end through the sore on the host's skin and can eject up to 1.5 million offspring into the water during the course of a day. These offspring cannot reinfect humans directly but, instead, carry out their life cycle in a species of microscopic aquatic crustacean. Humans are infected when they drink the water containing these tiny arthropods.

Evolutionary History

Modern nematodes possess two features that represent evolutionary milestones and are clear advantages over their predecessors: a complete digestive tract and a body cavity (pseudocoelom). A complete digestive tract permits food to pass through in one direction only—from mouth to anus—and opens up the possibility for different regions of the digestive tract to become specialized for different roles, vastly increasing digestive efficiency. Because flatworms and cnidarians have the limitation of only one digestive opening, they would not benefit from specialized regions in their digestive systems, because food moves around the entire gastrovascular cavity in a random manner. We see the first signs of digestive specialization in nematodes, which possess a muscular pharynx at the anterior end of the intestinal tract. Subsequent animal groups that we will examine have a much more extensive specialization of their digestive systems.

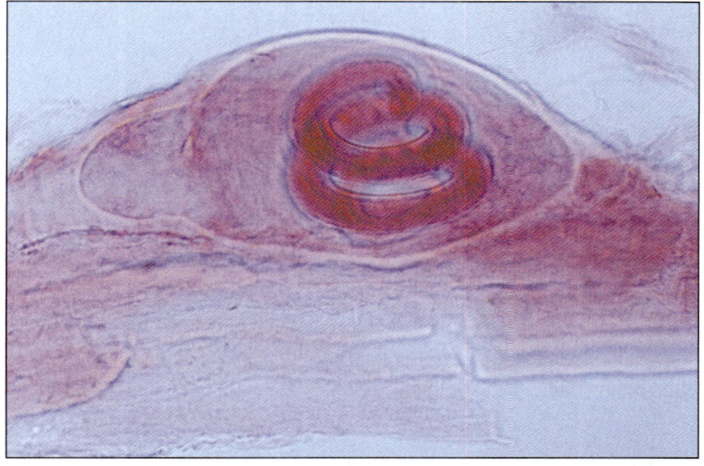

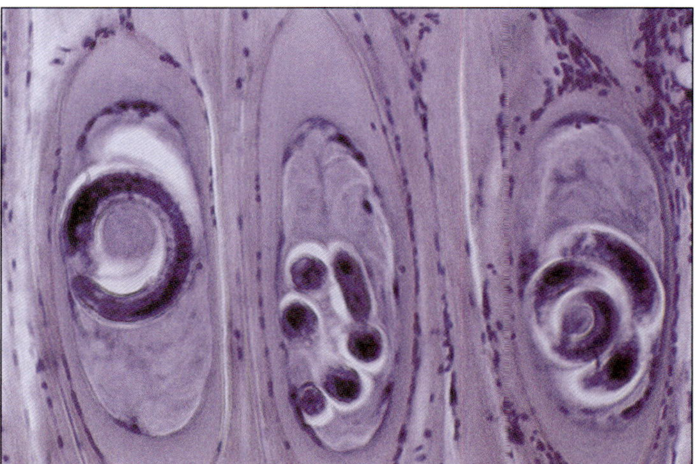

10.2 Larvae of the roundworm, *Trichinella spiralis*, the causative agent of trichinosis, reside in skeletal muscle cells of their host.

The evolution of a body cavity was a pivotal event in early animal evolution—so much so that this feature was retained by nearly every subsequent animal group and recruited for numerous functions. A **coelom** provides a space for the development of internal organs. Organs may be suspended within folds of tissue lining the coelom and can move independently of the outer body wall (for example, think about how your heart and stomach move independently of your body). The digestive tube can move food along its length independently of other body movements. Gonads have room to grow into the coelomic space during breeding seasons to increase reproductive potential. Coelomic fluid often acts as a crude circulatory medium, helping to transport nutrients, gases, and wastes. Nearby cells bathed in coelomic fluid may receive nutrients and oxygen while excreting wastes into the surrounding fluid.

The body cavity also serves as an enclosed, pressurized compartment of fluid that can act as a **hydrostatic skeleton**. To be effective, muscles have to pull against a structure that offers resistance. Many animals have skeletons, either a bony endoskeleton or a hard exoskeleton or shell. Early animals were soft bodied, yet a fluid-filled body cavity offered an incompressible structure that muscles could contract against effectively. Thus, we see significant improvements in locomotion in animal groups with coeloms, especially groups with both circular and longitudinal muscles that work in concert.

Nematodes, along with arthropods and seven other smaller animal phyla, are classified together in the Ecdysozoa (Fig. Appendix A.1), an assemblage that derives its name from a process all of its species undergo known as *ecdysis*—shedding or molting the tough, external body cuticle. Though Ecdysozoa includes only nine animal phyla, it contains more known species than all other animal, protist, fungus, and plant groups combined! Recent evidence from molecular research has led evolutionary biologists to generate drastically different phylogenies from those based on morphological and developmental properties alone, with respect to the placement of nematodes and the other invertebrate phyla. In our treatment, we have opted for the most current and widely accepted view in which roundworms and arthropods represent the descendants of one lineage of invertebrates—the Ecdysozoans—and flatworms, rotifers, molluscs, and annelids represent the descendants of another major invertebrate lineage—the Lophotrochozoans (Fig. 10.3).

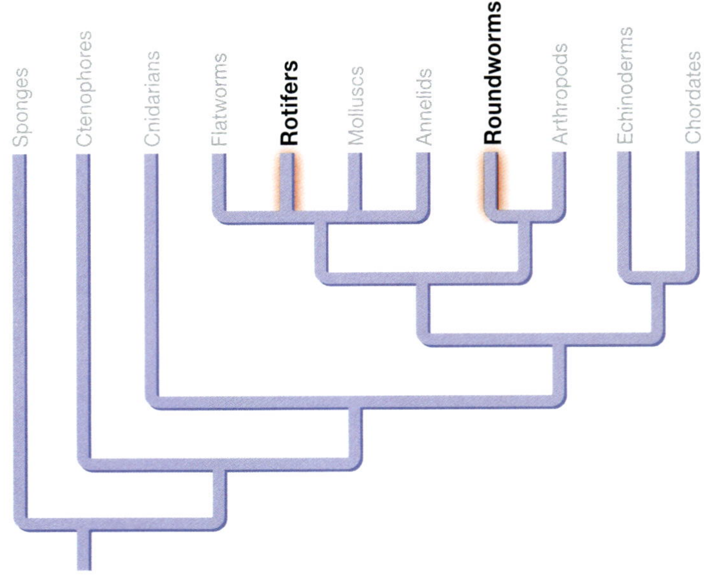

10

10.3 Modern roundworms and arthropods represent the descendants of one lineage of invertebrates—the Ecdysozoans—and flatworms, rotifers, molluscs, and annelids represent the descendants of another major invertebrate lineage—the Lophotrochozoans.

Exercise 10-1 Roundworms (Phylum Nematoda)

Materials Needed

- ❑ prepared dissection of *Ascaris lumbricoides*
- ❑ dissecting microscope
- ❑ latex gloves
- ❑ slide of *Ascaris* male and female c.s.
- ❑ compound microscope

Ascaris lumbricoides is a large, parasitic roundworm that infects humans and pigs. Estimates suggest that close to one-quarter of the world's population may be infected with this intestinal parasite, with infections being most common among immigrants and refugees. This parasite routinely spreads in areas where sanitation practices are poor, such as villages where inhabitants defecate near their dwellings, seeding the soil with eggs that may remain viable for years. Ingested eggs hatch in the small intestine of the host and the larvae burrow through the intestinal wall and are carried to the liver. From the liver, they migrate to the lungs, and later move up through the bronchi and trachea to the throat, where they are swallowed and carried back to the small intestine to complete their developmental cycle.

Be very careful when working with preserved nematodes. Nematode eggs are highly resilient to chemicals and may remain viable, even after the females have been preserved in formalin-based solutions. Though the chance of acquiring an infection from preserved laboratory specimens is remote, you still should exercise caution when handling them. Always wear gloves, keep your hands away from your eyes, nose, and mouth, and wash your hands afterward to avoid possible contamination.

Procedure

1 Obtain a prepared dissection of male and female specimens of *Ascaris lumbricoides*. (Your instructor should have set up several of these prior to your lab period.)

2 First determine the sex of each nematode. Males typically are smaller, have a curved posterior end, and have two small, spiny projections called **spicules** on the ventral surface near the anus, which are used to hold the female during copulation (Fig. 10.4).

3 After you have identified the sex of each specimen, differentiate the anterior and posterior ends. This is more apparent on the males because of their curved tails. The anterior end of both sexes is generally more tapered than the blunt posterior end. The **anus** is located on the ventral surface rather than at the terminal portion of the tail.

4 Draw male and female roundworms in the space provided below and label the anterior end, the posterior end, and the anus on each specimen.

5 Use a dissecting microscope to examine the anterior end of one specimen. In the space below, draw a sketch of the anterior end and label the triradiate lips surrounding the mouth. This is a distinguishing characteristic of the phylum; all nematodes possess three liplike plates arranged around the terminal mouth.

6 Find a section of the body wall where a piece of the thin cuticle has been dislodged, and examine it using your dissecting microscope. The cuticle is nonliving tissue composed of a transparent, proteinaceous substance. In parasitic species such as *A. lumbricoides*, this cuticle protects the animals from the digestive enzymes of the host. In terrestrial species, it protects the delicate tissues beneath from the abrasive forces encountered as roundworms burrow through the soil.

7 Draw a picture of the cuticle as it appears under the dissecting microscope in the space provided below.

8 After you have completed your examination of the external anatomy of each specimen, use Figures 10.4 and 10.5 and Table 10-1 to identify the internal anatomy of both sexes.

Ascaris Internal Anatomy

The digestive system of *Ascaris* is uncomplicated. As an intestinal parasite, its food already is partially digested by its host and the roundworm simply pumps in these nutrient-rich juices through the mouth using its muscular pharynx and transports this fluid to the long, thin **intestine** for nutrient absorption.

Roundworms possess a simple excretory system consisting of two **lateral lines** running the length of the body for the elimination of metabolic wastes. Careful inspection of the inner body wall should reveal two lateral lines on your specimen. **Dorsal** and **ventral nerve cords** are also present, but usually not visible on dissected specimens, because they are embedded within the longitudinal muscles of the body wall.

The remainder of the body cavity is occupied by reproductive organs. In both sexes the reproductive organs consist of extremely thin, coiled tubes that gradually expand in diameter until they reach the genital pore along the body wall. Males have a single tube that exits the body near the anus, and females have paired tubules that converge and exit the body closer to the anterior end (about one-third of the way down the body). In males this tube starts out as a fine, threadlike **testis,** where sperm production occurs. Sperm mature in the **vas deferens** and travel to the enlarged **seminal vesicle** for transport out of the body. In females, the fine, threadlike structures are the **ovaries.** Eggs are produced in the ovaries and stored in the **oviducts.** Fertilized eggs are stored in the enlarged, paired, branched **uterus** and are released from the body through the **vagina** and **genital pore.**

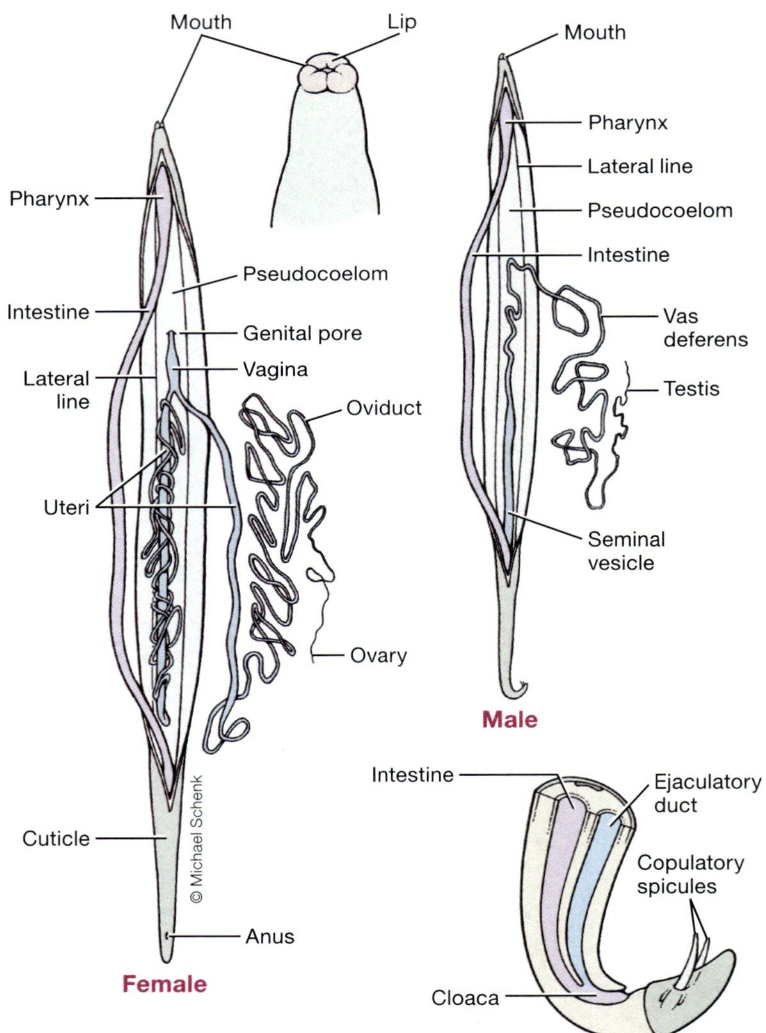

10.4 Anatomical features of the nematode, *Ascaris lumbricoides*.

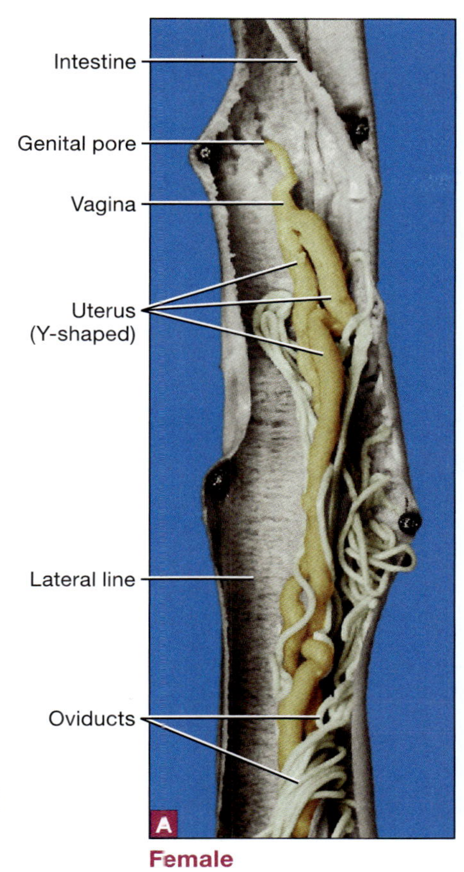

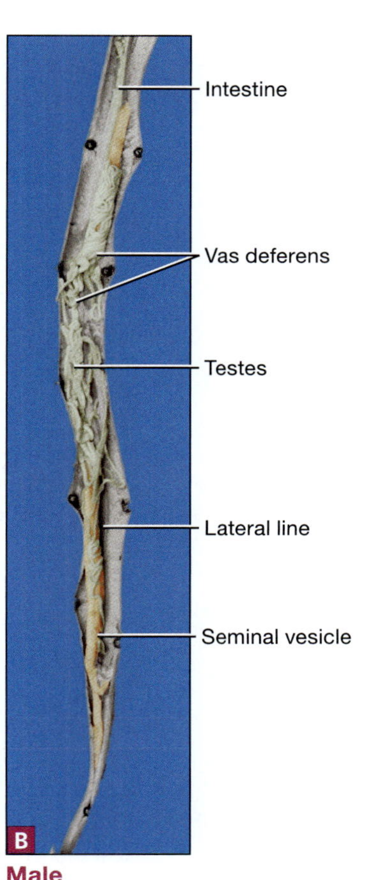

10.5 Anatomy of a dissected nematode, *Ascaris lumbricoides*: (**A**) female and (**B**) male depicted separately.

10

Table 10-1 Anatomy of a Roundworm (*Ascaris*)

Structure	Function
Mouth	Ingestion of food
Pharynx	Muscular region of digestive tract that "pumps" food through the mouth and into the intestine
Intestine	Ribbonlike digestive tract where absorption of nutrients occurs
Anus	Elimination of indigestible wastes (egestion)
Lateral lines	Longitudinal canals that function as the excretory system of the roundworm, releasing nitrogenous wastes in the form of ammonia and urea
Pseudocoelom	Body cavity lined on the inside by a layer of gastrodermis and on the outside by a layer of mesoderm
Testis (male)	Produces sperm
Vas deferens (male)	Stores mature sperm and transports them to seminal vesicle
Seminal vesicle (male)	Enlarged tube representing terminal portion of male reproductive tract, which transports mature sperm out of the nematode
Genital pore (female)	Point of entry for sperm and opening through which fertilized eggs are released from the body
Vagina (female)	Terminal portion of female reproductive tract, which receives sperm from males and directs eggs through genital pore
Branched uterus (female)	Site where developing eggs mature before being released
Oviduct (female)	Repository for eggs produced in ovary until fertilization
Ovary (female)	Produces eggs

9 After you have completed your examination of male and female dissected specimens, use a compound microscope to examine a prepared slide of cross sections (c.s.) of male and female *Ascaris* specimens (Fig. 10.6).

Ascaris Cross Section

The thick outer cuticle and underlying **epidermis** should be easily distinguishable on your slide. Beneath these layers are the longitudinal muscles described earlier. Embedded within this layer of muscles are the two lateral lines and the dorsal and ventral nerve cords. The nerve cords are depicted in Figure 10.6 at the twelve o'clock and six o'clock positions, and the lateral lines are located at the three o'clock and nine o'clock positions. A cross section of the thin, somewhat flattened intestine should be apparent in the pseudocoelom, with the rest of the body cavity occupied by reproductive organs. Differentiating the testis and the vas deferens (on males) or the ovaries, oviducts, and **uteri** (on females) is generally a matter of distinguishing between the different diameters of these structures. Testes and ovaries are always the smallest tubules in the body. The paired uteri are the most easily recognized because of their larger diameters and the fact that they are usually packed with eggs.

10

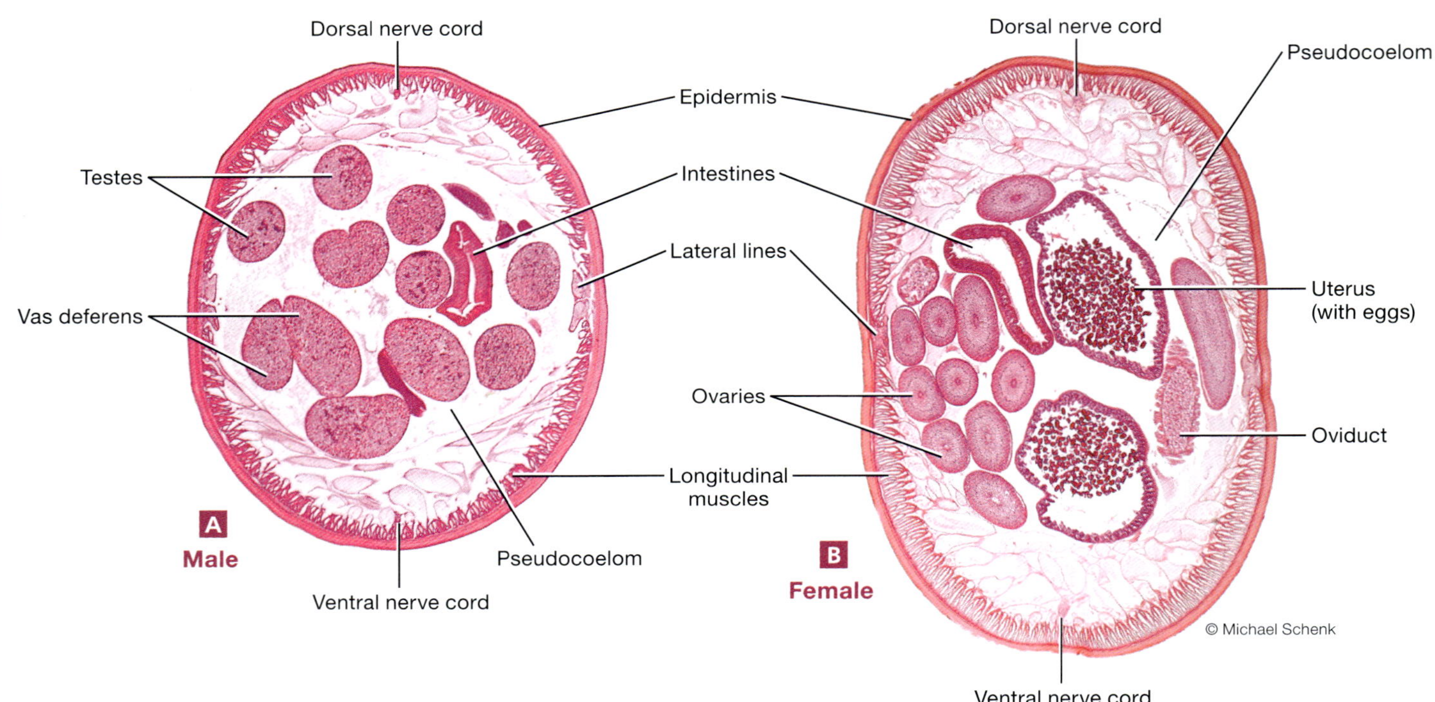

10.6 Transverse sections through the body of (**A**) male and (**B**) female nematodes.

1.1 List two external differences between male and female nematodes.

1.

2.

1.2 Is the intestine surrounded by an outer layer of muscle? (**HINT:** Think of the type of body cavity this animal possesses.)

1.3 Why do you suppose the digestive system of nematodes is rather unspecialized?

1.4 Do nematodes fit the pattern typical of parasites, in that a large portion of their body cavity is devoted to reproductive structures?

10

1.5 Through which structure are male gametes released?

1.6 Through which structure are females' fertilized eggs released?

Exercise 10-2 Investigating Locomotion in the Vinegar Eel

Materials Needed

- live vinegar eels (*Turbatrix aceti*)
- culture medium
- test tubes with stoppers
- water baths at 10°C, 15°C, and 20°C
- dissecting microscope

Like all invertebrates, roundworms are **ectotherms**, meaning they do not generate body heat internally. The label "cold-blooded" is not an appropriate term for describing ectothermic organisms, because many ectotherms actually sustain body temperatures well above that of most birds and mammals (the traditional "warm-blooded" creatures). Ectothermic simply means that the source of body heat for an organism is outside the body. Birds and mammals are **endotherms**, meaning that their body heat is generated from within.

It stands to reason, then, that an ectothermic animal such as a roundworm may be affected profoundly by changes in its surrounding environmental temperature. Such changes may affect its behavior, metabolic rate, growth rate, and even its basic biochemical pathways. These features make ectothermic animals, such as the vinegar eel, *Turbatrix aceti*, excellent models for elucidating the complex biological relationships between an animal and its thermal environment. Vinegar eels are small, free-living nematodes that are harmless to humans and feed on a microbial culture called mother of vinegar used to create vinegar (hence their common name).

Procedure

1 Place a few vinegar eels (*Turbatrix aceti*) into a clear test tube containing culture medium and insert the stopper.

2 Using the dissecting microscope, observe their pattern of locomotion. How would you describe this pattern?

3 Most nematodes display a characteristic undulating, whiplike pattern of movement. By focusing your attention on one small portion of the roundworm's body (for example, the head or tail), you will be able to count the number of undulations over a given time frame. Practice counting the number of undulations in your specimen over 30-second time intervals. Do this several times and record the number of undulations for each time interval below.

Time Interval #1: ________ (no. of undulations)

Time Interval #2: ________ (no. of undulations)

Time Interval #3: ________ (no. of undulations)

Time Interval #4: ________ (no. of undulations)

10

Check Your Progress

2.1 Is the number of undulations fairly consistent between trials?

2.2 What is the average undulation rate of vinegar eels under normal conditions? (Remember to express the rate as a function of time—that is, # undulations/minute.)

2.3 Based on your knowledge of nematode anatomy, what kinds of muscles are involved in producing these movements?

Now that you have a method for reliably quantifying the locomotive abilities of this animal, you will investigate the effects of a changing environmental parameter on locomotion in vinegar eels. You will subject vinegar eels to different environmental temperatures and record their undulation rates to see if temperature affects locomotion. Fifteen test tubes, each containing one vinegar eel, will be placed in water baths set at (or near) the following temperatures: 10°C, 15°C, and 20°C. You will record the undulation rate of each vinegar eel for 30 seconds (for a total of 15 recordings—five roundworms at each of three different temperatures). Before beginning the experimental protocol, complete the following section using the steps of the scientific method to approach this investigation systematically.

Using the Scientific Method

1 What is the general question that this experiment addresses?

2 Record the hypothesis you wish to test.

3 Identify the independent variable in this experiment.

> **Tip**
>
> The *independent variable* is the element that you change or manipulate and the *dependent variable* is the element that you measure to determine how it is affected by the changes you made to the independent variable.

10

4 Identify the dependent variable(s) in this experiment.

5 What are the levels of treatment (that is, levels of the independent variable)?

6 What served as the control treatment?

> **Tip**
>
> A *positive control* shows that the experiment is capable of producing a positive result when a phenomenon is expected, whereas a *negative control* is designed to demonstrate a lack of response or negative result when there should be no effect.

7 Identify several controlled variables.

Experimental Protocol

1 Work in groups of three for this portion of the experiment. Obtain 15 test tubes, each containing a single vinegar eel.

2 Label five test tubes 10°C, five tubes 15°C, and five tubes 20°C.

3 Place an appropriately labeled tube in each of the three water baths and let the vinegar eels habituate to their new thermal environment for 3–4 minutes.

4 Have each person in your group record the undulation rate of one vinegar eel for 30 seconds and record the data in Table 10-2.

5 Repeat steps 3 and 4 four more times, *using a new vinegar eel in a new test tube each time* for each of the three temperature settings (for a total of 15 recordings—five roundworms at each of the three different temperatures).

6 Record your data in Table 10-2.

Table 10-2 Locomotion in the Vinegar Eel (*Turbatrix aceti*)

Record the number of undulations per 30-second interval for five roundworms in each of the three temperature settings.

Specimen #	Temperature		
	10°C	**15°C**	**20°C**
1			
2			
3			
4			
5			
Average undulation rate (# undulations/30 sec.)			

Check Your Progress

3.1 Convert your results to the average number of undulations per minute for each of the three temperatures.

10°C = ________

15°C = ________

20°C = ________

3.2 Does environmental temperature affect locomotion in ectothermic animals such as vinegar eels?

3.3 Discuss the effects on locomotion that variation in the thermal environment of vinegar eels produces.

3.4 Based on your data, should you retain or reject your hypothesis?

10

Exercise 10-3 Rotifers (Phylum Rotifera)

Materials Needed

- ❑ live rotifer culture medium
- ❑ plastic dropper
- ❑ glass depression slides
- ❑ coverslips
- ❑ compound microscope

Rotifers (phylum Rotifera) represent a group of tiny invertebrates (0.05 mm to 2.0 mm in length) that commonly inhabit freshwater pools, ponds, and lakes, and are often found in abundance in waterlogged mosses and damp soil. Approximately 2,000 species have been described, most of which are found in freshwater with only a few marine species known to exist. Although rotifers are now not considered to be as closely related to roundworms as was once thought (Appendix Fig. A.1), we have included them for study in this chapter because of the many physical characteristics they share with roundworms. Like roundworms, rotifers are bilaterally symmetrical, have a pseudocoelom and a complete digestive tract with a separate mouth and anus, and are dioecious (separate sexes). Likewise, the fluid within the pseudocoelom serves as a hydrostatic skeleton that facilitates movements and distributes dissolved nutrients throughout the animal's body.

Two physical characteristics that distinguish members of this phylum from roundworms and other invertebrates are: (1) a ciliated crown (**corona**) that surrounds the anterior end of the body, and a pharyngeal apparatus known as the **mastax**, which bears teethlike structures that grasp and chew prey (in predatory species) or grind food particles (in suspension feeding species).

Rotifers are dioecious, with males generally being smaller than females, and most species reproduce sexually through internal fertilization. In some species the male inseminates his mate by stabbing her with his penis (through her body wall) and injecting sperm into her pseudocoelom—a process fittingly dubbed "hypodermic impregnation." Many freshwater species produce both rapidly hatching thin-shelled eggs and thick-shelled eggs that can lie dormant for long periods of time, withstanding low temperatures and desiccation. Other species consist of only females that produce more females from unfertilized diploid eggs, a type of asexual reproduction called **parthenogenesis**.

Parthenogenesis has been observed in several other invertebrate groups (nematodes, scorpions, aphids, wasps, and bees) as well as in many vertebrate groups (amphibians, lizards, fishes, and even birds). Such reproductive flexibility allows rotifer populations to grow rapidly when food is plentiful, to rebound after population crashes, and to survive other extreme changes in their environmental conditions.

Procedure

1 Place a drop of rotifer culture on a depression slide, cover with a coverslip, and examine with your compound microscope using low power.

2 Locate a specimen and then examine using medium power. Be sure to adjust the iris diaphragm and condenser settings to achieve optimum contrast and sharpness.

Rotifers have elongated, saclike bodies partitioned into three sections: a head, trunk, and foot; although different species may vary considerably in their overall shapes (Fig. 10.7). The anterior end is crowned with a ring of cilia (corona) and contains two eyespots. The trunk contains the major digestive, reproductive, and excretory organs. The foot possesses adhesive glands that secrete a type of chemical cement for anchoring the rotifer to the substrate for feeding. Rotifers swim by means of ciliary action of the corona and can crawl in a leechlike fashion, often alternating between these two means of locomotion as they search through organic debris and algae for suspended food particles. Food is swept into the mouth by ciliary action of the corona. Most species of rotifers have a large stomach and a short intestine leading to the anus, which is located at the base of the trunk on the ventral side of the animal.

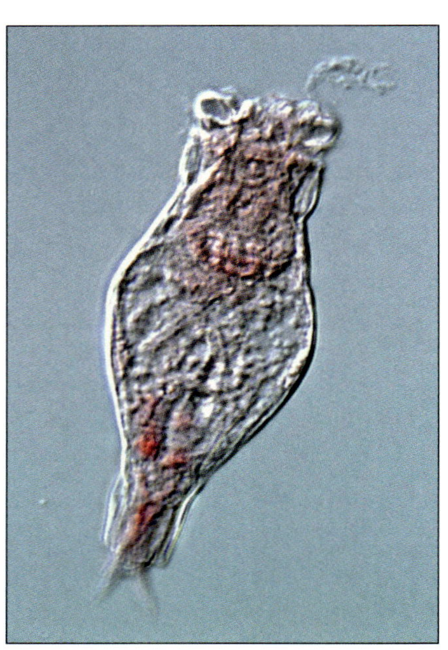

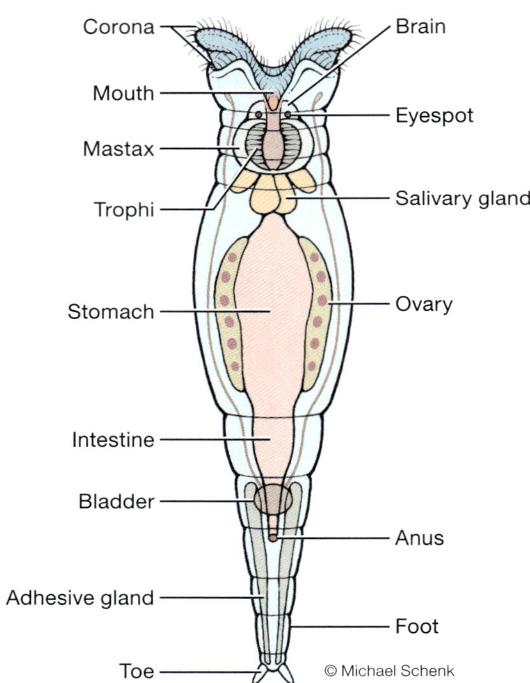

10.7 Anatomical features of the rotifer, *Philodina sp*.

3 Find a specimen on your slide in which you can easily see the corona, and carefully examine that structure. Can you detect the rhythmic beating of cilia on each of the two halves of the corona? Do the cilia on the two halves beat in the same direction, or in opposite directions?

4 Locate the mastax on your specimen and try to identify the teethlike jaws called trophi that protrude into the lumen of the digestive system.

5 Draw a sketch of your specimen in the space provided and label the following structures: corona, eyespots, mastax, stomach, intestine, and foot.

10

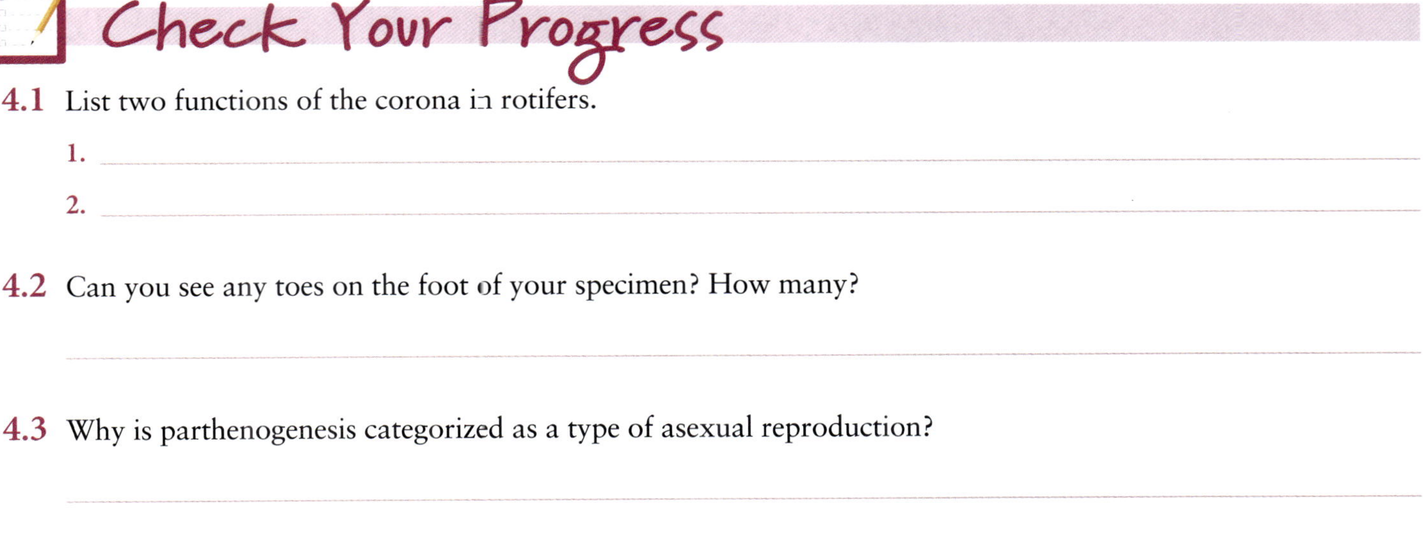

Check Your Progress

4.1 List two functions of the corona in rotifers.

1. ______

2. ______

4.2 Can you see any toes on the foot of your specimen? How many?

______

4.3 Why is parthenogenesis categorized as a type of asexual reproduction?

______

______

Notes

10

Name ______________________ Section ______________

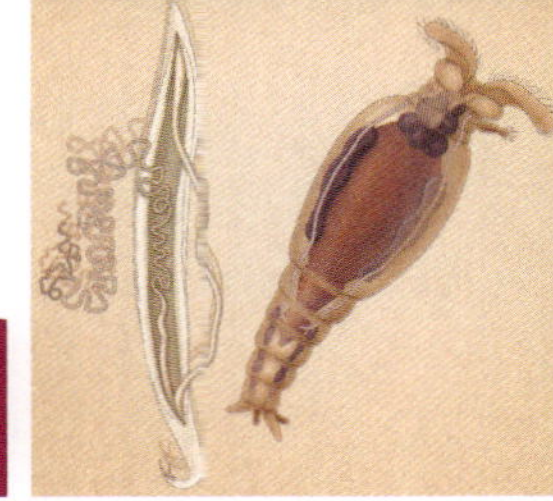

10 Questions for Review

1 Complete Summary Table A-1 in the Appendix, filling in the characteristics of roundworms in the appropriate row.

2 List two distinct evolutionary advancements of roundworms over flatworms.

3 What are the advantages of a digestive tract having a separate entrance and exit?

4 Contrast the nervous system of nematodes with that seen in flatworms in Chapter 9.

5 For a parasitic organism such as *Ascaris*, what would be the selective advantage of a cuticle?

6 For a free-living, terrestrial nematode, what would be the selective advantage of a cuticle?

7 In many parasitic species the sex ratio of males to females is skewed (not 1:1). Which sex do you suppose would tend to be more numerous? Why?

8 Explain the difference between a pseudocoelom and a true coelom.

9 Define ectotherm.

____

10 List two major features that nematodes and rotifers share in common *and* two that are found in rotifers but not in nematodes.

Nematodes and Rotifers:

1. ______________________________

2. ______________________________

Rotifers ONLY:

1. ______________________________

2. ______________________________

11 The hookworm, *Necator americanus*, which infects some 900 million people worldwide, may ingest more than 0.5 ml of human host blood daily. Given that an infection may number more than 1,000 individual hookworms, calculate the total volume of host blood that may be lost per day to a severe nematode infection.

12 Given that the total blood volume of the average adult human is 5 liters, calculate the percentage of total blood volume lost daily in the example above.

Name ______________________________ Section ______________

10 Laboratory Practical Challenge

The following evaluation is intended to replicate the method of assessment and type of questions you will be asked on lab practicals. If you have difficulty with these questions, you may benefit from additional review of the specimens provided in lab.

1 Identify the phylum of the specimen depicted below.

____

2 What is the function of the structure indicated by line A?

____

3 Identify the structure indicated by line B.

____

4 To what system does the structure indicated by line C belong?

____

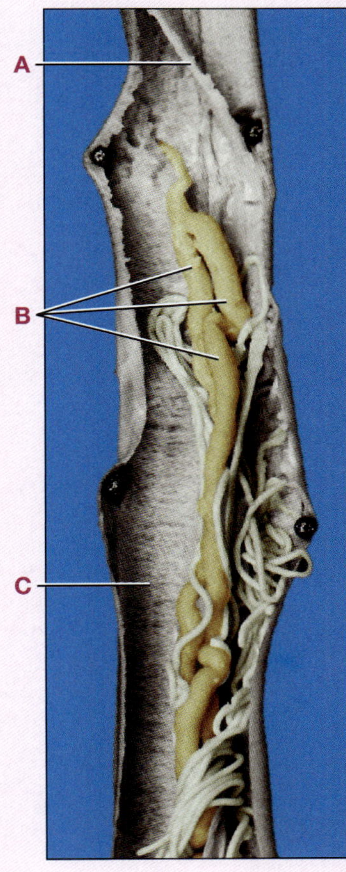

5 On the figure below, label the following structures: ventral nerve cord, lateral line, intestine, and uterus.

6 To what system does the structure indicated by line D belong?

7 Identify the type of tissue indicated by line E.

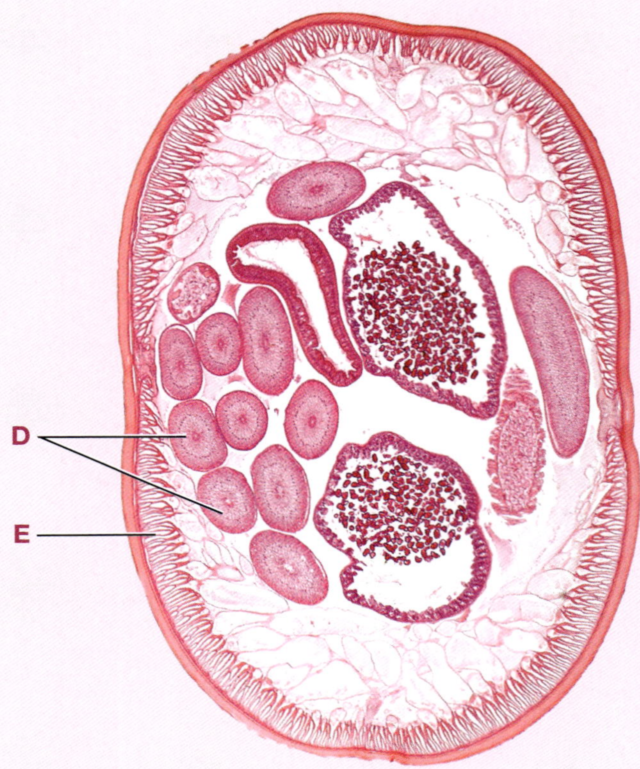

CHAPTER 11

Molluscs

After completing the exercises in this chapter, you should be able to:

1. List the distinguishing characteristics of the phylum Mollusca.
2. Understand the evolutionary position of molluscs on the phylogeny of major animal phyla.
3. Discuss the concept of adaptive radiation and explain how molluscs illustrate this phenomenon.
4. Identify the major anatomical features of a freshwater clam.
5. Explain the pattern of water flow through a freshwater clam and its significance.
6. Identify the major anatomical features of a squid.
7. Explain the basic differences between open and closed circulatory systems.
8. Compare and contrast the evolutionary modifications of filter-feeding bivalves with predatory cephalopods.
9. Discuss the concept of convergent evolution and explain how the mollusc and vertebrate eyes provide an example of this phenomenon.

The phylum Mollusca is a large and remarkably diverse collection of animals that includes clams, mussels, scallops, oysters, snails, slugs, octopuses, squid, and nautiluses. More than 93,000 described species of molluscs, and potentially 150,000 species, are in existence today. Despite the diversity within this phylum, the basic body plan of all molluscs is relatively similar. They are **bilaterally symmetrical**, unsegmented animals that have a true **coelom** and well-developed organ systems, and most species are **dioecious.**

Molluscs possess four major morphological features that distinguish them from other invertebrates:

1. a protective **shell** (reduced in some species),
2. a **mantle,**
3. a **visceral mass** that houses the major internal organs, and
4. a **foot** for locomotion.

The four basic body parts are modified in different ways in molluscs, and these morphological variations clearly illustrate how diversity can be achieved by simple alterations to a basic body plan. For example, the degree

11 of **cephalization** varies enormously within this phylum, ranging from a total lack of cephalization in clams and mussels to a well-developed brain and image-forming eyes in squid and octopuses. As a result, there is no single, typical mollusc. The chiton's body (class Polyplacophora), however, probably best resembles the basic mollusc body plan in its simplest and most ancestral form. This group of grazing, marine herbivores has bilaterally symmetrical bodies, covered by a dorsal shell consisting of eight overlapping plates (Fig. 11.1). The unsegmented, soft body is exposed on the ventral side, and the mantle, head, and foot are all clearly visible.

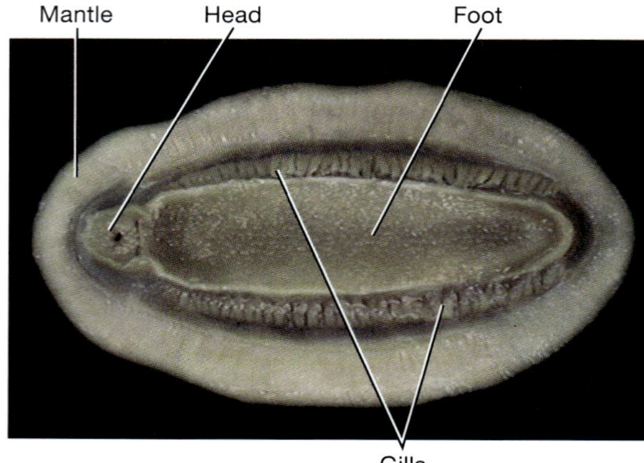

11.1 Modern chitons possess a fairly unspecialized, ancestral body plan depicting the four major features characteristic to the phylum.

Although zoologists have recognized a total of eight different classes of molluscs, more than 90% of the species described to date are grouped into just two classes:

1. Bivalvia (clams, oysters, scallops, and mussels), and
2. Gastropoda (snails, slugs, and nudibranchs).

Table 11-1 depicts the distinguishing characteristics of the five largest classes of molluscs.

Table 11-1 Five Largest Classes of Phylum Mollusca

Class and Representative Animals	Characteristics
Gastropoda (≈ 60,000 sp.) Snails, slugs, nudibranchs	Well-developed head with eyes and tentacles; body undergoes torsion during development; primarily marine, with some freshwater and terrestrial species
Bivalvia (≈ 20,000 sp.) Clams, oysters, mussels, scallops	Two-part, hinged shell; reduced head; marine or freshwater filter feeders
Polyplacophora (≈ 900 sp.) Chitons	Dorsal shell consists of eight overlapping plates; ventral body with head-foot; entirely marine
Scaphopoda (≈ 900 sp.) Tusk shells	Body enclosed in one-piece, curved, conical shell open at both ends; mouth with tentacles; head absent, entirely marine
Cephalopoda (≈ 786 sp.) Squids, octopuses, nautiluses	Foot modified into tentacles; shell greatly reduced; prominent head with well-developed eyes and beak; closed circulatory system; entirely marine, highly active predators

In this chapter you will study three highly divergent molluscs: a terrestrial snail, a freshwater clam, and a marine squid. Snails, slugs, and nudibranchs are members of the class Gastropoda—the largest class of molluscs, with species inhabiting marine, freshwater, and terrestrial environments. Most gastropods are sluggish, sedentary herbivores, but some are scavengers, and others, such as nudibranchs, are even carnivorous predators. Clams, oysters, scallops, and mussels are members of the class Bivalvia and are generally heavily shelled, stationary filter feeders that burrow beneath the sand and extend their siphons through the sand to filter water into their bodies and extract oxygen and nutrients. Squid are members of the class Cephalopoda, which also includes the octopus and nautilus. All cephalopods are highly active, visually oriented predators that hold their own in competition with some of the fiercest ocean predators. As you familiarize yourself with the anatomical structures of these organisms, keep in mind how each animal's anatomy has been modified to allow it to be adapted maximally to its particular lifestyle.

Evolutionary History

Molecular evidence suggests that there are three major groupings of bilaterally symmetrical animals: Lophotrochozoa, Ecdysozoa, and Deuterostomia (Fig. 11.2). Several of these animal groups appear in the fossil record as early as 570–560 million years ago (mya), and by the Cambrian explosion (535–525 mya) it appears that most of the major present-day animal phyla had arisen. Molluscs, along with several other invertebrate phyla, are classified together in one of these clades—the Lophotrochozoa, an assemblage that derives its name from two different features of its members. Molluscs (and annelids) go through a distinctive larval stage called the *trochophore*. Other lophotrochozoans develop a structure called a *lophophore*, a horseshoe-shaped crown of ciliated tentacles used in feeding. For a more complete phylogeny depicting the evolutionary relationships among the major animal phyla covered throughout this laboratory guide, review Figure A.1 in the Appendix.

Because of the extreme diversity of molluscan body forms, it may seem at first glance that these animals do not all belong in the same phylum. Such a vast range of morphological types within a group of organisms sharing a common lineage is an example of **adaptive radiation**—the evolution of numerous species from a common ancestor following migration into a new environment. Adaptive radiation often is accelerated when the new environment has few existing competitors, and the introduced organisms are able to disperse quickly into different niches. Over many generations these organisms become highly specialized to their particular surroundings, and divergent forms arise.

Molluscs branched off the main animal line more than 545 million years ago during a period when few other large organisms occupied the oceans. Competition for resources was relatively low, and there were numerous ecological niches that were either unfilled or easily conquered by these newcomers. Keep in mind that at this time no plants or animals were living on land; most life was confined to water. Actually, plant and animal life on land would not arrive for another 100 million years! Thus, today's living molluscs represent descendants of an ancient group of organisms—a group that has experienced hundreds of millions of years of gradual evolutionary change. In the ensuing time frame, molluscs have become adapted to virtually every type of freshwater and marine habitat, as well as numerous terrestrial habitats.

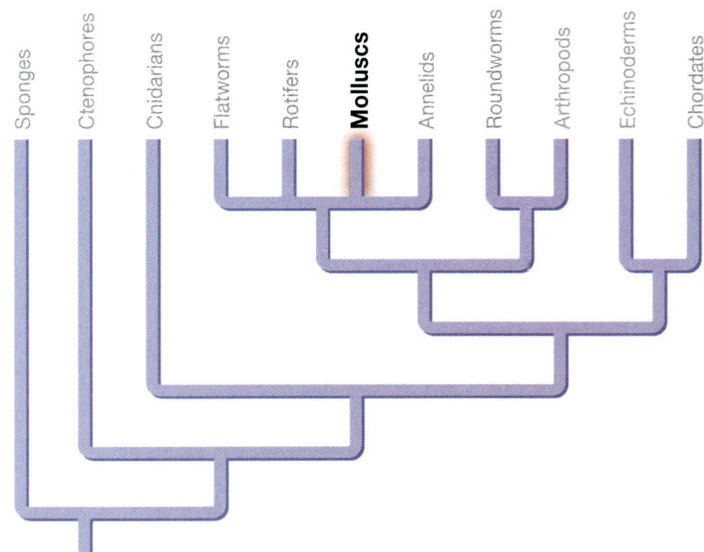

11.2 Molluscs are grouped with a collection of other invertebrate phyla into the Lophotrochozoan assemblage based on shared characteristics that indicate close evolutionary relationships.

Exercise 11-1

Snails, Slugs, and Nudibranchs (Class Gastropoda)

Materials Needed

- ❑ living snail (*Helix* or *Polygyra*)
- ❑ finger bowl
- ❑ glass plate
- ❑ blunt dissecting probe
- ❑ piece of lettuce
- ❑ piece of white paper
- ❑ sandpaper
- ❑ ruler
- ❑ cotton swab
- ❑ vinegar
- ❑ dissecting microscope

11.3 Snails represent one of the few groups of terrestrial molluscs.

Procedure

11

1 Obtain a live specimen of the common terrestrial snail, *Helix* or *Polygyra*.

2 Place the snail in a clean finger bowl and observe it with your dissecting microscope, using low light.

> **Note**
> *DO NOT use bright light, because it may injure the snail.*

Land snails have a well-defined head region with **eyes** located at the tips of two long, retractable ocular **tentacles**. The **mouth** is on the ventral side of the head, and a pair of sensory tentacles is located near the mouth, positioned toward the front of the head (Fig. 11.3).

3 Gently touch the tentacles of the snail with a blunt dissecting probe and observe the response.

4 Place a small piece of lettuce near the snail and see if it finds the food and begins to eat.

5 Gently transfer the snail to a clean glass plate and place the snail upside down over the finger bowl. Observe its underside with the dissecting microscope as the snail crawls across the glass. Notice the waves of muscular contraction along the snail's foot as it moves, and the slimy trail that it leaves behind in its path.

6 Place your snail on a piece of notebook paper and mark its starting place. Time the snail for 1 minute, observing where the snail moves. Measure how far the snail traveled in 1 minute and record that number below. Repeat this step two more times and average the three results.

Trial 1 = ______ cm/min. Trial 2 = ______ cm/min. Trial 3 = ______ cm/min.

Average = ______ cm/min.

7 Place your snail on a piece of coarse sandpaper and repeat Step 6. Record your data below.

Trial 1 = ______ cm/min. Trial 2 = ______ cm/min. Trial 3 = ______ cm/min.

Average = ______ cm/min.

8 Compare your data on snail crawling speeds with other students in your lab to get an idea of the range of speeds on the two different substrates.

9 Return your snail to the finger bowl. Dip a cotton swab in vinegar and hold it close to the snail's sensory tentacles, but do not make contact with the snail. Carefully observe the snail's response to the vinegar.

10 Place the cotton swab near other body parts to determine if the snail's reaction varies from its initial response.

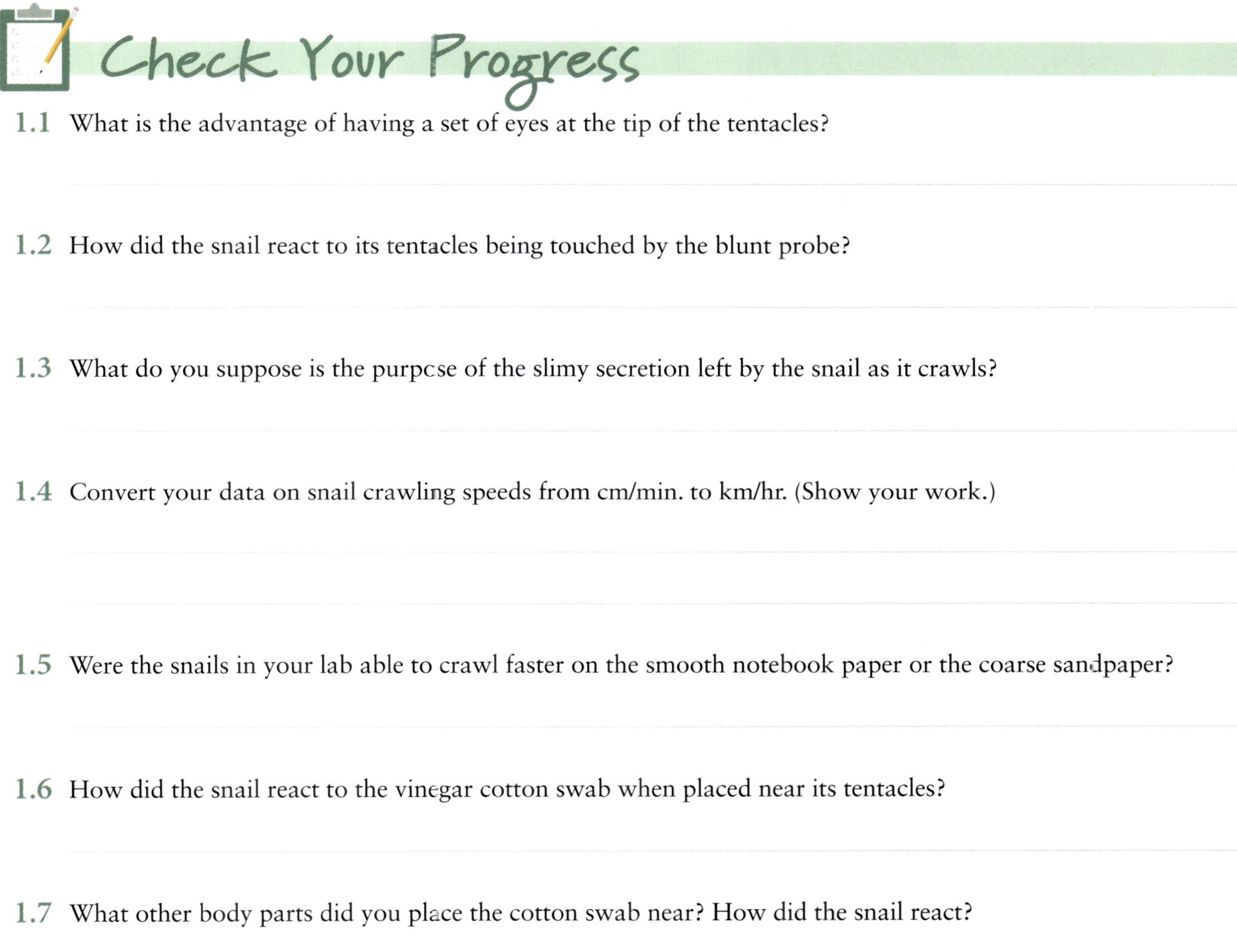

Check Your Progress

1.1 What is the advantage of having a set of eyes at the tip of the tentacles?

1.2 How did the snail react to its tentacles being touched by the blunt probe?

1.3 What do you suppose is the purpose of the slimy secretion left by the snail as it crawls?

1.4 Convert your data on snail crawling speeds from cm/min. to km/hr. (Show your work.)

1.5 Were the snails in your lab able to crawl faster on the smooth notebook paper or the coarse sandpaper?

1.6 How did the snail react to the vinegar cotton swab when placed near its tentacles?

1.7 What other body parts did you place the cotton swab near? How did the snail react?

Exercise 11-2 Clams and Mussels (Class Bivalvia)

Materials Needed

- ❑ preserved freshwater clam
- ❑ dissection tools
- ❑ dissecting pan
- ❑ latex gloves

11.4 External view of a bivalve shell: (**A**) dorsal view, and (**B**) left valve.

Procedure

1 Obtain a preserved specimen of the freshwater clam *Anodonta* or other bivalve.

2 Use Figure 11.4 to identify the anterior, posterior, dorsal, and ventral regions of the specimen, as well as the left and right valves. These areas will be important points of reference for identifying internal structures. Remember that bivalves are bilaterally symmetrical, so many of the internal structures are repeated on the left and right side of your specimen.

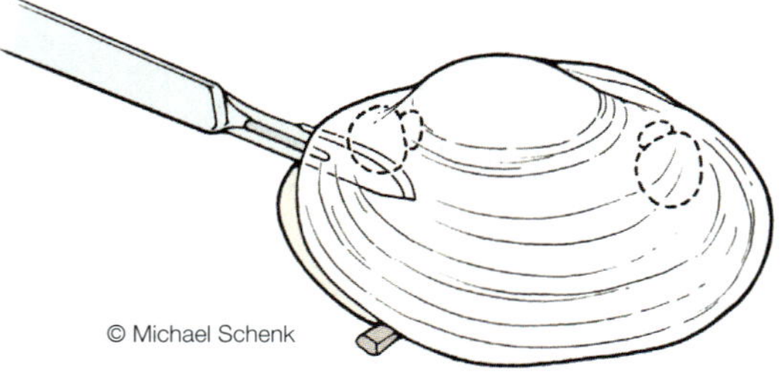

Dissection Diagram

1 Insert scalpel near wooden peg and carefully cut adductor muscle.

2 Repeat with other adductor muscle, being careful not to damage internal organs.

3 Gently peel mantle away from shell as you open specimen.

3 Review the Dissection Diagram before opening your specimen.

4 Your specimen should have a wooden peg in it to keep the valves partially separated. Carefully insert a scalpel in the space between the left valve and the mantle near the wooden peg, and slice through the nearby adductor muscle.

5 Repeat this procedure with the adductor muscle at the other end. This step requires a bit of feel to locate the second adductor muscle, because it usually is not visible through the small opening in the shell. Sliding the scalpel along the upper shell between the shell and the mantle tissue to locate the second adductor muscle will minimize the risk of damaging other internal organs.

6 To completely open your specimen, you may have to use a dissecting needle or probe to gently peel the thin, fleshy mantle away from the shell.

7 Once it is open, lay your clam on its right shell. For easier viewing of the internal structures, remove the left shell and the left mantle tissue.

Identify the larger posterior adductor muscle and the slightly smaller anterior adductor muscle that you cut through earlier to open your specimen (Fig. 11.5). These muscles allow the clam to close its shell and hold it closed with tremendous force. The remaining half of the thin, fleshy mantle should be evident along the inside of the right shell. Remember that the function of the mantle in molluscs is to secrete the protective shell. Just ventral to the posterior adductor muscle, you should be able to see two areas of tissue along the edge of the mantle that have a puckered, almost serrated appearance. These regions represent the incurrent and excurrent siphons, small separations of the mantle that bivalves use to bring in and expel water, respectively. When a bivalve closes its shell, both halves of the mantle are closed together, but small openings remain in these two regions, allowing water to pass in and out of the shell.

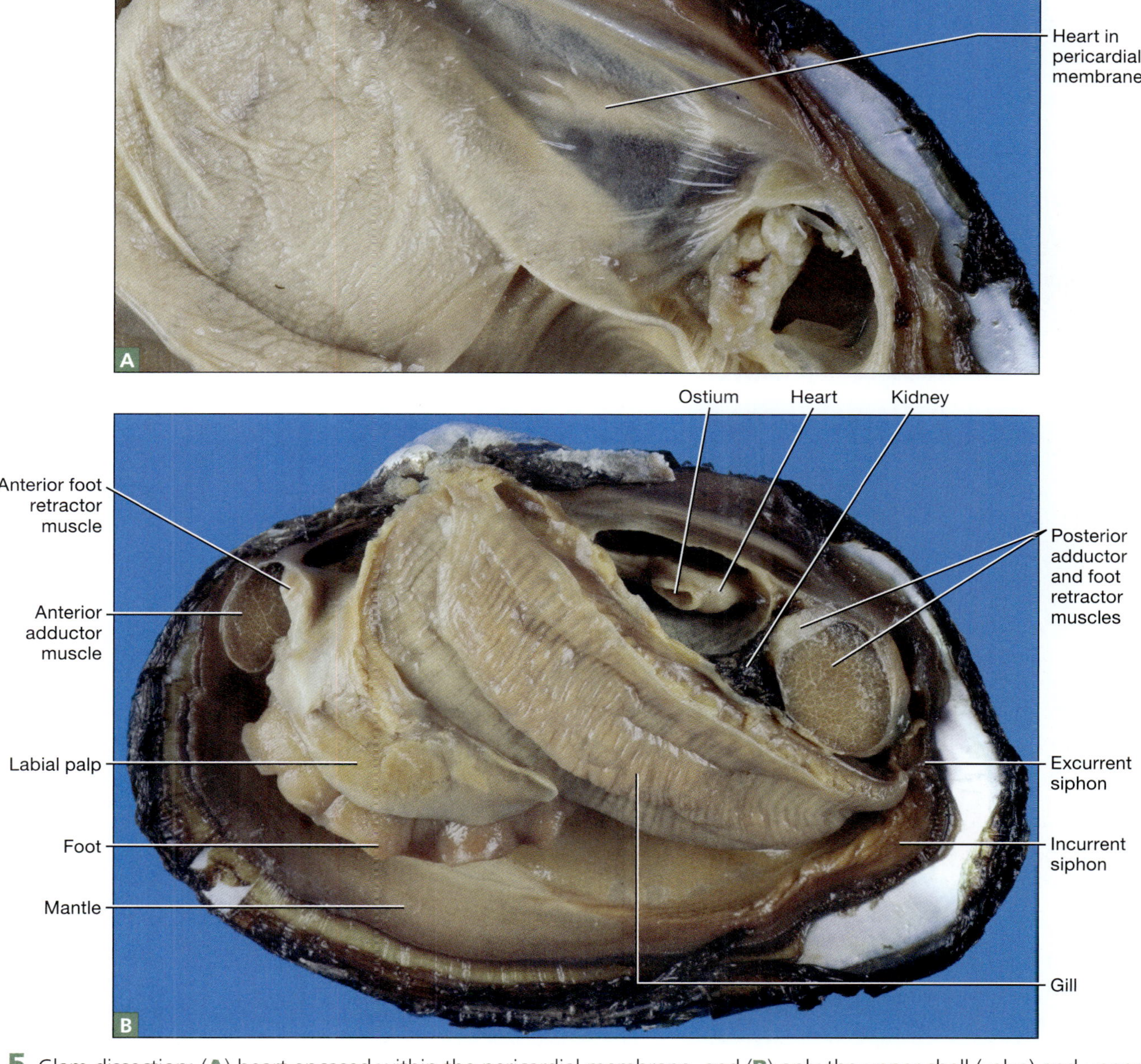

11.5 Clam dissection: (**A**) heart encased within the pericardial membrane, and (**B**) only the upper shell (valve) and upper mantle have been removed.

11

The foot, with its rippled border, and the somewhat swollen visceral mass are joined together in bivalves. In living clams, the foot is very elastic and flexible and may be extended through even a small slit in the shell for burrowing and locomotion. The **gills** are the large, striated flaps evident along each side of the body. Careful examination of each gill will reveal that it is a double fold of tissue with tiny ridges that increase surface area for respiration and filter feeding. Just anterior to the gills are the labial palps, fleshy folds of skin that collect food particles from the gills and transport them to the mouth. If you trace the groove that runs along the inside of each set of labial palps, you should be able to see where they converge along the midline of the clam's body. Though often difficult to expose in dissection, this is the region where the clam's mouth is located (Fig. 11.6B).

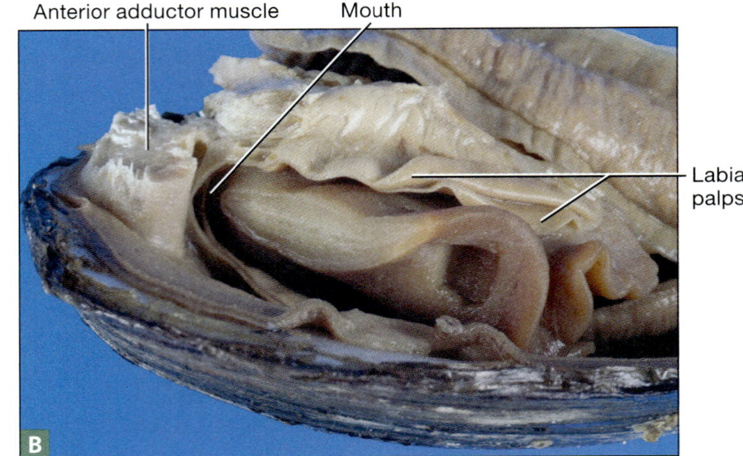

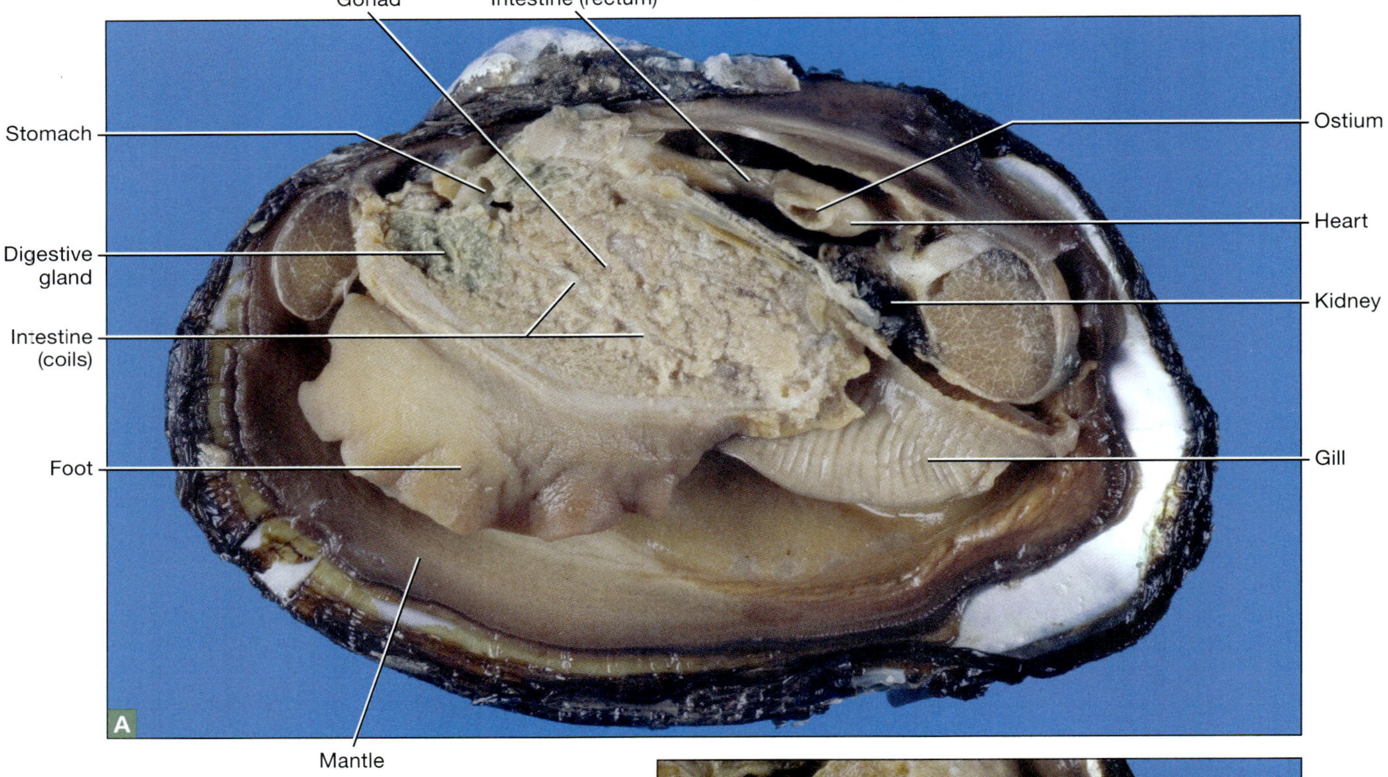

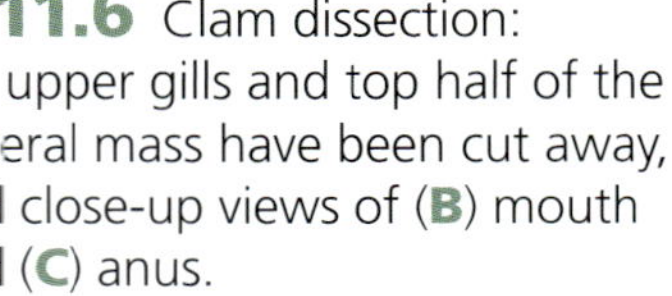

11.6 Clam dissection: (**A**) upper gills and top half of the visceral mass have been cut away, and close-up views of (**B**) mouth and (**C**) anus.

8 To expose the **heart**, you must carefully tease through the pericardial membrane (Fig. 11.5).

The pericardial membrane actually is mesodermal tissue that encloses the clam's coelom. In most molluscs the coelom is reduced to this small chamber surrounding only the heart, excretory organs, and in some other species, the gonads. Notice that the heart has openings. These openings, called **ostia** (singular = ostium), allow the heart to collect the oxygenated blood from the gills that returns via open sinuses around the heart. Remember that bivalves have an open circulatory system, meaning that their blood is not always confined within vessels.

An interesting feature of the architecture of the clam is that the heart actually sits around the terminal portion of the **intestine.** The bulk of the intestinal tract is located within the visceral mass. Follow the intestine posteriorly along the dorsal edge of the posterior adductor muscle to locate the **anus**, anchored to the side of the adductor muscle along the dorsal border of the excurrent siphon (Fig. 11.6C). Also, just ventral to the heart is a thin patch of dark tissue, the excretory organ in clams, known as the **kidney** (or nephridium).

9 Several major internal organs are located within the visceral mass. To dissect this structure, use your scalpel to make a *longitudinal incision* through the visceral mass, dividing it into two bilaterally symmetrical left and right halves (in the same plane that the shell opens).

10 Use Figures 11.6A and 11.7 to identify the internal organs of the visceral mass. You should be able to locate **gonad** tissue, coils of intestine, **digestive gland** tissue, and the **stomach.** Table 11-2 provides a review of the major organs of the bivalve covered in this dissection.

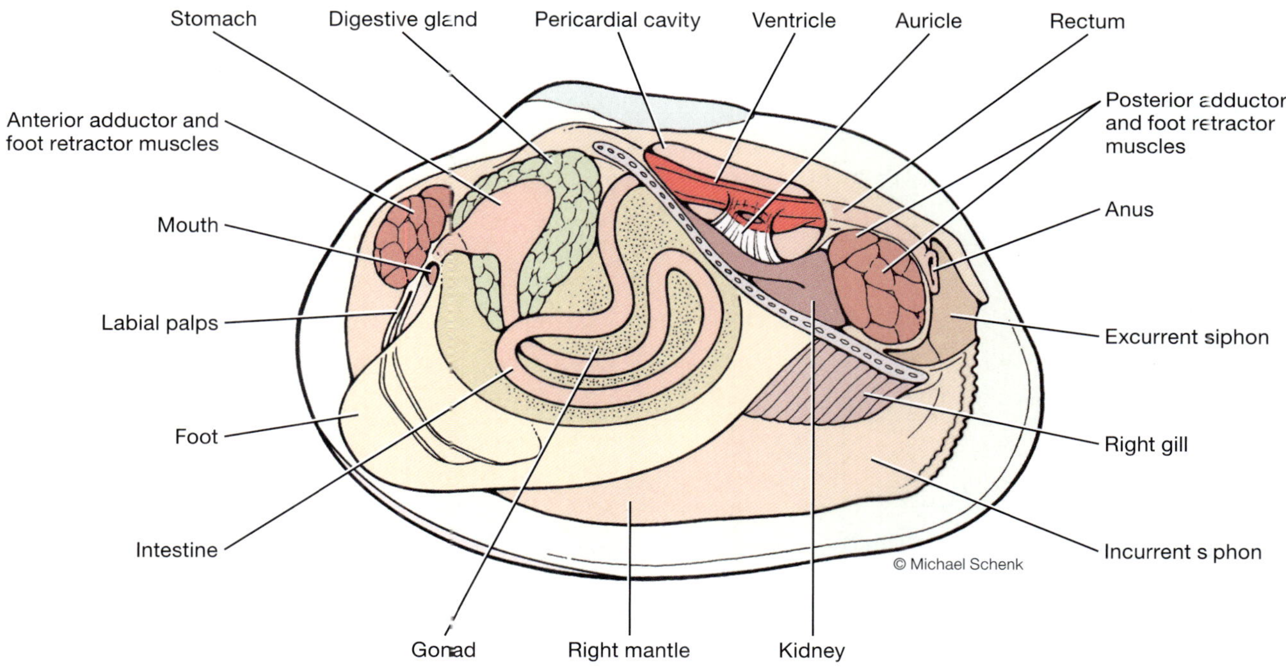

11.7 Internal anatomy of the freshwater clam.

11

Table 11-2 Anatomy of a Bivalve (Clam or Freshwater Mussel)

Structure	Function
Incurrent and excurrent siphons	Extendable, fleshy tubes that transport water into and out of the body
Gills	Used primarily for respiration and filter feeding; female freshwater mussels brood eggs in special gill pouches
Mantle	Thin membrane that secretes the shell
Shell	Hard outer covering that protects soft internal organs; composed of a mixture of calcium carbonate and protein
Foot	Muscular region adjacent to visceral mass; used for burrowing and locomotion
Visceral mass	Pouch that houses several major internal organs
Adductor muscles	Large, tubular muscles located at the anterior and posterior ends of the animal; close shell and hold valves tightly together
Labial palps	Fleshy folds of skin located near the mouth that collect food particles from the gills and transport them to the mouth
Mouth	Ingestion of food
Esophagus	Short tube through which food passes from mouth to stomach (rarely visible on dissection)
Stomach	Small chamber located within visceral mass for food storage
Digestive gland	Greenish, granular tissue that secretes digestive enzymes into stomach and intestine to assist in the chemical breakdown of food
Intestine	Coiled digestive tract where absorption of nutrients occurs
Anus	Elimination of indigestible wastes (egestion)
Gonad	Produces gametes for reproduction
Heart	Muscular portion of circulatory system that receives blood from the gills and pumps it through short arteries to neighboring tissues and organs
Kidney (nephridium)	Excretory organ of bivalve that filters nitrogenous wastes from the hemolymph and eliminates them from the body

Feeding and Reproduction

Clams and other bivalves depend on a constant flow of water through their bodies for acquiring oxygen and nutrients and for releasing gametes and wastes. As you examine your specimen, trace the flow of water through the body (Fig. 11.8). Water enters through the more ventral incurrent siphon and passes over the feathery gills. The gills extract oxygen and small, suspended food particles from the water while filtering out larger debris and releasing carbon dioxide into the water. The gills have a thin coating of mucus that traps food particles, allowing them to be passed by ciliary action toward the labial palps.

Water circulates dorsally through the mantle cavity and through the suprabranchial chambers within the gills and makes a 180° turn, passing along the dorsal aspect of the mantle cavity. Nitrogenous wastes are excreted by the kidney into the water as it passes. As the water leaves the clam through the more dorsal excurrent siphon, it passes directly past the anus, where wastes are eliminated from the digestive system and swiftly carried away from the animal.

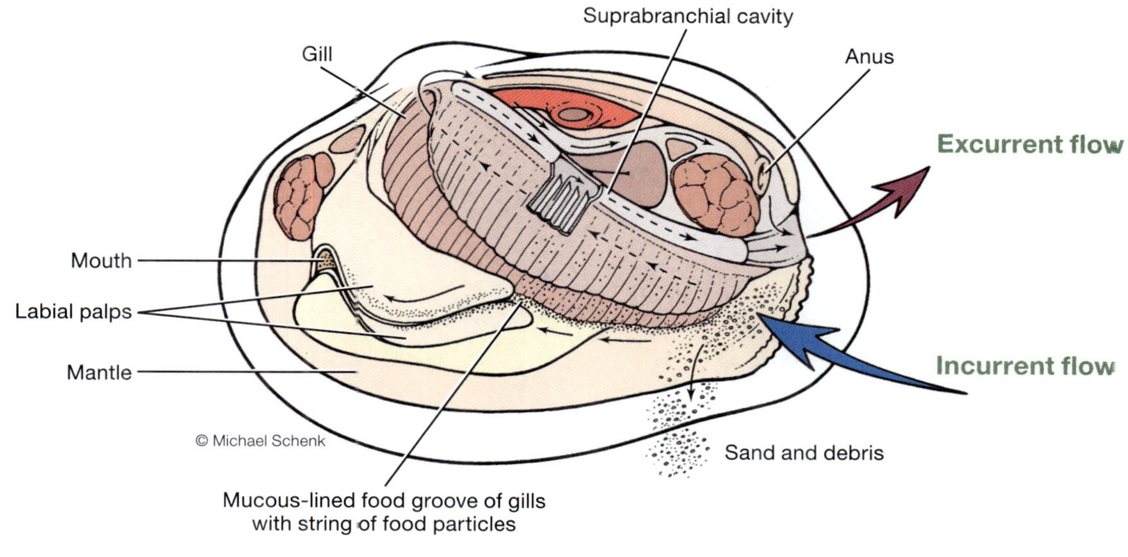

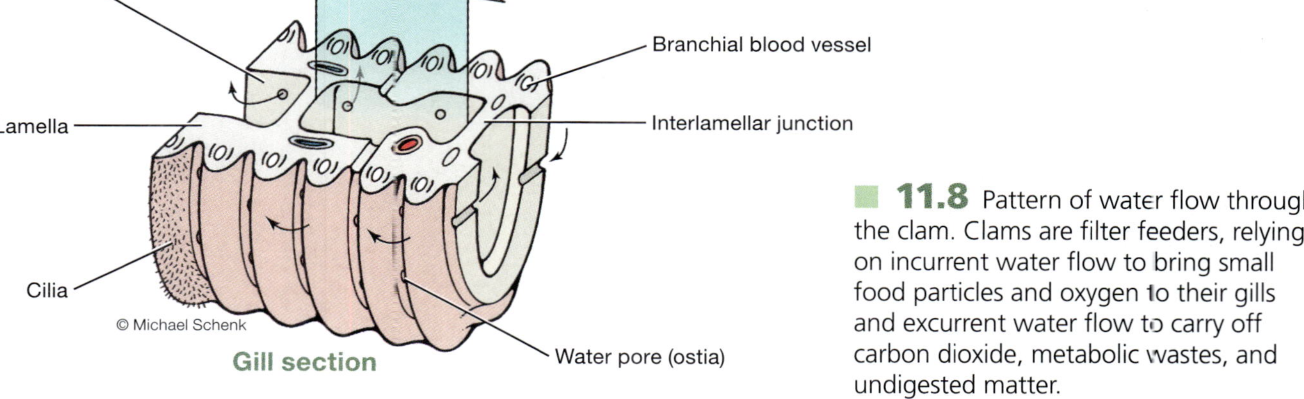

11.8 Pattern of water flow through the clam. Clams are filter feeders, relying on incurrent water flow to bring small food particles and oxygen to their gills and excurrent water flow to carry off carbon dioxide, metabolic wastes, and undigested matter.

11

In addition to respiration and food acquisition, the gills in female freshwater clams play an important role in reproduction. Remember that molluscs are dioecious, so the sexes are separate. Eggs are fertilized within special chambers in the female as sperm cells released from nearby males are brought in by water currents. Females brood fertilized eggs in special pouches in the gills until the eggs are ready to hatch.

The eggs develop into tiny larvae called glochidia that attach to the gills of certain fish species and act as external parasites, stockpiling the necessary nutrients from the host fish to complete their embryonic development. Later, the parasitic larvae detach from the gills of the host fish and settle to the seafloor to complete their transformation into free-living clams.

Circulation

Molluscs represent the first phylum we have studied whose members have a true circulatory system. Bivalves such as clams, mussels, and oysters have an **open circulatory system**—a system in which the blood is not always confined within a network of vessels. In bivalves, blood from tissues and major organs flows to the gills, where it is oxygenated and is directed passively back to the heart (Fig. 11.9). The oxygenated blood enters through openings in the heart called ostia and is pumped out of the heart through **arteries** to the mantle, foot, and visceral mass, where it empties into open sinuses in the tissues of these regions. Small **veins** collect the blood and return it to the gills.

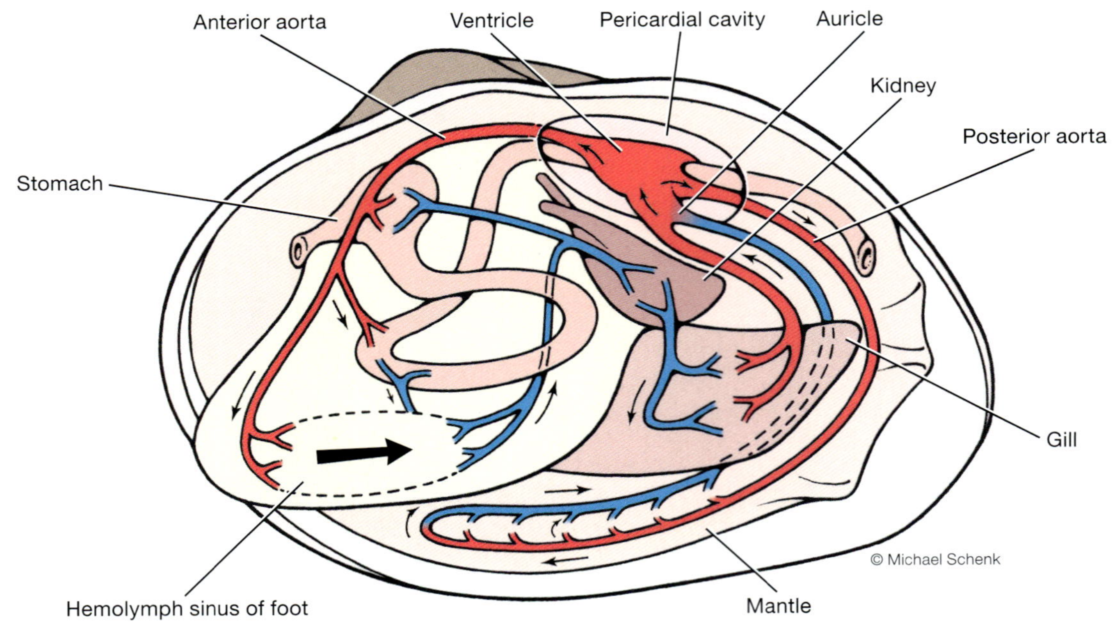

11.9 Schematic diagram of an open circulatory system as seen in clams.

Check Your Progress

11

2.1 List several features of the freshwater clam that enable it to thrive as a sedentary, aquatic organism.

2.2 What is the excretory organ of bivalves? Where in the body is it located?

2.3 Why do you suppose the adductor muscles are so well developed in bivalves?

2.4 Describe how the gills of a clam are an example of a multifunctional organ.

2.5 Discuss some of the disadvantages of an open circulatory system.

Exercise 11-3 Squid and Octopuses (Class Cephalopoda)

Materials Needed

- preserved squid
- dissection tools
- dissecting pan
- dissecting pins
- latex gloves

As their name implies, members of the class Cephalopoda have a modified "head-foot" with an array of prehensile tentacles and arms at one end of the body, and a visceral mass with short, protruding fins at the other end. The nautilus is the only living cephalopod that possesses an external shell; the shell is completely lacking in octopods and is reduced and internal in squid and cuttlefish. Although squid and cuttlefish typically use their muscular fins for leisurely locomotion, they also possess the ability to maneuver quickly by jet propulsion—rapidly expelling water from the mantle cavity through a tubular **siphon**. In fact, squid are among the fastest invertebrates on Earth, capable of reaching speeds up to 40 km/h (24 mph).

The evolutionary recruitment of the mantle cavity as the fluid reservoir for jet propulsion was incompatible with a hard external shell and explains why this feature is absent or highly reduced in most cephalopods. The consequence of this evolutionary trend is increased vulnerability to predation because of the lack of protection afforded by a shell. As a result, many octopods and squid developed alternative defense mechanisms.

The ability to change their skin color rapidly through the use of specialized epidermal cells called chromatophores allows them to blend in perfectly with their surroundings, making them even more effective predators. The chromatophores contain different-colored pigments and can contract or expand to produce a wide assortment of skin colors and patterns. Some species can become nearly translucent, and others can even glow in the dark! In addition, most cephalopods (except the nautilus) have an **ink sac** that can discharge a dark, cloudy liquid through the anus to confuse predators momentarily and provide them a few extra seconds for escape. For example, a squid may turn dark, squeeze ink out of its ink sac, then instantly turn lighter, or even clear, to confuse a predator and escape.

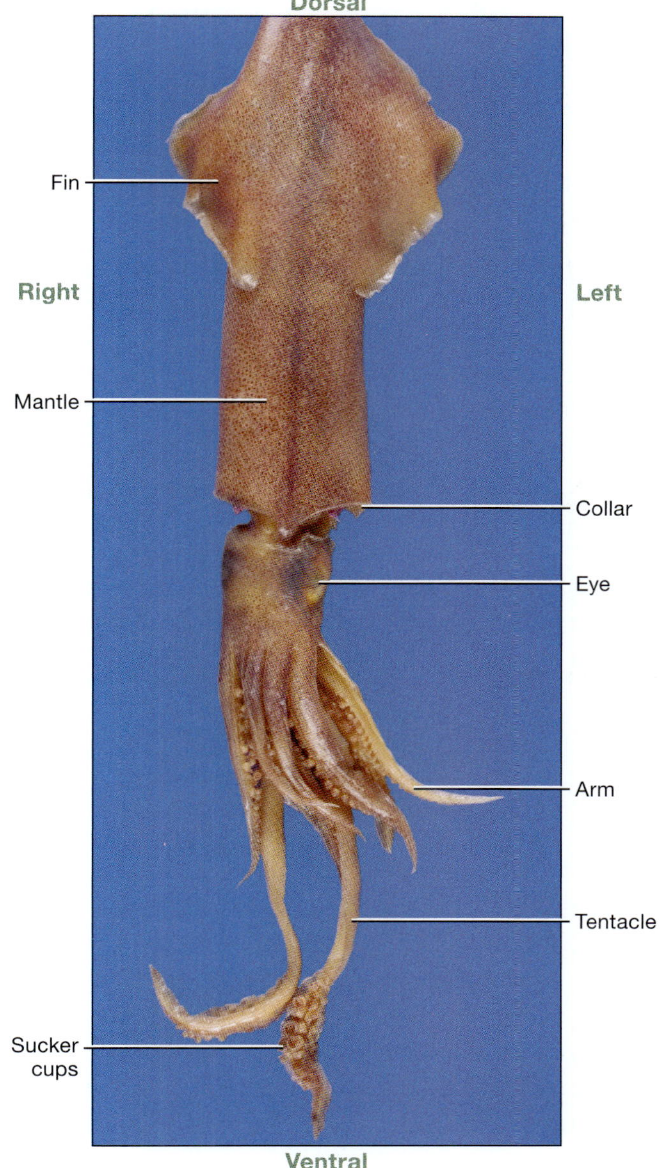

11.10 External anatomy of a squid, anterior view.

Procedure

1 Obtain a preserved squid and position it in your dissecting pan so that the side with the siphon is downward. This will position the anterior surface of the squid facing upward.

2 First, identify the anterior, posterior, dorsal, and ventral regions of your specimen, as well as the left and right sides of the body. Examine the external anatomy of the squid and identify the following structures: tentacles, arms, fins, siphon, mantle, eyes, and collar (Fig. 11.10 and Table 11-3).

Notice that squid have eight arms and two longer tentacles. The arms bear suckers along their entire length, but the two tentacles have suckers only at their distal ends. The remainder of each tentacle is a long, thin, elastic appendage. As squid get close to prey, they shoot out the long tentacles to capture their prey, pull it quickly toward the mouth, and use their shorter arms to ensnare it. They use their arms to rip apart prey and stuff it into the mouth, taking bites out of the prey with their sharp beak. The meat is ripped apart even more by the radula, a conveyor belt of tiny teeth inside the mouth that most mollusc species possess.

Table 11-3 External Anatomy of a Cephalopod (Squid)

Structure	Function
Collar	Fleshy border separating head-foot from visceral mass (mantle)
Eyes	Image-forming organs for detecting visual stimuli
Siphon	Hollow tube through which water is expelled from the mantle cavity at high velocity to propel the squid through the water
Mantle	Body tube encircling visceral mass forming a hollow chamber in which water is collected and used for propulsion
Arms	Shorter appendages (8) used to manipulate captured prey and act as a rudder for navigating while swimming
Tentacles	Long, extensible, prehensile appendages (2) for capturing prey
Fins	Triangular-shaped extensions of the dorsal end of the body tube that are used for leisurely swimming and for maneuvering during locomotion

3 Review the Dissection Diagram for suggested cuts to make on your specimen.

4 Flip your specimen so the posterior surface with the siphon is facing upward in your dissecting pan. Using scissors, make a shallow, longitudinal incision along one side of the body through the mantle, starting at the collar and extending to the dorsal end of the body tube past the fins toward the tip.

5 Repeat this incision on the other side of the body, bringing this cut all the way toward the dorsal tip of the body to join your first cut.

6 Use a blunt probe or teasing needle to carefully lift away the separated portion of the mantle from the underlying structures.

7 You may wish to use pins to keep the body tube open.

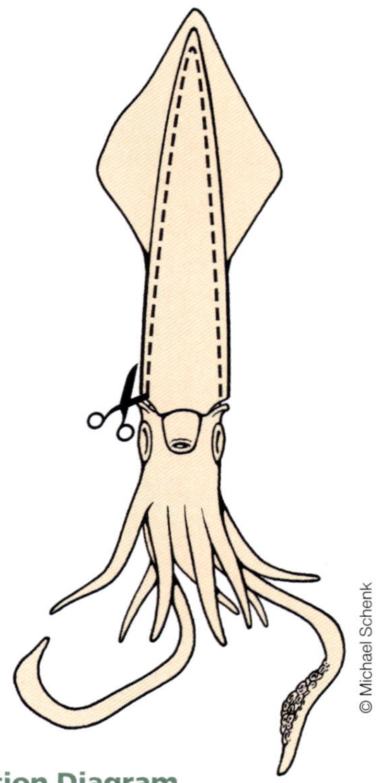

Dissection Diagram

1 Use scissors to make incisions along each side of the posterior surface of the body from the collar toward the dorsal tip of your specimen.

2 Gently tease away the separated portion of the mantle, being careful not to damage underlying structures.

3 Use pins as needed to spread the body open.

Reproduction

In male squid, sperm are produced in the single, large, white **testis** that lies dorsal to the hearts, and then pass into the ciliated funnel of the **vas deferens** for storage (Fig. 11.11). Using a specialized, convoluted and coiled tubule, known as the **spermatophoric gland**, the male packages sperm into a dense spiral covered with membranes and accessory structures called a **spermatophore**. Only about the size of a grain of rice (8–10 mm), each spermatophore contains 7–10 million spermatozoa! These spermatophores are stored in a special sac at the end of the spermatophoric gland until mating occurs. During copulation, the male uses a specially modified arm, the hectocotylus, to transfer spermatophore bundles from his **penis** to the inside of the female's mantle cavity. If you have a male squid, examine the arms. In mature males, the pedicels (stalks) of the proximal suckers will be noticeably longer and thinner on the hectocotylus than on the other seven arms. Often the female does not even wait for the male to remove his arm and rips it off in her haste to escape—really only a small loss for the male, because he will die shortly after mating anyway!

The spermatophore, stuck firmly to the inside wall of the female's mantle, is used to fertilize her eggs when she lays them. Female *Loligo* squid lay from 1,000 to 5,000 eggs in long strings stuck to the ocean floor. The eggs are produced in a single, large, yellowish **ovary**, which lies dorsal to the three hearts, and the eggs are released into the coelomic space, where they are picked up by the open, ciliated end of the **oviduct** known as the **ostium** (Fig. 11.12). Secretions released by two prominent glands, called the **nidamental glands**, and two smaller, oval glands beneath them, called the accessory nidamental glands, combine to produce a hard, foul-tasting capsule that encloses the eggs and deters predation.

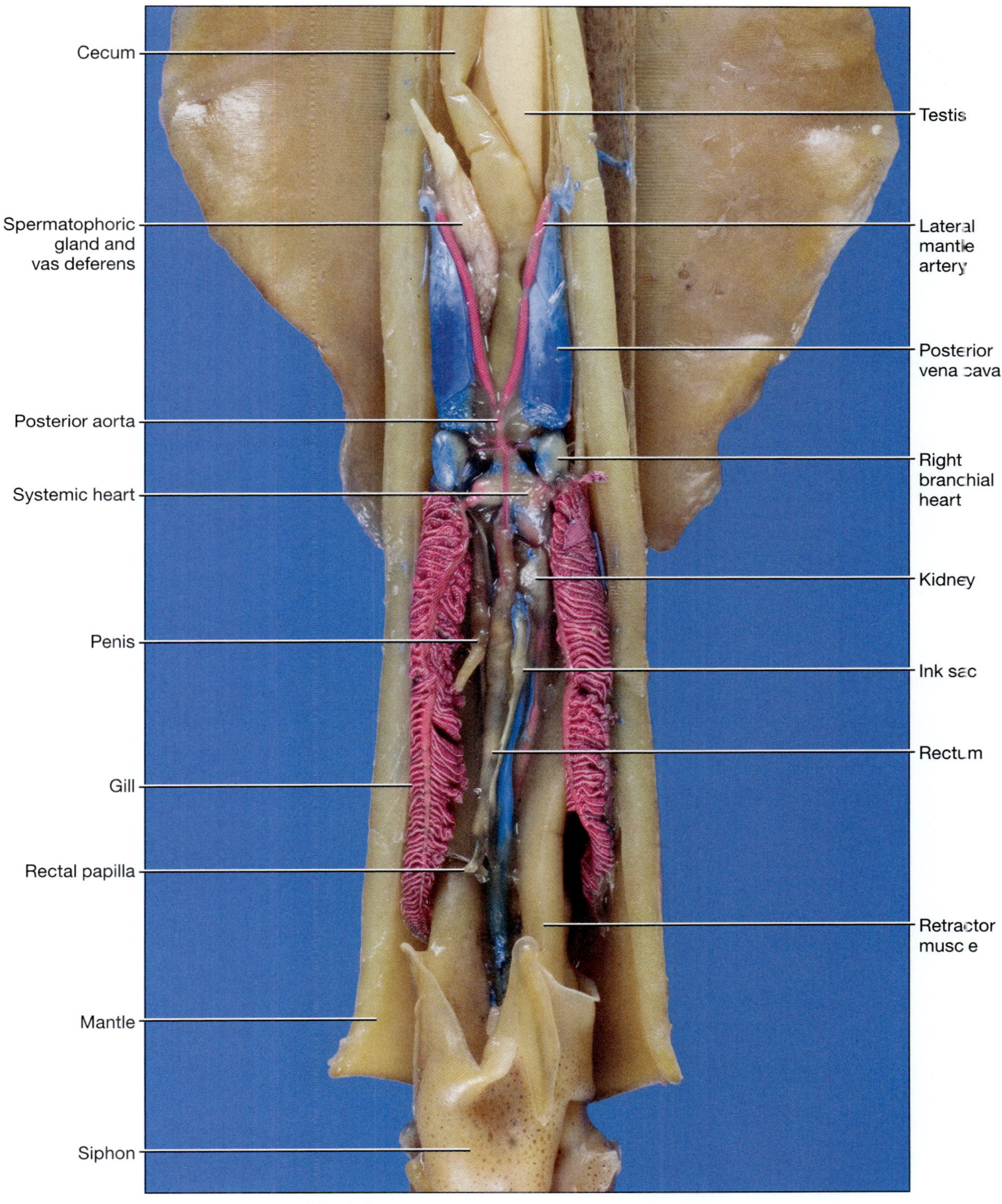

11.11 Dissection of the male squid, *Loligo*.

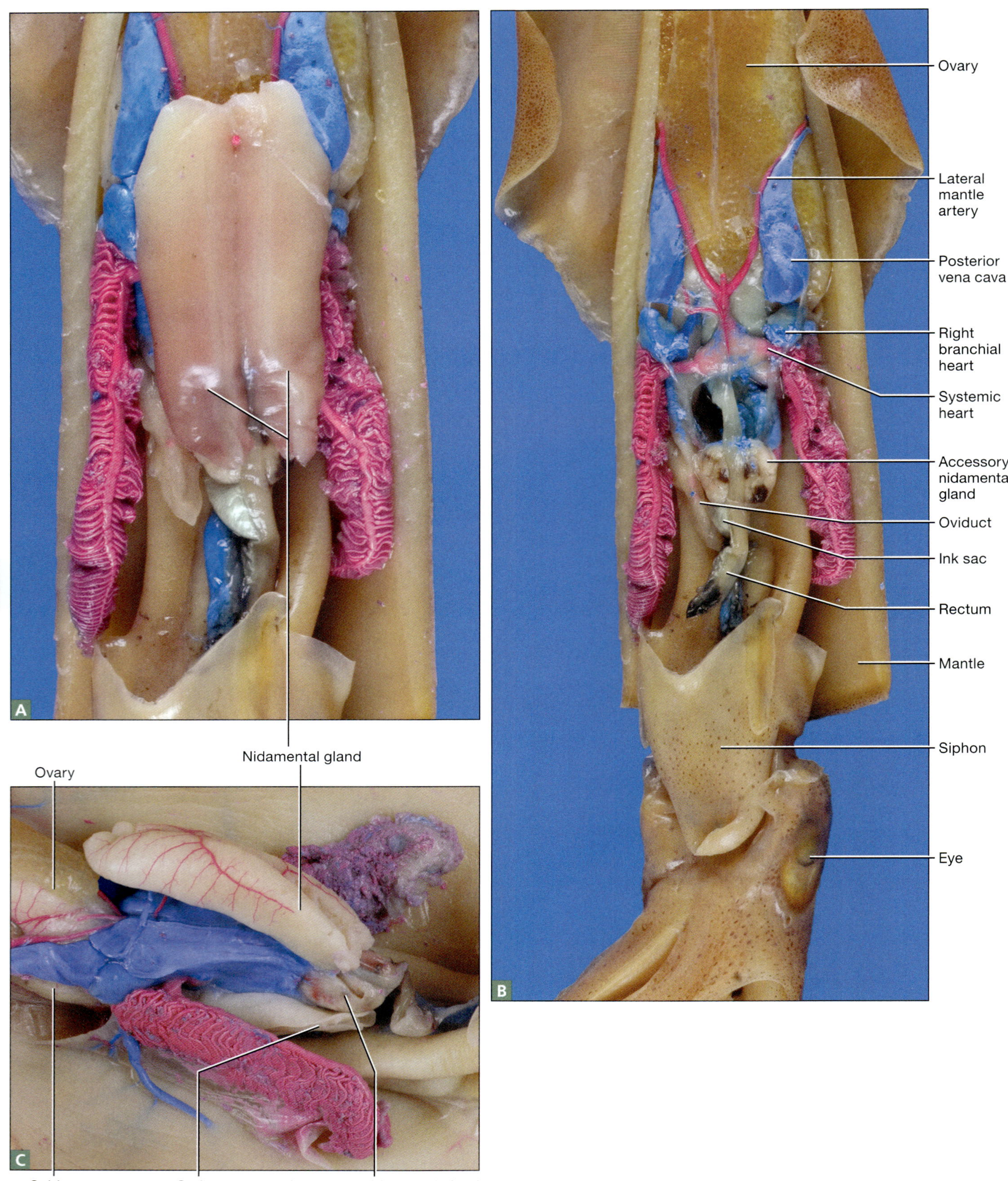

11.12 Dissection of female squid, *Loligo*: (**A**) nidamental glands in place, (**B**) nidamental glands removed to show underlying anatomy, and (**C**) close-up rotated 90° to depict the ostium of the oviduct.

As is the case with male squid, once the female mates and lays her eggs, she, too, usually dies. But, after several weeks of embryonic development, a new generation of baby squid, each about the size of a rice grain, hatches out and heads to deep water to feed for one or two years before returning to the shallow waters again to continue the cycle.

Digestion, Excretion, and Respiration

Food is swallowed and travels through a long, thin-walled **esophagus** into the small muscular stomach, where digestive juices from the **liver** and **pancreas** help to break it down. The mushy food then is passed into the large, thin-walled **cecum**, where it is absorbed into the squid's hemolymph (blood) and distributed throughout the body (Fig. 11.13). Wastes are passed along the **rectum**, through the anus, and exit the body through the siphon. The silvery, black ink sac, found along the distal portion of the rectum, opens by way of a small duct into the rectum, just proximal to the anus.

The excretory system of the squid is tied intricately to the circulatory system. The paired kidneys are more like sinuses than discrete organs, and are located near the systemic heart on either side of the intestine. They collect metabolic wastes from the major circulatory vessels in this region of the body and, because the squid lacks a urinary bladder, expel their waste products through the mantle funnel. The respiratory system of the squid consists of two large gills that oxygenate blood they receive from the branchial hearts and pass it to the centrally located systemic heart for distribution throughout the body (Table 11-4).

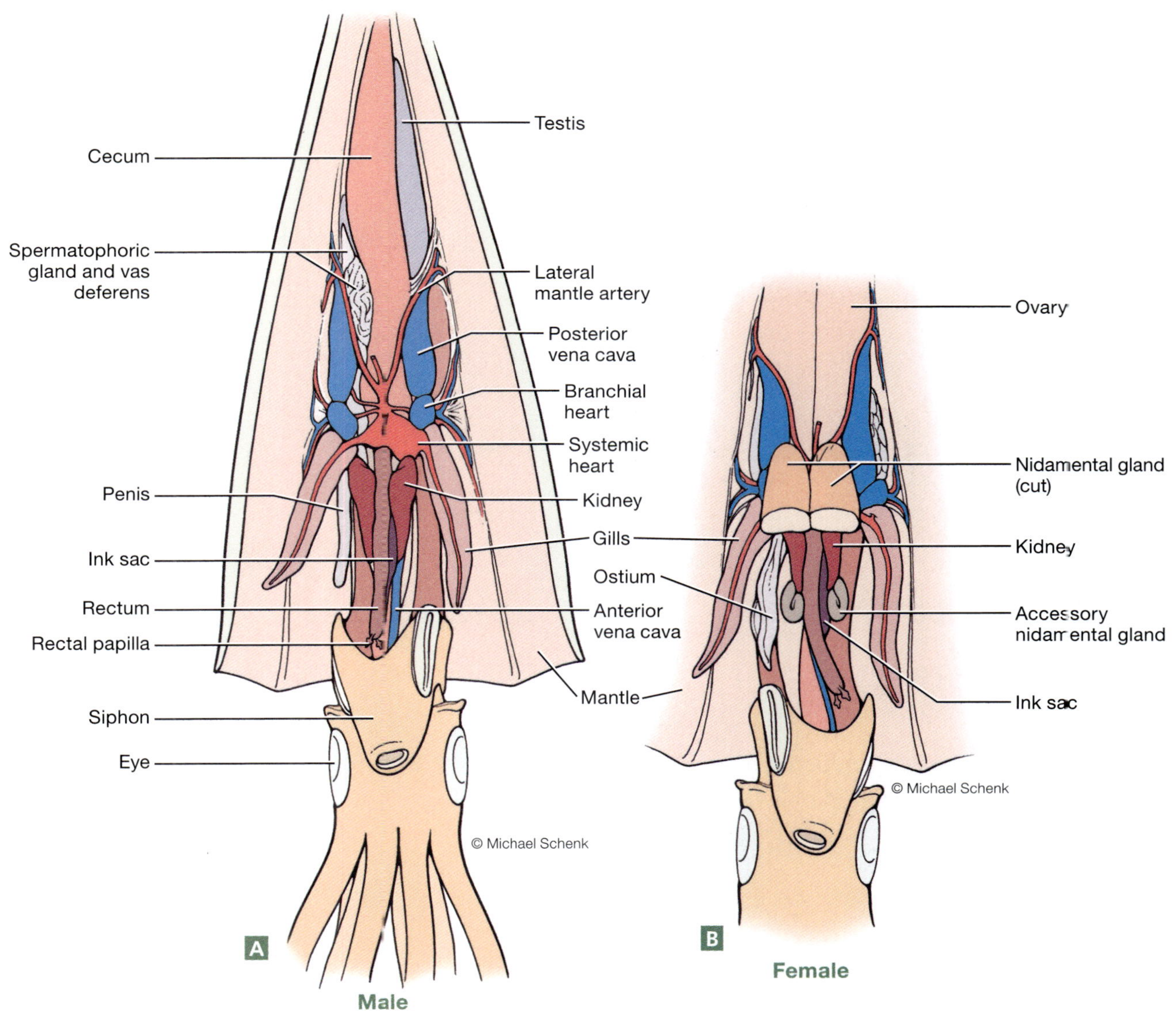

11.13 Internal anatomy of the (**A**) male, and (**B**) female squid, *Loligo*.

Table 11-4 Internal Anatomy of a Cephalopod (Squid)

Structure	Function
Gills	Paired, feathery organs used for respiration
Esophagus	Thin tube connecting the mouth to the stomach
Stomach	Small sac where food is stored and digested; digestion is entirely extracellular in cephalopods
Pancreas	Small, granular digestive gland that secretes enzymes into the stomach to assist in the breakdown of food
Liver	Large, elongated gland that releases secretions into the stomach to facilitate enzymatic digestion of food
Cecum	Large, thin-walled digestive chamber prominent along the dorsal end of the body
Rectum	Portion of digestive tract leading to anus
Ink sac	Large sac that opens into the rectum and secretes a dark brown or black fluid when the animal is alarmed
Anus	Terminal portion of digestive tract located near siphon
Testis (male)	Produces sperm; located in dorsal end of body tube
Vas deferens (male)	Open-ended chamber that receives and stores sperm
Spermatophoric gland (male)	Coiled, convoluted tubules that package sperm cells into spermatophore packets
Penis (male)	Muscular organ that gathers spermatophores and releases them outside the body where they are transferred by the male's hectocotylus to the female during mating
Ovary (female)	Site of egg production
Oviduct (female)	Open-ended chamber that receives eggs from the ovary
Nidamental glands (female)	Produce secretions that form an encasement around the strings of eggs to protect them
Retractor muscles	Long muscles that control the retraction of the siphon and the head
Kidneys	Paired excretory organs located near the systemic heart that filter nitrogenous wastes from the hemolymph and eliminate them from the body
Systemic heart	Large, muscular ventricle that receives oxygenated blood from the gills and pumps it throughout the body
Branchial hearts	Smaller, muscular chambers that receive deoxygenated blood from all parts of the body and pump blood to the gills

Check Your Progress

3.1 Is your squid specimen a male or female? Explain how you know.

3.2 Why are sensory structures more prevalent on cephalopods than on bivalves?

3.3 Put the following male reproductive structures in the proper order to represent the correct path of sperm: penis, testis, spermatophoric gland, vas deferens.

3.4 What are the functional differences between the gills of squid and the gills of clams?

Circulation

Most molluscs (clams, mussels, oysters, scallops, snails, and slugs) have an open circulatory system, meaning that the organ tissues are bathed in a pool of "blood" called **hemolymph**. Unique among molluscs, cephalopods have a much more efficient circulatory system, capable of supporting the comparatively high metabolic needs of swift-moving, active predators. The fast-paced life of a squid is possible partly because its hemolymph is pumped quickly around the body while remaining completely contained inside a network of veins, arteries, and capillaries in a **closed circulatory system** like that of humans.

At the center of this circulatory system are three hearts: the larger **systemic heart** pumps hemolymph around the body and to the brain, while two **branchial hearts**, one at the base of each gill, push hemolymph through the smaller capillaries of the gills (Fig. 11.14). Tremendous pressure is necessary to pump large volumes of blood through the

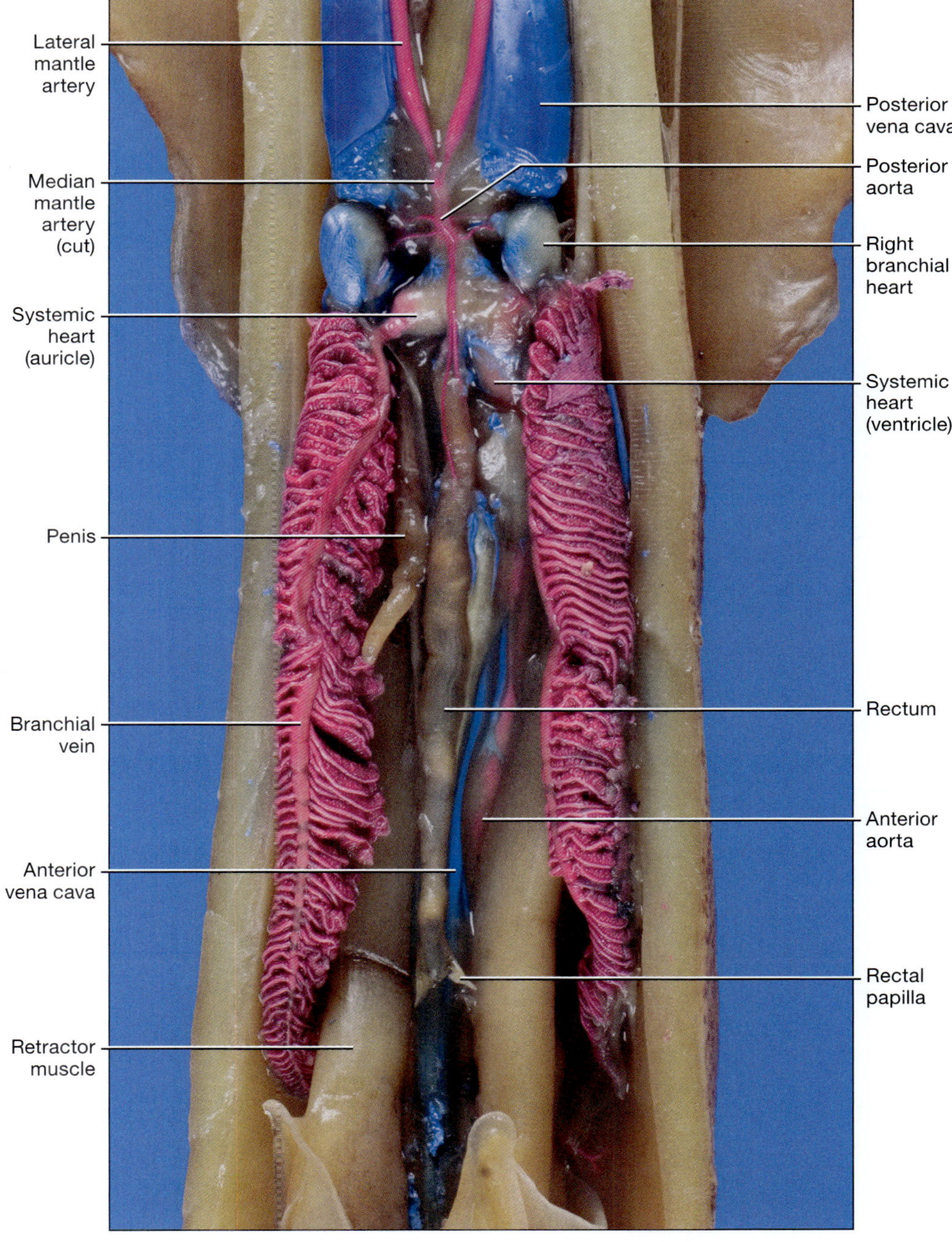

11.14 Major vessels and organs of the circulatory system in the squid.

11

extensive network of tiny vessels in the gills, and a dedicated heart for each gill provides the necessary blood pressure to accomplish this task. Whereas squid have evolved a separate heart to pump blood through each gill, vertebrates generate the high pressures necessary to pump blood through the gills or lungs with specialized chambers within a single heart. The selective forces of evolution often produce different solutions in different animal lineages to overcome similar problems.

The squid's systemic heart consists of a single chamber (ventricle) that receives oxygenated blood from the gills through the right and left branchial veins. The blood is sent out of the systemic heart through three vessels, the anterior aorta, the posterior aorta, and the smaller genital artery (Fig. 11.15). The posterior aorta quickly splits into three mantle arteries: a median mantle artery and two lateral mantle arteries. The major veins in squid are large and particularly prominent, especially in injected specimens. The anterior vena cava returns deoxygenated blood from the head and arms. It splits into two branches, with each branch leading to a separate branchial heart. The left and right posterior vena cavae are large vessels that collect deoxygenated blood from the dorsal portion of the mantle and route it to the branchial hearts. Each branchial heart pumps blood to its respective gill for oxygenation through a branchial artery.

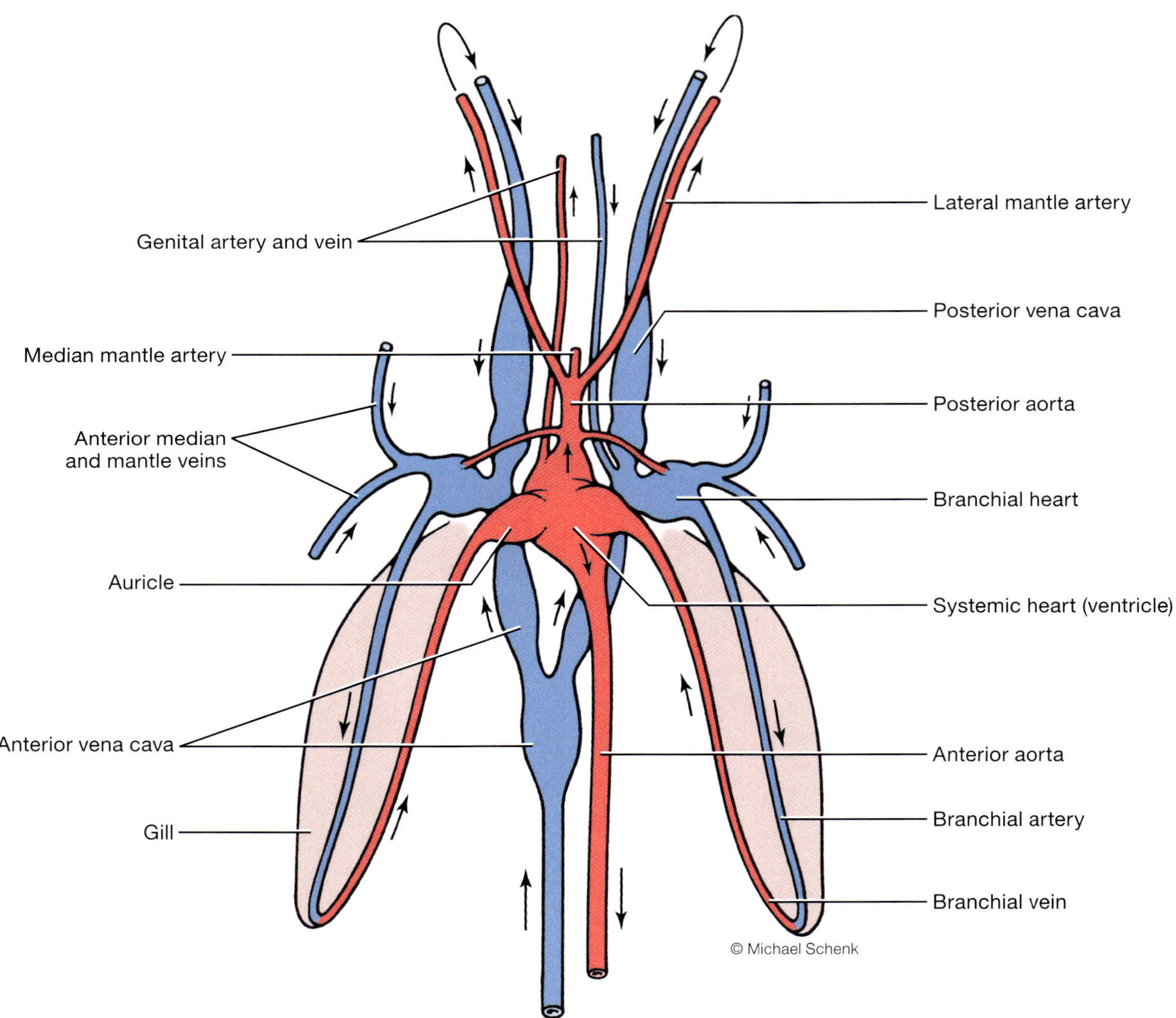

11.15 Pathway of blood through the major vessels and organs of the circulatory system in the squid.

The Cephalopod Eye

Perhaps the single most remarkable adaptation that squid possess is their well-developed eye. The striking similarity between cephalopod eyes and the eyes of vertebrates (birds, mammals, fishes, etc.) is a classic example of **convergent evolution** among animals; each group has independently evolved acute, image-forming eyes that are amazingly similar in structure. The eye of the squid contains a lens, cornea, iris, ciliary muscles, and a retina, just like the eyes of vertebrates. Though due to their independent evolution, a cephalopod's eye differs in a few important ways from a vertebrate's eye. Vertebrate eyes have an elastic lens and visual images are focused on the retina by altering the shape of the lens. Cephalopod eyes have a rigid lens and focus images on the retina by altering the distance between the lens and retina (just like in a camera).

Another major difference between the two eyes is in the way the photoreceptors in the retina receive light. In the vertebrate eye, the rods and cones are located at the very back of the retinal layer and point *away* from the pupil, forcing light to pass through the other accessory nerve cells before it is detected by the photoreceptors! Even though these other nerve cells are fairly transparent, some scattering of light rays occurs, resulting in a slight loss of visual acuity. This spatial positioning also places all of the accessory neurons *inside* the eye. To exit the eye, they come together in a large, cable-like nerve fiber (the optic nerve) and "push" the rods and cones aside to make a path through the back of the eye, thereby creating a blind spot in our visual field (Fig. 11.16A).

By contrast, photoreceptors in the cephalopod eye are positioned along the front of the retina, with the light-sensitive ends oriented *toward* the front of the eye, so light entering the pupil passes through the lens and stimulates the photoreceptors directly, without the scattering problem of the vertebrate eye. And because the accessory neurons are positioned *behind* the photoreceptors, the visual field of cephalopods has no blind spots (Fig. 11.16B). If you think about it, this is a much more logical blueprint for an eye than the architecture of the vertebrate eye!

This is an excellent illustration of how different evolutionary means may be employed to reach the same end—in this case a functional, image-forming eye. Remember—natural selection can operate only on natural variation in existing traits within populations, so the most logical design may not always be achievable. Without natural variability in a trait (for example, the elasticity of the lens), that trait will remain unchanged forever and other traits in which variability within the population does exist are modified to improve survival (for example, ciliary muscles that can move the lens back and forth to compensate for the inability of the lens to change shape).

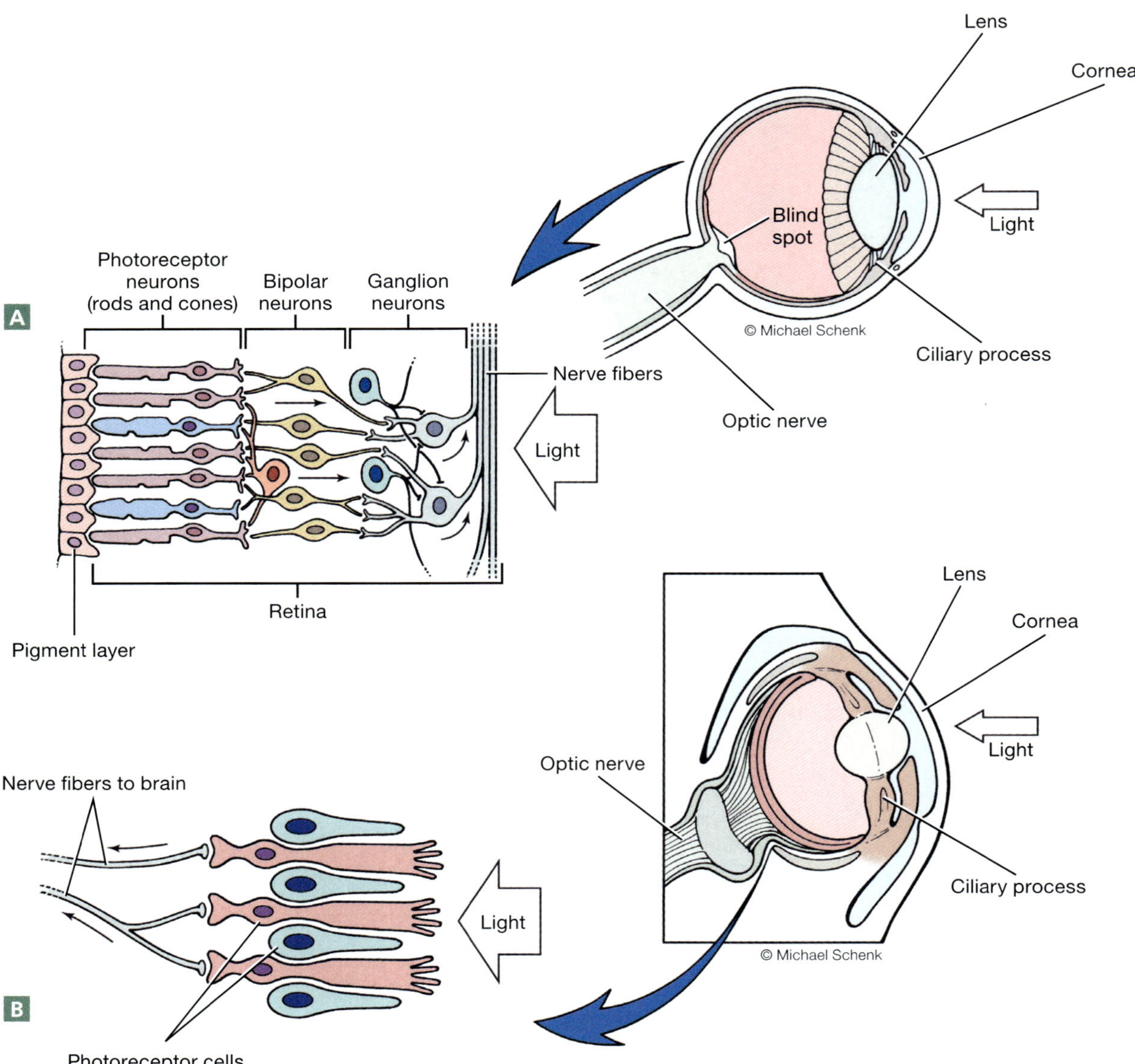

11.16 Comparison of internal anatomy of (**A**) vertebrate eye to (**B**) cephalopod eye, showing patterns of convergent evolution in distantly related groups. Small arrows depict direction of nerve impulses to the brain.

11

Name ______________________ Section ______________

11 Questions for Review

1 Complete Summary Table A-1 in the Appendix, filling in the characteristics of both bivalves (clams) and cephalopods (squid) in the appropriate rows.

2 What are the four major characteristic features of molluscs?

1. ______________________
2. ______________________
3. ______________________
4. ______________________

3 How does the foot of a bivalve differ from the foot of a cephalopod, or the foot of a snail? In what ways do these differences reflect specific adaptations for each animal to its unique lifestyle?

____

____

____

____

4 List at least three features of squid that contribute to their success as predators.

1. ______________________
2. ______________________
3. ______________________

5 What are some advantages of freshwater clams producing parasitic larvae that attach to the gills of fish?

____

____

____

6 Define *adaptive radiation*, and discuss how molluscs demonstrate this phenomenon.

7 Discuss how the circulatory system of cephalopods has been adapted to enable their active, predatory lifestyle.

8 Define *convergent evolution*, and discuss how the cephalopod eye and the vertebrate eye demonstrate this phenomenon.

Name ______________________ Section ______________

11 Laboratory Practical Challenge

The following evaluation is intended to replicate the method of assessment and type of questions you will be asked on lab practicals. If you have difficulty with these questions, you may benefit from additional review of the specimens provided in lab.

1 Identify the phylum and class of the specimen depicted below.

2 What is the function of the structure indicated by line A?

3 Identify the structure indicated by line B.

4 To what system does the structure indicated by line C belong?

11

5 On the figure below, label the following structures: gill, branchial heart, cecum.

6 To what system does the structure indicated by line D belong?

---

7 Identify the structure indicated by line E.

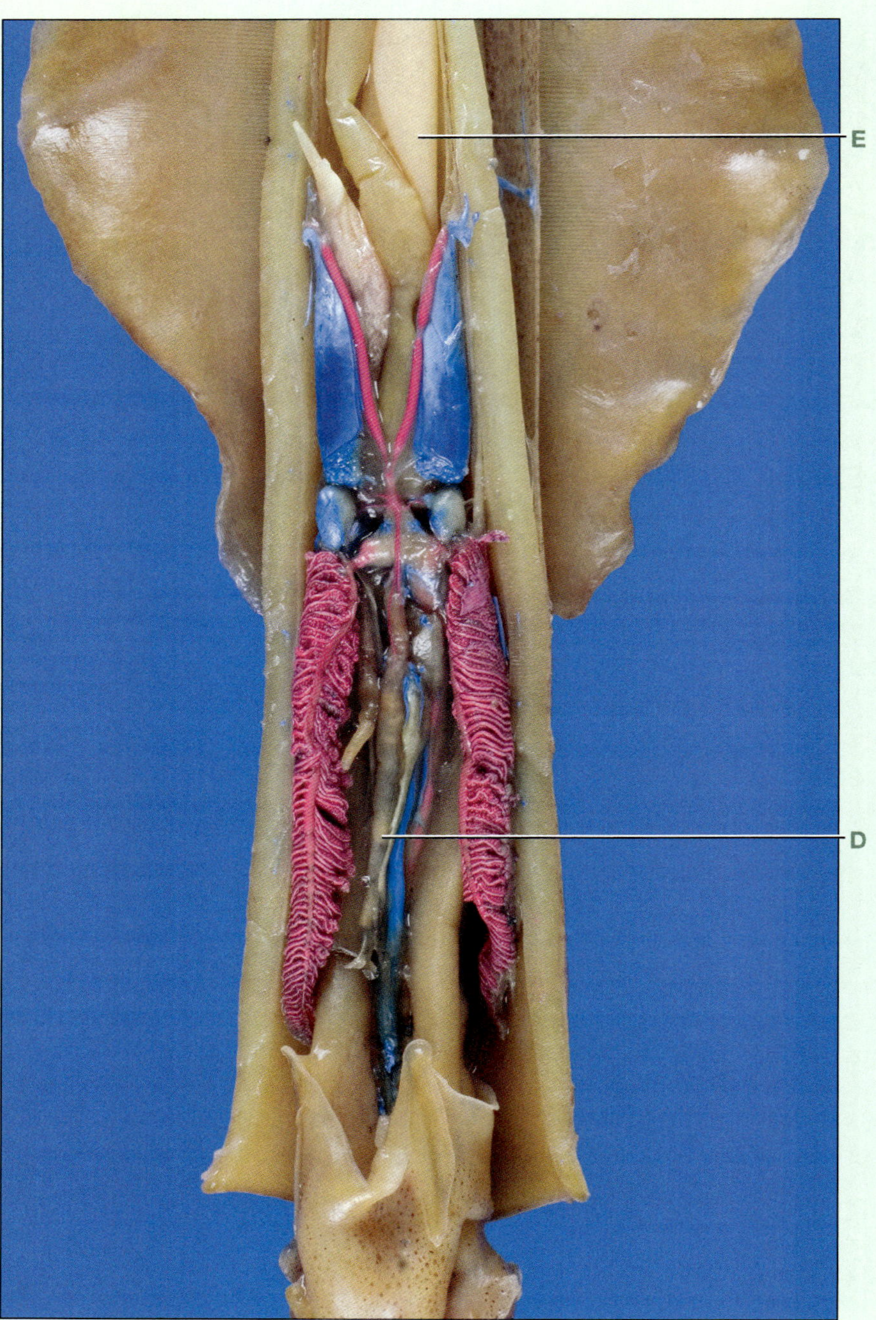

11

CHAPTER 12

Annelids

After completing the exercises in this chapter, you should be able to:

1 List the distinguishing characteristics of the phylum Annelida.

2 Identify specimens from the two major classes of annelids.

3 Understand the evolutionary position of annelids on the phylogeny of major animal phyla.

4 Discuss the evolutionary implications of segmentation and closed circulatory systems.

5 Identify the major anatomical features of sandworms, earthworms, and leeches.

6 Discuss the reproductive cycle of earthworms.

7 Characterize patterns of locomotion of different annelids.

The phylum Annelida contains more than 16,500 species of segmented worms. They occupy a wide variety of habitats, ranging from marine and freshwater areas to moist terrestrial locations, and they range in size from less than 1 millimeter to more than 3 meters in length. The phylum name is derived from the **segmentation** of their bodies (*annellus* means "little ring" in Latin). The body wall is segmented, and many organs and body parts show repetition inside.

One advantage of segmentation is that it facilitates locomotion. The coelom is divided into segments (partitioned internally by septa), and each segment has its own muscle groups. This arrangement allows the worm to elongate one part of its body while simultaneously shortening another part. The fluid within these segments creates a **hydrostatic skeleton**, providing a rigid structure against which the muscles contract.

Another advantage of segmentation is that it provides repeated body elements that may be modified in different ways to perform specialized functions relating to reproduction, feeding, locomotion, respiration, or excretion. For example, the **complete digestive system** of annelids contains many specialized subregions along the length of the body. Also, the anterior body segments have been modified for sensory and reproductive roles, and the posterior segments have been modified for nutrient absorption and excretion.

Other major characteristics of this phylum include a **closed circulatory system** consisting of multiple aortic arches (hearts), arteries, veins, and capillaries; a large, well-developed **coelom**; and **setae**, which are small, hair-like bristles used for locomotion. In fact, the degree of setal development is one distinguishing feature that biologists use to divide annelids into separate taxonomic groups. Annelids have a simple nervous system consisting of a rudimentary **dorsal brain** with two lobes and single **ventral nerve cord**. Respiration occurs through the skin, either in specialized regions, as in polychaetes, or across the entire epidermal surface, as in earthworms and leeches. Most annelids are **hermaphroditic**, meaning that an individual contains both male and female sex organs. They do not self-fertilize like many flatworms, however. Instead, two worms typically exchange sperm, simultaneously cross-fertilizing each other's eggs.

The taxonomic organization of this phylum has been under recent revison and many zoologists now recognize only two major classes of annelids:

1. Polycheata (clamworms, sandworms, fanworms, scaleworms, lugworms, and tubeworms)
2. Clitellata (earthworms, angleworms, blackworms, and leeches).

Formerly categorized in separate classes, leeches and oligochaetes (earthworms, blackworms, angleworms, etc.) are now considered as separate subclasses within the class Clitellata, based on the shared characteristic of having a visible clitellum during the reproductive phase.

In the following exercises you will examine representatives of these two classes from a combination of marine, freshwater, and terrestrial habitats. Although the notion of a segmented worm usually brings to mind images of the familiar terrestrial earthworm, the vast majority of annelid species occupy marine habitats and belong to the class Polychaeta. Most annelids are free-living. Leeches, however, are classified as semiparasitic because they may utilize host organisms for nutrition during part of their lives. As you examine the different members of this phylum, keep in mind the anatomical similarities that link them together and illustrate their common ancestry. But also pay close attention to the differences between each group that reflect specific adaptations to their particular lifestyles.

Evolutionary History

During the early evolution of animals 600 million years ago, the three primary habitats for animal life were the open sea, the surface of the seafloor, and the soft, muddy substrate beneath the seafloor surface. As early animals developed the ability for more complex movements some of these animals burrowed in the soft, seafloor mud. They were the segmented worms, as shown by fossil worm tunneling tracks dating back 540–650 million years.

What did the first annelids look like? From which ancestral group did they originate? Most evolutionary biologists agree on the answers to these questions. Modern polychaetes represent the closest living examples of what the ancestral annelids probably looked like. As the ancient members of this phylum moved into different habitats, populations slowly diverged into terrestrial and "semi-parasitic" groups, eventually giving rise to the oligochaetes and hirudinids. Molecular evidence suggests that annelids share a recent common evolutionary past with several other invertebrate phyla classified together into the Lophotrochozoa. Prior to the Cambrian explosion (535–525 mya) this assemblage is thought to have split from the Ecdysozoa, the line leading to roundworms, arthropods, and several other smaller invertebrate phyla (Fig. 12.1). For a more complete phylogeny depicting the evolutionary relationships among the major animal phyla covered throughout this laboratory guide, review Figure A.1 in the Appendix.

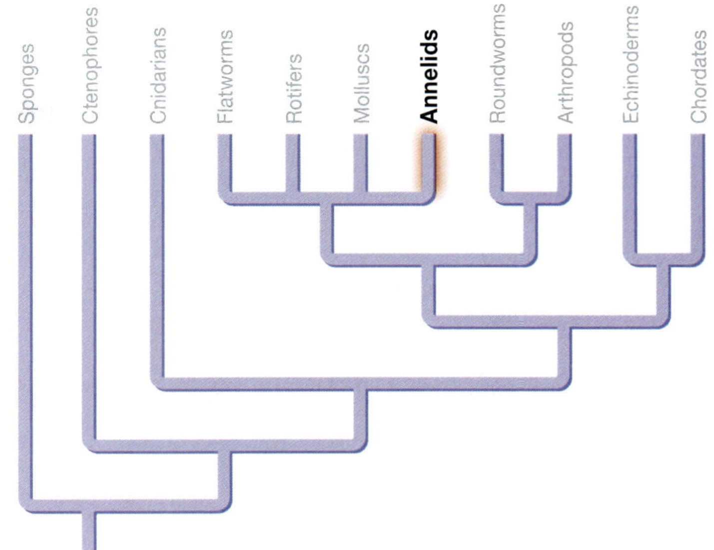

12.1 Annelids are grouped with a collection of other invertebrate phyla into the Lophotrochozoan assemblage based on shared characteristics that indicate close evolutionary relationships.

Exercise 12-1 Sandworms (Class Polychaeta)

12

Materials Needed

- preserved specimen of *Nereis*
- dissecting pan
- dissecting microscope
- latex gloves
- slide of *Nereis* c.s.
- compound microscope

The class Polychaeta (≈10,000 sp.) is a widely diverse group of marine segmented worms, including clamworms, sandworms, fanworms, scale-worms, lugworms, and tubeworms. The name *polychaeta* means "many bristles" and refers to the setae projecting from side flaps called **parapodia** that are a characteristic feature of this class. Parapodia have been modified in the different species as adaptations for different habitats and lifestyles. Species living in rocky environments have leg-like parapodia used in walking. In burrowing species, parapodia function as digging paddles. In some species, the parapodia are short and combine with peristaltic contractions of the body to move the worm through the mud. Filled with blood vessels, parapodia play an essential role in gas exchange by increasing the worm's surface area for diffusion of oxygen and carbon dioxide. Another characteristic feature that separates some polychaetes from their fellow annelids is their well-developed heads with specialized sense organs.

Some polychaetes build temporary burrows or permanent tubes of mud and mucous secretions, where they lead stationary lives filtering **plankton** and other suspended food particles from the water using feathery feeding tentacles—a process known as suspension feeding. Some species eat organic material that settles on the surface of the muddy substrate (detritus feeding). Other species are more aggressive, free-living predators that swim and scurry about in search of prey.

A typical example of a predatory polychaete is the sandworm, *Nereis*. Sandworms hide under rocks or in burrows during the day and emerge at night to search for food. Their slow crawling is carried out by means of the parapodia. Rapid crawling and swimming are achieved by undulation of the body. To capture prey, the sandworm quickly thrusts out of the mouth an eversible proboscis that is armed with piercing jaws that grab hold of the victim.

Procedure

1 Obtain a preserved sandworm (*Nereis*) and examine its external features (Fig. 12.2). Small **eyes,** two fleshy palps, and an array of thin **tentacles** are present on the head and are part of the sensory system of *Nereis*. On preserved specimens, the sharp jaws usually are retracted within the pharyngeal region. It may be possible to evert the jaws and pharynx by pulling on them with a small pair of forceps while pushing down on the head. Count the number of eyes and the number of tentacles present on the head of your specimen and record those values below.

# of eyes ______ # of tentacles ______

2 Examine the body of the sandworm. In polychaetes, the characteristic paired setae are located at the ends of feathery extensions called parapodia. Remember that these parapodia, while useful for locomotion, are highly vascular and, thus, are sites for gas exchange in this marine worm. How many parapodia per body segment are present on your specimen?

# of parapodia per segment ______

3 Examine a prepared slide of a cross section (c.s.) through the body of *Nereis*. Locate the following organs and regions depicted in Figure 12.3: dorsal blood vessel, intestine, coelom, parapodia, and setae (setae may not be visible on every slide). In the space below, describe the general shape of the lumen (inside) of the intestine. Make a note to compare this geometry to the lumen of the intestine in earthworms, which you will view shortly. Earthworms have evolved a structure that greatly increases the surface area of their intestine for more efficient digestion—a feature lacking in polychaetes.

Shape of intestinal lumen ______

12.2 Sandworm, (**A**) *Nereis*, a common marine polychaete, (**B**) parapodia with setae, (**C**) enlarged view of the head, and (**D**) head with pharynx everted to show jaws.

12

Dorsal blood vessel
Parapodia
Lumen of intestine
Coelom
Setae
A
B

12.3 Cross section through (**A**) the intestinal region of the sandworm, *Nereis*, and (**B**) magnified view depicting parapodia and setae.

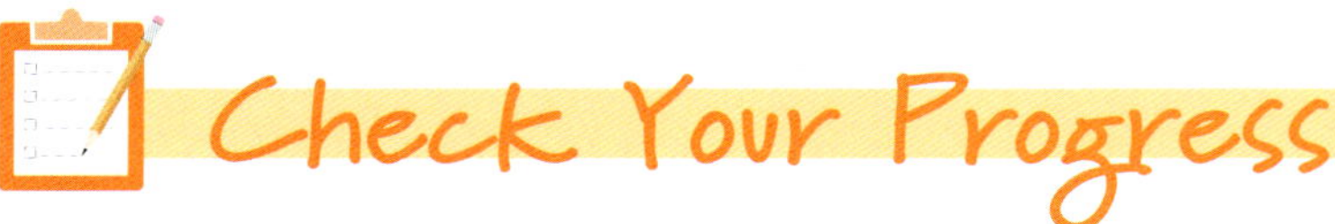

1.1 What kind of stimuli do you suppose the tentacles detect?

1.2 What features of sandworms would lead you to believe that they are free-living predators?

Exercise 12-2 Earthworms (Class Clitellata, Subclass Oligochaeta)

Materials Needed

- ❑ preserved specimen of *Lumbricus*
- ❑ slide of *Lumbricus* c.s.
- ❑ dissection tools
- ❑ dissecting pan
- ❑ dissecting pins
- ❑ latex gloves
- ❑ dissecting microscope
- ❑ compound microscope

Earthworm External Anatomy

Members of the class Clitellata, subclass Oligochaeta (≈3,000 sp.), such as earthworms, blackworms, and angleworms, are free-living worms that live in freshwater habitats or damp soil. Unlike polychaetes, oligochaetes lack parapodia and rely on their entire epithelial surface for gas exchange. Most possess short, bristly setae on each segment. Terrestrial members, such as the common earthworm, are well-suited for their fossorial (burrowing) lifestyle and possess many adaptive features for subterranean life.

Procedure

1 Obtain a preserved specimen of *Lumbricus terrestris*, the common earthworm, and examine its external anatomy. The anterior end features a small, slit-like **mouth**, covered by the fleshy **prostomium** (Fig. 12.4). Closely examine the prostomium on your specimen and speculate as to its function in the space provided below.

2 Run your fingers along the length of the body to feel the **setae**. Much like an athlete's cleats, these bristles help the earthworm grip the dirt and assist in crawling and burrowing. Describe what the setae feel like. Are they more or less spiny than those of the sandworm?

3 Locate the **clitellum**—a large band covering several segments about one-third of the way down the body from the mouth. The clitellum is used during reproduction for transferring sperm between individuals and in secreting a cocoon that contains the fertilized eggs.

4 At the posterior end of the body, locate the **anus**—the opening through which indigestible products are released from the digestive tract.

5 Careful inspection of the ventral surface of your worm with the aid of a dissecting microscope may allow you to see the openings to the reproductive organs of the worm. Most prominent are the swollen openings to each **vas deferens**, located on segment 15. Anterior to those, on the margins of segment 10 are the openings to the **seminal receptacles**. Each vas deferens releases sperm during copulation, and the seminal receptacles, as their name implies, simultaneously receive sperm (from another earthworm) during copulation. We will examine the reproductive cycle of the earthworm in greater detail later in the chapter.

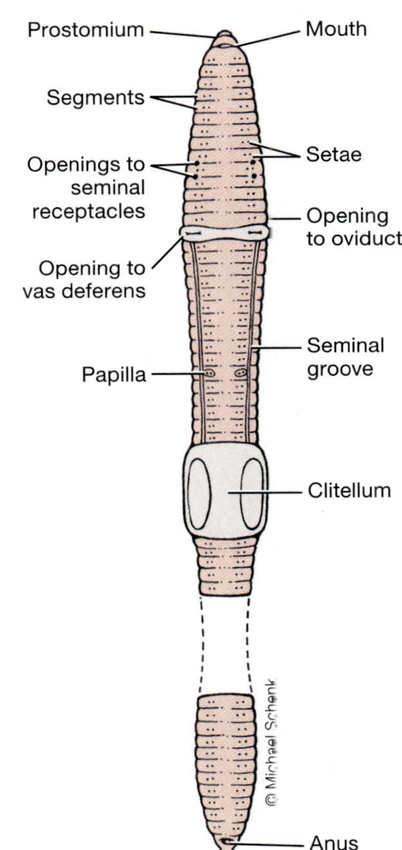

12.4 External anatomy of the common earthworm, *Lumbricus*, ventral body surface.

12

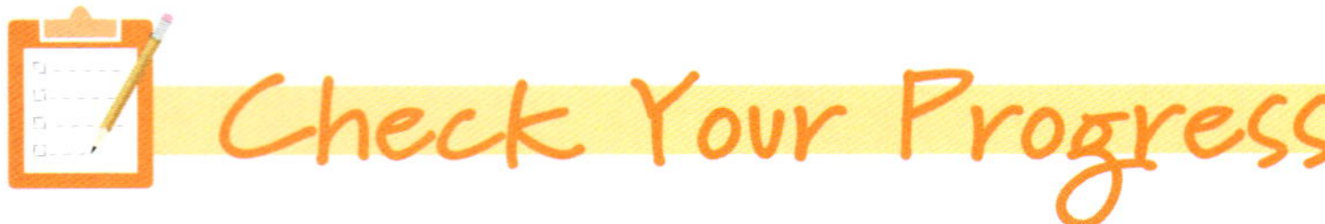

Check Your Progress

2.1 Do the setae feel different as you run your fingers in both directions along the body?

2.2 What does this tell you about an earthworm's ability to grip the soil?

2.3 What external features allow you to determine the anterior and posterior ends of your specimen? The dorsal and ventral surfaces?

2.4 What external features provide a hint that earthworms are hermaphroditic?

Earthworm Internal Anatomy

12

Procedure

1. Review the Dissection Diagram to assist you in opening your specimen.
2. Insert the tip of your dissecting scissors into the dorsal surface along the midline of the earthworm 5–10 segments posterior to the clitellum. Make a shallow incision, progressing anteriorly, snipping a segment or two at a time. Remember to keep the point of your scissors angled up toward the dorsal body wall to avoid damaging the internal organs.
3. Lengthen the incision all the way to the tip of the first segment (the prostomium).
4. Use pins to hold the body wall open as you identify the internal organs. **HINT:** If you plan to use a dissecting microscope to view the internal anatomy of your earthworm, pin your specimen near one side of your dissecting pan.
5. Table 12-1 provides a review of the major organs of the earthworm covered in this dissection.

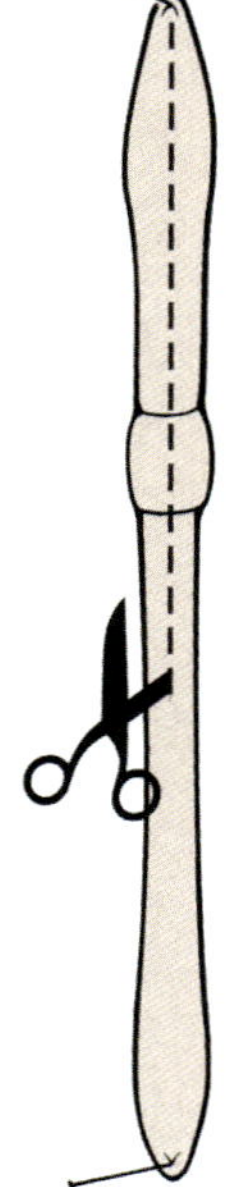

Dissection Diagram

1. Use small, sharp scissors to make a shallow incision along the dorsal midline of the body starting posterior to the clitellum and proceeding to the anterior tip of your specimen.
2. Gently open the specimen, using pins to spread the body.

The complete digestive system of the earthworm is the most prominent internal system and contains subregions specialized for different digestive functions—a resultant advantage of having a complete digestive system that moves food in only one direction. Soil is pulled into the mouth by pumping actions of the muscular **pharynx** (Figs. 12.5 and 12.6). The pharynx opens into the **esophagus**, which is usually hidden by the aortic arches and reproductive structures.

The esophagus empties into the thin-walled **crop**, where ingested soil is stored temporarily. The crop empties into the muscular **gizzard**, where mechanical breakdown occurs and organic matter is separated from the soil. This mixture then passes into the long **intestine**, where extracellular digestion and absorption of nutrients into the bloodstream occur. Undigested wastes are released through the anus. The efficiency of the intestine is improved by the increased absorptive surface area created by an inward fold in the intestinal wall known as the **typhlosole**. This fold is easily visible in a cross section of the earthworm.

Large, cream-colored **seminal vesicles** and smaller, cream-colored seminal receptacles are present in the anterior end between segments 9 and 15 and may obscure portions of the digestive and circulatory systems. The functions of these two organs will be covered, along with the other reproductive structures, later in this chapter.

All annelids possess a closed circulatory system consisting of a continuous network of veins, arteries, and capillary beds. Their blood contains hemoglobin, the same respiratory pigment found in our own blood cells. In earthworms, though, this pigment is not contained within cells but, rather, is dissolved directly in the blood plasma and occurs in much lower concentrations. Nonetheless, the presence of oxygen-binding molecules increases annelids' circulatory efficiency and allows them to live in stagnant mud on the ocean floor and in the soil of terrestrial environments where food is plentiful but oxygen is scarce.

Blood pressure in the vessels is maintained by the actions of five **aortic arches** ("hearts") located in segments 7–11. Dorsal and ventral blood vessels running along the length of the intestine are also visible. Blood is pumped anteriorly into the aortic arches by muscular contractions of the dorsal blood vessel and from there is sent through the ventral blood vessel to the cerebral ganglia, as well as posteriorly for distribution throughout the rest of the body.

Table 12-1 Internal Anatomy of the Earthworm, *Lumbricus terrestris*

Structure	Function
Mouth	Ingests soil
Pharynx	Muscular region of digestive system specialized for pumping in soil
Esophagus	Passageway between pharynx and crop
Crop	Thin-walled chamber where food is temporarily stored
Gizzard	Thick-walled, muscular chamber where soil is mechanically ground and usable organic materials are separated from indigestible materials
Intestine	Long tube occupying the majority of the body in which nutrients are absorbed into the bloodstream
Aortic arches ("hearts")	Specialized, muscular branches of the dorsal blood vessel that regulate blood pressure in the body
Dorsal blood vessel	Longitudinal blood vessel that returns blood to the aortic arches
Ventral blood vessel	Longitudinal blood vessel that distributes blood posteriorly to the body
Seminal vesicles	Cream-colored, lobed organs fastened ventrally, but extending dorsally around each side of the esophagus, that store maturing sperm
Testes (not visible)	Site of sperm production
Seminal receptacles	Ventrally located organs that receive sperm during copulation and store sperm until needed to fertilize eggs in cocoon
Ovaries	Site of egg production
Nephridia	Paired excretory organs found along the lateral margins of all but the most anterior and posterior segments; they release waste fluids out of the worm through small pores in the body wall
Septa	Thin, fleshy partitions between segments
Cerebral ganglia	Small, bilobed structure lying dorsal to the pharynx in segments 3 and 4; houses the majority of neural ganglia in the worm
Ventral nerve cord	Long, white "cord" located along the ventral surface of the body; contains large swellings of ganglia in each segment that handle the majority of coordination without intervention of the cerebral ganglia

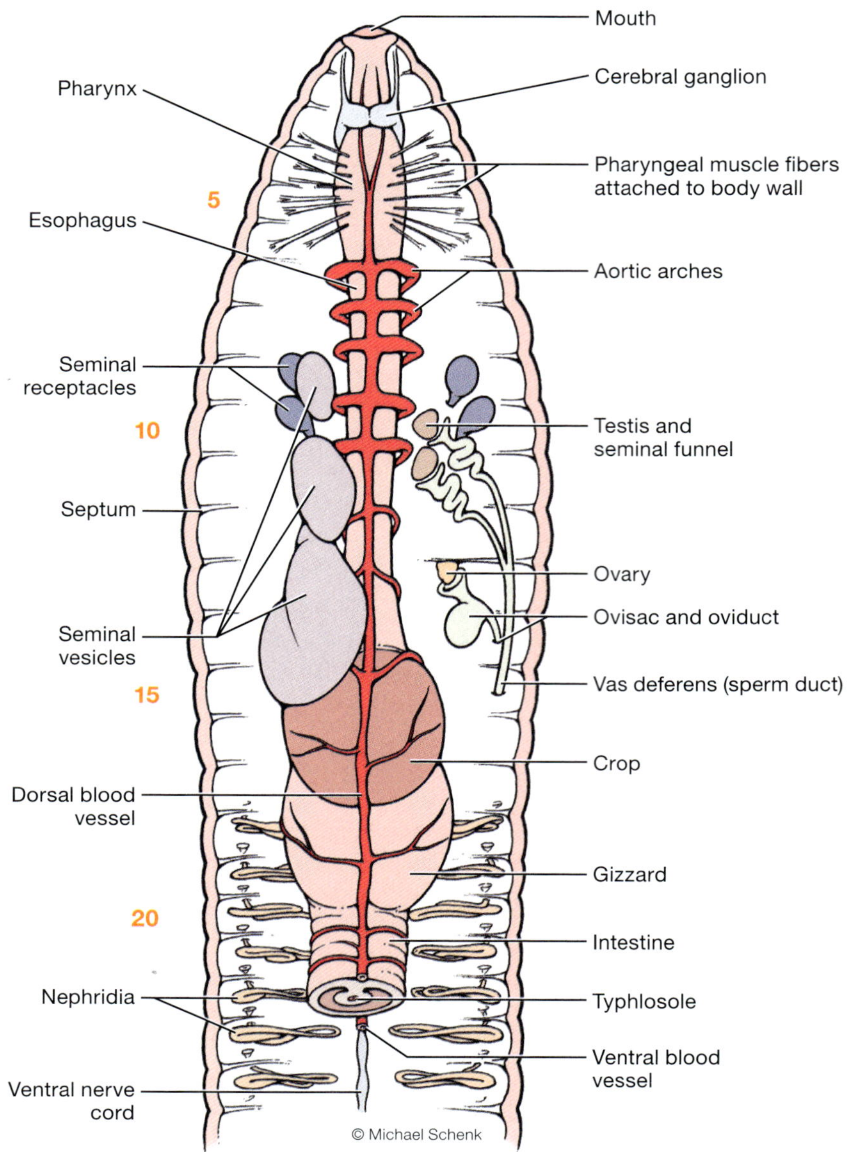

12.5 Internal anatomy of the anterior region of the earthworm, *Lumbricus*, dorsal view. Seminal vesicles have been removed on the right side to show underlying reproductive structures.

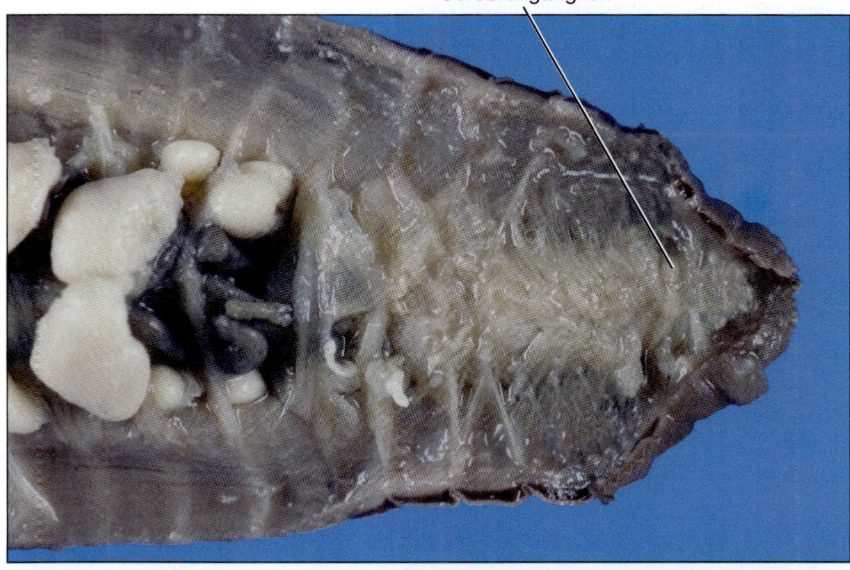

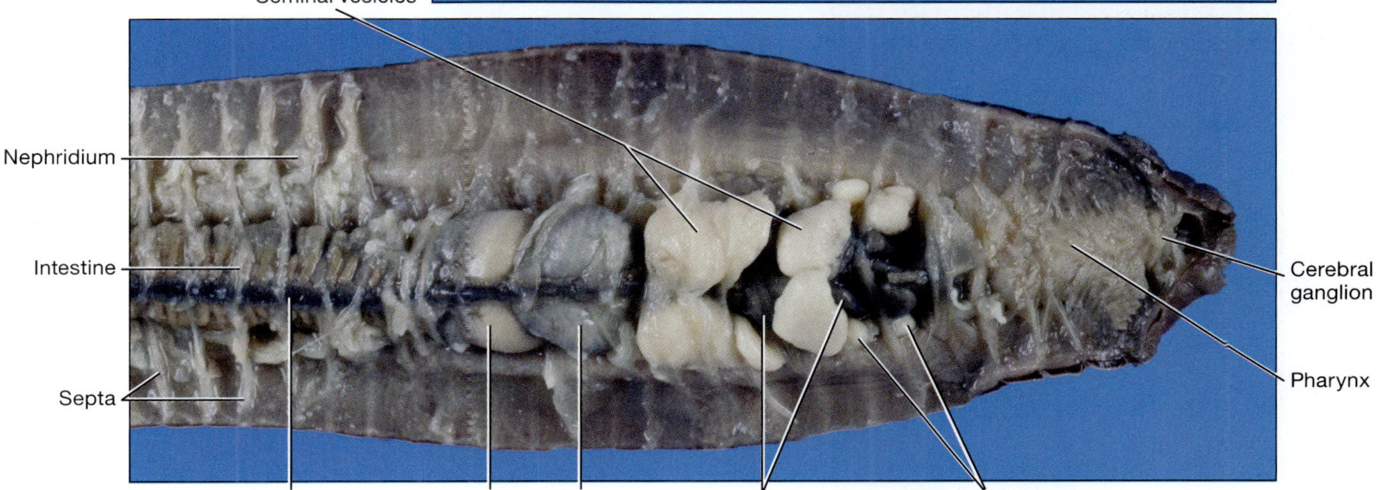

12.6 Dissection of anterior region of the common earthworm, *Lumbricus*. Inset photo shows close-up of the cerebral ganglia.

Excretion in earthworms is carried out in the **nephridia**—paired tubules in the coelomic space on either side of the intestine in the body segments posterior to the gizzard (Fig. 12.7). One end of the nephridium is an open ciliated funnel (nephrostome) that pulls coelomic fluid containing metabolic wastes into the bladder section of the tubule. Wastes pass through the tubule and exit through microscopic pores on the outer body wall (nephridiopores) while coelomic fluids are reabsorbed back into the body cavity. Notice that each pair of nephridia actually receive and filter the wastes from the adjacent anterior segment rather than from its own segment (Fig. 12.8).

The earthworm's nervous system consists of bilobed **cerebral ganglia** located at the anterior end of the animal, and a single ventral nerve cord (Figs. 12.7 and 12.8). As mentioned earlier, earthworms have no specialized respiratory structures. Carbon dioxide and oxygen diffuse freely across the moist surface of their skin directly into and out of the dorsal and ventral blood vessels lying just beneath the epidermis.

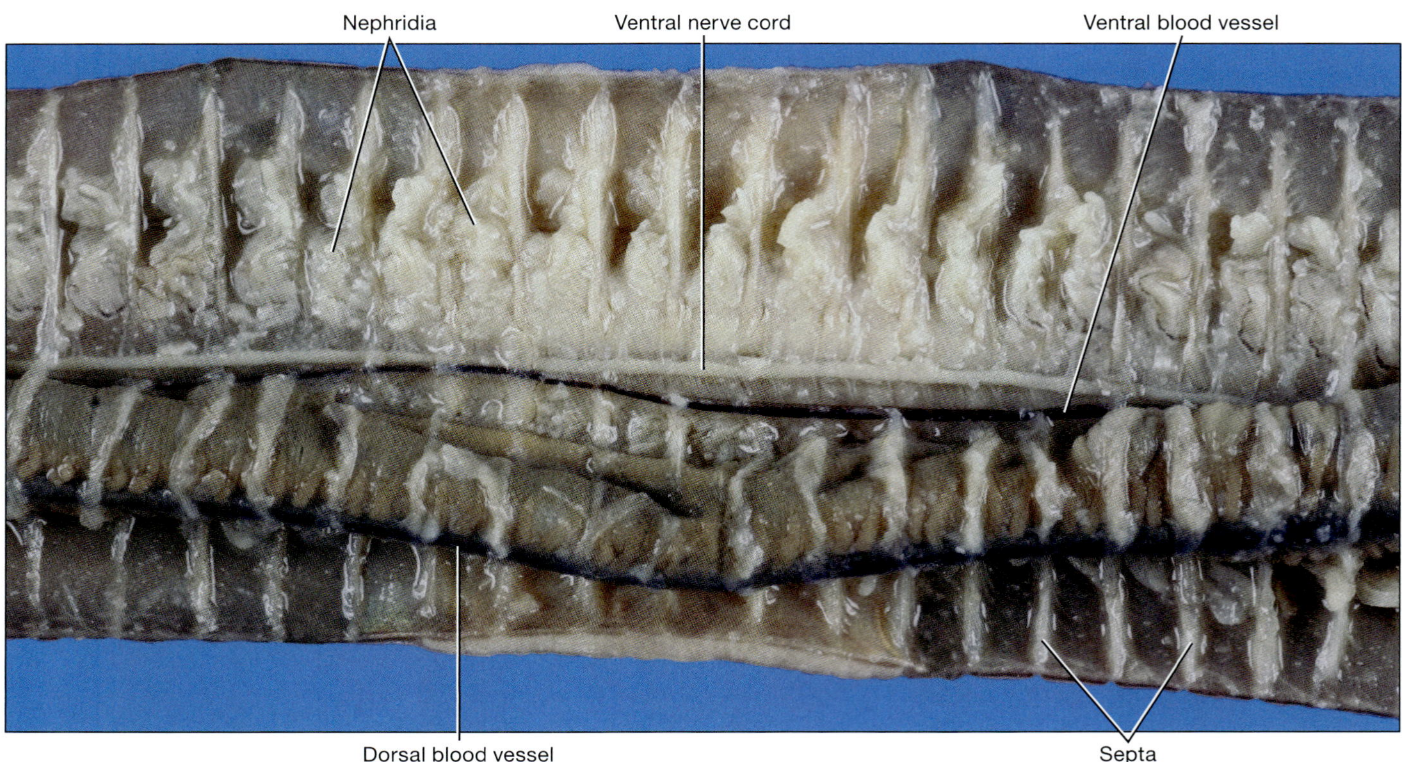

12.7 Dissection of the posterior region of the common earthworm, *Lumbricus*. Intestine has been reflected to the side to show underlying anatomy.

12

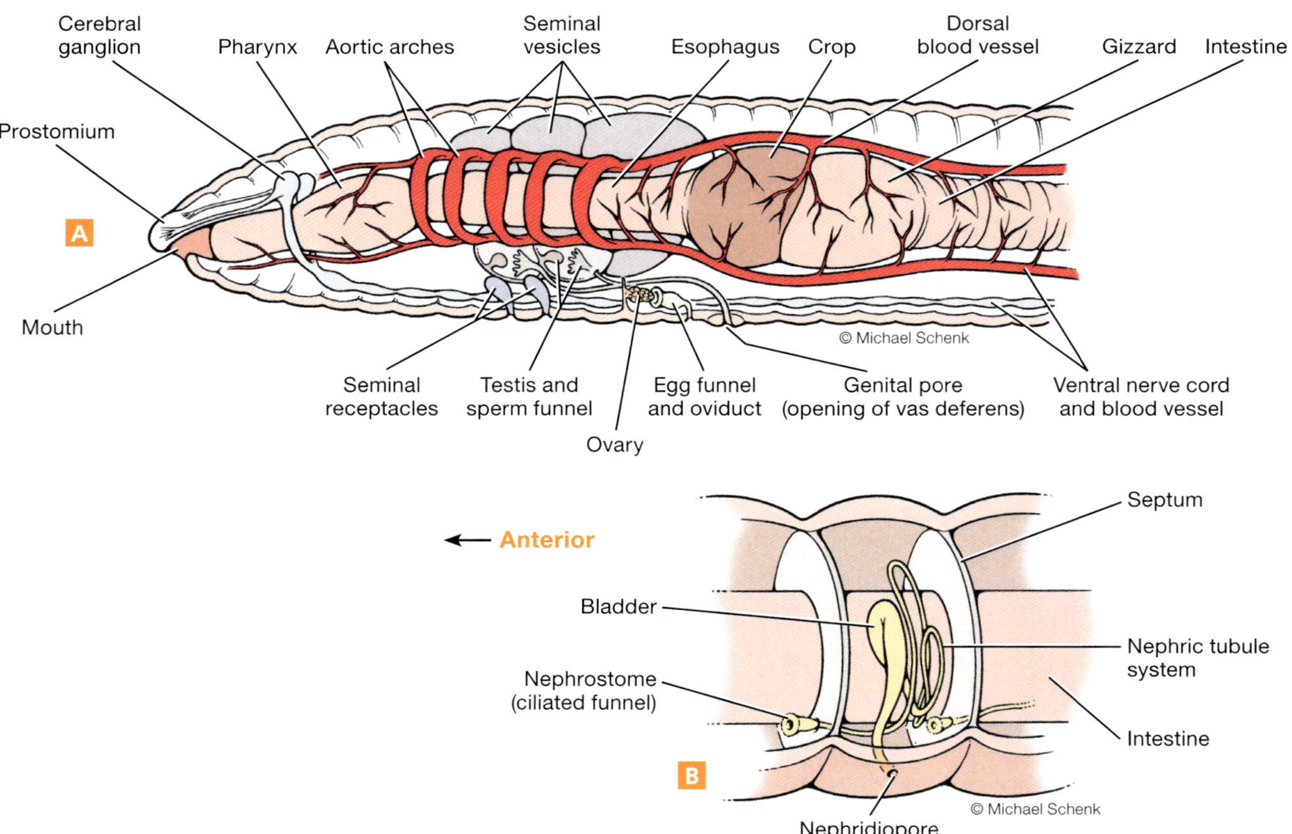

12.8 Internal anatomy of (A) anterior region of the earthworm, *Lumbricus*, lateral view, and (B) enlarged illustration of the nephridium, the excretory organ that removes metabolic wastes from the body.

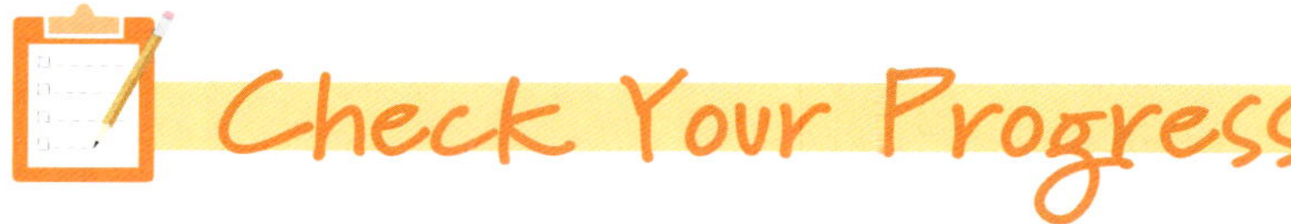

3.1 Are both male and female reproductive structures present in your specimen?

3.2 List several internal organs that are exemplary of segmentation in the earthworm.

3.3 Put the following digestive structures in the order in which food passes through them in the earthworm: intestine, esophagus, gizzard, anus, pharynx, mouth, crop.

Procedure

1 Next, examine a prepared slide of a cross section (c.s.) through the body of an earthworm using your compound microscope.

2 Locate the following organs and structures depicted in Figure 12.9: dorsal blood vessel, intestine, typhlosole, coelom, ventral nerve cord, epidermis, circular muscles, longitudinal muscles, nephridium, ventral blood vessel, and setae (setae may not be visible on every slide).

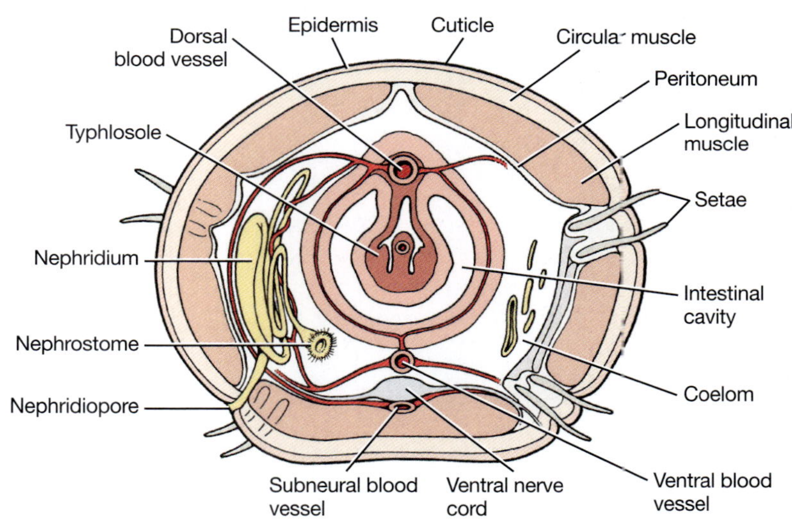

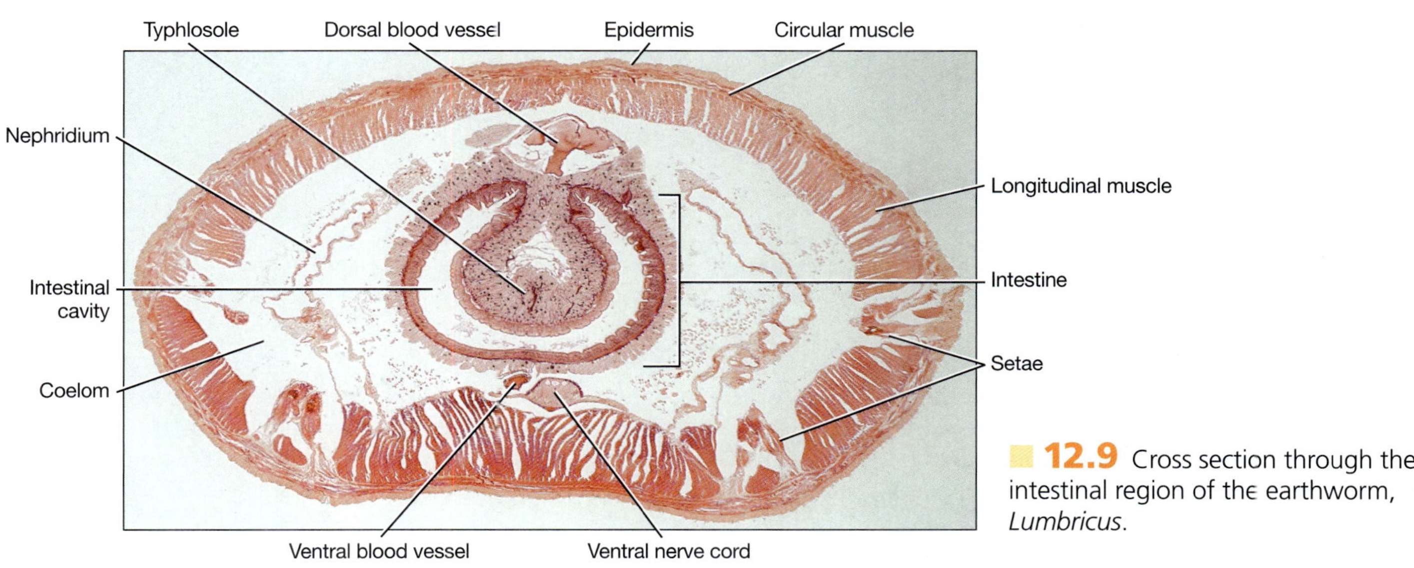

12.9 Cross section through the intestinal region of the earthworm, *Lumbricus*.

12

Earthworm Reproduction

Earthworms are **monoecious** (hermaphroditic), meaning that both male and female reproductive organs are found in the same animal. Mating occurs outside of their underground burrows at night, generally during warm, moist weather. Two worms align along their ventral surfaces with their heads pointing in opposite directions and secrete a slimy, mucous emission from each clitellum that holds them together during copulation (Fig. 12.10).

Sperm are discharged simultaneously from each worm and travel to the seminal receptacles of their mate along grooves in the ventral body surface. The worms separate, and each worm secretes a mucous band from its clitellum that forms a sticky cocoon that slides forward, picking up eggs from the egg sac and sperm from the seminal receptacles. Fertilization occurs within the cocoon, which slips off the anterior end of the worm and is deposited near the entrance to the worm's burrow. Eggs develop for two to three weeks before juvenile earthworms emerge.

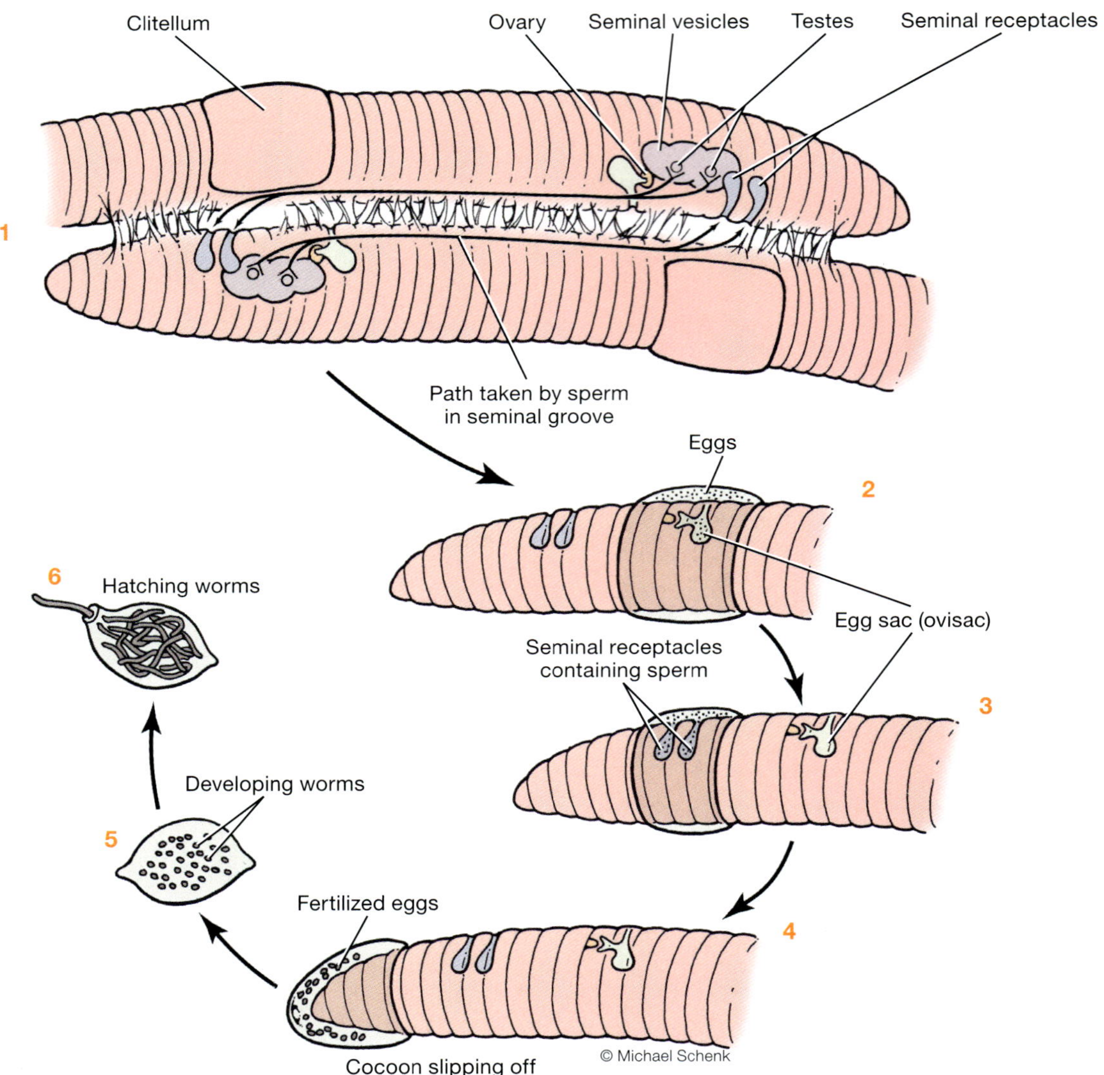

12.10 Copulation and formation of the egg cocoon in earthworms. (**1**) Earthworms align with heads pointing in opposite directions and exchange sperm. Worms separate and each worm secretes a mucous band from its clitellum that forms a cocoon and slides forward, (**2**) picking up eggs from the egg sac, and (**3**) then receiving sperm from the seminal receptacles. (**4**) Cocoon containing fertilized eggs slips off anterior end of worm and is deposited near entrance to worm's burrow. (**5**) Eggs develop for two to three weeks before (**6**) juvenile earthworms emerge.

Exercise 12-3 Leeches (Class Clitellata, Subclass Hirudinea)

Materials Needed

- living and preserved specimens of leeches
- living specimens of earthworms
- beaker
- pond water
- glass plates or petri dishes
- pan filled with moist sand or dirt
- prepared slide of *Hirudo* or other leech (w.m.)
- dissecting microscope

Leeches are members of the subclass Hirudinea (≈500 sp.) and represent an interesting assemblage of primarily freshwater annelids. Some leeches are scavengers. Other species prey on small, aquatic animals. Many leeches are ectoparasites of fish, and a few are bloodsucking ectoparasites of mammals. **Ectoparasites** are animals that attach themselves to the *outside* of the host organism to obtain their nourishment. Interestingly, medical practitioners have exploited this feature of leeches as far back as medieval times. Even today, doctors may attach leeches to patients to extract fluids that accumulate around injuries or to enhance the healing of incisions associated with surgeries. This may sound barbaric, but leeches are able to extract fluid much more efficiently and with less tissue damage than does hypodermic suction. These animals can quickly consume 5 to 10 times their original body weight in blood and body fluids!

Leeches do not crawl or burrow as other annelids do, but move either by swimming or "looping" like an inch worm. They lack setae and parapodia, and their external rings do not correspond to the pattern of internal segmentation. Like earthworms, leeches are hermaphroditic, and they possess a clitellum, but it is visible only during breeding seasons. Their most prominent and characteristic features are their suckers, used for attaching to and feeding off their hosts.

Procedure

1. Obtain a prepared slide of *Hirudo* or other leech species and examine it with a dissecting microscope.
2. Notice the prominent anterior and posterior **suckers** (Fig. 12.11). The mouth is located within the opening of the anterior sucker. The body is clearly segmented and you should be able to detect repeated elements within the body cavity—a defining feature of true segmentation.
3. You may be able to see outpockets of the intestine in each segment along with paired male and female reproductive structures.
4. Compare the internal anatomy of the leech with that of the sandworm and earthworm viewed earlier.
5. If living leeches are available in your laboratory, place one in a beaker of pond water and observe its movements.
6. Remove the leech from the beaker of water and place it on a wet glass plate or petri dish and observe its unique style of movement when out of the water. Describe this motion below.
7. Examine the living leech with a dissecting microscope to view its suckers more closely. Describe what you see inside each of the suckers.
8. If live earthworms are available in your laboratory, place one on your lab bench and observe its movements. By gently running your fingers along the sides of the earthworm's body you should be able to feel the bristly setae projecting from the lateral sides of the body. Unlike leeches, which lack setae, earthworms utilize these short bristles to obtain traction against their substrate as they crawl and burrow.
9. Experiment with different substrates to determine how they affect the ability of your earthworm to crawl. You may want to use a wet paper towel, a wet glass plate, a dry lab bench, or a shallow pan filled with dirt or moist sand.

12

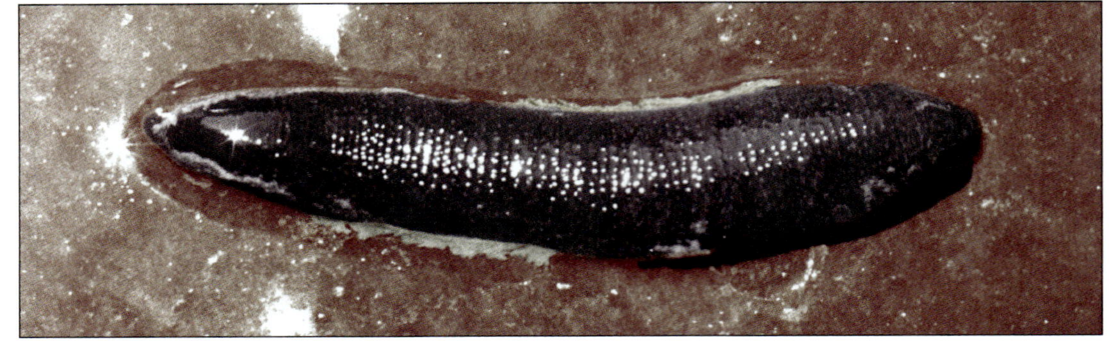

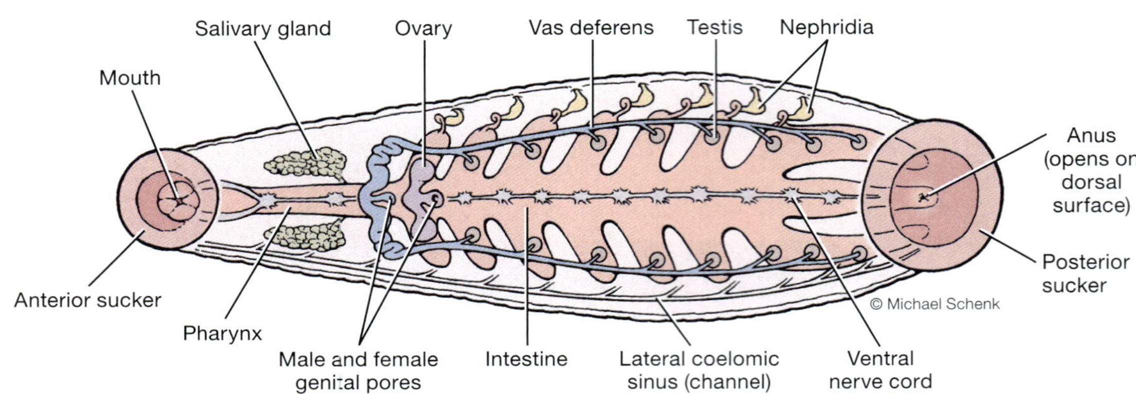

12.11 Leeches lack setae but have suckers for attaching and feeding off their hosts.

12

Check Your Progress

4.1 What anatomical similarities do you see among leeches, earthworms, and sandworms?

4.2 Do leeches possess setae? How do they move?

4.3 What is the function of the anterior sucker? The posterior sucker?

4.4 In what ways are the movements of leeches and earthworms similar? In what ways do their movements differ?

Exercise 12-4 Investigating Annelid Circulation and Locomotion

Materials Needed

- blackworms (*Lumbriculus variegatus*)
- springwater or distilled water
- petri dishes
- filter paper
- plastic dropper
- stopwatch or timer
- 1-inch long, straight human hairs (or paintbrush bristles)
- narrow rubber bands
- scissors
- applicator sticks
- ½-inch wide masking tape
- 5½ inch × ⅞ inch plastic weigh boats
- dissecting microscope

The freshwater oligochaete, *Lumbriculus variegatus* (often known as the blackworm or mudworm) is a common occupant of the shallow margins of ponds, marshes, and lakes throughout North America. Despite its wide distribution, its presence often goes unnoticed by the casual recreational user of such waters. Rarely exceeding 5 centimeters in length, this small, free-living annelid has proven to be easily obtained and cultured in the laboratory and can be used to illustrate a host of biological principles, including patterned regeneration of lost body parts, blood vessel pulsations, swimming and other locomotor reflexes, and giant nerve fiber action potentials.

Observation of Circulation

Annelids possess a closed circulatory system consisting of contractile vessels that channel blood through a contained series of arteries and veins. In this exercise you will see the closed circulatory system of the blackworm in action. In *L. variegatus* blood is pumped by rhythmic contractions of the dorsal blood vessel, moving the blood from the posterior end (where gas exchange occurs) toward the head. The blood is red and thus easy to distinguish from other body fluids due to the respiratory pigment **erythrocruorin**, a hemoglobin-like pigment that reversibly binds oxygen and carbon dioxide in much the same way that hemoglobin works in humans.

12

Procedure

1 Use a plastic dropper to place a blackworm on moist filter paper.

Note

Exposure to chlorinated water or soap residues will kill these worms quickly. Be sure you are using dechlorinated water or other invertebrate-safe culture and clean, chemical-free glassware and droppers.

2 Anterior and posterior regions are easy to distinguish because the head segments are blunter, wider, more maneuverable, and more darkly pigmented than posterior segments. In addition, the prostomium, muscular pharynx, and reproductive organs are contained within the first 8–10 anterior segments.

3 Position your specimen under the dissecting microscope and be sure you can differentiate the head and the posterior region.

4 You will notice two large blood vessels running longitudinally along the body. These are the dorsal and ventral blood vessels. The **dorsal blood vessel** is contractile and is responsible for the circulation of blood through the closed circulatory system. The **ventral blood vessel**, although large and easily visible, is not contractile.

5 Count the number of pulsations of the dorsal blood vessel for 1 minute, and record your finding below. Repeat this measurement four more times and average your results.

# of pulsations/minute ________ # of pulsations/minute ________

# of pulsations/minute ________ # of pulsations/minute ________

# of pulsations/minute ________ Average # of pulsations/minute ________

6 Closer inspection of the circulatory system should reveal numerous lateral branches of the dorsal blood vessel that transport blood to the organs and tissues within each segment.

7 In its natural setting, the blackworm uses its head to forage for aquatic vegetation and microorganisms while its posterior end projects upward and bends at a right angle, just breaking the surface tension of the water. This posture allows gases to be exchanged between the air and the pulsating dorsal blood vessel lying just beneath the body surface. Examine the dorsal blood vessel in the tail region of your specimen. Do you notice any anatomical or physiological adaptations in this region that would enhance gas exchange? Explain.

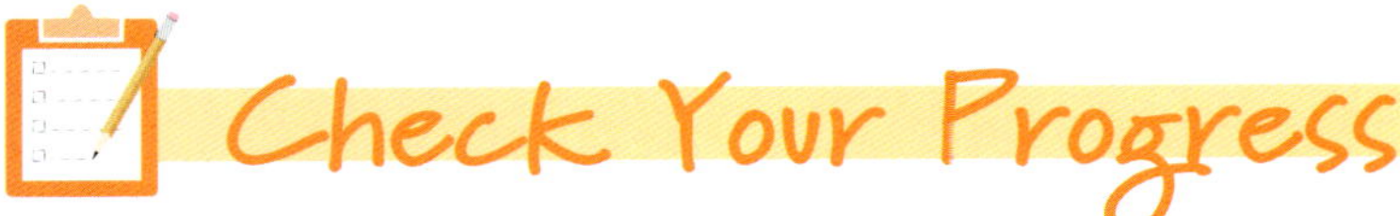

5.1 How did the pulsation rate of your specimen compare to others in your laboratory?

5.2 Collect pulsation rates from at least five other groups, and calculate a mean pulsation rate for blackworms in your laboratory.

12

5.3 What factors do you suspect could affect the pulsation rate of the dorsal blood vessel in blackworms?

Observation of Locomotion and Reflexes

No doubt you have seen an earthworm use peristaltic crawling to move across a sidewalk after a heavy rain or quickly recoil into its burrow when disturbed. Earlier in this chapter you had the opportunity to observe the locomotor patterns of earthworms and leeches. In this exercise you will see how blackworms utilize these same circular and longitudinal muscle layers to accomplish a surprisingly diverse array of acrobatic locomotor movements.

Procedure

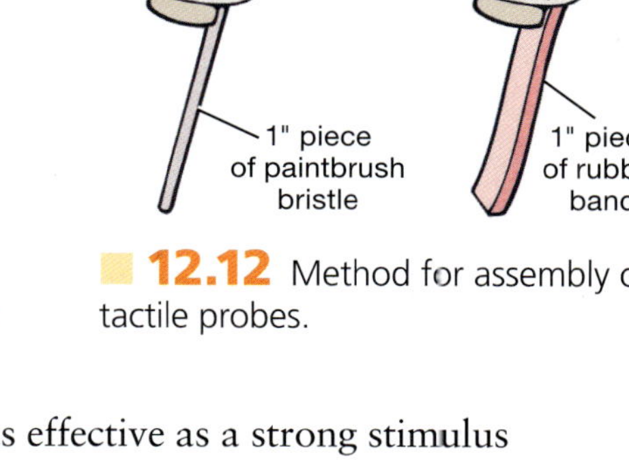

12.12 Method for assembly of tactile probes.

1. First, you will need to assemble a tactile probe. Cut a narrow rubber band into a 1-inch long, straight section and tape it to the end of a wooden applicator stick so approximately ½ inch of the rubber band protrudes past the end of the applicator stick (Fig. 12.12)
2. Fashion another tactile probe out of a 1-inch long, coarse, straight human hair or paint brush bristle using the same process as described in step 1 (Fig. 12.12).
3. Place a blackworm on moist filter paper.
4. Using the applicator stick with the rubber-band section, stroke the worm's posterior end lightly.
5. Record your observations in Table 12-2.
6. Using the same tactile probe, stroke the worm's head lightly, and record your observations.
7. Determine whether a weak stimulus (the tactile probe with the hair) is as effective as a strong stimulus (the tactile probe with the rubber band) in evoking crawling behavior.
8. In the space provided in Table 12-2, sketch the blackworm and label the anterior and posterior ends. On the sketch, depict all locations along the body where touching elicits *forward* crawling. Indicate where it fails to do so. Try touching at least six to eight different locations along the body, moving progressively forward from the posterior end to the anterior end. The result is a map of the touch sensory field for forward crawling.
9. Repeat this process to generate a touch sensory field map for rearward crawling by starting at the anterior end and working backward.

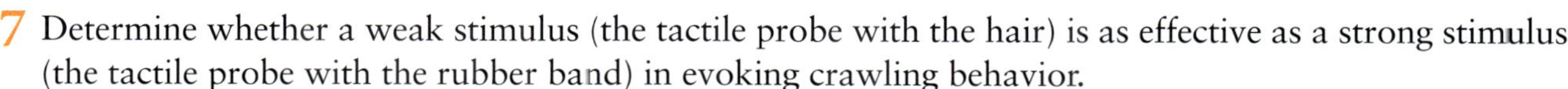

Table 12-2 Locomotor Properties of Blackworms in Moist, Terrestrial Environments

Treatment	Stimulus	Movement Evoked?	Direction of Movement	Description of Movement
Posterior end stroked	Strong			
Head stroked	Strong			
Posterior end stroked	Weak			
Head stroked	Weak			
Touch sensory field map for forward crawling				
Touch sensory field map for rearward crawling				

12

10 Construct two more tactile probes using 2-inch rubber bands and hair (or paint bristle) sections this time. Instead of leaving one end free, as before, form a loop with the rubber band and hair sections and tape the two free ends to the wooden applicator sticks. In each case, this will leave a loop approximately ½-inch long.

11 Transfer a fresh, unused worm to a plastic weigh boat containing 10–20 ml of springwater.

Note

For added strength, you may want to stack several weigh boats together. Clear petri dishes with white paper underneath may be used in place of weigh boats.

12 Using the tactile probe with the rubber loop, coax the blackworm to the center of the container.

13 After the blackworm has settled there, hold the tactile probe (with the rubber loop) at a 45° angle and quickly press the loop down on the blackworm's posterior end for a brief instant. If done properly, this should not injure or fragment the worm but, instead, induce swimming.

14 Record your observations in Table 12-3.

15 Repeat step 13, but this time press gently on the blackworm's head. Record your observations. In this case, the stimulus should not elicit swimming but, rather, a reversal response in which the blackworm quickly curls and then uncurls its body, causing a nearly 180° repositioning of its head and tail. Because blackworms cannot swim backward, reversal enables rapid repositioning of the head, presumably moving it away from the stimulus or threat and making possible subsequent forward swimming in the direction opposite the original stimulus.

16 As before, determine whether a weak stimulus (the probe with the hair loop) is as effective in evoking swimming or reversal behavior from both ends.

12

17 In the space provided in Table 12-3, sketch the blackworm, and label the anterior and posterior ends. On the sketch, depict all locations along the body where touching elicits *forward swimming*. Indicate where it fails to do so. As before, try to touch at least six to eight different locations along the body, moving progressively forward from the posterior end to anterior end to construct a map of the touch sensory field for forward swimming.

18 Repeat this process to generate a touch sensory field map for reversal behavior by starting at the head and working backward.

Table 12-3 Locomotor Properties of Blackworms in Underwater Environments

Treatment	Stimulus	Movement Evoked?	Direction of Movement	Description of Movement
Posterior end prodded	Strong			
Head prodded	Strong			
Posterior end prodded	Weak			
Head prodded	Weak			
Touch sensory field map for forward swimming				
Touch sensory field map for reversal behavior				

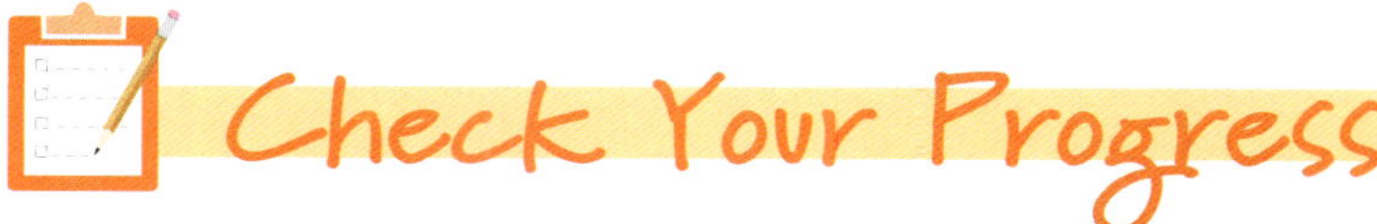

6.1 In both environments, were strong stimuli more effective than weak stimuli at evoking locomotor behaviors?

6.2 Did blackworms employ peristaltic movement for crawling? Is this more similar to locomotion in earthworms or in leeches?

6.3 Did blackworms employ peristaltic movement for swimming?

6.4 How would you describe the movements employed by blackworms for swimming?

12

Notes

12

Name ______________________ Section ______________

12 Questions for Review

1 Complete Summary Table A-1 in the Appendix, filling in the characteristics of annelids in the appropriate row.

2 List three major distinguishing features of annelids.

1. ______________________
2. ______________________
3. ______________________

3 What is a closed circulatory system?

____

4 How do the nephridia of annelids compare to the flame cells of flatworms?

____

5 Why must terrestrial annelids such as earthworms maintain a moist skin surface?

____

6 Match the taxonomic group with the correct organism or characteristic.

______ *Hirudo*	**a.** Class Polychaeta
______ *Lumbricus terrestris*	**b.** Subclass Oligochaeta
______ sandworms	**c.** Subclass Hirudinea
______ leeches	
______ free-living freshwater or terrestrial worms with setae	
______ ectoparasites with suckers for feeding; lack setae and parapodia	
______ marine worms with parapodia	

12

7 Describe the advancements in the specialization of the digestive tract of earthworms in comparison to the digestive tract of *Ascaris* (roundworm).

8 Discuss the difference in the role of the cuticle in the roundworm (*Ascaris*) and the earthworm (*Lumbricus*).

9 List two major features that roundworms and earthworms share in common *and* two that are found in earthworms but not in roundworms.

Roundworms and Earthworms:

1.

2.

Earthworms ONLY:

1.

2.

12

10 Discuss the differences between (1) fertilization in dioecious animals such as mammals, (2) self-fertilization in monoecious animals such as tapeworms, and (3) cross-fertilization in hermaphroditic (monoecious) animals such as earthworms.

Name ____________________ Section ____________

12 Laboratory Practical Challenge

The following evaluation is intended to replicate the method of assessment and type of questions you will be asked on lab practicals. If you have difficulty with these questions, you may benefit from additional review of the specimens provided in lab.

1 Identify the phylum and subclass of the specimen depicted below.

2 What is the function of the structure indicated by line A?

3 Identify the structure indicated by line B.

4 To what system does the structure indicated by line C belong?

12

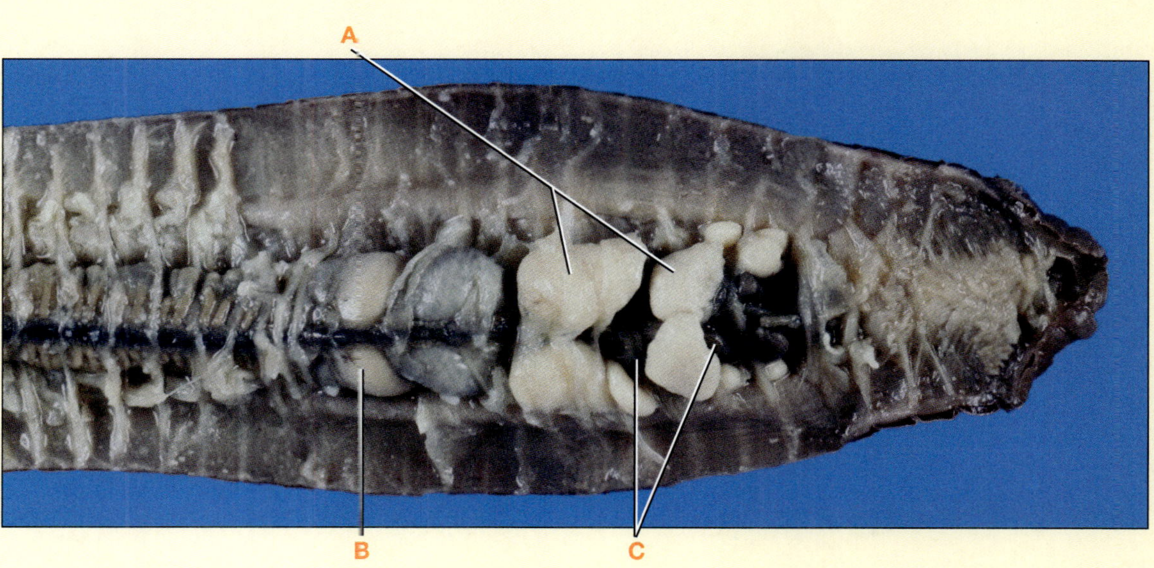

5 On the figure below, label the following structures: coelom, ventral blood vessel, intestine, and typhlosole.

6 To what system does the structure indicated by line D belong?

7 Identify the structure indicated by line E.

8 Identify the structure indicated by line F.

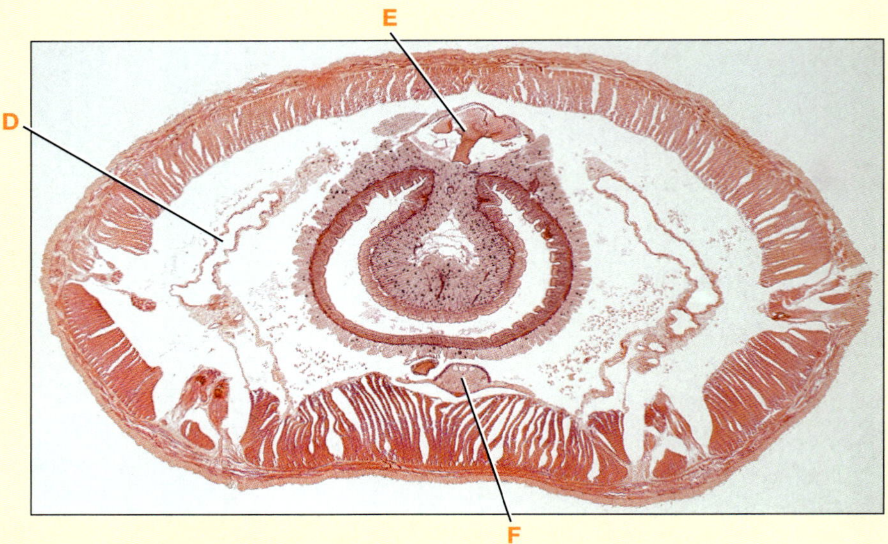

12

CHAPTER 13

Arthropods

After completing the exercises in this chapter, you should be able to:

1 List the distinguishing characteristics of the phylum Arthropoda.

2 Distinguish between the seven major classes of arthropods covered in this chapter.

3 Understand the evolutionary position of arthropods on the phylogeny of major animal phyla.

4 Discuss the evolutionary implications of jointed appendages and the ability to fly.

5 Identify the anatomy of a crayfish and be able to discuss the functions of each structure.

6 Identify the anatomy of a grasshopper and be able to discuss the functions of each structure.

7 Differentiate male and female crayfish and grasshoppers.

8 Categorize orientation behaviors in animals and discuss the evolutionary significance and possible origins of such behaviors.

The phylum Arthropoda is by far the largest in the animal kingdom, containing approximately 1.1 million named species and an estimated 10–30 million species overall! More than two-thirds of all named *animal* species on Earth are arthropods, and new species are literally being discovered every day, adding to the total of named species.

Despite their incredible diversity, all arthropods share two key characteristics in common: segmented bodies and jointed appendages. The latter trait is the inspiration for the phylum name. Although arthropod bodies are segmented like those of annelids, the individual segments often exist only during embryological development, and many eventually fuse into functional groups. The jointed appendages may be highly specialized, and their numbers often are reduced in more advanced members of this phylum. In arthropods such as the crayfish and grasshopper, jointed appendages have been modified into legs, pinchers (chelipeds), swimmerets, antennae, wings, and complex arrays of mouthparts (Fig. 13.1).

Arthropods also possess a hard **exoskeleton** made of chitin and protein that is secreted by and remains fused to the epidermis. Though the segments may appear to be separate, the exoskeleton is really a continuous structure with overlapping folds (Fig. 13.2). It is rigid in certain spots and flexible in others, allowing the body to bend and the appendages to move. The exoskeleton provides support for the body, provides an anchor for muscle attachment, protects the animal from predators, and impedes water loss. As the body grows larger inside, the old exoskeleton is shed periodically through a process called **ecdysis** (molting), in which the soft, new exoskeleton is secreted and fixed in place before the old shell is shed. You may remember from Chapter 10 that arthropods are grouped with roundworms in the lineage Ecdysozoa due to this shared trait (Fig. 13.3).

To circumvent the obvious logistical problem of having a larger exoskeleton inside a smaller one, the soft, new exoskeleton folds upon itself as it is being produced. After the old shell is shed and the body is free of its former constraints, the new, larger shell slowly expands to its final size and hardens in place. One downside to having a skeleton on the *outside* of the body is that it limits body size. The larger the body inside, the thicker and heavier the exoskeleton must be to support the body weight and bear the strain of muscular contractions. Large exoskeletons would simply be too heavy to move around. For this reason, terrestrial arthropods rarely are larger than a few inches long. Even aquatic arthropods, which gain some relief from this constraint because of the buoyancy of water, rarely grow larger than a few feet.

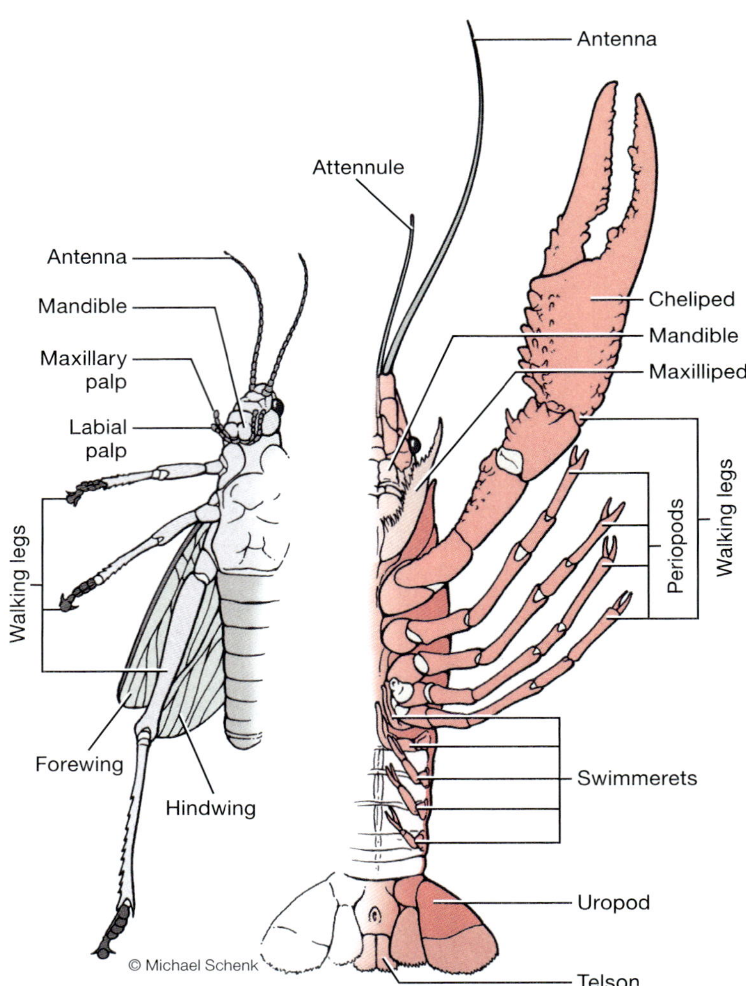

13.1 Jointed appendages of arthropods have been extensively modified over millions of years into highly specialized structures that best equip each species for its particular niche.

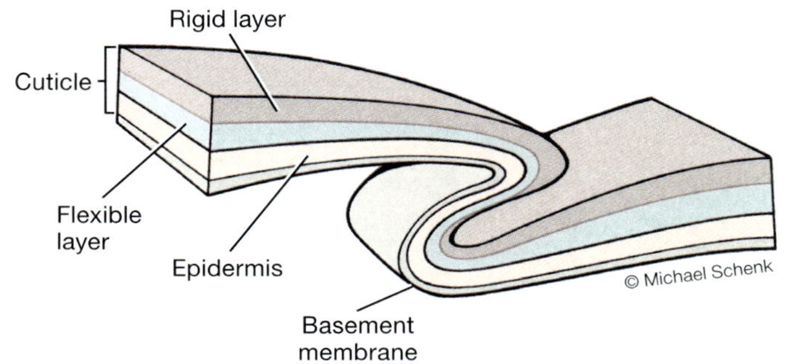

13.2 Segmented body plates are a characteristic of all arthropods.

13

In an animal encased in such a rigid suit of armor, the coelom can play no major role in locomotion, and in the course of arthropod evolution, the coelom has become greatly reduced. The main body cavity is instead a **hemocoel** (consisting of blood-filled sinuses in the tissues), comprising part of the **open circulatory system** that is characteristic of this group. The arthropod nervous system resembles that of annelids, consisting of a **dorsal brain** and a single **ventral nerve cord**. As in annelids, the ventral ganglia control much of the animal's behavior, and many arthropods can continue to carry out functions such as eating, moving, and copulation with the brain removed from the body. Most groups have specialized respiratory organs, varying from **gills** in crustaceans, to **book lungs** in arachnids, to **tracheae** and **spiracles** in insects. They also have specialized excretory organs for eliminating metabolic waste—the most common of which are the **Malpighian tubules** found in insects.

Providing a complete survey of the arthropod phylum in just one laboratory period would be an impossible task. Instead, we will examine specimens from seven major classes, comprising five different subphyla, and concentrate on the crayfish and grasshopper as representatives of a "typical" aquatic and terrestrial arthropod. The distinguishing characteristics of the seven major classes of arthropods presented in this chapter are summarized in Table 13-1. Three of these classes, Arachnida, Insecta, and Malacostraca, include well in excess of 95% of all arthropod species.

Evolutionary History

With the evolution of early protostomes, many major innovations in body structure had been established: bilateral symmetry, three primary germ layers, a true coelom, and segmentation in many groups. Early arthropods that sprang off the Ecdysozoan line took these existing traits and added two key elements: **jointed appendages** and an exoskeleton (Fig. 13.3). Jointed appendages came to be used for walking, flying, feeding, reproducing, and sensing the environment, while the exoskeleton provided protection, support, an anchor for muscles, and a barrier to minimize water loss.

As their numbers suggest, arthropods are arguably one of the most successful group of animals ever to occupy the planet. Part of their unsurpassed success stems from

Table 13-1 Major Subphyla and Selected Classes of Phylum Arthropoda

Subphylum	Class and Representative Animals	Characteristics
Trilobitomorpha	**Trilobita** (≈ 20,000 extinct sp.) Trilobites	All extinct; lived 542–251 mya; distinct head, thorax, and tail; body divided into three lobes by two longitudinal furrows; biramous appendages
Chelicerata	**Merostomata** (4 sp.) Horseshoe crabs	Lack antennae and mandibles; first pair of appendages are chelicerae, second pair of appendages are pedipalps, four pairs of walking legs; spiked telson and platelike gill opercula present on abdomen; marine
	Arachnida (≈ 74,000 sp.) Spiders, scorpions, ticks, mites, tarantulas	Lack antennae and mandibles; first pair of appendages are chelicerae, second pair of appendages are pedipalps, four pairs of walking legs; no abdominal appendages or telson; primarily terrestrial
Myriapoda	**Chilopoda** (≈ 2,500 sp.) Centipedes	Cylindrical body divided into head and segmented trunk; one pair of legs per body segment; one pair of antennae; poison fangs present; terrestrial
	Diplopoda (≈ 10,000 sp.) Millipedes	Cylindrical body divided into head and segmented trunk; two pairs of legs per body segment; one pair of antennae; poison fangs absent; terrestrial
Crustacea	**Malacostraca** (≈ 20,000 sp.) Crayfish, lobsters, crabs, shrimp, isopods	Two pairs of antennae, biramous appendages, one pair of mandibles, two pairs of maxillae; marine, freshwater, or terrestrial
Hexapoda	**Insecta** (≈ 1,100,000 sp.) Insects	Body divided into head, thorax, and abdomen; three pairs of legs on thorax; one pair of antennae; primarily terrestrial

the fact that they were the first animals to inhabit land, predating other familiar ancient animals such as dinosaurs by several hundred million years! During the evolution of arthropods, departure from the ancestral aquatic lifestyle favored the development of characteristics that permitted successful adaptations to the many ecological hurdles associated with terrestrial living, such as the need for stronger support systems, smaller body size, and reproductive, skeletal, digestive, excretory, and respiratory systems that conserved water. Also, arthropods were the first animal group to evolve the ability to fly, giving them access to a three-dimensional terrestrial landscape devoid of any other competitors.

Between 440 and 410 million years ago, arthropods gradually moved onto a previously unexploited terrestrial landscape that simultaneously was being populated by vascular plants, promoting relationships between plants and arthropods that still exist today. Virtually every kind of plant is eaten by one or more species of insect, and insects are vectors of pollen for most flowering plants. This close association accelerated adaptive radiation of arthropods throughout the terrestrial landscape. Since the beginning of their long evolutionary history, arthropods have spread through every terrestrial, aquatic, and aerial habitat imaginable and have had a profound impact on the evolution of numerous other species. For a more complete phylogeny depicting the evolutionary relationships among the major animal phyla covered throughout this laboratory guide, review Figure A.1 in the Appendix.

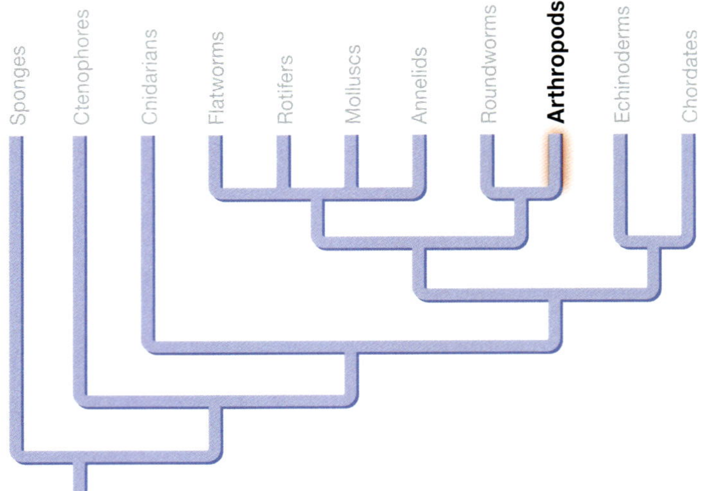

13.3 Arthropods are grouped with roundworms into the Ecdysozoan assemblage based on shared characteristics that indicate close evolutionary relationships.

13

Exercise 13-1 Trilobites (Subphylum Trilobitomorpha, Class Trilobita)

Trilobites were one of the most prominent marine arthropods of the Paleozoic era (570–250 mya) and have proven to be one of the most significant groups for understanding arthropod evolution. They first appeared during the Cambrian explosion and their extinction at the end of the Permian period more than 300 million years later is one of the defining events that scientists use to mark the end of the Paleozoic era. They are one of the most diverse groups of extinct animals preserved in the fossil record, with 10 recognized orders containing more than 20,000 named species. Everything we know about trilobites we have learned from fossils—no one has ever seen a living trilobite. Yet the incredible wealth of information scientists have gathered from studying fossilized trilobites is a testament to how valuable the study of fossils can be.

Trilobites are among the earliest arthropods to appear in the fossil record and thus provide scientists with direct evidence of what the first arthropods looked like. Surprisingly, many features of modern arthropods were already present in even the earliest trilobites. Fossils dating back more than 520 million years reveal that trilobites were marine animals with jointed appendages, chitinous exoskeletons, complex nervous systems, and highly flexible, segmented bodies divided into distinct regions (Fig. 13.4). They had complete digestive systems with specialized internal organs, compound eyes, single pairs of antennae, and they reproduced sexually. They crawled along the seafloor or within reefs, scavenging on organic debris or preying on smaller invertebrates. They dug into ocean sediments to hunt for food and to conceal themselves from predators. Fossilized trilobites exhibit an astonishing diversity in size and body morphology and species range in length from as small as 1 to 2 millimeters to 72 centimeters (nearly 2.5 feet!), with a typical size range of 3 to 10 centimeters (1.2 to 4 inches).

Because of their widespread distribution, vast numbers, spectacularly preserved details (often showing soft body parts), and species diversity, trilobite fossils have played a vital role in estimating the rate of speciation during the Cambrian explosion, were fundamental in developing the concept of punctuated equilibrium as a mechanism of evolution, and are used today by geologists as stratigraphic markers to date the age of rocks in which they are found.

13.4 Trilobites like this fossil of *Modicia typicalis* are extinct marine arthropods that lived throughout the Paleozoic era (570–250 mya).

Exercise 13-2 Horseshoe Crabs (Subphylum Chelicerata, Class Merostomata)

Materials Needed

- ❑ preserved horseshoe crab (*Limulus*)
- ❑ dissecting tools
- ❑ dissecting pan
- ❑ latex gloves

Horseshoe crabs represent a living relic among arthropods—modern animals that have remained virtually unchanged in the last 220 million years, and date back more than 425 million years in the fossil record. Along with arachnids, they are members of the subphylum Chelicerata; they lack antennae and mandibles and, instead, their first pair of appendages is modified into **chelicerae.** Horseshoe crabs are primarily nocturnal and swim on their backs in search of small molluscs and worms upon which to feed. They use their walking legs to scurry along the ocean floor and often nestle down in the sand to hide, leaving only their two compound eyes protruding above the sediment to keep a watchful eye for danger.

13

Procedure

1 Examine a preserved specimen of the horseshoe crab, *Limulus*. The leathery, horseshoe-shaped exoskeleton of this animal is divided into a **cephalothorax** (fused head and thorax) and an **abdomen** (Fig. 13.5).

2 On the dorsal side of the cephalothorax, locate the lateral **compound eyes** and the simple eyes. Their compound eyes are true, image-forming eyes, but their simple eyes can detect only shadows and differentiate between light and dark areas. Along its edge the abdomen has six pairs of spines and a long, spiked **telson** extending from the posterior edge.

3 Turn the horseshoe crab over and observe the ventral surfaces of the cephalothorax and abdomen. Notice that the cephalothorax contains several pairs of modified appendages, including chelicerae, pedipalps, and walking legs. The small, inward-folding chelicerae are used to hold and chew food. The abdomen also has numerous pairs of highly modified appendages—platelike structures called gill opercula that conceal the delicate **book gills** underneath. The beating of the gill opercula aids in swimming and aerates the book gills.

13

13.5 Horseshoe crab, *Limulus*, commonly found in shallow waters along the Atlantic coast: (**A**) dorsal surface, (**B**) ventral surface, and (**C**) illustration depicting arrangement of gill opercula and book gills. Modern horseshoe crabs are quite similar in appearance to ancient arthropods that lived hundreds of millions of years ago.

Check Your Progress

1.1 How many compound eyes do horseshoe crabs possess? How many simple eyes?

1.2 What do you suppose is the purpose or function of the telson?

1.3 How many pairs of chelicerae do horseshoe crabs possess? How many pairs of pedipalps? How many pairs of walking legs?

1.4 How many opercula and book gills are present on your specimen?

Exercise 13-3 Spiders, Scorpions, Ticks, and Mites (Subphylum Chelicerata, Class Arachnida)

Arachnids are another class of arthropods in the subphylum Chelicerata. Unlike the class Merostomata, however, which has only four extant species, the class Arachnida is an incredibly large group with more than 74,000 named species and displays tremendous diversity in appearance. Like all chelicerates, arachnids lack antennae and the cephalothorax bears a single pair of chelicerae, a single pair of pedipalps, and four pairs of walking legs. Yet, they exhibit enormous anatomical variation around these standard features. Spiders, tarantulas, daddy longlegs, ticks, mites, chiggers, and scorpions all belong to this class of arthropods (Fig. 13.6).

Arachnids are some of the oldest living arthropods and were among the first arthropods to inhabit land. Fossilized scorpions nearly identical to those alive today have been found in rock dating back to the Silurian period, 425 million years ago.

Most arachnids are carnivorous (with the exception of mites) and feed only on liquefied food, which they often digest externally by injecting powerful enzymes into their prey with their fangs (modified chelicerae). They have been portrayed inaccurately as sinister animals with a deadly reputation—generating unreasonable fear among humans for centuries. In reality, most arachnids are timid creatures, biting only in defense, and then usually only after provocation.

Nearly all arachnids are terrestrial and have specialized respiratory structures called book lungs for breathing. Small slits along the ventral surface of the abdomen allow air to enter a hollow compartment filled with numerous thin folds of the body wall that resemble pages of a book (hence the name). One side of each "page" is exposed to the air, and the inner layer of the page is bathed in hemolymph (blood), allowing gas exchange across the thin, moist surface.

Beach spider, *Argiope appensa*

Garden spider

Ticks, within the family Ixodidae: **A** dorsal view, and **B** ventral view

Rose-haired tarantula, *Grammostola cola*

Scorpion, *Pandinus*

13.6 Spiders, ticks, tarantulas, and scorpions are all members of the class Arachnida and possess four pairs of walking legs.

Exercise 13-4 Centipedes (Subphylum Myriapoda, Class Chilopoda)

Centipedes are carnivorous predators that feed on worms and insects and may be found hiding under rocks or logs or scurrying about the forest floor in search of prey. The body of these long slender animals is divided into a head and a trunk. They have one pair of legs per body segment, a single pair of antennae on the head, and a pair of appendages modified into poison fangs (Fig. 13.7). Used both for defense and also to subdue their prey, their poison often is quite toxic to humans, making the centipede bite extremely painful. Centipedes, along with millipedes, are classified in the subphylum Myriapoda. The term *myriapod* means "many-footed," referring to their multi-segmented trunks with paired jointed appendages on all trunk segments.

13.7 Garden centipede, *Lithobius*. Notice the single pair of legs per body segment characteristic of the class Chilopoda.

Exercise 13-5 Millipedes (Subphylum Myriapoda, Class Diplopoda)

Millipedes are slow, docile herbivores that feed on decaying vegetation. They are more secretive than centipedes and prefer dark, moist areas under logs, stones, or leaf litter. Like centipedes, their cylindrical body is divided into a head and a multi-segmented trunk (Fig. 13.8). Their characteristic feature is two pairs of legs per body segment—a condition resulting from the fusion of adjacent pairs of segments during development to produce double segments. They have a single pair of antennae but lack the poison fangs of centipedes and instead rely on foul-smelling, chemical secretions emitted through pores at the base of their legs to ward off potential predators. Rather timid in disposition, and without any other defenses, millipedes often coil into a ball when disturbed.

13

13.8 Sonoran desert millipede, *Orthoporus ornatus*. Notice the two pairs of legs per body segment characteristic of the class Diplopoda.

Exercise 13-6 Crustaceans (Subphylum Crustacea, Class Malacostraca)

Materials Needed

- ❑ preserved crayfish (*Cambarus*)
- ❑ dissecting tools
- ❑ dissecting pan
- ❑ dissecting microscope
- ❑ latex gloves
- ❑ watch glass or small petri dish

The crustaceans (subphylum Crustacea) belong to a large and diverse group of primarily aquatic arthropods that includes crayfish, lobsters, shrimp, crabs, pill bugs, barnacles, copepods, and krill (> 67,000 sp.) (Fig. 13.9). The largest class within this subphylum is the class Malacostraca, with more than 20,000 named species. The crayfish is one of the few freshwater members of this class and serves as an excellent model for studying the magnificent adaptations that aquatic arthropods have developed. More than 350 different species of crayfish are found in the United States, and they are of enormous ecological and economic importance.

Peppermint shrimp, *Lysmata wurdemanni*

Ghost crab, *Ocypode ceratophthalmus*

Freshwater crayfish, *Cambarus*

13.9 Shrimp, crabs, and crayfish belong to the class Malacostraca. Two pairs of antennae, biramous appendages, and a fused cephalothorax are distinguishing characteristics of this class.

Crayfish External Anatomy

Procedure

1 Obtain a preserved specimen of *Cambarus*, the freshwater crayfish. Examine the external features of your crayfish and refer to Table 13-2 as we discuss the anatomy.

The crayfish body is divided into two main regions, the cephalothorax and the abdomen, and appendages are present on both body regions. A large extent of fusion of segments occurs during embryonic development, and appendages become modified for various functions. Extending anteriorly from the cephalothorax is a pointed rostrum, and under the rostrum are a pair of stalked, compound eyes and two pairs of **antennae** (Fig. 13.10). The longer pair is true antennae, and the shorter, branched pair is called antennules. Both types contain sensory organs that function in touch and taste, but the antennules also provide information on equilibrium for the crayfish.

2 Lay your crayfish on its dorsal side to examine the mouthparts visible on the ventral surface (Fig. 13.11). One pair of mandibles, used for chewing food, two pairs of maxillae, used for holding food and bailing water from the gill chamber, and three pairs of maxillipeds, used for holding, tasting, and manipulating food, are all present near the mouth.

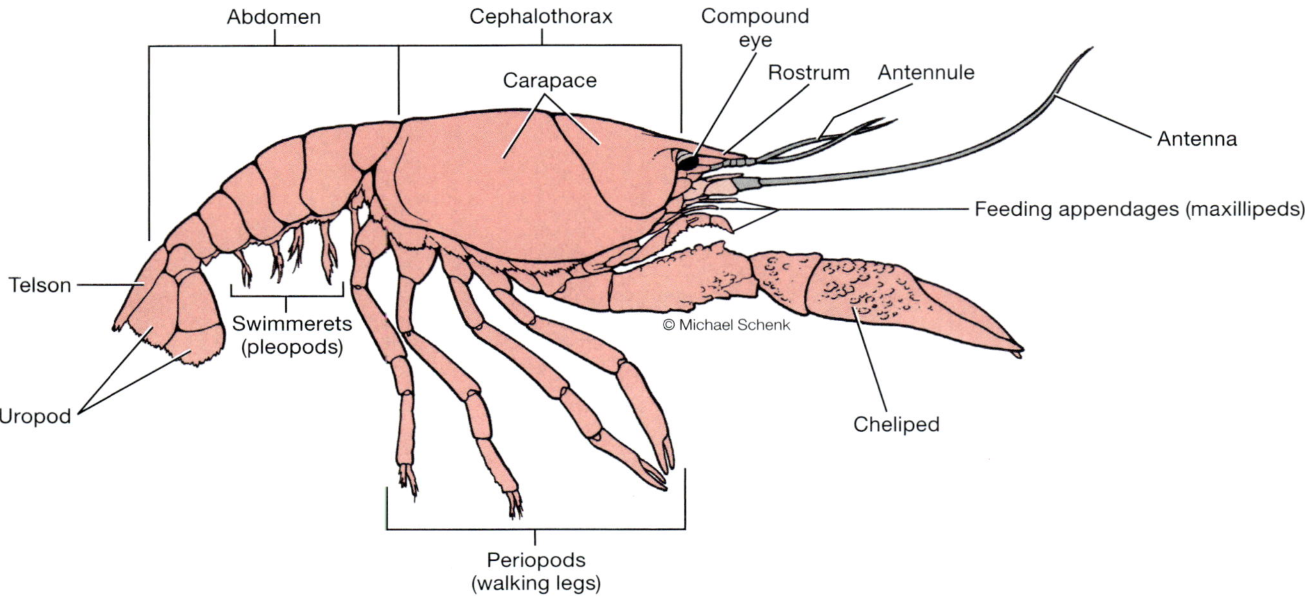

13.10 Common freshwater crayfish, *Cambarus*, lateral view.

3 Next locate the large pinchers, or **chelipeds** (Figs. 13.10 and 13.12). These hallmark appendages of crayfish are used both for defensive purposes and also for capturing and killing prey. Technically, they are considered the first pair of walking legs, even though they are not utilized for this function. Posterior to the chelipeds are several additional pairs of walking legs, also known as periopods, and appendages modified for swimming known as swimmerets, or pleopods. How many pairs of walking legs are present on your specimen (including the chelipeds)? How many pairs of swimmerets (pleopods)?

The pair of appendages on the last segment is flattened into two uropods. Along with the telson, they make up the tail fan that is used for swimming backward.

13.11 Common freshwater crayfish, *Cambarus*, ventral view of feeding appendages. The two pairs of maxillae are not identified in the photo.

4 With a pair of forceps, grasp a swimmeret at its base. While pulling on the swimmeret, shake it gently and pull it loose. Place the appendage on the stage of your dissecting microscope.

Notice that the distal end of each swimmeret is **biramous**, meaning it has two branches. Most of the crayfish appendages exhibit some degree of branching. The biramous nature of appendages is a specific characteristic of the subphylum Crustacea. In the male crayfish the first two pairs of swimmerets are modified for transferring sperm to the female during copulation. They are larger than the other swimmerets and are called copulatory swimmerets (Fig. 13.13). Sperm ducts may be visible at the base of the fifth pair of walking legs. Female swimmerets develop long, hairlike setae on the edges that attach eggs to the swimmerets during the egg-laying season. After the eggs are laid (via the **oviducts** at the base of the third pair of walking legs) and are attached to the swimmerets, they are brooded by the female until they hatch. The female's **seminal receptacle** is located between the fifth pair of walking legs, often obscured from view by the ends of the folded swimmerets (Fig. 13.13).

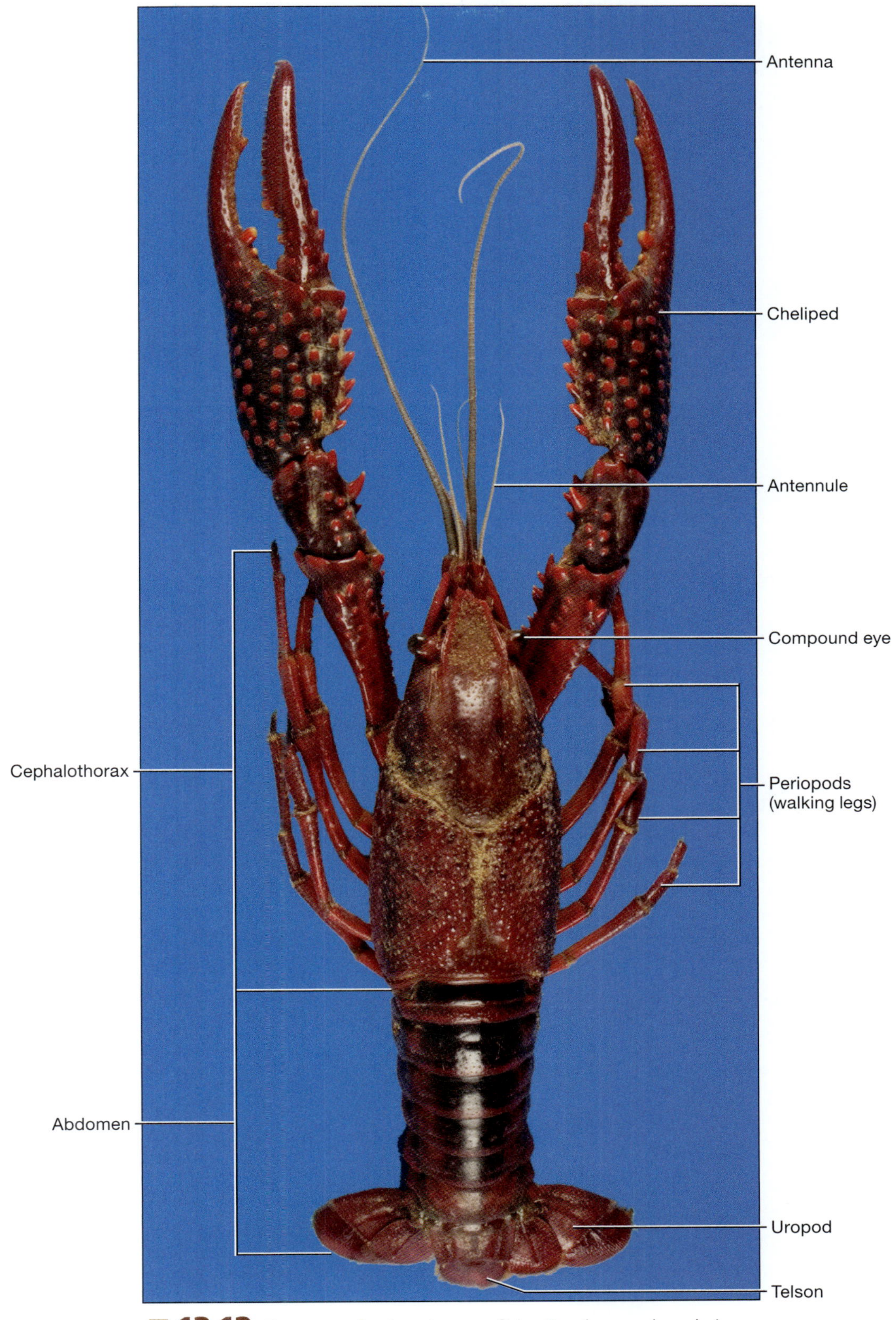

13.12 Common freshwater crayfish, *Cambarus*, dorsal view.

13

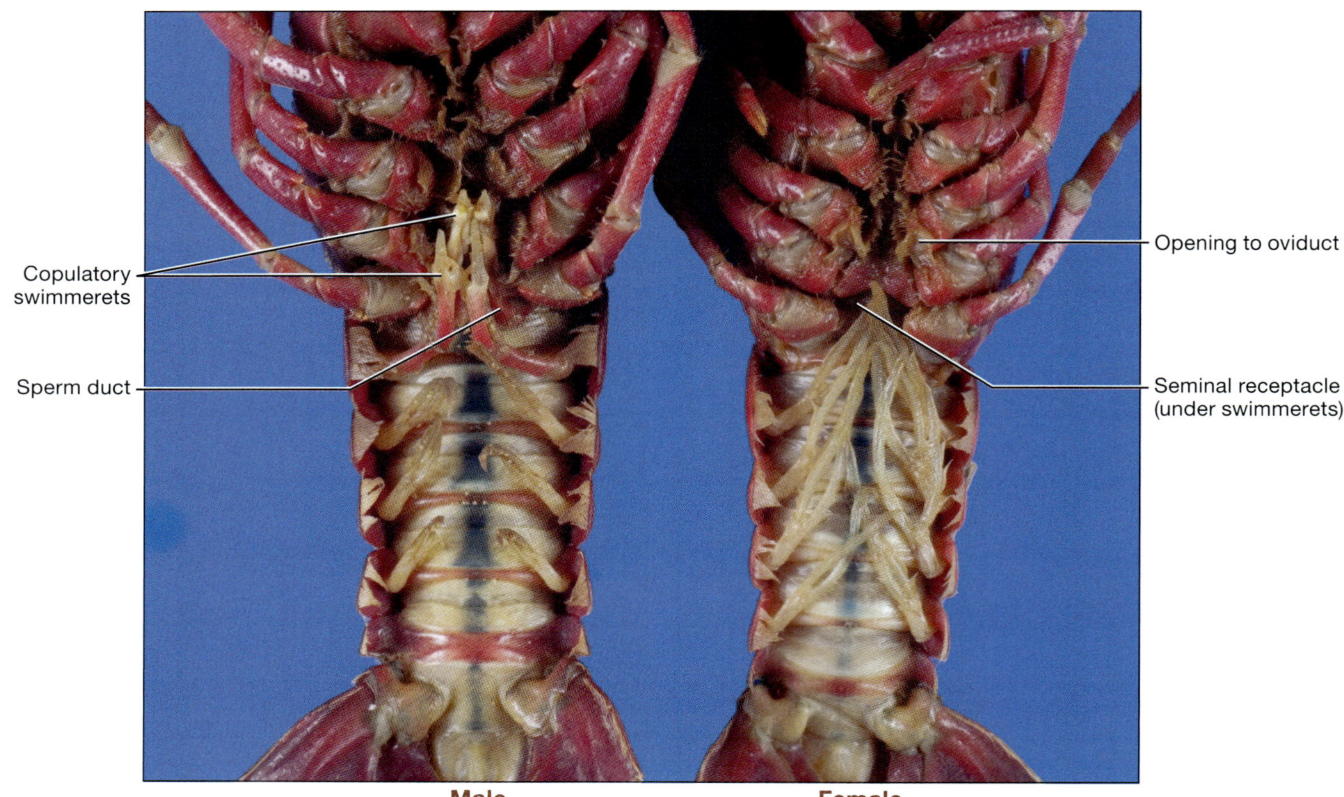

13.13 Male and female crayfish, ventral views. The first two pairs of swimmerets are greatly enlarged in the male for transferring sperm from the sperm duct to the female's seminal receptacle.

13

Table 13-2 External Anatomy of the Crayfish

Structure	Function
Rostrum	Pointed region between eyes demarcating anterior end of body
Compound eyes	Small, dark sense organs for detecting light and forming visual images
Antennae	Longer, paired appendages on head modified for chemosensory and tactile reception
Antennules	Shorter, paired appendages on head modified for chemosensory, tactile, and equilibrium reception
Maxillae	Two sets of short, paired appendages located on ventral surface near mouth used to hold food while the mandibles grind off bits and pieces to ingest
Maxillipeds	Three sets of paired appendages located on ventral surface near mouth used to manipulate food
Mandibles	Hard, chitinous mouthparts used to grind food into mouth
Mouth	Opening to digestive tract located on ventral surface of body
Chelipeds	Large pinchers used for grasping food and for defense
Walking legs (periopods)	Locomotion (walking on land and crawling across stream bottom)
Swimmerets (pleopods)	Abdominal appendages modified for swimming
Copulatory swimmerets (male)	Larger, club-shaped swimmerets used by male to transfer sperm during copulation
Uropod and telson (tail)	Broad, fan-shaped region of body used for rapid movement and for directional control during leisurely locomotion
Anus	Terminal point of digestive system located on ventral surface of telson

Check Your Progress

2.1 What advantage might mouthparts of various shapes provide a crayfish?

2.2 Are the antennae biramous? Are the antennules biramous?

2.3 What two "typical" arthropod body regions are fused into the cephalothorax?

2.4 Is your crayfish male or female? How do you know?

13

Crayfish Internal Anatomy

Procedure

1 To reveal the internal anatomy of your crayfish, you will have to make several cuts through the exoskeleton. Review the Dissection Diagram to assist you in this process.

Dissection Diagram

1 Use scissors to make incisions along each side of the dorsal surface of the cephalothorax starting from the base and angling toward the anterior tip of your specimen. Gently tease away the upper, separated portion of the cephalothorax, being careful not to damage underlying structures.

2 Make incisions along each side of the dorsal surface of the abdomen extending four to five segments posteriorly, and carefully remove the exoskeleton.

3 Make additional vertical incisions along each side of the head to separate each side of the cephalothorax from the body.

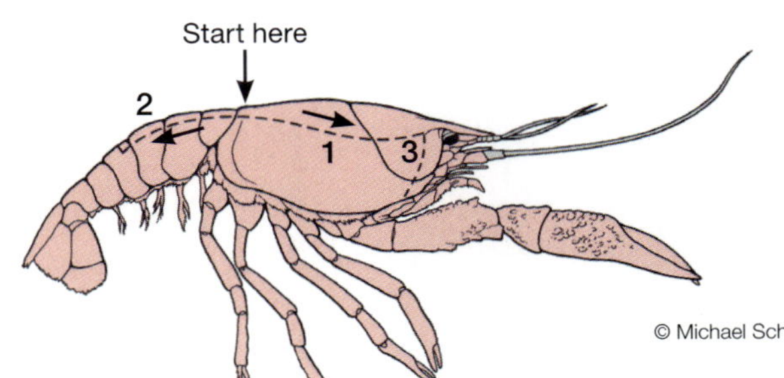

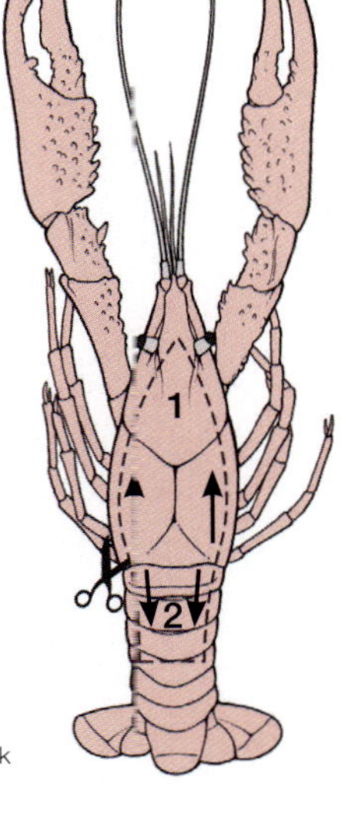

2 Using the pointed end of your dissecting scissors, make incisions along each side of the dorsal surface of the cephalothorax, starting from the posterior end of the cephalothorax and progressing rostrally, angling the incisions medially so they meet at the tip of the rostrum. Gently peel away the dorsal portion of the carapace, being careful not to pull away any of the internal organs that are attached to the underside of the exoskeleton.

3 Make additional incisions along each side of the dorsal surface of the abdomen, starting from the posterior end of the cephalothorax (your original incision point) and progressing posteriorly toward the tail. Extend these incisions four or five segments, and then cut laterally across the final segment. Carefully remove this portion of the exoskeleton covering the dorsal surface of the abdomen to expose the musculature of the tail and the blood vessels and digestive organs located within this region.

4 Finally, make vertical incisions along each side of the head, starting at the point where the cephalothorax and legs meet and extending the cuts behind each eye to the top of the head. Remove the remaining lateral sides of the carapace.

5 Use a probe to lift some of the feather-like gills away from the body. One of the first things you may notice is that the large, feathery gills are actually *external*. They reside between two protective pieces of exoskeleton—the outer lateral side of the carapace, which you removed, and a thinner, inner chitinous membrane. Pull on a gill with forceps, and you will notice that the gills are attached to the walking legs at their proximal juncture.

6 Remove a gill, place it in a small dish of water, and examine it using your dissecting microscope. Notice the central axis and the feathery gill filaments that extend from it.

7 Return to your crayfish and work from the dorsal surface downward (ventrally), identifying the internal organs and structures described in Table 13-3 as you go.

Table 13-3 Internal Anatomy of the Crayfish

Structure	Function
Gills	Respiration
Esophagus	Passageway between mouth and cardiac portion of the stomach
Cardiac chamber of stomach	Thick-walled, anterior portion of the stomach containing gastric mill—chitinous teeth that grind food into a liquefied mush
Pyloric chamber of stomach	Thin-walled chamber where chemical digestion of food occurs
Digestive glands	Accessory digestive organs that secrete enzymes into the pyloric stomach to facilitate chemical breakdown of food
Intestine	Long tube passing through abdominal region in which nutrients are absorbed into the bloodstream for delivery to the body tissues
Heart	Specialized, muscular chamber containing ostia (holes) to allow passive uptake of blood, which is delivered to the body tissues through arteries
Dorsal abdominal artery	Longitudinal blood vessel that distributes blood to the dorsal aspect of the abdomen
Green glands	Paired excretory organs found along the ventral margin of the head region; they release waste out of the crayfish through small pores in the ventral body wall
Testes (male)	Site of sperm production
Ovaries (female)	Site of egg production
Brain	Small, radiate structure lying dorsal to the green glands; houses the majority of neural ganglia in the crayfish
Circumesophageal nerves	Branches of the ventral nerve cord that bifurcate at the base of the brain and encircle the esophagus before merging along the ventral surface of the crayfish just posterior to the esophagus
Ventral nerve cord	Long, white "cord" located along the ventral surface of the body; contains large swellings of ganglia that handle the majority of coordination without intervention by the brain

8 Two sets of prominent muscles partially obscure some of the digestive organs and should be removed before proceeding. Use scissors to snip away the gastric muscles, which move the cardiac chamber of the stomach, and the mandibular muscles, which control the motion of the mandibles (Fig. 13.14A).

9 After differentiating the **cardiac chamber** and the **pyloric chamber** of the stomach, remove them from the crayfish and open the cardiac chamber to reveal the gastric mill (Fig. 13.14C).

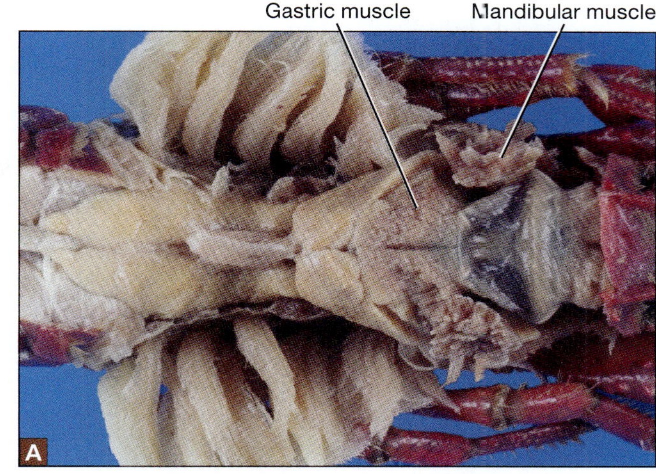

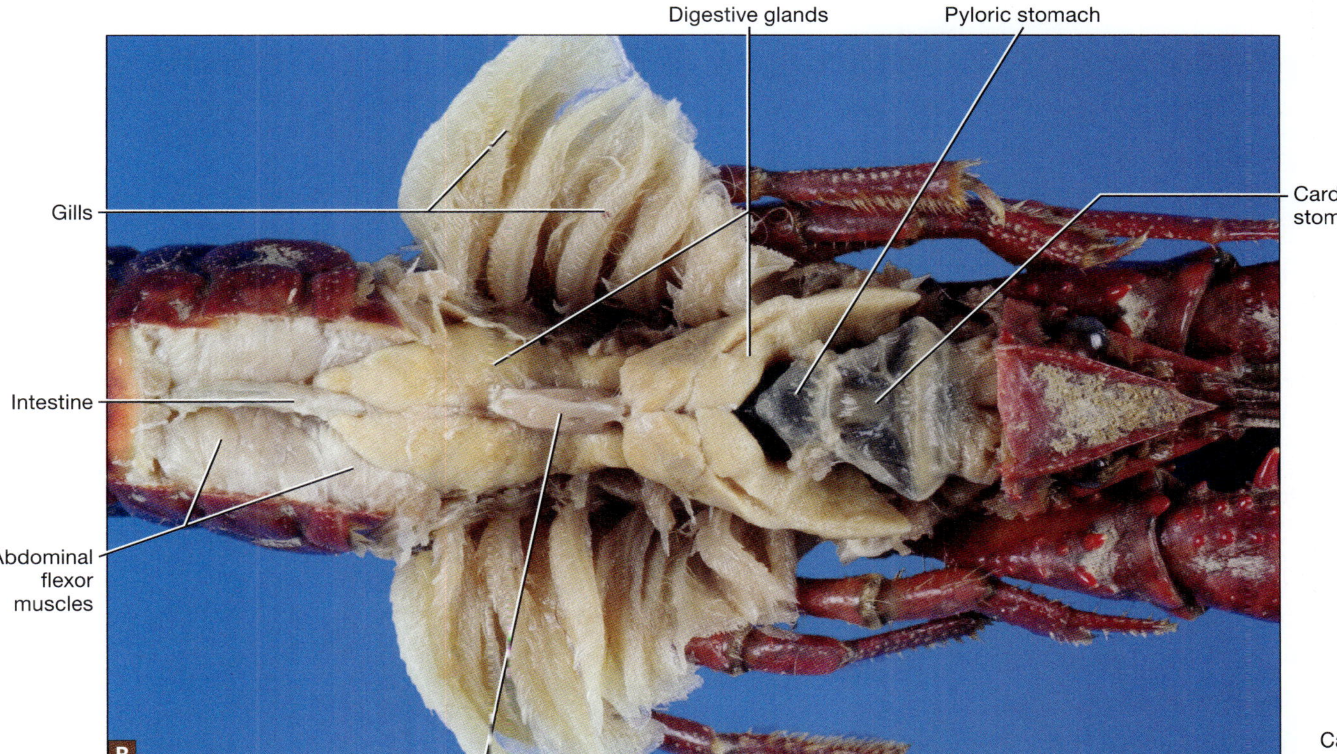

13.14 *Cambarus* dissection showing: (**A**) mandibular muscles and gastric muscles in place, (**B**) dorsal view with mandibular muscles and gastric muscles removed, and (**C**) close-up of gastric mill inside the cardiac stomach.

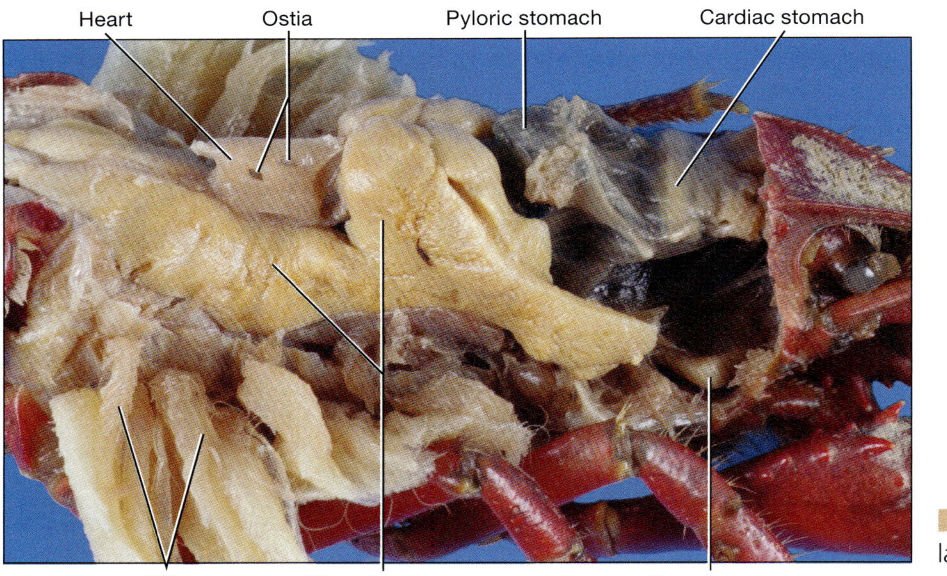

13.15 *Cambarus* dissection, lateral view.

The crayfish uses the chitinous teeth of the **gastric mill** to grind its food mechanically into smaller pieces for digestive enzymes secreted by the **digestive glands** to act upon as the food moves into the pyloric chamber of the stomach and then empties into the **intestine**, where nutrient absorption occurs.

The small, white, angular **heart**, with its numerous **ostia**, should be clearly visible, situated medially among the lobes of digestive gland tissue (Figs. 13.14–13.15). Arthropods have an open circulatory system, similar to that of bivalves. When the heart contracts, special valves prevent blood (hemolymph) from escaping through the ostia, and blood is directed out of the heart along arteries. Once the hemolymph reaches the tissues, it leaves the vessels and percolates through open spaces in the hemocoel. The blood returns to the heart by flowing to the pericardial space that surrounds the heart and re-enters the heart through the pores (ostia) on its surface.

13

10 Lift the heart gently with your forceps. Several arteries can be seen leaving the heart. Careful inspection of the dorsal surface of the intestine also may reveal the thin dorsal abdominal artery that runs along the length of the intestine.

Just ventral to the cardiac stomach and situated on each side of it are the **green glands**, the excretory organs of the crayfish (Figs. 13.15 and 13.17). These round glands are highly folded tubule systems that sit in the hemocoel. Hemolymph is filtered by the tubules, and concentrated urine is stored in small bladders and released to the outside through visible openings called nephropores, located on the ventral surface of the head at the base of each antenna.

11 Remove the heart and use a dissecting microscope to scan the dorsal aspect of the digestive glands for the small reproductive structures.

Ovaries on female specimens will be easier to spot than the extremely small **testes** of males, but careful examination of this region along the median plane of the dorsal surface of your crayfish should reveal the reproductive organs (Fig. 13.16). The size of the testes and ovaries and the presence of eggs within the ovaries depend upon the time of year the specimens were collected, because reproduction in this group is seasonal. Sperm are produced in the testes and travel along each **vas deferens** to a sperm duct, located on the animal's ventral surface at the base of the fifth pair of walking legs (fourth pair of periopods) (Fig. 13.16C). Sperm are transferred to and held in the seminal receptacle of the female, and fertilization occurs externally as she releases eggs from her oviducts.

12 To reveal the brain and ventral nerve cord, you must remove the stomach completely by cutting the short portion of esophagus between the mouth and the stomach.

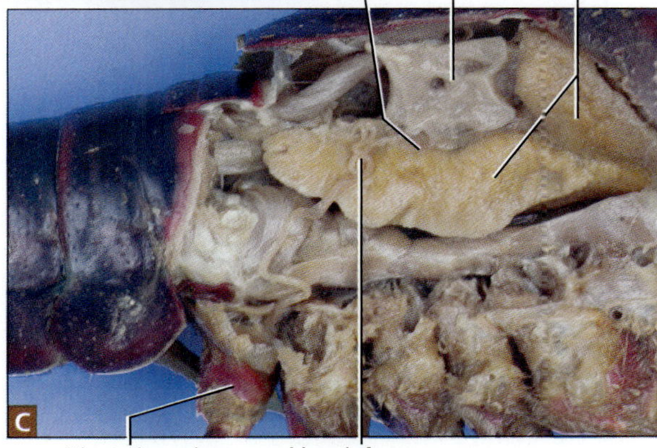

13.16 *Cambarus* dissection, dorsal view (**A** and **B**). The heart has been removed to show underlying (**A**) female and (**B**) male reproductive structures. The top portion of the cardiac stomach has been removed on the female to show the gastric mill within. Lateral view (**C**) of male showing the connection of the vas deferens to the sperm duct located at the base of the fifth walking leg (fourth periopod).

13

The brain is formed by fusion of ganglia from the first three segments. Leaving the brain are numerous nerve tracts leading to the antennae, antennules, and compound eyes, which gives the brain a stellate appearance (Fig. 13.17). Two large nerves exit the posterior margin of the brain encircling the esophagus; hence their name circumesophageal nerves. These two nerves come together behind the esophagus on the floor of the cephalothorax to form the subesophageal ganglion. This large ganglion is the result of the embryonic fusion of five pairs of thoracic and head ganglia. After the nerve cord leaves the subesophageal ganglion, it resembles the annelid nerve cord with its repetitive, segmented appearance.

13 You can trace the ventral nerve cord as it passes posteriorly from the subesophageal ganglion by carefully cutting away the flexor muscles in the abdomen (Fig. 13.18).

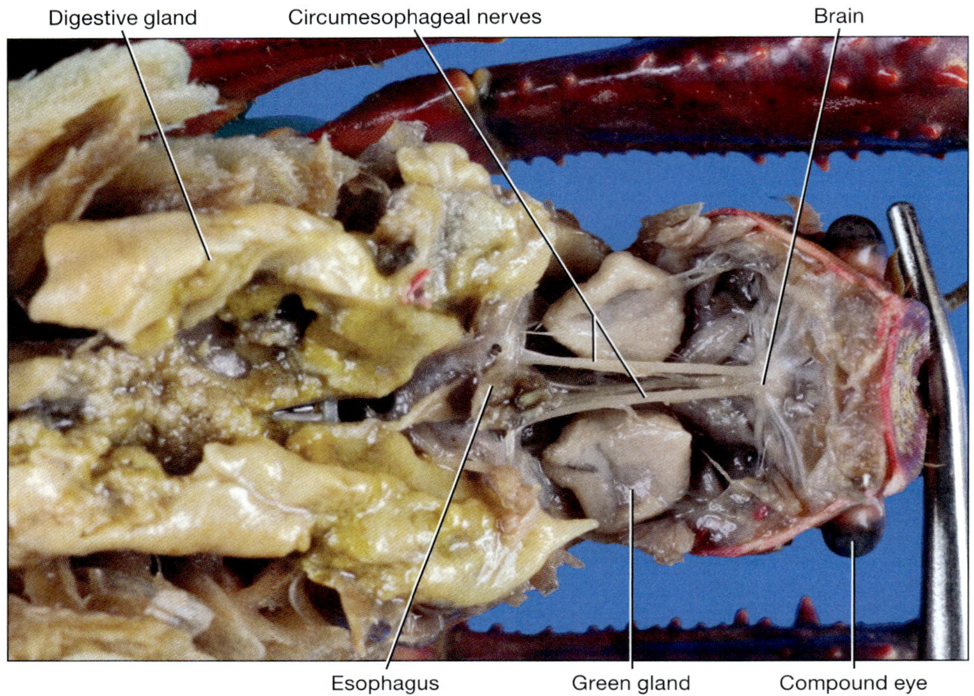

13.17 *Cambarus* dissection. Anterior region with stomach removed to show brain and green glands.

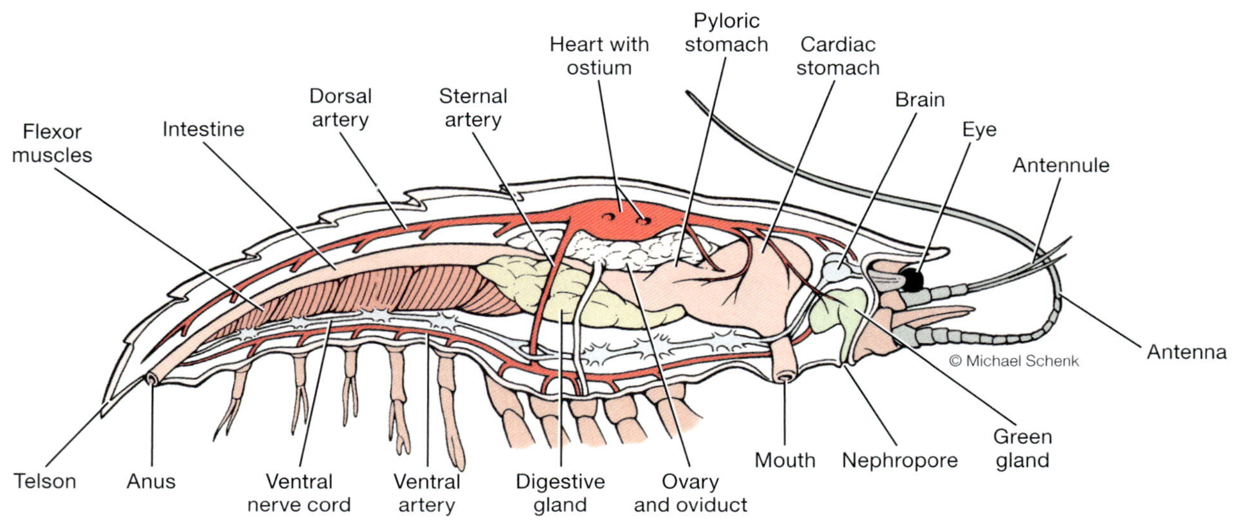

13.18 Internal anatomy of the female crayfish, lateral view.

Check Your Progress

3.1 What structure in the cardiac chamber of the stomach allows crayfish to digest their food mechanically?

3.2 What organ secretes enzymes into the pyloric chamber of the stomach to assist in the chemical digestion of food?

3.3 What type of circulatory system does the crayfish possess, open or closed? What anatomical evidence is present to support your answer?

3.4 Where on the body surface of the crayfish are the external openings for the excretory organs (green glands) located?

3.5 How do the relative positions of the brain and the nerve cord in crayfish compare to the positions of the brain and the nerve cord in the earthworm? Is this evidence of common ancestry?

13

Exercise 13-7 Insects (Subphylum Hexapoda, Class Insecta)

Materials Needed

- ❑ preserved grasshopper (*Romalea*)
- ❑ dissection tools
- ❑ dissecting pan
- ❑ dissecting pins
- ❑ latex gloves
- ❑ dissecting microscope

With nearly 1.1 million described species, the class Insecta (part of the subphylum Hexapoda) is by far the largest, most diverse, and most widespread class of arthropods. More than half of all named animal species are insects. The United States alone has more than 100,000 different named species of insects. Though their diversity is astonishing, the members of this group share several basic similarities (Fig. 13.19). All insects have three pairs of **uniramous** (unbranched) walking appendages, a single pair of antennae, and three distinct body regions: the head, thorax, and abdomen. Insects are primarily terrestrial creatures, and their evolution has been shaped by advances that have allowed them to thrive despite the unique set of challenges of a terrestrial environment, including:

1. stronger support systems and stronger walking appendages to overcome the forces of gravity,
2. waxy **cuticles** built to withstand the osmotic stresses of "dry" air, yet adapted for gas exchange,
3. the ability to fertilize eggs internally to prevent desiccation of exposed gametes,
4. specialized respiratory, excretory, and digestive structures modified to conserve water, and
5. appendages modified into **wings** to take advantage of aerial habitats.

Dragonfly, *Anax*

Seven-spot ladybird beetle, *Coccinella septempunctata*

Paper wasp, *Polistes*

Praying mantis, *Stagmomantis carolina*

13.19 All insects possess a single pair of antennae, three pairs of walking legs, and a body clearly divided into three main regions: head, thorax, and abdomen.

In the following exercise the grasshopper will serve as our representative example of an insect that displays all of these adaptations to terrestrial life.

Grasshopper External Anatomy

Procedure

1 Obtain a preserved specimen of *Romalea*, the common grasshopper, and examine its external features using Table 13-4 and Figure 13.20 to assist you in identification.

The insect body is divided into three main regions: the head, thorax, and abdomen. Notice the segmented nature of these regions and the many modified appendages in each of these areas. On the head you will find a single pair of antennae, one pair of compound eyes, and three smaller **ocelli** aligned in the shape of a triangle. The antennae are the sensory organs for touch, taste, and smell for the grasshopper. Surrounding the mouth is an array of specialized mouthparts used for chewing and manipulating food, including maxillary palps, labial palps, and mandibles. The thorax bears three pairs of walking legs and two pairs of wings. Along the sides of the abdomen, you should see small pores in each segment. These holes, called spiracles, represent the openings to the respiratory system of the grasshopper. Inside the body, the spiracles lead into small tubes, known as tracheae, through which air passes and gas exchange occurs.

2 To determine the sex of your grasshopper examine the posterior end of the abdomen. In addition to an **anus**, females possess an **ovipositor** ventral to the anus. The ovipositor is an opening bordered by two pairs of chitinous spikes that thrust into the soil and flex outward, creating a chamber in the soil in which the female deposits her eggs. Males lack an ovipositor (Fig. 13.20).

Table 13-4 External Anatomy of the Grasshopper

Structure	Function
Compound eyes	Paired, complex, image-forming photoreceptors composed of numerous ommatidia that create a fairly coarse-grained picture of their visual field
Ocellus (*pl.* = ocelli)	Simple photoreceptor consisting of a small cup backed by light-absorbing pigments; not capable of image formation
Antennae	Thin, paired appendages on head modified for chemosensory and tactile reception
Mouthparts	Multiple sets of hard, chitinous appendages used to chew food and manipulate food into mouth
Wings	Two sets of paired appendages modified for flight
Tympanum	Circular, pressure-sensitive membrane on each side of the thorax that detects sound waves; functions as an "ear drum"
Spiracles	External openings in abdomen that allow air flow into and out of tracheae
Ovipositor (females)	External reproductive opening bordered by pointy, chitinous teeth that penetrate the soil and create burrows for egg deposition

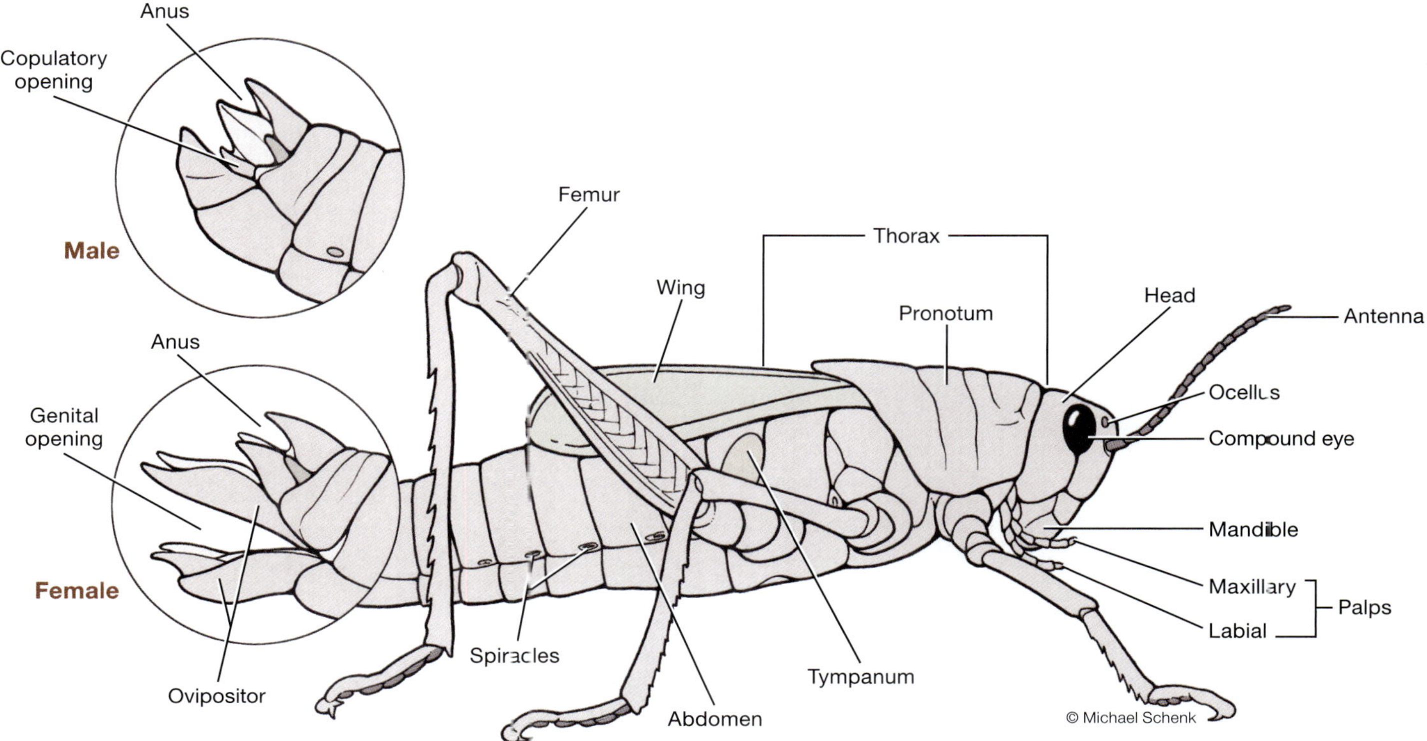

13.20 External anatomy of female grasshopper with comparison of the posterior regions of a male and female.

13

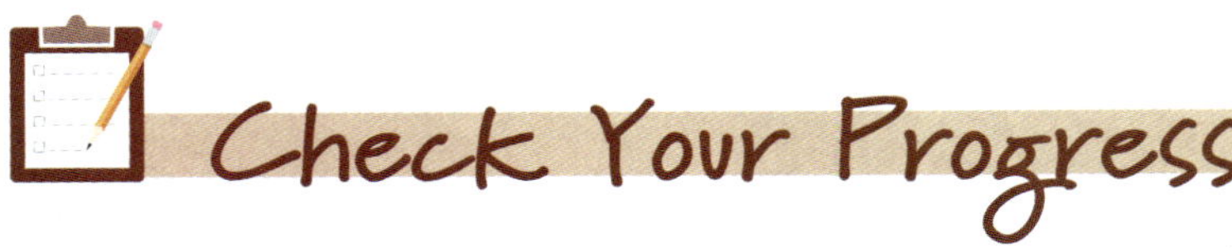

Check Your Progress

4.1 Are the grasshopper's antennae biramous? Are any of its appendages biramous?

4.2 How many major body regions do insects have?

4.3 How many pairs of walking appendages do insects have?

4.4 What other appendages are present for locomotion?

13

4.5 Is your grasshopper male or female? How do you know?

4.6 Why do the spiracles of the respiratory tract open to the outside of the body?

Grasshopper Internal Anatomy

Procedure

1 Review the Dissection Diagram to assist you in opening your specimen.

2 Using the pointed end of your dissecting scissors, make an incision along the dorsal margin of the grasshopper from the posterior end of the abdomen to the base of the head.

3 Gently peel apart the lateral portions of the thorax and abdomen, being careful not to pull away any of the internal organs that are attached to the underside of the exoskeleton.

4 Use pins to hold your specimen open, and secure it to your dissecting pan. Work from the dorsal surface downward (ventrally), identifying the organs and structures described in Table 13-5 as you proceed. **HINT:** If you plan to use a dissecting microscope to view the internal anatomy of your grasshopper, remember to pin your specimen near one side of your dissecting pan.

Dissection Diagram

1 Use scissors to make a single incision along the dorsal midline of the body from the anus to the base of the head.

2 Gently tease open the body, using pins to spread the body open.

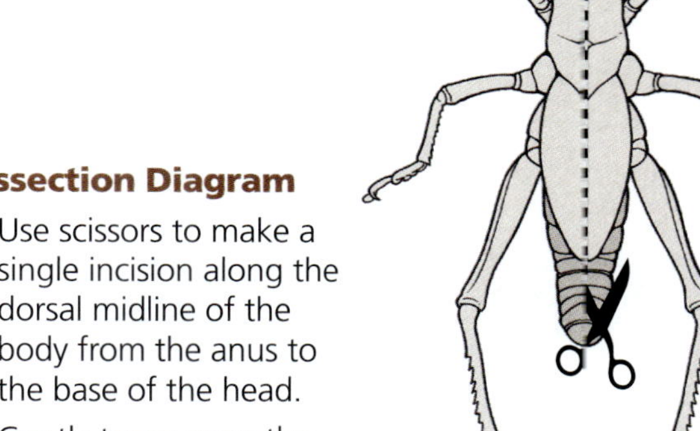

Table 13-5 Internal Anatomy of the Grasshopper

Structure	Function
Esophagus	Passageway between mouth and crop
Crop	Highly extensible, anterior portion of the digestive tract; serves as a storage compartment for food
Stomach	Chamber that receives semi-digested food from the crop and secretes enzymes for chemical digestion
Gastric ceca	Lobed, accessory digestive areas located at the juncture between the crop and stomach that facilitate chemical digestion
Intestine	Shortened tube passing through the abdomen in which nutrients are absorbed into the bloodstream
Rectum	Specialized swelling of the posterior portion of the digestive tract for efficient water reabsorption and conservation
Anus	Terminal portion of the digestive tract; regulates egestion of undigested food from the body
Ovaries (female)	Site of egg production and maturation; fertilization is internal and fertilized eggs are deposited in the soil
Testes (male)	Site of sperm production (may not be visible on specimen)
Heart (dorsal blood vessel)	Specialized muscular swellings of the dorsal blood vessel containing ostia (holes) to allow passive uptake of hemolymph that is delivered to the body tissues through arteries; veins do not exist in this open system
Malpighian tubules	Stringy, fibrous excretory organs scattered along the margins of the stomach and intestine; release metabolic waste into the digestive tract so excess water may be reabsorbed by the intestine and rectum and conserved in the body
Brain	Enlarged conglomeration of nervous tissue located in the head; processes sensory information and coordinates major body functions
Ventral nerve cord	Long, whitish "cord" located along the ventral surface of the body; contains large swellings of ganglia that handle some nervous coordination without intervention by the brain

If your specimen is a female, you immediately may notice the large ovary filled with eggs, which can often expand to nearly fill the abdominal cavity (Fig. 13.21A). Ovaries generally will be easier to spot than the smaller testes of males (Fig. 13.22), but careful examination of the dorsal surface along the median plane of the specimen should reveal the reproductive organs. The size of the testes and ovaries and the presence of eggs within the ovaries depend upon the time of year that the specimens were collected, because reproduction in grasshoppers is seasonal.

Notice the thin, filamentous tubules stretching between the internal organs and the spiracles (identified earlier) that open on the outside of the abdomen. These are tracheae—respiratory tubules of the grasshopper that conduct air flow from the outside environment directly to the tissues within the body.

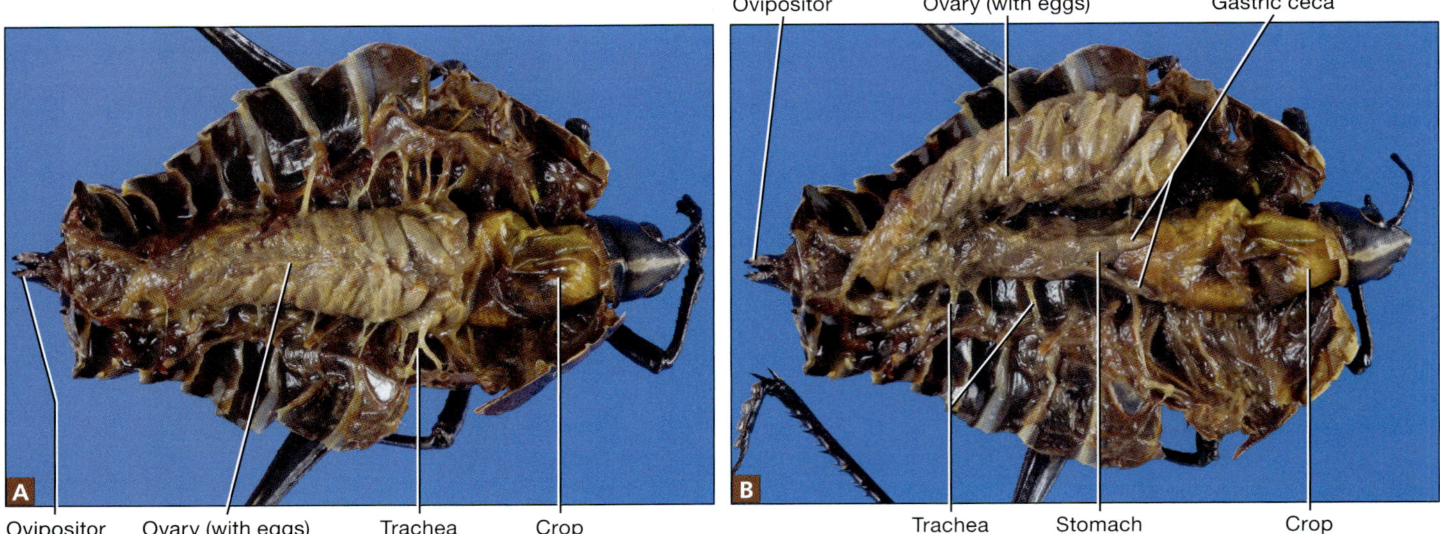

13.21 *Romalea* dissection, female: (**A**) ovaries with eggs left in place, and (**B**) ovaries with eggs reflected laterally to show underlying anatomy.

The first specialized region of the digestive system posterior to the mouth is the **esophagus**, which transfers food into a large storage organ known as the **crop**. Food passes from the crop into the **stomach**, where accessory organs, such as the **gastric ceca**, assist in the extracellular digestion of food (Figs. 13.21–13.23). Nutrient absorption occurs along the short length of intestine, and indigestible foodstuff passes into the enlarged **rectum**, an organ that plays a vital role in water conservation for this animal (Figs. 13.22 and 13.23). Remember—a major hurdle faced by terrestrial organisms is water loss. The chitinous exoskeleton plays a significant role in minimizing water loss through evaporation, and the rectum reabsorbs most of the water used in the digestive process, producing dry feces that are egested through the anus.

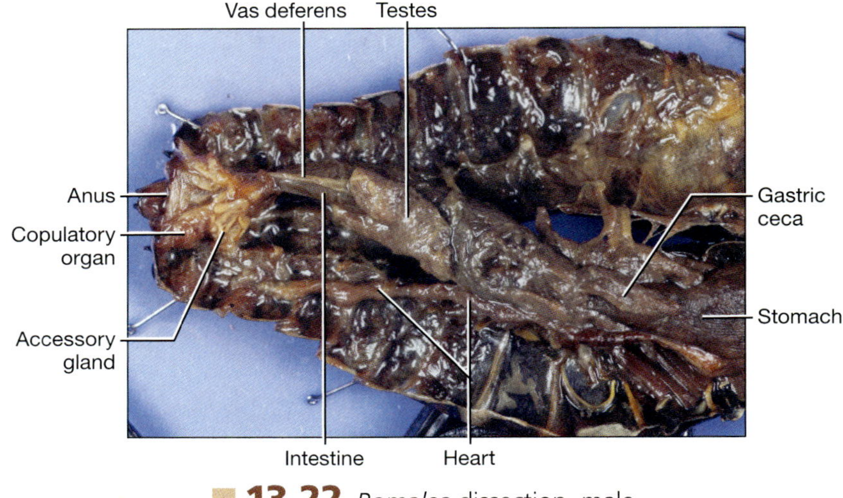

13.22 *Romalea* dissection, male.

The circulatory system of the grasshopper, like all arthropods, is an open system. Because oxygen is delivered to the grasshopper tissues by means of the highly branched tracheal system, the circulatory system of the grasshopper does not have to be particularly efficient. The only blood vessel is a dorsal blood vessel that runs the entire length of the animal. Because it has muscular swellings that help move hemolymph through the animal, this vessel sometimes is called a heart (Fig. 13.22). A series of holes called ostia allow blood to flow into the dorsal vessel.

The grasshopper has a system of filamentous, blind-ended tubules, the Malpighian tubules, to rid its body of metabolic wastes (Figs. 13.22 and 13.23). These tubules can be found along the lateral sides of the abdomen at the juncture of the stomach and intestine. Notice how they are connected to the anterior portion of the intestine. Malpighian tubules float freely in the coelomic cavity. They absorb metabolic wastes from the circulating hemolymph, concentrate the wastes (by pushing fluids back into the hemocoel), then deposit the concentrated waste in the gut to be eliminated with the feces.

Like most arthropods, insects have a dorsal brain and a ventral nerve cord (Fig. 13.23). Careful inspection of the nerve cord will reveal several enlarged areas along its length (**ganglia**) that coordinate information from the different regions of the body. The anterior region, the brain, has direct connections to the photoreceptors and antennae that you examined earlier.

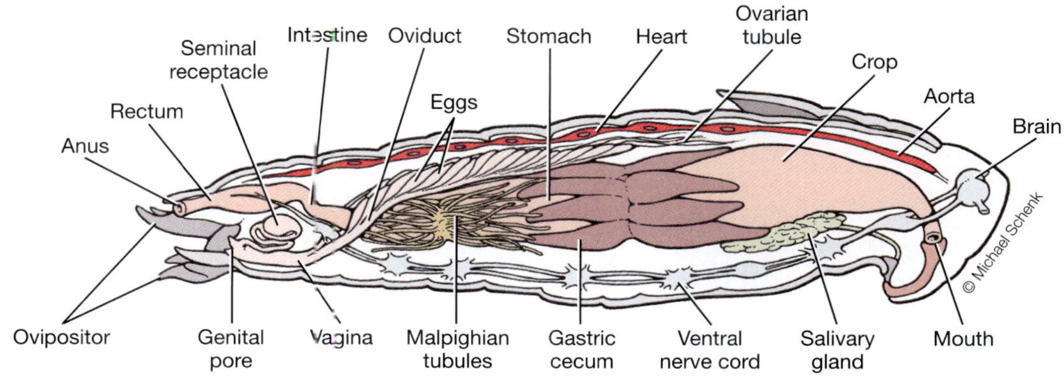

13.23 Female grasshopper, *Romalea*, internal anatomy.

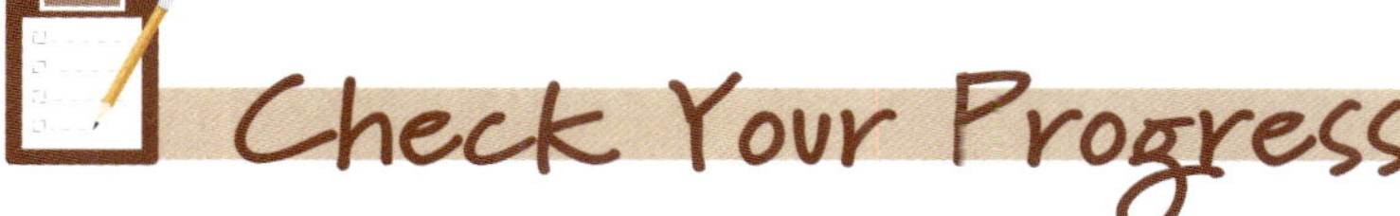

Check Your Progress

13

5.1 What digestive organ for food storage do grasshoppers possess that bears the same name in earthworms?

5.2 What two specialized areas of the digestive tract reflect adaptations to living in dry, terrestrial habitats?

5.3 What organ in grasshoppers performs the same function as the green glands in crayfish? In what way has this organ been adapted in grasshoppers to conserve water?

5.4 How does the position of the nerve cord in the body of grasshoppers compare to its position in crayfish?

5.5 What benefit do the gastric ceca provide for grasshoppers?

Exercise 13-8 Investigating Orientation Behaviors

Behavior can be defined generally as any overt response by an animal to a stimulus; thus, we can say that **orientation behaviors** refer to specific responses that involve positioning or movements by an animal.

A natural starting point in a behavioral study is to attempt to elucidate the nature of the stimulus that evokes a particular behavior. Thus, we typically observe the behavior first and then ask the question, "What causes that behavior?" Again we return to our earlier discussion of proximate versus ultimate behavioral causes from Chapter 8. If we are interested in the stimulus that elicited the behavior, then, we are looking for the **proximate cause** of the behavior.

You also may want to know *why* this behavior evolved in the first place and persisted over time in this species—the **ultimate cause**. For most behaviors, light, sound, gravity, temperature, moisture, or chemicals are likely candidates for potential proximate cues. Because many animals orient to stimuli that humans lack the ability to sense (for example, minute concentrations of chemicals, magnetic fields, ultraviolet light, or low-frequency vibrations), it is often not the stimulus most obvious to us that evokes the behavior we observe.

To effectively tease apart the stimuli causing a behavior, we first must be able to precisely describe the behavior we are observing. With respect to orientation behaviors, two main categories have been defined.

1. **Kinesis** is a general term for a random behavior that does not necessarily orient the animal toward a stimulus but, rather, alters its general rate of movement (orthokinesis) or degree of turning (klinokinesis). For example, terrestrial isopods respond to increases in humidity by decreasing their overall movements. This behavioral response tends to keep them in damper areas. As the humidity in their environment decreases, their rate of movement increases, presumably to carry them to a more favorable location.
2. In contrast to kinesis, a **taxis** is a directed behavior in which an animal orients or moves toward or away from a stimulus.

Both kineses and taxes may be *positive* (an increase in movement or movement toward the stimulus) or *negative* (a decrease in movement or movement away from the stimulus). Similarly a prefix may be attached to the taxis or kinesis to further describe the stimulus, as mentioned earlier with orthokinesis and klinokinesis. Other examples of descriptive prefixes are: photo- (light), chemo- (chemical), geo- (gravity), thigmo- (touch), and thermo- (heat).

Responses to Humidity and Light in Isopods

Isopods are the terrestrial crustaceans often referred to as sow bug, pill bug, or roly-poly. Common to wooded and urban settings alike, they generally are found in leaf litter or under other debris, where they feed on decaying organic matter. Isopods possess two pairs of antennae, one pair of compound eyes, and seven pairs of walking legs.

In the following experiment, modified petri dishes will be used to give these animals a choice between different environments that you will manipulate. In the first part of the experiment, you will examine the preferences of isopods for humid and arid environments. Then you will examine their preferences for light and dark environments. After you have determined their responses to moisture and light as separate stimuli, you will examine the interplay between these two stimuli to establish which stimulus is more influential.

Materials Needed

- ❑ terrestrial isopods (any species)
- ❑ petri dish "choice chambers"
- ❑ black cloth or construction paper
- ❑ plastic screen mesh
- ❑ paper towels or filter paper
- ❑ Drierite® desiccant
- ❑ gooseneck lamp with 60-watt bulb

Note

Keep the room lights low during these laboratory exercises. Stray light may cause unexpected results in these experiments.

Experimental Protocol

1 Working in groups of three or four students, obtain an isopod choice chamber that has been constructed by taping two petri dishes together, leaving a cutout passageway between the dishes.

2 Place a folded, moistened paper towel or filter paper in the bottom of one petri dish and sprinkle a thin, even layer of Drierite® (a commercial desiccant) in the bottom of the other half of the chamber (Fig. 13.24).

3 Position the plastic screen mesh over the bottom of the two dishes and place the lids securely on the chamber. The isopods will be crawling on the top surface of the screen mesh, so be sure there are no gaps at the edges of the chamber for escape. Rubber bands may be necessary to secure the lids tightly.

4 Allow the choice chamber to sit undisturbed for 5 minutes to allow the humidity level in each side of the chamber to stabilize.

5 After 5 minutes, introduce 10 isopods in each side of the chamber through the central holes and place rubber stoppers in the holes to close them.

6 Leave the chamber undisturbed for at least 10 minutes in a setting where it is exposed to low, uniform light, preferably away from the immediate area in which you are working.

7 While the animals are "selecting" their preferred location within the dishes, answer the following questions.

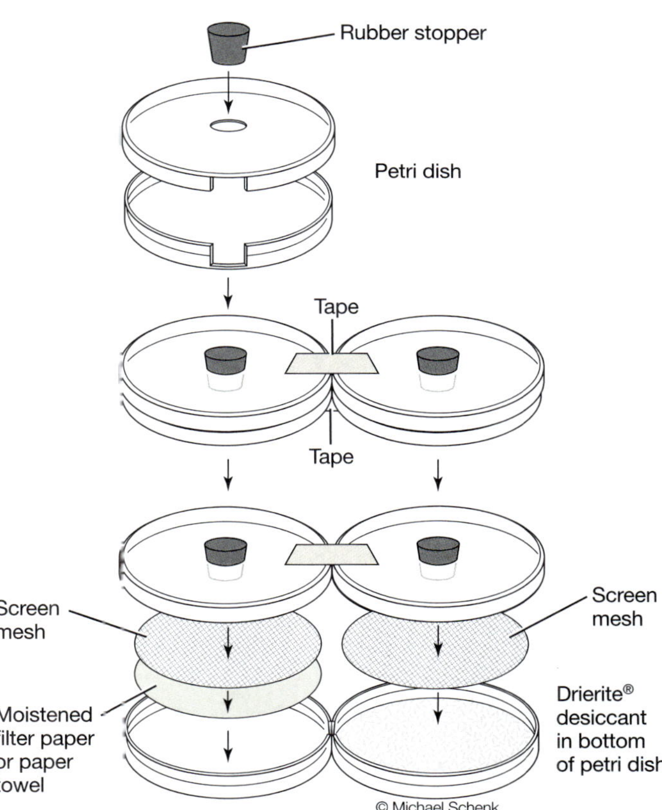

13.24 "Choice chamber" setup for testing isopod responses to humidity and light.

13

Using the Scientific Method

1 What is the general question that this experiment addresses?

2 Record the hypothesis you wish to test.

3 Identify the independent variable in this experiment.

Tip

The *independent variable* is the element that you change or manipulate and the *dependent variable* is the element that you measure to determine how it is affected by the changes you made to the independent variable.

4 Identify the dependent variable(s) in this experiment.

5 What are the levels of treatment (that is, levels of the independent variable)?

6 Did you set up a control treatment? If not, describe an appropriate control treatment for this experiment.

Tip

A *positive control* shows that the experiment is capable of producing a positive result when a phenomenon is expected, whereas a *negative control* is designed to demonstrate a lack of response or negative result when there should be no effect.

7 Identify several controlled variables.

8 After 10 minutes, return to your choice chamber and record the number of isopods in each side of the chamber. Record your findings in the first two rows of Table 13-6.

9 Remove the individuals from the chamber (your instructor may have a designated container for "used" isopods) and discard the Drierite®.

10 Now set up the choice chamber with moist, folded paper towels or filter paper in each side, replace the lids and rubber stoppers, and again let the dishes sit undisturbed for 5 minutes.

11 After 5 minutes, place 10 individuals in each dish, as before, then replace the stoppers and this time cover the top and sides of *one of the dishes* with black cloth or construction paper. Illuminate the uncovered side of the chamber with a gooseneck lamp containing a 60-watt bulb.

12 Let the choice chamber sit undisturbed for at least 10 minutes. While you are waiting, answer the following questions.

Using the Scientific Method

1 What is the general question being addressed by this experiment?

---

2 Record your hypothesis.

---

3 Identify the independent variable in this experiment.

---

4 Identify the dependent variable in this experiment.

---

5 What are the levels of treatment (that is, levels of the independent variable)?

13

13 After 10 minutes have passed, return to your choice chamber and record the number of isopods in each side of the chamber. Record your findings in the third and fourth rows of Table 13-6.

14 Finally, design and conduct an experiment to examine the interplay between these two stimuli. Your goal is to determine the preference of isopods between a brightly illuminated, humid environment and a dark, dry environment. Using the steps of the scientific method, answer the following questions, conduct your experiment, and record your findings in the final two rows of Table 13-6.

Using the Scientific Method

1 Record your hypothesis.

____

2 What are your two independent variables in this experiment?

____

Table 13-6 Responses to Humidity and Light in Terrestrial Isopods

Treatment	Number Found in Location
Dry	
Humid	
Light	
Dark	
Light, humid	
Dark, dry	

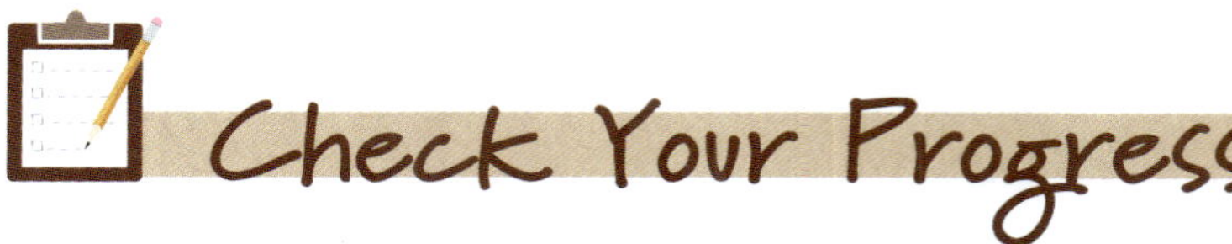

6.1 Could the isopods distinguish between the moist and the dry microhabitats?

6.2 Based on these findings, would you retain or reject your hypothesis?

6.3 Summarize your findings about the preference of isopods for humid versus arid microhabitats.

6.4 Explain how this behavior would be adaptive for these organisms.

6.5 Could the isopods distinguish between the light and dark regions?

13

6.6 Summarize your findings about the preference of isopods for light versus dark environments.

6.7 Explain how this behavior would be adaptive for these organisms.

6.8 Which stimulus—light intensity or relative humidity—is more influential in affecting isopods' habitat selection?

Phototaxis in Cladocercans

Often referred to as water fleas because of their superficial resemblance to the familiar arachnid associated with domestic pets, members of the genus *Daphnia* are actually tiny, aquatic crustaceans in the order Cladocerca commonly found in freshwater ponds and lakes (Fig. 13.25). Individuals possess branched antennae, a single, median compound eye, and flattened legs that are their primary respiratory organs. Like many other zooplankton, *Daphnia* are known to make daily vertical migrations within lakes in response to changing light intensity levels, moving up within the water column after sundown and returning to greater depths as day approaches. Presumably, these vertical migrations place the zooplankton within the upper, more productive parts of the aquatic ecosystem at a time when predation pressures and damaging illumination levels are lowest. In the following exercise you will determine the phototactic responses of *Daphnia magna* to changes in light source position and light intensity.

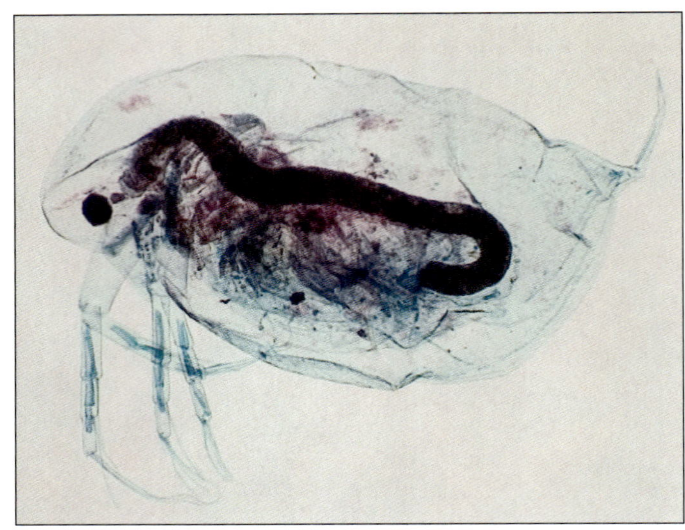

13.25 Cladocercan, *Daphnia*, is a common microscopic freshwater crustacean.

Experimental Protocol

13

1 Working in small groups of three or four students, obtain a 100 ml graduated cylinder containing a small population of *Daphnia*.

2 Suspend the cylinder by a clamp from a ring stand and use a gooseneck lamp with a 60-watt bulb to present light first from above for several minutes, noting *Daphnia's* response (Fig. 13.26). Use the ten 10-ml divisions on the cylinder to characterize the vertical distribution of the population.

3 To protect the organisms from excessive heat, use a petri dish filled with water as a heat shield. Suspend it between the cylinder and the lamp, using a beaker holder.

Materials Needed

- ❑ live *Daphnia magna*
- ❑ 100 ml graduated cylinders
- ❑ springwater
- ❑ fine mesh net or plastic droppers for dispensing *Daphnia*
- ❑ ring stands with clamps to hold cylinders
- ❑ beaker plates to hold heat shields
- ❑ medium-sized petri dishes (heat shields)
- ❑ black construction paper sleeves
- ❑ gooseneck lamps with 60-watt and 150-watt bulbs

4 Next, position the light below the cylinder and look for changes in the population's distribution. In the space provided below, describe the response of *Daphnia* to this low-intensity light. Would you characterize their behavior as positively phototactic?

With these initial observations in mind, you now will perform a series of experiments using different intensities of light. You will compare the vertical migrations of *Daphnia* under controlled conditions in five treatments: (1) no light, (2) low-intensity light from above, (3) low-intensity light from below, (4) high-intensity light from above, and (5) high-intensity light from below. Before beginning the experimental protocol, complete the following section using the steps of the scientific method to approach this investigation systematically.

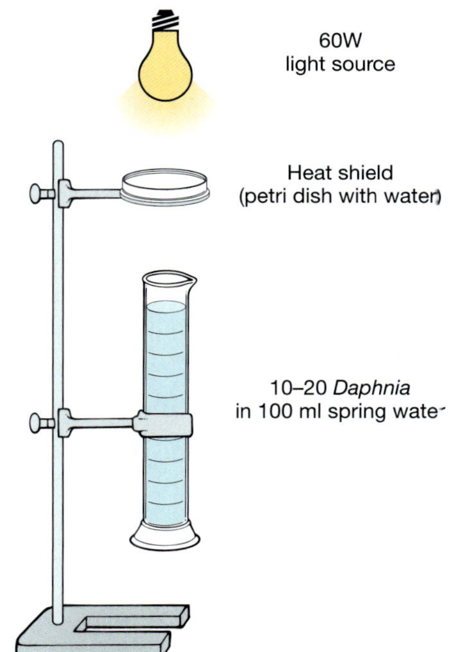

13.26 Experimental setup for gathering initial observations on cladocercan's (*Daphnia*) responses to light from above and below.

Using the Scientific Method

1 What is the general question being addressed by this experiment?

____

2 Record the hypothesis you wish to test.

____

3 Identify the independent variable in this experiment.

____

4 Identify the dependent variable in this experiment.

____

Tip

The *independent variable* is the element that you change or manipulate and the *dependent variable* is the element that you measure to determine how it is affected by the changes you made to the independent variable.

5 What are the levels of treatment (that is, levels of the independent variable)?

____

6 Explain how your initial observations constitute a control treatment.

____

7 Identify several controlled variables in this study.

____

Note

To expedite data collection, set up all five treatments simultaneously but in staggered fashion, a minute or so apart. Also, each group member should be responsible for counting *Daphnia* in a specific volume of the water column, because data collection will have to be completed quickly and accurately.

Experimental Protocol

1. Set up five 100 ml graduated cylinders with populations of *Daphnia* as described in steps 1 and 2 of the previous Experimental Protocol (page 280).
2. To eliminate side light in this study, place sleeves of black paper around the graduated cylinders. To create the "no light" treatment, apply additional pieces of black paper to the top and bottom of the designated cylinder (Fig. 13.27).
3. Use gooseneck lamps with 60-watt bulbs for the low-intensity light treatments and lamps with 150-watt bulbs for the high-intensity light treatments. Be certain that you use heat shields for these four treatments and that you position all four lights the same distance from their respective cylinders (Fig. 13.28).

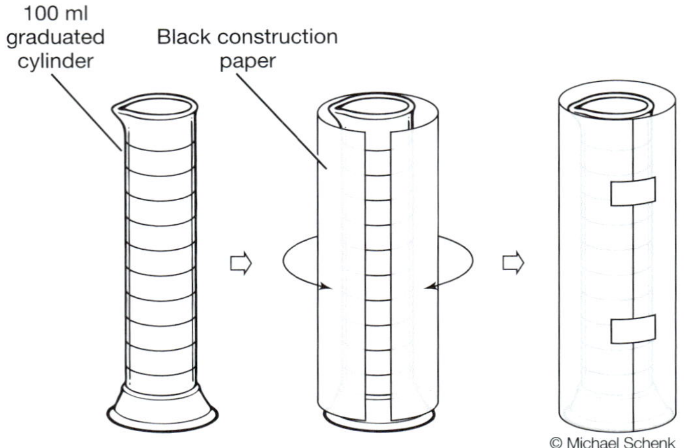

13.27 Procedure for outfitting 100 ml graduated cylinder with black paper sleeve to eliminate side light.

4. Allow the organisms 5 minutes to adjust their positions within the water column in each treatment.
5. After 5 minutes, remove the paper sleeve from each cylinder and quickly count the number of *Daphnia* in each 10 ml volume within the cylinder. Record your data in Table 13-7.

13

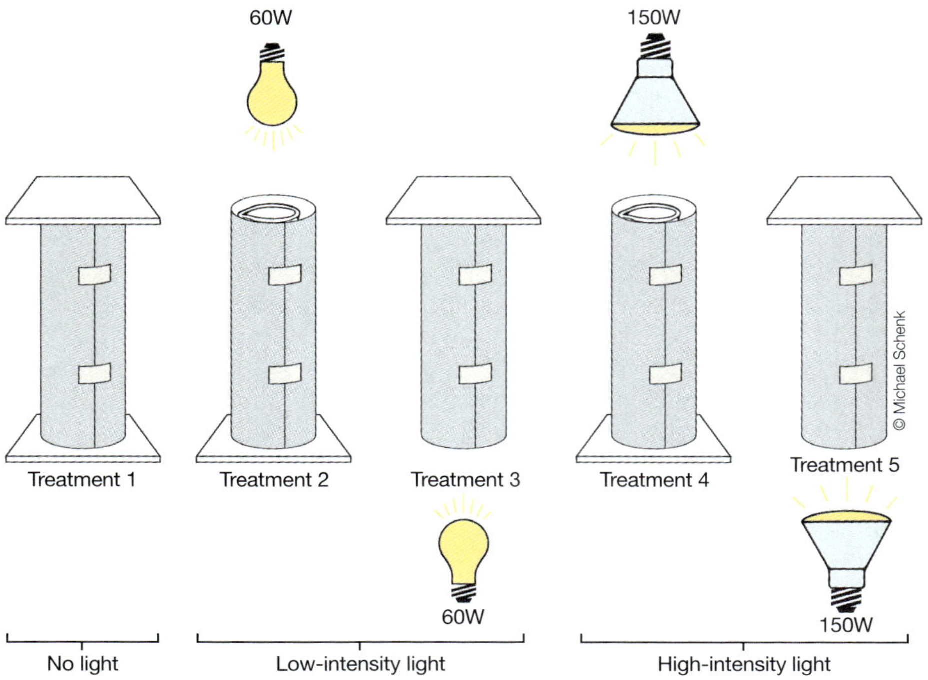

13.28 Experimental setup for testing cladocercan's (*Daphnia*) responses to both low-intensity and high-intensity light from above and below. Be certain that you use heat shields for the four treatments and that you position all four lights the same distance from their respective cylinders.

Table 13-7 Number of *Daphnia* in Each Volume of the Water Column after 5 Minutes of Exposure to Different Light Intensities

10 ml Volumes in Graduated Cylinder	No Light	Low-intensity Light		High-intensity Light	
		Light from Above	Light from Below	Light from Above	Light from Below
90–100					
80–90					
70–80					
60–70					
50–60					
40–50					
30–40					
20–30					
10–20					
0–10					

13

Check Your Progress

7.1 How does positioning the light from above and below allow you to tease apart the response to light from the response to gravity?

7.2 Did the response differ between the no-light, high-intensity light, and low-intensity light treatments? If so, how?

7.3 Based on these findings, would you retain or reject your hypothesis?

13

Name ______________________ Section ______________

13 Questions for Review

1 Complete Summary Table A-1 in the Appendix, filling in the characteristics of both malacostracans (crayfish) and insects (grasshopper) in the appropriate rows.

2 What phylum is considered to be the arthropod's closest relative? What major characteristic do arthropods share with this sister group?

____

____

3 List at least two advantages of an exoskeleton.

1. ______________________

2. ______________________

4 What disadvantages associated with having exoskeletons did arthropods have to overcome to allow natural selection to favor this trait? How did they circumvent these problems?

____

____

____

13

5 Which of the following respiratory structures IS NOT found in arthropods?

a. gills
b. trachea
c. lungs
d. book gills
e. spiracles

6 Match each organism to its correct class.

______ grasshopper	**a.** Class Merostomata
______ centipede	**b.** Class Arachnida
______ horseshoe crab	**c.** Class Chilopoda
______ millipede	**d.** Class Diplopoda
______ scorpion	**e.** Class Malacostraca
______ crayfish	**f.** Class Insecta
______ tick	
______ crab	

7 Arthropods have a/an ______ circulatory system and a/an ______ digestive system.

a. closed, complete

b. open, complete

c. closed, incomplete

d. open, incomplete

e. nonexistent, complete

8 List three major features that grasshoppers and crayfish share in common and three that are found in grasshoppers but not in crayfish.

Grasshoppers and Crayfish:

1. ______________________________

2. ______________________________

3. ______________________________

Grasshoppers ONLY:

1. ______________________________

2. ______________________________

3. ______________________________

9 Which class of arthropods contains the most species?

____

10 True or false? All arthropods have antennae.

____

13

11 Explain the distinction between a kinesis and a taxis in behavioral terms.

____

____

Name ______________________ Section ______________

13 Laboratory Practical Challenge

The following evaluation is intended to replicate the method of assessment and type of questions you will be asked on lab practicals. If you have difficulty with these questions, you may benefit from additional review of the specimens provided in lab.

1 Identify the phylum, subphylum, and class of the specimen depicted below.

2 What is the function of the structure indicated by line A?

3 Identify the structure indicated by line B.

4 Identify the structure indicated by line C.

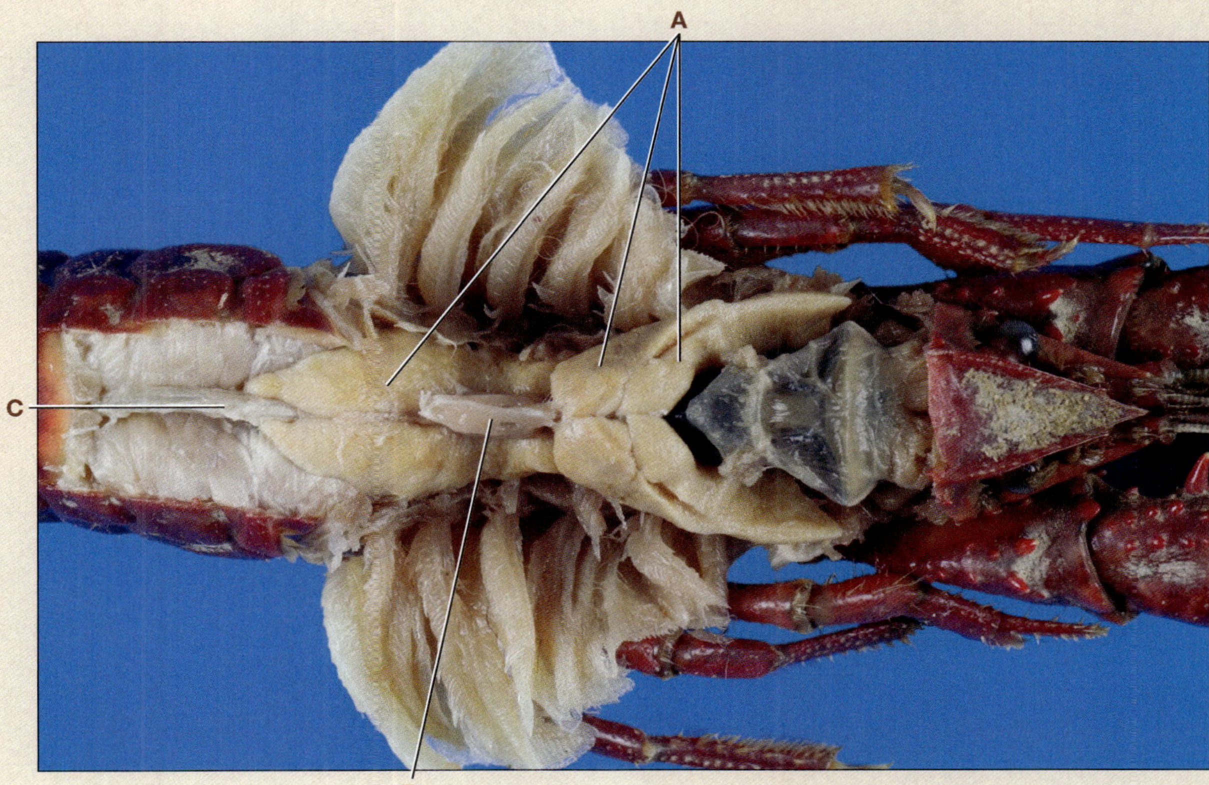

13

5 Identify the phylum, subphylum, and class of the specimen depicted below.

6 What is the function of the structure indicated by line D?

7 Identify the structure indicated by line E.

8 To what system do the structures indicated by line F belong?

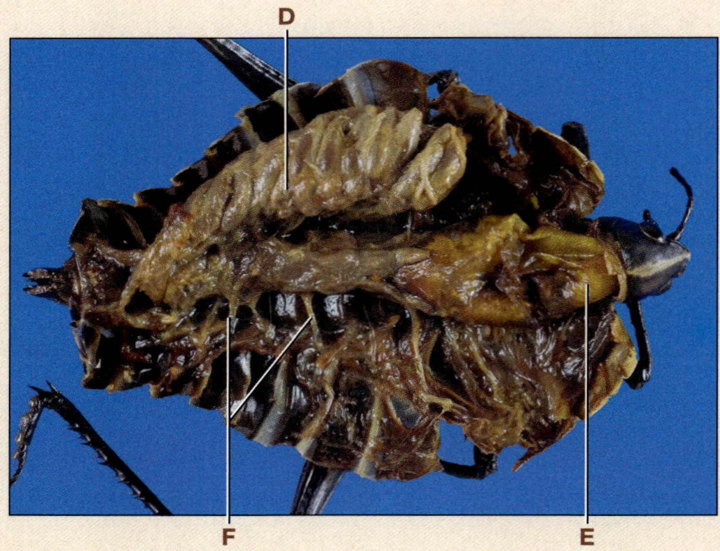

CHAPTER 14

Echinoderms

After completing the exercises in this chapter, you should be able to:

1 List the distinguishing characteristics of the phylum Echinodermata.

2 Identify specimens from the five major classes of echinoderms covered in this chapter.

3 Understand the evolutionary position of echinoderms on the phylogeny of major animal phyla.

4 Identify the internal and external anatomy of a sea star and discuss the functions of each structure.

5 Identify the internal and external anatomy of a sea cucumber and discuss the functions of each structure.

6 Characterize patterns of locomotion in sea stars and sea urchins.

Sea stars, brittle stars, sand dollars, sea urchins, sea cucumbers, feather stars, and sea lilies all belong to the phylum Echinodermata. Nearly all of the 6,425 described species comprising this phylum are marine organisms that vary in size from a few millimeters to more than a meter in length. Although their young are **bilaterally symmetrical**, the adults in this phylum typically display a pattern of **pentaradial symmetry** based on a five-point design. Radial symmetry and the absence of cephalization have evolved in the echinoderm line as adaptations to the primarily sedentary lifestyle these animals display. Echinoderms have no head or brain and, thus, no centralization of nervous control. Instead, they have a nerve ring with nerves radiating evenly around the body. Their skin covers a bony **endoskeleton** composed primarily of calcium carbonate **ossicles**. Tiny, elaborate structures called **dermal branchiae** (skin gills) are distributed along the surface of the epidermis. Dermal branchiae are thin outpockets of epidermis that allow gas exchange to occur between the outer seawater and inner coelomic fluid.

Other than these specialized regions of the epidermis, echinoderms have no true respiratory system—or circulatory system for that matter. Their relatively large **coelom** circulates gases and nutrients throughout the body. A part of the coelom has evolved into a complex system of tubules called the **water vascular system**. This system uses seawater to generate hydraulic pressure that powers their tube feet for locomotion and for capturing prey. Echinoderms are the only group of animals that possess a water vascular system. Five major classes of echinoderms are recognized, and their distinguishing characteristics are summarized in Table 14-1.

14

Evolutionary History

For nearly 300 million years echinoderms were among the most prevalent marine creatures in the oceans. An additional 13,000 species of echinoderms (including more than 20 extinct classes) are known only from the fossil record as a testament to their former reign. Because of their bony endoskeleton, echinoderms fossilized relatively well, providing an excellent record of their evolution dating back to the Cambrian period. Many groups, such as crinoids, date back as far as 600 million years!

Despite their unique appearance, echinoderms share a close evolutionary link to modern chordates (Fig. 14.1). Both phyla are believed to have diverged from a common lineage of bilaterally symmetrical ancestors nearly 600 million years ago, and since have grown apart in appearance. The pentaradial symmetry and lack of cephalization that sets them apart *physically* from chordates develops later, in the adult body, and evolved as an adaptation to their sessile existence. Evidence for their common ancestry with chordates can be seen best in the embryological pathways that both groups display. Echinoderms and chordates are both **deuterostomes**, meaning that

1. their mouths develop from the second embryonic opening,
2. they exhibit **indeterminate development**, and
3. embryonic cells divide by **radial cleavage**.

In deuterostomes the anus develops from the first embryonic opening, the blastopore. All other bilateral invertebrates that we have examined are **protostomes**, meaning that

Table 14-1 Five Classes of Phylum Echinodermata

Class and Representative Animals	Characteristics
Crinoidea (≈ 625 sp.) Sea lilies, feather stars	Stalk present and attached to aboral surface; mouth and anus both present on oral surface, branched arms with feathery pinnules; madreporite absent
Ophiuroidea (≈ 2,100 sp.) Brittle stars	Long, slender arms distinct from central disc; tube feet lack suckers and are not used for locomotion; no ambulacral grooves; madreporite on oral surface
Echinoidea (≈ 950 sp.) Sea urchins, sand dollars	Body spherical or disc-shaped and without arms; endoskeletal plates fused to form test; tube feet with suckers; well-developed, movable spines
Asteroidea (≈ 1,600 sp.) Sea stars	Broad arms indistinct from central disc; tube feet with suckers present in ambulacral grooves; anus and madreporite present on aboral surface
Holothuroidea (≈ 1,150 sp.) Sea cucumbers	Body soft and cylindrical and without arms; spines absent; tube feet with suckers; mouth surrounded by branching tentacles; madreporite internal

1. their mouths develop from the first embryonic opening (blastopore),
2. they exhibit **determinate development**, and
3. embryonic cells divide by **spiral cleavage**.

The fact that the protostome developmental pattern is so widespread among many different phyla suggests that it is the original pattern for animals and that it was characteristic of the common ancestor of all triploblastic animals.

The uniqueness of the deuterostome pattern suggests that it evolved once, in a common ancestor to the few phyla that exhibit it. This theory of a common deuterostome ancestor has gained additional support in recent years by comparisons of rRNA and other molecular research. For a more complete phylogeny depicting the position of echinoderms and the evolutionary relationships among the major animal phyla covered throughout this laboratory guide, review Figure A.1 in the Appendix.

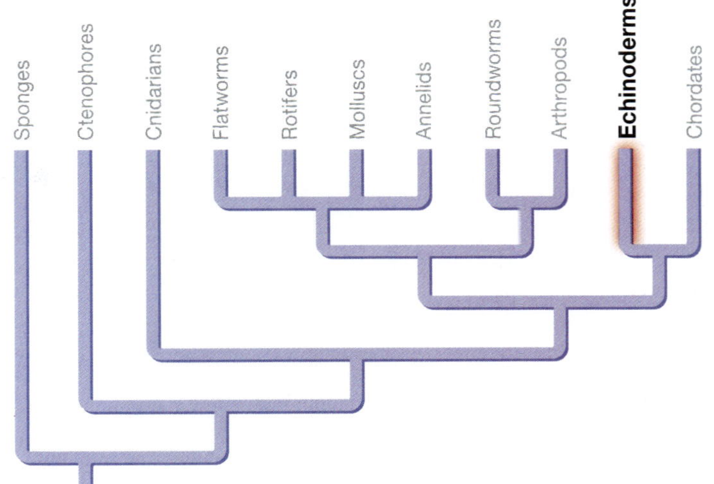

14.1 Echinoderms and chordates are considered sister taxa on the animal phylogeny because they share a common developmental pathway (deuterostome development) and other anatomical features (endoskeleton).

Exercise 14-1 Sea Lilies, Feather Stars (Class Crinoidea)

Crinoids include about 625 living species of sea lilies and feather stars. While they fit the basic characteristics of echinoderms, their anatomy has a much more primitive organization. Zoologists believe that sessile, stalked echinoderms similar to modern crinoids were the ancestors of all other echinoderms. More than 6,000 fossil species are known from this single class. The most obvious feature of a sea lily is the stalk that attaches it to the ocean floor or reef (Fig. 14.2). Small jointed appendages called cirri encircle the stalk and are used by crinoids to prop themselves or to move about the ocean floor.

The body wall is composed of an outer layer of articulating, calcareous ossicles, covered by a thin, poorly developed epidermis. The calyx contains the digestive and reproductive organs. Radiating from the calyx are five flexible, highly branched arms with feathery extensions called pinnules. The pinnules contain sticky mucus to capture small food particles, which are carried along the arms toward the mouth (filter feeding). In crinoids both the mouth and anus are located on the **oral** surface—the surface that points upward and away from the stalk. Another primitive feature of crinoids is the lack of a **madreporite**. Instead of seawater, the water vascular system is filled entirely with coelomic fluid.

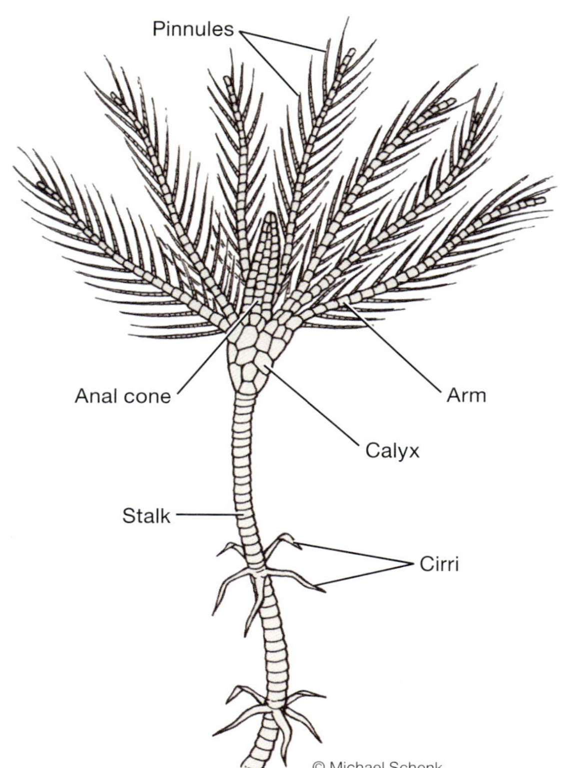

Red-stalked crinoid

14.2 Sea lilies, members of the class Crinoidea, represent the most ancestral body form of modern echinoderms.

Exercise 14-2 Brittle Stars (Class Ophiuroidea)

14

With more than 2,000 named species, brittle stars comprise the most abundant and widespread class of echinoderms, the class Ophiuroidea. Their most characteristic feature is the five long, slender arms with which they rapidly scurry about the ocean floor and will readily cast off if grabbed by a predator—resulting in the common name of "brittle" stars (Fig. 14.3). In members of this class, the arms are distinctly demarcated from the central disc—a distinguishing feature that sets them apart from sea stars in the class Asteroidea. Brittle stars' tube feet lack suckers and are not used for locomotion. Instead they push and pull themselves along with their long, flexible arms.

Despite their abundance, their secretive nature makes them one of the less commonly encountered echinoderms. They usually hide under rocks or in crevices during the day and come out under the cover of darkness to feed. They are filter feeders and use their tube feet to capture suspended microplankton and organic matter and to direct food toward the mouth. Another unique feature of this class is the location of the madreporite on the oral surface next to the mouth.

Green brittle star, *Ophiarachna incrassata*

14.3 Brittle stars, class Ophiuroidea, have long, slender arms that lack suckers on their tube feet.

Exercise 14-3 Sea Urchins, Sand Dollars (Class Echinoidea)

Sea urchins are members of the class Echinoidea, which also includes sand dollars and the lesser-known heart urchins, totaling around 950 species. The most distinguishing feature of sea urchins is their long, rigid, movable **spines**, used primarily in locomotion and defense (Fig. 14.4). The spines are attached at their bases by "ball-and-socket" joints to small muscles that coordinate locomotion with the extensible tube feet. In some species of urchins the spines exude toxins and serve as both a chemical and a physical defense system, allowing them to wedge securely into crevices between rocks and ward off predators with painful stings.

In most species the ossicles underneath the epidermis are fused together, forming large plates that constitute the solid, inflexible **test** that also characterizes members of this class. As an urchin or sand dollar grows, the test enlarges in all directions along sutures, much the way a child's skull grows inside its scalp.

Sand dollars are essentially "flattened" sea urchins, with all of the same internal structures organized inside a compressed test—an evolutionary consequence of their particular niche as echinoderms that burrow beneath the soft sands and slowly creep through sediments in search of food (Fig. 14.4). They have very short, velvety spines covered in cilia that allow them to move across the seabed.

Echinoids are herbivores that use their sharp teeth to scrape algae and plant material from rocks. Highly extensible **tube feet** occur in rows along the body, and small **pedicellariae** are present for cleaning the body surface of parasites and foreign debris. Respiration occurs by means of tiny skin gills (dermal branchiae) located in rows along the body.

The water vascular system of echinoids follows the typical echinoderm plan, with a madreporite, stone canal, ring canal, and lateral canals, the latter being fused to the inner surface of the test along rows that correspond to the arrangement of tube feet protruding through the body wall. Sea urchins and sand dollars are **dioecious** (as you may remember from the fertilization exercises in Chapter 4), and sperm or eggs are released through the genital pores on the **aboral** surface. Fertilization is external in echinoids.

Green sea urchin, *Strongylocentrotus droebachiensis*

Sand dollars, *Echinarachnius parma*

14.4 Sea urchins and sand dollars, class Echinoidea, lack arms altogether and instead have movable spines that facilitate locomotion.

Exercise 14-4 Sea Stars (Class Asteroidea)

Materials Needed
- ❏ preserved sea star (*Asterias*)
- ❏ dissecting tools
- ❏ dissecting pan
- ❏ dissecting microscope
- ❏ latex gloves

Sea stars, often erroneously referred to as "starfish," belong to the class Asteroidea, a group of slow-moving, carnivorous predators that methodically stalk their prey—usually even slower moving molluscs—along rocky coastlines and coral reefs. The undersides of their arms are covered with tube feet, bearing suckers that are used for both locomotion and prey capture. They have a tenacious grip and many species can pry open the halves of a fully closed clam or mussel just enough to evert the cardiac stomach and insert it into the gap in the shell to begin digesting their prey externally! The approximately 1,600 living species of sea stars make them the second largest group of echinoderms (Fig. 14.5).

Sea stars (starfish), *Asterias*

Sea star (starfish), *Asterias*

14.5 Representative sea stars, class Asteroidea.

Sea Star External Anatomy

Procedure

1 Obtain a preserved specimen of *Asterias* or another genus of sea star, and place it in your dissecting pan. Examine the external features of your specimen and refer to Table 14-2 as you review the anatomy. Even though most echinoderms are dioecious, sea stars have no external features to distinguish the sexes.

2 Your first task is to distinguish the oral surface from the aboral surface. The oral surface is the "underside" of the animal, containing the mouth and hundreds of tube feet embedded in grooves along the arms. The aboral surface bears the madreporite—the off-center, external opening to the water vascular system (Fig. 14.6). The anus is located in the center of the body on the aboral surface but usually is too small to be visible with the naked eye.

Table 14-2 External Anatomy of the Sea Star

Structure	Function
Spines	Calcareous projections for protection and support
Pedicellariae	Pincerlike structures believed to kill small organisms that might settle on body surfaces, thus keeping the epidermis free of parasites and algae
Dermal branchiae	Gas exchange and excretion through simple diffusion
Mouth	External opening to cardiac stomach (through a short esophagus)
Anus	Regulates egestion of undigested food (feces) from the body
Madreporite	Porous entrance to the water vascular system that serves as both pressure regulator and simple filter
Ambulacral grooves	Radiate from the mouth to the tip of each arm and house the tube feet
Tube feet	Locomotion and prey capture

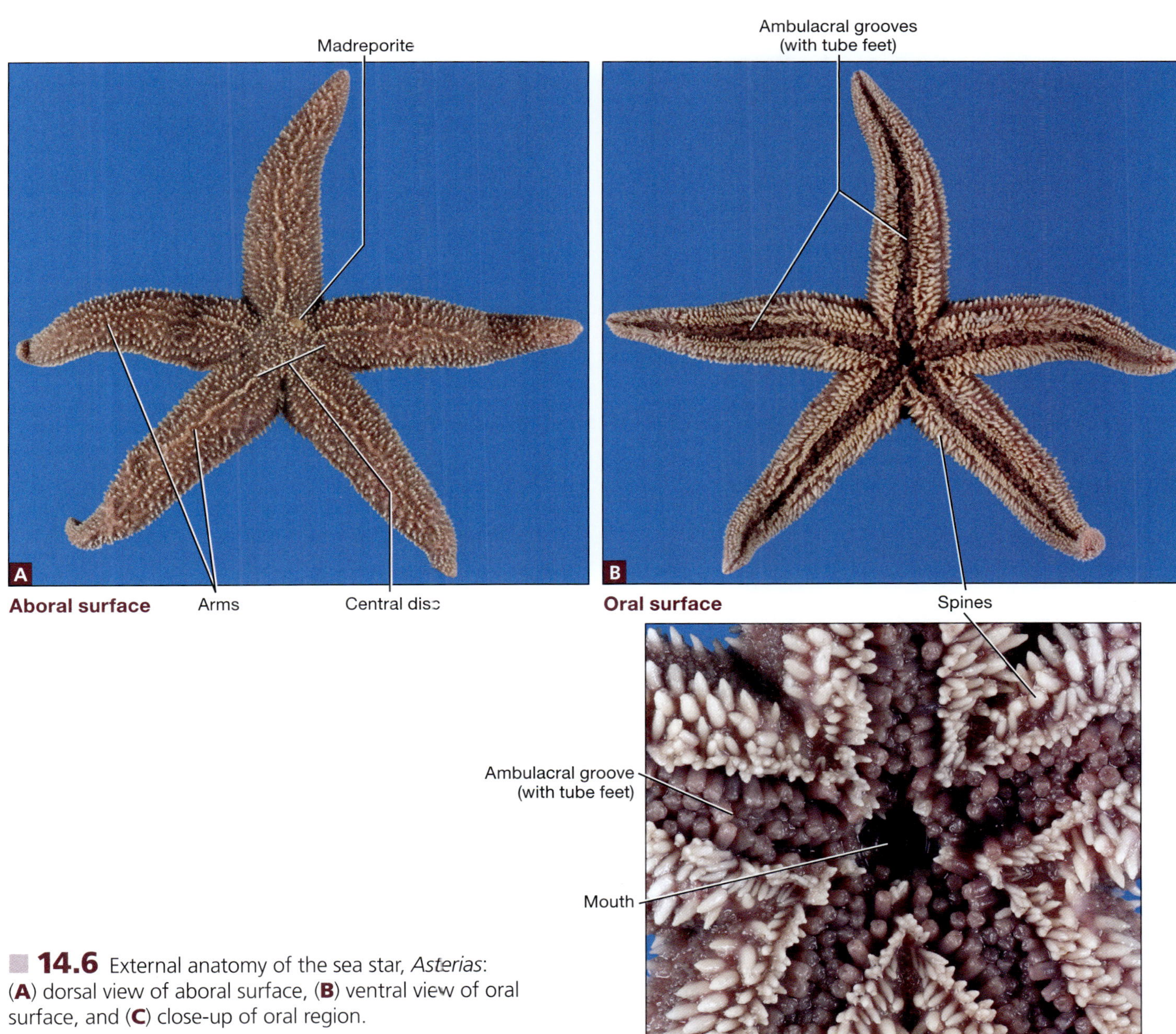

14.6 External anatomy of the sea star, *Asterias*: (**A**) dorsal view of aboral surface, (**B**) ventral view of oral surface, and (**C**) close-up of oral region.

3 Using a dissecting microscope, scan the epidermis on the aboral surface for small, pincerlike pedicellariae, which the sea star uses to cleanse its skin surface of parasites and foreign debris; and for dermal branchiae, which aid in gas exchange and excretion (by simple diffusion). In the space provided, draw a picture of the pedicellariae as they appear on your specimen.

Check Your Progress

1.1 Are the spines of sea stars movable?

1.2 Were you able to locate pedicellariae on the epidermis? Dermal branchiae?

1.3 How do the arms and central disc of sea stars differ from those of brittle stars discussed earlier?

14

Sea Star Internal Anatomy

To reveal the internal anatomy of your sea star, you will make several cuts through the epidermis. Review the Dissection Diagram to assist you in this process.

Procedure

1 Use a sharp pair of dissecting scissors to make a shallow incision on the aboral side through the epidermis encircling the central disc, leaving the madreporite in place.

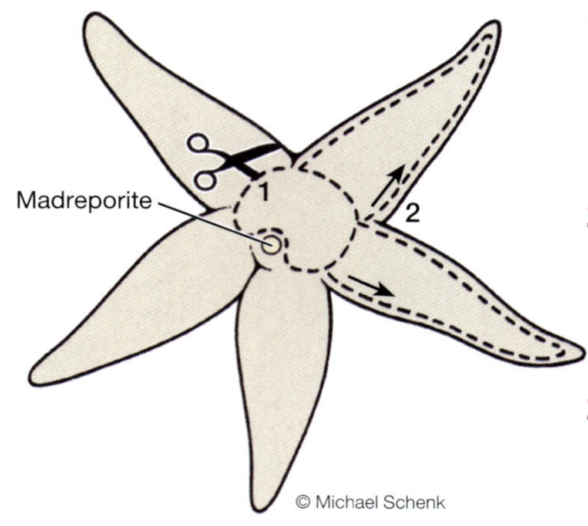

Dissection Diagram

1 Use scissors to make a shallow incision through the epidermis encircling the central disc, leaving the madreporite in place.

2 Make additional incisions along the outer margins of one or more arms, extending from original circular incision distally along arm.

3 Gently tease away the separated portions of the epidermis, being careful not to damage underlying structures.

2 Make additional incisions along the outer margins of one or more arms, extending from the original circular incision distally along each arm.

3 *Do not* carelessly rip away the epidermis; rather, gently tease it apart from the underlying tissue with a dissecting needle or blunt probe, especially in the region of the central disc. The underlying organs lie very close to the inner surface of the epidermis and actually adhere to it in some places. Sloppy technique here may ruin underlying organs!

4 The internal anatomy of each arm is identical, so you should work from the aboral surface toward the oral surface, identifying each structure or organ as you proceed (Table 14-3).

Digestive and Reproductive Systems

A challenge faced by sea stars during their developmental transformation from a bilateral juvenile into a radial adult is the reorganization and compression of the digestive system. Because of their flattened body shape and the limited distance between the mouth and the anus, the coelomic spaces in the arms have been recruited to house the bulk of the digestive system.

The mouth, located on the oral side, leads through a short esophagus to the muscular **cardiac stomach** that fills most of the central disc. Connected to the cardiac stomach is the thin-walled, **pyloric stomach** (Figs. 14.7 and 14.8). Pairs of large **digestive glands** extending into each arm connect to the pyloric stomach by thin pyloric ducts. Passing aborally from the pyloric stomach is a short **intestine**. Attached to the intestine are small, branched, rectal ceca that increase the absorptive surface area of the intestine. Undigested wastes are eliminated from the body through the **anus**, which opens on the aboral surface.

When a sea star captures a bivalve, it locates the opening of the shell and attaches numerous tube feet to each valve. The arms pull tightly on the two valves, eventually causing the bivalve's muscles to fatigue and allow the sea star to pry apart the shell just enough to create a thin opening. The sea star everts its cardiac stomach through its mouth and wedges it into the small slit in the bivalve's shell. Powerful enzymes from the digestive glands pour onto the body of the bivalve, quickly digesting the bivalve's soft tissues, and the partially digested fluid is drawn through the cardiac stomach into the pyloric stomach, where digestion continues. Once feeding is complete, the sea star pulls its stomach back through its mouth into its original position in the coelomic cavity.

The reproductive system of the sea star consists of pairs of **gonads** housed along the interior, oral surfaces of each arm, underneath the digestive glands (Figs. 14.7 and 14.8). At the point where the arm joins the central disc, each gonad is attached to the side of the arm by a small duct that delivers sperm or eggs outside the body. Sea stars, like all echinoderms, are dioecious, but the sexes are difficult to differentiate through dissection due to their lack of distinguishing external features. Fertilization is external; thus, males and females must coordinate the timing of their reproductive efforts to ensure fertilization of the eggs.

Table 14-3 Internal Anatomy of the Sea Star

Structure	Function
Cardiac stomach	Can be everted through the mouth to envelope prey; site of initiation of digestion
Pyloric stomach	Receives secretions of digestive glands for chemical digestion
Digestive glands	Secrete digestive enzymes for breakdown of food; play a major role in absorption and storage of food materials
Rectal ceca	Small, branched "trees" that increase the surface area of the intestine for more efficient nutrient absorption
Gonads	Produce gametes for reproduction
Ring canal	Portion of water vascular system encircling the mouth
Radial canals	Portions of water vascular system emanating from the ring canal and leading into each arm of the sea star
Stone canal	Portion of water vascular system leading from the madreporite to the ring cana
Madreporite	Porous entrance to the water vascular system that serves as both pressure regulator and simple filter
Ampullae	Provide hydraulic pressure for movement of the tube feet
Ossicles of endoskeleton	Support

14

14.7 *Asterias* dissection.

14

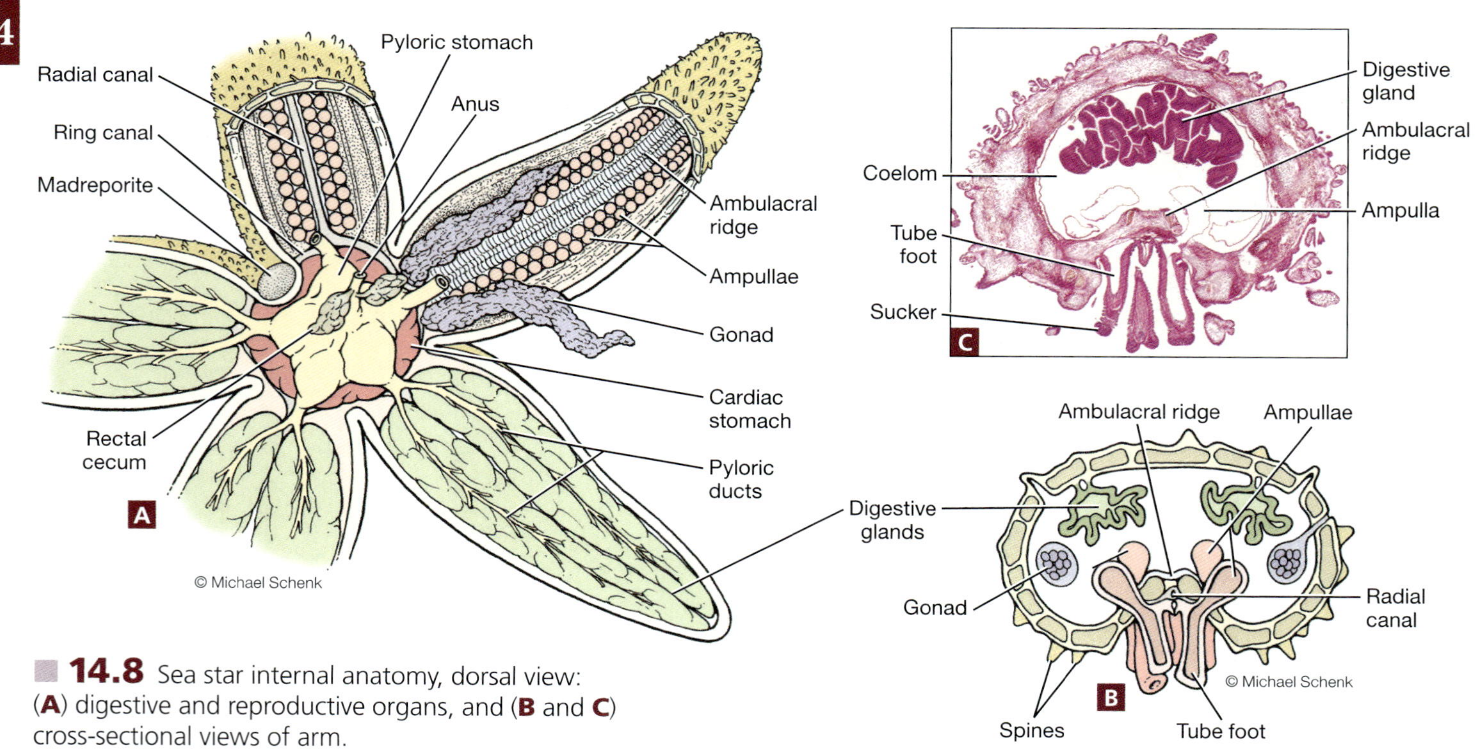

14.8 Sea star internal anatomy, dorsal view: (**A**) digestive and reproductive organs, and (**B** and **C**) cross-sectional views of arm.

5 To view the water vascular system, you must remove the components of the digestive and reproductive systems from an arm and the central disc.

Water Vascular System

The water vascular system of the sea star is fairly representative of echinoderms in general. It consists of a series of internal canals that branch from a centralized **ring canal** into primary **radial canals** that terminate into the hundreds of tube feet that line the **ambulacral grooves** along the oral surfaces of the arms (Figs. 14.8–14.10). Seawater enters this system through the madreporite and is kept under pressure by muscular contractions by the sea star, affecting an entire repertoire of behaviors, including locomotion, reproduction, defense, and prey capture.

The tube feet are connected to **ampullae** that contain seawater. When small valves along the ambulacral ridge are closed, contraction of the muscles in an ampulla forces water into the tube foot, causing it to elongate lengthwise. Postural muscles in the tube foot allow for lateral movements of the tube foot, while contractions of retractor muscles shorten the tube foot and force the water back into the ampulla. Remarkably, coordination of the hundreds of tube feet is accomplished by the nerve ring and radial nerves without any centralized control center!

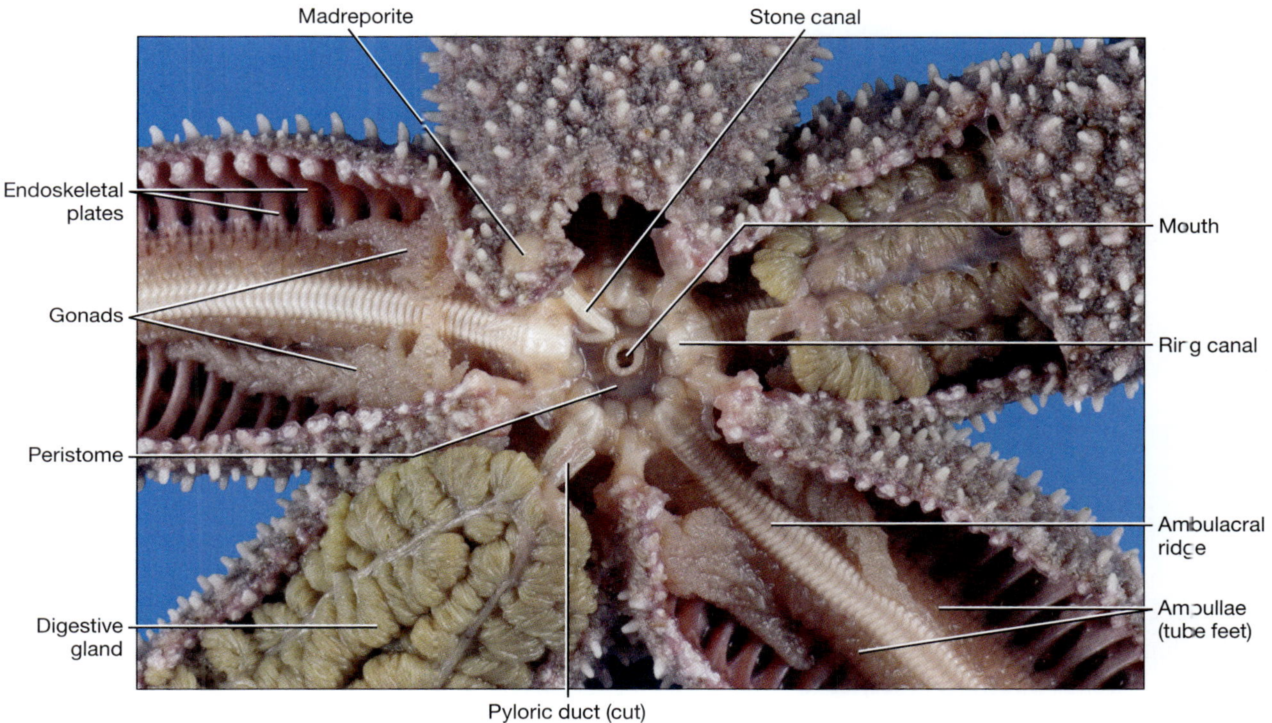

14.9 *Asterias* dissection. Cardiac and pyloric stomachs removed to show water vascular system.

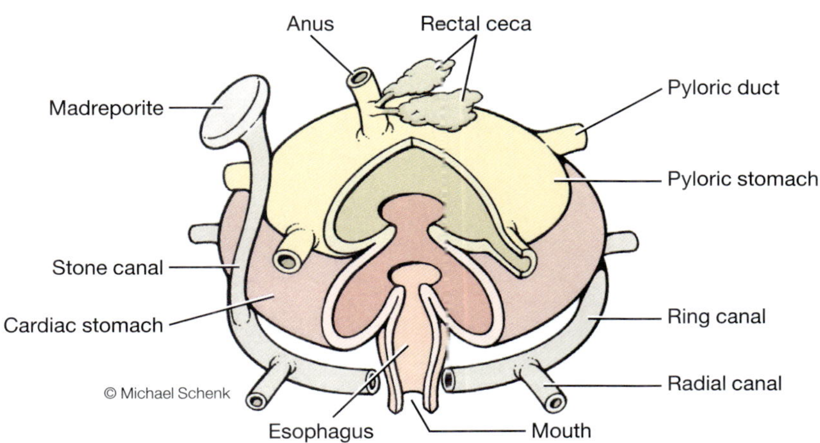

14.10 Cutaway view of the digestive system and water vascular system of the sea star.

14

Check Your Progress

2.1 What role does the madreporite play in the water vascular system of sea stars?

2.2 Does the sea star possess a heart or circulatory system? How do you suppose nutrients and oxygen are distributed to the body tissues?

2.3 Does the sea star possess an excretory system? How do you suppose metabolic wastes are eliminated from the body?

2.4 Why are ossicles classified as an endoskeleton?

14

Exercise 14-5 Sea Cucumbers (Class Holothuroidea)

Materials Needed

- ❏ preserved sea cucumber
- ❏ dissecting tools
- ❏ dissecting pan
- ❏ latex gloves

Sea cucumbers belong to the class Holothuroidea. This class contains approximately 1,150 species of cylindrical, soft-bodied, sluggish, filter-feeding echinoderms (Fig. 14.11). Sea cucumbers superficially resemble elongated urchins with their spines removed and their ossicles reduced to microscopic structures embedded deep within the body wall. Their tough, leathery epidermis lacks spines, pedicellariae, and dermal branchiae.

Procedure

1. Obtain a preserved sea cucumber and place it in your dissecting pan.
2. Examine the external features of your sea cucumber and refer to Table 14-4 as you review the anatomy. Notice the warty, leathery texture of its skin. Tube feet occur sparsely along the body wall and are arranged in five longitudinal rows, corresponding to the pentaradial symmetry characteristic of all echinoderms. Near the mouth, the tube feet are

14.11 Sea cucumbers are members of the class Holothuroidea.

modified into feathery **tentacles** that sea cucumbers use to capture suspended food particles from the surrounding water (Fig. 14.12). Sea cucumbers sometimes are categorized as deposit feeders, because they are also able to extract organic matter from the fine muddy deposits of the deep seafloor.

3 To view the internal anatomy of your specimen use dissecting scissors to make a longitudinal incision along the body wall from the tentacles toward the anus.

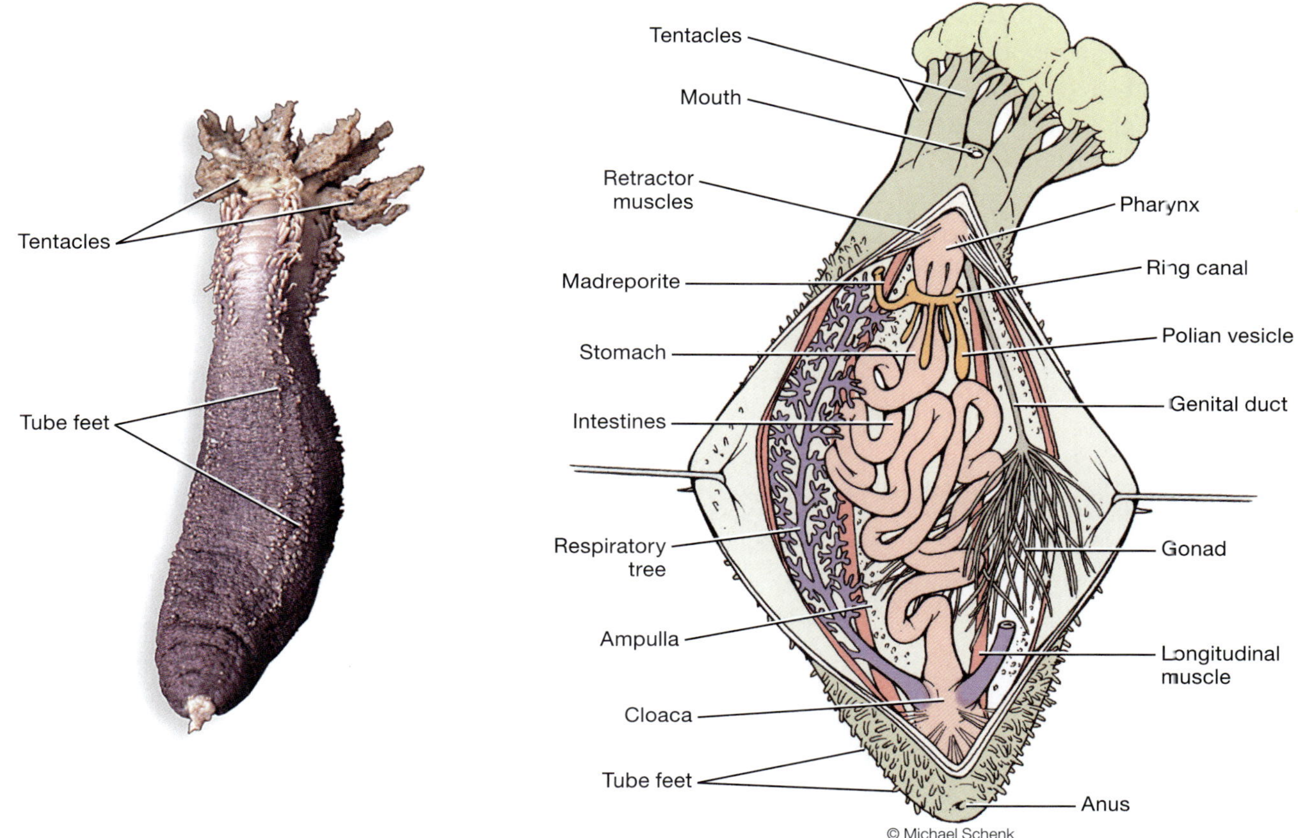

14.12 Sea cucumber, *Cucumaria*, external and internal anatomy.

Table 14-4 Anatomy of the Sea Cucumber

Structure	Function
Tentacles	Food collection
Ring canal	Portion of water vascular system encircling the junction of pharynx and stomach
Ampullae	Provide hydraulic pressure for movement of the tube feet
Pharynx	Transports food from mouth to stomach
Madreporite	Entrance to the water vascular system that serves as a pressure regulator; floats freely within the coelomic cavity and is not connected to the outer body wall
Stomach	Extremely reduced in holothurians; site of food storage and initiation of digestion
Intestine	Final digestion of food received from stomach and subsequent nutrient absorption
Gonad	Produces gametes for reproduction
Tube feet	Locomotion
Respiratory trees	Gas exchange
Cloaca	Pumps a ventilating current of seawater into and out of respiratory trees
Anus	Regulates egestion of undigested food (feces) from the body

The digestive system of sea cucumbers contains a muscular **pharynx**, small stomach, and a greatly elongated intestine leading to an anus (Fig. 14.12). The water vascular system follows the typical echinoderm pattern, except for the madreporite, which lacks a direct connection to the outside of the body. Instead, it floats freely in the large coelomic cavity; thus, the water vascular system of sea cucumbers uses coelomic fluid rather than seawater as its pressure-generating medium.

Sea cucumbers, like most echinoderms, lack a circulatory system and use coelomic fluid as the primary circulatory medium as well. A single, filamentous gonad is present near the oral end of the body, and gametes are released through a gonopore near the base of the tentacles. Unique paired structures, found only in holothurians, are the **respiratory trees**—highly branched outgrowths of the hindgut (Fig. 14.12). Typically, two of these structures are present and are connected to the **cloaca**. Seawater is pumped into and out of the respiratory trees by muscular contractions of the anus, and gas exchange occurs across the thin tissues of the respiratory trees as water passes over them.

No discussion of sea cucumbers would be complete without mentioning their amazing ability to eviscerate their internal organs. When sufficiently disturbed or provoked, the muscular body wall quickly contracts and some portion of the viscera is forced out through the anus, essentially turning the sea cucumber inside out! In some species this is limited to expulsion of the respiratory trees, but others exhibit true evisceration and the entire contents of the body cavity are expelled. Eventually the lost body parts are re-formed, reflecting the astounding regenerative capabilities of echinoderms!

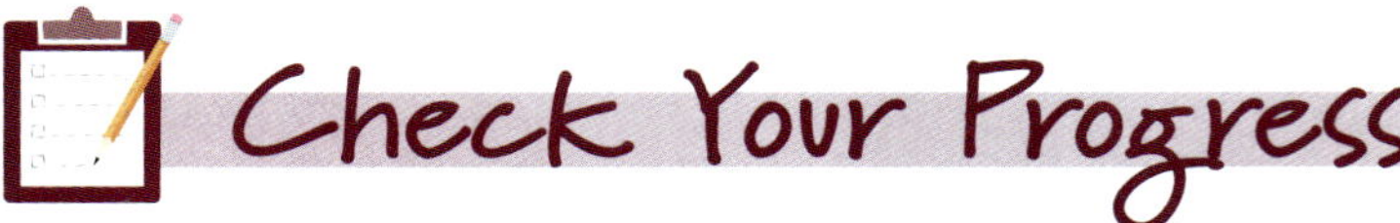

3.1 What method of food acquisition do sea cucumbers employ?

3.2 Because sea cucumbers have lost the sharp spines and hard endoskeleton on which most echinoderms rely for protection, how do they defend themselves?

14

3.3 What characteristics do sea cucumbers possess that would cause zoologists to group them in the same phylum with other echinoderms?

Exercise 14-6 Investigating Locomotion in Sea Stars and Sea Urchins

Materials Needed
- ❑ live sea stars and sea urchins
- ❑ 2 large finger bowls
- ❑ seawater

If living sea stars and sea urchins are available in your laboratory, you can perform some simple exercises to characterize the different methods of locomotion that these two groups use. Despite the simple, decentralized nature of the echinoderm nervous system, they clearly are capable of sophisticated, coordinated behaviors and movements. Evidence even supports the contention that a limited amount of learning is possible in some echinoderms!

Procedure

1 First obtain a living sea urchin and place it in a large finger bowl filled with seawater.

2 Observe its pattern of locomotion and answer the following questions.

Are the spines utilized in locomotion?

Are the tube feet utilized in locomotion?

3 Gently roll the sea urchin over on its "side." Be careful not to pull too hard on it as you dislodge the urchin from the glass container or you will rip off its tube feet!

Can the sea urchin move as effectively on its "side" as in its former orientation, or does it display a tendency to right itself to its former position?

Do you suppose that sea urchins can move fairly easily in any orientation?

4 Now obtain a living sea star and place it in another large finger bowl containing seawater.

5 Observe its pattern of locomotion and answer the following questions.

Are the spines utilized in locomotion?

Are the tube feet utilized in locomotion? Do the tube feet move in random fashion or in smooth, coordinated waves?

Do the arms move appreciably during locomotion?

6 Gently flip the sea star over onto its aboral surface, again being careful to dislodge it *gently* from the bottom of the container.

Can the sea star move as effectively on its aboral side as in its former orientation?

Did the sea star "right itself" to its former position with the oral surface facing downward?

What does this tell you about a sea star's ability to sense directional stimuli such as gravity?

Do you suppose that sea stars can move with equal ease in any orientation?

7 Once again, gently flip the sea star over onto its aboral side. This time, as it rights itself, pay close attention to which of its arms it uses in the righting response.

8 Repeat this flipping process several more times—each time identifying which arm (or arms) the sea star uses to accomplish the righting response. You may find it helpful to place a small identifying mark on the arm or arms that you wish to watch, especially if you cannot distinguish them from the other arms.

Does the sea star favor one arm over the others when righting itself?

What does this tell you about a sea star's ability to distinguish among its arms and coordinate their movements?

How is this similar to dominant "handedness" in humans?

14

Name ____________________ Section ____________

14 Questions for Review

1 Complete Summary Table A-1 in the Appendix, filling in the characteristics of echinoderms in the appropriate row.

2 Given that modern echinoderms and chordates look nothing alike, what factors have led zoologists to propose a close evolutionary relationship between these two phyla?

3 Echinodermata is the first phylum you have examined whose members possess true endoskeletons. What are some advantages that an endoskeleton provides?

4 What is the primary function of the water vascular system?

5 What is unique about the madreporite of sea cucumbers (class Holothuroidea)?

6 Which of the following structures in echinoderms *does not* play a role in respiration (gas exchange)?

a. dermal branchiae

b. respiratory trees

c. coelomic fluid

d. pedicellariae

e. all play a role in respiration

7 Match the class with the correct organism or characteristic.

______	*Asterias*	**a.** Class Crinoidea
______	sea lilies, feather stars	**b.** Class Ophiuroidea
______	sea cucumbers	**c.** Class Echinoidea
______	brittle stars	**d.** Class Asteroidea
______	soft, cylindrical body with leathery skin	**e.** Class Holothuroidea
______	body contains stalk; mouth and anus on oral surface	
______	body plates fused into solid test with long, movable spines	

8 Briefly explain how radial symmetry and the lack of cephalization usually associated with it can have adaptive value for sessile animals and be favored by natural selection.

____

____

9 Fill in Table 14-5 with the appropriate characteristics for each class of echinoderms covered to compare the adaptations to different lifestyles among members of these groups.

Table 14-5 Comparison of Major Characteristics among Echinoderm Classes

	Class				
Characteristic	**Crinoidea**	**Ophiuroidea**	**Echinoidea**	**Asteroidea**	**Holothuroidea**
Shape of arms					
Tube feet (present or absent; lack or possess suckers)					
Ossicles (well developed, reduced, etc.)					
Feeding method					
Location of madreporite					
Spine structure					

Name ______________________ Section ______________

14 Laboratory Practical Challenge

The following evaluation is intended to replicate the method of assessment and type of questions you will be asked on lab practicals. If you have difficulty with these questions, you may benefit from additional review of the specimens provided in lab.

1 Identify the phylum and class of the specimen depicted below.

____

2 What is the function of the structure indicated by line A?

____

3 Identify the structure indicated by line B.

____

4 To what system does the structure indicated by line C belong?

____

14

5 On the figure below, label the following structures: gonad, respiratory tree, mouth, pharynx, and anus.

6 To what system does the structure indicated by line D belong?

---

7 Identify the structure indicated by line E.

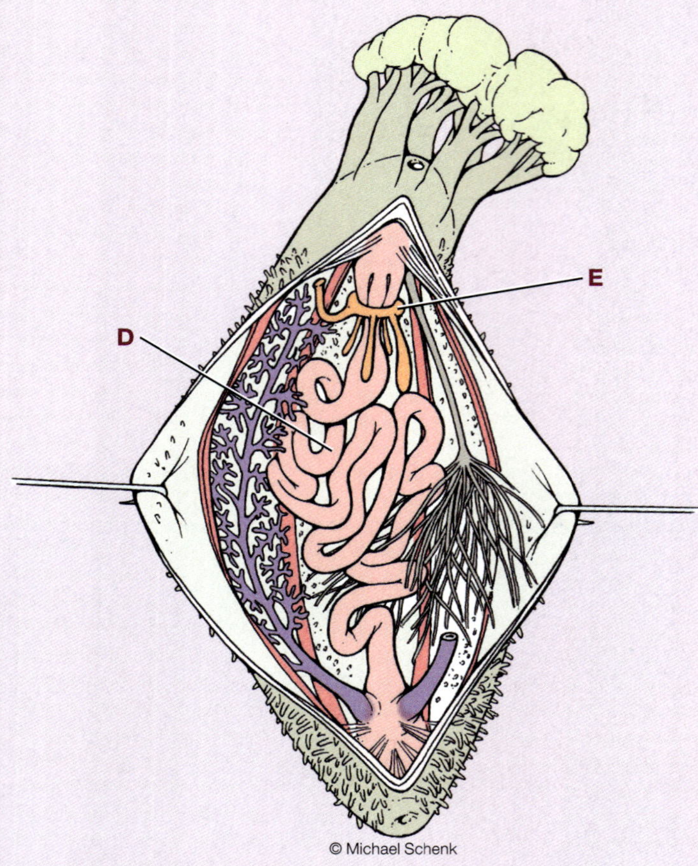

14

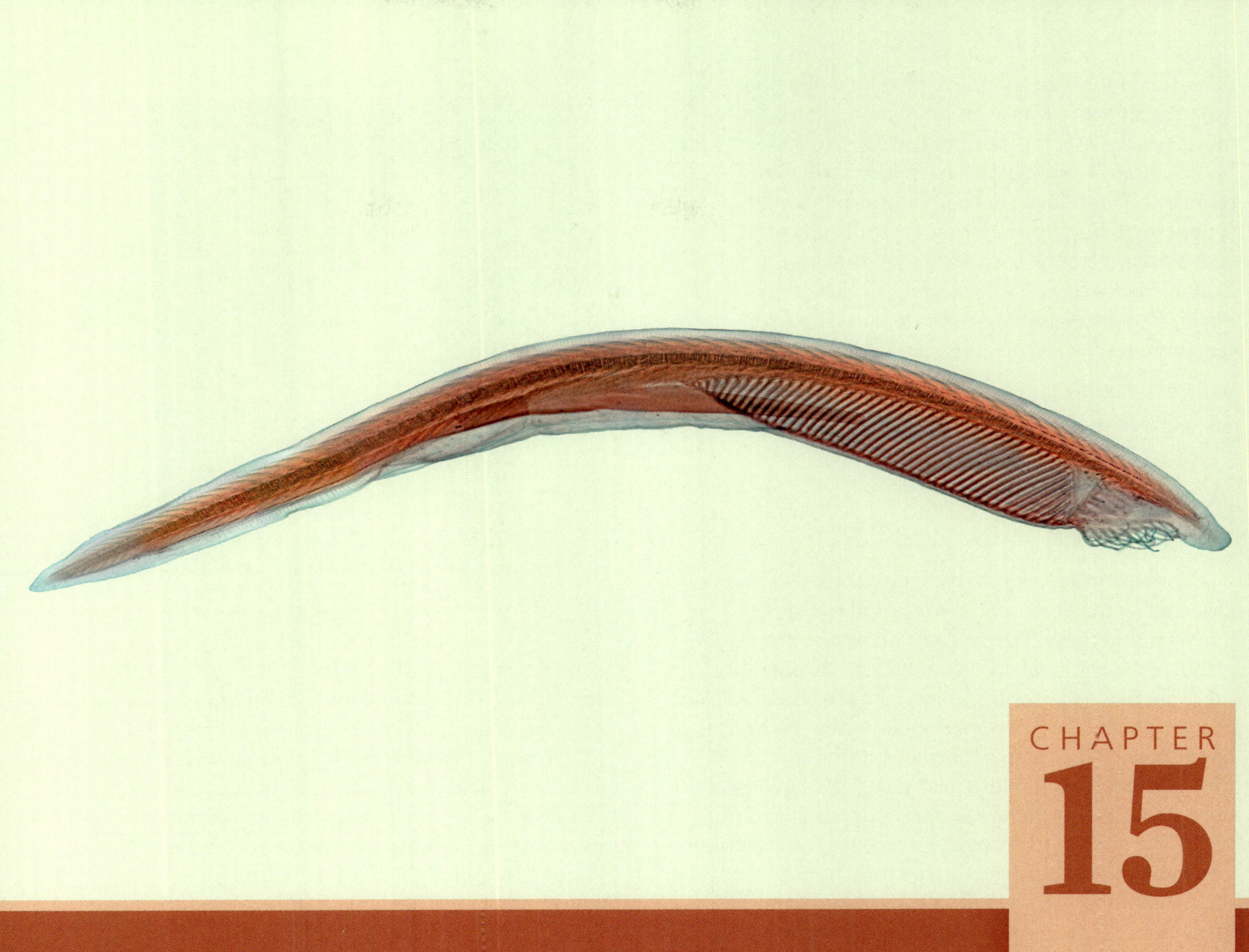

CHAPTER 15

Tunicates and Cephalochordates

After completing the exercises in this chapter, you should be able to:

1 List the five distinguishing characteristics of the phylum Chordata.

2 List the major characteristics of tunicates and cephalochordates and distinguish between members of these two subphyla.

3 Understand the evolutionary position of tunicates and cephalochordates on the phylogeny of chordates.

4 Identify the internal and external structures of the tunicate, discuss their functions in the body, and compare them to homologous structures in other chordates.

5 Identify the internal and external structures of the lancelet, discuss their functions in the body, and compare them to homologous structures in other chordates.

Chordata is a large, familiar phylum containing a tremendously diverse group of more than 55,000 living species. We are chordates, as are birds, dogs, lizards, frogs, lampreys, fishes, sharks, turtles, and elephants. Chordates are triploblastic, bilaterally symmetric deuterostomes that typically have an endoskeleton and closed circulatory system—characteristics they share with many other animal phyla. Five distinguishing characteristics, however, set chordates apart from all other animal groups:

1. a **notochord,**
2. a **dorsal, hollow nerve cord,**
3. **pharyngeal gill slits,**
4. a **postanal tail**, and
5. an **endostyle**, or thyroid gland.

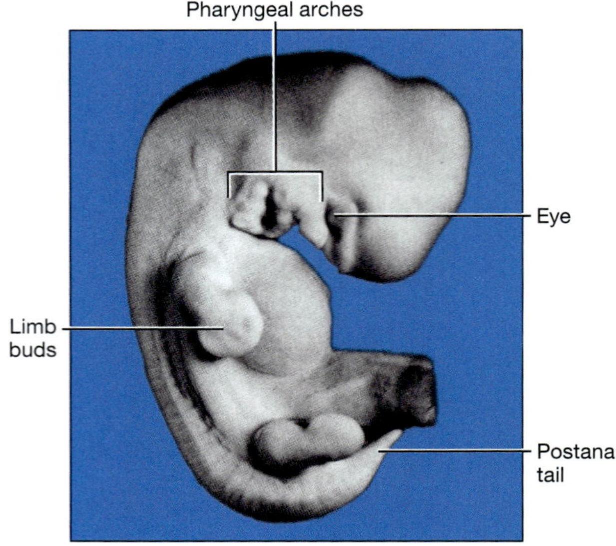

15.1 Pharyngeal gill arches and a postanal tail can easily be seen in this seven-week-old human embryo.
Photo courtesy of the Carnegie Institution of Washington, DC.

Until recently, biologists recognized only four distinct chordate characteristics. New research has shown, however, that the endostyle, or its derivative, the thyroid gland, is present in all chordates and is not found in any other animals. All chordates have all five of these features at some time in their lives, often only during embryological development. For example, human embryos have all five characteristics but lose their pharyngeal gill arches, notochord, and postanal tail during development (Fig. 15.1). The notochord becomes modified into the vertebral column, and the pharyngeal arches form parts of the jaw, the hyoid bone, the three middle-ear bones, and the Eustachian tubes connecting the middle ears to the oral cavity.

Chordates are a **monophyletic** group, with all descendants having a common evolutionary ancestor, so many of the morphological features of the members of its subgroups are **homologous**. Homologous structures are inherited structures in different species that are similar because of their common ancestry. The five chordate characteristics listed above are all examples of homologous structures. As you proceed through the next several chapters, keep this in mind and attempt to identify the many homologies in each subgroup we study.

Because of our extensive knowledge of these groups and their evolutionary relevance to humans, the remainder of this book is devoted to exploring the homologies and intricate differences among the major subphyla and classes of chordates. In this chapter you will examine two subphyla within the phylum Chordata: the subphylum Urochordata (also known as Tunicata), which includes tunicates; and the subphylum Cephalochordata (also known as Acrania), which includes lancelets. Studying these two groups of chordates first gives us a reference point to learn more about what early chordates were like and the evolutionary line that led to the third subphylum of chordates: Vertebrata. These three subphyla, their major classes, and their distinguishing characteristics are summarized in Table 15-1.

Table 15-1 Major Subphyla and Selected Classes of Phylum Chordata

Subphylum	Class and Representative Animals	Characteristics
Urochordata[1] (Tunicata)	**Ascidiacea** (≈ 1,600 sp.) Tunicates	Notochord and nerve cord present only in larvae; adults are sessile; segmentation absent; body enclosed in tunic; entirely marine
Cephalochordata[2] (Acrania)	**Cephalochordata** (≈ 29 sp.) Lancelets	Notochord and nerve cord present throughout life; segmentation present; lance-shaped and without head; entirely marine
Vertebrata	**Petromyzontida** (≈ 35 sp.) Lampreys	Enlarged brain enclosed in cranium; nerve cord surrounded by rudimentary vertebrae, jaws absent; paired fins absent; notochord present throughout life
	Chondrichthyes (≈ 750 sp.) Sharks, rays, skates	Jaws present; cartilaginous endoskeleton; gill opercula absent; placoid scales; claspers present; internal fertilization; mostly marine
	Actinopterygii (≈ 27,000 sp.) Ray-finned fishes	Jaws present; bony endoskeleton; fins with bony rays; gill opercula present; swim bladder present; external fertilization; both marine and freshwater
	Amphibia (≈ 4,800 sp.) Frogs, toads, salamanders	Moist skin without scales; unshelled eggs; undergo larval stage prior to becoming adult; tetrapods with four digits on forelimbs; gas exchange via lungs and/or moist skin; external fertilization; ectothermic
	Reptilia (≈ 8,000 sp.) Turtles, lizards, snakes, alligators, crocodiles	Dry skin with keratinized scales; amniotic egg usually with leathery shell; no larval stage; tetrapods with five digits on forelimbs, claws present; gas exchange via lungs; internal fertilization; ectothermic
	Aves[3] (≈ 8,800 sp.) Birds	Front pair of limbs usually modified for flight; feathers present; horny beak without teeth; hard-shelled amniotic egg; endothermic
	Mammalia (≈ 4,800 sp.) Mammals	Body covered with hair; mammary glands present; single lower jaw bone (mandible); embryos develop in uterus with placental attachment; complete secondary palate; endothermic

[1] The name Tunicata has been suggested as a replacement for the subphylum name Urochordata.

[2] There are currently taxonomic discrepancies about whether the name Cephalochordata should be the subphylum or the class name for this group. The alternative subphylum name of Acrania also is considered valid by the ITIS (Integrated Taxonomic Information System). Source: www.itis.gov.

[3] Many zoologists now consider Aves to be a subclass of Reptilia

15

Evolutionary History

The first chordates arose from a deuterostome lineage related to echinoderms around 570 million years ago (mya), during the same geologic era when early molluscs, annelids, and echinoderms were diversifying in the oceans (Fig. 15.2). During this relatively brief evolutionary period (of 40 million years) most of the 34 animal phyla that we recognize today came on the scene. So significant was this time period in animal evolution that it often is referred to as the Cambrian explosion.

Earlier theories about the first chordates hypothesized that they were sessile creatures similar to the adult tunicates you will study in this chapter, and that the free-swimming chordates evolved from their motile larvae, which developed functional gonads before undergoing metamorphosis and becoming sessile adults. More recent fossil evidence and molecular data suggest that the sessile tunicates represent a derived body form and that free-swimming adult forms actually were the ancestral condition. Thus, our current thinking is that early chordates somewhat resembled the motile larvae of tunicates or the adults of modern cephalochordates such as the lancelet.

The next evolutionary step, the origin of true vertebrates, occurred around 530 mya. More than 300 fossils unearthed in China in 1999 have revealed a series of animals that are intermediates between cephalochordates and true vertebrates. One of these small creatures had some vertebrate characteristics, such as paired **eyes** and an enlarged **brain**, along with a notochord, **pharynx**, and dorsal nerve cord, but it lacked a cranium, ears, and other important vertebrate characteristics. These early protovertebrates were larger and more highly motile than modern cephalochordates, and they most likely were predators rather than filter feeders. Still, these transitional forms appear to provide a direct link between cephalochordates such as the lancelet and the first true vertebrates. For a more complete phylogeny depicting the evolutionary relationships among the chordates covered in the remainder of this laboratory guide, review Figure A.2 in the Appendix.

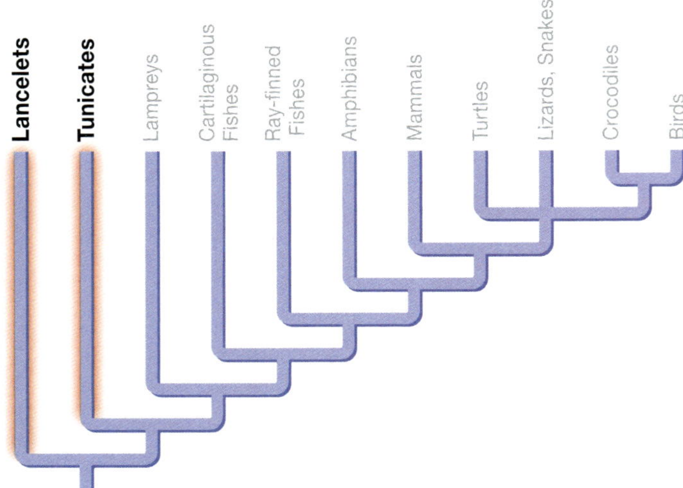

15.2 Modern-day lancelets are among the simplest animals classified as chordates and retain all five chordate characteristics in the adult stage. Tunicates, though now thought to represent a more recent offshoot from the main chordate line, possess all five distinguishing chordate characteristics only in their larval forms. Their sessile lifestyle is believed to be a derived condition, and the subsequent loss of these characteristics is considered an adaptation to this sedentary lifestyle.

Exercise 15-1 Tunicates (Subphylum Urochordata)

Materials Needed

- preserved specimen of tunicate adult
- prepared slide of tunicate larva w.m.
- dissecting microscope
- dissecting pan or finger bowl
- compound microscope

Early chordates, though now thought to be motile, were most likely suspension feeders with pharyngeal pouches and slits that allowed them to take in water through the mouth and filter small food particles into the gut. Because of their relatively primitive design, modern tunicates give us a glimpse into the past and allow us to hypothesize about the anatomy and physiology of the earliest chordates.

The subphylum Urochordata (or Tunicata) includes about 1,600 species of marine, bag-like, filter-feeding animals. The name "tunicate" is derived from the tough, fibrous coating (tunic) that houses the body of these primarily sessile animals (Fig. 15.3). They often live in large colonies and attach to rocks or other hard surfaces along the ocean floor. Tunicates commonly are called sea squirts—a name that was coined from the observation that they will squirt bursts of water forcefully from their excurrent siphons when agitated. Were it not for the anatomy of their larval forms, no one would think that a creature as bizarre as this could be a close relative to vertebrates.

15.3 Tunicates are often called "sea squirts" because of their ability to discharge bursts of water from their excurrent siphon.

Adult Tunicates

Procedure

1 Obtain a preserved specimen of *Molgula*, *Ciona*, or other species of adult tunicate, and place it in your dissecting pan or finger bowl. Refer to Table 15.2 as we discuss the anatomy of your specimen.

Adult tunicates are primarily sessile filter feeders that pull water through the **incurrent siphon** into the pharynx where suspended organic particles are filtered out by the pharyngeal slits (Fig. 15.4). As was the case with the sponges we studied earlier, tunicates are extremely effective filter feeders; even small ones can filter hundreds of liters of water per day and remove well over 95% of the bacteria and other organic matter suspended in it. Sticky mucus, secreted by the endostyle, coats the walls of the pharynx, and ciliary action propels the food into the **stomach**, while this filtered water passes into the **atrium** before exiting the animal through the **excurrent siphon.** Nutrient absorption occurs along the small length of **intestine**, and waste products are eliminated through the **anus**, which also empties its contents into the excurrent siphon.

Tunicates possess an **open circulatory system**—one that is largely without vessels. Blood flows through large sinuses or channels in the tissues. They have a small, contractile **heart** and a series of rudimentary vessels that innervate the pharynx for gas exchange and delivery of oxygenated blood to the other body tissues.

Tunicates have no specialized excretory organs. Instead, they deposit nitrogenous wastes as solid crystals of guanine or uric acid in the tunic. Individuals often look like they have been dusted with a fine layer of white dots—a result of the accumulation of these granules in the tunic. The dorsal nerve cord, once prominent in the larval stage, degenerates in the adult to a single nerve **ganglion** located between the two siphons.

All tunicates are **hermaphroditic**, and often are able to self-fertilize. **Gonads** can develop just about anywhere within the animal's atrium, and empty their gametes into the atrium near the excurrent siphon. Embryos resembling tadpoles exit the parent's excurrent siphon and swim for a short period before selecting a substrate.

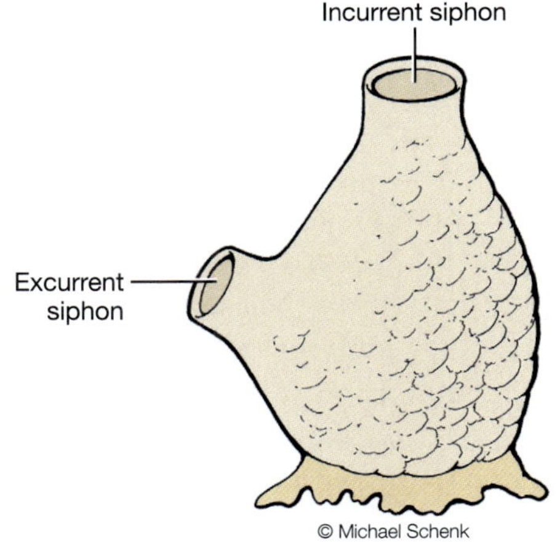

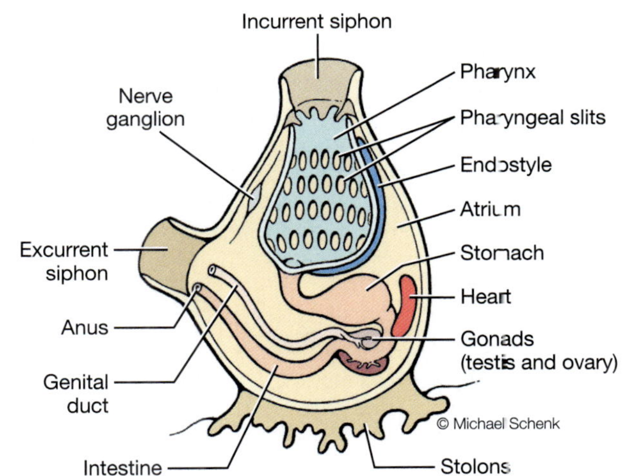

15.4 Adult tunicates are sessile, filter-feeding, bag-like organisms with incurrent and excurrent openings.

15

Larval Tunicates

Procedure

1 Obtain a prepared slide of a tunicate larva w.m. and examine it under low power using your compound microscope.

The tadpole-like larvae of tunicates display their chordate characteristics distinctly, and you should be able to easily identify the notochord, dorsal nerve cord, pharyngeal slits, postanal tail, and perhaps the endostyle (Fig. 15.5).

Unlike the adult form, larval tunicates are free-swimming, though they do not eat during the short time they exist in this phase. Instead, they swim for only a few days before settling and attaching to the ocean floor. The larvae glue themselves, nose first, to the substrate, using secretions of their adhesive discs. Once firmly attached, they undergo an exceptionally rapid and complex metamorphosis into the sessile, adult form. Epidermal cells covering the animal's surface quickly contract, and within a period of just 10–15 seconds, completely crush the tail, its inner notochord, nerve cord, and muscles, along with the brain and sensory structures (ocellus and statocyst) into a mass of tissue. Over the next few hours the internal viscera rotate, the two siphons enlarge and differentiate, and a single nerve ganglion develops from the remnants of the larval nervous system.

Table 15-2 Anatomy of the Tunicate

Structure	Function
Incurrent siphon	Water enters this opening, bringing fresh oxygen and suspended food particles
Excurrent siphon	Water exits this opening, carrying wastes and carbon dioxide out of the animal
Pharynx	Receives incoming water from mouth, entraps food in a film of mucus, and is responsible for gas exchange
Endostyle	Secretes mucus that forms sticky coating of pharynx to trap food particles
Pharyngeal slits	Allow for outflow of water from pharynx into atrium
Atrium	Body chamber that receives water passed from pharynx through pharyngeal slits
Stomach	Primary digestive organ
Intestine	Site of nutrient absorption
Anus	Regulates egestion of undigested food (feces) from the body
Stolons	Holdfasts that allow animal to anchor itself firmly to substrate
Heart	Pumps blood through vessels and sinuses of open circulatory system
Adhesive discs	Used by larvae to attach to substrate prior to metamorphosis
Notochord	Provides skeletal support for anchoring of swimming muscles in the larval form
Dorsal nerve cord	Handles nervous coordination in the larval form

15

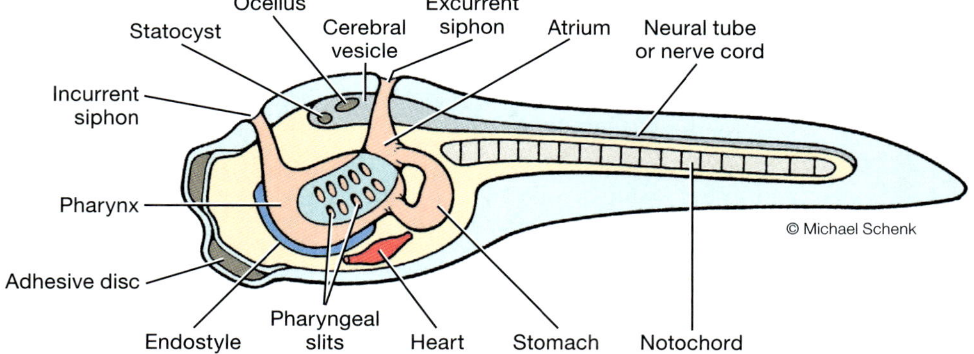

15.5 Tunicate larvae possess all five distinguishing chordate characteristics.

Exercise 15-2 Lancelets (Subphylum Cephalochordata)

Materials Needed

- ❑ prepared slide of lancelet w.m.
- ❑ prepared slide of lancelet c.s.
- ❑ compound microscope

The lancelet (or amphioxus) is a common member of the subphylum Cephalochordata. Lancelets display several intermediate characteristics between invertebrates and vertebrates, making them a useful organism for understanding the evolutionary transition in the ancestral line of chordates. All five chordate characteristics are present in the adult stage, yet lancelets lack the vertebral column and cranium of vertebrates. This small subphylum contains just 29 species of small, laterally flattened, fishlike filter feeders that commonly are found in shallow marine and brackish waters.

Lancelets spend most of their time partially buried in the sand with only their heads protruding above the sediment. They do possess the ability to swim by what best could be described as an erratic, convulsive, twitching motion. Nonetheless, their ability to swim is tantamount to dispersal and mating.

Procedure

1 Obtain a prepared slide of a whole mount (w.m.) of a lancelet and view it using low power on your compound microscope.

2 Begin by identifying the structures at the anterior end of your specimen and progress posteriorly, using Figures 15.6–15.8 and Table 15-3 to assist you.

Locomotion in this group is achieved by alternating muscular contractions of the segmented **myomeres** against the fairly rigid notochord (Fig. 15.6). As myomeres along one side of the body contract, opposing myomeres along the other side of the body relax and the body bends. These lateral undulations are enhanced by the presence of small dorsal and ventral fins that aid in steering and maneuvering. The same motions employed for swimming also double as the lancelet's mechanism for burrowing into the soft sand for protection and feeding. During feeding, a current of water is driven into the pharynx by cilia on the **oral cirri** and **wheel organ**, bringing with it suspended food particles and plankton that are filtered across the pharyngeal gills (Figs. 15.7 and 15.8).

As small food particles are trapped by the coating of mucus on the **gill bars**, the water passes through the gill slits into the atrium and flows posteriorly along the body, eventually exiting the body through the ventral **atriopore**. At the same time, the trapped food particles are transported along the epibranchial groove of the pharynx into the intestine for digestion and nutrient absorption.

15

Digestion initially is extracellular but is completed intracellularly in the walls of the intestine and the small, lateral outpocket of the intestine known as the **hepatic cecum**. In addition to digestion, the hepatic cecum functions in lipid and glycogen storage and is believed to be the evolutionary precursor of the vertebrate liver. Indigestible food particles pass through the intestine and exit the body through the anus.

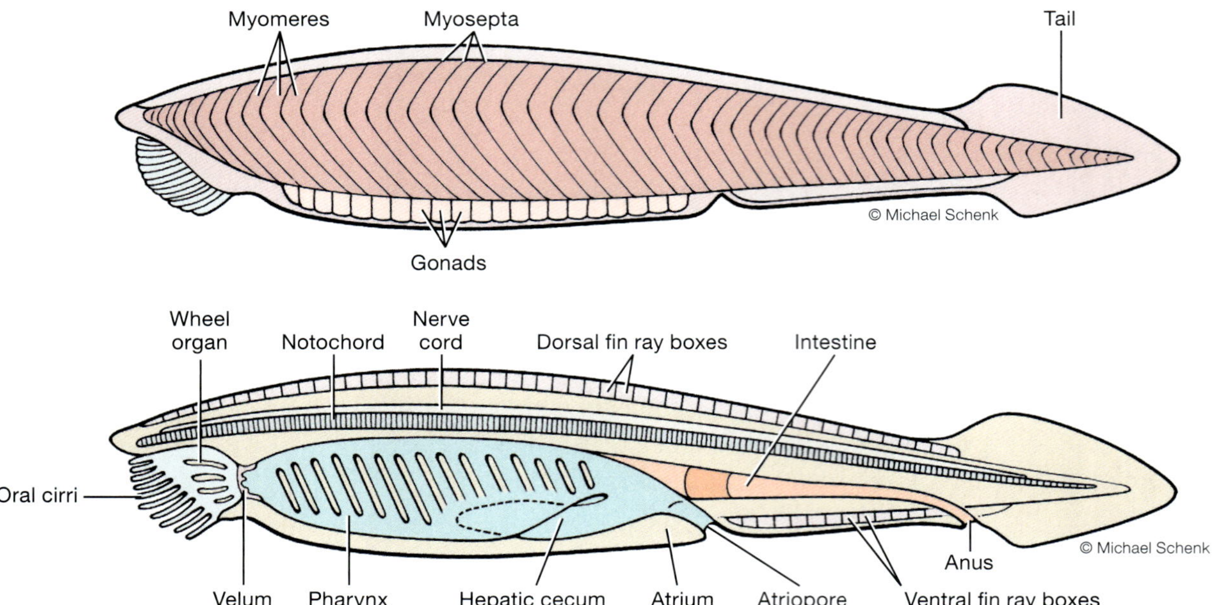

15.6 Adult lancelet, *Branchiostoma*, basic anatomy. All five chordate characteristics are easily identified in the adult stage.

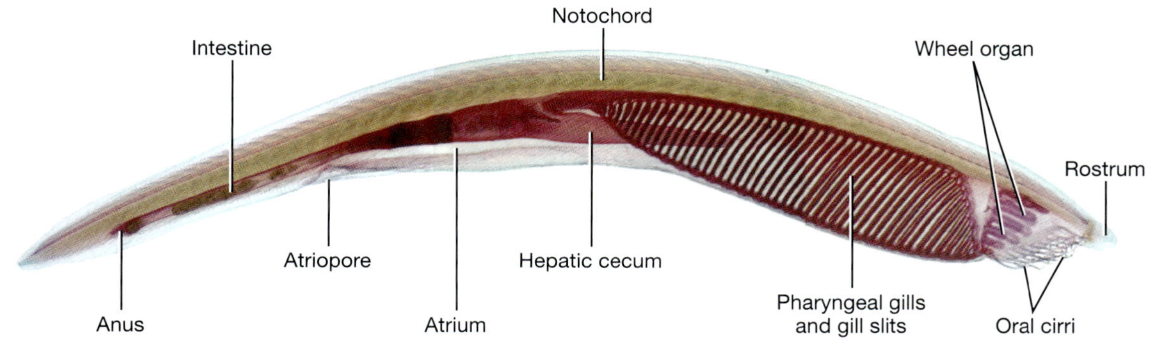

15.7 Adult lancelet, *Branchiostoma*, stained preparation, lateral view.

15

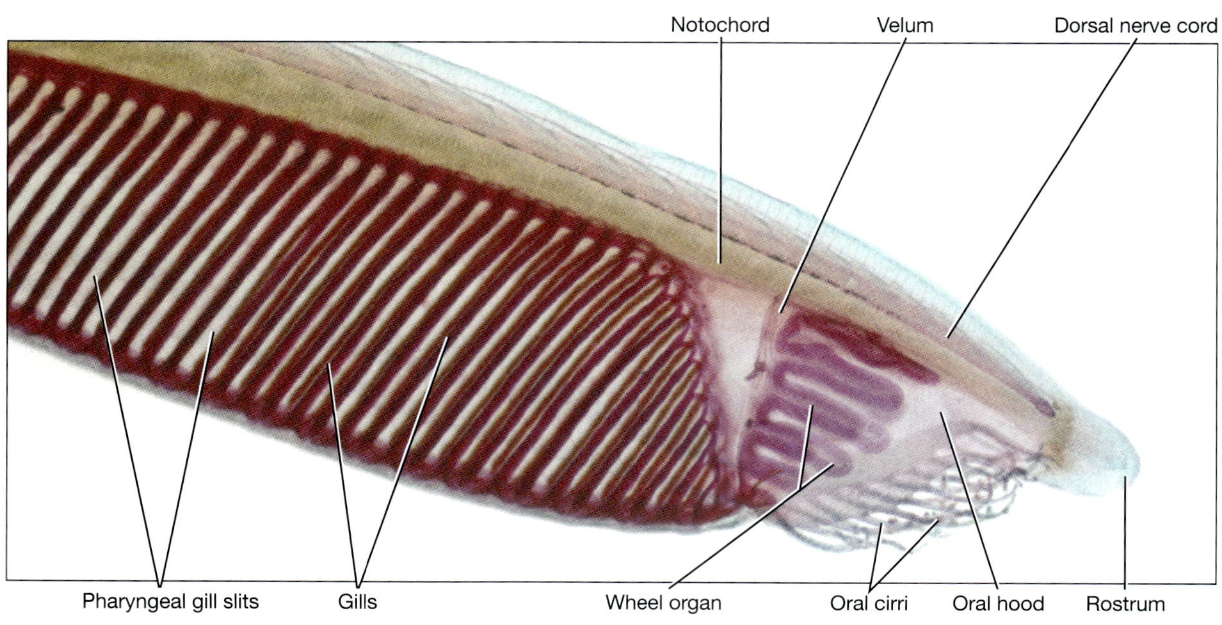

15.8 Lancelet, *Branchiostoma*, anterior region.

Table 15-3 Anatomy of the Lancelet

Structure	Function
Myomeres	Provide muscular movements for swimming and burrowing
Notochord	Provides skeletal support for anchoring myomeres to provide lateral body movements
Dorsal nerve cord	Handles majority of nervous coordination without intervention of the brain
Rostrum	Anterior projection shielding entrance to the mouth dorsally
Wheel organ	Lined with cilia, which produce a current of water that brings food into the mouth
Oral cirri	Act as a strainer to exclude larger particles from the mouth (filter feeding)
Pharyngeal gills	Have thin mucous coating that traps ingested food particles; also contain blood vessels supplementing gas exchange for respiration
Gill slits	Allow for outflow of water from pharynx between gill bars for food capture and respiration
Velum	Transverse partition encircling mouth
Pharynx	Receives incoming water from mouth, entraps food in a film of mucus, and is responsible for some gas exchange
Hepatic cecum	Lateral outpocket of intestine responsible for intracellular digestion of small food particles and lipid and glycogen storage
Intestine	Site of enzymatic digestion of larger food particles
Atrium	Ventral body chamber that receives water passed from pharynx through gill slits
Atriopore	Discharges water from the atrium to the external environment
Anus	Regulates egestion of undigested food (feces) from the body
Dorsal aortae	Paired blood vessels that join behind the pharynx to form the median dorsal aorta, which carries nutrient-laden blood posteriorly to the body tissues and intestine
Nephridia	Ciliated ducts that transport metabolic wastes from the dorsal portions of the coelom to the atrium
Testes (male)	Produce sperm for external fertilization
Ovaries (female)	Produce eggs for external fertilization

3 Next, obtain a prepared slide of a cross section (c.s.) of the lancelet and view it using low power on your compound microscope.

4 Use Figure 15.9 to help you identify the anatomical structures visible on the cross section.

Like most chordates, lancelets have a **closed circulatory system** consisting of a series of closed vessels for blood flow. They have no heart, but a full complement of arteries and veins exists. The paired dorsal **aortae** and single ventral aorta are often visible on cross sections through the middle of the body. The blood contains no respiratory pigments and thus is colorless, functioning primarily in nutrient distribution rather than in gas exchange and transport. The pharyngeal gills play a minor role in respiration, but the bulk of gas exchange occurs across thin flaps of highly vascular tissue along the ventral surface of the body wall.

Excretion is accomplished through segmented **nephridia** akin to those seen in annelids. Metabolic waste is collected in the nephridia and channeled to the atriopore for elimination from the body. In lancelets the sexes are separate but structurally similar. They possess from 25 to 38 pairs of serially arranged gonads along the ventral region of the body wall in the atrium, lateral to the pharynx. A hallmark of chordates is the presence of a hollow nerve cord along the dorsal surface of the body. Aside from this feature, lancelets have little else in the way of a central nervous system. Cephalization is absent; instead, segmentally arranged nerves handle the majority of muscular coordination along the body axis.

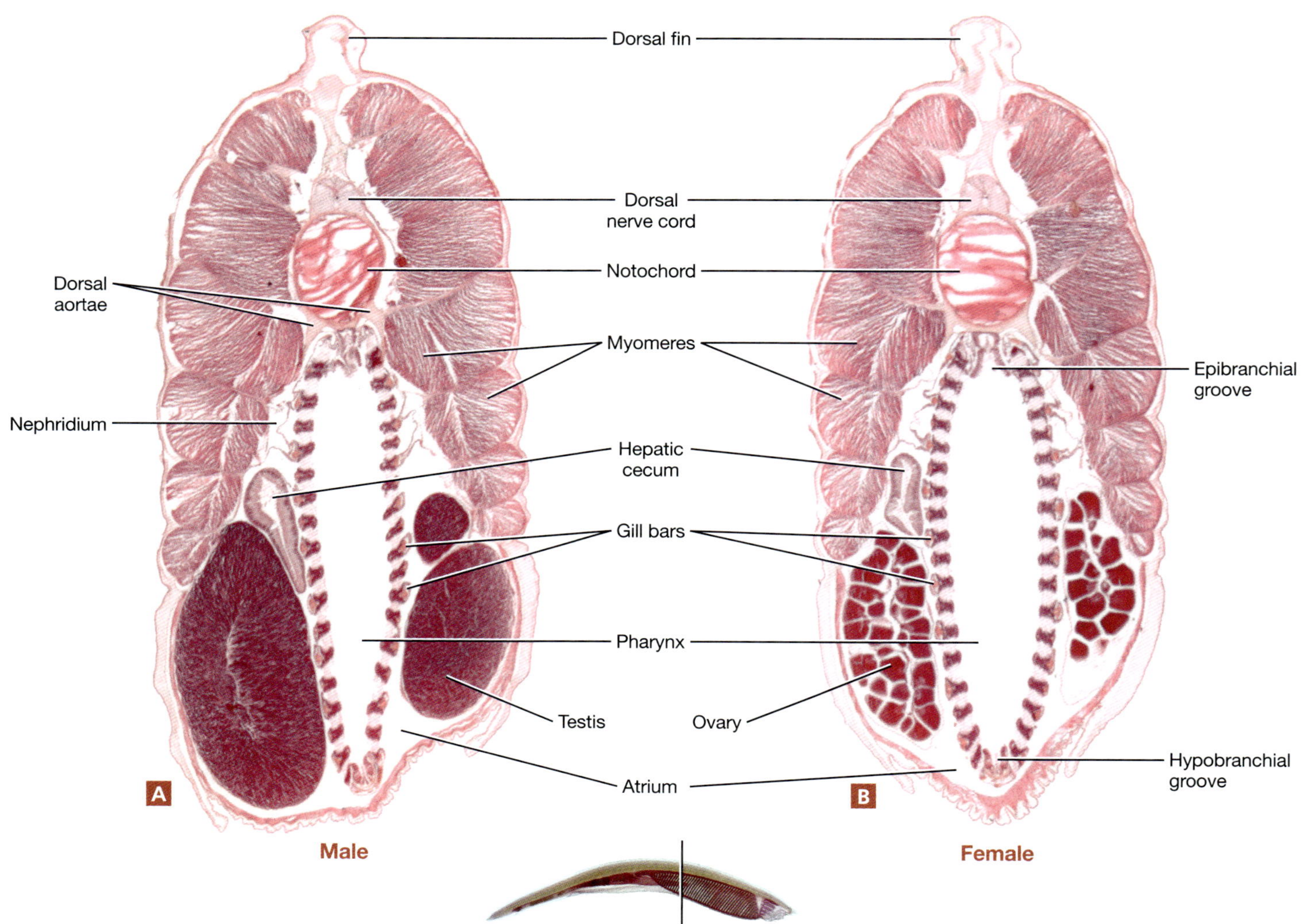

15.9 Transverse sections through the pharyngeal regions of (**A**) male, and (**B**) female lancelets.

15

Name ______________________ Section ______________

15 Questions for Review

1 Complete Summary Table A-1 in the Appendix, filling in the characteristics of both tunicates and lancelets in the appropriate rows.

2 List the five characteristics common to all chordates.

1. ______________________
2. ______________________
3. ______________________
4. ______________________
5. ______________________

3 What structures common to all chordates do adult tunicates possess? Why are tunicates considered chordates?

4 List two functions of the pharynx in tunicates.

5 Which structures in the lancelet exhibit segmentation?

6 What other animals have you encountered that use filter feeding to get food?

7 Describe two aspects of the circulatory system of lancelets that have evolved in more complex chordate groups, such as mammals.

8 How does gas exchange occur in the lancelet?

9 In lancelets, water that has filtered across the gill bars and has picked up metabolic waste products from the nephridia exits the body through the ________________, while indigestible food is eliminated from the body through the ________________.

10 Why are tunicates and lancelets useful for understanding the evolutionary changes that have occurred during chordate evolution?

15

Name ______________________ Section ______________

15 Laboratory Practical Challenge

The following evaluation is intended to replicate the method of assessment and type of questions you will be asked on lab practicals. If you have difficulty with these questions, you may benefit from additional review of the specimens provided in lab.

1 Identify the phylum and subphylum of the specimen depicted below.

____

2 What is the function of the structure indicated by line A?

____

3 Identify the structure indicated by line B.

____

4 To what system does the structure indicated by line C belong?

____

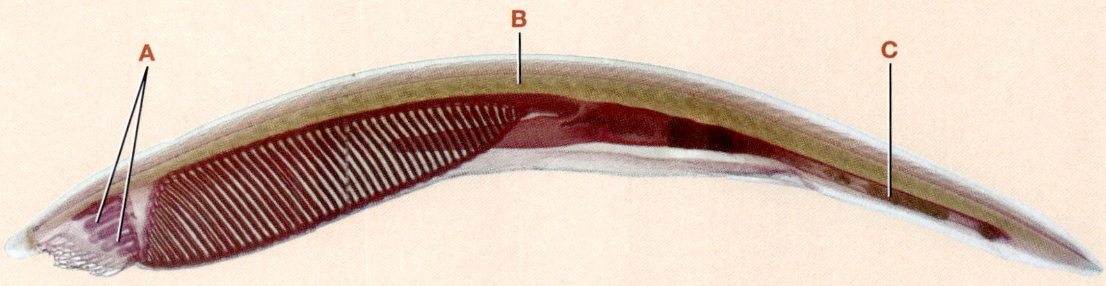

15

5 On the figure below, label the following structures: gonad, dorsal nerve cord, notochord.

6 What is the function of the structure indicated by line D?

7 Identify the structure indicated by line E.

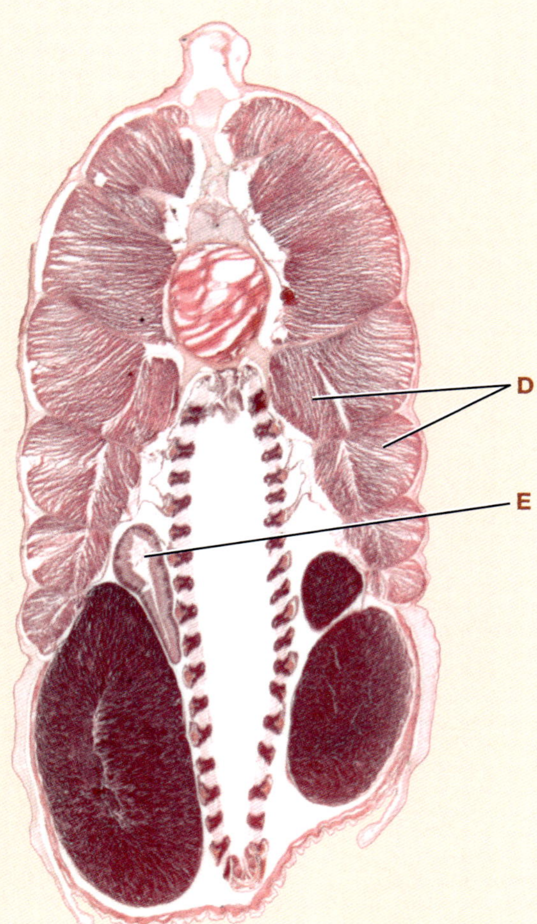

15

CHAPTER 16

Jawless Fishes

After completing the exercises in this chapter, you should be able to:

1 List the distinguishing characteristics of the class Petromyzontida.

2 Understand the evolutionary position of jawless fishes on the phylogeny of chordates.

3 Discuss significant adaptations of jawless fishes in the evolution of early vertebrates.

4 Identify the major external features and internal organs of an adult lamprey, discuss their functions, and compare them to homologous structures in other vertebrates.

5 Identify the major external features and internal organs of a larval lamprey, discuss their functions, and compare them to structures in adult lancelets.

Jawless fishes, such as lampreys and hagfishes, are the most primitive of living fishes and are the only living descendants of an early vertebrate group called ostracoderms. They represent a key position on the evolutionary tree of chordates—they are the first true **vertebrates** (subphylum Vertebrata) we have studied. Though plentiful during the Paleozoic era, jawless fishes have since been far outnumbered by jawed vertebrates such as sharks, ray-finned fishes, lobe-finned fishes, amphibians, reptiles, birds, and mammals. Today there are only around 70 species of hagfishes and 35 species of lampreys. These two groups of jawless fishes may appear morphologically similar at first glance, but are sufficiently different from each other that zoologists have assigned them to separate taxonomic classes.

Lampreys belong to the class Petromyzontida, and are characterized by a smooth, scaleless, fusiform body and a sucker-like buccal funnel for attaching to the host organism. Most species are **ectoparasites** that feed on the blood of other fishes by clamping onto the host's body and using their rasping tongues to scrape through the outer body wall of their host. They are the first chordates we have studied with a distinct head, a true cranium containing a tripartite brain, and sense organs specialized for taste, smell, hearing, and sight. These structures enabled early chordates to process a wealth of sensory stimuli and to coordinate more complex movements and feeding behaviors, giving them a distinct advantage over their predecessors. Though dorsal fins are present, lampreys lack the paired fins of later fishes. Adults possess a cartilaginous skeleton with rudimentary vertebrae, but the notochord persists in the adult as the primary axial support element.

Today lampreys are distributed worldwide, with some species living in freshwater lakes and others in oceans, but all lampreys return to freshwater rivers or streams to reproduce. The sexes are separate (dioecious) and fertilization is external—females create shallow nests on the stream bottom to deposit their eggs, and males release sperm over the eggs to fertilize them. Adults die shortly after spawning.

Evolutionary History

The fossil record now contains a diverse collection of early chordate fossils that depicts the evolutionary transistion to early vertebrates. The first true vertebrates appeared in the ancient seas during the Cambrian period around 530 mya. These small fish, less than a foot long, lacked jaws and teeth and were bottom-dwellers that sucked up small food particles from the ocean floor. They wriggled about haphazardly without fins to guide them and only a primitive tail to propel them. Some of these early vertebrates show fossilized remains of ear capsules, eye capsules, and skull parts, indicating they were true craniates. They had well-formed brains, and more advanced body musculature than cephalochordates.

During the Cambrian period, a lineage of early craniates gave rise to vertebrates with even more elaborate skeletons and more complex nervous systems than those of their predecessors. These heavily armored, jawless fishes were collectively known as ostracoderms and they excelled at capturing food and avoiding predators. Many species developed primitive fins to help them swim, and massive shields of bony armor for protection. Bony structures called fin rays stiffened their fins and provided thrust and steering control when swimming after prey or away from predators. Increased musculature associated with the gill slits permitted the development of more efficient gas exchange systems. Hearts with at least two chambers, red blood cells containing hemoglobin, and kidneys to filter metabolic wastes enabled higher metabolic rates. These adaptations supported faster swimming, greater endurance, and enhanced maneuverability and allowed ostracoderms to thrive during the Silurian and Devonian periods (444–360 mya), eventually radiating into nearly 40 different families of jawless fishes.

Their reign ended when lighter-bodied fishes with more streamlined bodies, paired fins, and movable jaws began their gradual conquest of the oceans. Today, only two living offshoots from these ancient jawless ancestors remain—hagfishes and lampreys (Fig. 16.1). For a more complete phylogeny depicting the evolutionary relationships among the chordates covered in the remainder of this laboratory guide, review Figure A.2 in the Appendix.

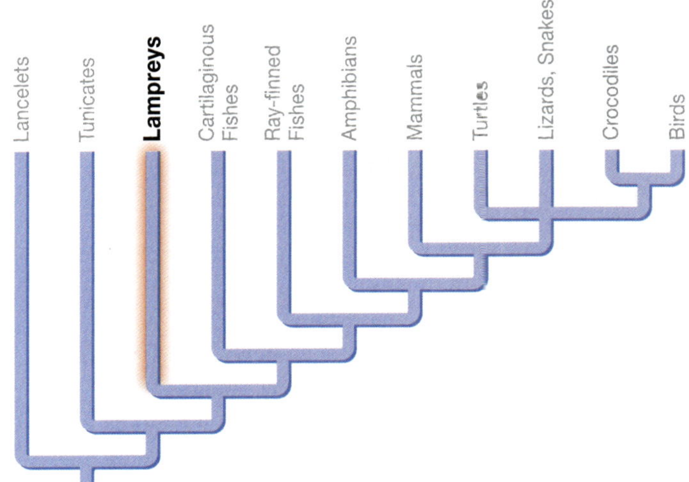

16.1 Lampreys are one of only two groups of living jawless fishes descended from an ancient lineage of jawless vertebrates that inhabited the oceans and lakes more than 500 million years ago. They represent one of the least complex groups of true vertebrates, possessing a cranium, partitioned brain, and a rudimentary vertebral column.

Exercise 16-1 Lampreys (Class Petromyzontida)

External Anatomy

Materials Needed

- preserved lamprey, *Petromyzon marinus*
- sharp kitchen knife
- dissecting tools
- dissecting pan
- latex gloves
- slides of lamprey larvae w.m. and c.s.
- dissecting microscope

Procedure

1. Obtain a preserved lamprey (*Petromyzon marinus* or other species) and examine its external features.
2. Use Figures 16.2 and 16.3 to assist you in identifying the external anatomy of your specimen.
3. As you examine your specimen, try to draw comparisons between the external features of the lamprey and the lancelet and with the external features of more familiar aquatic vertebrates such as sharks and bony fishes.

As you pass your hand over the lamprey's streamlined, fusiform body you will notice that its tough skin is smooth and scaleless—in sharp contrast to its ostracoderm ancestors whose heavily armored bodies were encased in bony plates. Many of the ancestral features of ancient jawless fishes were lost over the past 500 million years, and modern lampreys represent a mosaic of ancestral and derived characteristics. Several characteristics, such as the oral area and respiratory system, represent specializations for their parasitic lifestyle.

The paired pectoral and pelvic fins of more familiar fishes are not found in lampreys, and instead, only dorsal and caudal fins are present. The anterior end of the body has a pair of lateral, image-forming **eyes** and a single, median **nostril** on the dorsal surface of the head (Fig. 16.2). Pores along the sides of the head represent the lamprey's **lateral line** system,

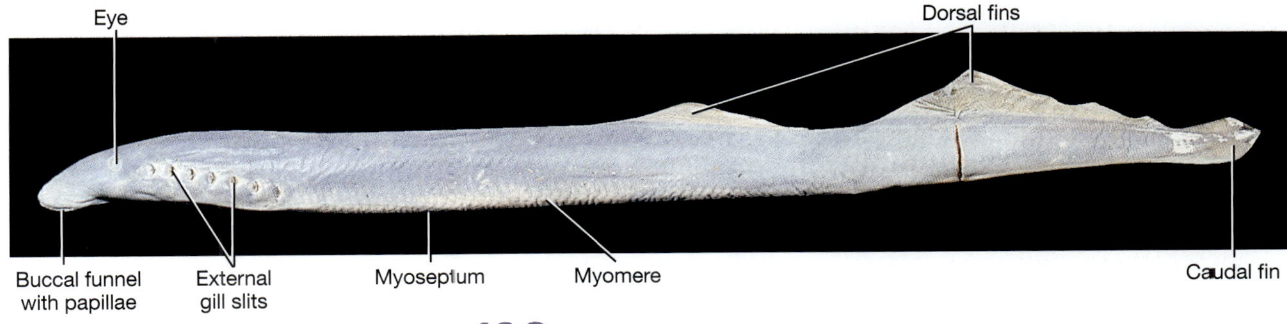

16.2 Lamprey, external anatomy.

a characteristic of nearly all fishes. Receptors embedded within these pores detect vibrations and water currents in the animal's surroundings. In lampreys, these receptors are open to the exterior and are not protected within canals as they are in bony fishes. Seven pairs of external **gill slits** open along the lateral aspect of the head and are used for respiration.

The oral region of the lamprey displays its unique adaptations for parasitic feeding (Fig. 16.3). The terminal mouth lies within a circular, buccal funnel fringed by small, fingerlike sensory buccal papillae. Inside the mouth are toothlike organs and a tongue with sharp, keratinized teeth used for rasping holes in the outer body wall of the host. The lamprey's horny, toothlike organs are of epidermal origin and thus are not homologous with true vertebrate teeth, which are derived from mesoderm and ectoderm. The lamprey uses its muscular buccal funnel to create a strong suction, allowing it to remain firmly attached to its host, consuming blood and body fluids as its host swims about.

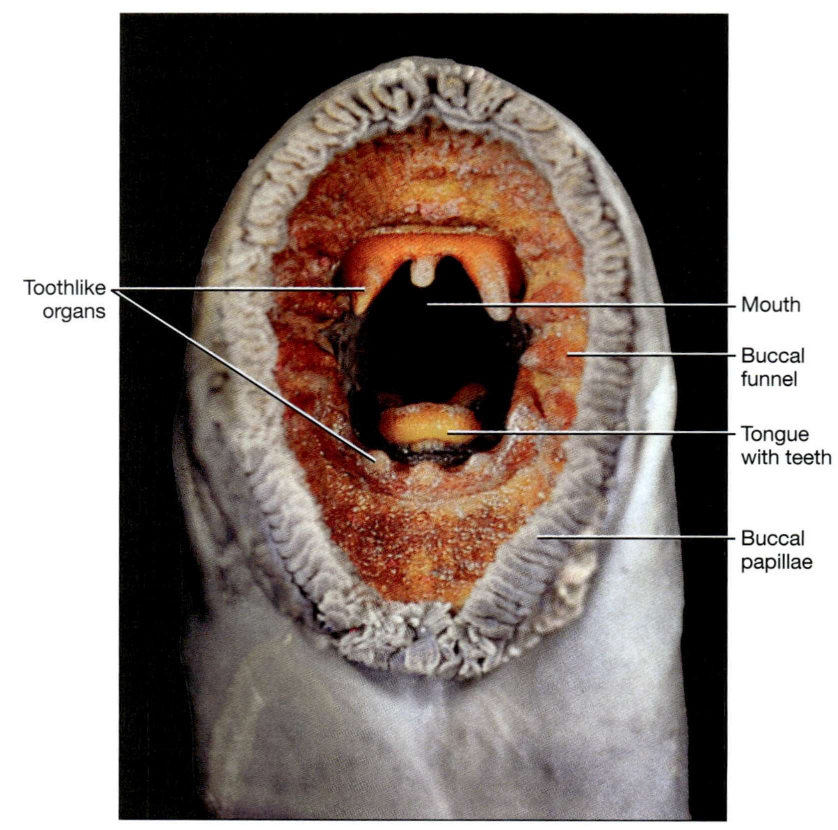

16.3 Lamprey, oral region, ventral view.

Check Your Progress

1.1 What features of the lamprey's head represent homologies to other vertebrate groups?

16

1.2 What external adaptations for a parasitic lifestyle does the lamprey exhibit?

1.3 What is the function of the lateral line system?

1.4 Why do you suppose the lateral line system has been retained in most aquatic vertebrates, but lost in reptiles, birds, and mammals?

Internal Anatomy

To effectively study the internal anatomy of the lamprey you must employ a different approach to dissection than the usual method of making a ventral incision along the midline of the body. Instead you will utilize a combination of transverse (cross) sections and midsagittal sections through the body that better preserves the relationships between internal organs and vessels and gives you a clearer picture of the lamprey's internal anatomy. Your specimens may be injected with latex to simplify identification of the blood vessels. If your specimen has been double-injected with latex, arteries will appear red and veins will appear blue. Single-injected specimens will have red arteries.

Your instructor may have prepared sagittal sections of lampreys as demonstrations in advance of your lab. Check with your instructor to see if prepared sagittal sections are available for you to view. Alternatively, some teams in your lab may be instructed to prepare midsagittal sections of their specimens and other teams may be instructed to prepare transverse sections. Be sure you are clear about your assignment before you begin cutting your specimen. Regardless of which section you are asked to make, you will need to study both types of sections to gain a complete understanding of the lamprey's anatomy.

Procedure

1 Obtain a lamprey that has been bisected along its midsagittal plane, or use a sharp kitchen knife to bisect your specimen into lateral halves.

2 To make the cut yourself, have your partner hold the lamprey securely while you begin your cut at the dorsal edge of the sucker and carefully slice completely through the body along the middorsal line to produce two equal halves. Employ a sawing motion with the knife as you extend your cut posteriorly. If done properly, the cut should pass through the middle of the nostril, pineal eye, and dorsal fins.

3 To prepare transverse sections for study, you will make three separate cuts through the anterior region of the body as follows:

a. Through the lateral eyes and dorsal pineal eye

b. Through the fourth gill slit

c. Through the body approximately 0.5 cm posterior to the seventh gill slit

4 Use Table 16-1 and Figures 16.4–16.6 to assist you in identifying the internal anatomy of your lamprey.

Digestive System and Sensory Structures

Like all vertebrates, lampreys have a complete digestive system beginning with a mouth at the anterior end of the body and an anus near the posterior end. The buccal papillae encircling the buccal funnel perform a sensory role and enhance the suction forces necessary for the lamprey to remain attached to its host (Figs. 16.3 and 16.4). The muscular tongue is equipped with horny teeth that the lamprey uses to rasp through the body wall of the host to create seeping wounds, and the paired oral glands secrete a powerful anticoagulating agent that prevents the host's blood from clotting while the lamprey feeds.

Because lampreys lead a parasitic lifestyle and ingest blood and body fluids that are nutrient-rich and require little additional digestion, the remainder of the digestive system is much simpler than that of other nonparasitic vertebrates. Ingested fluids pass through the **esophagus** to a fairly unspecialized **intestine** where nutrient absorption into the bloodstream occurs (Fig. 16.5). Lampreys lack a stomach, but they do have a **liver** that produces digestive compounds that are released into the intestine to increase digestive efficiency. Indigestible material is passed out of the body through an **anus** located within the lamprey's ventral **cloaca**.

Sensory structures in the lamprey are concentrated at the anterior end of the body—a condition known as **cephalization**. In addition to the external sensory structures discussed earlier (eyes, lateral lines, nostril, buccal papillae), lampreys have several sensory structures embedded in the head that are only visible through dissection. The external nostril terminates in an **olfactory sac** containing a thin layer of sensory epithelium that allows the lamprey to "smell" dissolved chemicals in the surrounding water. Respiratory movements of the pharynx compress and expand a small reservoir known as the **hypophyseal sac**, helping to circulate a current of water into and out of the nostril and over the olfactory epithelium. The **pineal organ**, located on the dorsal surface of the head just posterior to the nostril, is a photosensitive "eye" that can detect changes in illumination and communicate directly with the brain to regulate internal activities of the lamprey that are dependent on diurnal and seasonal cycles.

Table 16-1 Internal Anatomy of the Lamprey

Structure	Function
Olfactory sac	Enlarged, internal chamber of the nostril containing a sensory epithelium that allows the lamprey to "smell" dissolved chemicals in the surrounding water
Pineal organ	Photosensory organ located posterior to the nostril; controls many of the lamprey's activities that are dependent on photoperiod
Eye	Image-forming photoreceptive organs that lack eyelids; each eye contains a lens and a pigmented retinal layer homologous to those present in other vertebrates
Brain	Center of neural control; partitioned into a cerebrum, cerebellum, and medulla oblongata
Notochord	Primary axial support element for the body
Spinal cord	Posterior extension of the brain that lies dorsal to the notochord and contains spinal nerves
Hypophyseal sac	Terminal region of the nostril that circulates water through the nostril and over the olfactory epithelium to facilitate olfaction
Tongue (with teeth)	Muscular organ in the oral cavity that rasps through the body wall of the host to create seeping wounds for the lamprey to feed
Oral gland	Secretes anticoagulating agent during feeding that prevents host's blood from clotting and promotes blood flow
Gill filaments	Highly vascular structures for gas exchange with the bloodstream
Branchial pouch	Respiratory chamber that houses the gill filaments and receives water from the external gill slits
Heart	Two-chambered organ (one atrium, one ventricle) that pumps deoxygenated blood to the gills for oxygenation and from the gills throughout the various organ systems of the body
Sinus venosus	Thin-walled sac where deoxygenated blood from the body collects before being passed into the atrium of the heart
Transverse septum	Partition that separates the coelom into the pericardial cavity and the pleuroperitoneal cavity
Dorsal aorta	Channels oxygenated blood from the gills to the posterior regions of the body
Ventral aorta	Channels blood from the heart to the gills for oxygenation
Esophagus	Transports food to intestine
Intestine	Primary site of nutrient absorption
Liver	Accessory digestive organ that produces digestive compounds and detoxifies many constituents of the absorbed digested compounds
Kidneys	Paired, flattened excretory organs that filter nitrogenous wastes from the blood

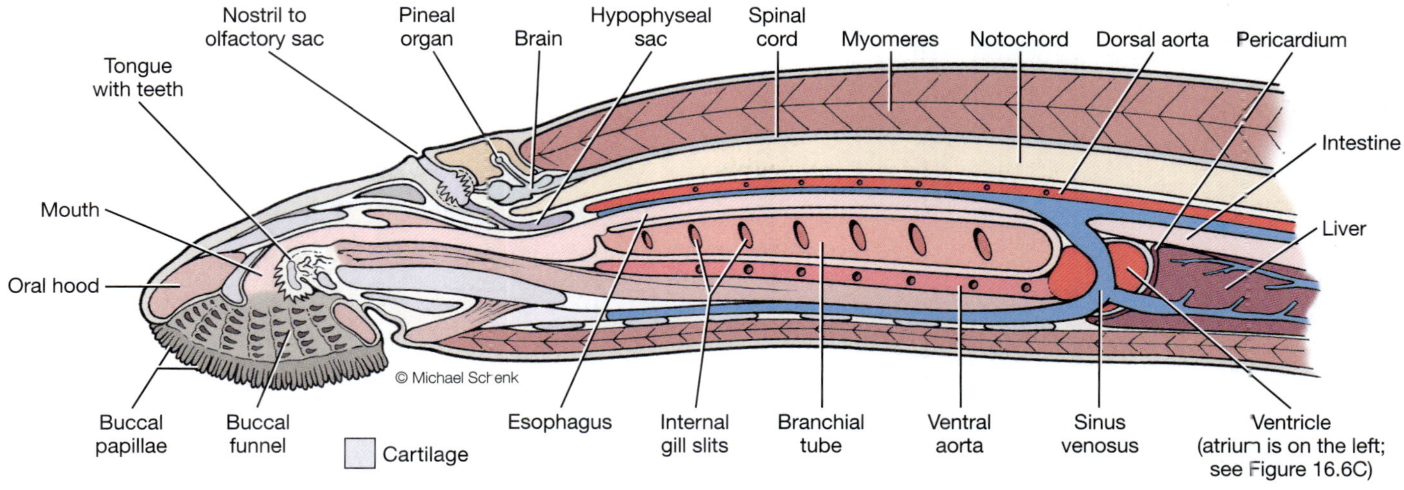

16.4 Lamprey, anterior anatomy, lateral view.

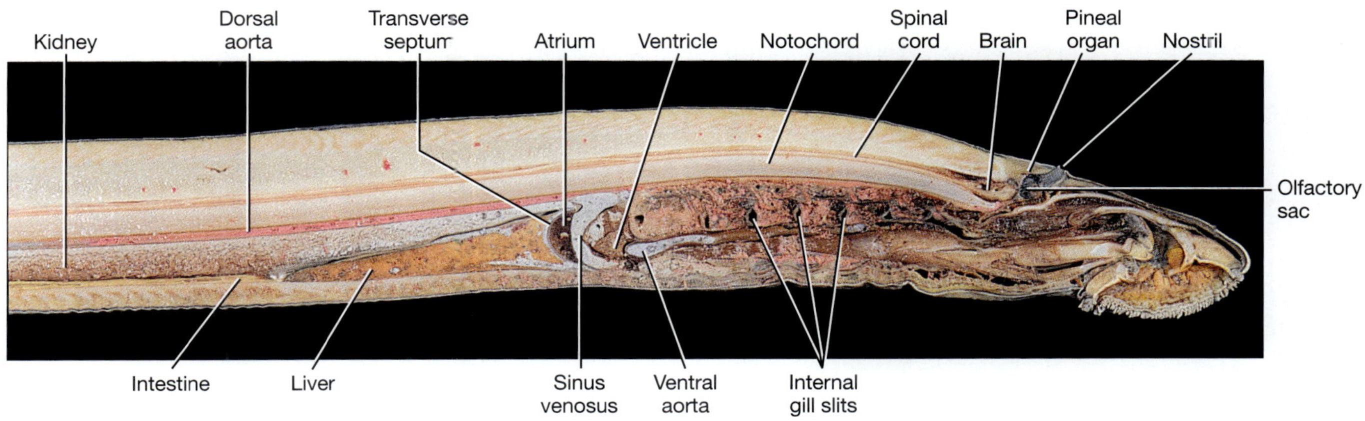

16.5 Lamprey, anterior end, sagittal section.

Circulatory and Respiratory Systems

Lampreys have a closed, single-circuit circulatory system sending blood from the heart to the gill filaments and then on to the rest of the body, before returning it to the **heart.** The heart has two primary chambers, an **atrium** and a thicker-walled **ventricle**, along with a posterior tubular sinus venosus that lies between the atrium and ventricle and collects deoxygenated blood from the veins in the body (Figs. 16.5–16.6C). Blood is pumped from the sinus venosus into the atrium, and from the atrium into the ventricle. From the ventricle, blood is pumped into the **ventral aorta** and is routed to the **gill filaments** through afferent branchial arteries for oxygenation in capillary beds within the gills. Oxygenated blood leaves the gills through efferent branchial arteries and is channeled to the anterior parts of the body through carotid arteries and to the posterior regions of the body through the **dorsal aorta** (Fig. 16.6B). Metabolic wastes are filtered from the bloodstream by a pair of long, laterally compressed **kidneys** located on either side of the intestine and eliminated from the body through a urogenital papilla within the cloaca.

The respiratory system consists of seven pairs of branchial pouches that connect to a modified pharynx known as the branchial tube (Fig. 16.6B). The separation of the pharynx from the food passage enables a feeding lamprey to pump water into and out of the gill pouches through the external gill slits while its mouth maintains tight suction on the host's body. When the lamprey is not feeding, water flow follows the typical pattern seen in other aquatic vertebrates, entering the body through the mouth and passing over the gill filaments and out through the external gill slits.

16

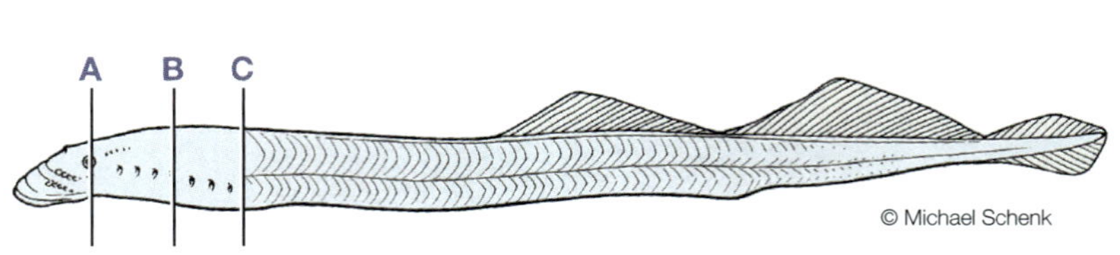

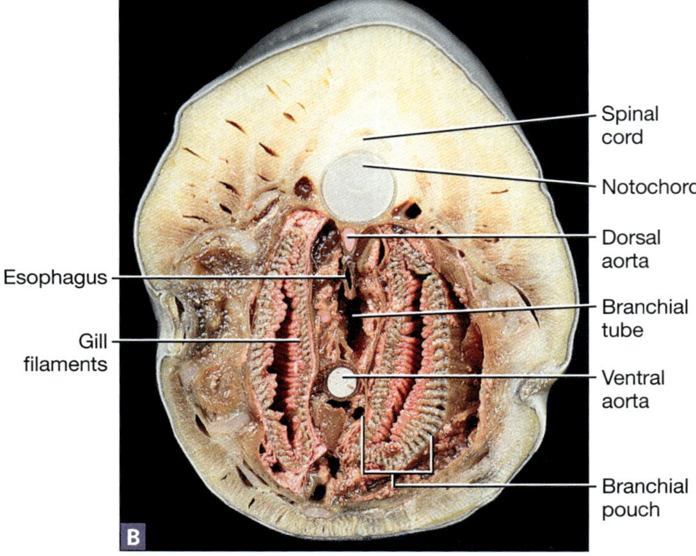

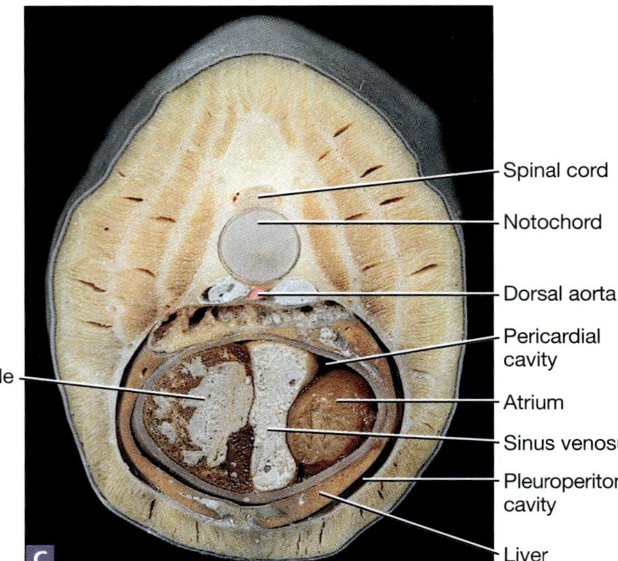

16.6 Lamprey, transverse sections: (**A**) through the lateral and pineal eyes; (**B**) through the fourth gill slit; and (**C**) 0.5 cm posterior to the seventh gill slit.

Check Your Progress

16

2.1 How do the oral glands improve the lamprey's ability to feed?

2.2 Why would it be problematic for a feeding lamprey to pass water through its mouth to the gills?

2.3 List several examples of cephalization in lampreys.

2.4 What is meant by the phrase "single-circuit circulatory system"?

Larval Lamprey

Procedure

1 Obtain prepared slides of a whole mount (w.m.) and cross section (c.s.) of a lamprey larva and examine them using the dissecting microscope. If you have several cross sections on your slide, choose one through the pharyngeal region of the larva that captures the gill filaments.

2 Use Figures 16.7 and 16.8 to assist you in identifying the anatomy of your specimens.

Larval lamprey hatch from eggs on the bottoms of freshwater streams or rivers and begin a developmental stage that lasts five or six years. The lamprey larva is so different in appearance from the adult that zoologists once believed it was a different kind of animal and gave it the name *Ammocoetes*. Even after the mistake was realized this name persisted, and today you will commonly hear the larval lamprey referred to as an ammocoete larva. The larval lamprey bears a remarkable resemblance to the lancelet you studied in Chapter 15, both in its anatomy and its lifestyle. However, larval lampreys and adult lancelets differ in a number of significant ways, and it is essential that you be able to distinguish members of these two groups.

Like lancelets, lamprey larvae are filter feeders that burrow into the sandy streambed leaving only their **oral hoods** protruding. Muscular contractions of the pharynx generate a current of water that brings dissolved oxygen and nutrients into the oral hood. Large particles are strained out by the **oral lobes** as the stream of water passes the **velum** and enters the pharynx (Fig. 16.7). You may recall that cephalochordates relied on ciliary movements to generate water currents for filter feeding. The active use of the pharynx as a muscular pump to bring in food and dissolved oxygen represents a significant step away from the feeding mechanism seen in the lancelet.

The lamprey larva also possesses lateral eyes and a distinct tripartite brain that is subdivided into regions that, in the adult, will become the cerebrum, cerebellum, and medulla oblongata. The **subpharyngeal gland** lies ventral to the pharynx and represents a homologue to the **endostyle** of cephalochordates and urochordates. In the lamprey, the subpharyngeal gland begins to take on some of the capabilities of the organ we know of in humans as the thyroid gland—functions such as fixing iodine and producing thyroid-like hormones. Because larval lampreys are not sexually mature, there are no gonads present in the coelomic cavity, as you observed in adult lancelets. A single dorsal aorta is present in the lamprey larva, in contrast to the paired dorsal aortae seen in lancelets. Other organs found in the lamprey larva that are not present in lancelets are a liver, a gallbladder, and a heart, though the gallbladder degenerates upon metamorphosis of the larva into an adult lamprey.

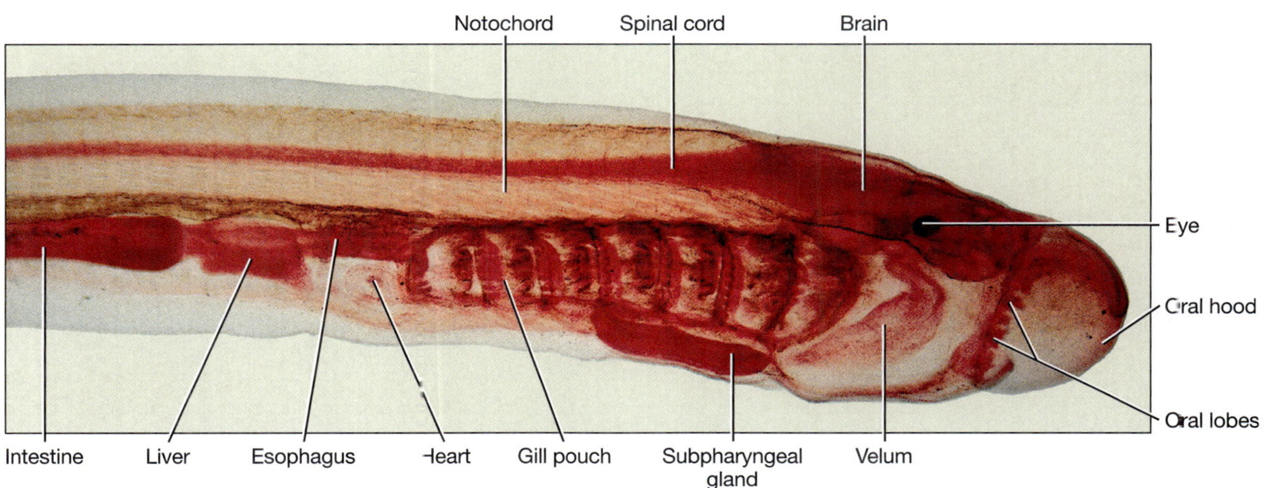

16.7 Lamprey ammocoete larva, anterior end, lateral view.

The remainder of the larvae's anatomy is generally reminiscent of the lancelet. A dorsal nerve cord (spinal cord) sits atop the larger notochord that forms the primary axial support element for the body (Fig. 16.8). Thick myomeres border the notochord and spinal cord on either side and are utilized for locomotion in much the same way they are in lancelets, producing erratic, convulsive, twitching motions to aid in dispersal and burrowing.

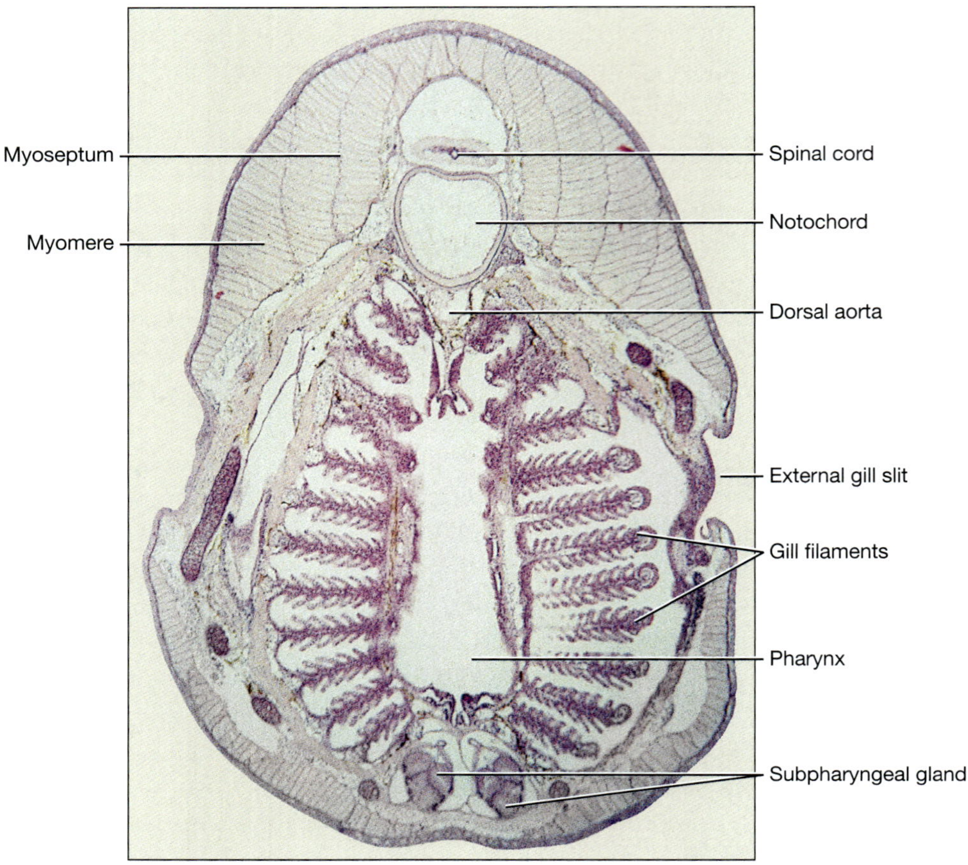

16.8 Lamprey ammocoete larva, transverse section through the pharynx.

16

Check Your Progress

3.1 List three features of the lamprey larva that are shared with lancelets.

3.2 List three features of the lamprey larva that are not shared with lancelets.

3.3 List the five characteristic chordate features that are visible in the lamprey larva.

Name ______________________ Section ______________

16 Questions for Review

1 List at least four adaptations of early ostracoderms that allowed them to become so successful during the Silurian and Devonian periods.

1. ______________________

2. ______________________

3. ______________________

4. ______________________

2 Discuss how the adult lamprey's digestive system reflects adaptations for its parasitic lifestyle.

____

____

____

3 Match the structures on the left with their function on the right.

	Structure		Function
____	olfactory sac	**a.**	circulates water into and out of nostril
____	pineal organ	**b.**	primary axial support element for body
____	hypophyseal sac	**c.**	secretes anticoagulating agent
____	buccal papillae	**d.**	chamber containing sensory epithelium for "smell"
____	notochord	**e.**	fingerlike sensory projections around mouth
____	oral gland	**f.**	photosensory organ that responds to changes in light

16

4 List the three regions of the heart in lampreys in the order that blood flows through them, starting with the return of deoxygenated blood from the body.

1. ______________________

2. ______________________

3. ______________________

5 Describe how the process of respiration in the lamprey larva differs from that seen in lancelets.

16

Name ______________________ Section ______________

16 Laboratory Practical Challenge

The following evaluation is intended to replicate the method of assessment and type of questions you will be asked on lab practicals. If you have difficulty with these questions, you may benefit from additional review of the specimens provided in lab.

1 Identify the phylum, subphylum, and class of the specimen depicted below.

____

2 What is the function of the structure indicated by line A?

____

3 Identify the structure indicated by line B.

____

4 To what system does the structure indicated by line C belong?

____

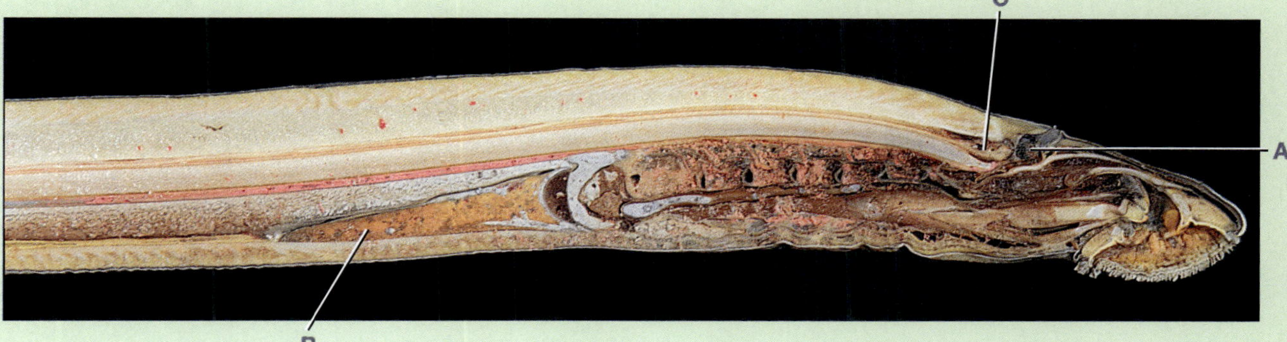

16

5 What is the function of the structure indicated by line D?

6 Identify the structure indicated by line E.

7 To what system does the structure indicated by line F belong?

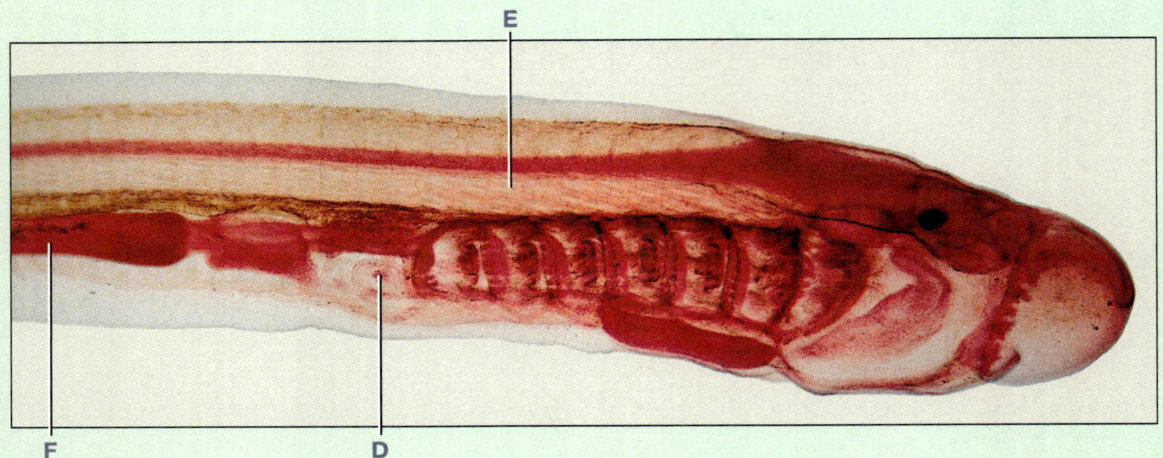

16

Cartilaginous Fishes

After completing the exercises in this chapter, you should be able to:

1 List the distinguishing characteristics of the class Chondrichthyes.

2 Understand the evolutionary position of cartilaginous fishes on the phylogeny of chordates.

3 Discuss the significance of jaws and paired fins in the evolution of early vertebrates.

4 Identify the major external features of a shark and compare them to homologous structures in other vertebrates.

5 Identify the major internal organs in a shark, discuss their functions in the body, and relate them to homologous organs in other vertebrates.

All fishes belong to the subphylum Vertebrata and exemplify organisms that are highly adapted to aquatic environments. Jawless fishes, such as lampreys and hagfish, are the most primitive of living fishes and are grouped in the superclass Agnatha. Jawed, cartilaginous fishes, including sharks, rays, and skates, are in the class Chondrichthyes (Fig. 17.1). Presently, named species of cartilaginous fishes number about 1,000—considerably fewer than what once roamed the oceans.

Modern cartilaginous fishes possess many of the familiar vertebrate characteristics such as a bilaterally symmetrical body, a well-developed head bearing sensory organs and a terminal mouth with jaws, a braincase (cranium), a vertebral column enclosing the dorsal nerve cord, and a closed circulatory system with a chambered heart.

The cartilaginous skeleton characteristic of this class of fishes is a derived feature; their ancestors had bony skeletons. Yet, even with a lighter skeleton and the increased buoyancy provided by oil stores in a huge liver, cartilaginous fishes are denser than water and will sink if they stop swimming. Continual swimming (or at least continual water movement) is also necessary to maintain water flow through the exposed **gill slits** and across the gills. The heterocercal tail and angled pectoral fins of most cartilaginous fishes provide natural lift to help them maintain their vertical position in the water column.

Their impressive combination of a sleek, **fusiform** body, strong jaws armed with sharp teeth, powerful swimming musculature, and well-developed sensory organs has secured them a place among the most successful aquatic predators to ever swim the oceans. Their keen olfactory organs allow them to detect prey at distances of more than a kilometer; they are able to sense low-frequency vibrations using their **lateral line** system, and they can even detect the weak electrical fields of living creatures with electroreceptors on the head (ampullae of Lorenzini).

Evolutionary History

Over a period of tens of millions of years early vertebrates made the gradual transition from filter-feeding bottom-dwellers to more agile jawless fishes with enhanced muscular, skeletal, circulatory, respiratory, and nervous systems. Yet these early advancements only represented the first chapter in the success story that vertebrates would experience over the next 500 million years on Earth.

The next major breakthroughs in vertebrate evolution were jaws and paired fins. Hinged jaws enabled vertebrates to grip food firmly. Paired fins, along with the tail, enabled them to maneuver more precisely while swimming. With their jaws and paired fins, many of these early fishes made the transition from bottom-feeders to active predators and eventually replaced most jawless fishes by the end of the Devonian period (416–360 mya). The upper and lower jaws of vertebrates evolved directly from modifications to the first two pairs of pharyngeal gill arches and their associated musculature. Once functional jaws were established, the remaining gill slits no longer were required for filter feeding and became dedicated to gas exchange with the external environment.

Two early groups of jawed fishes were the placoderms—armored fishes with paired fins—and the cartilaginous fishes of the class Chondrichthyes.

Blue spotted stingray, *Taeniura lymma*

Nurse shark, *Ginglymostoma cirratum*

17.1 Modern stingrays and sharks are the remaining descendants of cartilaginous fishes that roamed the oceans more than 400 million years ago.

Though placoderms ultimately became extinct, cartilaginous fishes thrived in the Devonian seas. Over time, their heavy dermal armor was reduced significantly, their bony skeletons were transformed into lighter, cartilaginous frames, and they diverged into shark-like and ray-like forms that have changed only slightly over the ensuing ages (Fig. 17.2).

Also abundant and highly successful throughout the Devonian period were two other groups of jawed fishes, the acanthoidians and bony fishes. Acanthoidians were armored fishes with paired spines and pectoral and pelvic fins and eventually became extinct during the Permian period (299–250 mya). The bony fishes, however, would undergo a monumental diversification and radiation that would continue for the next 250 million years and mark them as the most successful group of aquatic vertebrates ever to live. For a more complete phylogeny depicting the evolutionary relationships among the chordates covered in the remainder of this laboratory guide, review Figure A.2 in the Appendix.

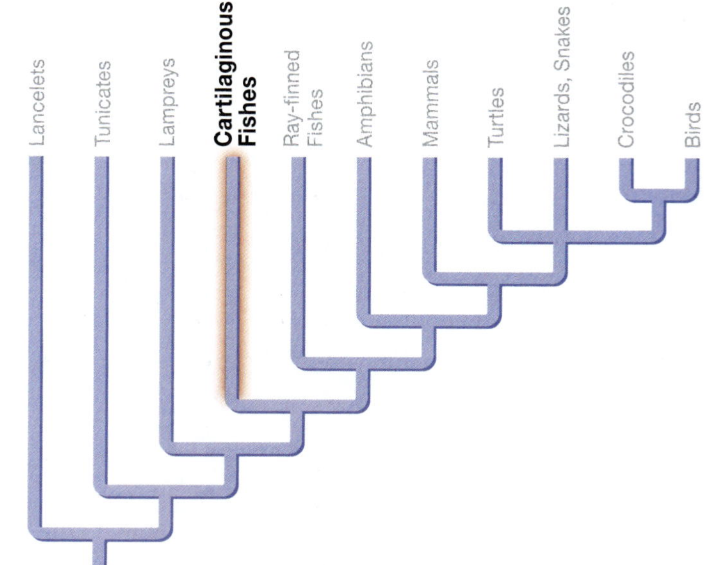

17.2 Cartilaginous fishes and bony fishes first appeared approximately 400 million years ago with innovations such as paired fins, jaws, and mineralized skeletons, representing a significant divergence from their vertebrate predecessors.

Exercise 17-1 Sharks, Rays, and Skates (Class Chondrichthyes)

External Anatomy

Materials Needed

- ❑ preserved dogfish shark, *Squalus*
- ❑ dissecting tools
- ❑ dissecting pan
- ❑ latex gloves

Procedure

1. Obtain a preserved dogfish shark and examine its external features.
2. Use Figure 17.3 and Table 17-1 to assist you in identifying the external anatomy of your shark.
3. As you examine your specimen, try to draw comparisons between the many homologous structures that are shared by other familiar vertebrates.

Notice that sharks have a streamlined, fusiform body shape designed to minimize drag during swimming. Their heads are flattened and wedge-shaped, a feature that generates lift at the anterior end of the body as they move through the water. The ventral side of the head bears paired **nostrils** and a subterminal **mouth** equipped with numerous rows of razor-sharp teeth to slice through prey, tearing off large chunks of flesh. The mouth also brings in water that is passed through the gills to extract oxygen. The dorsal surface of the head bears two large **eyes** with upper and lower immovable eyelids, two **spiracles** that serve as modified gill slits that open into the pharynx, and five pairs of external gill slits that allow water to exit the body. You may recall from Chapter 16 that jawless fish such as the lamprey had seven pairs of external gill slits. As the ancestors of modern fishes evolved jaws, the skeletal rods from two pairs of gill slits were recruited to form the bony elements of the jaws and the result was a reduction in the number of functional gill slits. The position of the spiracles on the dorsal surface allows the more flattened rays and skates, which spend a large portion of their time lying on the ocean floor, to draw in "clean" water without the possibility of damage to their delicate gills, which sand and debris might produce if water were inhaled through the ventral mouth.

Sharks possess a number of flattened fins, including two dorsal fins, an anal fin, and paired pectoral and pelvic fins, as well as a heterocercal tail, a type of caudal fin that has a large dorsal lobe and a much smaller ventral lobe.

17

Collectively, the fins are used to propel the animal through the water, as well as balance the shark against destabilizing forces generated when swimming that would otherwise cause the shark to veer off course. In males, the pelvic fins have claspers, modified structures that assist in reproduction (Fig. 17.3B).

During mating, male and female sharks align their bodies side by side and the male shark inserts one of his claspers into the cloacal opening of the female. A groove along the dorsal surface of his clasper channels sperm into her cloacal opening. In sharks, fertilization is internal, and the young develop inside the female, enclosed in eggs and without any placental connection to the mother. Males also possess a urogenital papilla through which sperm and urine pass, visible within the cloacal opening. Females instead have a urinary papilla through which only urine passes (Fig. 17.3C).

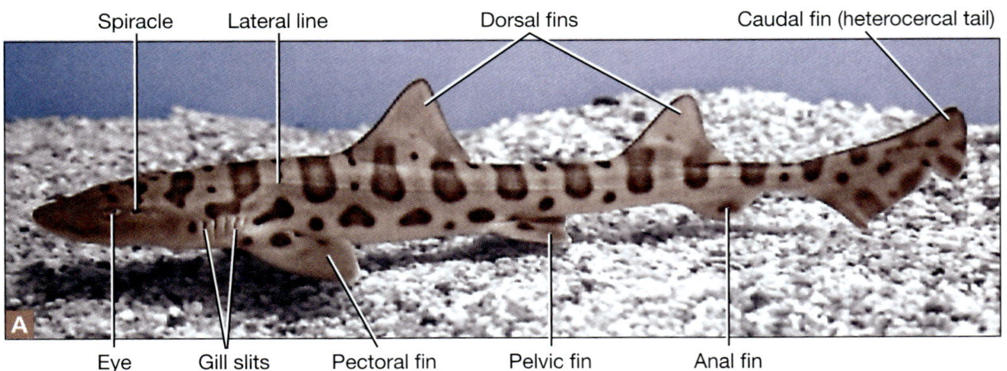

17.3 External anatomy: (**A**) lateral view of the leopard shark, *Triakis*; and ventral views of (**B**) male and (**C**) female dogfish sharks, *Squalus*.

A faint line running along the length of the body marks the position of the lateral line system found in sharks, bony fishes, and some aquatic amphibians. This extensive network of pores arranged over the surface of the head and along the body contains highly sensitive sensory cells that function as both mechanoreceptors, allowing the shark to detect water currents and pressure changes produced by nearby objects, and electroreceptors, capable of detecting the minute electrical currents produced across the cell membranes of animals, such as fish and invertebrates—their typical prey. Sharks employ the lateral line system both as a navigational tool and as a resource for detecting prey in especially dark, murky waters where vision is limited.

Table 17-1 External Anatomy of the Shark

Structure	Function
Eyes	Large, image-forming sight organs with immovable eyelids
Nostrils	Paired openings on the ventral surface of the head leading to olfactory receptors
Mouth	Subterminal opening to digestive tract; lined with razor-sharp teeth
Spiracles	Paired openings on the dorsal surface of the head that function as modified gill slits allowing water to be drawn into the pharynx and across the gills
External gill slits	Five pairs of openings along the sides of the head that allow the release of water
Pectoral fins	Steering and braking while swimming and maintenance of dorsal-ventral orientation
Pelvic fins	Steering while swimming
Anal fin	Steering while swimming
Caudal fin	Provides thrust and acts as a rudder while swimming
Dorsal fins	Steering and maintenance of dorsal-ventral orientation while swimming
Abdominal pores	Paired openings near the cloaca; the function of which is unclear
Lateral lines	Specialized sensory organs that detect vibrations and current disturbances as well as weak electrical fields in the water; used for navigation and prey capture

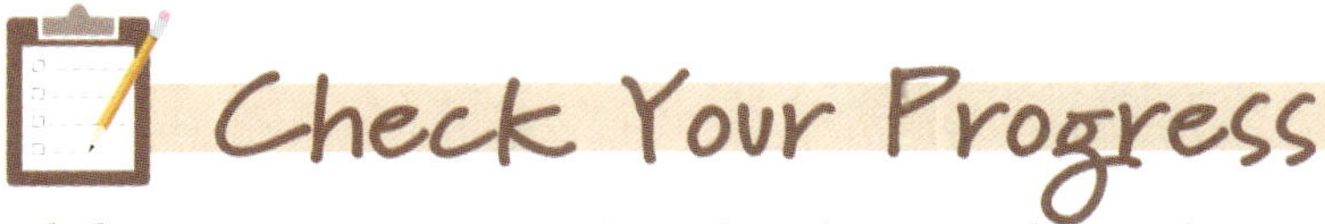

1.1 Is your specimen male or female? How do you know?

1.2 What features of the head represent homologies to other vertebrate groups?

1.3 What is the function of the spiracles?

1.4 Discuss the evolutionary significance of jawless fishes having seven pairs of external gill slits and sharks having only five pairs.

1.5 What is the function of the lateral line system?

1.6 Why do you suppose the lateral line system has been retained in fishes and some aquatic vertebrates, but lost in reptiles, birds, and mammals?

Digestive System

Procedure

1 To open your specimen and view the internal anatomy, review the Dissection Diagram for suggested cuts to make along the ventral surface of the body.

2 Insert your scissors in the slit along the ventral surface of the body wall made during the injection process, and extend this incision anteriorly to the base of the mouth, then posteriorly to the cloaca.

3 Next, make a transverse incision across the body just in front of the pelvic fins, and another transverse incision just in front of the pectoral fins.

4 Finally, make longitudinal incisions from the outer corners of the mouth posteriorly to intersect the transverse cut made anterior to the pectoral fins.

5 Use the following text, Table 17-2, and Figures 17.4–17.6 to assist you in identifying the digestive organs and other internal structures of your shark.

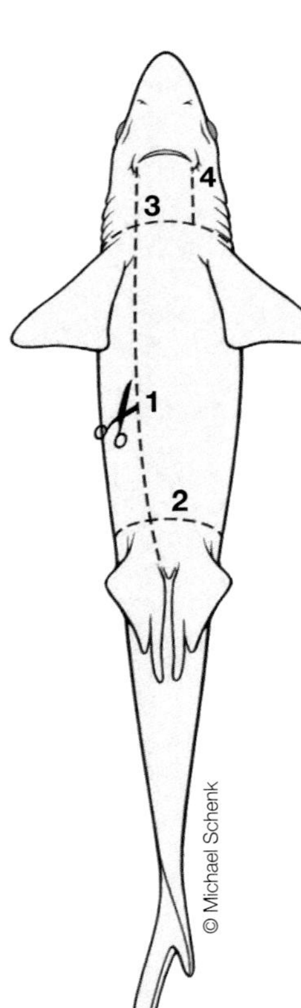

Dissection Diagram

1 Insert your scissors in the slit in the body wall made during the injection process, and extend this incision anteriorly to the corner of the mouth, then posteriorly to the cloaca.

2 Next make a transverse incision across the body just in front of the pelvic fins, and

3 another just in front of the pectoral fins.

4 Finally, make longitudinal incisions from the outer corners of the mouth posteriorly to intersect the transverse cut made anterior to the pectoral fins.

Two accessory digestive organs, the large, multilobed liver and the glandular pancreas, are present in the shark. Food does not pass directly into accessory digestive organs, but these organs do play a vital role in the digestive process by secreting digestive enzymes to help break down food. The **liver** secretes bile, which is stored in the **gallbladder** and released into the **duodenum**, where it aids in emulsifying fats (Fig. 17.4). The liver also detoxifies many constituents of the absorbed digested compounds and functions in lipid and glycogen storage. One reason the liver in sharks is relatively large is that it functions as an additional aid to buoyancy. Cartilaginous fishes lack a swim bladder and, instead, use the liver to store large reserves of oil, allowing them to lighten their overall density and approach neutral buoyancy. After identifying the liver, reflect it dorsally (or remove it) to view the underlying digestive organs.

The **pancreas** in sharks is a multilobed organ as well, and is embedded in the mesenteries of the stomach and intestine. The pancreas functions both as an accessory digestive organ, producing digestive enzymes that are released into the duodenum, and as an endocrine organ, producing hormones that help control blood-glucose levels.

Sharks do not chew their food; instead, they tear chunks from their prey and swallow them whole. These large bites of food pass first into the **esophagus** and then into the **stomach** (Fig. 17.5). Notice that the esophagus and the stomach almost seem to be a continuous structure with equally large diameters. This arrangement allows large fragments of food to be swallowed easily and pass into the stomach without difficulty. The stomach makes a J-shaped turn and passes food into the duodenum, the first portion of the intestine (Fig. 17.6).

Table 17-2 Internal Anatomy of the Shark

Structure	Function
Liver	Produces bile and detoxifies many constituents of the absorbed digested compounds
Gallbladder	Stores bile produced by the liver
Pancreas	Produces digestive enzymes and delivers them to the duodenum, produces hormones that control blood-glucose levels
Esophagus	Transports food to the stomach
Stomach	Site of food storage and initiation of digestion
Duodenum	Receives secretions from the liver and pancreas for further breakdown of food from the stomach
Ileum (valvular intestine)	Primary site of digestion where most of the absorption of nutrients into the bloodstream occurs
Cloaca	Common chamber for the release of urine, feces, and semen
Gills	Highly vascular structures for gas exchange with the bloodstream
Spleen	Elongate organ that stores blood and recycles worn-out red blood cells
Heart	Two-chambered organ (one atrium, one ventricle) that pumps deoxygenated blood to the gills for oxygenation and from there throughout the various organ systems of the body
Sinus venosus	Thin-walled sac where deoxygenated blood from the body collects before being passed into the atrium of the heart
Conus arteriosus	Contractile portion of the heart that receives blood from the ventricle and sends it to the gills for oxygenation via the ventral aorta
Kidneys	Paired excretory organs that filter nitrogenous wastes from the blood
Mesonephric ducts	Passageway leading from the kidney to the cloaca; transports urine, and in males, semen
Testes (males)	Paired reproductive organs that produce sperm
Epididymis (males)	Paired reproductive organs that store sperm
Ovaries (females)	Paired reproductive organs that produce eggs
Ostium tubae (females)	Common opening to the oviducts through which eggs released from the ovaries pass
Oviducts (females)	Paired tubules where fertilization occurs
Shell glands (females)	Small, oval glands that secrete a thin membrane around fertilized eggs
Uterus (females)	Large, paired horns of the reproductive tract where embryonic development occurs

Inside, the stomach contains **rugae**, longitudinal folds along the lining of the stomach that permit the stomach to expand to accommodate large meals. During the course of vertebrate evolution, strong selective pressures have promoted more efficient digestive tracts that were able to extract more nutrients from food. This meant increasing the absorptive surface area of the intestinal tract and slowing the speed at which food passes through the intestinal tract. Sharks accomplish this in a unique way: The **ileum** of the small intestine contains a **spiral valve** that both increases the absorptive surface area and greatly slows the passage of food, ensuring longer exposure to digestive enzymes and increased opportunity for nutrient absorption by the intestinal walls. Indigestible remnants pass through a short **colon** and out of the body through the **cloaca**.

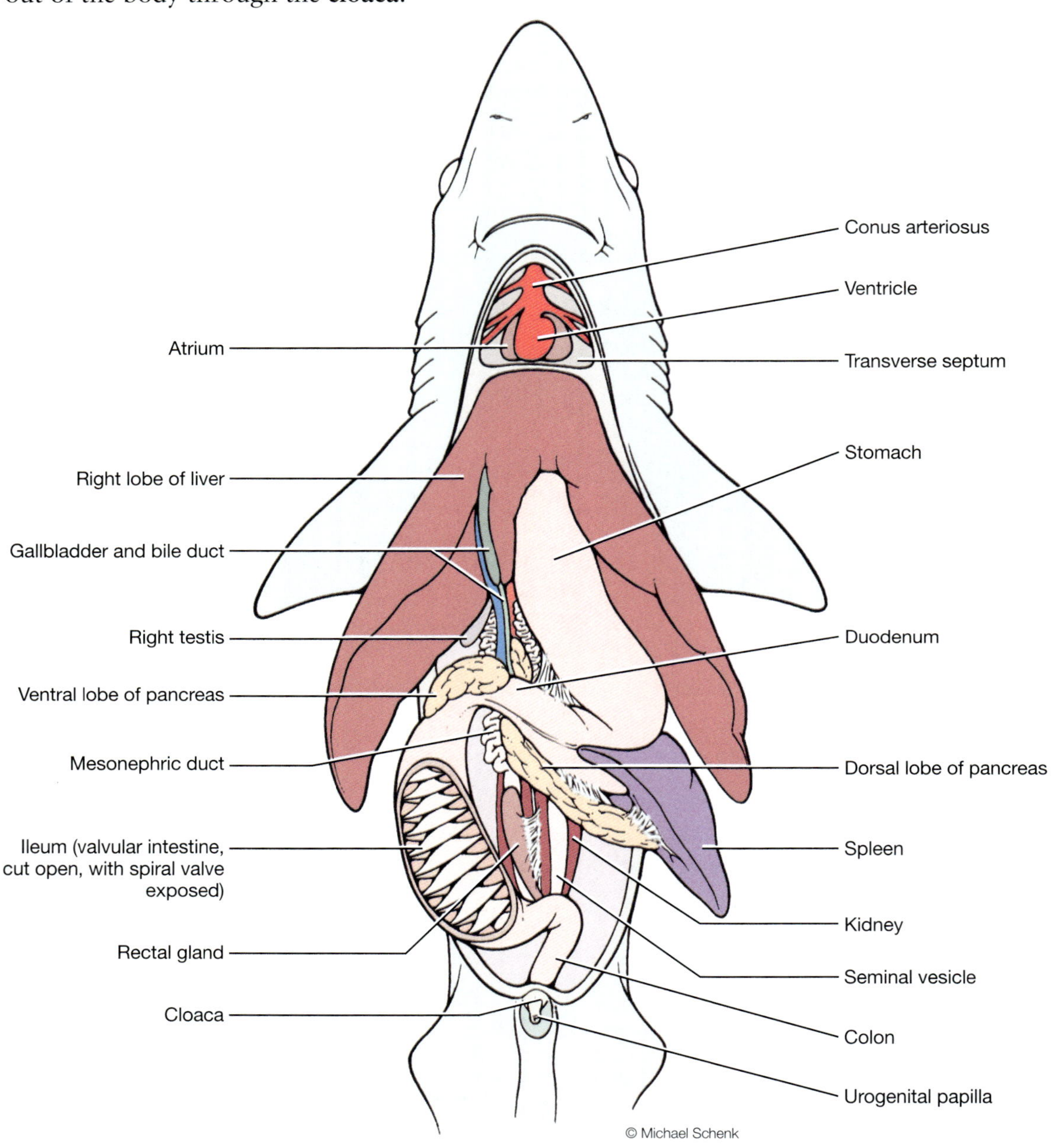

17.4 Dogfish shark, *Squalus*, internal anatomy.

17

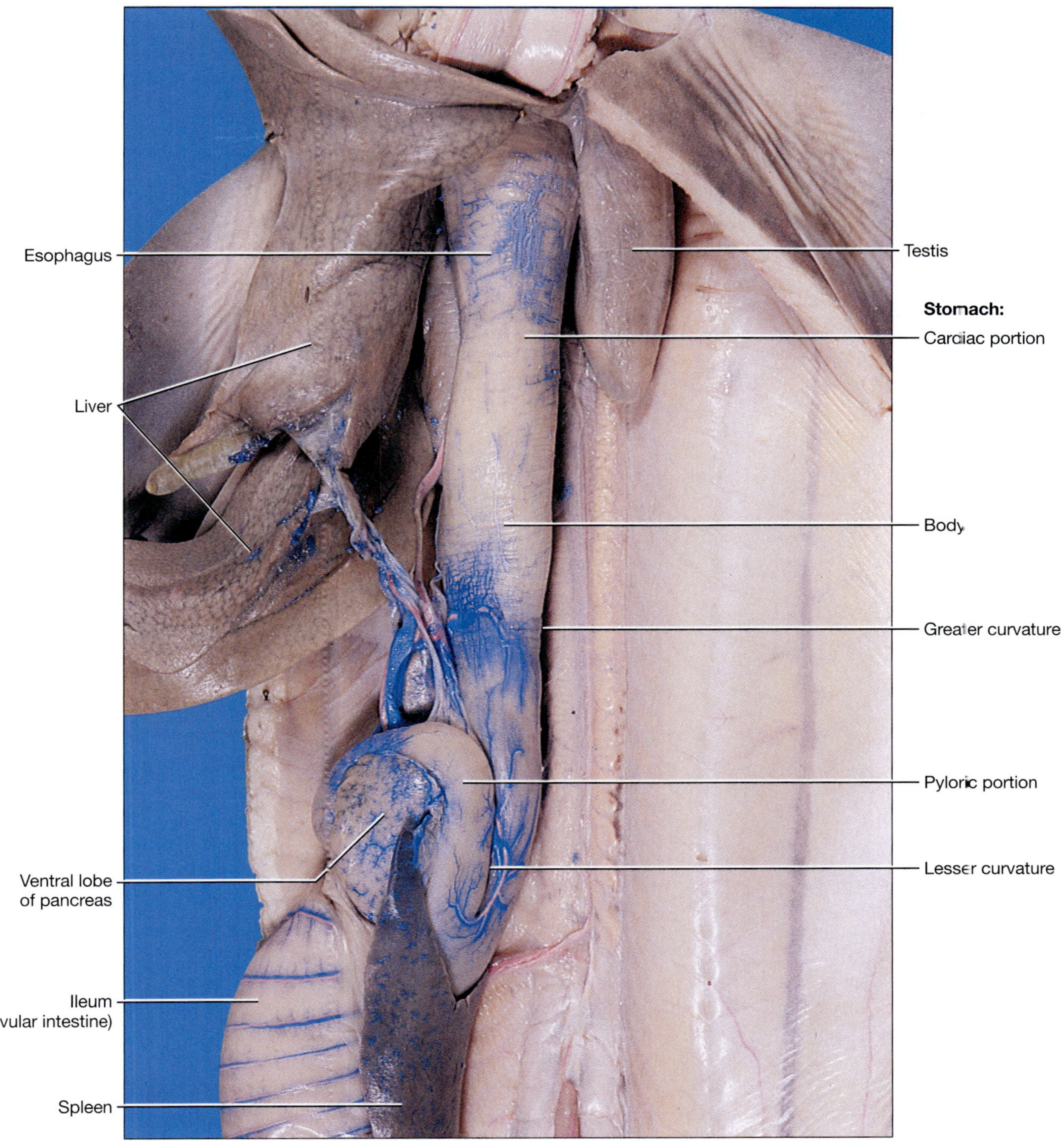

17.5 Dogfish shark, *Squalus*, esophagus and stomach.

17

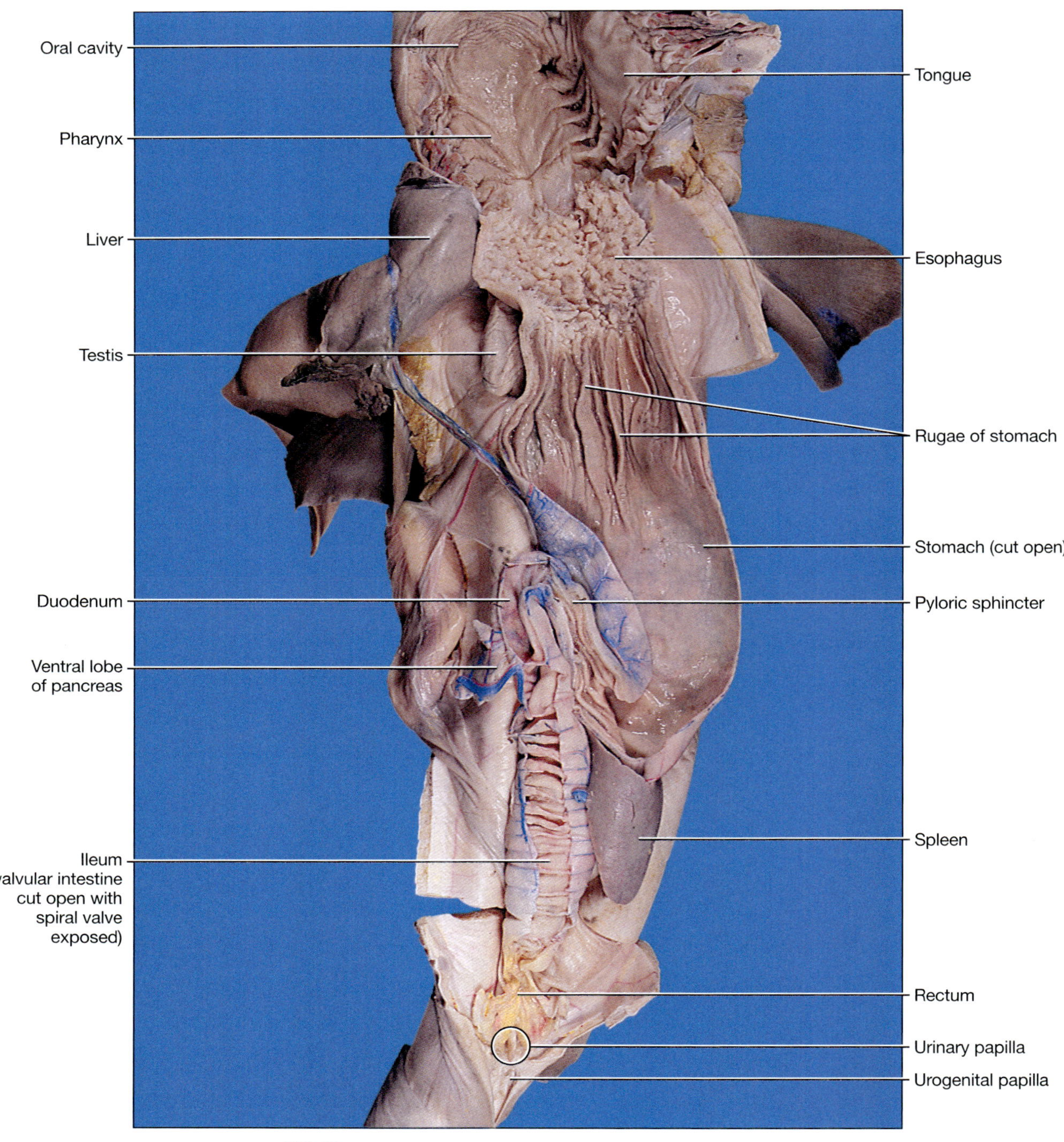

17.6 Dogfish shark, *Squalus*, dissected digestive canal.

Circulatory and Respiratory Systems

The circulatory pathway of cartilaginous fishes is a single-circuit system sending blood from the heart to the gills and then on to the rest of the body, before returning it to the heart. Though we simplistically think of the hearts of cartilaginous and bony fishes as two-chambered structures composed of a single **atrium** and a single, thick-walled **ventricle**, there are actually four contractile compartments that deliver blood to the body (Fig. 17.7). Presumably, the ancestral condition in vertebrates consisted of four linearly arranged chambers in the following order: sinus venosus, atrium, ventricle, and conus arteriosus. These four chambers have been retained in cartilaginous fishes, bony fishes, and most amphibians, though they no longer are arranged linearly in these groups. Instead, the heart folded into somewhat of an S-shaped configuration and became more compact over time. Reptiles, birds, and mammals incorporated the sinus venosus and conus arteriosus into the walls of the heart while partitioning the atrium and ventricle into paired atria and ventricles through the development of internal septa.

In sharks, deoxygenated blood returns from the body tissues and collects in the **sinus venosus**, a thin-walled sac located on the dorsal surface of the heart. Blood is pumped into the atrium, and then into the more muscular ventricle. From the ventricle, blood is pumped into the **conus arteriosus** and then passes into the **ventral aorta** and then through five pairs of afferent branchial arteries to the gills (Fig. 17.8).

Oxygenation of the blood occurs in the **gills**, and blood then travels through four efferent branchial arteries to their convergence at the **dorsal aorta**. The dorsal aorta diverges into numerous smaller arteries that distribute oxygenated blood to the capillary beds of the body tissues, where gases, nutrients, and metabolic wastes are exchanged (Fig. 17.9). Deoxygenated blood returns through the venous system, converging into two large posterior cardinal sinuses before draining into the sinus venosus to start the circuit again (Fig. 17.10).

Note

The accompanying illustrations of arterial and venous circulation in the shark (Figs. 17.9 and 17.10) provide a detailed depiction of the major blood vessels for reference purposes and may be beyond the scope that you are required to know. Check with your instructor to see which arteries and veins you will be expected to learn.

Cartilaginous fishes lack bones and, thus, naturally lack bone marrow, which is the primary source of red blood cell production in bony vertebrates. Instead, the **spleen** is the sole organ that produces red blood cells in sharks, rays, and skates. Additionally, the spleen filters the blood, removing worn-out red blood cells, harmful bacteria, and viruses in much the same way as other lymph organs do.

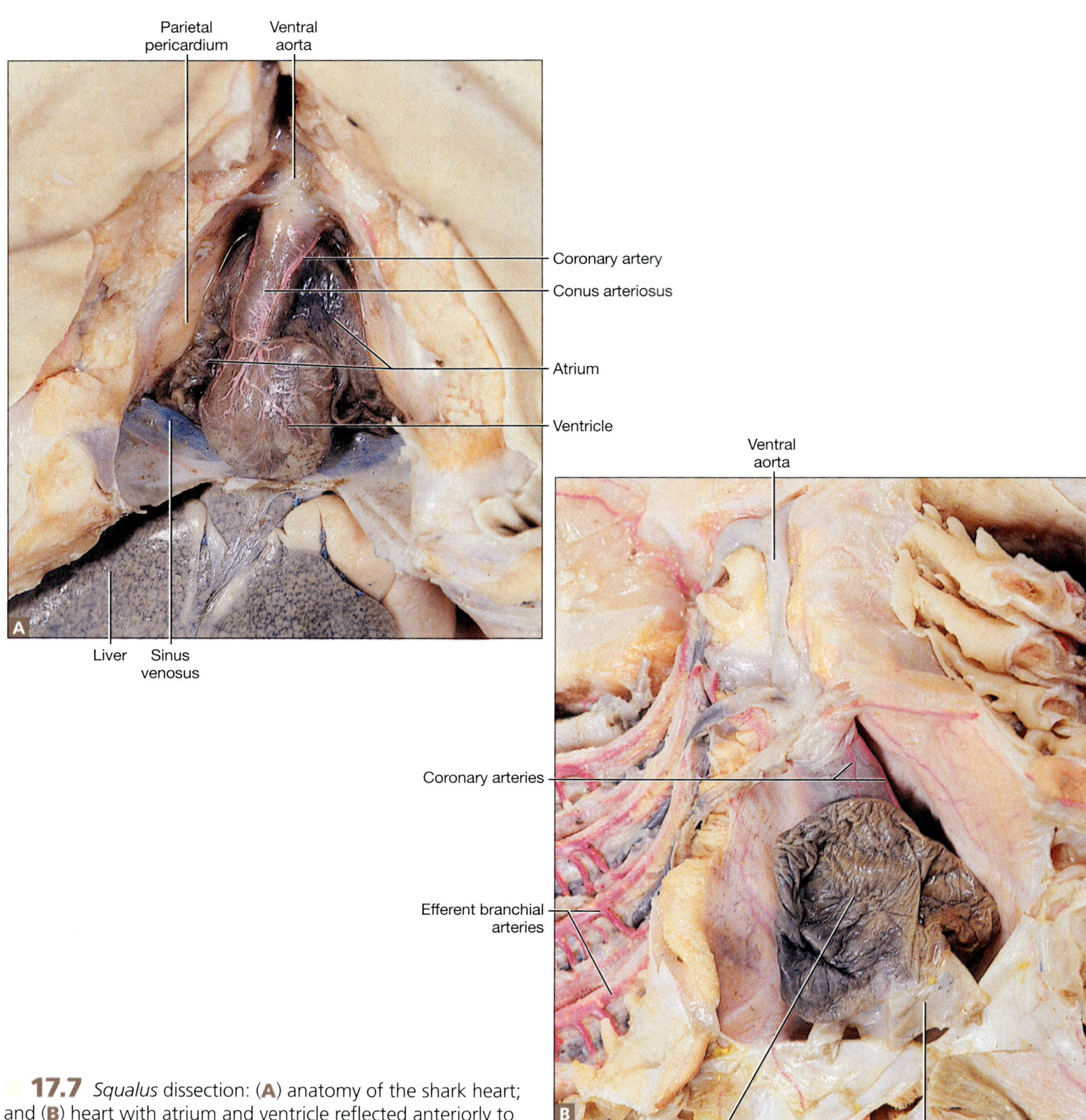

17.7 *Squalus* dissection: (**A**) anatomy of the shark heart; and (**B**) heart with atrium and ventricle reflected anteriorly to reveal the sinus venosus.

17

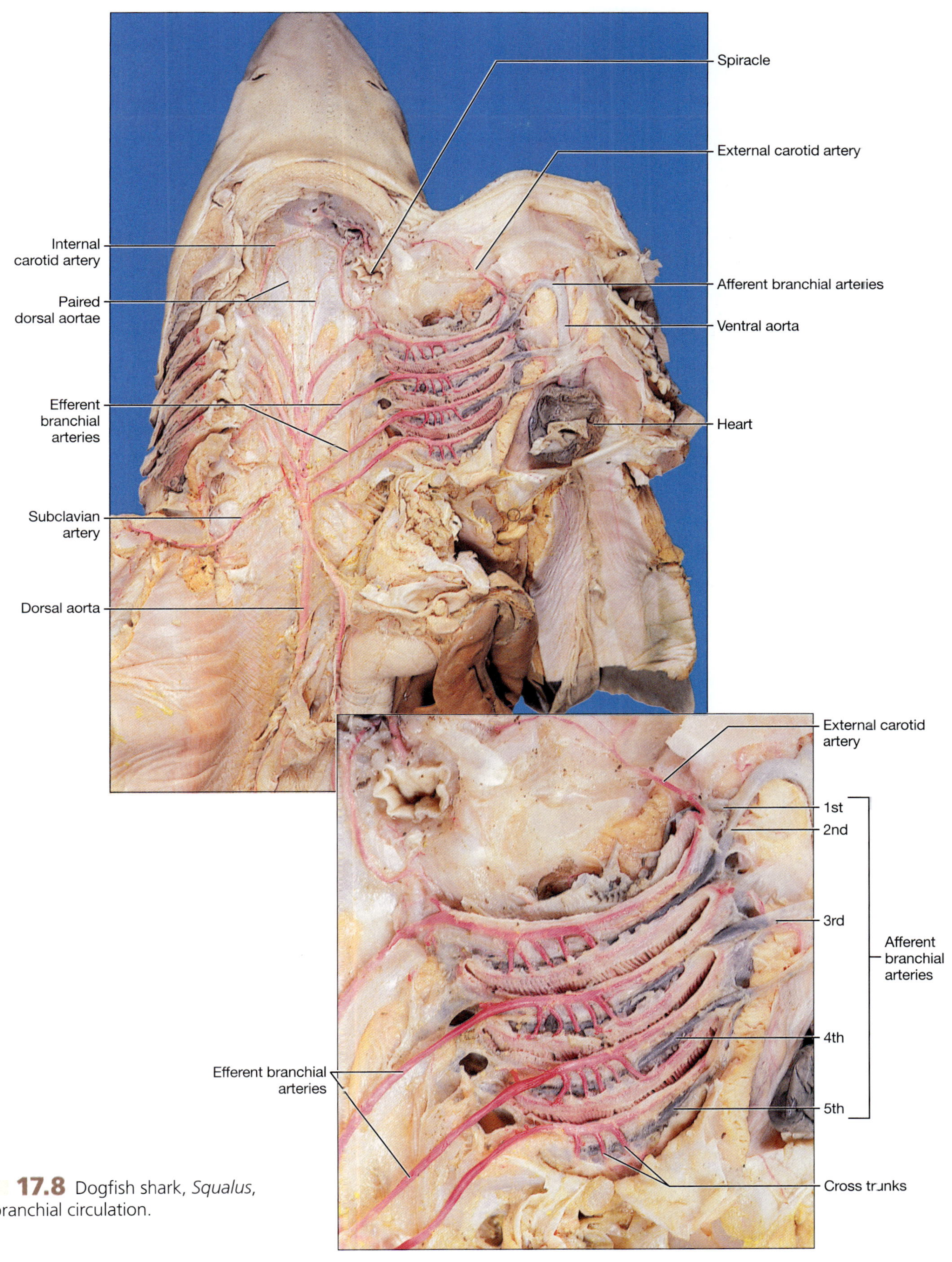

17.8 Dogfish shark, *Squalus*, branchial circulation.

17

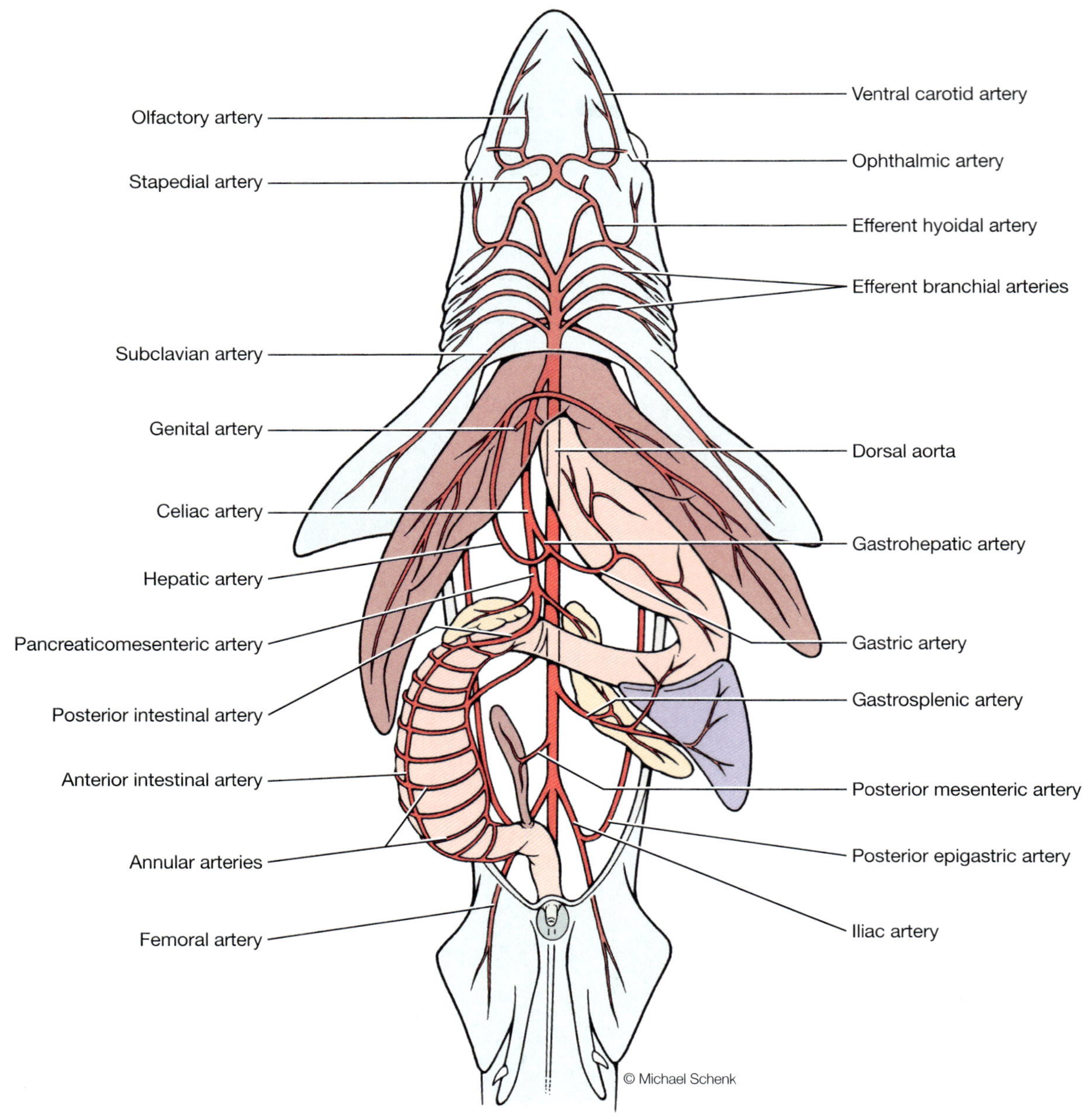

17

17.9 Dogfish shark, *Squalus*, arterial circulation depicted to show extensive networking of vessels to distribute oxygenated blood throughout the body.

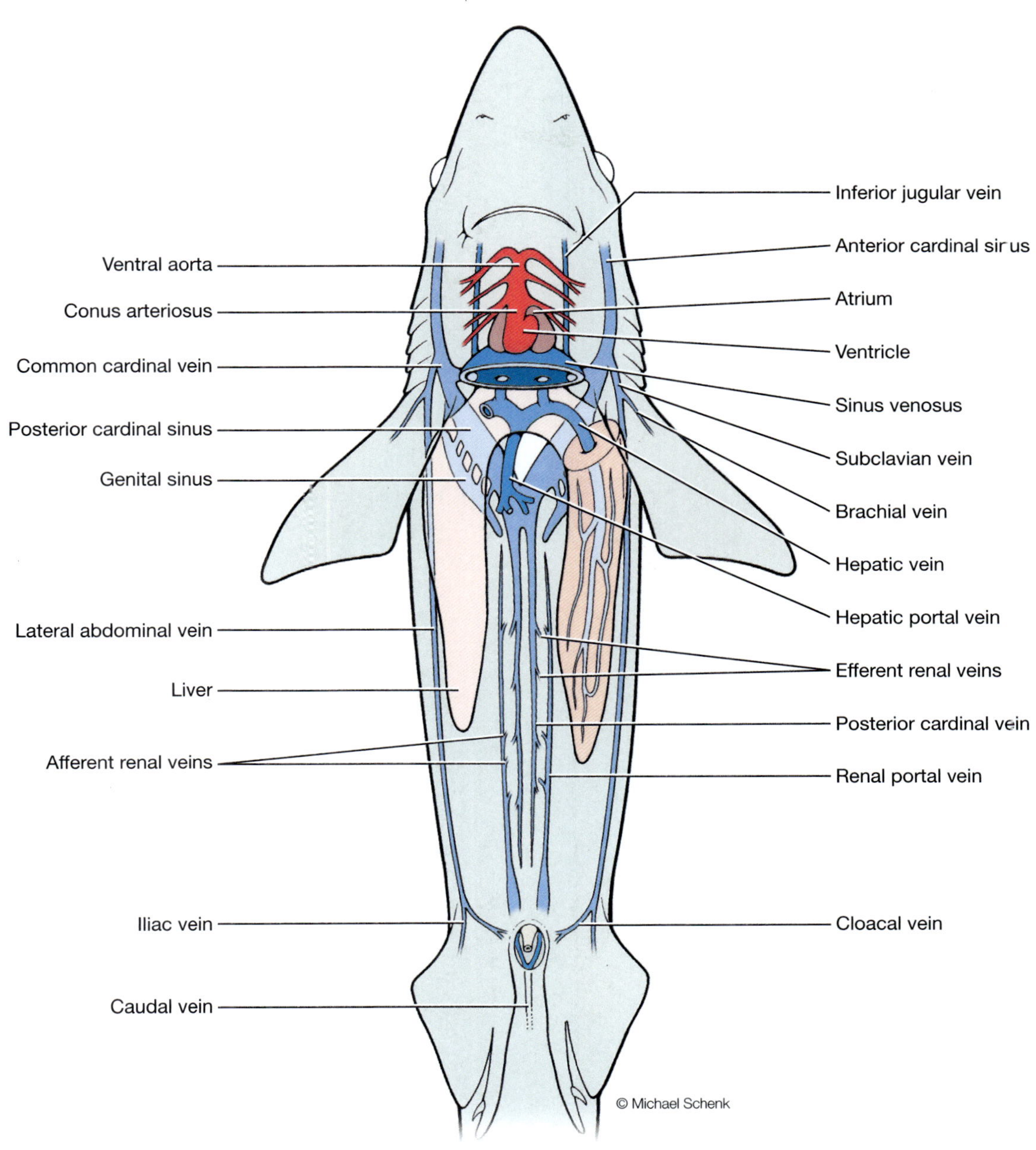

17.10 Dogfish shark, *Squalus*, venous circulation depicted to show extensive networking of vessels to return deoxygenated blood to the heart from all regions of the body.

17

Excretory and Reproductive Systems

In cartilaginous fishes, bony fishes, and many other vertebrates, the excretory and reproductive systems share many organs and often are combined into a single system commonly called the urogenital system. In both sexes of sharks, the paired **kidneys** are long, brownish organs that lie along the dorsal body wall on either side of the dorsal aorta. In males, filtrate flows from each kidney along a mesonephric duct and through the urogenital papilla into the cloaca for discharge from the body (Fig. 17.11). The mesonephric duct also carries sperm from the testes to the cloaca during copulation. In females, the mesonephric duct is very short and originates from the posterior end of the kidney, making it more difficult to locate.

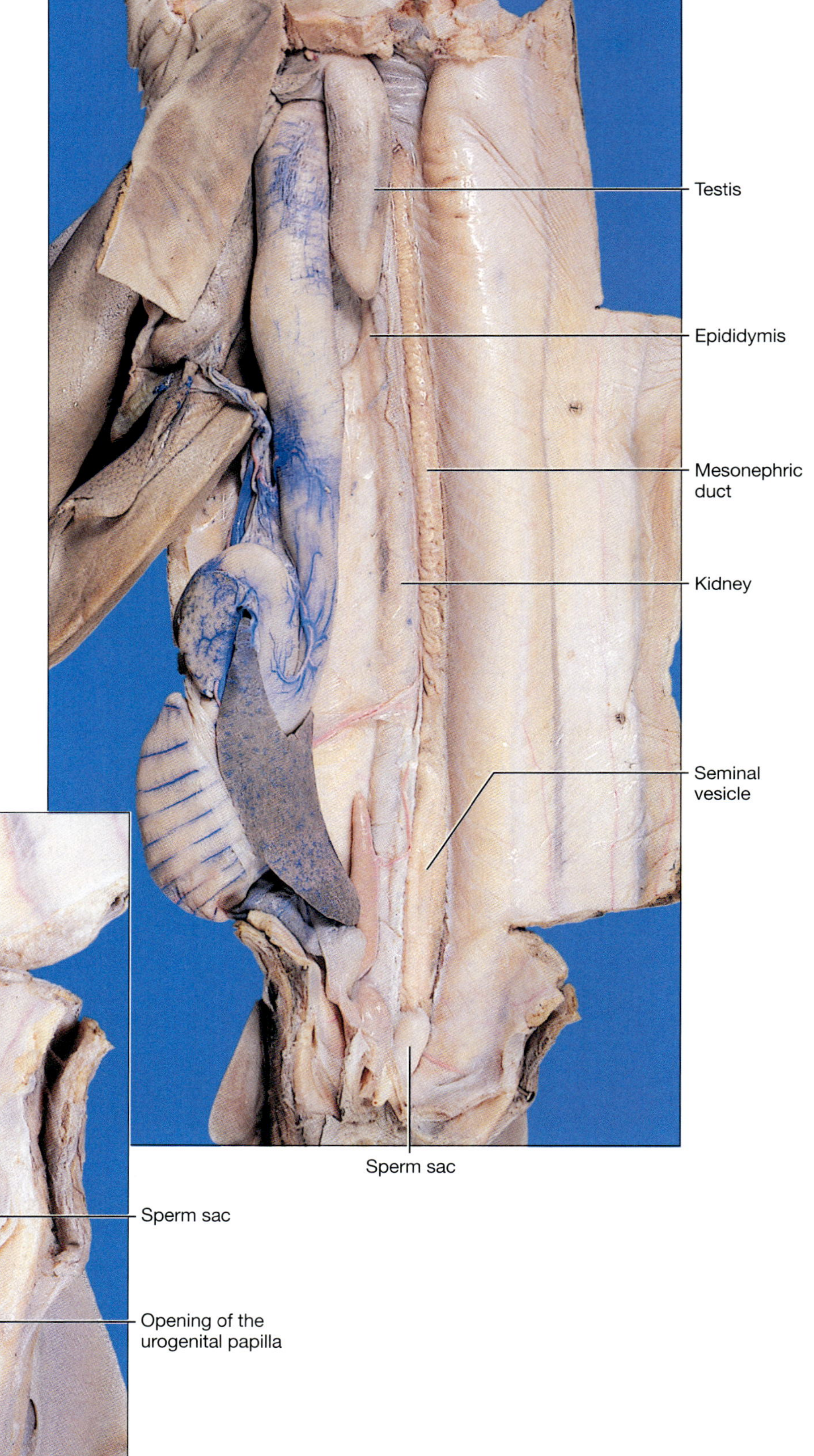

17.11 Dogfish shark, *Squalus*, male reproductive system.

In males, sperm are produced in paired **testes** that lie along the dorsal body wall on either side of the dorsal aorta at the anterior end of each kidney. Mature sperm pass through a system of minute tubules known as efferent ductules into each **epididymis**, where they are stored (Fig. 17.11). During copulation, sperm are discharged along the mesonephric duct and out of the body through the urogenital papilla that opens into the cloaca.

In females, eggs are produced in the paired **ovaries**, which lie against the dorsal body wall on either side of the esophagus (Fig. 17.12). Mature eggs leave the ovaries and pass into the abdominal cavity, where they are drawn into the oviducts through a common opening, the ostium tubae. Fertilization occurs internally in the **oviducts**, and the zygotes pass through shell glands to receive thin, membranous shells before being transported to the paired uteri. In most species of sharks, multiple young develop within each **uterus**, encased in their thin membranes. Just prior to birth, the pups break free from their eggs while still in the uterus and are born live (Fig. 17.13).

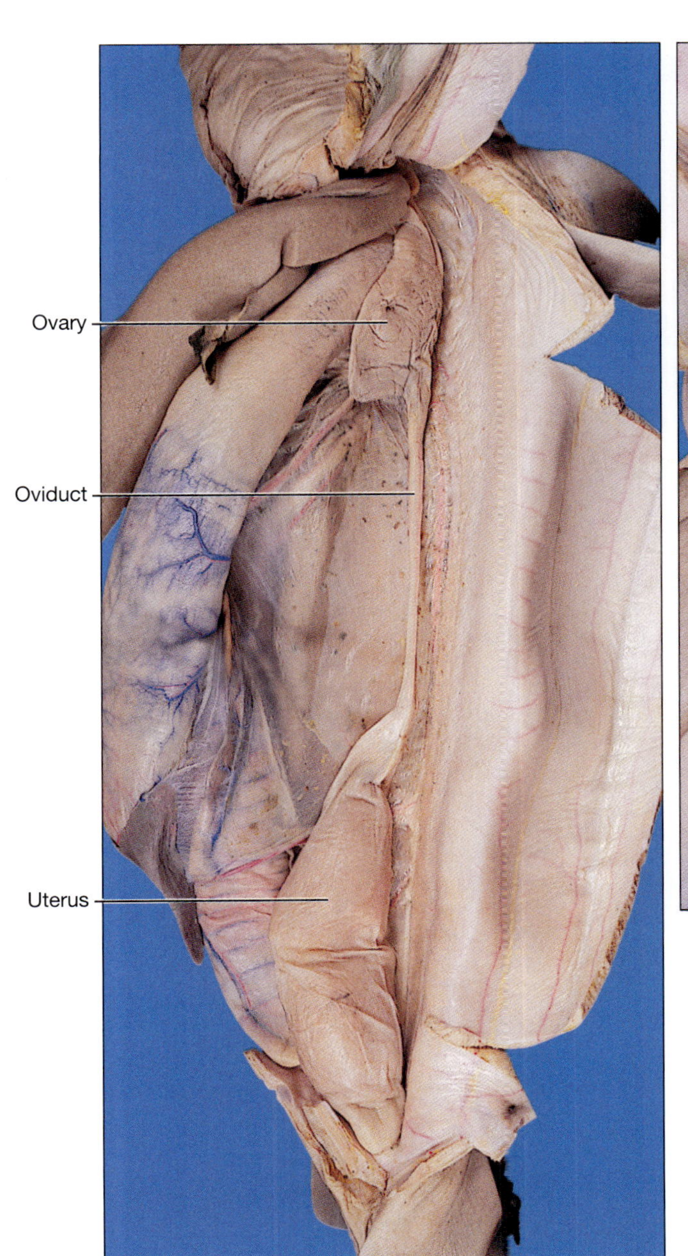

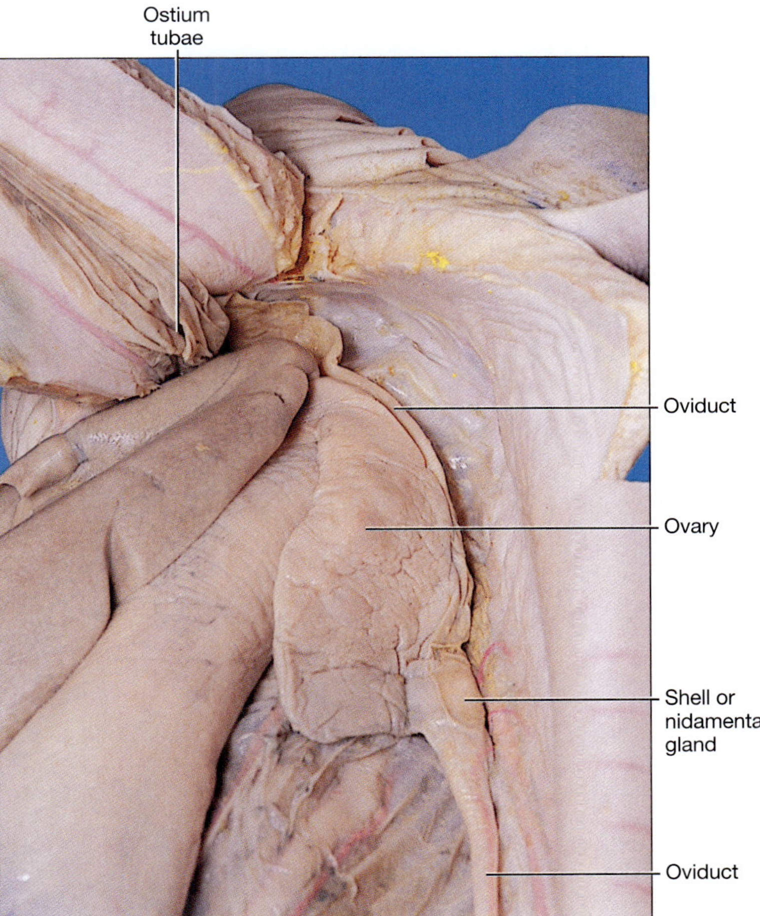

17.12 Dogfish shark, *Squalus*, female reproductive system.

17

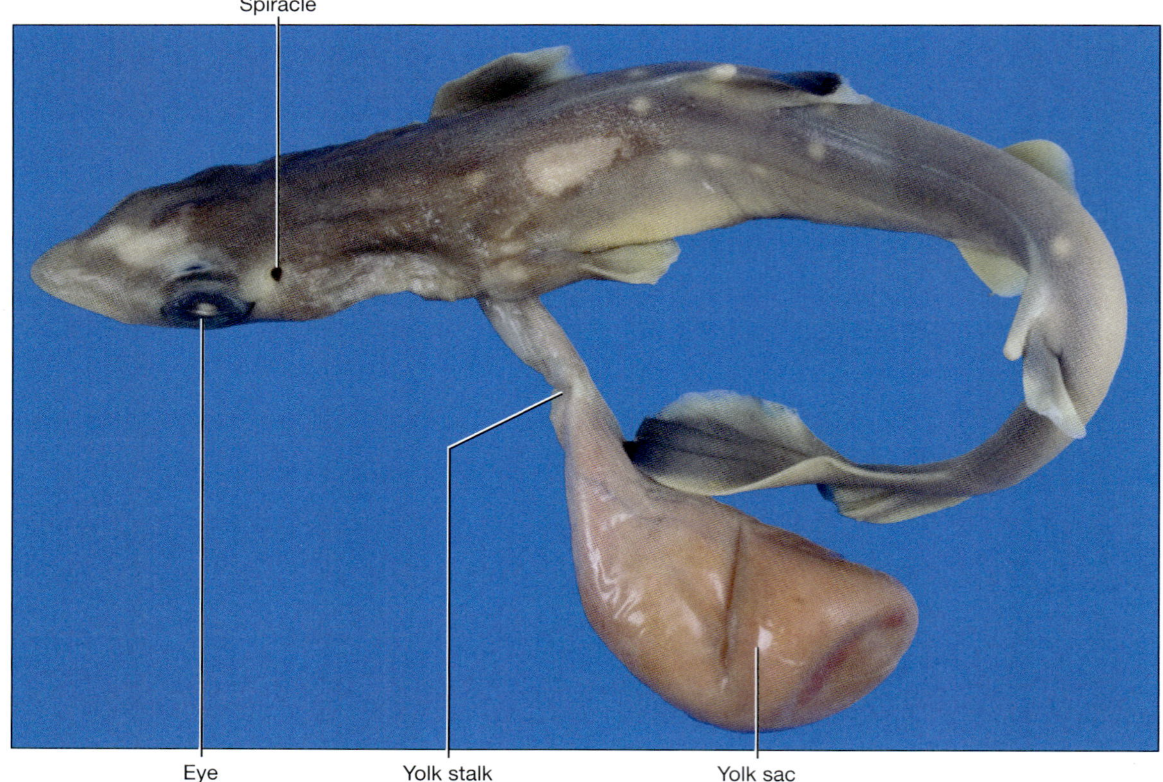

17.13 Dogfish shark pup prior to birth. The pup remains attached to the yolk sac until it has developed enough to survive on its own, at which time it will break free of its thin, membranous shell and leave the mother's uterus.

17

Name ______________________ Section ______________

17 Questions for Review

1 During the evolution of vertebrates, where did jaws come from and what advantages did they offer?

2 Describe how the spiracles are functionally different from the other gill slits in cartilaginous fishes.

3 What organ in sharks helps them remain more buoyant in water? How does it aid in buoyancy?

4 Discuss how the digestive system of the shark has been modified to increase its overall efficiency.

5 What organ functions as the primary source of red blood cell production in sharks?

6 List the four contractile regions of the heart in cartilaginous fishes in the order that blood flows through them, starting with the return of deoxygenated blood from the body.

1. ______
2. ______
3. ______
4. ______

7 Contrast the function of the mesonephric duct in male and female sharks.

8 Discuss why sharks must continuously swim (or remain in areas with moving water currents).

Name ______________________ Section ______________

17 Laboratory Practical Challenge

The following evaluation is intended to replicate the method of assessment and type of questions you will be asked on lab practicals. If you have difficulty with these questions, you may benefit from additional review of the specimens provided in lab.

1 Identify the phylum, subphylum, and class of the specimen depicted below.

____

2 What is the function of the structure indicated by line A?

____

3 Identify the structure indicated by line B.

____

4 Identify the structure indicated by line C.

____

5 What is the function of the structure indicated by line D?

6 To what system does the structure indicated by line E belong?

7 Identify the structure indicated by line F.

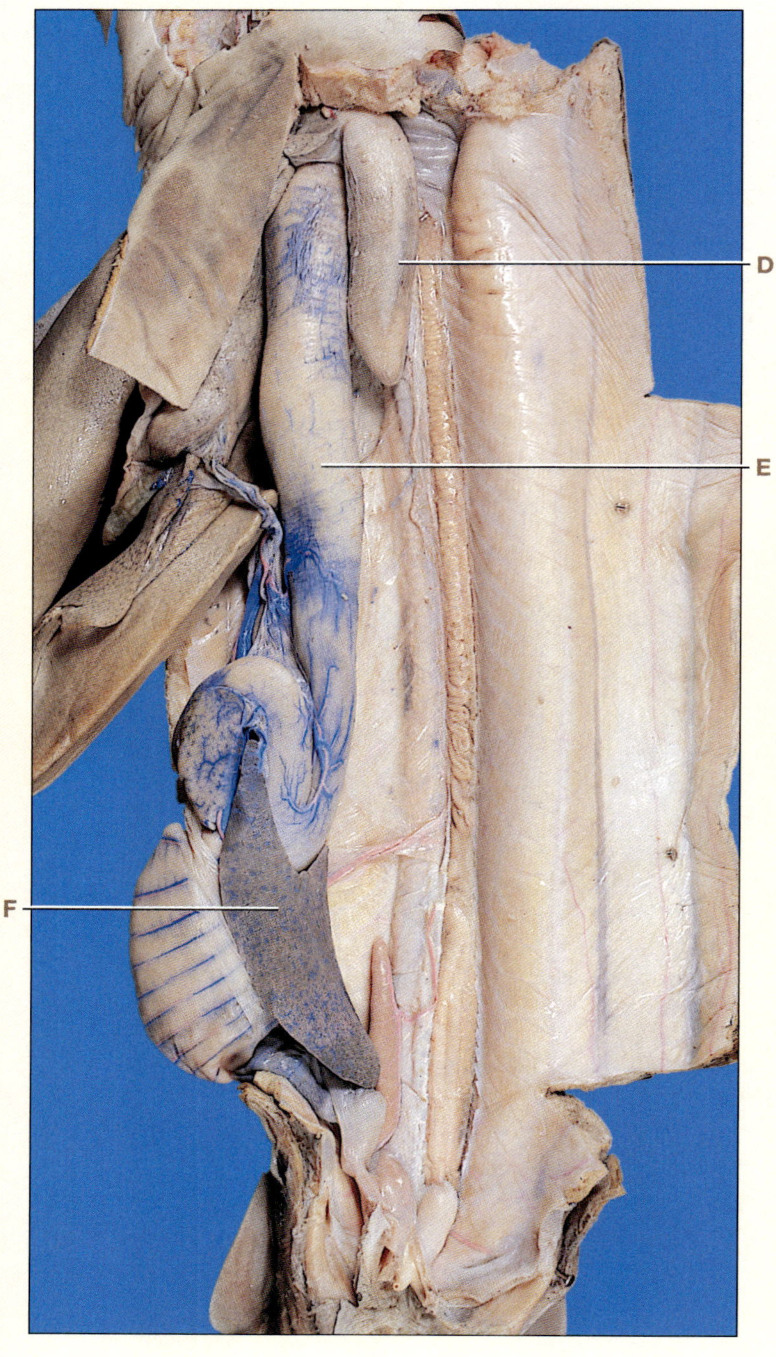

17

CHAPTER 18

Ray-finned Fishes

After completing the exercises in this chapter, you should be able to:

1. List the distinguishing characteristics of the class Actinopterygii.
2. Understand the evolutionary position of ray-finned fishes on the phylogeny of chordates.
3. Discuss the significance of bony fins, gill opercula, and the swim bladder in the evolution of early vertebrates.
4. Identify the major skeletal features of a ray-finned fish and compare them to homologous bones in other vertebrates.
5. Identify the major external features of a ray-finned fish and compare them to homologous structures in other vertebrates.
6. Identify the major internal organs in a ray-finned fish, discuss their functions in the body, and relate them to homologous organs in other vertebrates.

This chapter focuses on the subgroup of bony fishes within the class Actinopterygii. This single class of vertebrates contains more than 27,000 living species—more than half of all known vertebrates, more than all jawless fish, cartilaginous fish, amphibian, reptile, bird, and mammal species combined! The class Actinopterygii includes everything from freshwater species such as bass, trout, perch, and minnows to marine species such as flounder, grouper, tarpon, sailfish, and even sea horses (Fig. 18.1).

The remarkable success of the ray-finned fishes has resulted from several significant anatomical adaptations brought about by the unique selective pressures of the aquatic world in which they have been evolving for hundreds of millions of years. Faced with a medium that contains less than 1/20 as much oxygen as air, the **gills** of modern fishes have become the most effective respiratory structures in the animal kingdom. The efficiency of their gills is increased by the presence of an **operculum**, a movable flap covering each gill that can pump water over the gills, allowing a fish to breathe even while remaining stationary.

Unlike sharks, which will sink without continual muscular effort, bony fish are equipped with a **swim bladder** that allows them to achieve neutral buoyancy in the water column and provides an additional means of gas exchange. Their paired fins are supported by parallel bony rays, giving them increased support and stiffness, along with the ability to be folded against the body for maneuvering. These features are homologies shared with other vertebrate groups that have been modified in fishes to best equip them to cope with the challenges they face underwater.

Evolutionary History

18

Although bony fishes appear slightly earlier than cartilaginous fishes in the fossil record, both groups began their diversification around the same time, during the Devonian period (416–360 mya). Unlike sharks, however, bony fishes evolved in freshwater habitats and their ancestors had rudimentary lungs—a feature retained by the lobe-finned fishes and lungfishes. These paired air sacs were connected to the back of the throat and could be inflated with air to regulate the fish's buoyancy in the water column.

By the end of the Devonian period, the bony fishes had diverged into two major groups: the lobe-finned fishes and lungfishes (class Sarcopterygii), and the ray-finned fishes (class Actinopterygii). Although only a few genera of lobe-finned fishes and lungfishes still survive today, the ray-finned fishes underwent a major adaptive radiation that gave rise to the remainder of modern fishes (Fig. 18.2). Changes in the fins of bony fishes increased stability, maneuverability, and speed. The dorsal fin of bony fishes transitioned from a heavy, fixed keel to a thinner, more

Jack mackerel, *Trachurus declivis*

Leafy seadragon, *Phycodurus eques*

Lionfish, family Scorpaenidae

18.1 With more species than all other classes of vertebrates combined, the ray-finned fishes display remarkable diversity in their adaptations to the basic "fish" body plan.

flexible and foldable fin. The departure from the asymmetrical heterocercal tail in favor of a symmetrical homocercal tail balanced muscular contractions of the upper and lower tail surfaces, helping to generate faster swimming. Changes in jaw musculature and jaw suspension allowed rapid jaw protrusion and flaring of the opercula to create a highly sophisticated suction device for capturing prey. As gills and opercula became more efficient, the paired lungs were replaced by a swim bladder for regulating buoyancy and counteracting the heavier density of a bony skeleton.

Modern ray-finned fishes have become the most plentiful class of vertebrates, with more than 27,000 named species. Although most lineages of lungfishes died out, these groups were dominant predators in shallow freshwater habitats during the Devonian period. They also were of monumental importance to another branch in vertebrate evolution: The common ancestor of amphibians and all other tetrapods was a group of Devonian lungfishes. For a more complete phylogeny depicting the evolutionary relationships among the chordates covered in the remainder of this laboratory guide, review Figure A.2 in the Appendix.

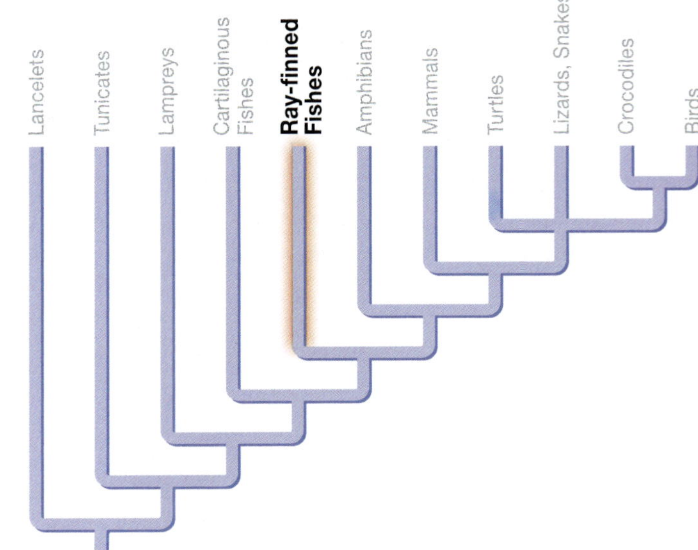

18.2 Ray-finned fishes and cartilaginous fishes first appeared approximately 400 million years ago with innovations such as paired fins, jaws, skulls, and multichambered hearts, representing a significant divergence from their vertebrate predecessors.

Exercise 18-1 Ray-finned Fishes (Class Actinopterygii)

Materials Needed

- ❑ mounted bony fish skeleton
- ❑ preserved perch
- ❑ dissecting tools
- ❑ dissecting pan
- ❑ latex gloves
- ❑ compound microscope
- ❑ clean glass slides
- ❑ coverslips

Skeletal System

Procedure

1. Obtain a mounted, articulated skeleton of a ray-finned fish such as a perch.
2. Examine it carefully, using Figure 18.3 to assist you in identifying the skeletal elements present.

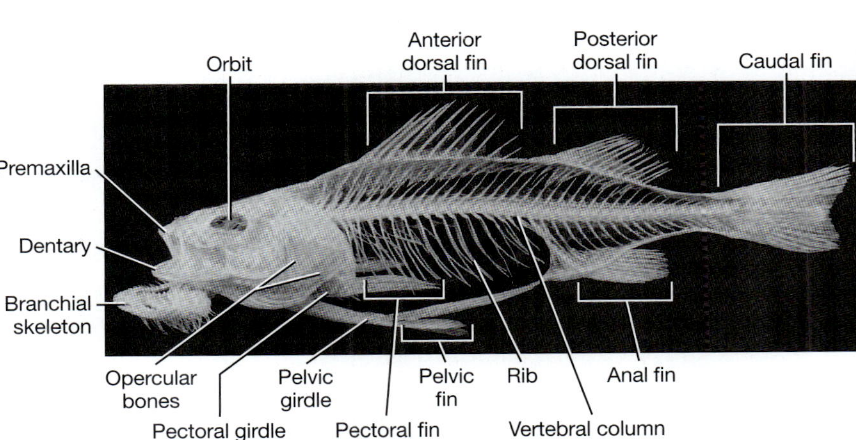

18.3 Skeleton of a ray-finned fish (perch). Notice the parallel, bony rays that lend support to each fin.

Notice that the skull is very heavily ossified and consists of both a **dermal exoskeleton** and a bony **endoskeleton** to provide the rigid, protective encasement for the brain and sensory organs. The dermal exoskeleton of ray-finned fishes is comprised of bony structures (fin rays and scales) that are derived from a process referred to as intramembranous ossification, in contrast to dermal bone found in the bony endoskeleton, which is deposited in layers by osteocytes. The jaws of carnivorous ray-finned fish, such as perch, have extensive hinging that allows them to open widely enough to accommodate large prey. Some fish are capable of swallowing prey that are nearly as large as their own bodies! Food items are not chewed in the mouth but, rather, are swallowed whole by most fish; thus, the teeth generally are small, numerous, sharp, and function mainly in preventing prey from escaping as they are swallowed.

Check Your Progress

1.1 What evidence of segmentation do you see in the perch skeleton?

1.2 Are all of the fins supported by bony elements?

1.3 Does the vertebral column extend all the way to the caudal fin?

1.4 Where is the skeleton reduced in its degree of coverage?

1.5 List two bones that are homologous with other vertebrates.

External Anatomy

18

Procedure

1. Obtain a preserved ray-finned fish such as a perch and examine its external features. Perch are common inhabitants of freshwater lakes, ponds, and occasionally streams, and are close relatives of bass, crappie, and other sunfish.
2. Use Table 18-1 and Figure 18.4 to assist you in identifying the external anatomy of your fish.
3. As you examine your specimen, try to draw comparisons between the many structures that are shared by other familiar vertebrates.

Notice that ray-finned fish have a streamlined, **fusiform** body shape designed to minimize drag during swimming. The body is thickest about one-third of the way back from the head and it tapers in both directions. The large head bears two **external nares** (nostrils), two large **eyes** that lack eyelids, a terminal **mouth** equipped with small teeth, and a bony operculum covering the feathery gills on each side of the head. Locate the **anus** on the ventral surface of the body anterior to the anal fin. In addition to an anus, female perch possess a single **urogenital pore** for the release of eggs and metabolic wastes. Males have a separate **genital pore** and urinary pore, in addition to an anus.

4 Remove a **scale** from your specimen and view a wet mount of the scale using a compound microscope set on low power. Notice the numerous concentric ridges on the scale. Fish scales grow continuously throughout the fish's life, and these ridges actually represent growth rings. As the fish ages, new material is deposited around the margin of each scale, forming the pattern you see. For this reason, scientists can determine the age of a fish by counting these growth rings.

5 In the space provided below sketch a scale from your fish and estimate the age of the fish from the number of rings. Each ring corresponds to one year of growth.

Table 18-1 External Anatomy of the Ray-finned Fish

Structure	Function
Eyes	Large, image-forming sight organs; lacking eyelids
External nares (nostrils)	Paired openings in dorsum of head leading to olfactory receptors
Mandible	Lower jawbone bearing teeth for prey capture
Maxilla	Upper jawbone, fused to skull, bearing teeth for prey capture
Opercula	Paired bony flaps that cover the gills on either side of the head, attached anteriorly and dorsally but open posteriorly and ventrally for the release of water
Pectoral fins	Steering and braking while swimming and maintenance of dorsal-ventral orientation while suspended
Pelvic fins	Steering while swimming
Anal fin	Steering while swimming
Caudal fin	Provides thrust and acts as a rudder while swimming
Dorsal fin	Steering and maintenance of dorsal-ventral orientation while swimming
Urogenital pore (female)	Common external opening for the release of urine and reproductive products; males have separate urinary and genital openings
Anus	Posterior opening of the digestive tract through which undigested food particles (feces) are eliminated from the body
Lateral lines	Specialized sensory organs that detect vibrations and current directions in the water

Operculum
Lateral line
Anterior dorsal fin
Posterior dorsal fin
Eye
External nares
Premaxilla
Mandible
Maxilla
Caudal fin
Anal fin
Pectoral fin
Pelvic fin
Ovary
Rectum
Anus
Urogenital pore
Anal fin

18.4 White perch, *Morone americana*, external anatomy. Inset photo shows close-up of the urogenital pore and anus of a female.

Check Your Progress

2.1 Is your specimen male or female? How do you know?

2.2 Does most of a fish's power for movement come from the tail or from other fins?

2.3 What features of the head represent homologies to other vertebrate groups?

2.4 Which fins would be used to slow the fish after a quick burst of speed?

2.5 What is the function of the lateral lines along the fish's body?

Digestive System

To reveal the internal anatomy of your perch, you will make several cuts through the body wall and ribs. Review the Dissection Diagram to assist you in this process.

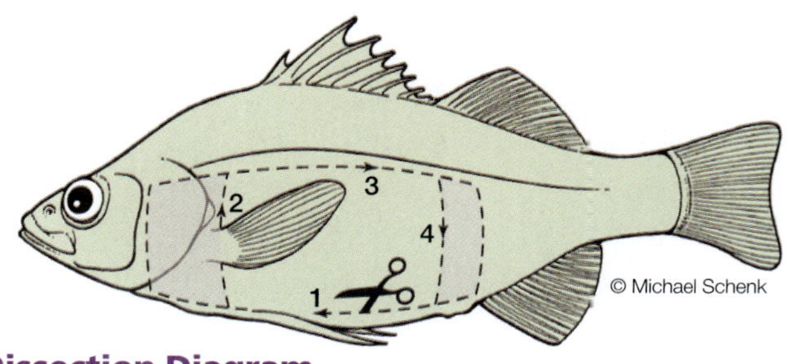

Dissection Diagram

1 Using scissors, make an incision along the ventral surface of the specimen, starting just in front of the anus and progressing anteriorly.

2 Extend this incision dorsally behind the left pectoral fin,

3 then posteriorly along the spine, and

4 then ventrally to intersect the original incision point. Carefully remove the rectangular section of tissue. Extend this rectangular opening anteriorly and posteriorly as needed to reveal additional underlying structures.

Procedure

1 First, use scissors to make a shallow, longitudinal incision along the ventral midline of the body starting just anterior to the anus. Extend your incision anteriorly through the pelvic girdle to just behind the pectoral fin.

2 Extend this incision dorsally behind the pectoral fin.

3 Then extend it posteriorly along the spine.

4 Finally, extend the incision ventrally to intersect the original incision point.

5 Carefully remove the rectangular flap of tissue and identify the internal organs depicted in Figures 18.5–18.6 and listed in Table 18-2.

6 Extend this rectangular opening anteriorly and posteriorly as needed to reveal additional underlying structures.

The large, whitish liver is one of the most prominent organs in the anterior region of the body cavity. Along with the **pancreas** and **gallbladder**, the **liver** is an accessory digestive organ. Remember—food does not pass directly into accessory digestive organs, but they play a vital role in the digestive process by secreting digestive compounds to help break down food.

7 After identifying the liver, reflect it dorsally (or remove it) to view the underlying digestive organs.

Food enters the mouth and passes through the **esophagus** into the large, pouch-like **stomach.** The fish stomach is extremely elastic and may stretch to nearly fill the coelomic cavity when large prey is swallowed. Three fingerlike projections of the small intestine, known as **pyloric ceca**, provide additional surface area for digestion. The S-shaped **small intestine** is bound tightly to its adjacent organs by mesentery that must be teased apart carefully to view the length of the digestive tract. Chemical digestion and nutrient absorption occur along the length of the small intestine. Indigestible matter passes along the short **rectum** to the anus for discharge from the body.

The swim bladder is a long, shiny, thin-walled sac that fills a large portion of the dorsal aspect of the body cavity (Fig. 18.6). It is filled with gas from the blood (mostly oxygen) and serves as both an accessory respiratory organ and a buoyancy organ. A fish can regulate the volume of gas in the swim bladder to control its buoyancy, allowing it to remain neutrally buoyant as it moves from shallower, less-dense water, to deeper, denser water.

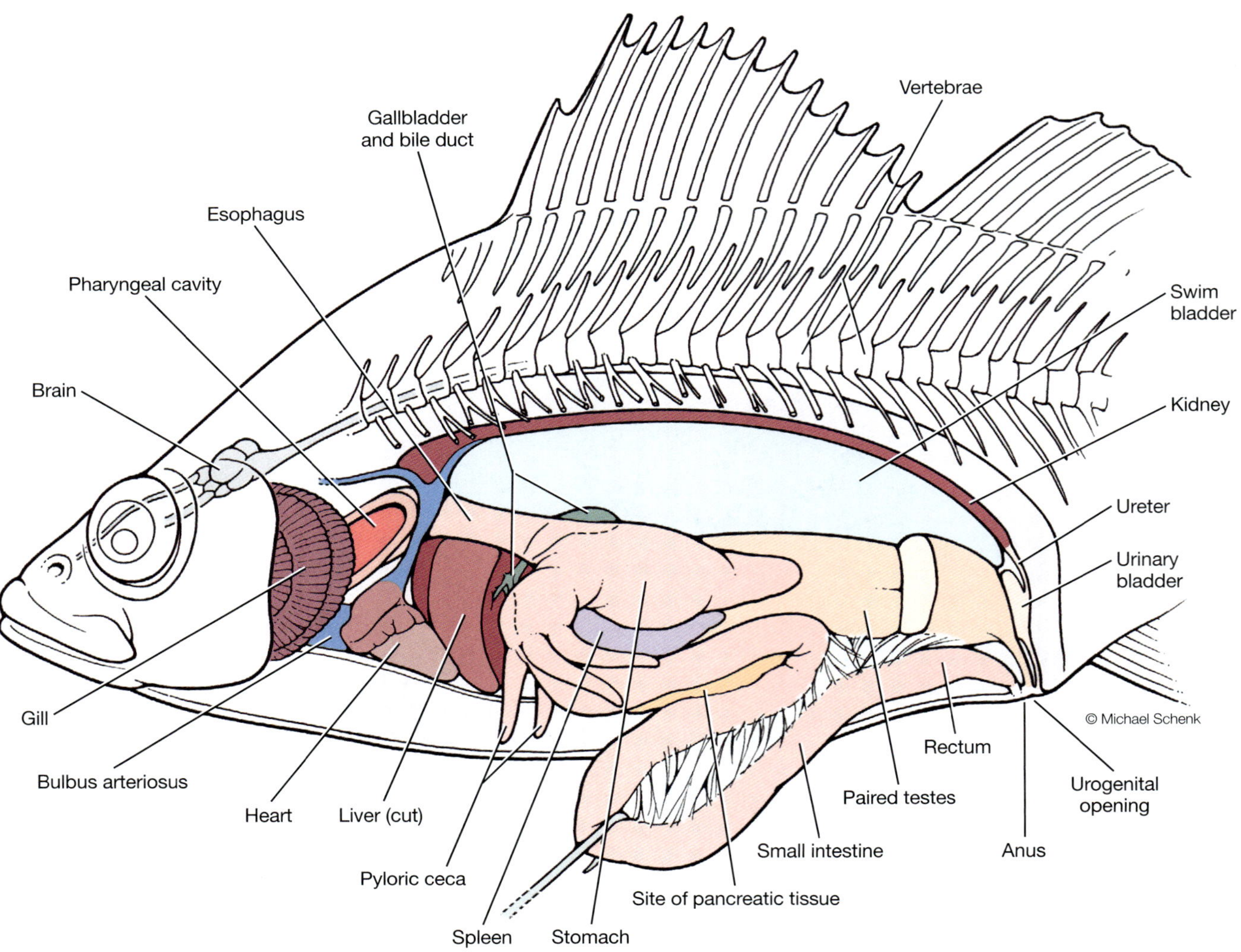

18.5 Male ray-finned fish, internal anatomy. The liver and testis are depicted as if they have been "cut," to reveal the underlying organs.

18

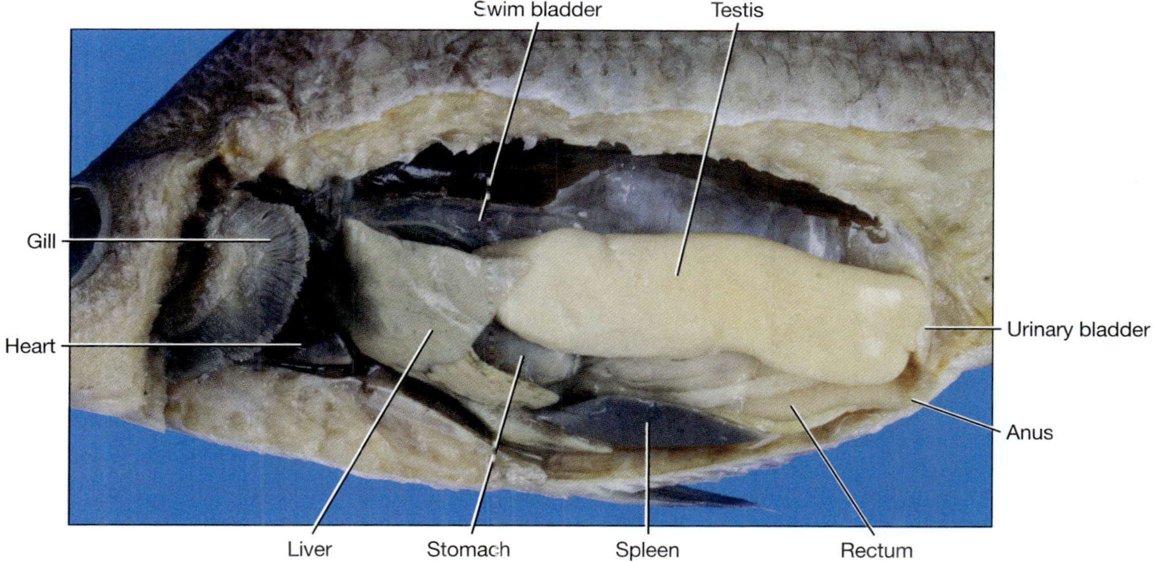

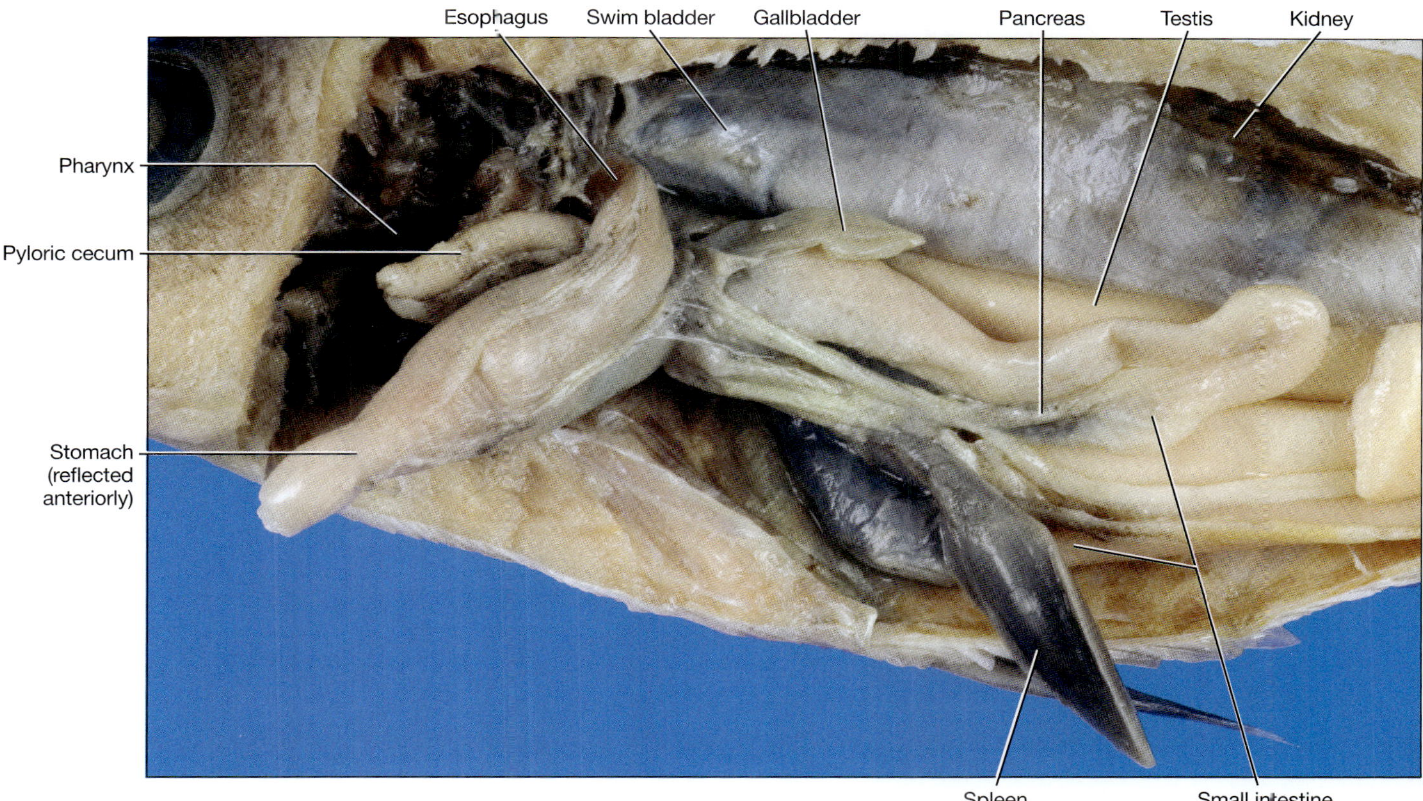

18

18.6 Male white perch, *Morone americana*, internal anatomy. The liver and testis have been partially removed, and the stomach has been reflected anteriorly to reveal underlying organs.

Table 18-2 Internal Anatomy of the Ray-finned Fish

Structure	Function
Gills	Contain capillary beds for gas exchange for respiration
Heart	Two-chambered organ (one atrium, one ventricle) that pumps deoxygenated blood to the gills for oxygenation and from there throughout the various organ systems of the body
Liver	Large whitish organ that detoxifies many constituents of the absorbed digested compounds and functions in lipid and glycogen storage
Gallbladder	Stores bile produced by the liver
Spleen	Elongate organ that stores blood and recycles worn-out red blood cells
Pyloric ceca	Three short pouches extending laterally from the small intestine near its juncture with the stomach that increase digestive surface area of the intestine
Stomach	Site of food storage and initiation of digestion
Small intestine	Digestion of food received from the stomach and nutrient absorption into the bloodstream
Rectum	Terminal portion of intestinal tract that leads to the anus
Testes (male)	Paired organs that produce sperm for transport through the sperm ducts and release through the genital pore for external fertilization
Ovary (female)	Single (fused) organ that produces eggs for transport through the short oviduct and release through the urogenital pore for external fertilization
Anus	Regulates egestion of undigested food (feces) from the body
Urinary bladder	Storage organ for ammonia prior to elimination through urogenital opening
Vertebrae	Columnar units of the vertebral column that provide support, house the dorsal nerve cord, and articulate with the ribs
Swim bladder	Hollow, gas-filled sac that serves as a buoyancy organ
Ureters	Paired tubes that transport kidney filtrate (ammonia) to the urinary bladder
Kidneys	Filter nitrogenous wastes from the blood

Excretory and Reproductive Systems

Ray-finned fishes are **dioecious** (sexes are separate) but often difficult to sex using external features alone. If your specimen is a male, you may have already noticed one of the two long, rectangular, cream-colored **testes** that occupy a large majority of the posterior coelomic cavity, especially during the breeding season (Fig. 18.6). Sperm are produced in the paired testes and are released through a separate dedicated genital pore. Female perch possess a single **ovary**, posterior to the stomach, that varies in size depending on the season (Fig. 18.7). Eggs are produced in the ovary and travel along a short **oviduct** before being released from the body through the urogenital pore (Fig. 18.8). Fertilization in ray-finned fishes is external.

The **kidneys** of ray-finned fish such as the perch are paired masses of tissue that lie along the dorsal body wall next to the vertebrae and extend the entire length of the abdomen. To view them, you must remove the swim bladder. Metabolic wastes are filtered from the blood by the kidneys and are concentrated into ammonia and stored in a small **urinary bladder** before being released from the body, either through a dedicated urinary pore (in males) or through a common urogenital pore (in females) (Fig. 18.8).

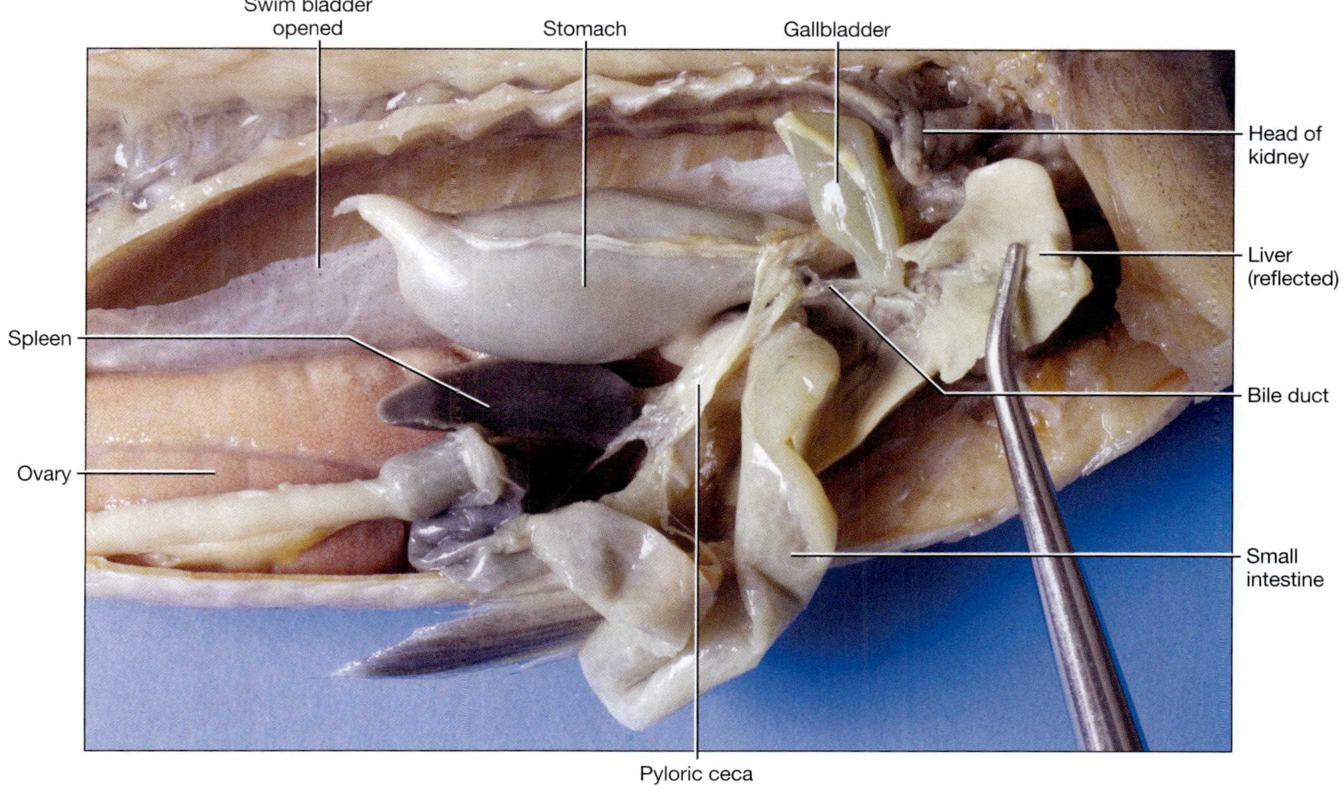

18.7 Female white perch, *Morone americana*, internal anatomy. Right side of specimen is depicted for better exposure of kidney, gallbladder, and liver.

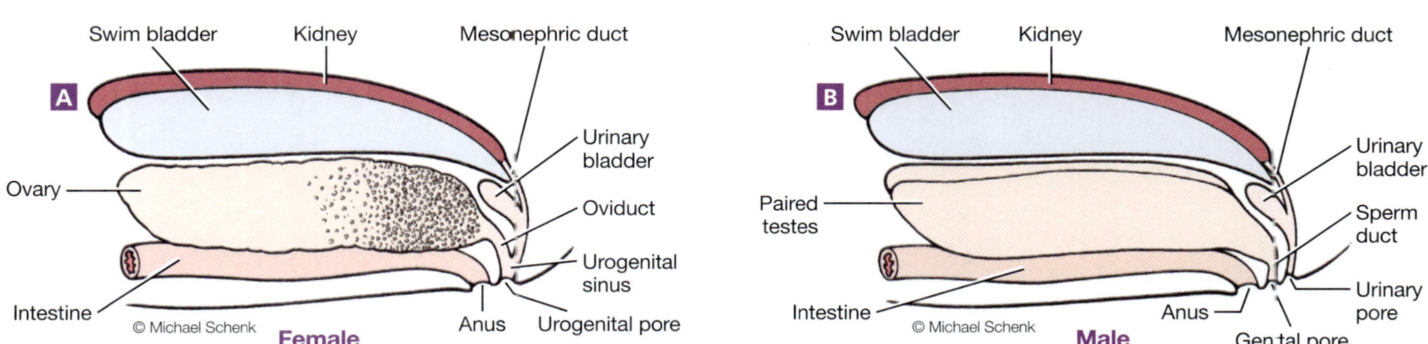

18.8 Urogenital openings of (**A**) female, and (**B**) male ray-finned fish. Females have two openings: an anus and a urogenital opening for both reproductive and urinary products. Males have three separate openings: an anus, a genital pore, and a urinary pore.

18

Circulatory and Respiratory Systems

Procedure

1 To reveal the heart, you may have to extend your midventral incision farther anteriorly.

The heart lies ventral to the gills and anterior to the liver and is enclosed within a pericardial cavity. Although we commonly refer to fish as having a two-chambered heart, as was the case with sharks, there are two specialized regions in addition to a larger, thin-walled atrium and a smaller, thick-walled ventricle (Fig. 18.9). Deoxygenated blood returning from the body tissues collects in the **sinus venosus**, passes into the **atrium**, is routed into the **ventricle**, is forced into the **bulbus arteriosus** by contractions of the ventricle, and passes through the ventral aorta to the gills. The sinus venosus and bulbus arteriosus function together to maintain an even flow of blood into the heart and into the gills.

Fish blood contains the respiratory pigment hemoglobin (found in all vertebrates), which increases the oxygen-carrying capacity of the blood and gives the blood its familiar red color. Oxygenation of the blood occurs in the gills, and the blood then travels throughout the rest of the body before returning to the heart (Fig. 18.10). This type of circulatory pattern is referred to as a single-circuit system, meaning the blood leaving the heart makes a single path through the gills and all body organs before returning to the heart. The evolution of lungs in tetrapods dramatically altered the architecture of the heart and circulatory pathways and has resulted in a double-circuit system with separate pulmonary and systemic circuits.

In ray-finned fishes, the **spleen** is the organ that produces red blood cells and filters the blood, removing worn-out red blood cells, harmful bacteria, and viruses in much the same way other lymph organs do.

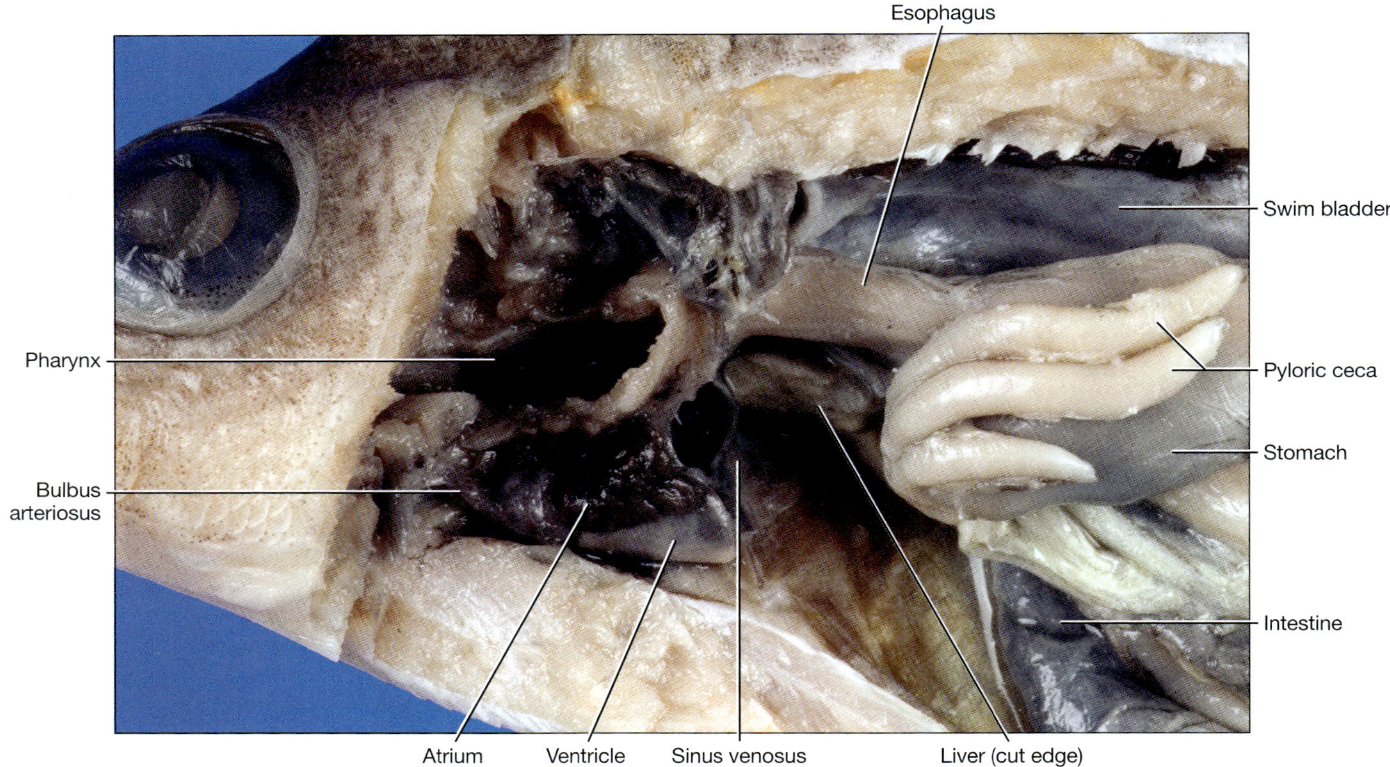

18.9 White perch, anterior region, dissected to show heart and associated anatomy (gills have been removed).

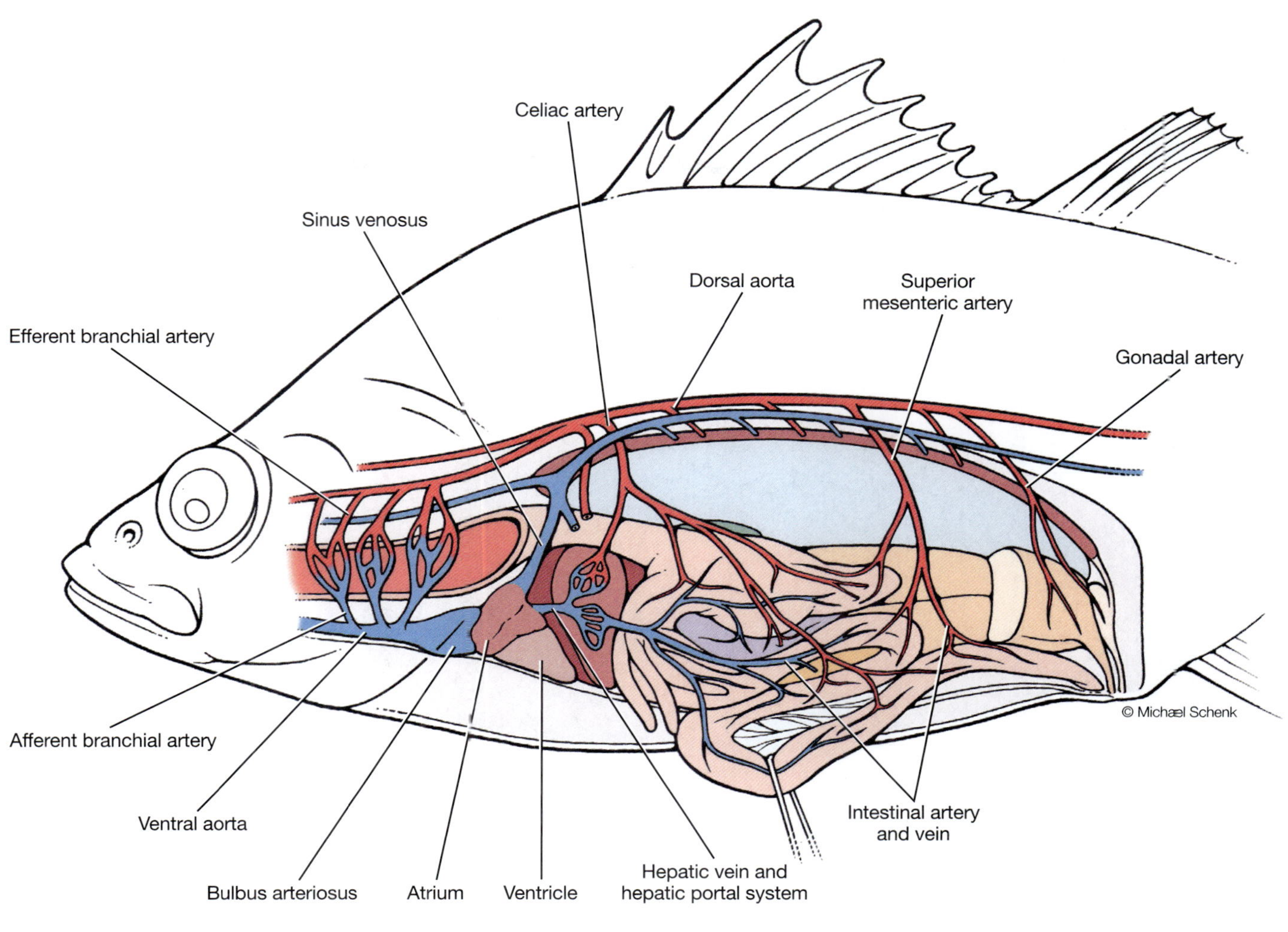

18.10 Ray-finned fish, major circulatory vessels.

2 To view the respiratory structures of your perch, carefully cut away the bony operculum from one side of the body to expose the gills.

Four gills are present beneath each operculum, and each gill consists of numerous **gill filaments** that extend posteriorly from the gill arch (Fig. 18.11). These highly vascular filaments provide a large surface area for contact with the water, to facilitate gas exchange.

3 Remove one of the gills and examine it closely. Notice the hard, serrated structures projecting anteriorly. These structures are **gill rakers**, which protect the gill apparatus and prevent the passage of coarse debris across the delicate gill filaments.

In addition to the operculum, the fundamental structure of the gill filament and the arrangement of gill filaments along the gill in ray-finned fishes differ markedly from those found in sharks and other cartilaginous fishes. Bony elements (rather than cartilaginous arches) support the perch's gills, and the pattern of blood flow through the gill filament is more complex in ray-finned fishes, allowing for more efficient exchange of oxygen and carbon dioxide between capillaries and water.

18

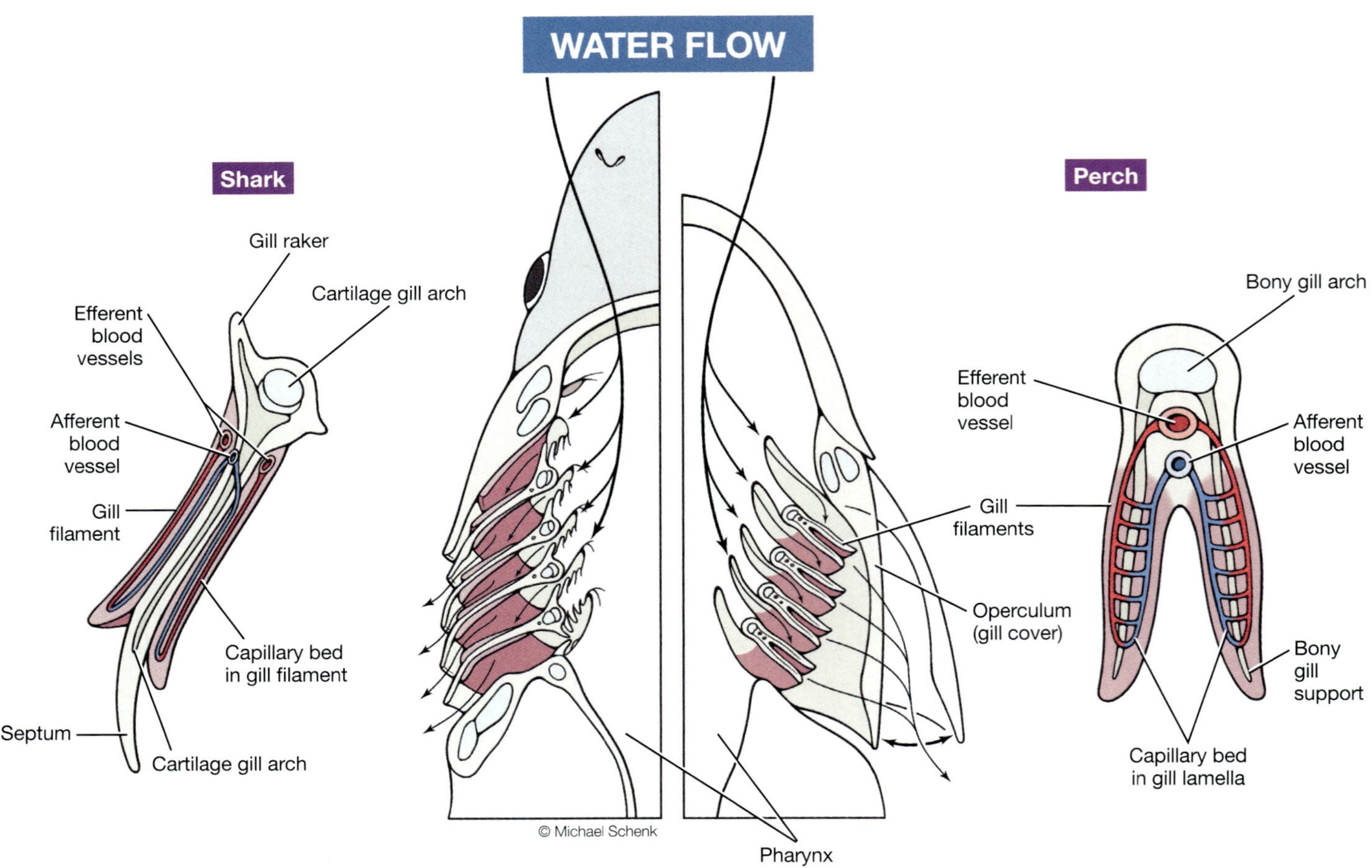

18.11 Comparison of the respiratory structures of a shark and typical ray-finned fish. In both groups, afferent blood vessels (blue) bring deoxygenated blood to the gill filaments, and efferent blood vessels (red) channel oxygenated blood away from the gill filaments.

18

Name ______________________________ Section ______________

18 Questions for Review

1 Opercula are absent in cartilaginous and jawless fishes and represent an evolutionary advancement seen in bony fish. How do opercula allow a bony fish to pass water over its gills while the fish remains stationary? Why would this be viewed as an adaptation?

2 The swim bladder is another evolutionary advancement in bony fish. What advantages do fish with swim bladders have?

3 What anatomical evidence supports the contention that fish have keen senses of vision and smell?

4 Match the structures on the left with the systems to which they belong on the right. (Some letters will be used more than once.)

_____ gills	**a.** reproductive
_____ liver	**b.** excretory
_____ pyloric ceca	**c.** digestive
_____ kidney	**d.** respiratory
_____ gonad	**e.** circulatory
_____ ureter	
_____ spleen	
_____ urinary bladder	

18

5 Name the four specialized regions of the perch heart in the order blood passes through them.

1. ______

2. ______

3. ______

4. ______

6 Explain what is meant by single-circuit circulatory system.

______

______

______

Name ______________________ Section ______________

18 Laboratory Practical Challenge

The following evaluation is intended to replicate the method of assessment and type of questions you will be asked on lab practicals. If you have difficulty with these questions, you may benefit from additional review of the specimens provided in lab.

1 Identify the phylum, subphylum, and class of the specimen depicted below.

____

2 What is the function of the structure indicated by line A?

____

3 Identify the structure indicated by line B.

____

4 To what system does the structure indicated by line C belong?

5 What is the function of the structure indicated by line D?

6 Identify the structure indicated by line E.

7 Identify the structure indicated by line F.

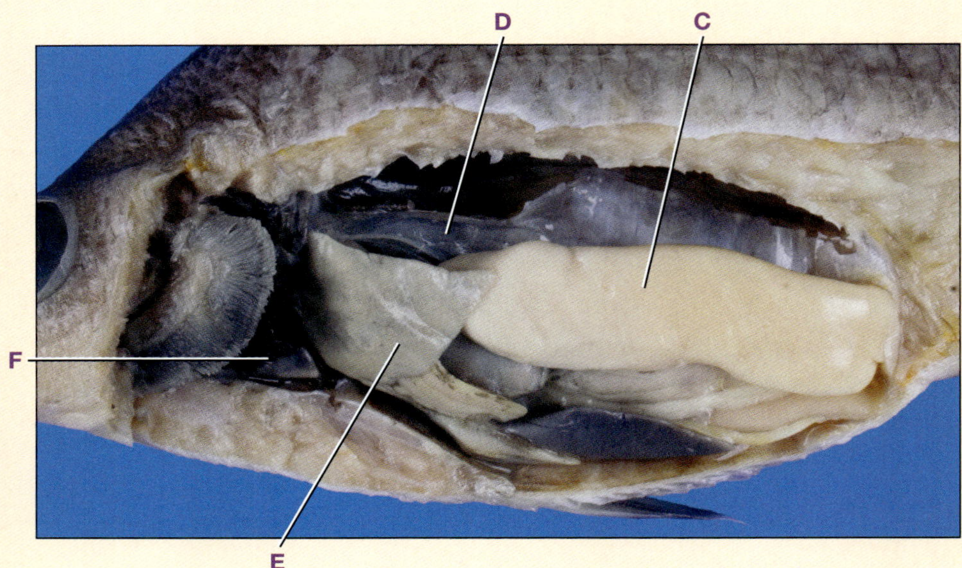

18

CHAPTER 19

Amphibians

After completing the exercises in this chapter, you should be able to:

1 List the distinguishing characteristics of the class Amphibia.

2 Understand the evolutionary position of amphibians on the phylogeny of chordates.

3 Discuss the significance of lungs and paired bony limbs in the evolution of vertebrates.

4 Identify the major external features of the frog and compare them to homologous structures in other vertebrates.

5 Identify the major skeletal features of the frog and compare them to homologous bones in other vertebrates.

6 Identify the major muscles of the frog and compare their morphology and functions to homologous muscles in other vertebrates.

7 Identify the major internal organs in the frog, discuss their functions in the body, and relate them to homologous organs in other vertebrates.

One of the most significant events in vertebrate evolution was the gradual movement of small groups of early vertebrates from water onto land. Amphibians tell a fabulous story about the characteristics that were present in these early vertebrates to allow such a successful shift. In many ways, modern amphibians represent an evolutionary transition between fishes and terrestrial vertebrates (reptiles, birds, and mammals) and have a unique blend of characteristics that are well suited for life in and out of the water. As aquatic larvae, many possess the fusiform body shape and external gills of fish and have long, laterally compressed tails to assist in swimming. Yet, adults have a thinner, lightweight skull and four jointed limbs—**tetrapods**—rather than fins for locomotion.

Lungs are present in most groups, and the closed circulatory system is powered by a three-chambered heart that partially separates oxygenated and deoxygenated blood. The thin skin of amphibians lacks scales or other keratinized tissue necessary to retard evaporative water loss, leaving most amphibians extremely susceptible to dehydration. Many species also rely on the skin as a supplementary organ for respiration. As a result, most amphibians occupy damp, humid habitats that allow them to keep their skin sufficiently moist to prevent dehydration and facilitate gas exchange. Amphibians also rely on water for reproduction. They exhibit external fertilization, and their eggs lack the membranes necessary to prevent desiccation and, thus, must be deposited in moist or aquatic areas.

Just over 6,000 living amphibian species belong to the class Amphibia within the subphylum Vertebrata (Fig. 19.1). In this chapter you will examine the anatomy of a frog as a representative amphibian. Keep in mind that no one amphibian could accurately represent the many variations of body styles and characteristics within the class Amphibia, but frogs do possess most of the typical amphibian features. In addition, they display many unusual features that reflect specific adaptations to their own unique lifestyle.

Evolutionary History

Lobe-finned fishes from the class Sarcopterygii were thought to be extinct until 1938, when a commercial fisherman brought up in a fishing net a living relic known as the coelacanth. Since then, more than 200 specimens of these large fish have been found in the deep waters off the coasts of Africa and Indonesia. Since their discovery, they have been studied extensively and have been postulated as the ancestor of modern tetrapods. Their paired pelvic fins are supported by bones that include direct homologues to the femur, tibia, fibula, ankle, and toe bones, and their paired pectoral fins are supported by bones that include direct homologues to the humerus, radius, ulna, wrist, and finger bones (Fig. 19.2). During the 1980s and 1990s additional morphological research and comparisons of mitochondrial DNA of another group of lobe-finned Sarcopterygians, known as lungfishes, provided additional evidence that lobe-finned fishes were the ancestors of modern tetrapods.

So how did these early fishes make the transition onto land? Remember that early bony fishes of the Devonian period already had paired **air sacs** connected to the back of the throat that could be inflated with air and function as rudimentary lungs. Fishes that also had sturdier fins, supported by jointed, bony elements with sophisticated

African tree frog

Eastern newt, *Notophthalmus viridescens*

19.1 Frogs and salamanders display typical adult amphibian features: moist skin, four limbs, and lungs.

fin musculature, would have been more successful at traversing the dry stretches of land between isolated pools of water that were common during the seasonal droughts of the late Devonian period. These individuals would have had a tremendous advantage for survival under those conditions. During this same period, terrestrial plants and terrestrial arthropods had become established, so any large vertebrates that could survive on land would have enjoyed a landscape rich with resources and devoid of any major competition for food.

The earliest amphibian fossils found in rock are from Greenland, dating back 370 mya, and are of a creature known as *Ichthyostega*. This was a well-built animal with four sturdy limbs supported by heavy bones, yet it was quite fishlike in overall appearance and probably spent most of its life in water. Numerous other fossils that span an entire period of 10 million years (380–370 mya) clearly document the intermediate forms and evolutionary transitions between the early lobe-finned fishes of the Devonian period and ancient amphibians.

One of the more recent, and more significant of these "transitional" fossils, was unearthed in 2004 in the frigid, barren Canadian tundra of Ellesmere Island. Now the home to polar bears, caribou, and arctic wolves, this portion of North America was actually situated along the equator 375 million years ago, when a large fishlike tetrapod sloshed around the shallow streambeds that dotted an almost tropical landscape. Named *Tiktaalik* by the team of paleontologists who discovered it, this creature had a combination of fishlike characteristics (scales, fins, gills, and lungs) and tetrapod characteristics (a neck, ribs, flexible forelimbs, and a flat skull with eyes on top) (Fig. 19.3). *Tiktaalik* (the name means large, shallow-water fish) documents key steps in the vertebrate transition from water to land, and its features suggest that key tetrapod traits, such as wrists, ribs, and a flexible neck, were ancestral to the tetrapod lineage.

During the Carboniferous period (360–299 mya), amphibians underwent a population explosion, thriving in the wet, low-lying tropical swamplands that covered

Coelacanth, *Latimeria chalumnae*

African lungfish, *Neoceratodus forsteri*

19.2 Coelacanths and lungfishes are living lobe-finned fishes in the class Sarcopterygii and are thought to be direct descendants of a lineage of lobe-finned fishes that gave rise to modern tetrapods.

19.3 Artist reconstruction of *Tiktaalik roseae*, a fishlike tetrapod with a flattened head and body that lived 375 million years ago. This large aquatic predator represents an important intermediate step in the evolutionary transition from fish to animals that walked on land.

19

a large portion of the terrestrial landscape. Some amphibian groups left the marshes and migrated to the dry upland areas. These groups developed bony plates and armor and grew to incredible sizes, some as large as small horses.

By the Permian period (299–250 mya), amphibians had reached the peak of their success with more than 40 families, most of which were either semiaquatic or completely terrestrial. A mass extinction at the end of the Permian period, however, wiped out more than 95% of all existing species, signifying the end of the reign of amphibians and the rise of a new land vertebrate—reptiles. Amphibians became relegated to entirely aquatic niches, and their numbers continued to dwindle to only two families (frogs and salamanders), from which all modern amphibians descended (Fig. 19.4).

Amphibians would have to wait another 150 million years, until the end of the Cretaceous period and the corresponding mass extinction of the dinosaurs, before they would make a resurgence. Over the past 65 million years, amphibians have regained their numbers slowly and currently exist in 37 families worldwide. For a more complete phylogeny depicting the evolutionary relationships among the chordates covered in the remainder of this laboratory guide, review Figure A.2 in the Appendix.

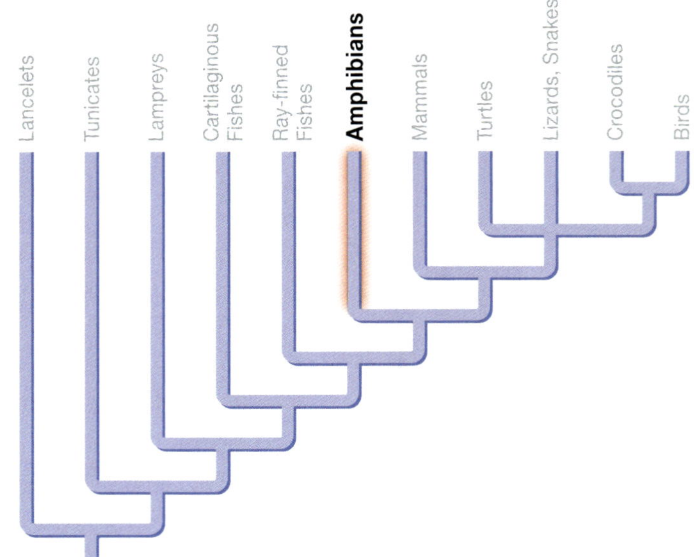

19.4 Some 370 million years ago, the first amphibians added four sturdy limbs, lungs, and a three-chambered heart to the basic vertebrate plan, allowing them to live part of their life cycle on land.

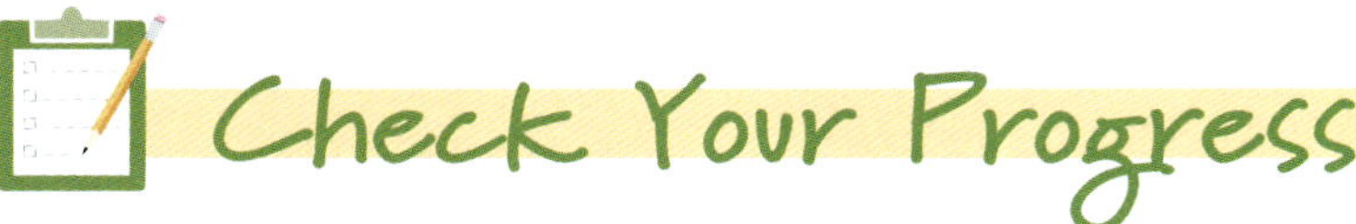

Check Your Progress

1.1 What adaptations do modern amphibians have to land? To water?

1.2 Describe the environmental conditions and the traits selected for during the Devonian period.

19

Exercise 19-1 Frogs (Order Anura)

Materials Needed
- ❑ mounted frog skeleton
- ❑ preserved frog
- ❑ dissecting tools
- ❑ dissecting pan
- ❑ dissecting pins
- ❑ latex gloves

Frogs and toads constitute the order Anura, the most numerous and diverse group of amphibians, with more than 20 recognized families containing more than 5,420 species. They occupy habitats ranging from tropical rain forests, high mountain regions, and temperate woodlands and meadows, to deserts and savannas. Though they can be found on most islands and all continents except Antarctica, more than 80% of all known frog species are concentrated in tropical rain forests—habitats that are being destroyed at an alarming rate.

External Anatomy

Procedure

1 Obtain a preserved frog specimen.

2 Examine its external features, using Figure 19.5 to assist you in identifying the anatomy.

The primary features that distinguish anurans from other amphibians are their compact bodies and powerful hind limbs, well-adapted for jumping, and the absence of a tail in the adult stage. Their **eyes** and **nostrils** are positioned on top of the head to allow frogs to breathe and watch while sitting in water, with the rest of the body hidden from view beneath the surface. Their eyes generally are large and are inclined slightly forward to increase depth perception. Frogs have a blind spot in their visual field directly in front of the snout and must turn their heads slightly to one side to see small prey that are close. Movable eyelids protect the eyes from dust and soil and help keep them moist.

Next to each eye is a prominent eardrum called a tympanic membrane. Amphibians are the most ancestral vertebrates to have a middle ear cavity for conducting sound vibrations from the eardrum to the inner ear. Small movements of the tympanic membrane are amplified and transferred by the middle ear bones to fluid-filled capsules in the inner ear, containing hair cells that bend from the pressure of the sound waves and generate nerve impulses that are sent to the brain where they are interpreted as sounds.

19.5 Leopard frog, *Rana pipiens*, external features.

Skeletal System

Procedure

1 Obtain a mounted, articulated skeleton of a frog.

2 Examine it carefully, using Figure 19.5 to assist you in identifying the skeletal elements present.

Many of the bones seen in the frog represent homologies to familiar bones present in most mammals, including humans (for example, the humerus, scapula, femur, and atlas), providing evidence of their common ancestry. Many specializations in the frog skeleton, however, reflect adaptations necessary to withstand the forces generated while jumping great distances and landing safely.

The pelvic girdle is reinforced by the paired ilia (singular = ilium), the central urostyle, and the ischium, which serves as a strong anchor for the attachment of each femur and the corresponding musculature. Notice how the vertebral column is compressed into just 10 **vertebrae**. The tenth vertebra is the long urostyle, which develops from the fusion of the last several vertebrae to form a strong, rigid base for bearing the forces necessary to propel the frog into a jump and to support its landing. Likewise, in the forelimbs and hind limbs some bones are fused during development to increase strength in these areas of the body. The radius and ulna fuse to form a common radioulna, and the tibia and fibula fuse to form the tibiofibula (Fig. 19.6). The tarsals of each hind limb are elongated and articulate with the tibiofibula and metatarsals to increase flexibility in the distal portion of the hind limbs, enhancing both swimming and jumping ability.

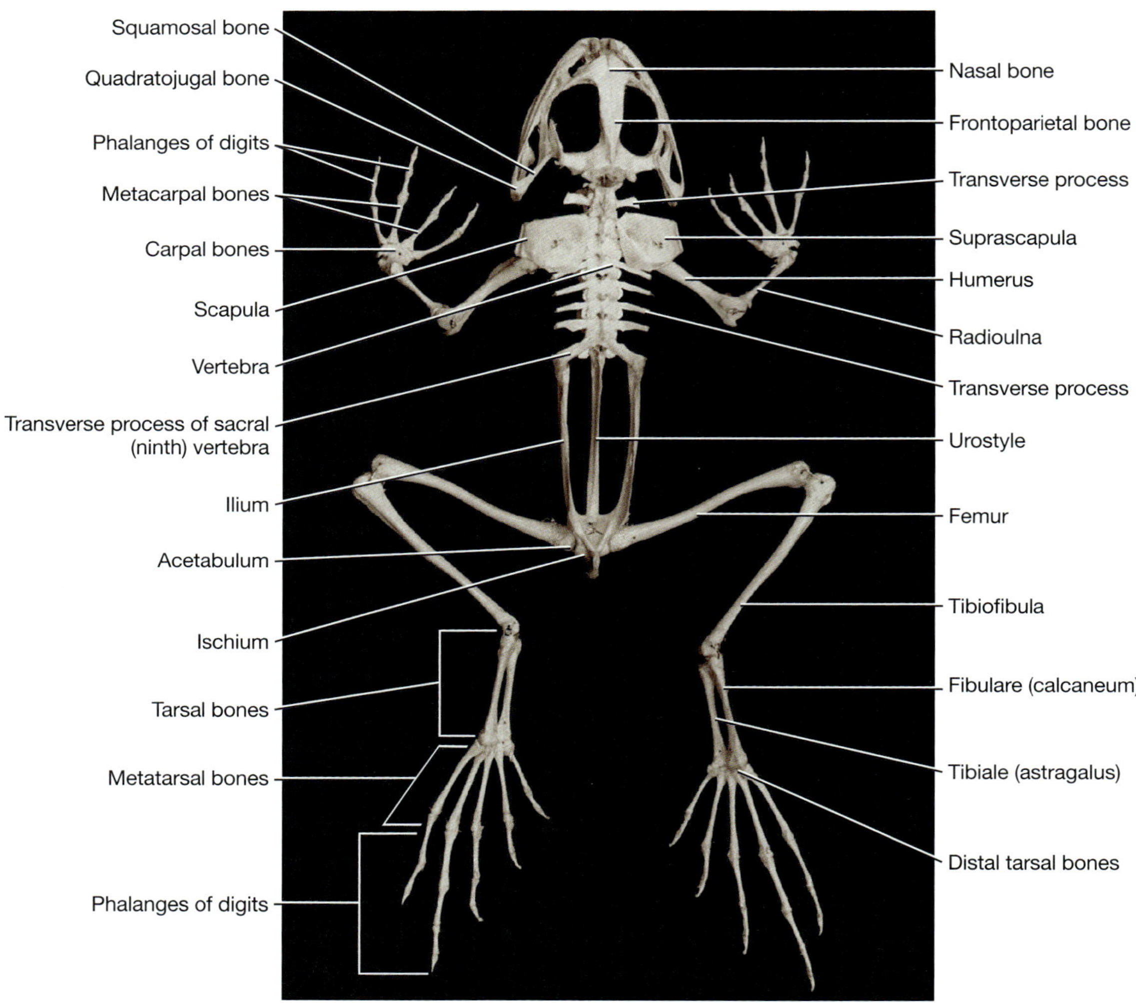

19.6 Frog, skeletal anatomy, dorsal view.

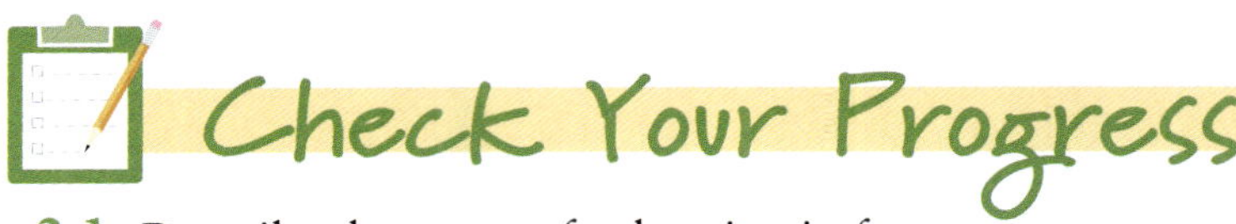

2.1 Describe the system for hearing in frogs.

2.2 What evidence of segmentation do you see in the frog skeleton?

2.3 How many toes are present on the forelimbs? On the hind limbs? How does this compare to mammals?

2.4 What bony remnant of the postanal tail that was present in the tadpole stage remains in the adult?

2.5 Are ribs present in the frog?

2.6 List five bones in the frog that are directly homologous to bones in humans.

1.
2.
3.
4.
5.

Musculature

With frogs having more than 200 muscles, a comprehensive study of their musculature would entail many hours of tedious dissection and is beyond the scope of this laboratory manual. Instead, we will confine our study of the muscular system to the major superficial muscles of the body surface. Because this is our first in-depth investigation of muscles, some background information should make your observations more productive. 19

All skeletal muscles have a fixed end, the **origin**, and a movable end, the **insertion**. Most skeletal muscles taper from the thicker **muscle belly** in the middle to thinner, tough, white tendons on the ends, which anchor the muscle to bones or other muscles. The direction in which a muscle exerts force—its **action**—also plays a role in its shape and where it inserts and originates. A muscle that **adducts** moves a limb toward the midline of the body. Conversely, a muscle that **abducts** moves a limb away from the midline of the body. Some muscles are **flexors** and bend one part of the body toward another part, and other muscles are **extensors** and straighten or extend a body part.

Because muscles can contract in only one direction, and thus perform only unidirectional movements, they generally are arranged in **antagonistic** pairs or groups; that is, they work in opposition to one another. Muscles usually are grouped closely together, so it is often difficult to tell where one muscle ends and another begins. Thus, you will have to pay careful attention to the orientation of the muscle fibers. Often, this will give you clues as to where two muscles cross or abut.

To observe the musculature of your frog, you first must remove the skin completely from your specimen. Refer to the Dissection Diagram to assist you in this process.

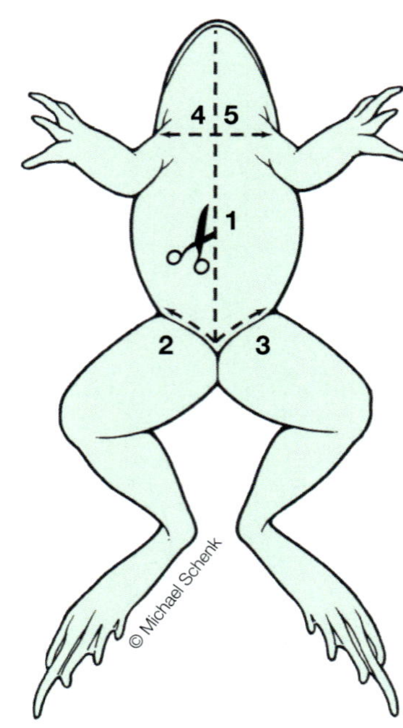

Dissection Diagram

Using scissors, make a midventral incision through the skin starting at the base of the abdomen (**1**) and extend it anteriorly to the jaw. Next, make lateral abdominal incisions (**2–3**) and then lateral thoracic incisions (**4–5**) to create flaps that will allow access to the internal structures.

Procedure

1 Place the frog on its dorsal surface in your dissecting pan and use scissors to make an incision through the skin (but *not* through the body wall!) along the midventral line from the cloaca anteriorly to the tip of the jaw.

2 Begin to separate the skin from the underlying muscles carefully, using a blunt probe where necessary.

3 Make two transverse incisions through the skin around the body, one just in front of the hind limbs and the other just in front of the forelimbs.

4 Use a blunt probe to carefully tease the skin away from the muscles around the entire body. In most cases you should be able to "peel" away the skin with your fingers.

5 Using a similar technique, remove the skin from the forelimbs and hind limbs of your specimen. Continue until you have exposed the muscles of your specimen completely.

6 Use Tables 19-1–19-3 and Figures 19.7–19.9 to identify the major superficial muscles and their actions in the frog.

Table 19-1 Ventral Musculature of the Frog

Body Region	Muscle Name	Action
Neck, shoulders, and abdomen	Mylohyoid	Assists in swallowing and breathing
	Coracoradialis	Flexes forearm
	Pectoralis major	Adducts forelimb
	Deltoid	Flexes shoulder and adducts forelimb
	Cutaneous pectoris	Compresses subcutaneous lymph sacs
	Rectus abdominus	Compresses abdomen and flexes trunk
	External oblique	Compresses abdomen and flexes trunk
Forelimb	Anconeus	Extends forearm
	Extensor carpi radialis	Extends wrist and digits
	Extensor digitorum communis longus	Extends hand and digits
	Flexor carpi radialis	Flexes hand and digits
	Flexor carpi ulnaris	Flexes hand and digits
	Palmaris longus	Flexes digits

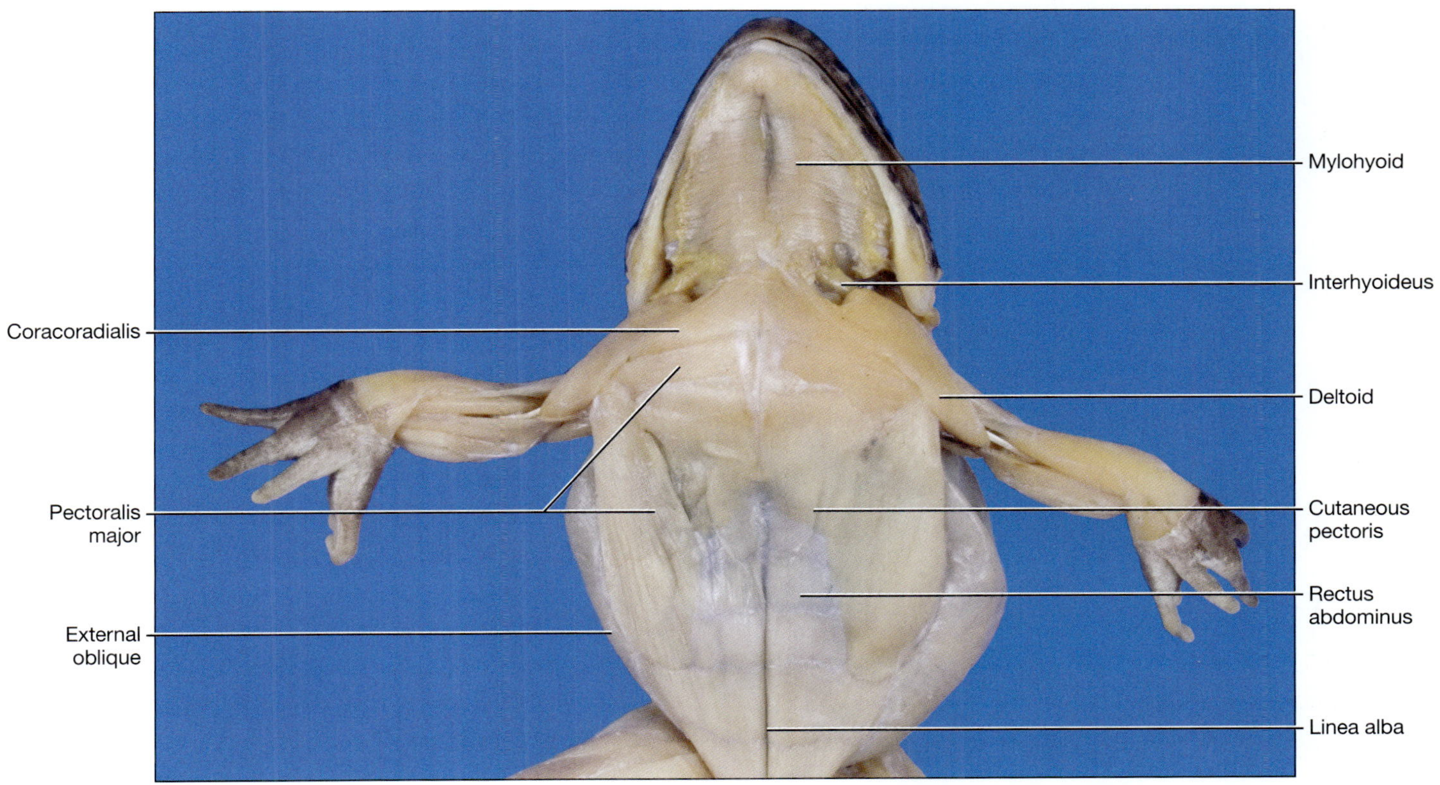

Extensor digitorum communis longus
Extensor carpi radialis
Deltoid
Mylohyoid
Interhyoideus
Coracoradialis
Radius
Pectoralis major
Flexor carpi radialis
Flexor carpi ulnaris
Cutaneous pectoris
Palmaris longus
Anconeus

19.7 Frog, ventral musculature, thoracic region and forelimb.

Table 19-2 Ventral Musculature of the Frog

Body Region	Muscle Name	Action
Pelvis, thigh, and hind limb	**Adductor magnus**	Adducts thigh
	Pectineus	Adducts thigh
	Adductor longus	Adducts thigh
	Sartorius	Flexes hind limb and adducts thigh
	Gracilis major and minor	Flexes hind limb and retracts and adducts thigh
	Semitendinosus	Flexes hind limb and retracts thigh
	Semimembranosus	Flexes hind limb
	Triceps femoris	Extends hind limb
	Tibialis posterior	Extends hind limb

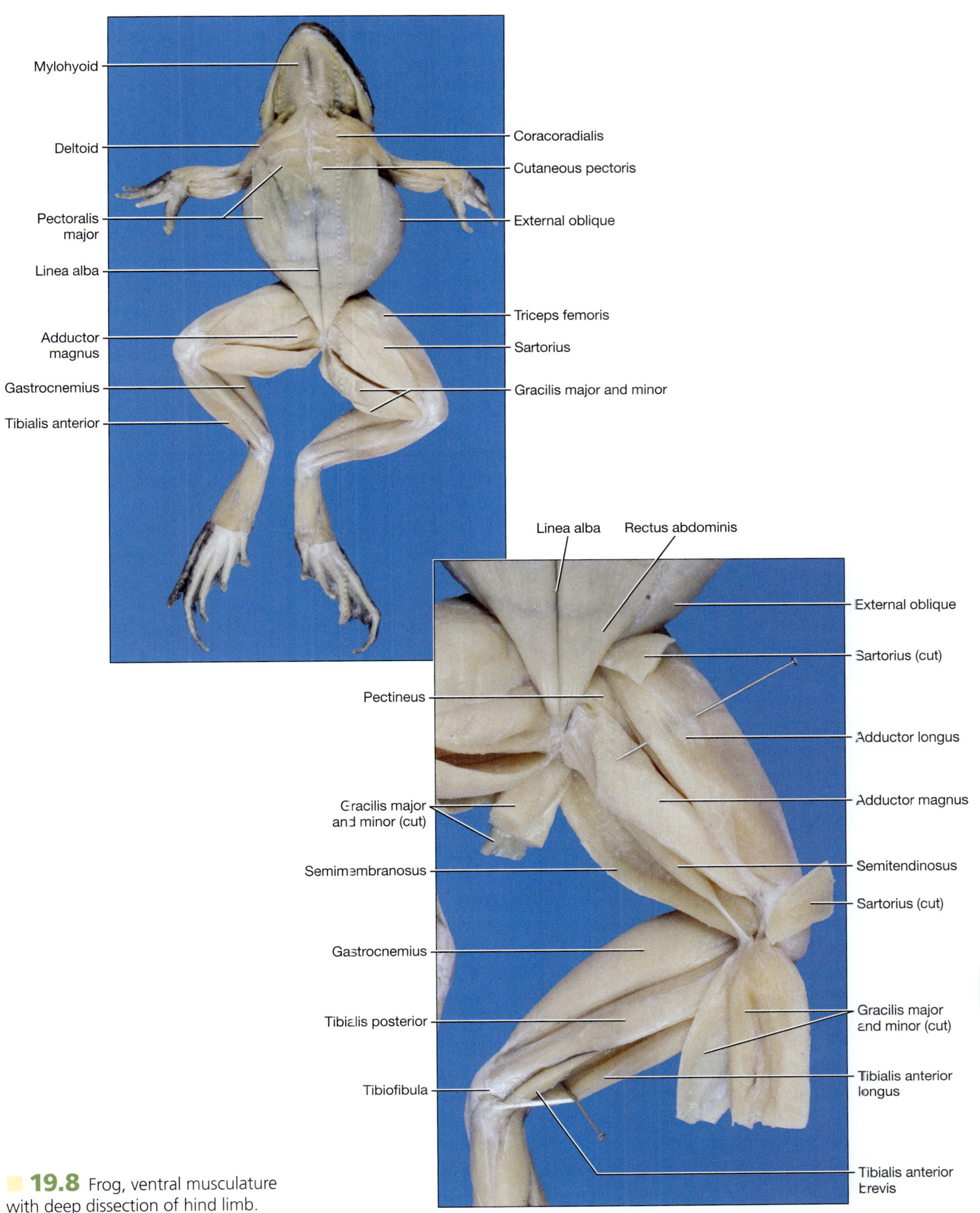

19.8 Frog, ventral musculature with deep dissection of hind limb.

Table 19-3 Dorsal Musculature of the Frog

Body Region	Muscle Name	Action
Neck, shoulders, and back	Temporalis	Flexes head and lifts mandible
	Dorsalis scapulae	Abducts humerus
	Latissimus dorsi	Abducts humerus
	Anconeus	Extends forearm
	Iliolumbar	Braces the back and extends pelvic girdle
	Longissimus dorsi	Braces the back and extends pelvic girdle
	Coccygeoiliacus	Braces the back and extends pelvic girdle
Pelvis and thigh	Gluteus	Rotates thigh forward
	Iliacus internus	Moves thigh forward
	Adductor longus	Adducts thigh
	Adductor magnus	Adducts thigh
	Piriformis	Retracts thigh
	Iliofibularis	Flexes hind limb
	Gracilis minor	Flexes hind limb
	Semimembranosus	Flexes hind limb
	Triceps femoris	Extends the lower hind limb
	Semitendinosus	Flexes lower hind limb and retracts thigh
Shank and hind foot	Peroneus	Flexes hind foot
	Abductor brevis dorsalis	Abducts digits of hind foot
	Flexor digitorum brevis	Flexes digits of hind foot
	Gastrocnemius	Extends hind foot

Check Your Progress

3.1 Name one antagonistic pair of muscles and state its function.

3.2 Why are antagonistic pairs necessary?

19

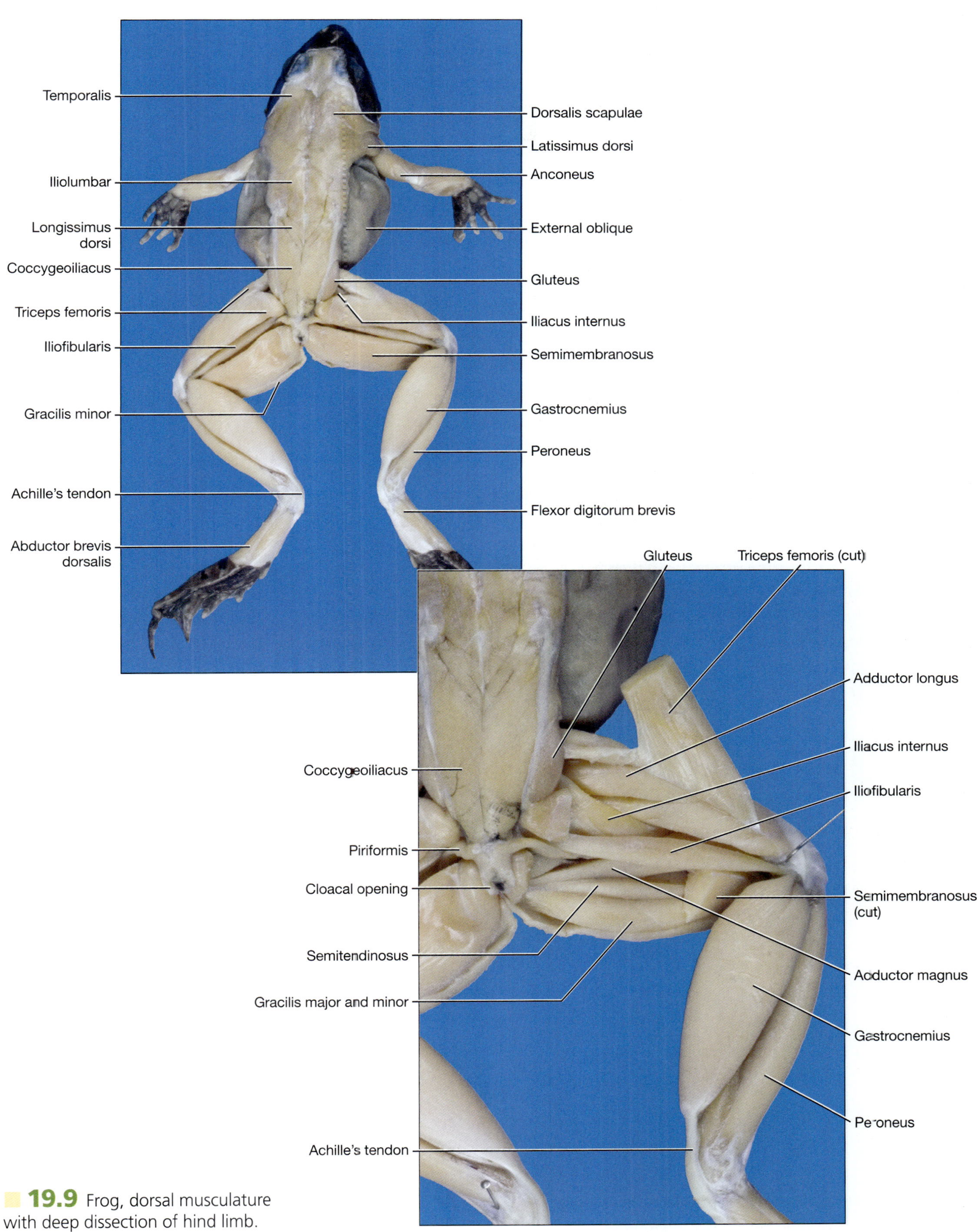

19.9 Frog, dorsal musculature with deep dissection of hind limb.

Digestive System

Procedure

1 Place your frog on its dorsal surface in your dissecting pan.

2 Using scissors, make a shallow, longitudinal incision along the midventral line through the body wall from the cloaca anteriorly to the tip of the jaw.

3 Make two transverse incisions along the sides of the body, just behind the forelimbs and just in front of the hind limbs.

4 Spread apart these flaps and pin them to the dissecting tray to view the internal organs of your specimen.

5 Use Table 19-4 and Figures 19.10–19.12 to identify the internal organs and their corresponding functions in the body.

Depending on the time of year your specimen was collected, a large portion of the abdominal cavity may be filled with bright orange or yellow, fingerlike structures known as **fat bodies** (Fig. 19.10). The fat bodies actually are attached to the kidneys and serve as repositories for lipid stores that the body draws upon when food is scarce or during the breeding season when energy requirements are higher. The fat bodies may have to be removed to gain access to the underlying anatomy of your specimen. Be careful not to remove the testes (if you have a male frog) in the process of removing the fat bodies.

Frogs capture prey in their mouths with the aid of their long, sticky tongues. The mouth is wide to accommodate large prey items, but is not equipped for chewing; thus, prey is swallowed whole. Food items pass down the **esophagus** and digestion begins in the **stomach**. The large, multilobed **liver** produces bile that is stored in the **gallbladder** and released into the **bile duct**. In the leopard frog, the gallbladder is a small, round, green bulb located under the lobes of the liver. Bile, along with digestive enzymes from the **pancreas**, travels into the **duodenum**, the anterior portion of the small intestine. Nutrients are absorbed throughout the remainder of the small intestine (collectively known as the jejunoileum), and water and ion absorption take place in the **large intestine** (Figs. 19.11 and 19.12). Indigestible food items exit the body through the **cloaca**.

Table 19-4 Digestive Anatomy of the Frog

Structure	Function
Fat bodies	Repositories for lipid reserves
Esophagus	Transports food to the stomach
Stomach	Site of food storage and initiation of digestion
Liver	Produces bile and detoxifies many constituents of the absorbed digested compounds
Gallbladder	Stores bile produced by the liver
Bile duct	Transports secretions from the liver and the pancreas to the duodenum
Pancreas	Produces digestive enzymes and delivers them through the common bile duct to the duodenum
Duodenum	Receives secretions from the liver and the pancreas through the common bile duct for further breakdown of food from the stomach
Jejunoileum	Site of completion of digestion, where most of the absorption of nutrients into the bloodstream occurs
Large intestine	Site of absorption of water as well as certain vitamins and ions
Cloaca	Common chamber for the release of urine, feces, and gametes

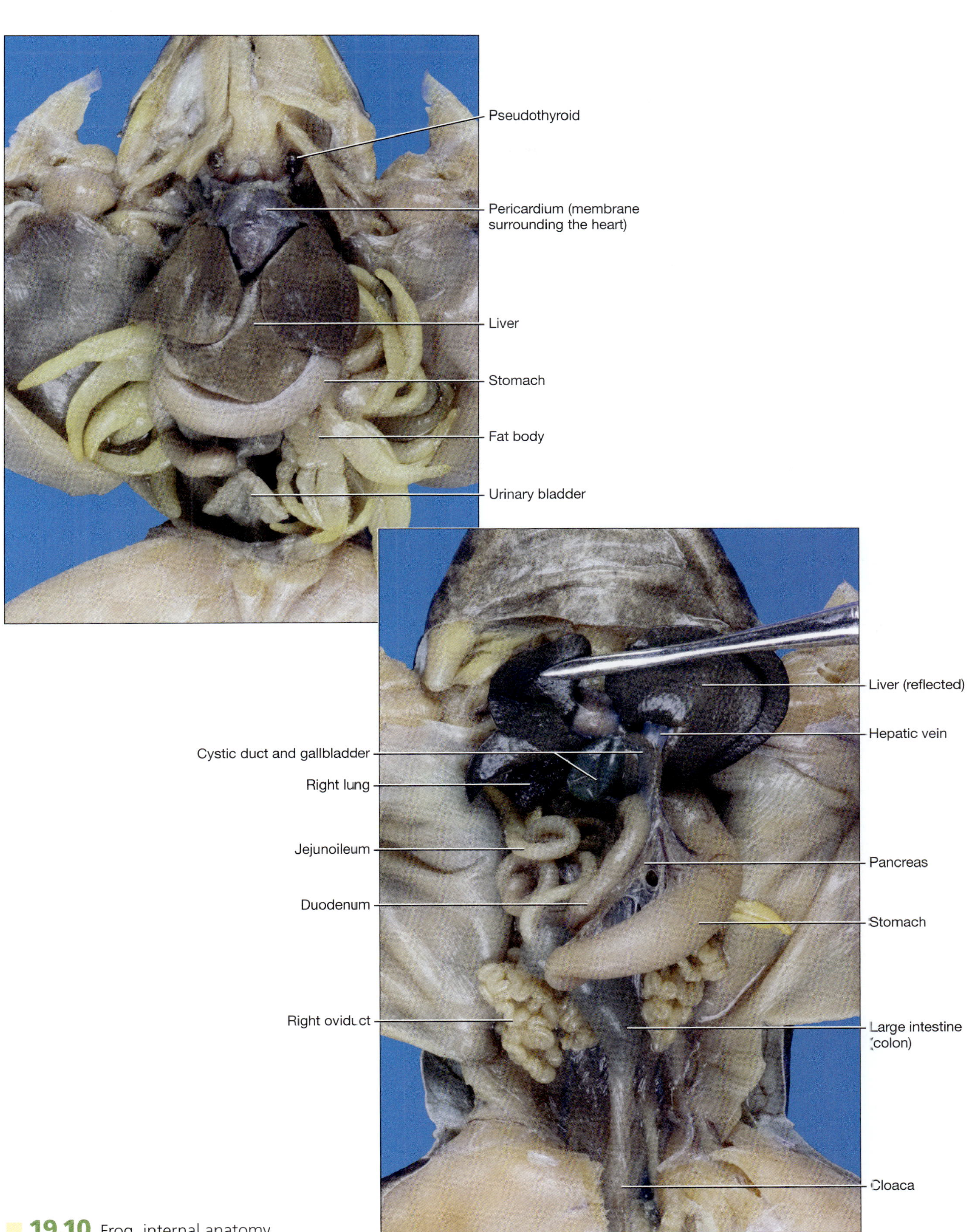

19.10 Frog, internal anatomy.

19

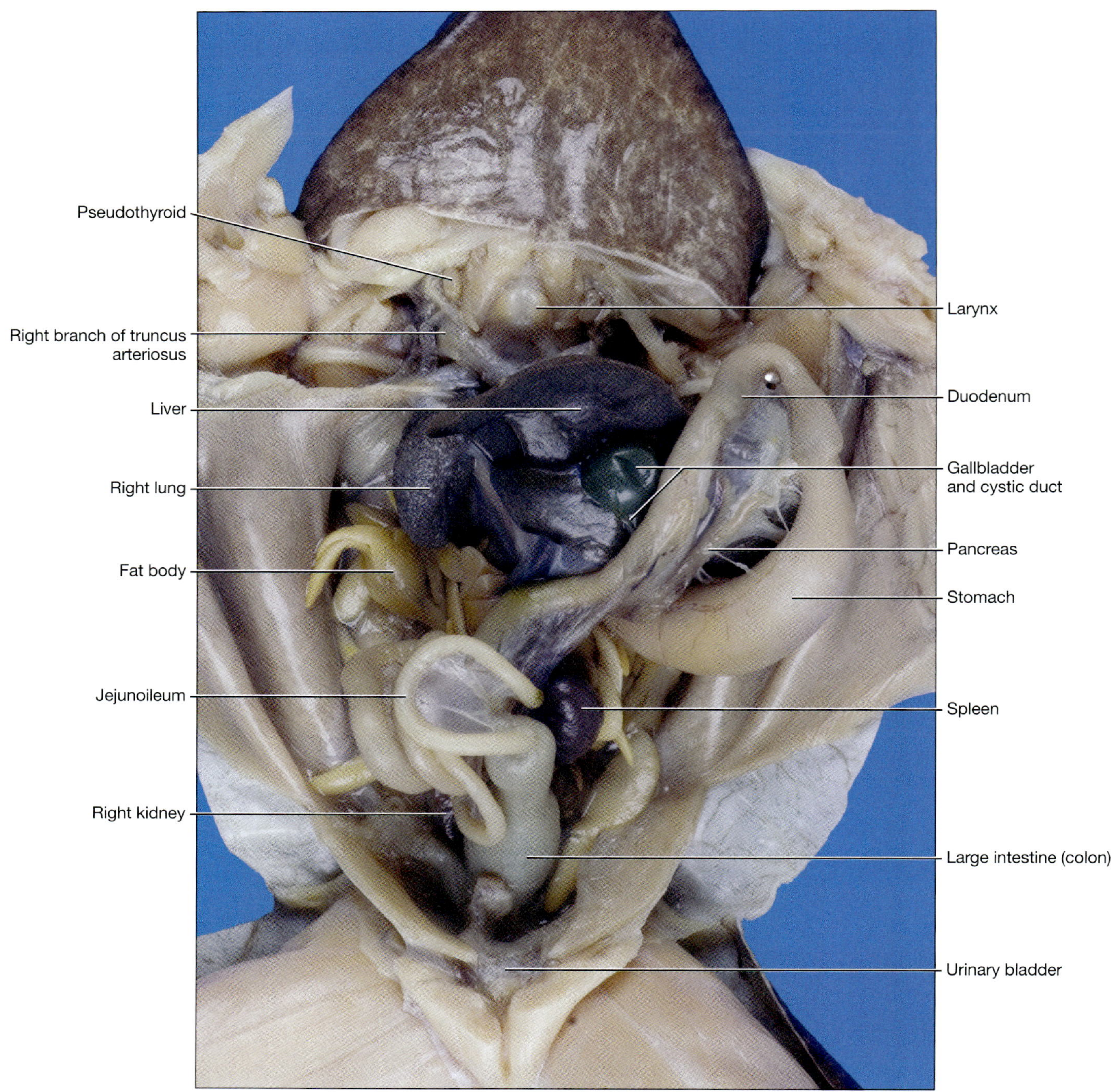

19.11 Frog, internal anatomy.

19

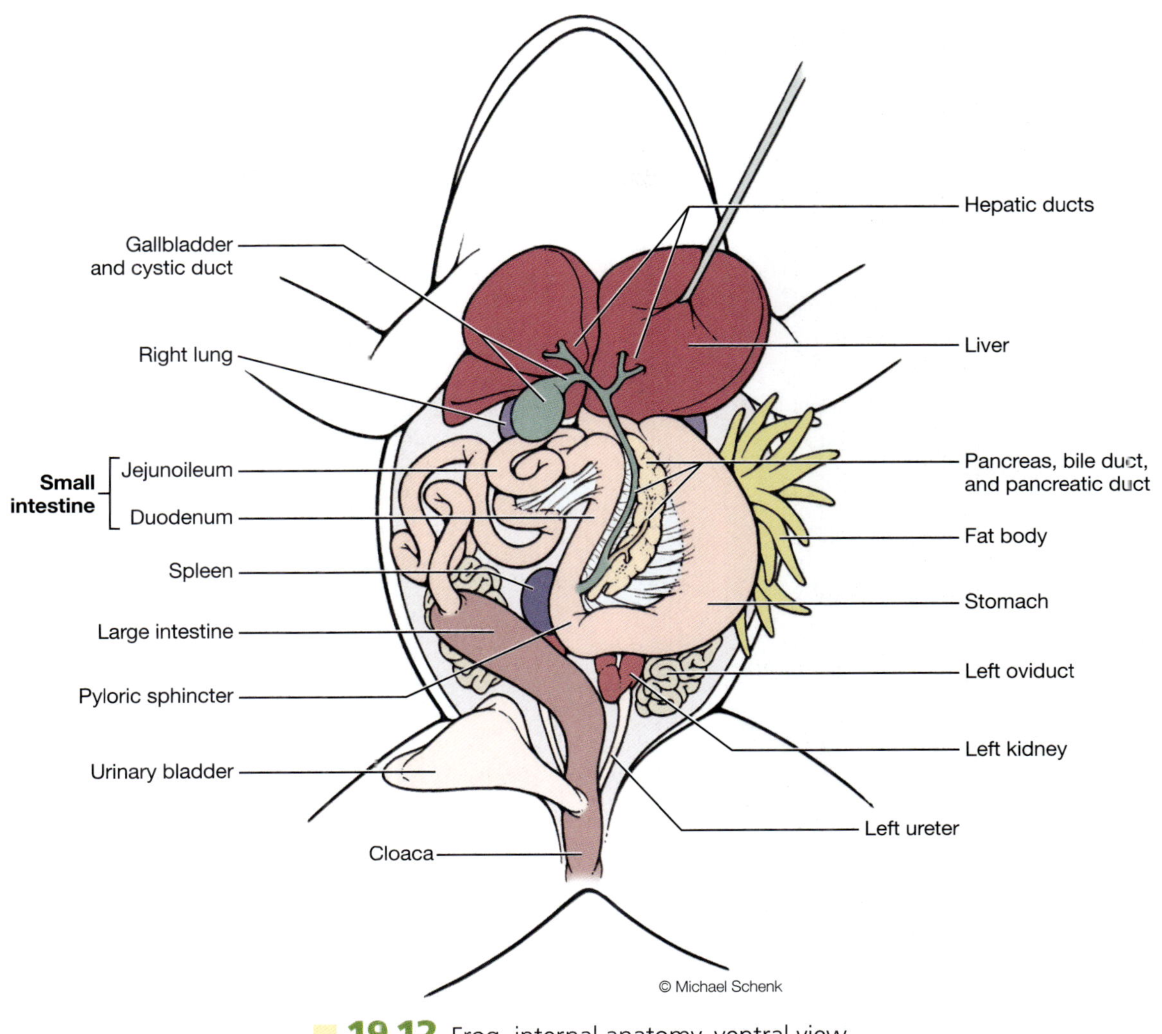

19.12 Frog, internal anatomy, ventral view.

Circulatory and Respiratory Systems

Small, paired lungs behind the heart and liver serve as the primary organs of respiration in leopard frogs (Fig. 19.13). Amphibians lack a diaphragm, so each breath must be "swallowed" and forced into the lungs, using muscles in the throat region. Most amphibians supplement gas exchange through vascular patches of skin that must remain moist to be effective. Certain desert-dwelling toad species even have the ability to absorb water through these patches simply by sitting on an area of damp rock or soil.

The evolutionary adaptations necessary to shift successfully from gill-breathing to lung-breathing altered dramatically the architecture of the heart and the circulatory pathways of amphibians. Two major challenges that early terrestrial animals faced were the development of a good internal vascular surface for gas exchange and, more importantly, an effective mechanism to move air. As a result, modern amphibians represent a transitional state between fishes and fully terrestrial vertebrates. Amphibians are the first group to possess a double-circuit system. Separate pulmonary and systemic branches route blood to the lungs and to the rest of the body, respectively, allowing each circuit to operate under different pressures.

The small, three-chambered heart, encased in a pericardial membrane, is the driving force behind the closed circulatory system in amphibians. The heart has been divided partially between these two circuits with left and right **atria** to collect blood from the pulmonary and systemic circuits, respectively, but oxygenated and deoxygenated blood are not kept completely separated in the common **ventricle**.

As blood leaves the heart, the **spiral valve** inside the conus arteriosus serves as a partition to route oxygenated and deoxygenated blood to their appropriate circuits and allows only a partial mixing of the blood (Fig. 19.13).

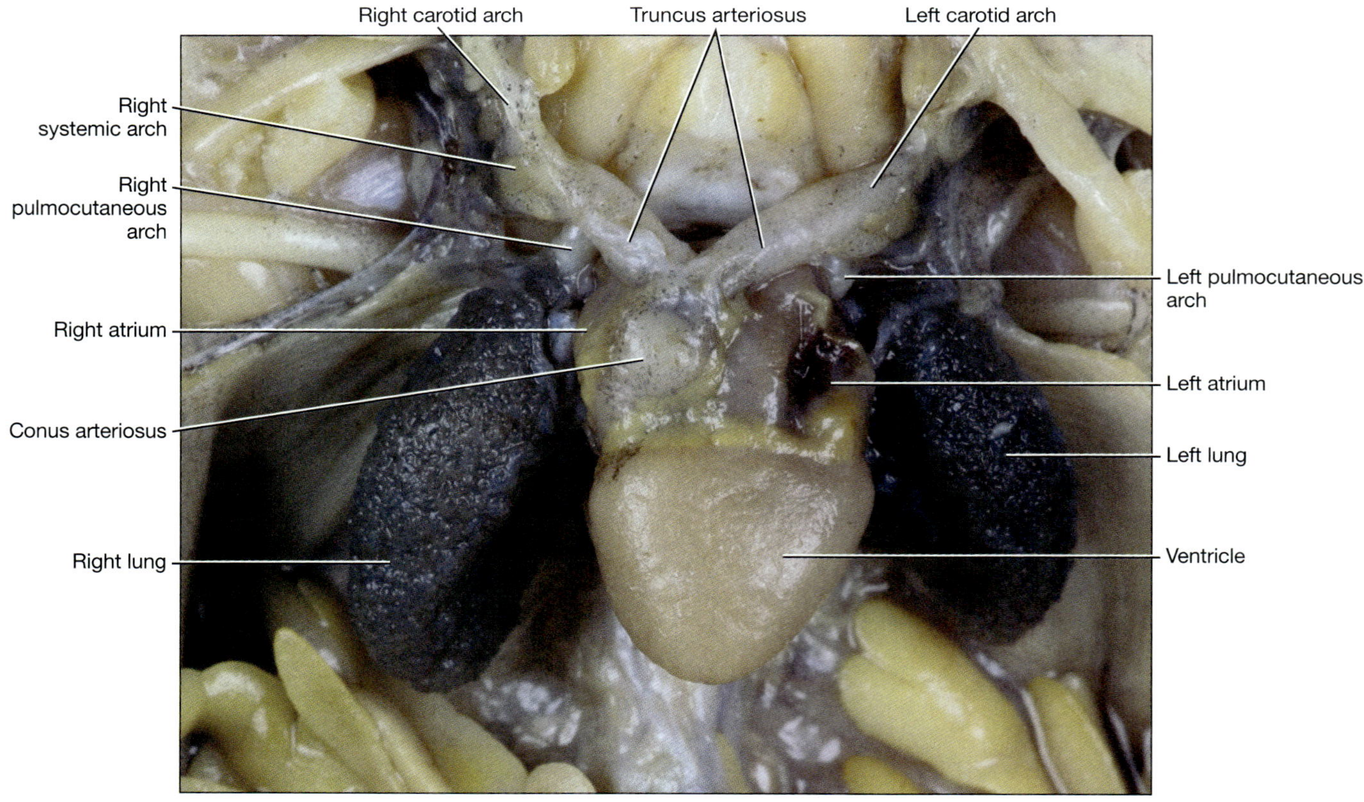

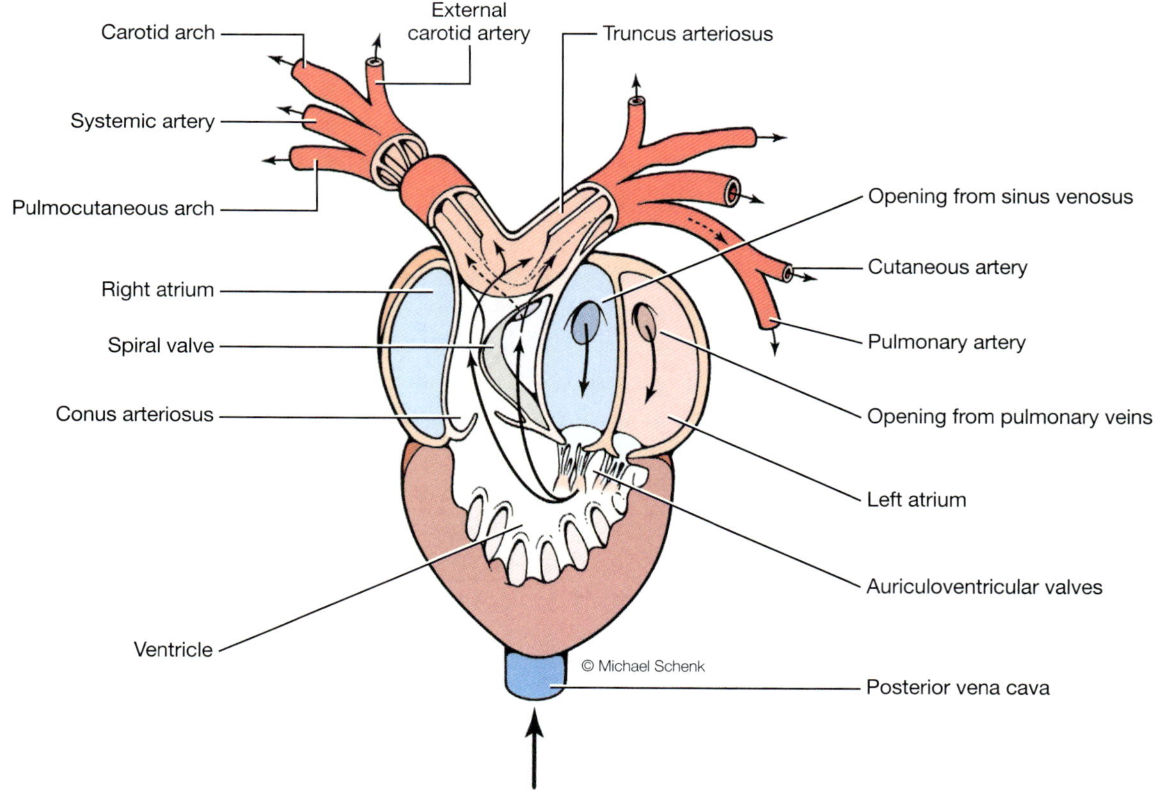

19.13 Frog, heart, lungs, and associated vessels.

19

As a result, oxygenated blood is pumped out of the ventricle of the heart through the conus arteriosus, then through the truncus arteriosus to the body, while deoxygenated blood travels through the pulmocutaneous arches to the lungs (Fig. 19.14). Deoxygenated blood from the body collects in an area on the dorsal surface of the heart formed by the convergence of three large veins (the two anterior vena cavae and the single posterior vena cava) called the sinus venosus and returns to right atrium of the heart (Fig. 19.15).

Note

The accompanying illustrations of arterial and venous circulation in the frog (Figs. 19.14 and 19.15) provide a detailed depiction of the major blood vessels for reference purposes and may be beyond the scope that you are required to know. Check with your instructor to see which arteries and veins you will be expected to learn.

In the typical circulatory pathway, arteries carry blood directly from the heart to capillary beds in the body tissues and veins carry blood directly from the capillary beds in these tissues to the heart. In a portal system, blood flows from capillary beds in the tissues through portal veins to a *second* set of capillary beds, before returning to the heart. Like fishes, amphibians possess two portal systems that interrupt this return of blood to the heart, providing further evidence of their close evolutionary ties to the fishes.

These two portal systems in amphibians are called the hepatic portal system and the renal portal system. In the **hepatic portal system**, blood from the capillary beds of the small and large intestines, the spleen, the pancreas, and the stomach is diverted to the capillary beds of the liver by the hepatic portal vein before entering the posterior vena cava and returning to the heart (Fig. 19.15). This extra step allows blood from the stomach and intestines to be filtered of its sugars and toxins by the liver before the blood is sent to the rest of the body. Also, hormones produced by the pancreas can be directed to their target organ, the liver, without the delay and diluting effect of having to travel through the entire circulatory system. In the case of the **renal portal system**, blood returning from the hind limbs is diverted to a network of capillaries in the kidneys, before entering the posterior vena cava and returning to the heart (Fig. 19.15).

A small, round, reddish-brown spleen is present in frogs, embedded in the mesentery near the stomach and large intestine (Fig. 19.11). The **spleen** is a multifunctional organ of the circulatory system that stores blood, recycles worn-out red blood cells, and produces lymphocytes (white blood cells).

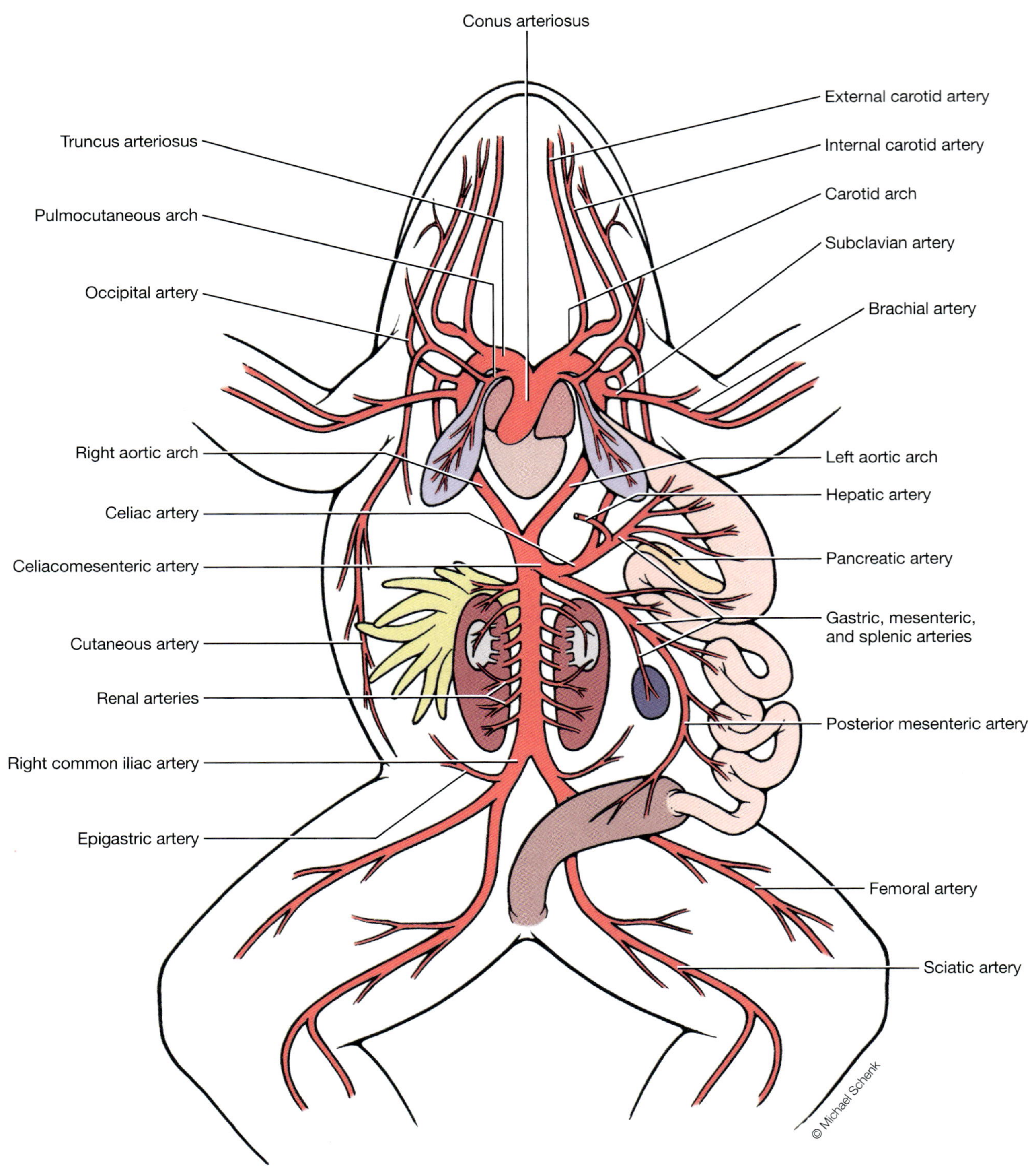

19

19.14 Frog, arterial system depicted to show extensive networking of vessels to distribute oxygenated blood throughout the body.

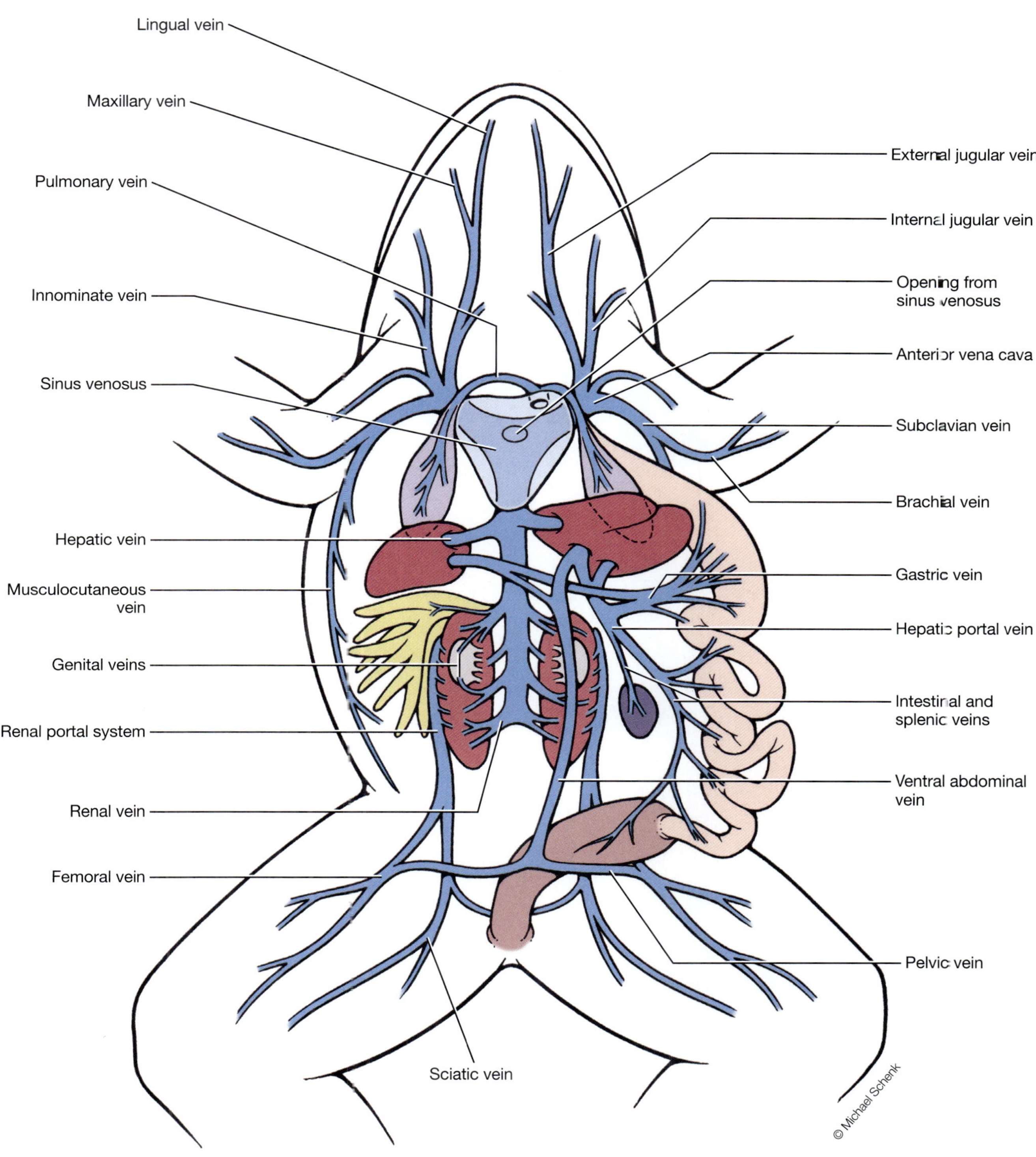

19.15 Frog, venous system depicted to show extensive networking of vessels to return deoxygenated blood to the heart from all regions of the body.

4.1 Why do you suppose the fat bodies of frogs would be larger during the breeding season?

4.2 Describe how a single-circuit and double-circuit circulatory system differ.

4.3 What adaptation have amphibians developed to help maintain separation of oxygenated and deoxygenated blood within the heart?

Urogenital System

Procedure

1 Carefully remove the liver and other digestive organs of your specimen by cutting the esophagus and the terminal portion of the large intestine where it enters the cloaca.

2 If you have not done so already, remove the fat bodies from one or both sides of the body, being careful not to remove the testes in male specimens.

3 Use Table 19-5 and Figures 19.16–19.18 to assist you in identifying the structures of the urogenital system of your frog.

Male frogs possess a pair of testes. Each is attached to the ventral surface of a kidney by **vasa efferentia**, and also may be attached to the fat bodies on that side of the body (Figs. 19.16 and 19.18A). Sperm are transported through the vasa efferentia to the **kidneys**, then through the **ureters** to the cloaca for release from the body. Vestigial oviducts are present in the males of many frog species, including leopard frogs. Do not let their presence confuse the identification of your specimen's sex.

In female frogs, eggs are produced in the paired **ovaries** and are released into the **coelom** when mature (Figs. 19.17 and 19.18B). The eggs are captured by the funnel-shaped openings—the **ostia**—of each **oviduct** and are transported to the corresponding **uterus** on that side of the body. As the eggs pass along the length of the oviducts, they are encased with several layers of gelatinous material secreted by the oviductal walls. In frogs, the uterus is simply the enlarged, terminal portion of the oviduct that temporarily stores unfertilized eggs prior to mating. During mating, the female frog releases her eggs through the cloaca into the water while the male frog simultaneously releases a cloud of sperm cells over the eggs to complete the act of external fertilization.

In both sexes, the kidneys are flat, elongated organs that lie on either side of the posterior vena cava dorsal to the intestinal tract. As described in the previous section, "Circulatory and Respiratory Systems," the kidneys have several arterial and venous attachments to the circulatory system along their length for blood filtration. The kidneys filter metabolic wastes from the blood and pass them through the paired ureters into the cloaca and then to the **urinary bladder** for storage. Frogs and toads are exceptionally adept at conserving water, and many species are able to reabsorb water directly from urine stored in their urinary bladders during prolonged dry periods. Urine passes out of the frog through the cloaca, a common chamber for urine, feces, and gametes (Fig. 19.18).

Table 19-5 Urogenital Anatomy of the Frog

Structure	Function
Adrenal glands	Produce hormones that regulate blood pressure and metabolism
Kidneys	Paired organs that filter nitrogenous wastes from the blood
Ureters	Transport urine from the kidneys to the urinary bladder
Ureter apertures	Paired orifices through which urine from the ureters enters the cloaca before passing into the urinary bladder for storage
Cloaca	Common chamber for collecting materials from the digestive, excretory, and reproductive systems prior to their discharge from the body
Urinary bladder	Site of storage for urine prior to discharge from the cloaca
Vasa efferentia (male)	Small ducts that transport sperm from the testes to the kidneys
Seminal vesicles (male)	Contribute seminal fluid for sperm entering cloaca to assist in dispersion and insemination
Testes (male)	Paired organs that produce sperm for transport via the vasa efferentia through the kidneys to the ureters and discharge through the cloaca
Uteri (female)	Site of storage of unfertilized eggs prior to discharge through the cloaca
Ovaries (female)	Organs that produce eggs for transport through the oviducts to the uteri and discharge through the cloaca
Oviducts (female)	Paired tubes that transport eggs from the ovaries to the uteri and secrete a jellylike coating for protection of the eggs
Ostia (female) *sing.* = ostium	Openings in the anterior ends of the oviducts through which eggs released from the ovaries into the coelomic cavity enter the oviducts

19.16 Male frog, urogenital system: (**A**) depicts fat bodies and testes in place; (**B**) fat bodies and testis removed from animal's left side to reveal underlying anatomy.

19

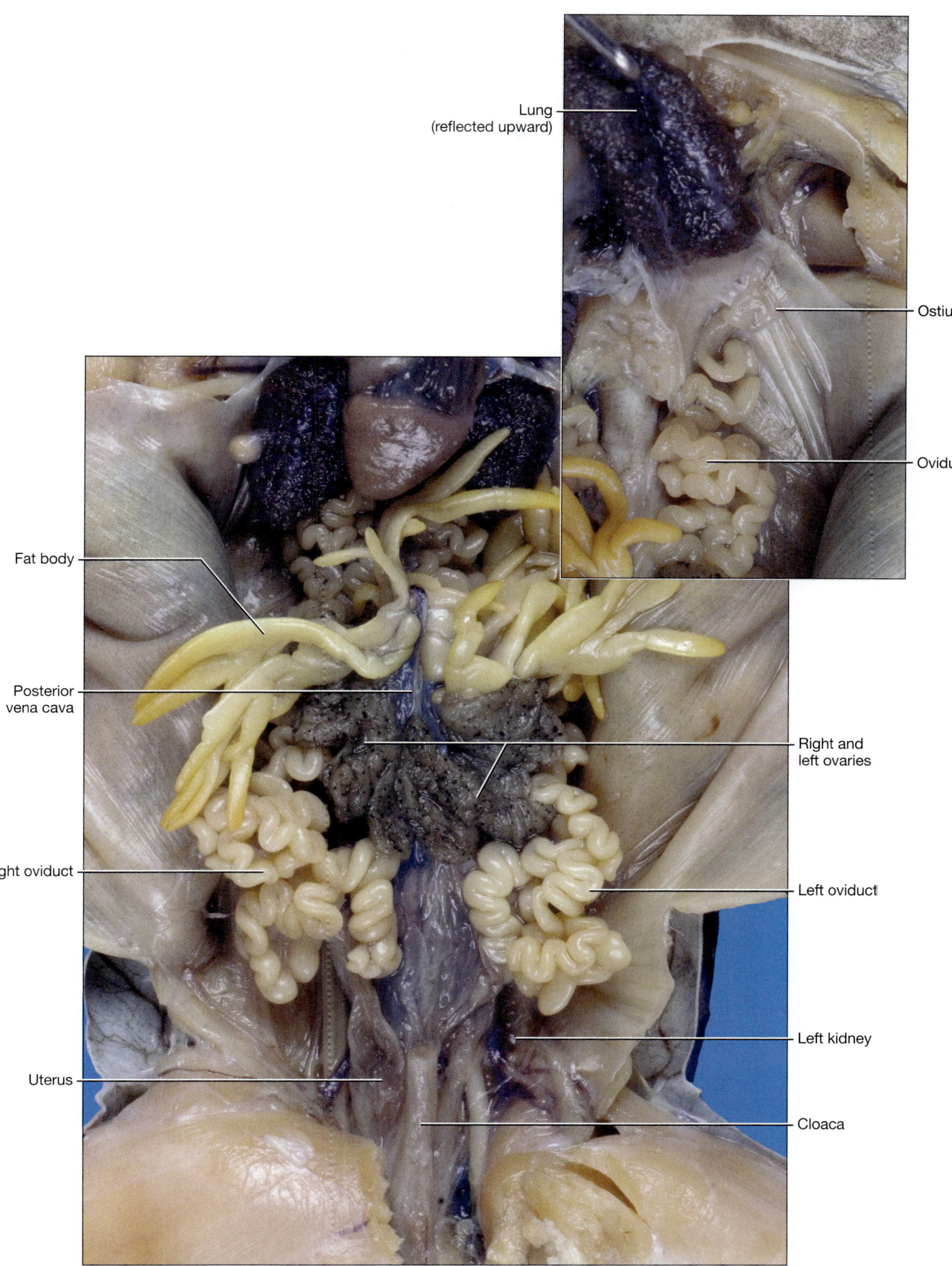

19.17 Female frog, urogenital system. Inset shows ostium of the oviduct.

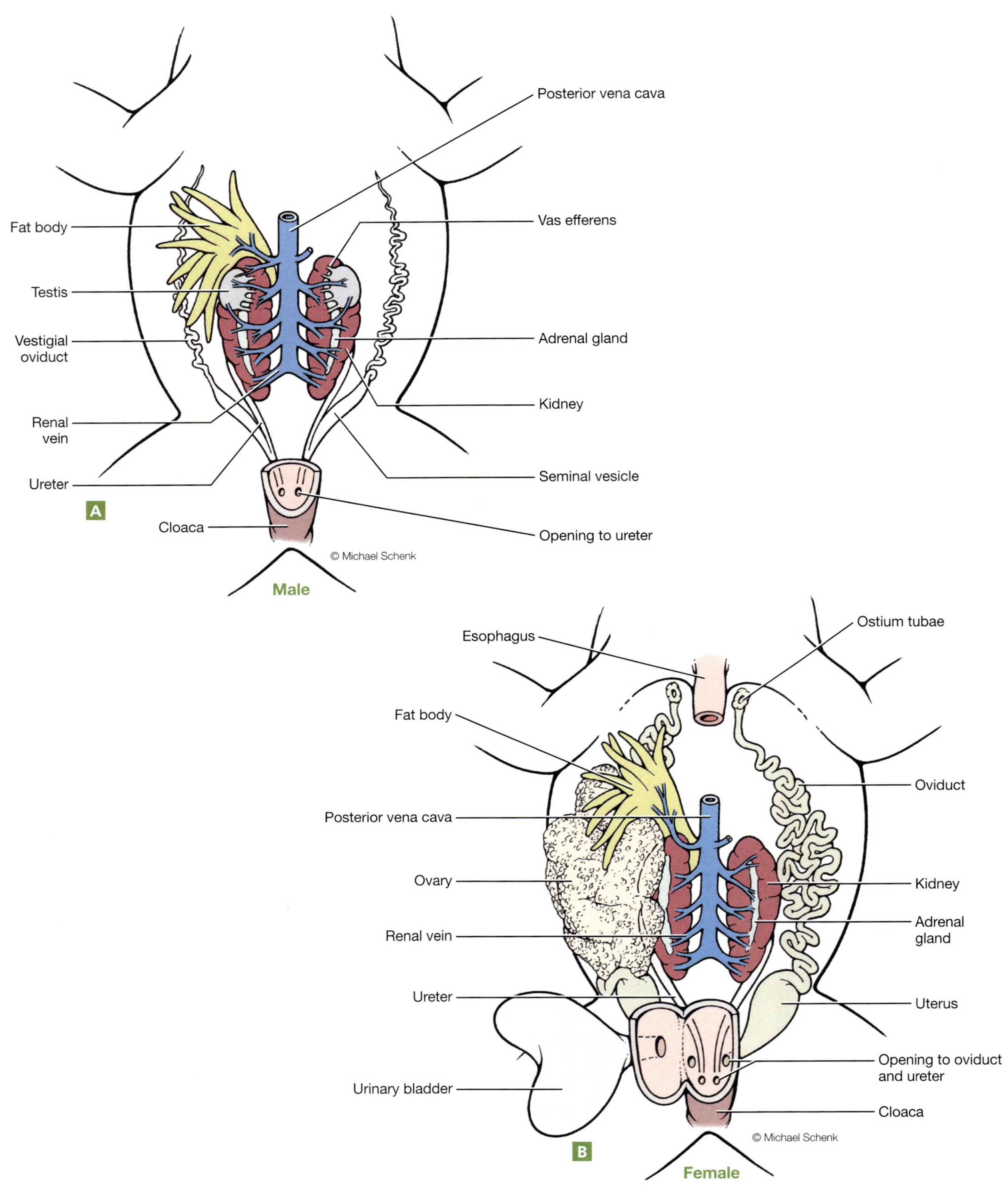

19.18 Comparison of the urogenital systems of (**A**) male, and (**B**) female frog. Urinary bladder not depicted on male.

Name ____________________ Section ____________

19 Questions for Review

1 Which features of the skeletal system n frogs represent specific adaptations for jumping?

2 Which two accessory digestive organs release compounds into the duodenum through the bile duct?

3 What is the functional difference between a cloaca and an anus?

4 How many chambers does a frog heart have? Is the circulatory system open or closed?

5 Describe how a portal system differs from a regular circulatory pathway.

6 What organ, in addition to the lungs, do frogs use to supplement gas exchange?

7 Which one of the following organs is *NOT* part of the female frog's reproductive pathway?

a. ureter

b. ovary

c. oviduct

d. uterus

e. cloaca

8 List all of the organs in the male frog that are common to both the excretory *AND* reproductive systems.

9 Describe three defining characteristics of amphibians.

1.

2.

3.

10 Describe evidence supporting the placement of amphibians in the phylogeny of chordates.

19

Name ______________________ Section ______________

19 Laboratory Practical Challenge

The following evaluation is intended to replicate the method of assessment and type of questions you will be asked on lab practicals. If you have difficulty with these questions, you may benefit from additional review of the specimens provided in lab.

1 Identify the structure indicated by line A.

____

2 What is the function of the structure indicated by line B?

____

3 To what system does the structure indicated by line C belong?

____

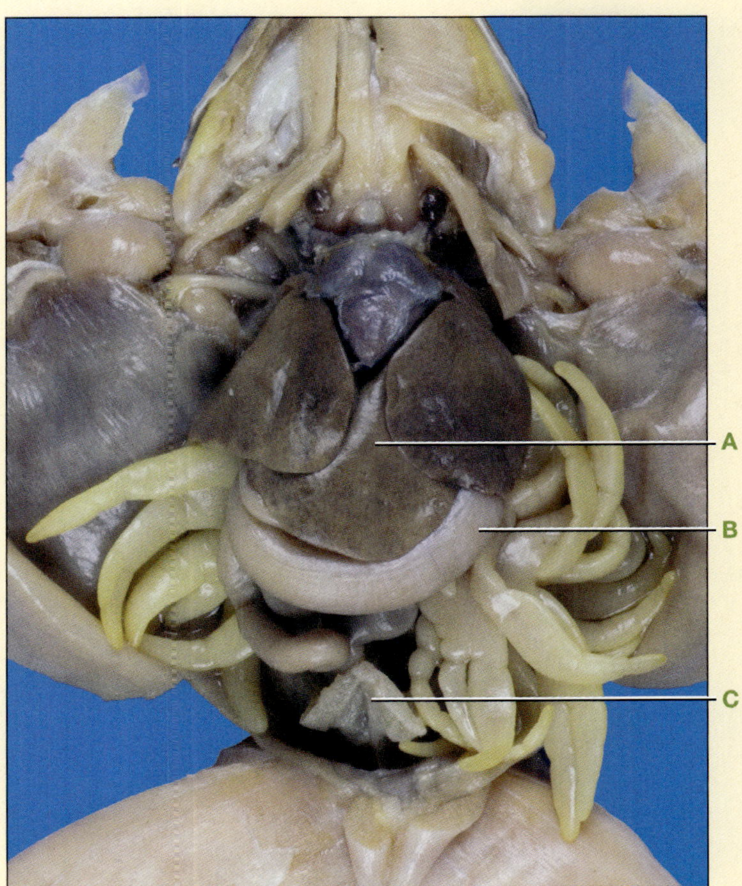

19

4 To what system does the structure indicated by line D belong?

5 What is the function of the structure indicated by line E?

6 Identify the structure indicated by line F.

7 Identify the structure indicated by line G.

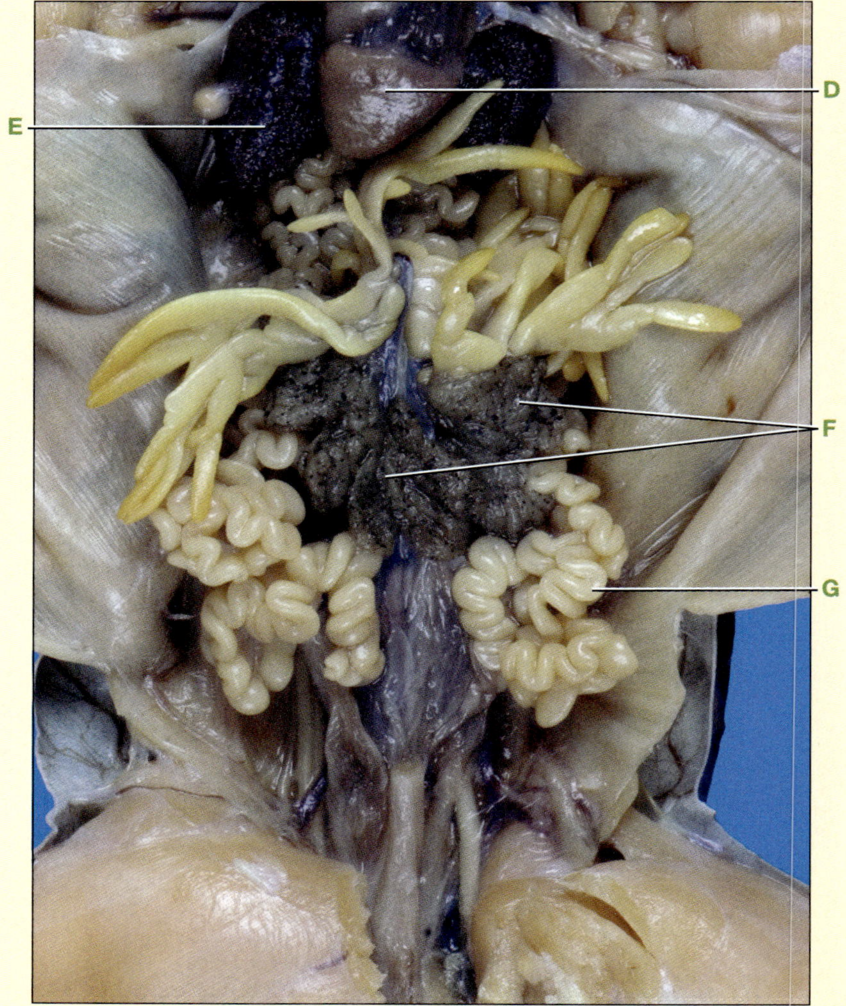

19

CHAPTER 20 Reptiles

After completing the exercises in this chapter, you should be able to:

1 List the distinguishing characteristics of the class Reptilia.

2 Understand the evolutionary position of reptiles on the phylogeny of chordates.

3 Discuss the significance of the amniotic egg and scales in the evolution of terrestrial vertebrates.

4 Identify the major external features and internal organs of a turtle, discuss their functions in the body, and compare them to homologous structures in other vertebrates.

5 Identify the major external features and internal organs of a snake, discuss their functions in the body, and compare them to homologous structures in other vertebrates.

6 Contrast the unique characteristics of alligators and crocodiles from those of other reptiles.

In this chapter you will examine in detail two of the 9,500 species in the class Reptilia, a group of animals often thought of as cold, creepy, primitive creatures on the way out, superseded by birds and mammals. Reptiles, however, are in many respects much more successful than often is thought (Fig. 20.1). In habitats such as deserts, reptiles are the dominant vertebrate group as a result of one huge advantage they hold over birds and mammals: Being less dependent on maintaining a constant body temperature, they can survive on a fraction of the food that birds and mammals require, and thus are able to thrive in harsh environments where food is scarce and temperatures fluctuate dramatically.

Reptiles are classified as **ectotherms**—animals that derive the bulk of their internal body heat from external sources. Fish, amphibians, and invertebrates all fall into this category as well. Only birds and mammals are true **endotherms**—generating most of their body heat from within.

All reptiles are characterized by dry skin composed of overlapping, **keratinized scales**. In turtles, some of these scales have been modified into the bony **scutes**, or plates, of the shell. The tough outer skin of reptiles provides an excellent barrier against dehydration—a formidable concern for all terrestrial animals. Unfortunately this type of skin allows for limited growth and must be replaced as the animal grows. As a result, reptiles shed their skin periodically, through a process known as **ecdysis** (molting).

Another feature that has contributed to the success of reptiles in terrestrial environments is the evolution of the **amniotic egg**, sometimes referred to as the cleiodoic egg. This type of egg has a tough, sometimes leathery, outer shell that is impermeable to water loss but permits gas exchange. For such an impenetrable barrier to be placed around an egg *before* it leaves the female's body, the egg must be fertilized before a shell is constructed around it. Thus, mechanisms of **internal fertilization** arose in reptiles and, with them, **copulatory organs**. The development of the amniotic egg and internal fertilization enabled reptiles to be the first truly terrestrial animals that do not depend on water for reproduction. Reptiles

Hawaiian green turtle, *Chelonia mydas*

Ornate box turtle, *Terrapene ornata*

California king snake, *Lampropeltis getula*

Side-blotched lizard, *Uta stansburiana*

20.1 Turtles, snakes, and lizards are representatives of the class Reptilia. Hatched from amniotic eggs, they display direct development and possess dry, scaly skin and copulatory organs.

also have **direct development** with no larval stage—another difference from amphibians.

The turtle and snake have been selected as study specimens for this chapter because, in one sense, they represent extremes at opposite ends of the evolutionary history of reptiles, yet they share many typical reptilian characteristics. Turtles are the most ancient surviving lineage of reptiles, appearing almost 300 mya, and have remained virtually unchanged for the past 200 million years. Snakes emerged as a more recent offshoot from the main branch of reptiles around 170 mya and brought many highly specialized features.

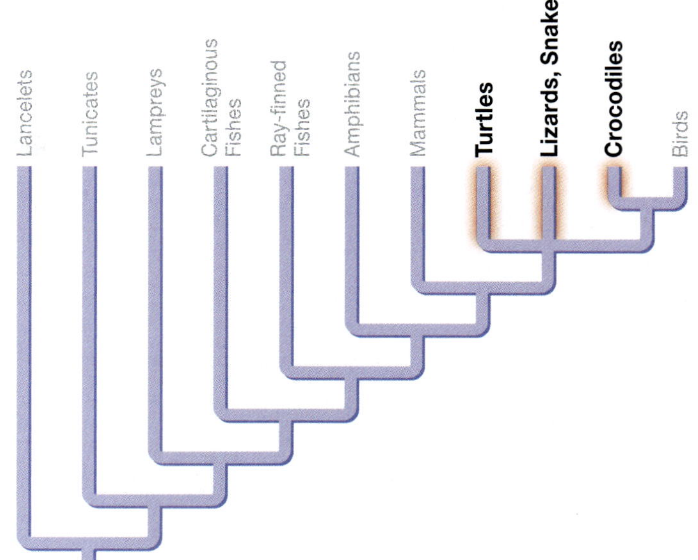

20.2 Reptiles covered in this chapter represent a paraphyletic group—meaning the class Reptilia, as we define it, does not include all the descendants of its ancestor. Birds are still currently assigned to a separate class (Aves) but are direct evolutionary descendants of ancient reptiles (dinosaurs).

Evolutionary History

By the end of the Permian period, 250 mya, most amphibian families had become extinct and were being replaced slowly by reptiles as the dominant terrestrial vertebrates (Fig. 20.2). The first reptiles actually had been around since the Carboniferous period (360–299 mya), and these early reptiles underwent a major adaptive radiation around 290 mya, diverging into two major branches: the sauropsids and the synapsids. Sauropsid reptiles became subdivided into the anapsids and the diapsids. These terms referred to the number of temporal openings in the skull behind the orbit. Synapsids have a single temporal opening behind the orbit, diapsids have two temporal openings behind the orbit, and anapsids lack any openings behind the orbit (Fig. 20.3). The number and distribution of openings in the sides of the skull affected the anchoring of head and jaw musculature and, thus, diet and feeding habits, and played a significant role in the evolutionary history of these groups.

Synapsid forms were the most ancestral and gave rise to a branch of reptiles known as the therapsids (doglike carnivores) with high metabolic rates and eventually fur—the lineage that eventually gave rise to modern mammals. Anapsid forms gave rise to modern turtles. Diapsid forms gave rise to the remainder of the reptilian lineage, including the dinosaurs, crocodiles, snakes, lizards, and birds. Including birds as a subgroup of reptiles may seem odd, but birds and crocodiles both share more in common with dinosaurs and each other than they do with other modern reptiles.

The key to the success of reptiles was undoubtedly the amniotic egg. Breaking the tether to aquatic reproduction by enclosing the embryo in its own watery environment with membranes to prevent dehydration allowed reptiles to flourish as the global climate continued to dry over the ensuing millennia. Amphibians never became fully terrestrial because their eggs had to be laid in water to avoid desiccation. Reptiles also developed more advanced respiratory mechanisms to aid in breathing. Amphibians gulp air and use their throat musculature to *push* air into their lungs, limiting their breathing capacity to the volume of the mouth and throat. In reptiles the thoracic musculature developed to expand the rib cage and allow air to be *pulled* into the lungs by negative pressure, permitting them to take advantage of the full capacity of their lungs. Two additional characteristics that early reptiles developed were dry, scaly skin that resisted dehydration and internal fertilization, increasing their success on land.

For more than 200 million years, reptiles were the dominant land vertebrates until 65 mya, when a huge asteroid roughly six miles in diameter slammed into the region of North America that we now recognize as the Yucatan Peninsula of Mexico. It carved out a crater more than 100 miles in diameter. The resulting explosion and shockwaves from this collision vaporized the asteroid and sent fire and debris miles into the atmosphere, landing as far away as modern-day New York City. During this same time period, unrelated, enormous volcanic eruptions were occurring simultaneously on the far side of the globe in what is now India, spewing ash and toxic gases into the atmosphere and burying that region of the planet under hundreds of feet of volcanic rock.

The global climate changes that ensued in the wake of this catastrophic duet brought about one of the great worldwide mass extinctions in the history of our planet, in which an estimated 76% of all plant and animal species perished. Yet, life survived. Among the creatures that were spared were small, insectivorous mammals that had been relegated to nocturnal, arboreal lifestyles during the reign of the dinosaurs. When the dust settled, these inconspicuous, mouse-sized mammals would diversify and spread throughout the landscape into the numerous niches vacated by dinosaurs and ultimately give rise to the lineage of modern mammals. For a more complete phylogeny depicting the evolutionary relationships among the chordates covered in the remainder of this laboratory guide, review Figure A.2 in the Appendix.

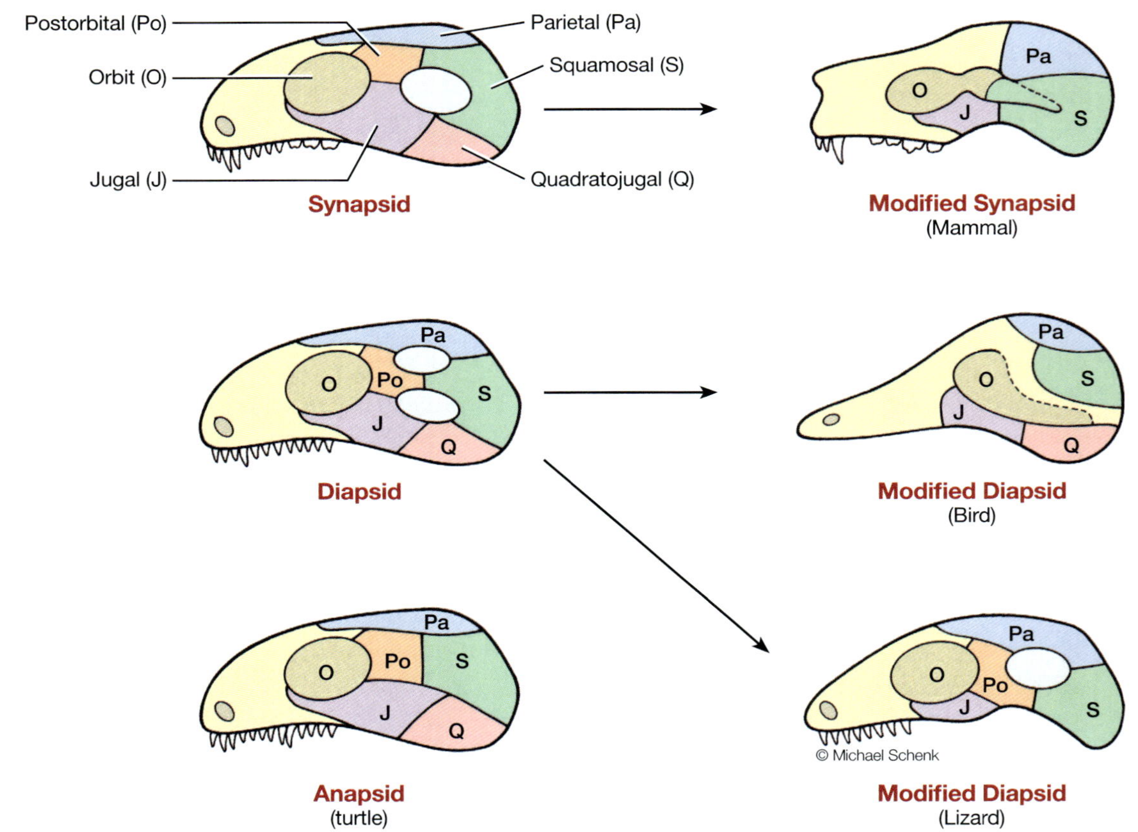

20.3 Evolution of temporal openings. Modern mammals evolved from synapsid reptiles. Modern snakes, lizards, crocodiles, and birds evolved from diapsid reptiles, and modern turtles evolved from anapsid reptiles.

Exercise 20-1 Turtles and Tortoises (Order Testudines)

Materials Needed

- ❑ mounted turtle skeleton
- ❑ preserved turtle
- ❑ assortment of turtle shells/ specimens (optional)
- ❑ dissecting tools
- ❑ dissecting pan
- ❑ latex gloves

Approximately 300 living species of turtles and tortoises constitute the order Testudines. Though slow and methodical in their gait, turtles have spread successfully throughout the globe into temperate and tropical regions on all continents except Antarctica. Individuals of some species may live upward of 120 years, and adults range in size from only a few inches long and a few ounces to more than 6 feet in length. The leatherback sea turtle is the largest species, with recorded weights over 1,500 pounds! Lacking the speed and ability to catch fast-moving prey, most species of turtles graze on vegetation, scavenge on decaying animal and plant matter, or capture slow-moving animals such as insects, earthworms, and molluscs.

External Anatomy

Procedure

1 Obtain a preserved turtle and place it on its ventral surface in your dissecting pan.

2 Examine the paired forelimbs and hind limbs. Your specimen is an aquatic species that has partially webbed feet to aid in swimming. Long claws also are present at the tips of the toes, for digging and crawling.

3 Examine the head.

Eyes, with movable eyelids, and ears are present, although turtles have no external ear openings (Fig. 20.4A). Paired **nostrils** are also present at the end of the snout. The tip of the mouth is covered by a sharp, epidermally derived projection called the beak. Modern turtles lack teeth but are capable of tearing apart their food into bite-sized pieces, using their strong neck muscles and sharp beaks.

Notice that space exists inside the shell at the base of the neck and each leg for the head and limbs to be withdrawn into the shell. This feature affords extra protection for the turtle when it is confronted with danger. In contrast to modern turtles, fossil evidence indicates that the earliest turtles had teeth and were unable to retract their heads into their shells.

The most prominent feature of any turtle is, without doubt, its **shell**. No other vertebrate has evolved an armor quite like the turtle's shell. So successful was this protective covering that it quickly became the cornerstone of turtle architecture, with all other anatomical adaptations being built around it. The shell is composed of an upper section, the **carapace**, and a lower section, the **plastron**, that are fused along the lateral margins into one solid structure. In many species the plastron has one or two transverse **hinges**, allowing the ends of the plastron to open and close. Turtles with hinged shells are able to seal themselves tightly inside their shells when threatened. An equally important benefit of a shell that closes tightly is to prevent evaporative water loss and dehydration—supported by the observation that most terrestrial turtles can close their shells tightly, but few aquatic species have hinged shells.

4 Examine the plastron. The hinges (when present) are located at the sutures along the anterior and posterior borders of the abdominal scales.

5 Press against the ends of the plastron to determine the location of the hinges.

The **scales** (scutes) of the carapace and plastron are composed of epidermal tissue covering hard, fused bony plates underneath (Fig. 20.4). The names of the scutes generally correspond to the region of the body they cover (for example, vertebral scales, pectoral scales, and abdominal scales). Biologists use the number, arrangement, and color patterns of these scutes to identify different turtle species.

Although a turtle sheds its skin as it grows, its shell remains in place throughout the animal's life and, thus, bears the scars and signs of age that accumulate through the years. As was the case with fish scales, close examination of the scutes reveals faint, concentric growth rings that give an idea of a turtle's relative age. In terrestrial species that do not shed their scutes, each ring represents one year's growth. Many aquatic species shed their scutes regularly, and within a few years the impressions of the earlier growth rings fade, making age estimates unreliable.

6 If other turtle specimens are available, compare the number and arrangement of scutes on the plastron and carapace, as well as the location and number of hinges on the plastron.

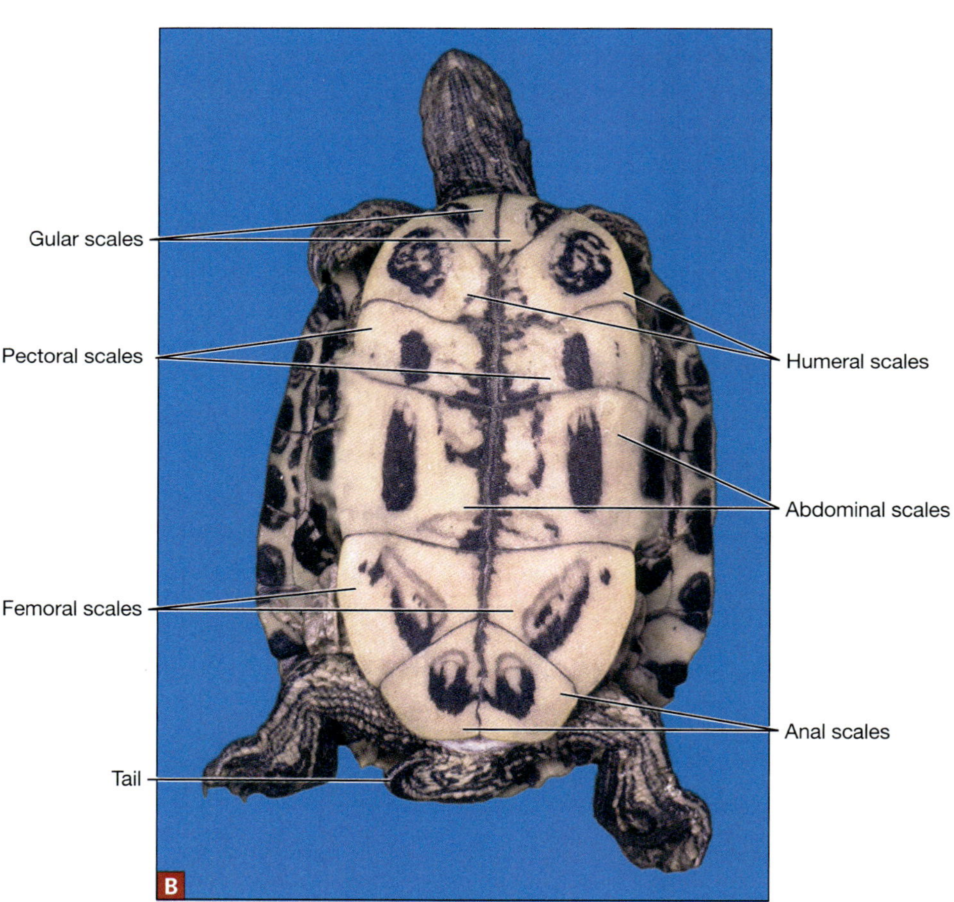

20.4 Red-eared slider, *Trachemys scripta*, external anatomy: (**A**) dorsal view, and (**B**) ventral view.

20

Check Your Progress

1.1 Count the growth rings on one of the costal scales (identified in Fig. 20.4) on your specimen and estimate the turtle's age.

1.2 How many toes are present on the forelimbs? On the hind limbs? How does this compare to amphibians? To mammals?

1.3 What external features of the turtle are homologous in mammals?

Skeletal System

Procedure

1. Obtain a mounted, articulated skeleton of a turtle.
2. Examine it carefully, using Figure 20.5 to assist you in identifying the skeletal elements present.

As was the case with the frog skeleton, many of the bones seen in the turtle represent homologies to familiar bones present in most mammals, including humans (for example, the humerus, scapula, femur, tibia, and fibula), providing evidence of their common ancestry. But many specializations in the frog skeleton reflect adaptations necessary to withstand the forces generated while jumping great distances and landing safely. In contrast, the turtle skeleton has been modified around the architectural cornerstone of the body—the shell. Notice that the dorsal surfaces of the pectoral and pelvic girdles, as well as the dorsal ribs and many of the **vertebrae**, have been fused to the shell to increase the strength of these connections. The pectoral girdle, pelvic girdle, and limb bones also have been reinforced to support the additional weight of the shell. The bones of the neck, forelimbs, and hind limbs articulate in a way that enables them to be withdrawn into the shell. Ventral ribs are absent, replaced by the plastron covering the ventral surface of the body. You may remember that turtles are anapsid reptiles, meaning that they have no additional openings in the temporal portion of the skull behind the orbit, and this characteristic should be evident by examining the turtle's skull.

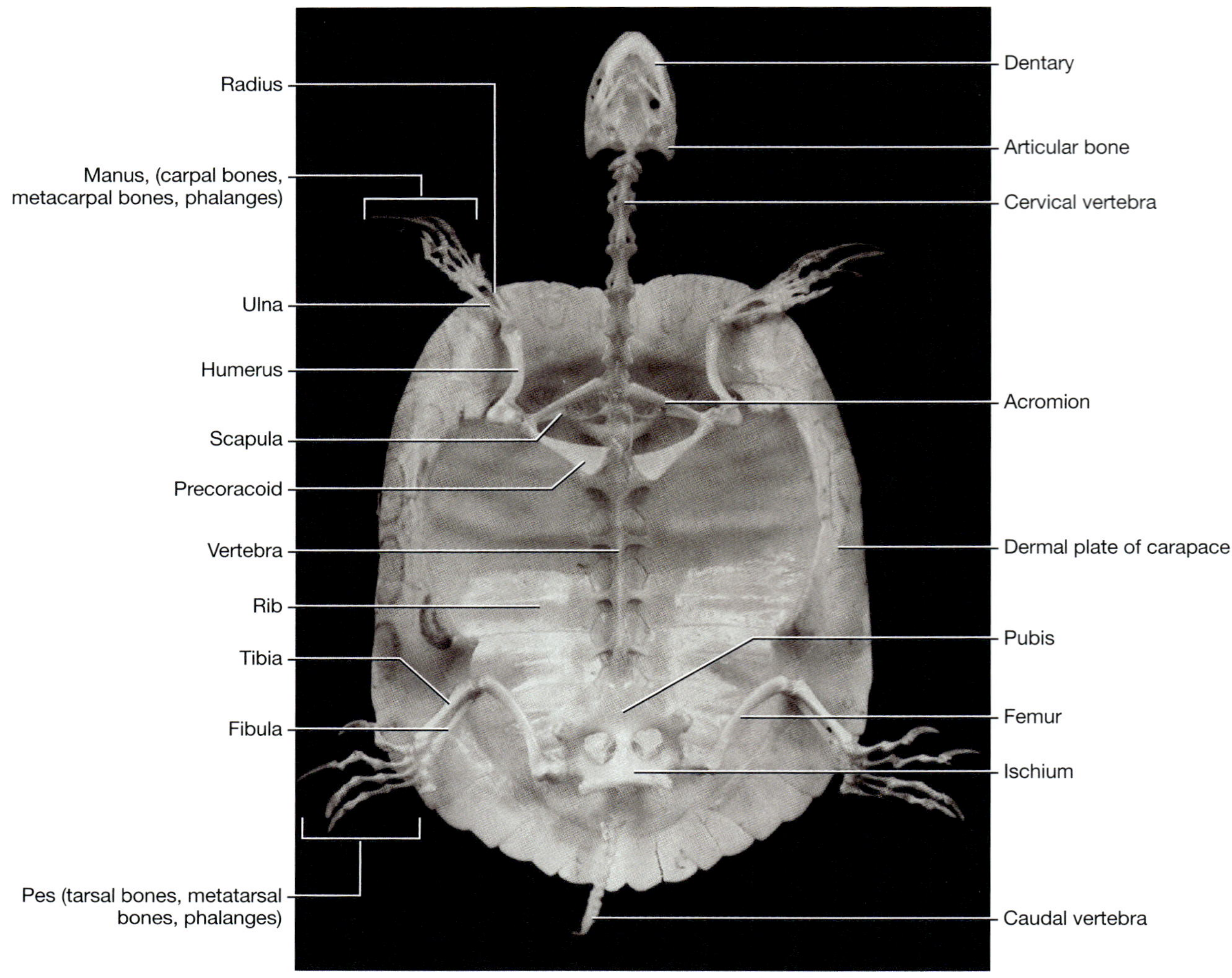

20.5 Turtle skeletal anatomy, ventral view.

Check Your Progress

2.1 List five bones in the turtle that are homologous to bones in other vertebrates (such as humans).

2.2 Of what materials is a turtle's shell constructed?

20

2.3 Are turtles synapsid, diapsid, or anapsid reptiles? How do you know?

Digestive System

Procedure

1 Lay your turtle on its carapace with the plastron facing you.

2 Opening a turtle is a bit like opening a tin can in the middle without losing any of its contents! You must carefully separate the plastron from the carapace and other soft tissues along its entire margin.

3 Preserved specimens usually are shipped with the lateral margins of the shell already cut. All you should have to do to open your specimen is use scissors or a scalpel to separate the plastron from the soft tissues of the neck and legs.

4 When the plastron has been removed, you will notice that the heart is encased within a pericardial membrane, and the internal organs are encased within a visceral peritoneum. Both of these membranes should be removed carefully before proceeding.

5 Use Table 20-1 and Figures 20.6–20.8 to guide you through the dissection and identification of the major internal organs of your specimen.

The **esophagus** in most reptiles is equipped with several longitudinal folds to permit great distensibility for swallowing large food items. The **stomach**, found beneath the left lobe of the liver, resembles that of a frog in many respects, as do the duodenum and convoluted small intestine. The **liver** is large and prominent in turtles, occupying a significant portion of the abdominal cavity (Fig. 20.5A). The **gallbladder** may be found underneath the right lobe of the liver. As in other vertebrates, the gallbladder stores bile produced by the liver and releases it into the duodenum. A long, slender **pancreas** is located along the anterior margin of the duodenum, where it releases digestive enzymes to assist the digestive process (Fig. 20.7). A **cecum** is present in turtles (and some snakes) for fermentation. This large, thin-walled pouch demarcates the beginning of the large intestine and contains anaerobic bacteria that break down cellulose and other plant materials. The digested products, devoid of their nutrients, empty into the **cloaca** and are expelled from the body through the cloacal opening (Fig. 20.8).

The cloaca of reptiles is divided internally into three chambers (as in birds):

1. the **coprodeum**, which receives fecal matter directly from the intestine,
2. the middle chamber, the **urodeum**, which receives reproductive and urinary products, and
3. the last chamber, the **proctodeum**, which acts as a general collecting area for digestive and excretory wastes.

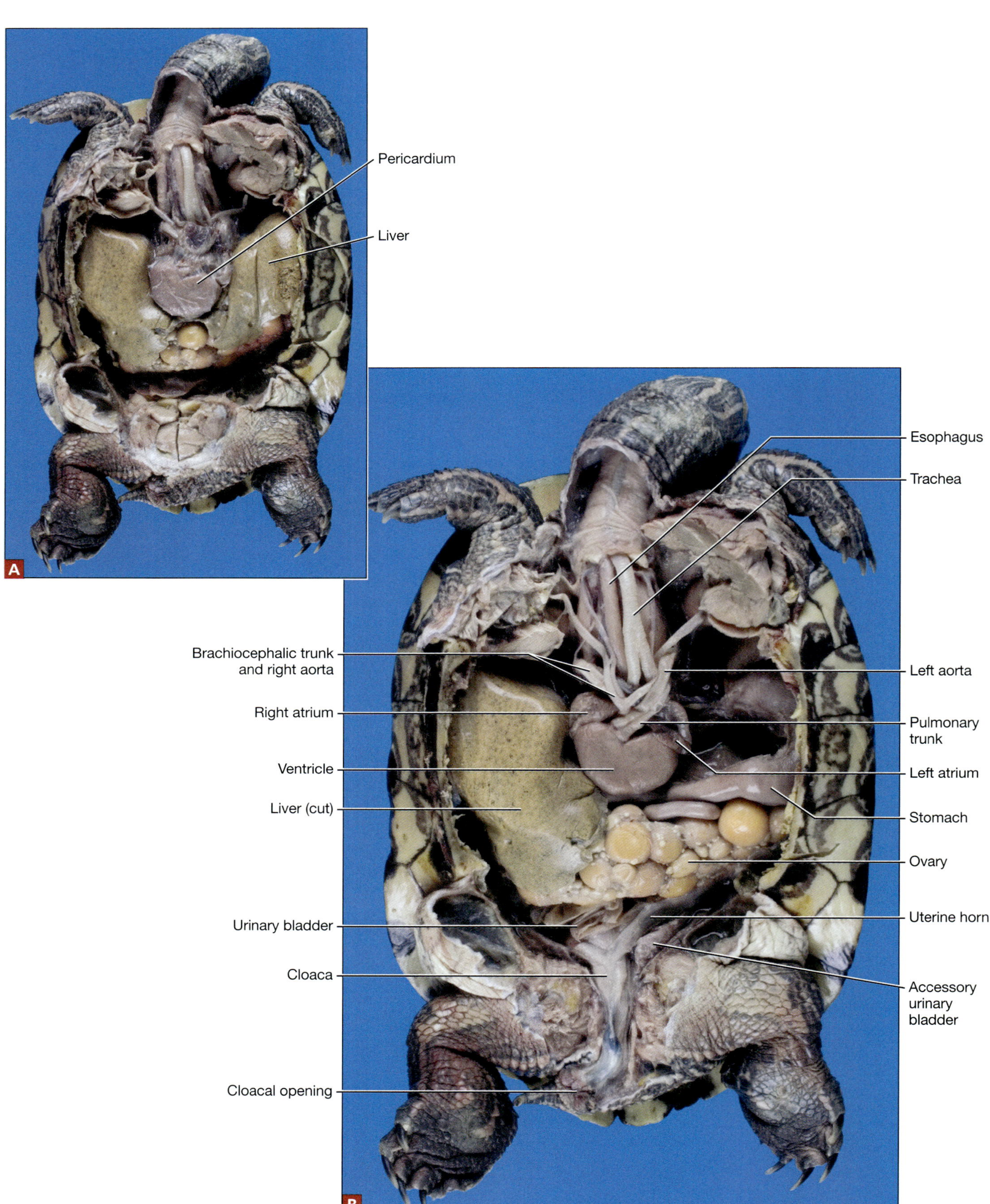

20.6 Red-eared slider, *Trachemys scripta*, internal anatomy: (**A**) ventral view, visceral peritoneum removed; and (**B**) ventral view, pericardial membrane, visceral peritoneum, and left lobe of liver removed.

20

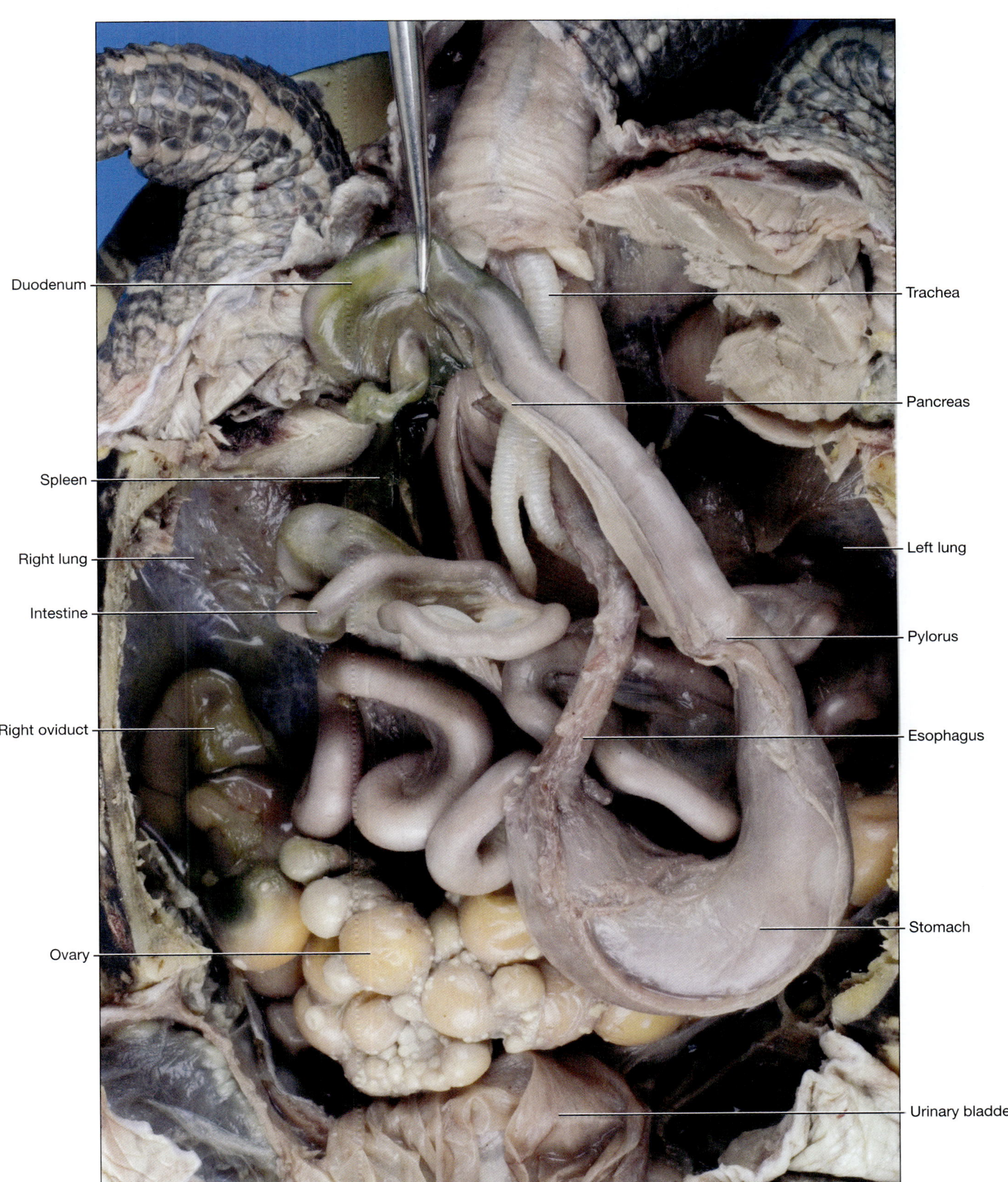

20

20.7 Red-eared slider, *Trachemys scripta*, underlying internal anatomy (liver removed).

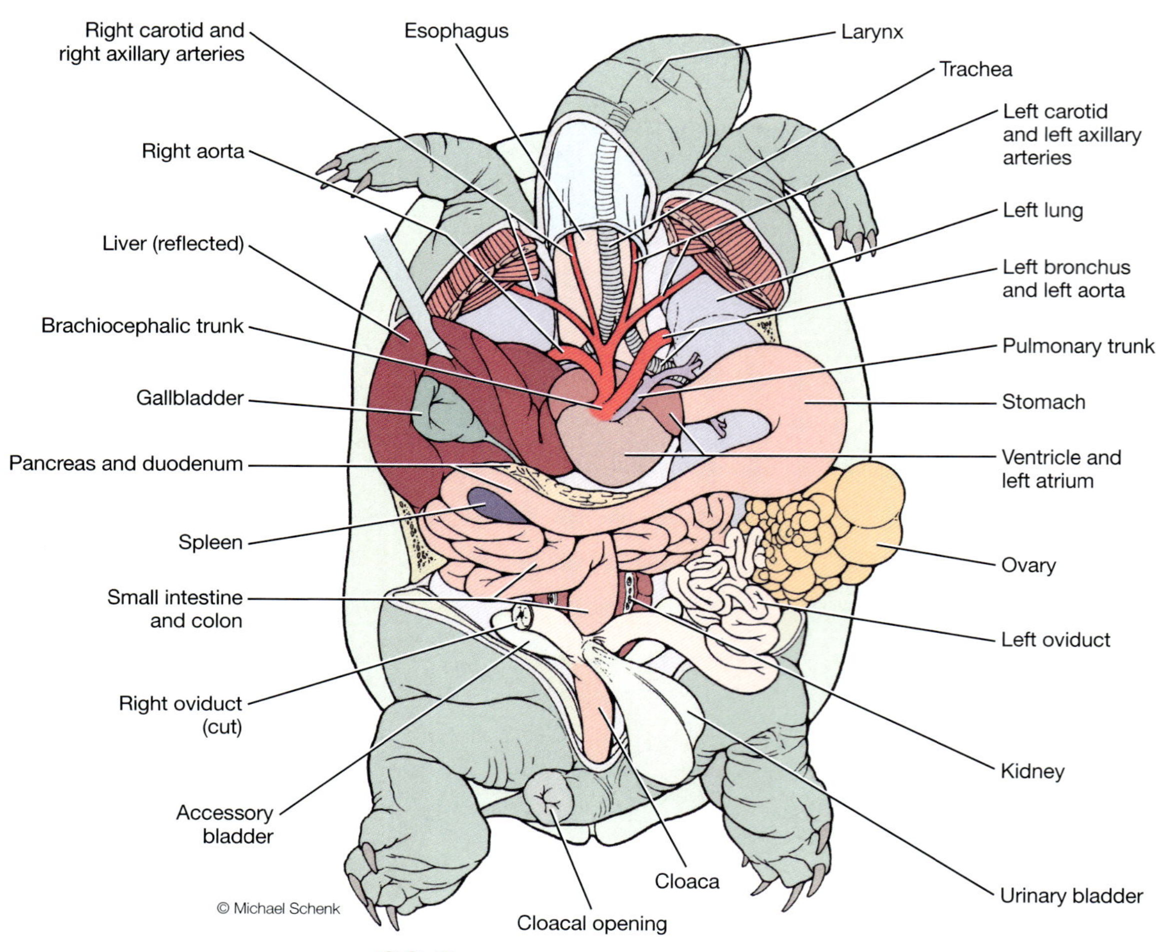

20.8 Turtle, generalized internal anatomy.

Table 20-1 Internal Anatomy of the Turtle

Structure	Function
Trachea	Conducts air to and from the lungs during respiration
Thyroid gland	Controls metabolism, growth rates, blood calcium levels, and shedding cycle
Liver	Produces bile, converts glucose to glycogen for storage, detoxifies many constituents of the absorbed digested compounds
Gallbladder	Stores bile produced by the liver
Heart	Three-chambered organ (two atria, one ventricle) that receives oxygenated blood from the lungs and pumps it via the arteries throughout the body
Lungs	Paired, highly vascular organs for respiration
Stomach	Site of food storage and initiation of digestion
Large intestine	Responsible for reabsorption of water and electrolytes; transports feces to coprodeum via peristalsis
Cecum	Large, thin-walled pouch in herbivorous species demarcating the beginning of the large intestine; contains anaerobic bacteria responsible for fermentation of cellulose and other plant materials
Pancreas	Produces digestive enzymes and delivers them through the pancreatic duct to the duodenum; also produces a suite of hormones for the endocrine system
Spleen	Stores blood, recycles worn-out red blood cells, produces lymphocytes
Kidneys	Filter blood (creating urine) and are responsible for osmoregulation
Testes (male)	Produce sperm
Ovaries (female)	Produce eggs

Circulatory and Respiratory Systems

Reptiles have a three-chambered **heart** that is somewhat more sophisticated than the amphibian heart in its internal design and, despite its single ventricle, basically functions as a four-chambered heart, keeping oxygenated and deoxygenated blood sufficiently separated. Gone from this circulatory plan are the conus arteriosus (or bulbus arteriosus) and sinus venosus present in fishes and amphibians. With two separate atria and a partially divided ventricle, the reptilian heart can generate sufficient contractile pressure to push blood through its double circuit, first to the lungs and back to the heart, then throughout the rest of the body.

Three large arterial trunks extend from the anterior surface of the heart: the pulmonary trunk, the left aorta, and the right aorta (Fig. 20.9). The pulmonary trunk receives blood from the right side of the ventricle and routes it to the lungs for oxygenation. The left aorta carries oxygenated blood from the left side of the ventricle to the abdominal organs. The right aorta also receives oxygenated blood from the left side of the ventricle but sends it to the head, neck, and pectoral regions via the brachiocephalic trunk, which branches immediately before the right aorta bends posteriorly to route blood to the pelvic regions via the dorsal aorta.

The flexibility of a single ventricle and separate aortae permits controlled mixing of pulmonary and systemic blood and allows blood to be shunted preferentially to the critical organs (brain, heart, forelimbs, and hind limbs) while diving—giving reptile hearts a huge advantage over the less-flexible, four-chambered heart of birds and mammals. Less-critical processes such as digestion can be suspended temporarily during long dives, as the left aorta delivers poorly oxygenated blood to those regions. A **spleen,** located near the pancreas (Fig. 20.8), is present to filter the blood of worn-out blood cells.

The respiratory system of turtles follows the typical vertebrate plan with a single **trachea** branching into two primary **bronchi**, one leading to each **lung** (Fig. 20.10). Two functional lungs are present, and in turtles and crocodiles the cartilaginous rings completely surround the trachea, preventing collapse of this structure during respiration or feeding. Reptiles (with the exception of crocodiles) lack a diaphragm, and most snakes and lizards use the intercostal muscles of the ribs to breathe. But breathing with a shell surrounding the body poses a unique problem for turtles. Instead of using their intercostal muscles, turtles utilize muscles that expand and contract the forelimb and hind-limb pockets to create the negative pressure necessary to ventilate the lungs. Abdominal muscles that force visceral organs against the lungs aid in expiration.

For some turtle species, lungs are not the only respiratory organ; they also have specialized patches of highly vascular skin lining the throat and cloaca through which gas exchange can occur underwater. Most aquatic turtles are extremely tolerant of low oxygen levels. How long a turtle can stay submerged depends on the species, the ambient temperature, and the amount of dissolved oxygen in the water. Red-eared sliders can survive while submerged in water fully saturated with oxygen for more than 24 hours, and sea turtles are capable of surviving indefinitely in these conditions.

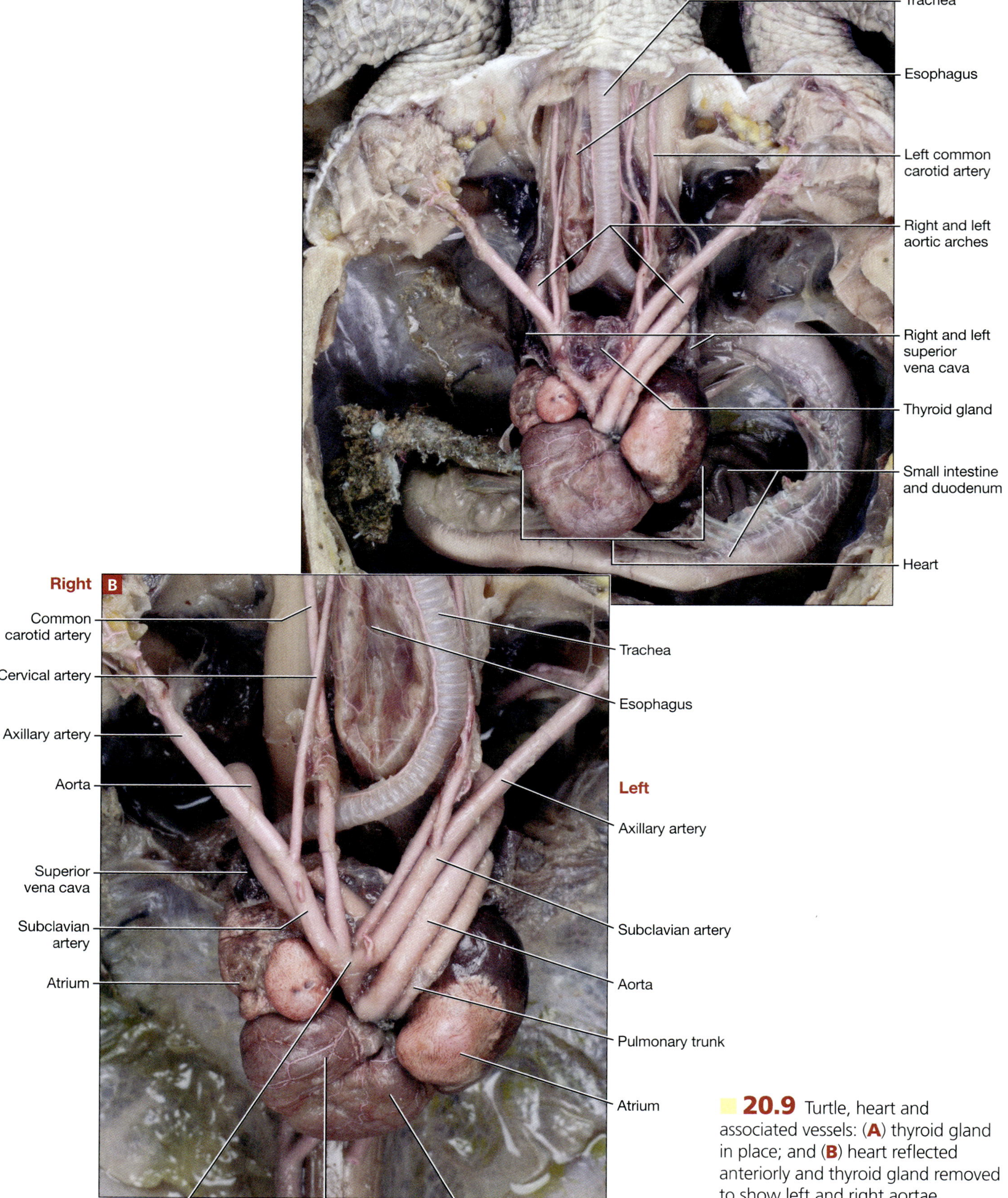

20.9 Turtle, heart and associated vessels: (**A**) thyroid gland in place; and (**B**) heart reflected anteriorly and thyroid gland removed to show left and right aortae.

20

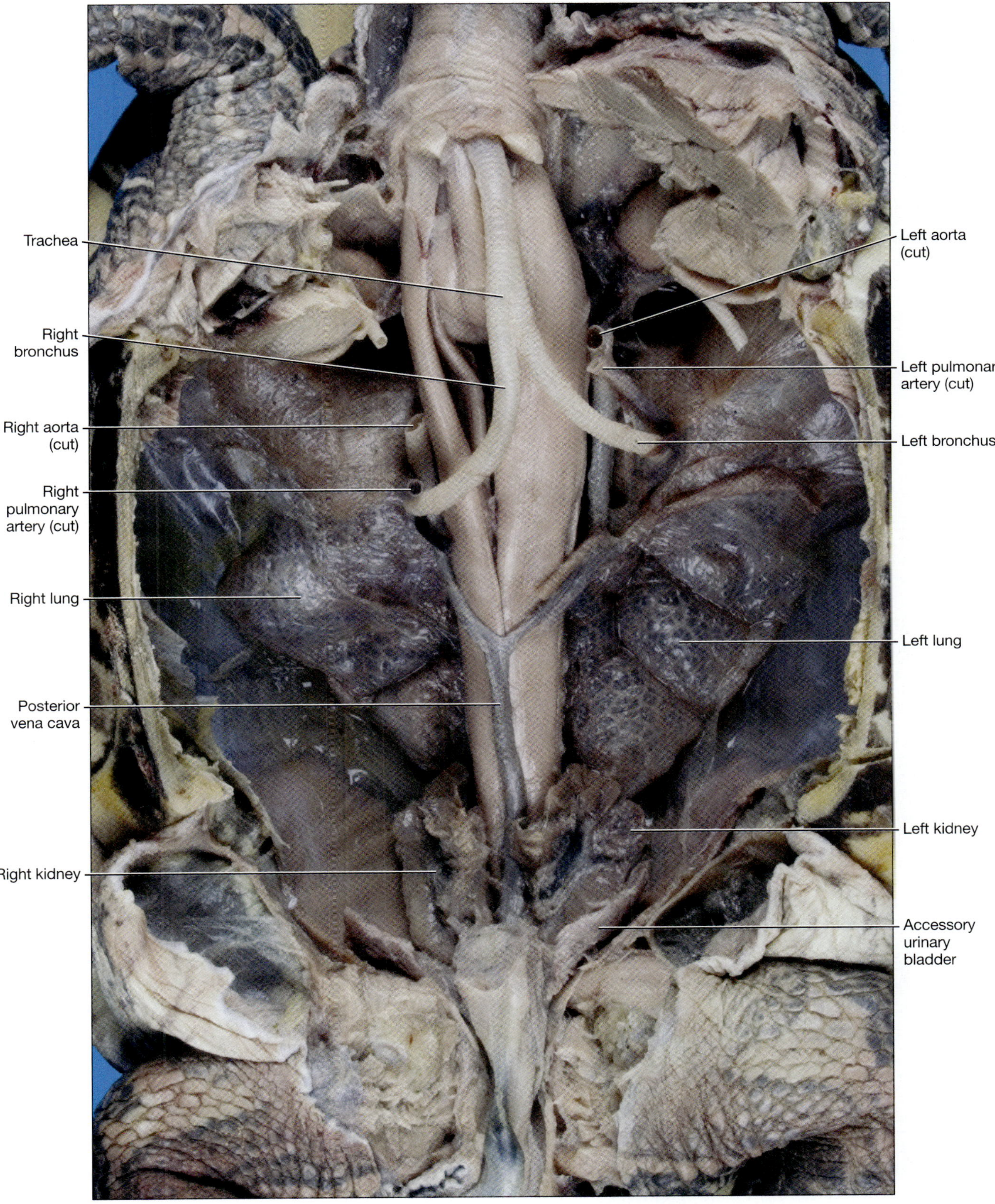

20.10 Turtle, respiratory and urogenital structures, heart and major organs of the digestive system removed.

20

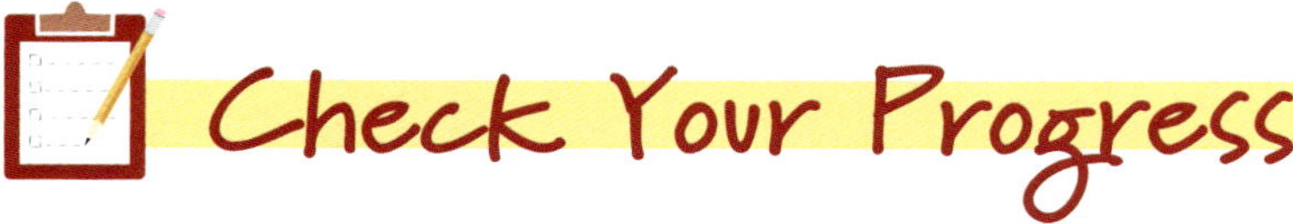

3.1 Would you expect the cecum to be relatively large or small in herbivorous turtles?

3.2 How many chambers does a turtle's heart have?

3.3 How many aortae do turtles have?

3.4 How many functional lungs do turtles possess?

3.5 How are turtles able to stay submerged beneath water for such long periods of time?

3.6 Explain blood shunting and the advantage to a three-chambered heart for aquatic turtles.

Urogenital System

Reptiles have a metanephric **kidney** that is more efficient than fish and amphibian kidneys and is more adept at conserving water. The paired kidneys typically are offset in the turtle, with the right kidney in front of the left (Fig. 20.10). Urine enters the urodeum (of the colon) from the **ureters** and then flows *backward* into the bladder for storage. Turtles have large **urinary bladders** and accessory bladders, and can recirculate urine through their bodies to reuse the water in it.

The location of gonads varies slightly in males and females. In females, the **ovaries** are attached to the posterior side of the lungs near the rear of the shell (Fig. 20.11A). As eggs mature in the ovaries, they pass into the **oviducts**, where the eggs are fertilized and subsequently encased in their protective, leathery shells by **shell glands**. In males, the **testes** are attached to the kidneys (Fig. 20.11B). Male turtles and tortoises possess a single copulatory organ, the **penis**, which is housed in the front portion of the cloaca within the base of the tail and everts into the proctodeum during copulation. During copulation the male mounts the female's shell from behind and locks his hind legs underneath her shell. To achieve intromission, the male often stands on his hind legs while balancing his plastron against the female's carapace. In many species, such as the eastern box turtle, the male's plastron has a concave portion that fits against the convex curvature of the female's carapace, making this balancing act somewhat less awkward.

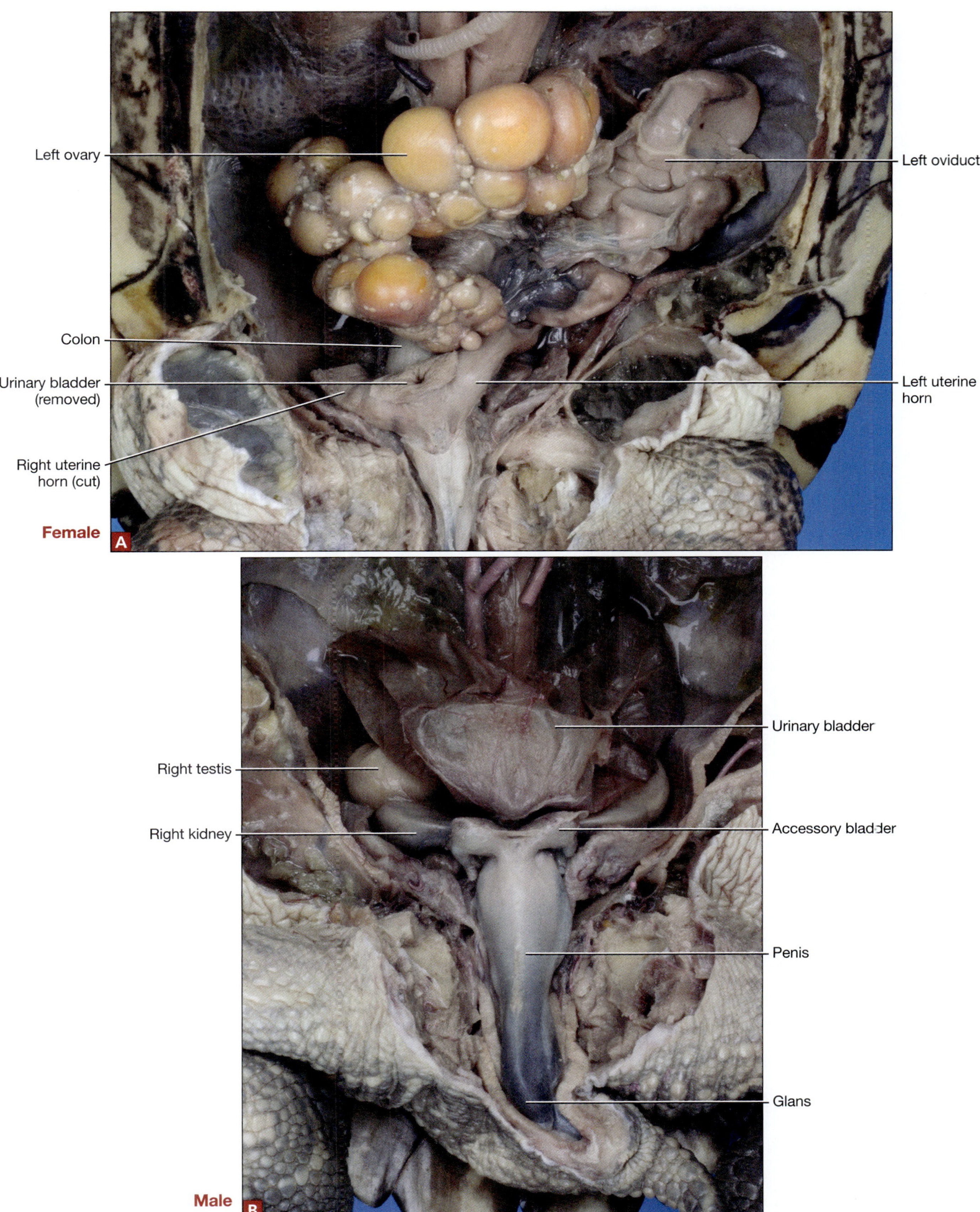

20.11 Turtle, urogenital anatomy: (**A**) female, right ovary and uterine horn removed to show underlying structures; and (**B**) male turtle.

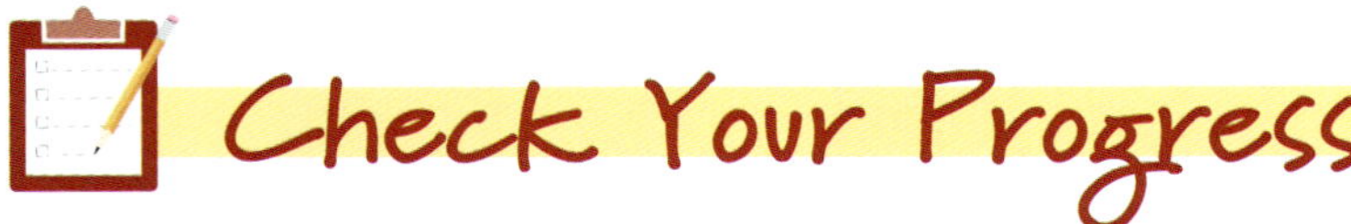

4.1 Are turtles dioecious? How do you know?

4.2 How is the turtle's reproductive system modified for increasing the survival of the eggs on land?

4.3 How does the turtle's urinary system help it survive in dry environments?

Exercise 20-2 Snakes and Lizards (Order Squamata)

Materials Needed

- ❑ mounted snake skeleton
- ❑ preserved snake
- ❑ dissecting tools
- ❑ dissecting pan
- ❑ latex gloves

Approximately 5,600 species of lizards and 3,400 species of snakes constitute the order Squamata, the largest and most diverse order of reptiles. Both groups have been highly successful and have worldwide distribution, with the exception of the Arctic and Antarctic regions. Lizards probably are the most familiar of the reptiles, and certainly the most typical in their body plan. In comparison, snakes have a "strangeness" that has made them an often feared and misunderstood group of reptiles. Further evidence of their mystique lies in the fact that in many cultures, for thousands of years, snakes have been revered as magical creatures with special medicinal and sometimes spiritual powers, and often are important religious symbols. In reality, snakes are a simple and fascinating assemblage of creatures that exemplify the extent to which selective pressures can drive evolution. The serpentine body plan has been extensively shaped by natural selection, becoming extremely elongated with organ reduction and rearrangement and the ability to swallow and digest incredibly large prey. To say that snakes are simply lizards without legs is to grossly underestimate their true uniqueness.

External Anatomy

Procedure

1. Obtain a preserved snake specimen.
2. Examine it carefully, using Figure 20.12 to assist you in identifying the external features of the head.

Some may see the lack of legs in snakes as a handicap, but the fact that snakes have lost their legs over the course of evolution actually has improved their ability to move. Some snakes, such as racers and coachwhips, can easily outpace humans over difficult terrain. Eyelids in snakes are also absent; instead, a transparent scale called the **spectacle** protects each eye. External ear openings are absent as well, and though they are unable to hear higher-frequency sounds, most snakes respond to low-frequency ground vibrations.

What snakes lack in their ability to hear they make up for many times over in their exceptional ability to smell. Snakes have a unique sensory structure called the **vomeronasal organ** (Jacobson's organ), which has paired openings in the roof of the mouth. As the forked tongue flicks, scent molecules are picked up on each fork and deposited into separate openings of the vomeronasal organ. The nerve fibers from each opening remain separate all the way to the brain and are interpreted as left and right olfactory fields, giving snakes the ability to smell in stereo! By simply probing the air or substrate with their tongues, they are able to determine both the direction and the proximity of nearby objects by detecting minute differences in the relative concentrations of odor molecules on each fork of the tongue.

Some species have pit receptors along the margins of the jaw to detect infrared radiation, allowing them to sense prey that are warmer than their surroundings, such as lizards, birds, mammals, or even other snakes (Fig. 20.12). These organs are the most sensitive heat receptors known, able to sense changes of less than 0.001°C! With them, a snake can distinguish both the direction of an object and its precise distance, allowing it to hunt successfully even in complete darkness.

20.12 In addition to eyes and nostrils, some snakes, like this green tree python, have pit receptors located along the sides of the mouth that detect heat—an adaptation that enhances their ability to prey on vertebrates that are typically warmer than their surroundings.

All snakes are carnivorous predators that either subdue their quarry by constriction or immobilize it with venom. Snakes possess teeth, but do not chew their prey. Nonvenomous snakes have four rows of upper teeth and two rows of lower teeth that are recurved and angle toward the back of the mouth to facilitate swallowing. In venomous snakes a pair of maxillary teeth usually is modified into hollow, surgically sharp fangs to deliver their toxic venom deep into the tissue of their victims.

Skeletal System

Procedure

1 Obtain a mounted, articulated skeleton of a snake.

2 Examine it carefully, using Figure 20.13 to assist you in identifying the skeletal elements.

Unlike most vertebrates that have rigid skulls, snake skulls have movable joints in the upper and lower jaw bones, the quadrate bones, and even the front of the braincase. Likewise, the two mandibles of the lower jaw are not fused together at the front of the mouth, allowing them to spread apart. These features enable snakes to open their mouths extremely wide to swallow prey that often is several times larger than their body diameter. The body of the snake has been lengthened by the addition of extra vertebrae to the spinal column, with anywhere from 200 to 400 vertebrae being common in many species (Fig. 20.13).

Each vertebra has a pair of curved ribs attached to it, and the vertebrae articulate with each other at no fewer than 10 surfaces, creating a strong, supple column. Ventral ribs are absent in snakes. And although forelimbs and hind limbs are absent in all modern snake species, Old World snakes such as boas and pythons retain vestiges of their tetrapod ancestry: Vestigial pelvic girdles are still present in many of these snakes, hidden beneath the skin. Males of these species have a pair of small, pointed spurs that protrude through the body on either side of the cloaca, the only externally visible remnants of the hind limbs that their ancestors once had.

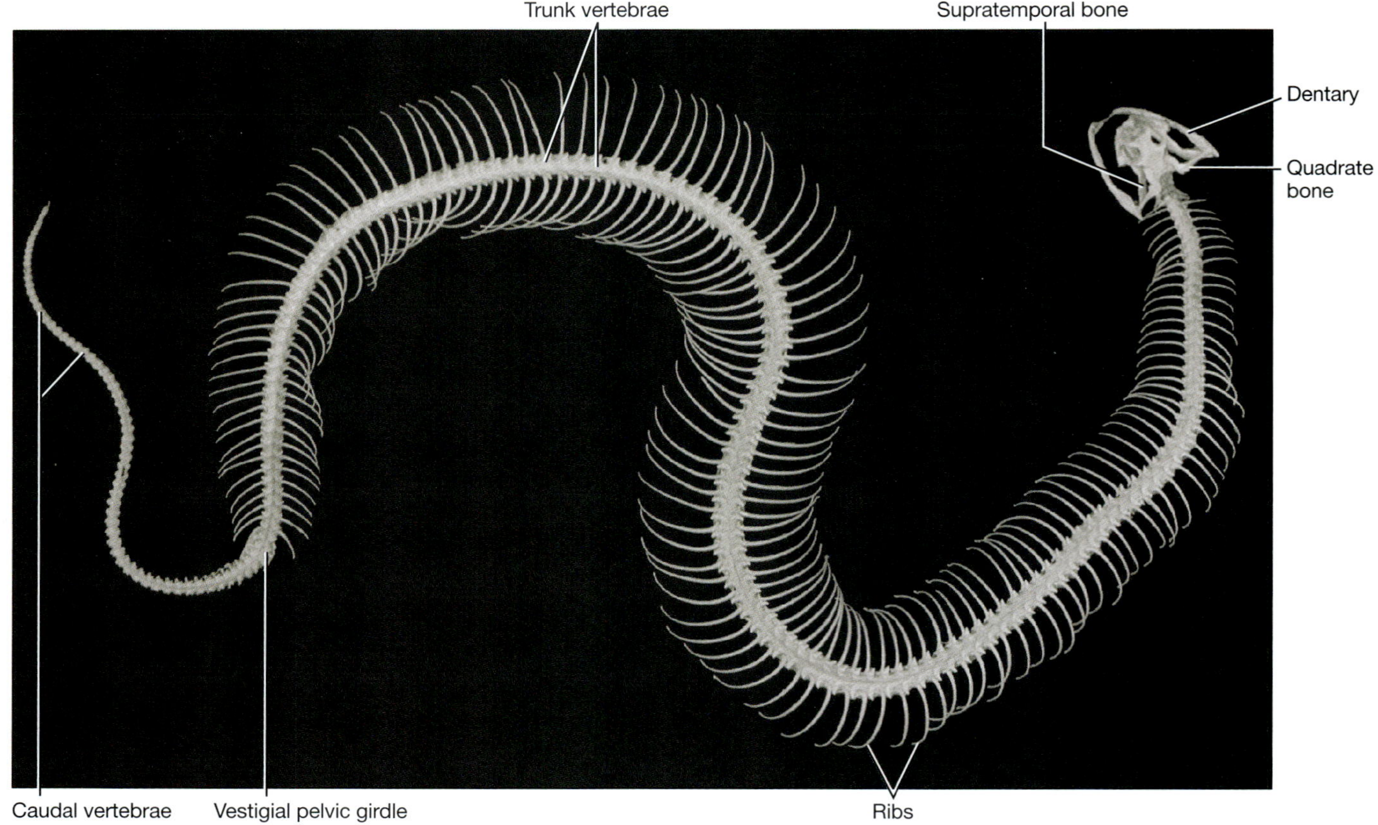

20.13 Snake (python) skeleton, dorsal view.

Check Your Progress

5.1 Which of the following senses do you think a rattlesnake (a pit viper) would rely on most to detect its prey? (Check all that apply).

_______ sight

_______ smell

_______ sound

_______ temperature

5.2 What anatomical modifications does the snake's skull have that permit swallowing large prey?

____

20

5.3 Did the ancestors of snakes have legs? What evidence in modern snakes supports your answer?

____

____

Digestive System

Procedure

1 To open your specimen and view the internal anatomy, review the Dissection Diagram for suggested cuts to make.

2 Use dissecting scissors to make parallel, longitudinal incisions along each side of the ventral surface of the body, starting at the base of the head and progressing posteriorly along the entire length of the body, stopping just anterior to the cloacal opening.

Dissection Diagram
Using scissors make parallel incisions at the base of the neck on the ventral surface of your specimen and extend these incisions along the lateral edges of the ventral body surface to the cloacal opening.

3 Make transverse incisions across the body at the ends of these incisions to connect them and carefully remove the skin from the ventral surface of your specimen.

4 Use Table 20-2 and Figures 20.14–20.19 to help you identify the internal organs in the snake.

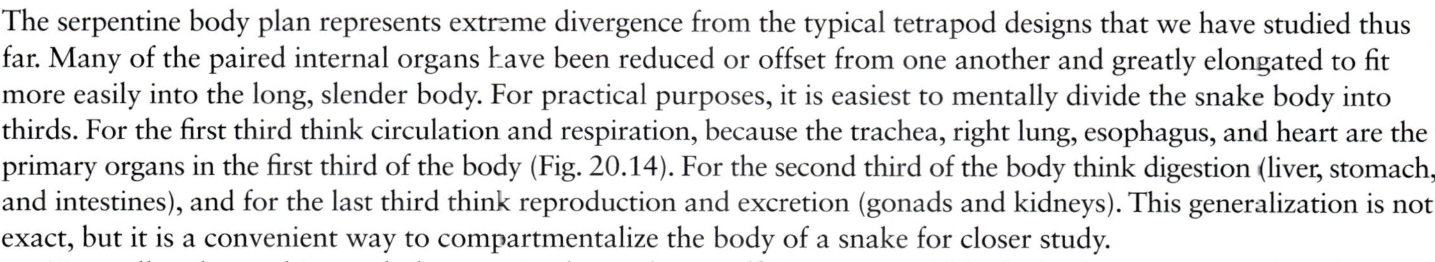

The serpentine body plan represents extreme divergence from the typical tetrapod designs that we have studied thus far. Many of the paired internal organs have been reduced or offset from one another and greatly elongated to fit more easily into the long, slender body. For practical purposes, it is easiest to mentally divide the snake body into thirds. For the first third think circulation and respiration, because the trachea, right lung, esophagus, and heart are the primary organs in the first third of the body (Fig. 20.14). For the second third of the body think digestion (liver, stomach, and intestines), and for the last third think reproduction and excretion (gonads and kidneys). This generalization is not exact, but it is a convenient way to compartmentalize the body of a snake for closer study.

To swallow large objects whole, an animal must have sufficient space within the body to accommodate the prey, so it should be no surprise that snakes have an incredibly long and highly expandable esophagus. The esophagus leads to the stomach, where digestion begins (Figs. 20.14–20.15). When a snake swallows extremely large prey, the first section of the prey swallowed (usually the head) reaches the stomach and the digestive process begins while the snake is slowly swallowing the remainder of the prey. Often the snake is prevented from completely swallowing its prey until a significant portion of its victim has been digested to create enough room in the esophagus for the remainder of the prey to be swallowed.

From the stomach on, the digestive tract of snakes follows the typical reptilian format. As in turtles, a large liver produces bile, detoxifies many constituents of the absorbed digested components, and converts glucose to glycogen for storage. A gallbladder and pancreas also are present in snakes, found near the posterior tip of the liver. The long, coiled **intestine** occupies much of the second third of the body and is the primary site for nutrient absorption into the bloodstream. Food then passes into the **large intestine**, which is responsible for reabsorption of water and electrolytes. Indigestible wastes pass into the coprodeum and are expelled from the body through the cloaca.

Table 20-2 Internal Anatomy of the Snake

Structure	Function
Trachea	Conducts air to and from the lung during respiration
Heart	Three-chambered organ (two atria, one ventricle) that receives oxygenated blood from the lung and pumps it via the arteries throughout the body
Lung	Single, highly vascular organ for respiration; the right lung is functional while the left lung is reduced to a vestigial remnant
Liver	Produces bile, converts glucose to glycogen for storage, detoxifies many constituents of the absorbed digested compounds
Gallbladder	Stores bile produced by the liver
Stomach	Site of food storage and initiation of digestion
Spleen	Stores blood, recycles worn-out red blood cells, produces lymphocytes
Pancreas	Produces digestive enzymes and delivers them through the pancreatic duct to the duodenum; also produces a suite of hormones for the endocrine system
Large intestine	Responsible for reabsorption of water and electrolytes; transports feces to coprodeum via peristalsis
Testes	Produce sperm
Vasa deferentia	Transport sperm to cloaca
Ovaries	Produce eggs
Oviducts	Capture eggs and transport them to the shell gland (uterus)
Kidneys	Filter blood (creating urine) and are responsible for osmoregulation
Ureters	Transport urine to cloaca
Cloaca	Common chamber for the release of urine, feces, and gametes

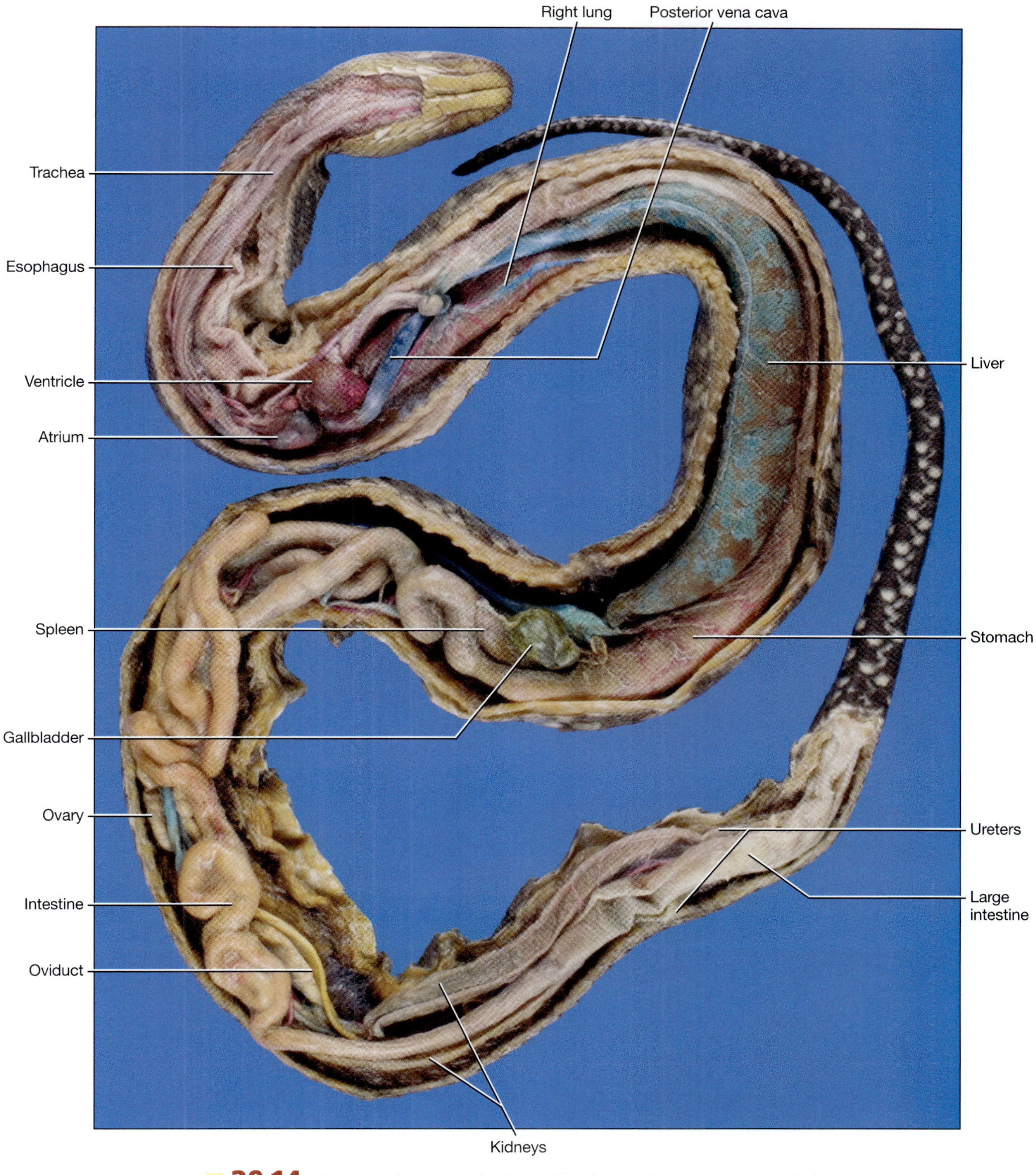

20.14 Water snake, *Nerodia rhombifer*, internal anatomy, ventral view.

20

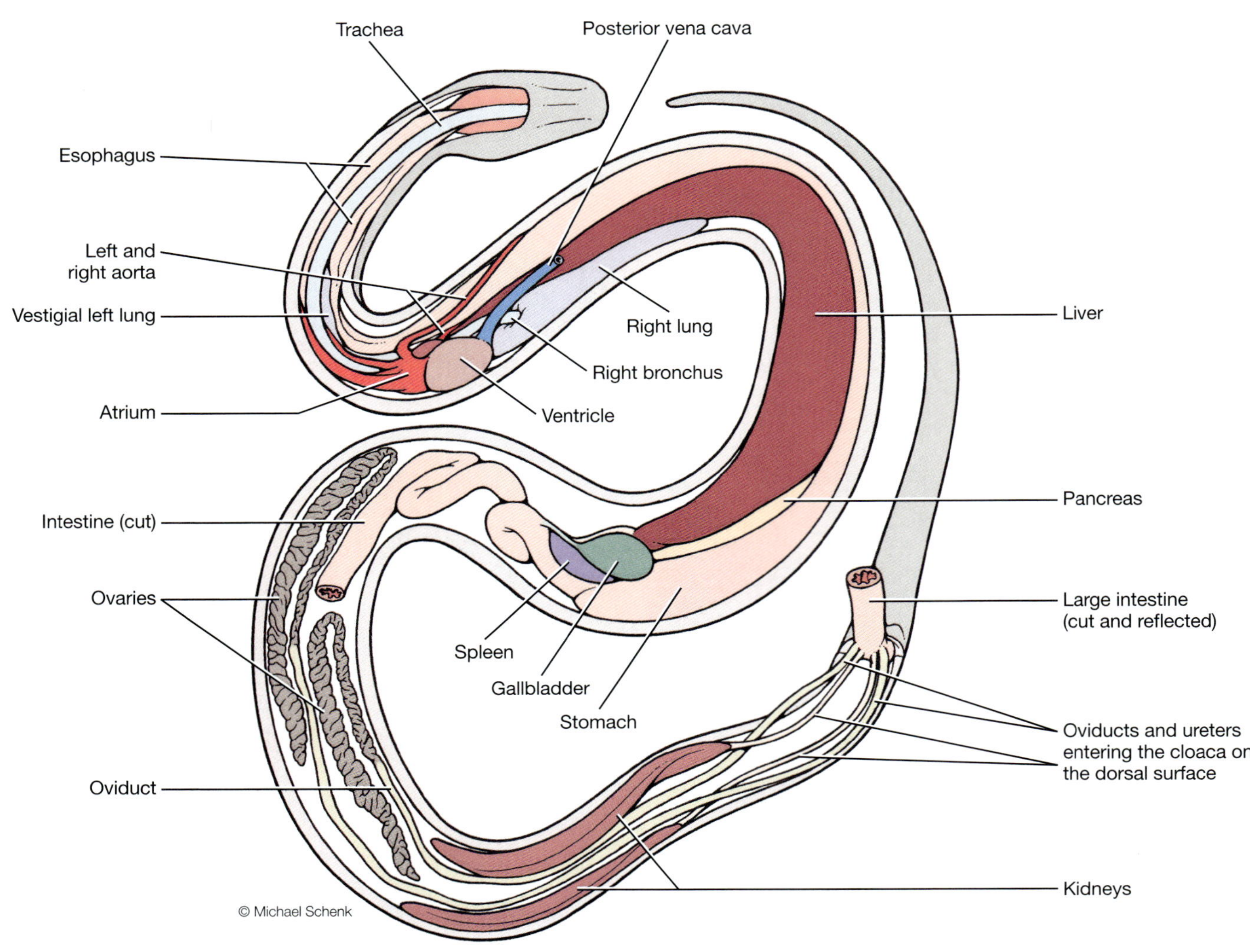

20.15 Snake, generalized anatomy.

Circulatory and Respiratory Systems

The respiratory system of snakes is unique among vertebrates. The long trachea contains cartilaginous rings like those present in most terrestrial vertebrates, but the rings are incomplete and do not entirely surround the trachea (Fig. 20.16). This permits the trachea to collapse somewhat to allow large food items to pass through the expandable esophagus more easily. Snakes have only one functional lung, the right lung (Fig. 20.16). In most species the left bronchus terminates in a vestigial, nonfunctional lung. In some water snakes this nonfunctional lung is modified into an air sac that enhances their buoyancy. Like most reptiles, snakes lack a diaphragm and rely on contractions of intercostal muscles between the ribs to ventilate the lung.

Circulation follows the typical pattern seen in reptiles, with a three-chambered heart powering a double-circuit system that delivers blood first to the lung for oxygenation through a pulmonary circuit and then through a systemic circuit to the body tissues (Fig. 20.17). The spleen is attached superficially by connective tissue to the gallbladder and pancreas and functions to filter the blood and recycle old red blood cells.

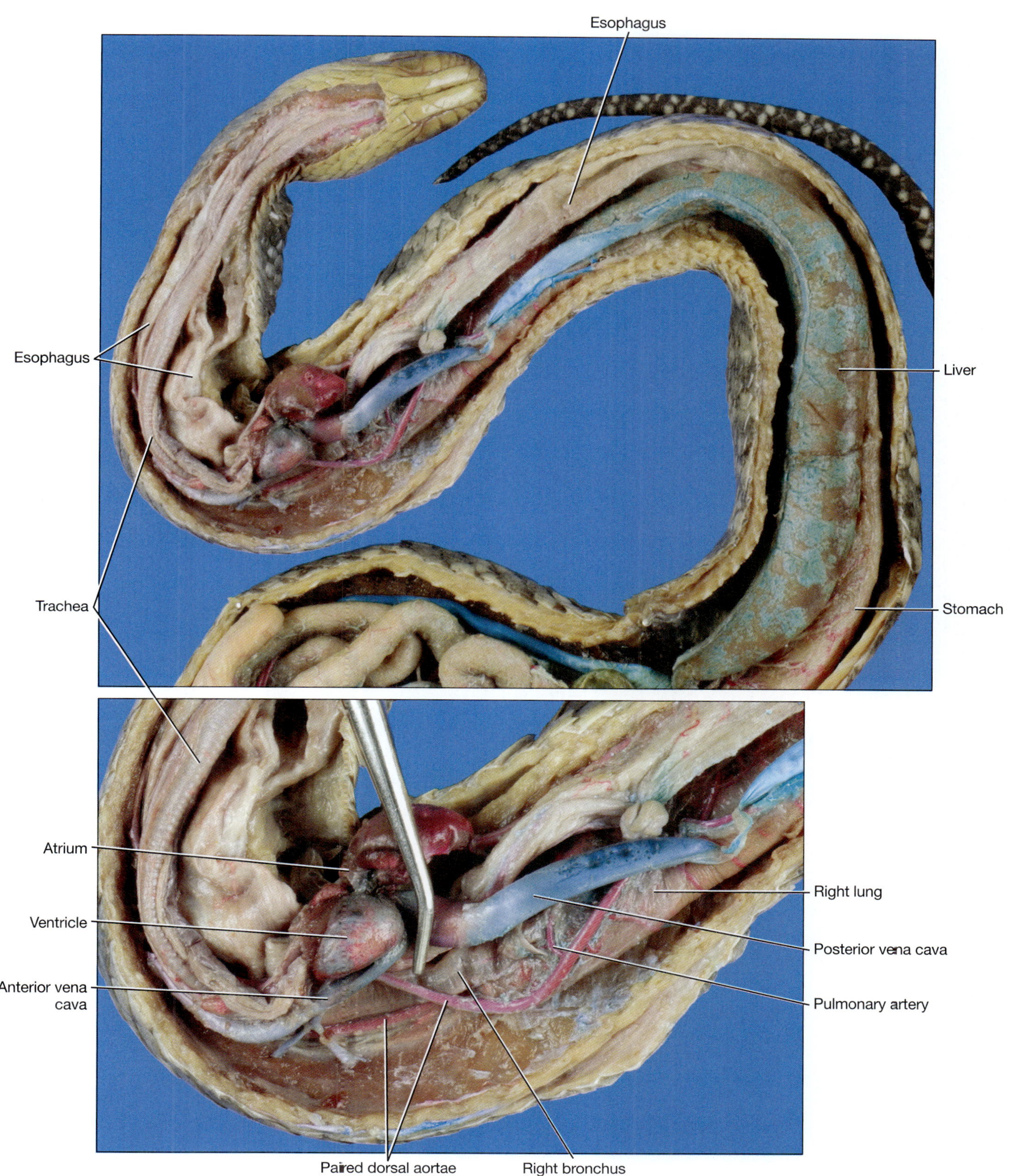

20.16 Water snake, *Nerodia rhombifer*, anterior half of body, ventral view.

20

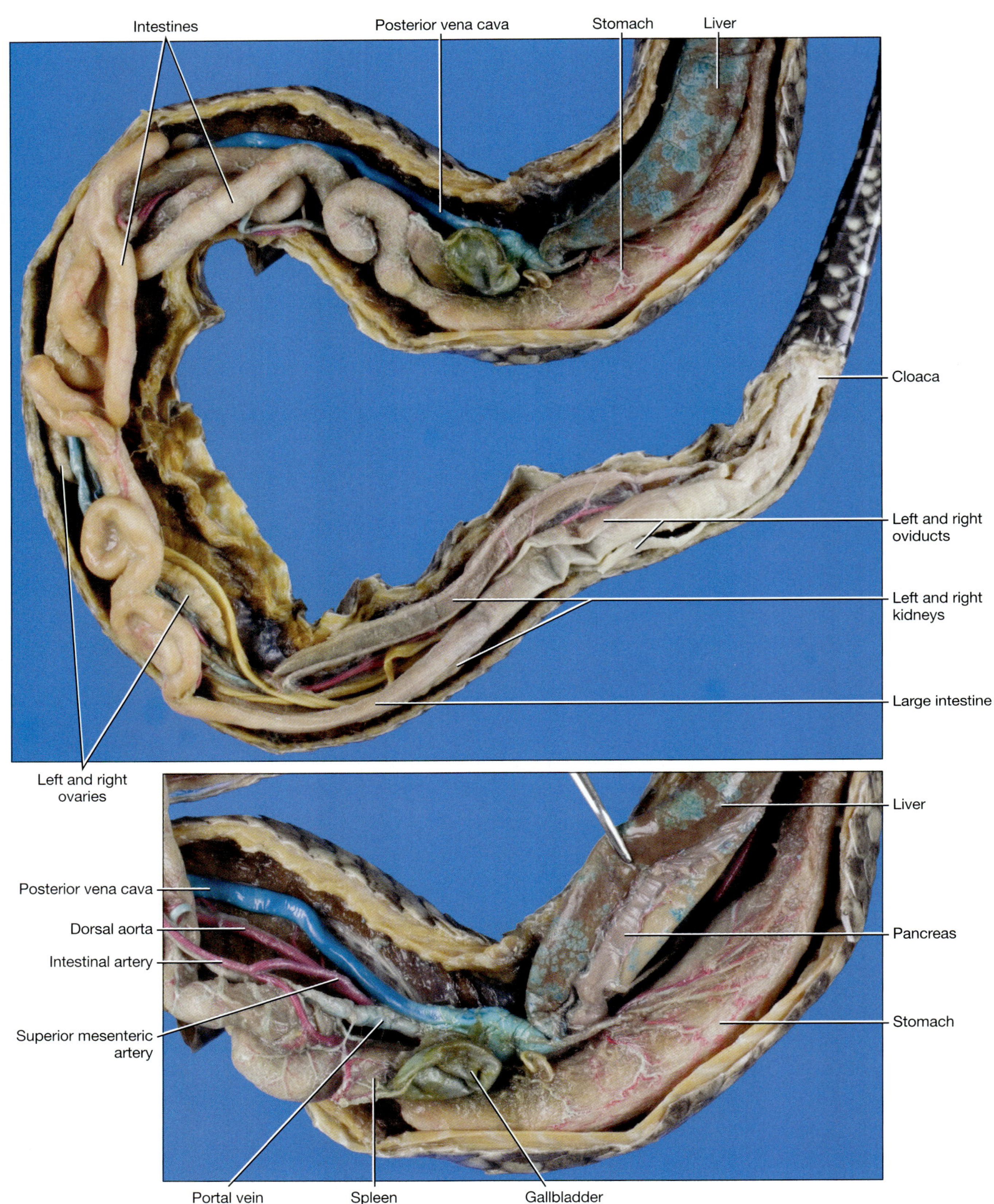

20

20.17 Water snake, *Nerodia rhombifer*, posterior half of body, ventral view.

Urogenital System

Snakes lack a urinary bladder; thus, urine storage is minimal and is confined to the small volume of fluid that the paired ureters can hold. As in turtles and lizards, urine flows from the ureters into the urodeum and out of the body through the cloaca. Paired gonads are located along the margins of the kidneys and generally are offset from one another, with the right gonad situated closer to the head than the left gonad. Females possess elongated ovaries that produce eggs, which are transported to the oviducts (Fig. 20.18). Fertilization in snakes is internal, and the fertilized eggs are shelled and deposited in the soil beneath rocks, logs, or leaf litter. Males have paired testes that lie anterior to the kidneys and transport sperm along the **vasa deferentia** (singular = vas deferens) (Fig. 20.19).

Male snakes and lizards possess paired copulatory organs called **hemipenes**. Each hemipenis is attached directly to a vas deferens and receives sperm only from the testis on that side of the body. During mating, only a single hemipenis is used, but snakes (and lizards) often alternate between the left and right hemipenis when successive matings occur within a short time interval. This behavior maximizes the number of viable sperm that can be released with each mating attempt.

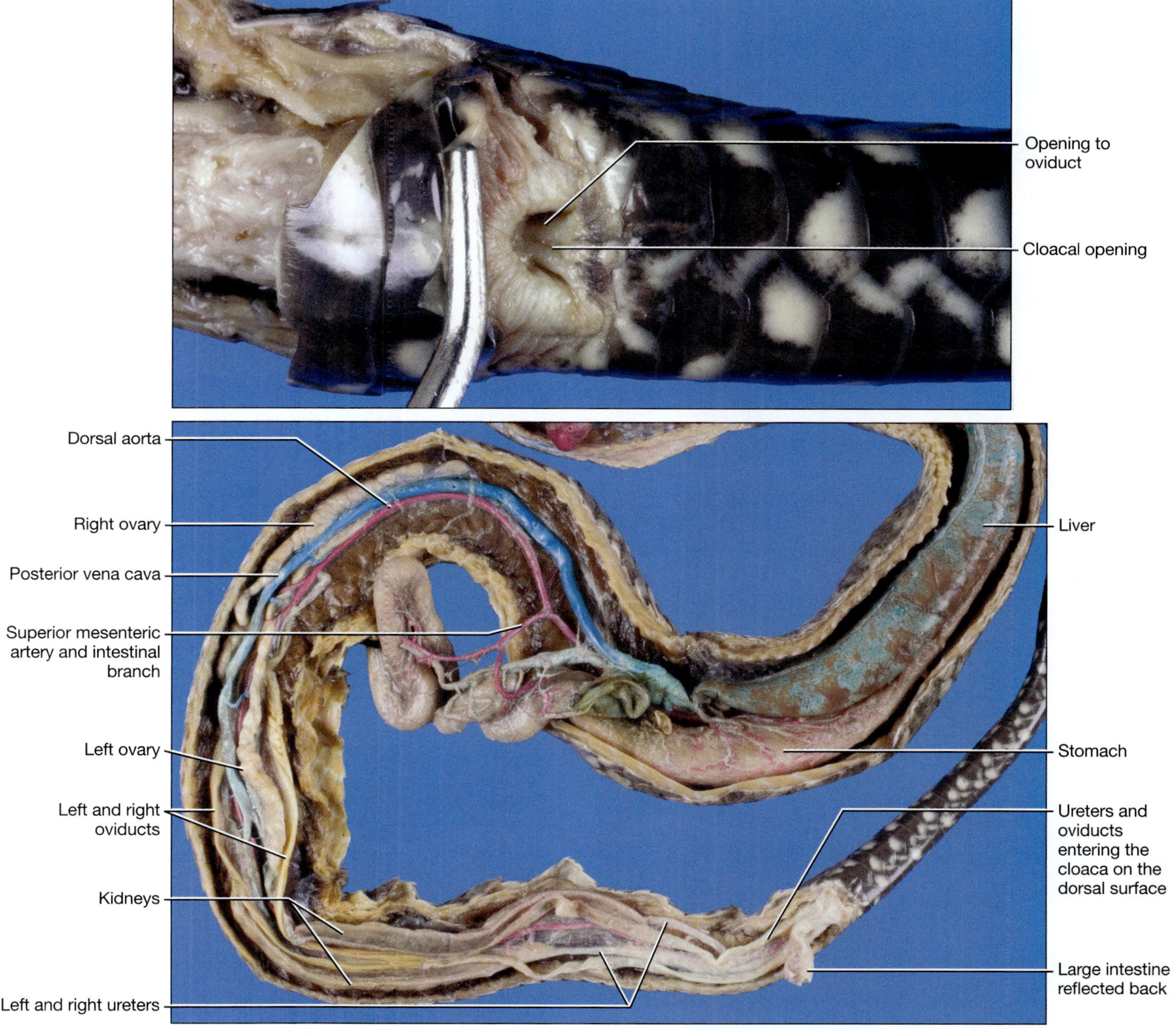

20.18 Female water snake, *Nerodia rhombifer*, urogenital system.

20

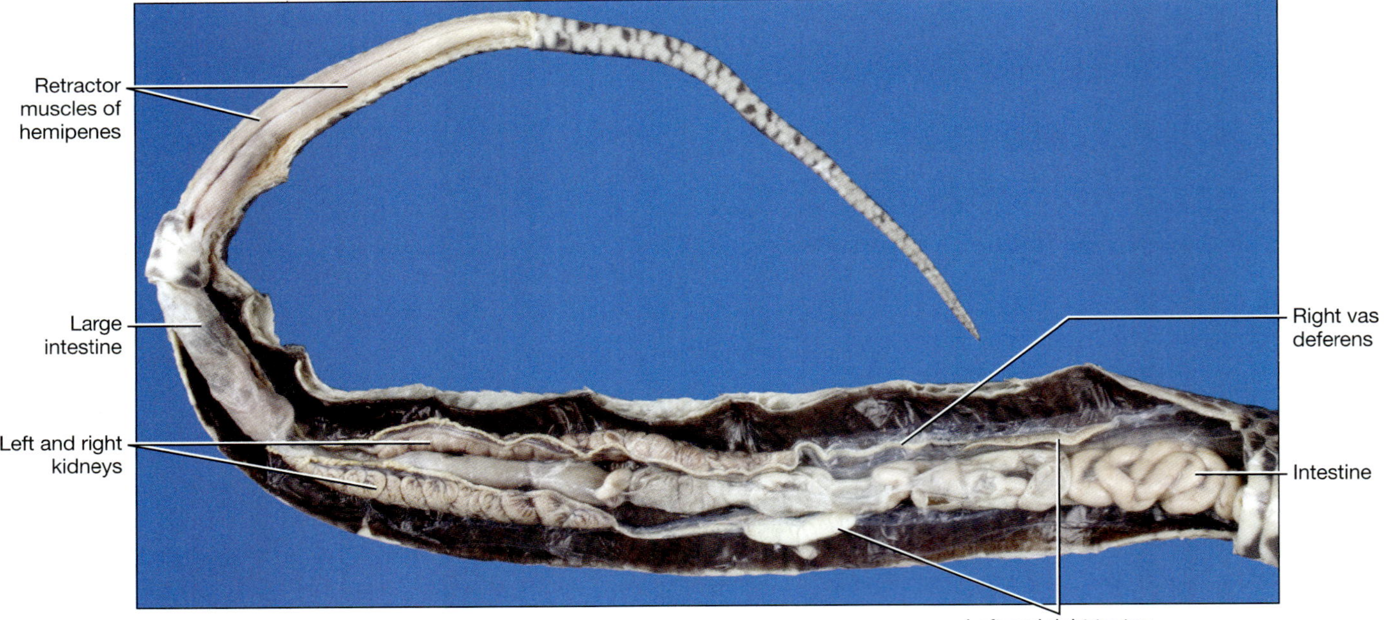

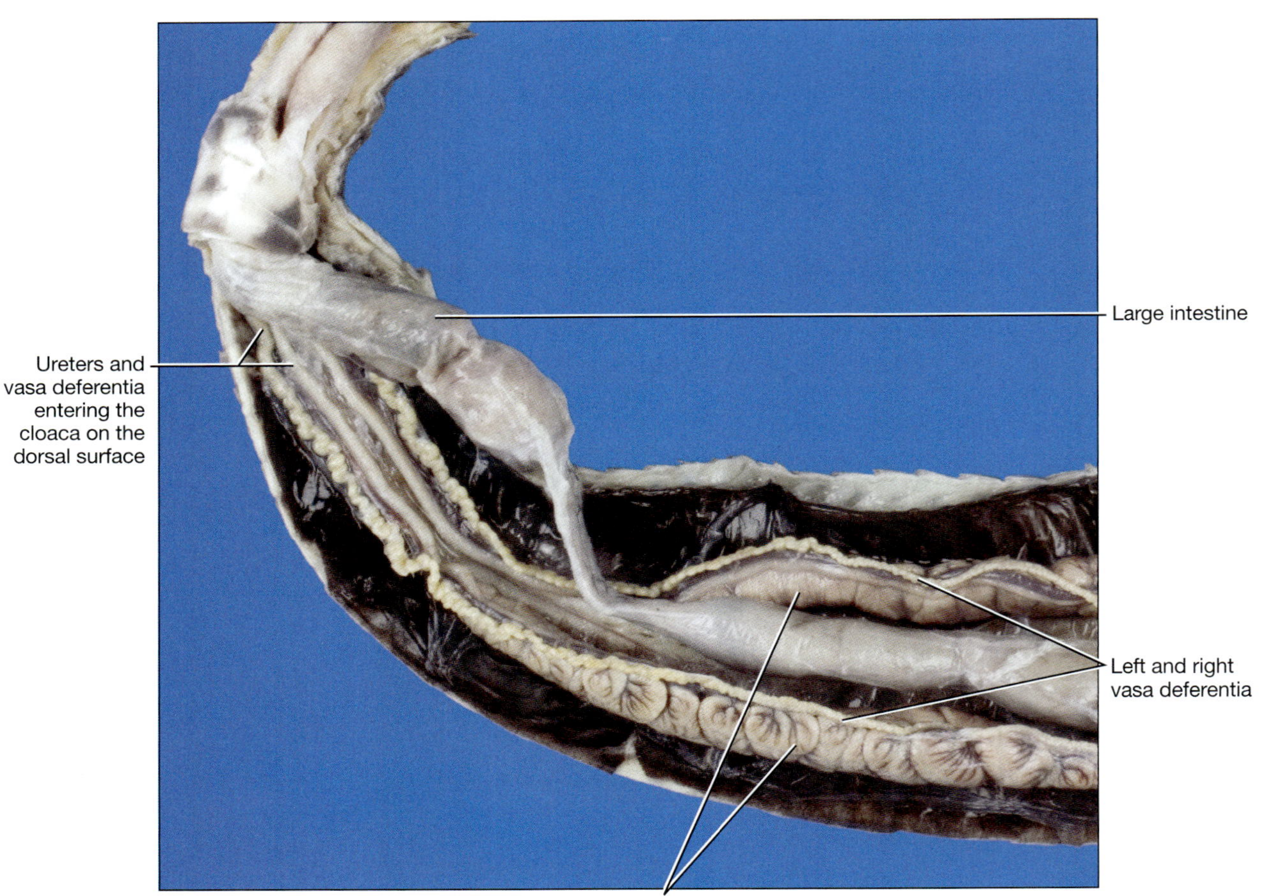

20

20.19 Male water snake, *Nerodia rhombifer*, urogenital system.

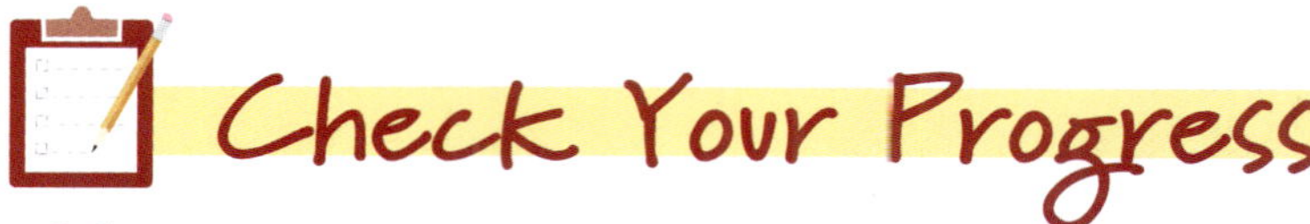

6.1 How many functional lungs do snakes possess?

6.2 Do you suppose the ancestors of modern snakes had more than this number of functional lungs? What evidence supports your answer?

6.3 What anatomical anomaly explains why snakes often urinate if you hold them vertically?

6.4 How many copulatory organs does a male snake possess? How does this compare to turtles? To lizards?

Exercise 20-3 Alligators and Crocodiles (Order Crocodilia)

Alligators and crocodiles are true living relics, having changed very little over the last 200 million years. Though they once existed in far greater numbers, only 23 species are living today, and more than half of these species are listed as threatened because of exploitation and destruction of their natural habitat by humans. Crocodilians live in equatorial and temperate regions worldwide, and adults range in size from 5 to 25 feet in length. Alligators, such as the American alligator, can be distinguished from true crocodiles by the fourth mandibular tooth, which fits into the upper jaw and cannot be seen when the jaw is closed (Fig. 20.20). In crocodiles this tooth protrudes upward from the lower jaw and is visible when the mouth is closed.

20.20 Modern alligators, such as the American alligator, *Alligator mississippiensis*, are direct descendants of dinosaurs and share many anatomical similarities with birds, their closest living relatives.

Alligators and crocodiles are unlike any other reptile group. They are classified as reptiles in the order Crocodilia, but they resemble birds far more than they do other living reptiles, and they are modern birds' closest living relatives (Fig. 20.2). Birds and crocodilians both are direct descendants of dinosaurs and have many features in common. Female crocodilians build mound nests and show a high degree of parental care for their eggs during incubation and for their young after hatching. They guard their nests fiercely, and mothers often pick up partially hatched eggs and gently crush them in their mouths to free the young. In some species, females transport newly hatched offspring in their mouths to specially excavated pools—nursery areas in the quiet backwaters of swamps—where they will protect the young for several months. In the American alligator, parental care may last as long as one to two years.

Alligators and crocodiles are the only living reptiles to communicate through vocalizations. Males bellow loudly during the breeding season, and the young produce high-pitched grunts shortly after hatching, which trigger the mother to excavate the nest to uncover her newly hatched offspring. Like birds, they have a true cerebral cortex, giving them the capacity for more learned behaviors as opposed to purely instinctive behaviors, and they have a four-chambered heart.

Unlike other reptiles, crocodilians have a diaphragm. Though it performs the same function, it differs from the type of diaphragm found in mammals. It is not muscular and is attached to the liver instead. Backward movement of the liver and diaphragm serve to ventilate the lungs. Crocodilians have a complete secondary palate that closes the glottis and allows them to feed while submerged without the risk of swallowing water and drowning. Much like their dinosaur ancestors, their legs are positioned underneath the body, giving them a "higher gait" than modern lizards and bringing the belly off the ground when walking and running.

An interesting problem that crocodiles have overcome involves their circulatory pathway. Other reptiles have a three-chambered heart with an incomplete septum that permits controlled mixing of pulmonary and systemic blood, as well as two aortae; the left one routes blood to the abdominal organs and the right one sends blood to the head, forelimbs, and hind limbs (Fig. 20.21B). This allows partially oxygenated blood to be shunted preferentially to the abdominal organs while oxygen-rich blood is sent to the head and limbs—keeping these crucial areas of the body well-supplied during long underwater dives.

The four-chambered heart of birds and mammals is advantageous for endothermic animals that must maintain high metabolic rates and correspondingly high blood pressures. Complete separation of the pulmonary and systemic circuits in this system allows for different pressures to be generated on each side of the heart—low pressure to send blood to the lungs across delicate respiratory membranes, and high pressure to route blood to the rest of the body tissues and organs (Fig. 20.21A). But crocodiles inherited a four-chambered heart from their dinosaur ancestors, as did birds. Yet, modern crocodiles are ectothermic and benefit from being able to spend long periods of time underwater, waiting to ambush their prey.

Over time, these selective pressures have modified the circulatory pathway of alligators and crocodiles. In a sense, they re-evolved a shunting system similar to that seen in other modern reptiles (Fig. 20.21C). Though their ventricles remain divided by a complete septum, crocodiles are able to mix pulmonary and systemic blood through a small opening between the right aorta and left aorta, giving them the same physiological flexibility that other aquatic reptiles have and demonstrating how natural selection must work within the confines of inherited traits to produce adaptations that better equip animals to their environments.

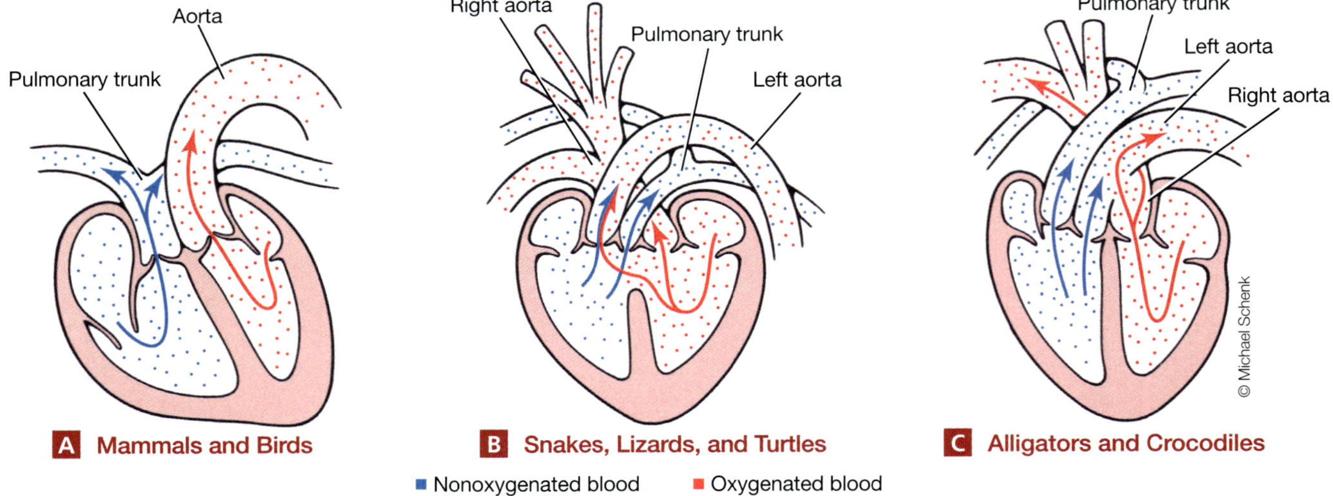

20.21 (**A**) Mammals and birds independently evolved four-chambered hearts that completely separate oxygenated blood from deoxygenated blood. (**B**) Modern snakes, lizards, and turtles have a three-chambered heart with an incomplete septum that allows controlled mixing of pulmonary and systemic blood. (**C**) Crocodiles inherited the four-chambered heart from dinosaurs, as did birds, but re-evolved a shunting system to allow controlled mixing of pulmonary and systemic blood.

Name ________________________ Section ____________

20 Questions for Review

1 Name four characteristic features of reptiles that have contributed directly to their success in terrestrial habitats.

1. ____________________

2. ____________________

3. ____________________

4. ____________________

2 Define ectothermic.

____

3 Match the type of skull with each group of animals that possesses it. (Some letters will be used more than once.)

______ turtles	**a.** synapsid
______ birds	**b.** diapsid
______ mammals	**c.** anapsid
______ lizards	
______ snakes	
______ crocodiles	

4 Explain how the reptile heart (turtle) differs from the heart of an amphibian (frog).

____

____

____

5 Describe five ways in which alligators and crocodiles differ from other reptiles we have studied.

1. ____________________

2. ____________________

3. ____________________

4. ____________________

5. ____________________

6 Contrast the breathing mechanisms in turtles, snakes, and crocodiles.

7 Fill in Table 20-3 with the appropriate characteristics for each group of reptiles, to compare the adaptations to different body shapes and lifestyles between turtles and snakes.

Table 20-3 Comparison of Anatomical Characteristics between Turtles and Snakes

	Turtle	Snake
Number of limbs		
Eyelids (present or absent)		
Teeth (present or absent)		
Number of functional lungs		
Tracheal rings (complete or partial)		
Diaphragm (present or absent)		
Urinary bladder (present or absent)		
Location of urine storage		
Number of copulatory organs in male		

20

Name ______________________ Section ______________

20 Laboratory Practical Challenge

The following evaluation is intended to replicate the method of assessment and type of questions you will be asked on lab practicals. If you have difficulty with these questions, you may benefit from additional review of the specimens provided in lab.

1 Identify the phylum, subphylum, and class of the specimen depicted below.

____

2 What is the function of the structure indicated by line A?

____

3 Identify the structure indicated by line B.

____

4 Identify the structure indicated by line C.

____

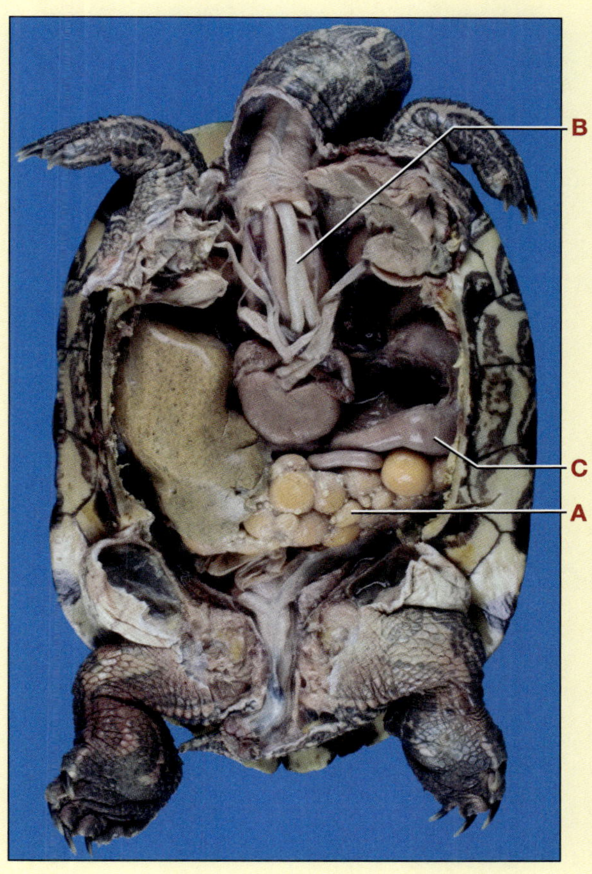

20

5 What is the function of the structure indicated by line D?

6 Identify the structure indicated by line E.

7 To what system do the structures indicated by line F belong?

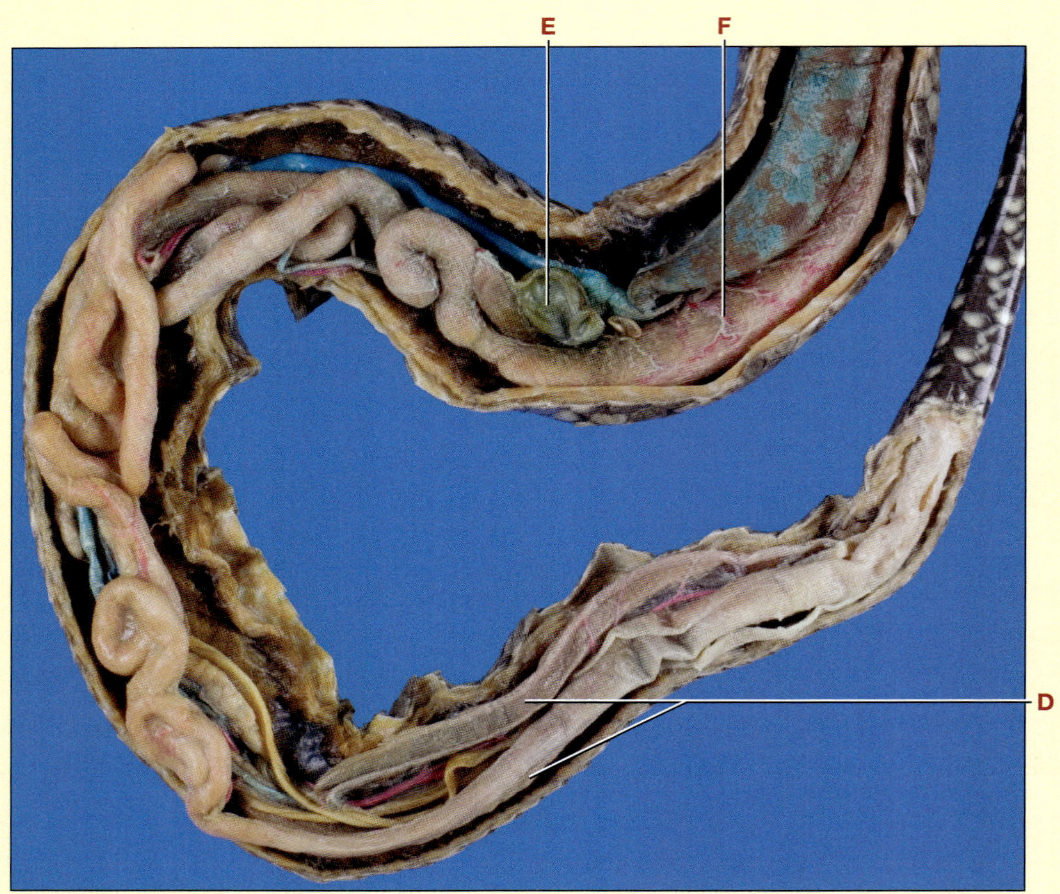

20

CHAPTER 21

Birds

After completing the exercises in this chapter, you should be able to:

1 List the distinguishing characteristics of the class Aves.

2 Understand the evolutionary position of birds on the phylogeny of chordates.

3 Discuss the significance of feathers in the evolution of homeothermic vertebrates.

4 Identify the major skeletal features of the pigeon and compare them to homologous bones in other vertebrates.

5 Identify the major internal organs in the pigeon, discuss their functions in the body, and relate them to homologous organs in other vertebrates.

6 Understand the basic structure of feathers.

Birds are the most numerous of all *terrestrial* vertebrate groups. The class Aves contains nearly 10,000 different species, ranging from tiny hummingbirds that weigh less than 2 grams to ostriches that weigh more than 120 kg, and wandering albatrosses with 2-meter wingspans (Fig. 21.1). Birds and mammals are the only living groups of **endothermic** animals, which means that they generate most of their body heat from within. And because their internal core body temperatures do not fluctuate much from their set points, birds and mammals often are referred to as **homeothermic** organisms.

The first birds diverged from the main reptilian line around 150 million years ago. Yet, in that relatively brief time span, they have diversified and adapted to fill many niches that reptiles and mammals were not equipped to exploit.

As birds evolved, the demands of flight shaped their anatomy and selected for characteristics that were both light and streamlined. The hallmark of birds is their magnificent feathers, which provide protection, insulation, and a lightweight, streamlined surface area for flight. Their hollow bones reduce weight and show marked fusion, especially in areas that must bear the forces generated during flight and landing. Birds lack urinary bladders and, instead of storing watery urine for prolonged periods of time, they concentrate nitrogenous wastes into a dry, solid compound (**uric acid**) to avoid carrying additional water weight. The paired ovaries and oviducts, characteristic of most female vertebrates, have been reduced to only one of each in birds, saving weight and freeing up coelomic space, and they remain small except during the breeding season.

To maintain the higher metabolic rates required for endothermy and energetically expensive aerobic endeavors such as flight, birds evolved an extremely efficient combination of circulatory and respiratory systems. They have a four-chambered heart containing separate pulmonary and systemic circuits (much like the hearts of crocodiles and mammals), and their internal organs are intermeshed with air sacs connected to the lungs, permitting a unidirectional flow of air through their lungs. This combination of features allows more oxygen to dissolve into the blood and to be routed more efficiently to the major flight muscles.

Navigation during flight requires excellent vision, and birds generally have large, well-developed **eyes** and correspondingly large visual centers in the brain. External ear openings are present but are hidden beneath the plumage of the head. The narrow jaws form a lightweight beak that lacks teeth and superficially resembles the beaks of many turtles, though beak shapes have taken on a variety of shapes and sizes suited to different diets and behaviors over the course of avian evolution.

Evolutionary History

Although today we would have no difficulty distinguishing a bird and a reptile, this would not have been so obvious in the past, because these two groups share a common

Great blue heron, *Ardea herodias*

Magellanic penguins, *Spheniscus magellanicus*

American robin, *Turdus migratorius*

21.1 Birds such as herons, penguins, and robins all belong to the class Aves.

evolutionary lineage. In fact, the class Aves, as it is historically known, is contained within the evolutionary line of all reptiles—making modern birds a highly specialized group of reptiles (actually living descendants of dinosaurs) (Fig. 21.2).

The evolutionary success of birds is probably attributed most appropriately to their development of feathers—modified scales inherited from their reptilian ancestors—both because of their ability to provide lift for flight and for their insulating properties. If damaged, feathers can be replaced easily, unlike the fleshy wings of bats or the membranous wings of insects. Fossil evidence seems to suggest that feathers evolved first as a means of insulation and were utilized by much later groups for the purposes of flight.

The first true birds appear in the fossil record in rock dating back 147 mya. The famous fossil of *Archaeopteryx*, first found in a Bavarian limestone quarry in 1861, set the world of paleontology ablaze as the first "dinosaur" fossil to have imprints of feathers. *Archaeopteryx* differed from modern birds in many ways. It had teeth, a long bony tail, solid bones, no enlarged breastbone, and dinosaur-like forelimbs. Yet, its well-developed feathers and ability to fly, or at least glide, were unmistakably birdlike (Fig. 21.3).

During the Cretaceous period (146–65 mya), a diverse array of birds evolved and shared the landscape with the other dinosaurs of that time. Because feathers and the delicate, hollow bones of birds do not fossilize particularly well, our understanding of the evolutionary relationships among these early birds is still incomplete. Comparisons of the DNA among living birds, however, have allowed us to piece together an idea of the evolutionary relationships among modern birds. The large, flightless birds such as the ostrich, kiwi, and emu appear to be the most ancient and dinosaur-like of modern birds. Waterfowl, such as ducks, geese, and swans of the order Anseriformes, probably evolved next during the early Cretaceous, followed by owls, woodpeckers, parrots, and swifts. The songbirds—representing the largest order of birds (Passeriformes)—most likely evolved around 100 mya, but the most specialized orders of birds, such as shorebirds, birds of prey, flamingos, and penguins, did not appear until late in the Cretaceous period. Today, biologists recognize 30 orders of living birds with 166 families distributed among these orders. For a more complete phylogeny depicting the evolutionary relationships among the chordates covered in the remainder of this laboratory guide, review Figure A.2 in the Appendix.

21.2 The first true birds appear approximately 150 million years ago in the fossil record. Their feathers are one of the defining characteristics that separate modern birds from crocodiles and other vertebrates.

21.3 Fossil of *Archaeopteryx*, a crow-sized flying reptile with feathers that lived during the late Jurassic period (195–135 mya). This specimen, preserved in remarkable detail in limestone from Bavaria, Germany, provides direct evidence that birds evolved from dinosaurs.

21

Exercise 21-1 Birds (Class Aves)

Skeletal System

Materials Needed

- ❑ mounted bird skeleton
- ❑ preserved pigeon
- ❑ dissecting tools
- ❑ dissecting pan
- ❑ latex gloves
- ❑ dissecting microscope

Procedure

1 Obtain a mounted, articulated skeleton of a pigeon.

2 Examine it carefully, using Figure 21.4 to assist you in identifying the skeletal elements present.

Many adaptations for flight are apparent in the skeleton of birds, the primary adaptation being hollow bones. Birds have extremely thin, lightweight bones, many of which bear pneumatic extensions from the lungs. Thus, a bird's skeleton constitutes a much smaller percentage of the animal's total body weight than does a mammal's skeleton. Notice the long, articulated vertebrae of the neck. They are organized to provide extreme mobility and flexibility of the head so it remains fixed in position during flight while the torso oscillates freely with each wing beat. A flexible neck also aids in feeding and preening.

In sharp contrast to the long neck, the vertebrae of the trunk are compressed into a strong, fused group of bones that serves as a solid attachment for the large muscles of the pectoral and pelvic regions. The **synsacrum** represents 13 posterior vertebrae fused to bear the animal's weight on its two legs, and the **pygostyle** represents several fused caudal vertebrae that support the tail feathers (Fig. 21.4). Along the midventral line of the torso, the sternum has a prominent **keel** for attachment of the large pectoral muscles that drive the wings during flight. Most of the bones of the legs and wings are homologous with limb bones in other vertebrates, though significant modifications of the basic tetrapod blueprint have been made during the course of avian evolution.

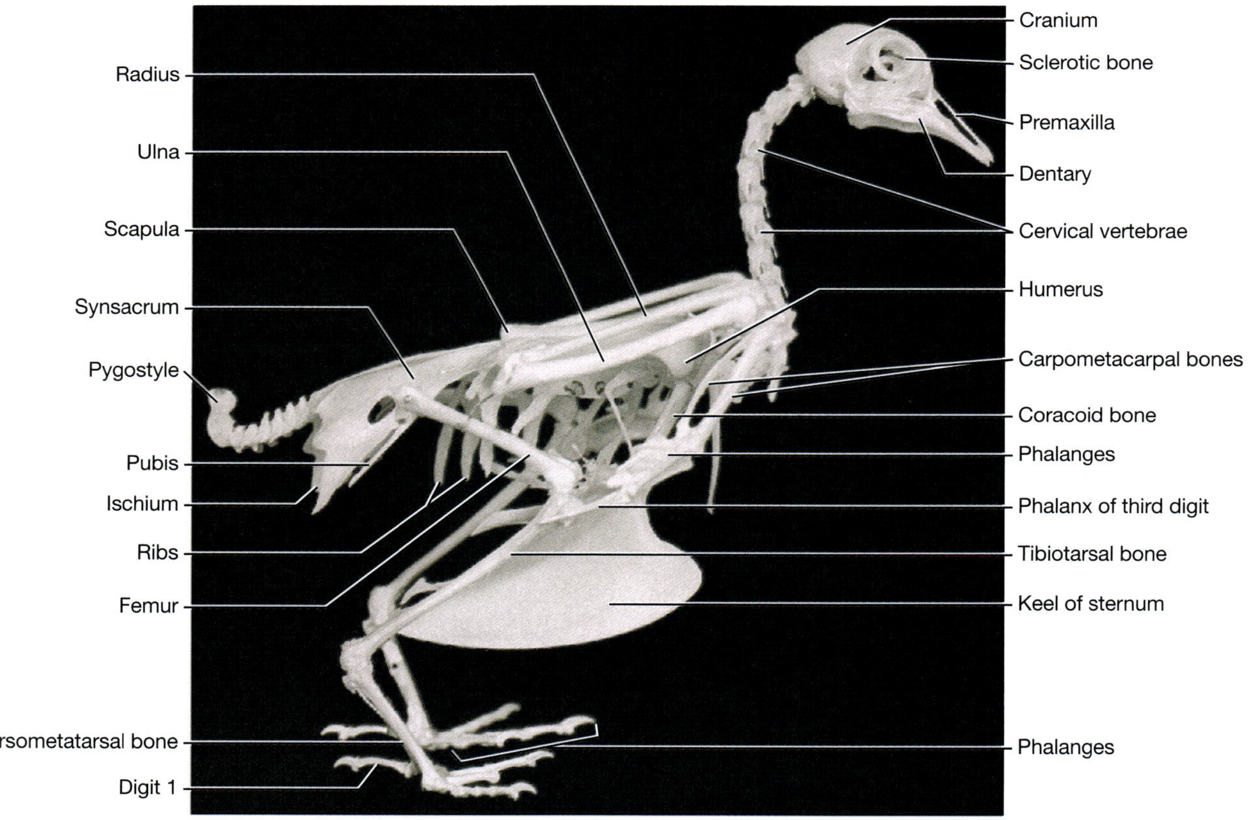

21.4 Pigeon, *Columba*, skeleton.

With respect to the legs, the fifth toe has been lost and the first toe is pointed posteriorly, acting as a "prop" on the ground and allowing birds to perch on tree limbs. The many bones of each ankle are fused into a single tarsometatarsal bone that acts as a shock absorber during landings. The location and articulation of each femur with the pelvic girdle helps to shift the weight of the animal forward, keeping its center of gravity over the feet. Each wing bears three short fingers and a fused carpometacarpal bone.

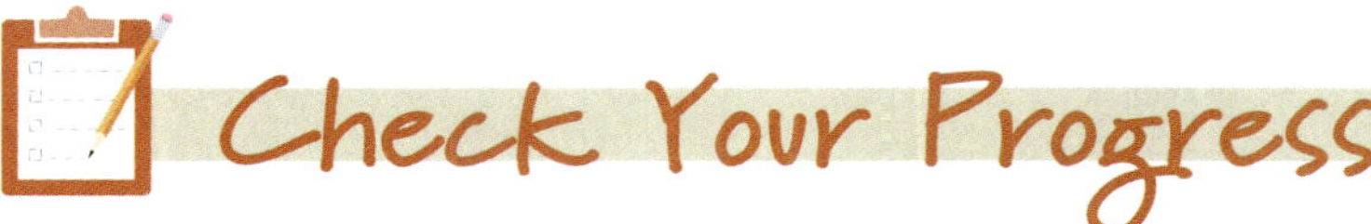

Check Your Progress

1.1 What common feature of bird bones has made flight possible?

1.2 List five bones in the pigeon that are directly homologous to bones in humans.

1.3 List five bones in the pigeon that do not have direct homologues in humans.

1.4 What are three adaptations of birds for flying?

Respiratory System

Note

Dissection of the air sacs should be performed before dissection of the visceral organ systems using specially prepared specimens with latex injection of the air sacs and spaces. Check with your instructor to see if a demonstration specimen is available with the respiratory system already dissected.

Procedure

1 Position your specimen on its back with its ventral surface facing you, and use scissors to make a longitudinal incision along the midventral line of your specimen from the base of the neck to the cloaca, as depicted in the Dissection Diagram.

2 Use a blunt probe to separate the skin and feathers from the underlying musculature of the breast and abdomen, being careful not to damage the crop, which lies just beneath the skin at the base of the neck.

© Michael Schenk

Dissection Diagram
Using scissors, make a longitudinal incision along the midventral line of your specimen from the base of the neck to the cloaca. Use a blunt probe to separate the skin and feathers from the underlying musculature of the breast and abdomen, being careful not to damage the crop, which lies just beneath the skin at the base of the neck.

21

3 Next, use a scalpel to cut transversely through the middle of the pectoral muscle and underlying muscles. Remove the pectoral musculature from the sternum.

4 Use scissors to cut through the keel of the sternum and separate the underlying membranes from the bone.

5 Make an incision along the midline of the abdominal wall to open the abdominal cavity.

6 Use scissors or bone cutters to cut through each rib near its articulation with the vertebrae. You may have to remove one of the wings completely to gain access to the lungs and air sacs.

7 Use Figures 21.5–21.8 and Table 21-1 to assist you in identifying the internal structures of your specimen.

Most birds possess external nares (nostrils) for breathing and inspired air travels through the larynx and then down the **trachea**. The **larynx** opens and closes the glottis during the breathing cycle and prevents foreign material from entering the lower respiratory tract. Like reptiles and mammals, birds employ a negative-pressure system to generate air flow, meaning they inhale air by suction. Unlike reptiles and mammals, birds have air sacs and a unique lung ventilation mechanism that permits one-way flow of air over their gas exchange surfaces, resulting in more efficient (and thus smaller) lungs than those of mammals of comparable body size. The air sacs constitute about 80% of the total volume of the respiratory system in birds and may completely surround the heart, liver, kidneys, testes, ovaries, and intestines (Fig. 21.5).

Although air flows into and out of the trachea and nostrils tidally, air flows through the lungs and air sacs in a nearly unidirectional path during both inhalation and exhalation. The breathing process requires two complete cycles of inhalation and exhalation for air to pass all the way through the respiratory system, but the benefits of this two-cycle process are twofold:

1. No dead air space exists in the avian respiratory system.
2. "Fresh" air is passing over some portion of the respiratory tract at all times—with every inhalation and exhalation.

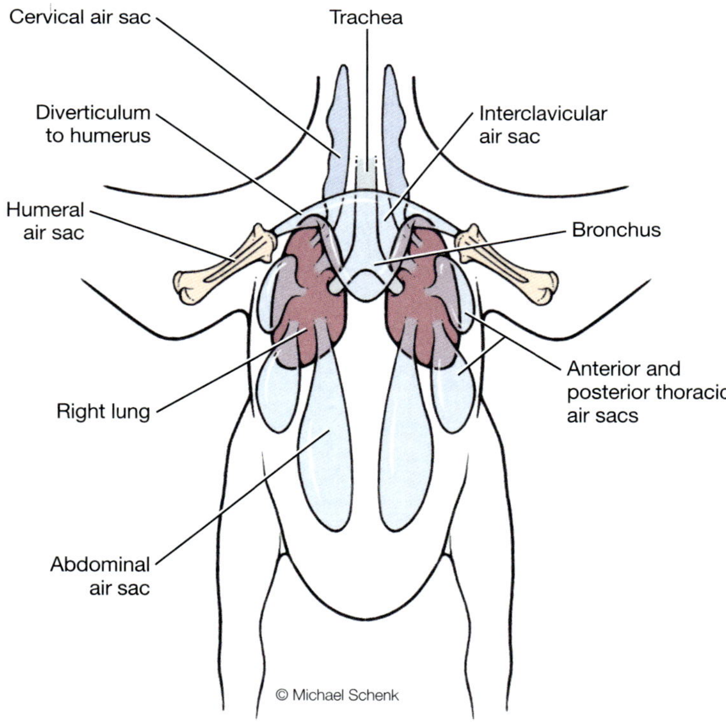

21.5 Avian respiratory system depicted to show the relationship of the thoracic and abdominal air sacs to the lungs and trachea.

21

When a bird inhales, air bypasses the lungs and travels first to a set of abdominal air sacs. Upon the first exhalation, the abdominal air sacs contract, pushing air into the lungs. With the second inhalation, air passes through the lungs and fills a set of thoracic air sacs; when the second exhalation occurs, the thoracic air sacs contract, sending the air out of the body. Of course this cycle happens continuously, so with each inhalation and exhalation, there are essentially two separate volumes of air (one from the current inhalation and one from the previous inhalation) that are moving through different portions of the respiratory system.

In mammals the trachea bifurcates more than 20 times, yielding more than a million tubules that terminate in the thin-walled alveoli where gas exchange occurs. In birds, however, the trachea divides only a few times, forming two main bronchi that lead to the lungs and then branch into only a few dozen secondary and tertiary bronchi within the lungs. Generally, each air sac has one connection to a primary or secondary bronchus, but it may have several indirect connections to tertiary bronchi. Gas exchange occurs along tiny air capillaries that emanate from each bronchus.

In mammals the entire lung expands and contracts with assistance from the diaphragm during each inhalation and exhalation. Birds lack a diaphragm and, thus, rely on movements of the ribs and sternum to bring in and expel air, much like lizards and snakes. Collectively, the unique anatomy and bizarre physiology of the avian respiratory system make it one of the most complicated and poorly understood respiratory systems among vertebrates.

Digestive System

Because birds lack teeth, mechanical digestion of food does not begin in the mouth; rather, food items are swallowed whole. The tongue lacks taste buds and is used primarily to manipulate food into the mouth and down the **esophagus** into the crop. The **crop** stores food and regulates the passage of food into the gizzard (Figs. 21.6–21.8). Food passes from the crop to the gizzard via a short tube known as the **proventriculus**. This digestive passageway mixes peptic enzymes with the food to begin the process of chemical digestion. The **gizzard** thoroughly pulverizes and churns the food before sending it into the intestinal tract. A **liver** and a **pancreas** are present and release their respective digestive compounds through multiple ducts into the **small intestine** where nutrients are absorbed into the bloodstream. The small intestine is partitioned into three distinct subregions: the duodenum, the jejunum, and the ileum.

A pair of vestigial **ceca** (singular = cecum) are present at the juncture of the ileum and large intestine, and the short length of **large intestine** finishes the digestive process by reabsorbing water, vitamins, and electrolytes into the bloodstream and transports feces to the coprodeum of the **cloaca**. As in reptiles, the cloaca is partitioned into three semi-distinct chambers: the **coprodeum**, the **urodeum**, and the **proctodeum**.

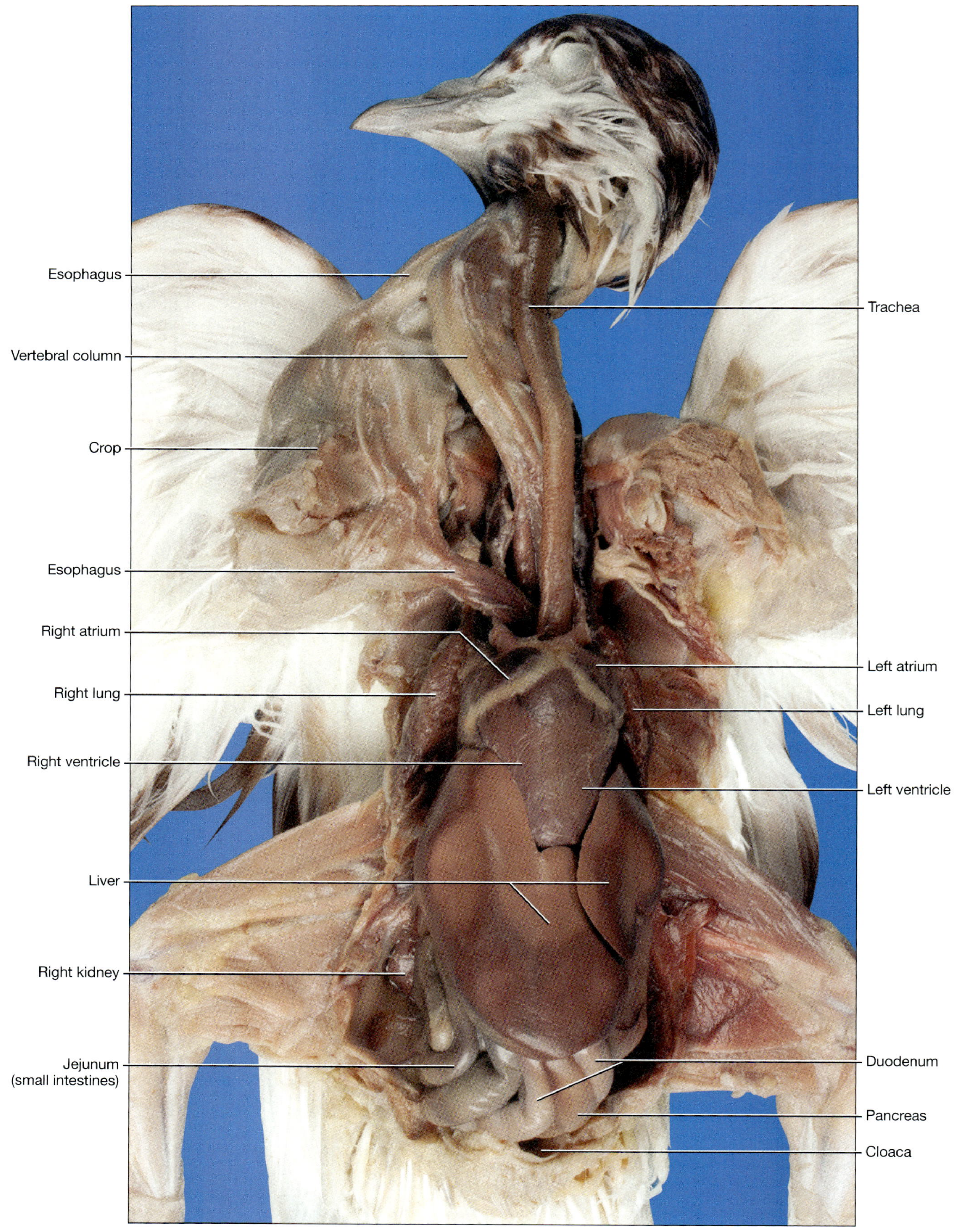

21.6 Pigeon dissection: ventral view of viscera with crop displaced to the animal's right.

21

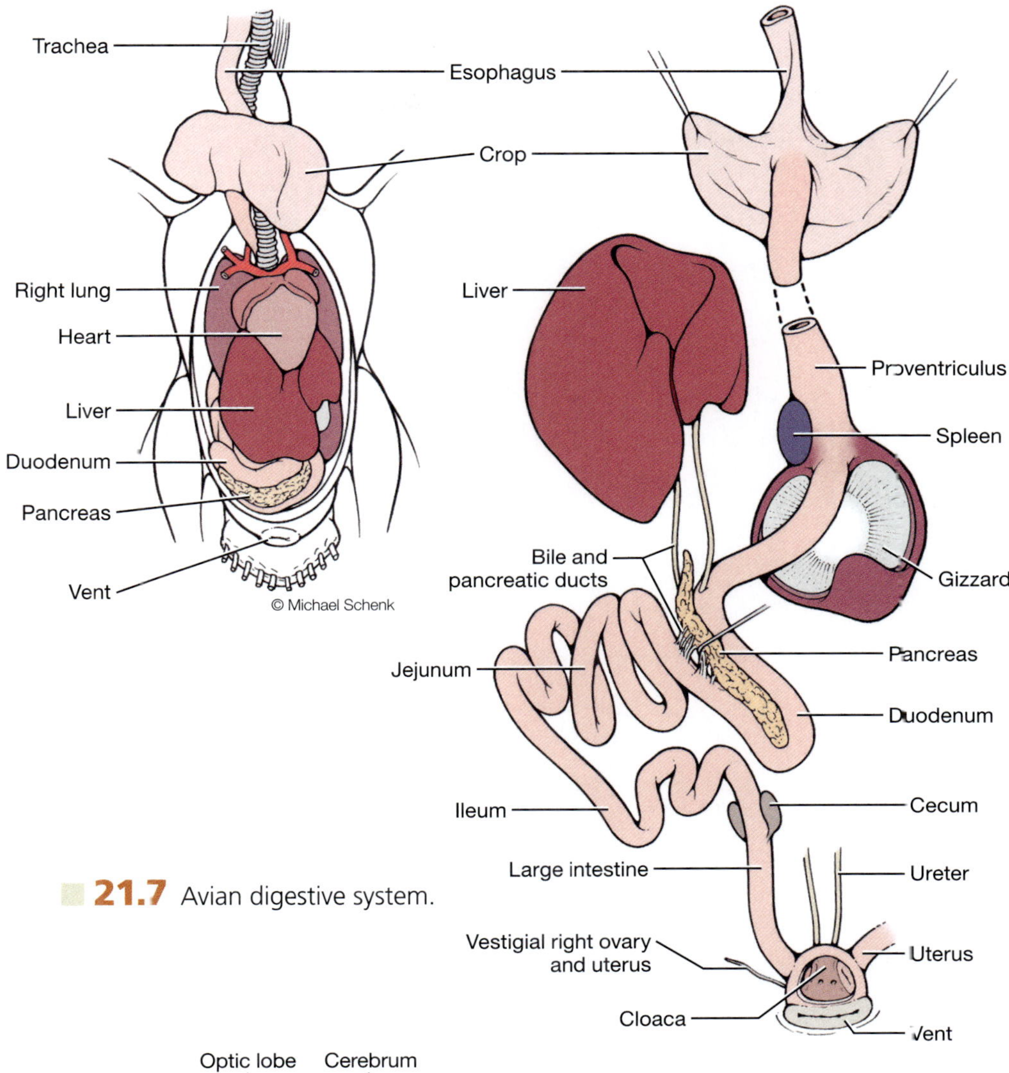

21.7 Avian digestive system.

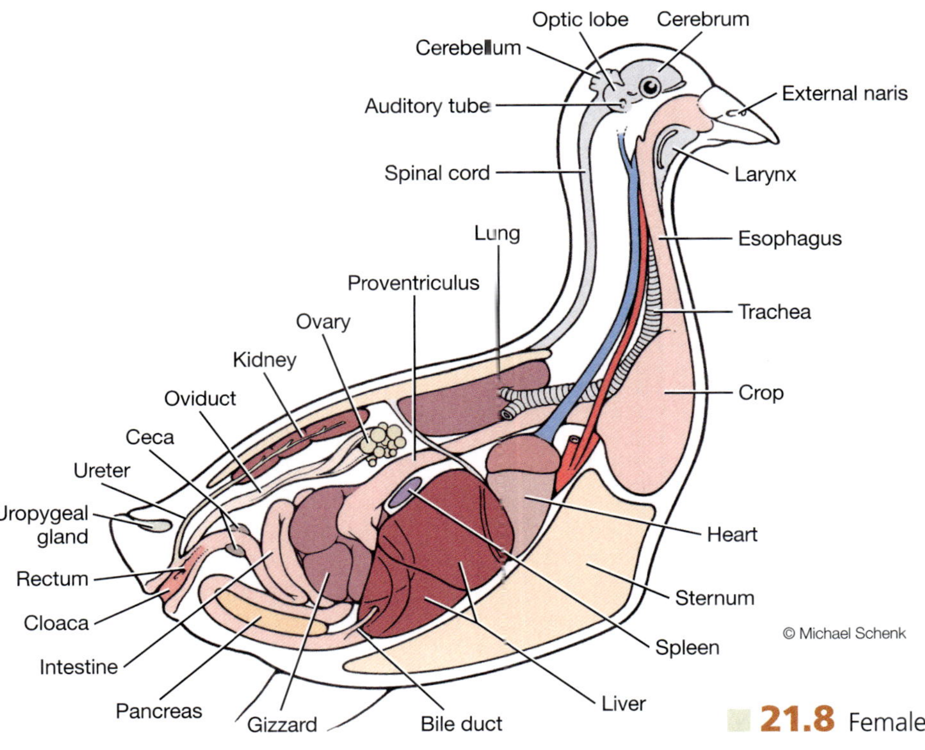

21.8 Female pigeon, internal anatomy.

Table 21-1 Internal Anatomy of the Pigeon

Structure	Function
Larynx	Opens and closes glottis during respiratory cycle and prevents foreign material from entering the lower respiratory tract
Trachea	Conducts air to and from the lungs during respiration
Lungs	Paired organs of respiration connected to air sacs permitting unidirectional flow of air
Heart	Four-chambered organ (two atria, two ventricles) that receives oxygenated blood from the lungs and pumps it via the arteries throughout the body
Esophagus	Transports food to the crop
Crop	Stores ingested food to await passage to gizzard; found in seed- and grain-eating species
Proventriculus	Mixes food with peptic enzymes for digestion
Gizzard	Thick-walled, muscular pouch that pulverizes and churns food prior to its passage into the intestine; larger in seed-eating species
Pancreas	Produces digestive enzymes that are released into the small intestine
Large intestine	Continued digestion of food particles; transports feces to coprodeum of cloaca via peristalsis
Cloaca	Common chamber for collection of materials from the digestive, excretory, and reproductive systems prior to their discharge from the body; partitioned into a coprodeum, urodeum, and proctodeum as in reptiles
Uropygeal gland	Secretes oily compound that coats feathers, preventing them from getting wet
Vasa deferentia (male)	Transport sperm from the testes to urodeum of cloaca
Testes (male)	Produce sperm
Ovary (female)	Single organ (left only) that produces eggs
Ureters	Transport uric acid mixture from the kidneys to the cloaca
Kidneys	Paired organs that filter nitrogenous wastes from the blood and produce dry uric acid with little water that is transported to the urodeum of the cloaca, where it mixes with feces as it is released from the body
Auditory tube	Channels sound waves to auditory receptors; similar to the ear canal of humans
Cerebellum	Primarily a reflex center for integration of skeletal muscle movements; responsible for coordination and balance
Optic lobes	Process visual information from the eyes
Cerebrum	Largest portion of the brain; interprets sensory impulses and coordinates voluntary movements

Circulatory System

Higher blood pressures and rapid, efficient transport of blood gases are vital to animals with high metabolic rates. The physiological demands associated with endothermy and flight shaped the circulatory system of birds to have major advancements over their reptilian and amphibian predecessors. Birds (and mammals) have a double-circuit system, with separate pulmonary and systemic pathways, made possible by the complete separation of oxygenated and deoxygenated blood that the four-chambered heart provides (Fig. 21.9). In birds however, the right side of the heart (specifically the right ventricle) pumps blood to the body tissues and organs, and the left ventricle of the heart sends blood to the lungs. Likewise, the right atrium receives oxygenated blood from the lungs, and the left atrium collects deoxygenated blood from the body. This reverse arrangement is a function of the independent evolution of these otherwise similar circulatory systems in mammals and birds.

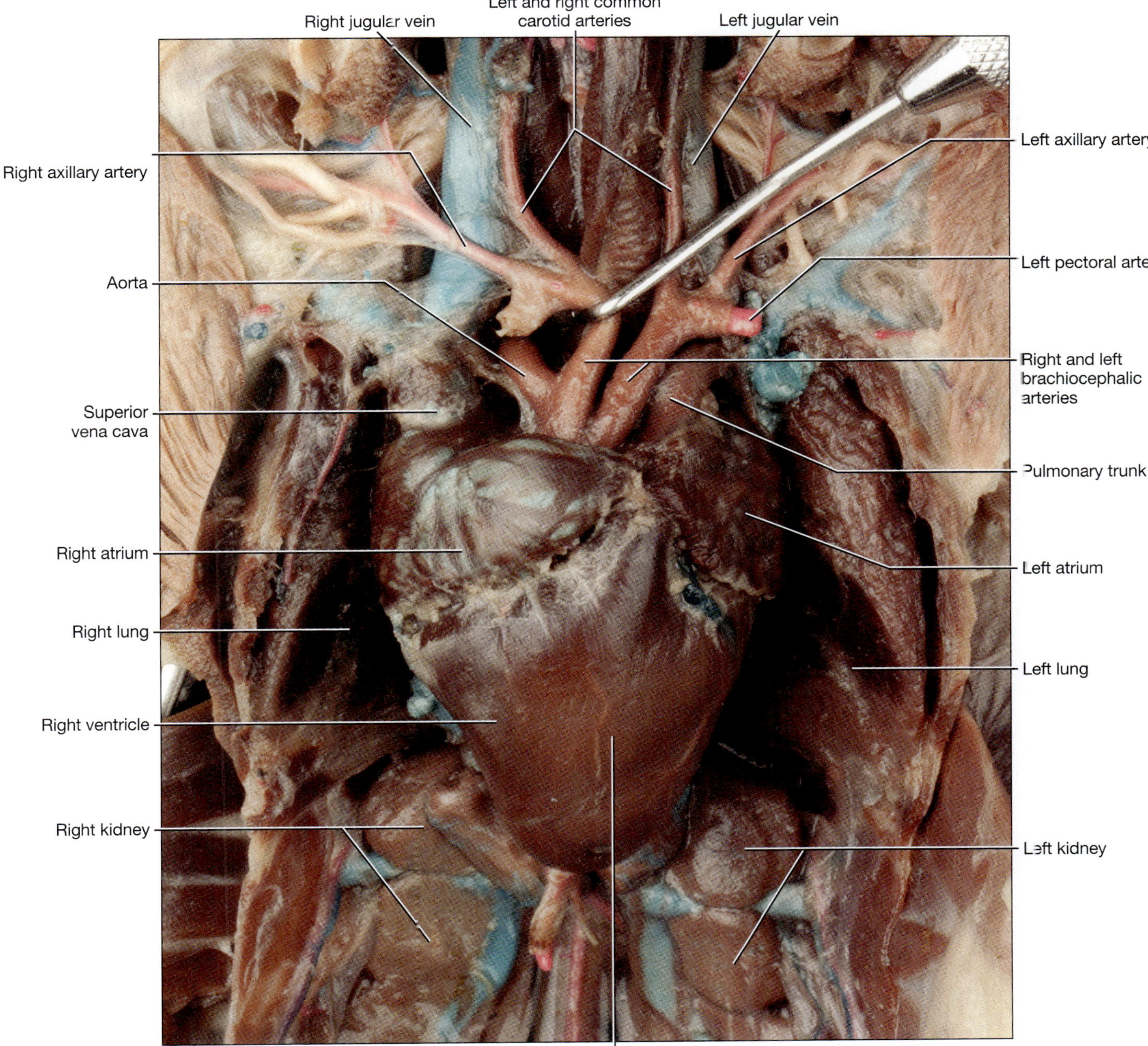

21.9 Pigeon, ventral view of thoracic region showing four-chambered heart and major circulatory vessels.

Reproductive and Excretory Systems

Female reproductive organs in birds are limited to the left side (Figs. 21.10 and 21.11A). The right ovary and oviduct degenerate during development, occasionally leaving an oviductal remnant attached to the cloaca. Mature ova produced in the left **ovary** are released at daily intervals into the coelomic space and are swept into the nearby opening of the left **oviduct** known as the **ostium**. Fertilization of each ovum typically occurs near the ostium soon after the ovum enters the oviduct. As the fertilized egg (or zygote) moves along the length of the oviduct toward the cloaca, glandular secretions of the oviductal walls first produce a layer of albumin around the zygote, followed next by egg and shell membranes, and finally more albumin and shell pigments. Eggs pass through the urodeum and are laid through the female's cloacal opening (or vent). Depending on the particular species of bird, the process of egg production takes about 24 hours from ovulation to laying. Unlike in mammals, the sex of the offspring in birds is determined by chromosomes in the egg rather than in the sperm.

In male birds, both **testes** are present and increase in size substantially as the breeding season approaches (up to 1,000-fold in some species!). The testes are located in the abdomen along either side of the dorsal aorta, usually surrounded by air sacs (Fig. 21.11B). Movement of air through the air sacs is presumably responsible for cooling the testes to maximize sperm production (as optimal sperm production occurs at temperatures slightly lower than the average internal body temperature of most birds). Sperm travel along the paired **vasa deferentia** to the urodeum, from which they are released from the body. Pigeons lack a penis, as do most birds; the male deposits sperm in the female by pressing his cloaca tightly against hers as he ejaculates. A few bird species, such as chickens, turkeys, ducks, and geese have a penis that channels sperm directly into the urodeum of the female.

Excretion in birds is accomplished by the **kidneys** and the **salt glands**. Avian kidneys have some features that resemble reptilian kidneys and other features that are more similar to the kidneys of mammals. The kidneys are paired and the **ureters** that drain the kidneys open directly into the urodeum of the cloaca (Fig. 21.10). No urinary bladder is present in birds. Instead, nitrogenous wastes are excreted as uric acid (as in many reptiles), and these concentrated wastes mix with fecal matter from the coprodeum before being voided from the body.

In addition to the kidneys, all birds possess paired salt glands, embedded in the orbits of the eyes. In marine birds these glands are especially well developed and may account for more than 90% of the salts excreted by the body. The viscous salty fluid, composed principally of sodium chloride, is released through small secretory pores located near the eyes or along the beak.

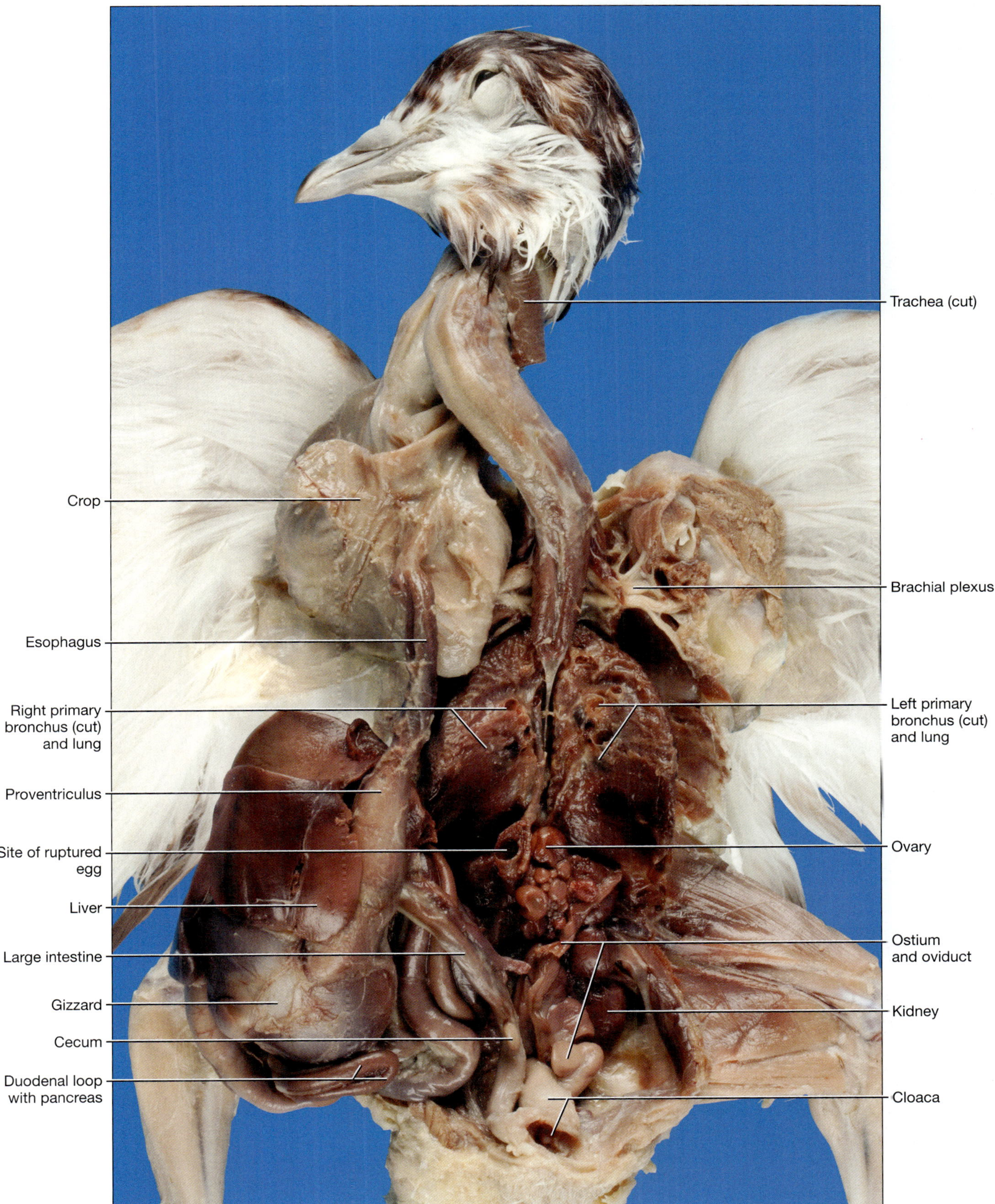

21.10 Female pigeon dissection: ventral view of viscera, alimentary canal displaced to the side to reveal underlying reproductive and urogenital anatomy.

21

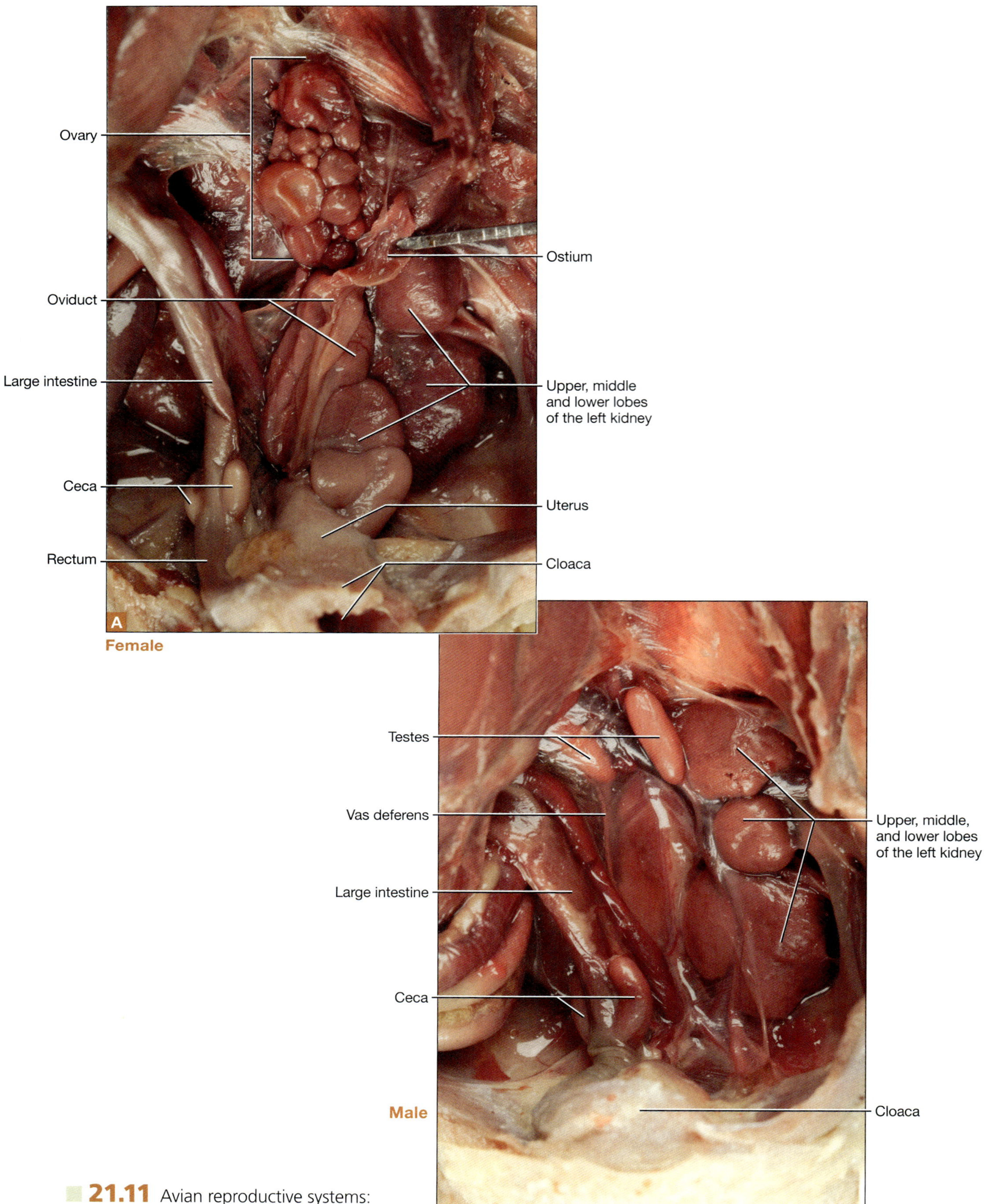

21

21.11 Avian reproductive systems: (**A**) female pigeon, and (**B**) male pigeon.

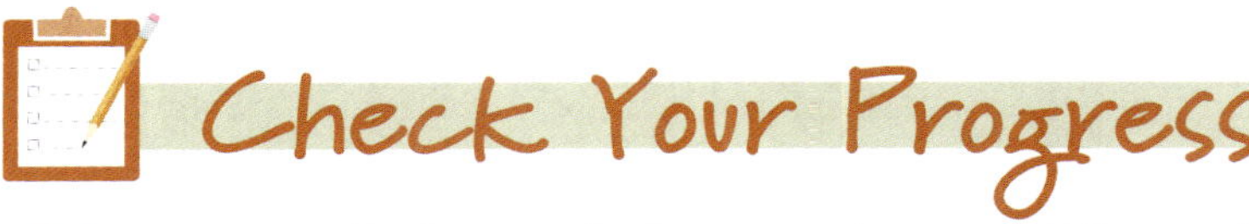

2.1 Functionally, how is the larynx in humans different than in birds?

______

2.2 Number the following organs in the order that represents the correct path of food through the digestive tract of a bird:

_____ cloacal opening

_____ small intestine

_____ crop

_____ proventriculus

_____ mouth

_____ large intestine

_____ esophagus

_____ gizzard

_____ coprodeum

2.3 Which side of the female reproductive tract is functional in birds?

______

2.4 Do birds possess a urinary bladder? How is this adaptive?

______

______

Exercise 21-2 A Closer Look at Feathers

Materials Needed

- ❑ contour feather
- ❑ down and filoplume feathers (optional)
- ❑ dissecting microscope

Birds possess several different types of feathers, including contour feathers used for flight, soft down feathers found beneath the contour feathers for insulation, and filoplume feathers with a thin shaft and small tuft on the end that play a sensory role. Feathers begin their development in much the same manner as a reptile's scales, as small protrusions of the epidermis. But rather than flattening out into a scale, they roll into a cylindrical bud that becomes enclosed in a protective sheath that will split open as the feather nears the end of its growth. Mature feathers actually are dead tissue, much like a human hair, and are molted periodically as new feathers grow to replace them.

Procedure

1 Obtain a contour flight feather and examine its basic structure.

This type of feather is composed of a long shaft, or rachis, that contains parallel barbs forming two vanes, one on each side of the shaft (Fig. 21.12). The side of the feather with the shorter barbs represents the leading edge of the feather and is designed to produce lift as the feather is swept through the air. The feather is anchored to its follicle in the skin by the hollow calamus, or quill, as it often is called.

2 Examine the feather using a dissecting microscope.

Under low magnification you should be able to see that each barb bears numerous hooklets and barbules that overlap to interlock the barbs in much the same way that Velcro® works. With the wear and tear of a bird's daily activities, these hooklets and barbules frequently are pulled apart, compromising the feather's aerodynamic properties. When a bird runs its feathers through its beak while preening, it is rezipping the hooklets and barbules of its feathers together to restore its feathers to their proper condition.

3 If any other types of feathers are available for study, such as filoplume or down feathers, examine them under the dissecting microscope and compare them to the contour feather.

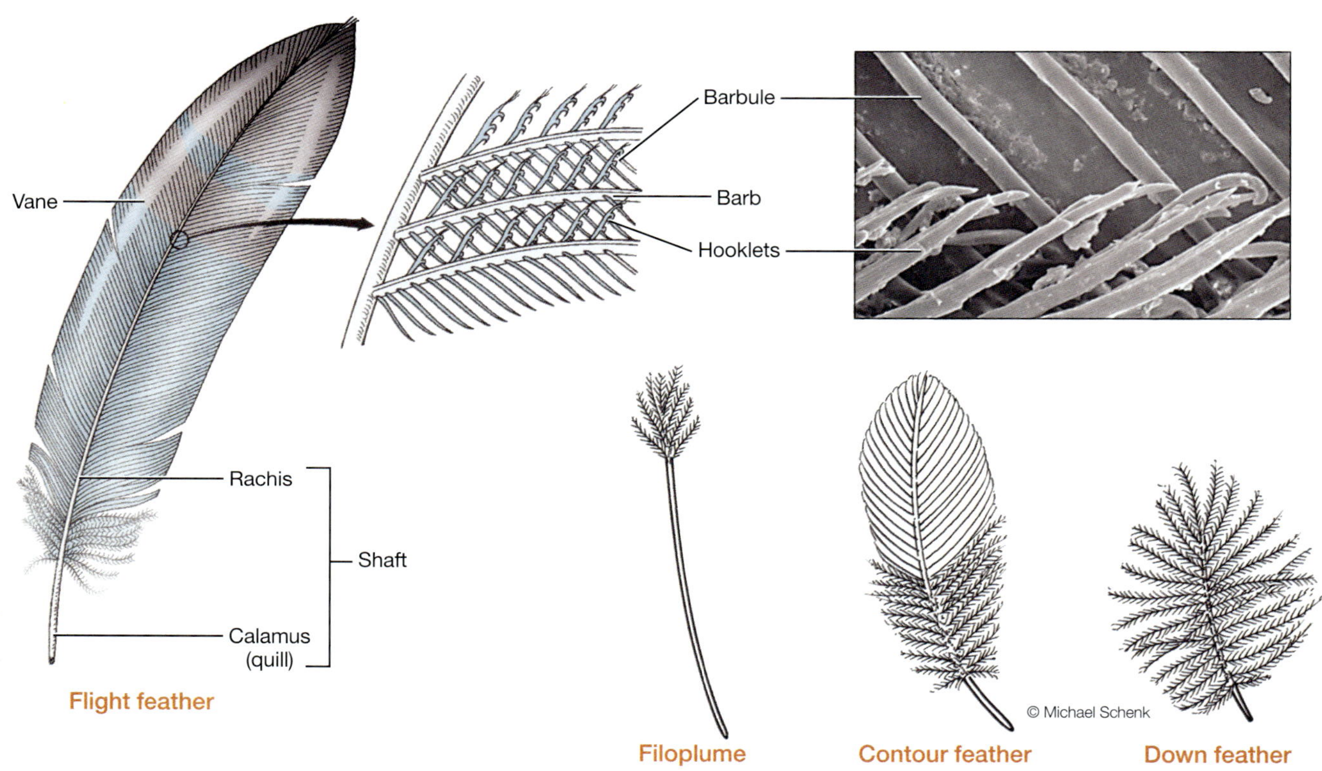

21.12 Anatomy of avian feathers: flight feather, filoplume, contour feather, and down feather.

21

Name ______________________________ Section ______________

21 Questions for Review

1 Name at least two structures in the avian respiratory system that are not present in mammals.

2 What unusual excretory organ do birds possess on their heads?

3 What organ in birds secretes an oily compound to waterproof the feathers?

4 How many chambers does a bird's heart possess? Is this more similar to the heart of reptiles or to that of mammals?

5 Describe respiration in birds, noting the advantages of air sacs.

6 What anatomical features of birds permit them to maintain constant and relatively high internal body temperatures?

7 Identify several anatomical similarities between birds and reptiles that reflect the common evolutionary lineage of these two groups.

8 List at least three avian developments that make flight possible and efficient.

9 What are the two primary functions of feathers?

Name ______________________ Section ______________

21 Laboratory Practical Challenge

The following evaluation is intended to replicate the method of assessment and type of questions you will be asked on lab practicals. If you have difficulty with these questions, you may benefit from additional review of the specimens provided in lab.

1 What is the function of the structure indicated by line A?

____

2 Identify the structure indicated by line B.

____

3 Identify the structure indicated by line C.

____

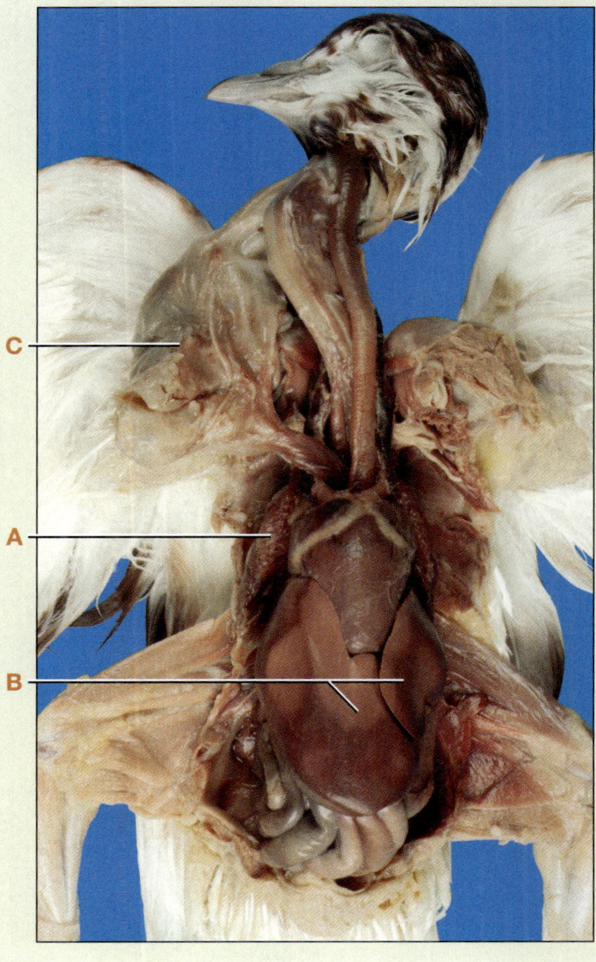

21

4 What is the function of the structure indicated by line D?

5 Identify the structure indicated by line E.

6 To what system do the structures indicated by line F belong?

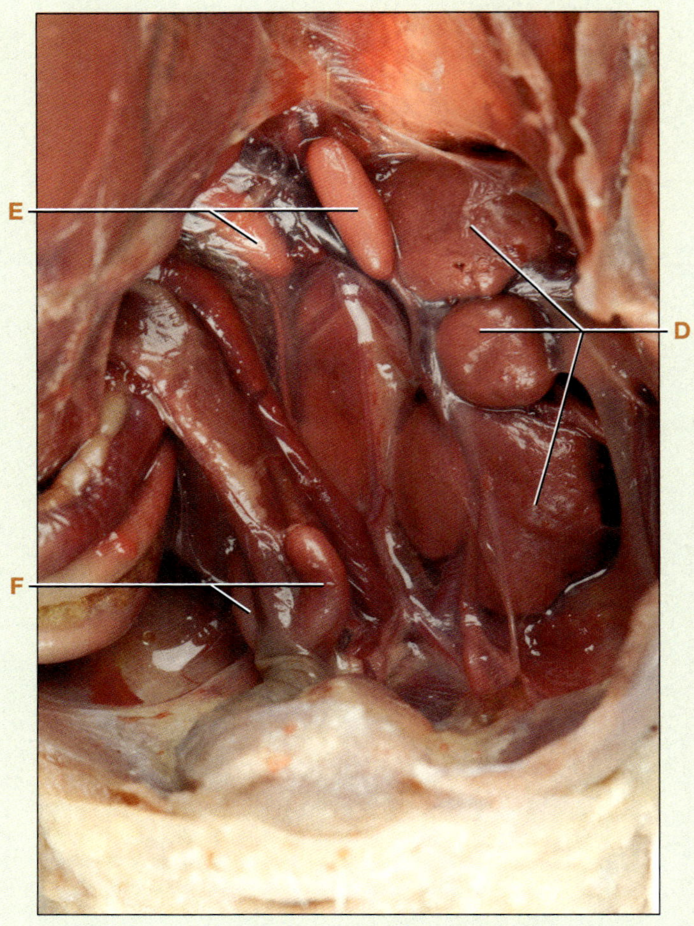

CHAPTER 22

Mammals

After completing the exercises in this chapter, you should be able to:

1. List the distinguishing characteristics of the class Mammalia.
2. Understand the evolutionary position of mammals on the phylogeny of chordates.
3. Discuss the significance of hair, mammary glands, a complete secondary palate, and large cerebral cortex in the evolution of homeothermic vertebrates such as mammals.
4. Identify the principal divisions of the mammalian skeleton and the major bones of those regions.
5. Identify the major anatomical structures and internal organs of the fetal pig, discuss their functions in the body, and compare them to homologous structures of other vertebrates.
6. Identify the major anatomical structures and internal organs of the rat, discuss their functions in the body, and compare them to homologous structures of other vertebrates.
7. Identify the principal regions of the mammalian brain and discuss their functions.
8. Identify the components of the mammalian eye and discuss their functions.
9. Identify the principal regions of the mammalian heart and discuss the pathway of blood through the heart.

Modern mammals display tremendous diversity in size, ranging from small bats weighing only 1.5 grams to giant whales topping the scales in excess of 90,000 kg, with a heart weighing tons and arteries large enough that a human could swim through them! Mammals live on all seven continents and occupy nearly every type of habitat on Earth. Most mammals are terrestrial, although whales and porpoises are highly adapted mammals designed for aquatic living.

Though outwardly diverse, the nearly 5,400 species within the class Mammalia possess a suite of common characteristics that collectively sets them apart from all other chordates (Fig. 22.1). Mammals are **endothermic, homeothermic** organisms that have epidermally derived hair covering their bodies for insulation. All except the spiny anteater (echidna) and platypus are **viviparous,** meaning that they give birth to live young. The male reproductive organ, the penis, permits **internal fertilization,** and the embryo is retained within the uterus of the female for the duration of fetal development. Although mammals produce relatively few young, they invest considerable time and effort in caring for them. In females, modified sweat glands known as **mammary glands** produce and secrete milk to nourish their young.

Most mammalian groups have **fetal membranes** (allantois, chorion, and amnion), and the fetus receives nutrients and oxygen from its mother through **placental** attachment. Mammals are characterized by heterodont dentition—teeth that differ structurally to accommodate foods in different ways. Perhaps the most universal, yet most subtle, characteristic of mammals is that each side of the lower jaw is composed of a single bone, the dentary bone. A muscular **diaphragm** is present, separating the thoracic and abdominal cavities and assisting with

Malaysian fruit bat, *Pteropus hypomelanus*

Koala, *Phascolarctos cinereus*

Bottlenose dolphin, *Tursiops truncatus*

Orangutan, *Pongo pygmaeus*

22.1 Mammals have diversified in size and form over the past 100 million years to adapt to a wide array of ecological niches.

ventilation of the lungs, and a four-chambered heart with separate pulmonary and systemic circuits keeps oxygenated and deoxygenated blood apart.

A characteristic of mammals that played a major role in their evolutionary rise to the intellectual pinnacle of the animal world is the relatively large **cerebrum** present in most groups. This feature, coupled with superior sensory structures, gives mammals the ability to detect and respond to stimuli, modify their behaviors, and communicate with other members in their group.

Evolutionary History

The earliest mammals probably arose sometime around 190 million years ago (mya) from synapsid reptiles and were contemporaries of the dinosaurs for 125 million years. During much of that time, mammals were small, secretive, nocturnal creatures that lived in trees and ate insects, occupying ecological niches that dinosaurs could not dominate. Around 65 mya, as the reign of the dinosaurs came to an abrupt end, mammalian evolution exploded with diversification driven by adaptations to newly opened niches vacated during the massive Cretaceous extinctions (Fig. 22.2). By the end of the Cretaceous period, mammals had diverged into three major groups—egg-laying mammals (monotremes), pouched mammals (marsupials), and placental mammals—and reached their greatest diversity 15 mya during the Tertiary period.

Today, only one order of egg-laying mammals remains (Monotremata), containing just four species of echidnas and one species of duck-billed platypus. Seven orders of marsupials continue to thrive in a few select places on Earth, primarily in regions where they have little competition from placental mammals (for example, Australia). The most successful branch of mammals has been the placental mammals, with 18 currently recognized orders and more than 5,000 living species worldwide.

As the mammalian lineages diverged from reptiles, they developed numerous adaptations that have been keys to their success. Morphological changes positioning the limbs beneath the body allowed mammals to have a more upright posture than reptiles, increasing their stability and improving their gait. Modifications to the jaw and associated musculature and the progression from homodont dentition to heterodont dentition with specialized, differentiated teeth improved their feeding efficiency. The advent of a complete secondary palate, separating the oral and nasal cavities, allowed mammals to eat and breathe at the same time.

22.2 The first mammals arose nearly 200 million years ago, branching off the vertebrate line that eventually led to modern reptiles and birds many millions of years before those groups diverged.

Hair evolved as an effective insulation to maintain the higher inner body temperatures that mammals generated through endothermy. To offset the chance of overheating from a thick coat of fur covering the body, sweat glands evolved to aid in thermoregulation. Modifications to certain patches of sweat glands led to the development of mammary glands to provide nourishment for the young, which lengthened the period of parental care over that of reptiles.

Changes in the breathing mechanism, such as the loss of abdominal ribs and the evolution of a muscular diaphragm, improved mammals' respiratory capacity. Two of the bones forming the lower jaw joint (the articular and quadrate bones) migrated into the middle ear, became reduced in size, and linked with the stapes to produce a three-bone structure, providing mammals with much better amplification of sound waves and, thus, better hearing than their reptilian contemporaries. This left mammals with a single lower jaw bone on each side (the dentary bone) to support the teeth. An especially large cerebral cortex equipped mammals with a more sophisticated ability to sense stimuli, process information, and communicate.

Since their origins over 190 million years ago, mammals have become one of the most biologically differentiated groups in the animal kingdom. One group of mammals in particular—humans—unquestionably has been the most impactful animal species on Earth. Life began on our planet some 3.5 billion years ago, and animal life began roughly 700 million years ago, yet the first true hominids did not appear until 7 million years ago, and *Homo sapiens* have been around only for the last 200,000 years—less than 0.03% of the time that animal life has existed—a mere blink of the eye in geologic time. Despite our existence for only a tiny fraction of history, no other mammal—or animal, for that matter—has been able to understand its planet and its origins, or been able to influence life on this planet in such a profound manner. For good and for bad!

Exercise 22-1 Skeletal System

Materials Needed

- ❑ preserved mounted cat skeleton

The skeletal system of vertebrates plays an important role in supporting the body and holding animals upright, yet the skeleton must allow for flexibility so animals can perform a wide array of motions. Thus, while the skeletal system is composed of many individual calcified bones that are quite rigid, the many different kinds of joints connecting these bones permit movement. Here we provide a detailed analysis of the cat skeleton, because it is a popular option for teaching the osteology of a mammal with an anatomy comparable to that of humans. Because all mammals share a common ancestry, you will be able to see many bones in the cat that are homologous to humans as well as the other vertebrates that we have studied previously.

Procedure

1 Obtain a mounted, articulated skeleton of a cat.

2 Examine it carefully, using Figure 22.3 to assist you in identifying the major skeletal elements present.

The mammalian skeleton is composed of two major regions: the axial skeleton and the appendicular skeleton. The **axial skeleton** consists of the skull, vertebral column, and rib cage, and forms the longitudinal axis of the body. The **appendicular skeleton** consists of the bones of the forelimbs and hind limbs, as well as the bones that attach the limbs to the axial skeleton, the pectoral girdle, and the pelvic girdle (Fig. 22.3).

The Axial Skeleton

The lower jaw in mammals is composed of a fused pair of single dentary bones called the mandible (Fig. 22.3). This one characteristic separates mammals from all other classes of vertebrates (such as fish, reptiles, and birds). All of the teeth of the lower jaw are anchored in the mandible, and many **foramina** are present for innervation of the teeth, lips, and gums. The actual braincase is composed of several bones that are fused together to form immovable connections known as sutures.

The vertebral column has two basic purposes:

1. to protect the delicate spinal cord that passes through it, and
2. to permit flexibility, support, and anchor points for muscle attachments.

Thus, all vertebrae have the same basic morphology, with minor modifications depending on their specific location along the length of the spine.

The seven cervical vertebrae constitute the most cranial portion of the vertebral column. The skull joins the vertebral column at the first cervical vertebra, known as the atlas. This is a highly specialized vertebra designed to fit precisely into the convex bulges in the base of the skull, the occipital condyles. The second cervical vertebra is the axis, which articulates with the atlas to form a pivot point for full rotation of the head. The remaining cervical vertebrae are similar in morphology to one another.

The thoracic region of the cat is composed of 13 thoracic vertebrae (12 in humans). On the articulated skeleton, notice the many ribs enclosing the chest region that extend from these vertebrae. These ribs provide protection and support for the heart and delicate lungs that lie inside the thoracic cavity. Caudal to the thoracic vertebrae are the seven lumbar vertebrae (five in humans). These are the largest of the vertebrae and have no true ribs extending from them. During embryonic development a special group of three vertebrae fuse together in the cat (five fuse together in humans) to form the sacrum, an especially strong region that supports the pelvic girdle and hind limbs. Finally, the caudal vertebrae continue from the base of the sacrum to the tip of the tail. Human caudal vertebrae are less numerous, minimally functional, and often fused. Collectively, they are referred to as the coccyx in humans. In mammals with long tails, the caudal vertebrae play an important role in locomotion, maneuverability, and balance.

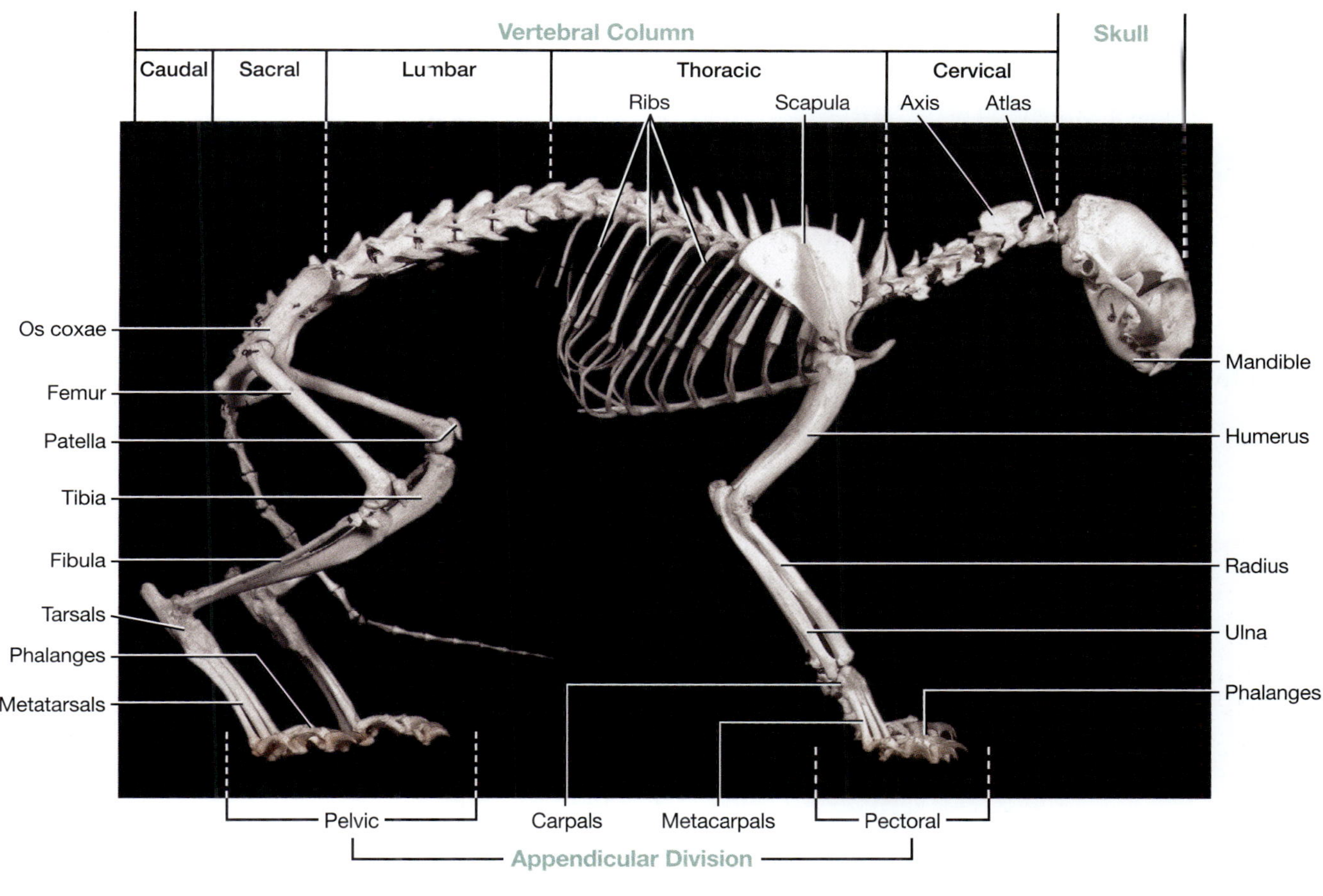

22.3 Cat, skeletal anatomy.

The Appendicular Skeleton

The scapula, or shoulder blade, forms the base of the forelimb (Fig. 22.3). This flattened, triangular bone is not actually attached to the axial skeleton; rather, it floats in a cavity created by the muscle layers surrounding this region. These muscles hold the scapula tightly in place but permit the flexible, fluid running motion characteristic of many quadrupedal mammals. The distal end of the scapula, however, is attached to the head of the humerus, the single bone that constitutes the proximal portion of the forelimb. The head of the humerus articulates with the scapula, and the distal end of the humerus articulates with the radius and ulna, the two bones that make up the distal portion of the forelimb. The radius is the smaller of the two and articulates proximally with the humerus and distally with the ulna and a large carpal bone (the scapholunate).

The seven bones of the wrist are known as the carpals, and the five metacarpals and phalanges make up the forefoot (or manus). A unique feature in the cat that is not present in humans is the presence of retractable claws on the distal phalanges. The claws are withdrawn into sheaths when they are not in use. Retractable claws represent a significant evolutionary adaptation of the feline family for capturing and grasping large, fast-moving prey.

The os coxae are paired bones fused together to form the pelvis, the basis of the pelvic girdle in mammals (Fig. 22.3). The femur is the long, proximal hind limb bone that articulates with the os coxa. The tibia and fibula are the more distal hind limb bones, with the tibia being the larger of the two. The seven ankle bones of the hind foot, or pes, are collectively called the tarsals, and the remaining bones of the pes are the metatarsals and phalanges.

The large bone in the hind foot that forms the slight bulge in the back of the hind limb in mammals is the calcaneus bone. This is homologous to our heel bone. The first digit in the hind foot (corresponding to our big toe) is greatly reduced in most quadrupedal mammals. As a result, animals such as the cat have four primary phalanges. Cats display **digitigrade** locomotion, meaning that they walk on the tips of their digits, or phalanges. Human locomotion is classified as **plantigrade**, which means that we walk on the soles of our feet; our body weight is supported primarily by our metatarsals and tarsals rather than by our phalanges.

Check Your Progress

1.1 What single skeletal feature distinguishes mammals from all other vertebrates?

1.2 List five bones that are directly homologous in birds and mammals.

1.
2.
3.
4.
5.

1.3 What other mammals can you think of that display digitigrade locomotion?

1.4 What other mammals can you think of that display plantigrade locomotion?

Exercise 22-2 Fetal Pig (Order Artiodactyla)

Materials Needed

- ❑ preserved fetal pig (double- or triple-injected recommended)
- ❑ dissecting tools
- ❑ dissecting pan
- ❑ dissecting pins
- ❑ latex gloves

Pigs (*Sus scrofa*) are members of the order Artiodactyla, the "even-toed" hooved mammals, along with antelopes, deer, cattle, sheep, goats, and hippopotamuses. The fetuses used for dissection come from mature female sows that, when slaughtered for their meat, are discovered to contain unborn young. They are not raised specifically for dissection purposes; therefore, their use as dissection specimens does not promote unnecessary animal deaths for scientific purposes. The gestation period of these animals typically is 114 days (16 weeks). The pigs used for dissection generally are between 80 and 100 days old (8–12 inches in length) and possess fully developed organs and organ systems suitable for study.

External Anatomy

Procedure

1 Obtain a preserved specimen of a fetal pig and use Figure 22.4 to identify the external features of your specimen.

2 Notice the sensory organs concentrated around the head—**eyes**, which are not yet open; ears; **external nares**, or nostrils, for breathing and for sensing chemicals dissolved in the air; and **vibrissae** on the chin and around the snout (vibrissae are not depicted in Fig. 22.4).

These organs all play a collective role in the pig's ability to sense and respond to stimuli in its environment. The external coverings of the pig eye consist of upper and lower eyelids and a reduced nictitating membrane, which moves laterally from the medial corner of the eye.

The vibrissae (commonly called whiskers) are used for tactile sensations. The base of each vibrissa is attached to a sensory nerve that is triggered by air movements or physical contact to the whisker. Other vibrissae are located above the eyes and on the cheeks and chin. Collectively, vibrissae provide important sensory cues for spatial orientation.

3 Identify the **umbilical cord** protruding from the ventral side of the abdomen (Fig. 22.4).

This structure carries nutrient- and oxygen-rich blood to the fetus and removes excess metabolic waste products and carbon dioxide from the fetal system. Males and females both possess an **anus**, located just ventral to the base of the tail, and both sexes have **mammary papillae** around the abdominal region near the umbilical cord. In females these will develop into the mammary glands and will be used to secrete milk during lactation for the newborn young. Although males also possess mammary papillae, these structures do not provide any known function.

4 You should be able to determine the sex of your pig using external features. In addition, you are expected to be familiar with the external structures that are unique to each sex, so you should work closely with another group that has a pig of the opposite sex.

Females have a **urogenital opening** ventral to the anus, near the base of the tail (Fig. 22.4B). This represents the opening to the reproductive pathway and serves as a channel for the release of excretory products (urine) from the body. This opening may be slightly obscured by the genital papilla, a short projection that develops into the clitoris in the adult female. The clitoris is a direct homologue to the male penis and plays a similar role in sexual sensation, sending information about sexual stimulation to the brain.

Males are identified most easily by the location of the urogenital opening on the ventral surface just caudal to the umbilical cord (Fig. 22.4C). This is the opening of the urethra, which releases excretory products (urine) and semen in the adult pig. The penis is not fully developed in the fetus and still is embedded in the tissues of the abdomen along the ventral abdominal surface. If your male pig is old enough, you may see a **scrotum** near the anus. As development proceeds, the **testes**, which originally form deep inside the abdominal cavity near the kidneys, migrate caudally and eventually descend into the scrotum.

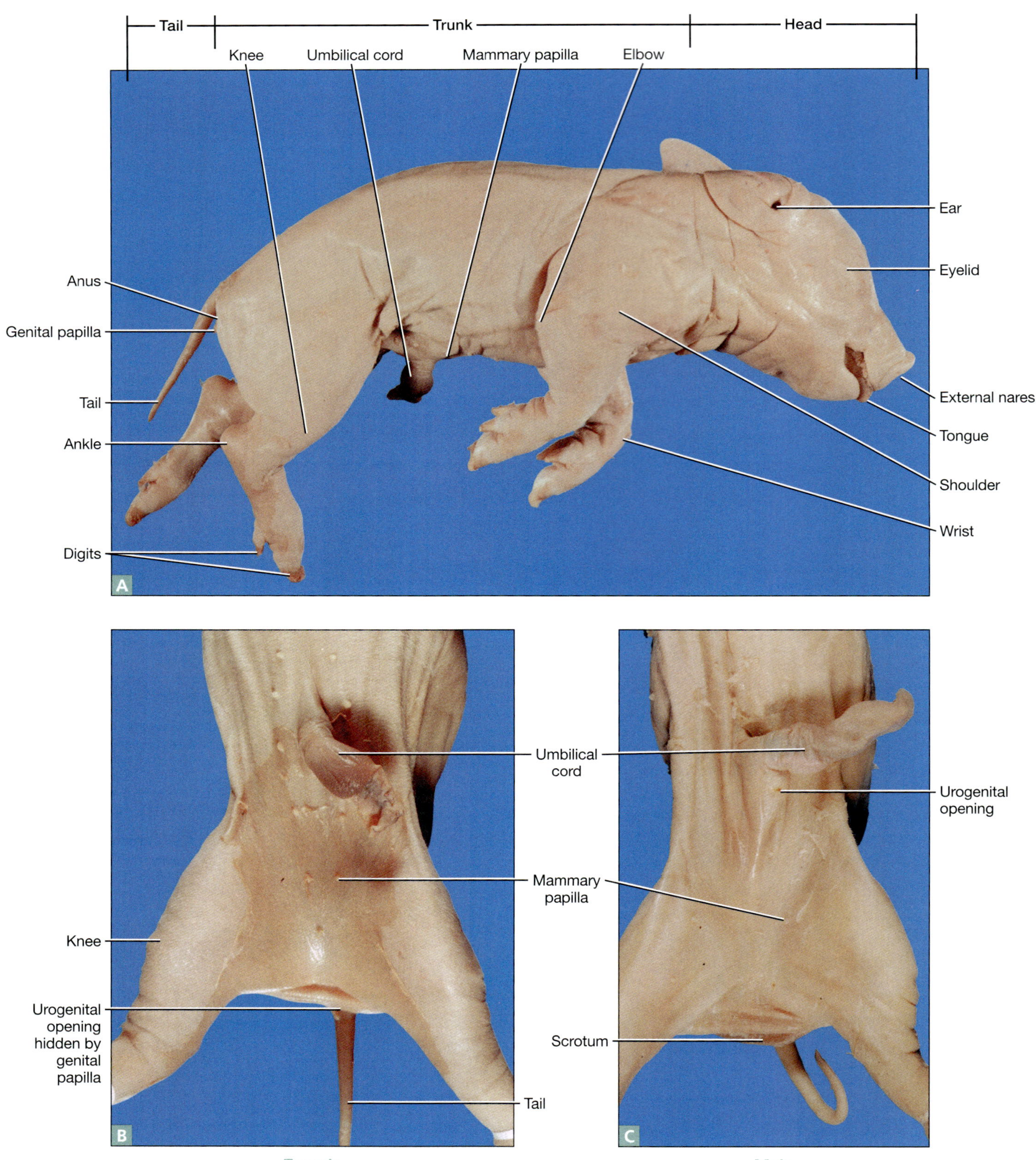

22.4 Fetal pig, external anatomy: (A) lateral view; and ventral views of (B) female, and (C) male specimens.

22

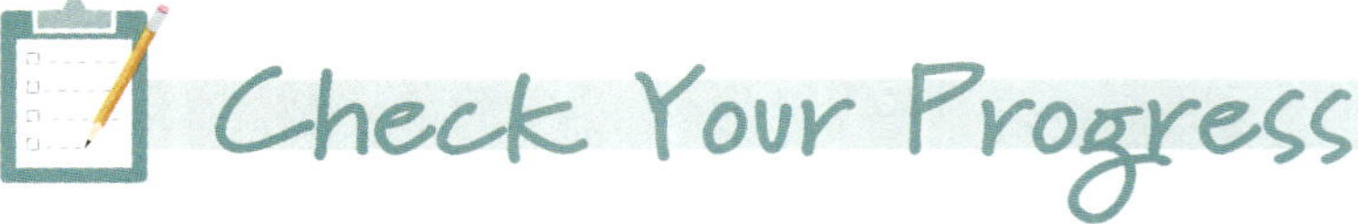

2.1 Is your pig male or female? How do you know?

2.2 What external features of the fetal pig are characteristic of all mammals?

2.3 What sensory organs are present on the head?

Musculature

Muscles are designed with one basic purpose in mind—movement. Muscles work to either move an animal through its environment or move substances through an animal. Vertebrates have three basic types of muscle tissue: (1) skeletal muscle and (2) cardiac muscle, both of which possess striated fibers, and (3) smooth muscle, sometimes called visceral muscle. Some of these muscles, such as skeletal muscle, can be controlled voluntarily by the animal. Others, such as cardiac and many smooth muscles, produce involuntary actions that are regulated by the autonomic nervous system.

All skeletal muscles have a fixed end, the **origin**, and a movable end, the **insertion**. Most skeletal muscles taper from a thicker **muscle belly** in the middle to thinner, tough, white tendons on the ends that anchor the muscle to bones or other muscles. The direction in which a muscle exerts force—its **action**—also plays a role in its shape and where it inserts and originates. A muscle that **adducts** moves a limb toward the midline of the body. Conversely, a muscle that **abducts** moves a limb away from the midline of the body.

Some muscles are **flexors**, bending one part of the body toward another part. Other muscles are **extensors**, straightening or extending a body part. Because muscles can contract in only one direction, and thus perform only unidirectional movements, they generally are arranged in antagonistic pairs or groups; that is, they work in opposition to one another. Because muscles usually are grouped closely together, it often is difficult to determine where one muscle ends and another begins. Thus, you will have to pay careful attention to the orientation of the muscle fibers. Often this will give you clues as to where two muscles cross or abut.

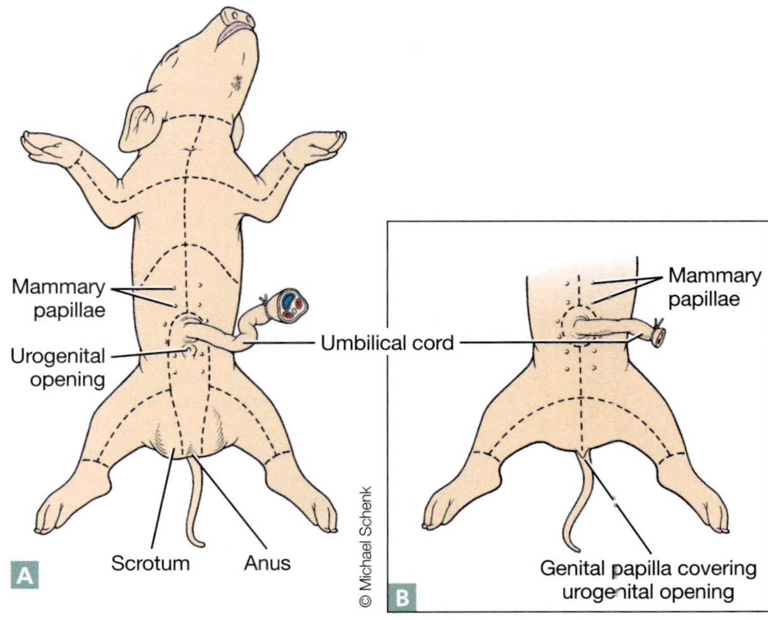

Dissection Diagram
Insert a finger (or blunt probe) in the opening created by the injection process, and free the skin from the underlying fascia along the midline of the body. Use scissors to make the illustrated incisions through the skin of a (**A**) male or (**B**) female specimen, and use a blunt probe to tease the skin away from the muscles. Do not cut completely through the body wall if you intend to study the musculature.

Procedure

1 To observe the musculature of the pig, you must first remove all of the skin from the external body surface. Place your pig on its dorsal surface in the dissecting pan and insert a finger or blunt probe in the opening created by the injection process, to free the skin from the underlying fascia along the midline of the body.

2 Separate the skin from the underlying muscles carefully, using a blunt probe where necessary.

3 Work laterally from the midline, using the Dissection Diagram as a guide for the subsequent incisions to make.

4 Use a blunt probe to tease the skin away from the muscles carefully. In most cases you should be able to peel away the skin with your fingers.

5 Continue until you have completely exposed the muscles of your specimen.

6 Use Tables 22-1–22-2 and Figures 22.5–22.6 to identify the selected muscles and their actions in the pig.

Table 22-1 Superficial Musculature of the Lateral Aspect of the Fetal Pig

Body Region	Muscle Name	Action
Neck	Cleidomastoid	Moves forelimb cranially
	Cleidooccipitalis	Moves forelimb cranially
Pectoral region	Supraspinatus	Extends shoulder
	Deltoid	Flexes shoulder; abducts forelimb
	Trapezius	Elevates scapula and draws scapula laterally
	Triceps	Extends forelimb
	Brachialis	Flexes forelimb
	Pectoralis profundus	Adducts forelimb
	External oblique	Compresses abdomen and flexes trunk
Forelimb	Extensor carpi radialis	Extends and flexes wrist
	Extensor digitorum lateralis	Extends fourth digit
	Extensor digiti V	Extends fifth digit
	Extensor digitorum communis	Extends joints of principal digits
	Ulnaris lateralis	Extends and flexes wrist
	Flexor digitorum profundus	Flexes wrist and digits
Hind limb	Gluteus medius	Abducts and extends thigh
	Gluteus superficialis	Abducts thigh
	Tensor fascia latae	Extends hind limb
	Semitendinosus	Extends thigh and flexes hind limb
	Biceps femoris	Abducts thigh; flexes hind limb
	Semimembranosus	Extends hip and adducts hind limb
	Gastrocnemius	Extends hind foot
	Extensor digitorum quarti and quinti	Extend fourth and fifth digits
	Peroneus longus and tertius	Flex hind foot
	Tibialis anterior	Flexes hind foot

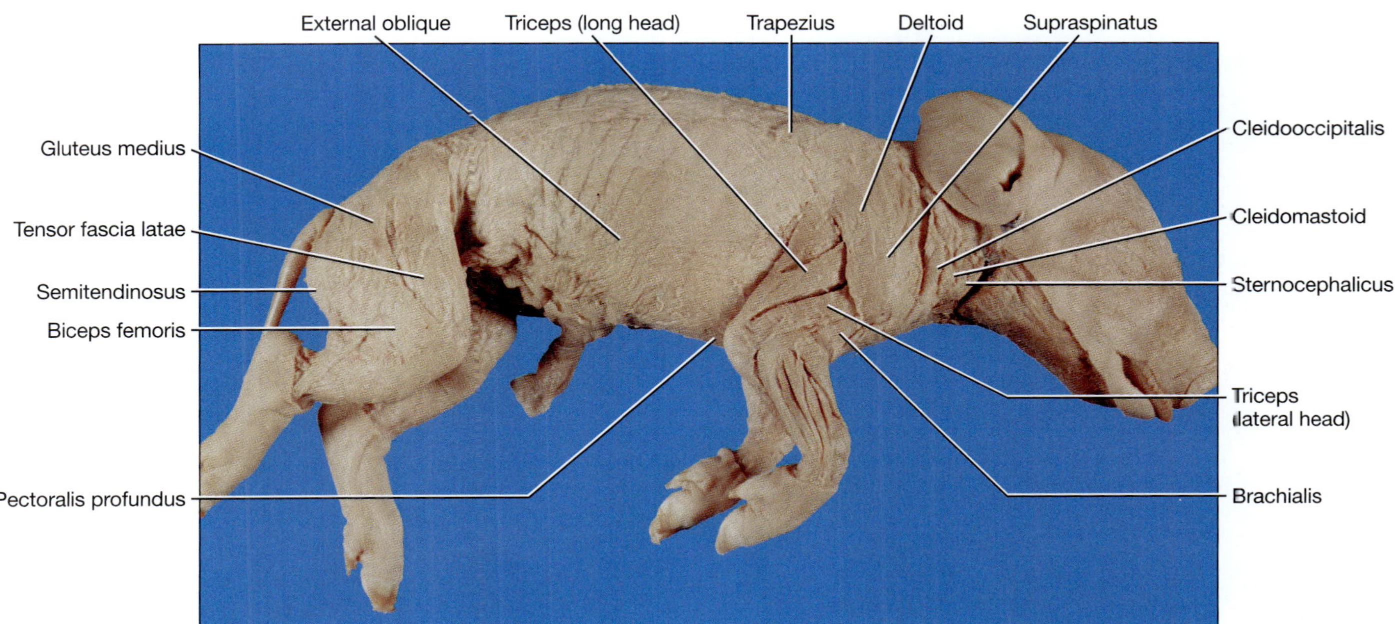

Gluteus medius
Tensor fascia latae
Gluteus superficialis
Semitendinosus
Biceps femoris
Semi-membranosus
Gastrocnemius
Extensor digitorum quarti and quinti
Peroneus longus
Peroneus tertius
Tibialis anterior (cranial)

Supraspinatus
Deltoid
Triceps (long head and lateral head)
Brachialis
Extensor carpi radialis
Extensor digitorum lateralis
Extensor digiti V
Extensor digitorum communis
Flexor digitorum profundus
Ulnaris lateralis
Extensor carpi oblique

22

22.5 Fetal pig: superficial muscles, lateral aspect.

Table 22-2 Superficial and Selected Deep Musculature of the Medial Aspect of the Fetal Pig

Body Region	Muscle Name	Action
Neck	Mylohyoid	Raises floor of mouth
	Digastric	Depresses mandible
	Masseter	Elevates mandible
	Stylohyoid	Raises hyoid bone; draws base of tongue and larynx caudally
	Omohyoid	Draws hyoid bone caudally; retracts roof of tongue
	Sternohyoid	Pulls hyoid bone and tongue caudally
	Sternomastoid	Flexes head
	Thyrohyoid	Moves hyoid bone caudally and dorsally
	Sternothyroid	Moves larynx caudally
Pectoral region	Brachiocephalicus	Moves forelimb cranially
	Pectoralis superficialis	Adducts forelimb
	Pectoralis profundus	Adducts forelimb
Hind Limb	Psoas major	Flexes hip and rotates thigh
	Iliacus	Flexes hip and rotates thigh
	Rectus femoris	Extends hind limb
	Vastus medialis	Extends hind limb
	Sartorius	Extends hind limb; adducts thigh
	Gracilis	Adducts thigh
	Semitendinosus	Extends thigh and flexes hind limb
	Pectineus	Flexes hip and adducts thigh
	Adductor	Adducts thigh
	Tibialis anterior	Flexes hind foot

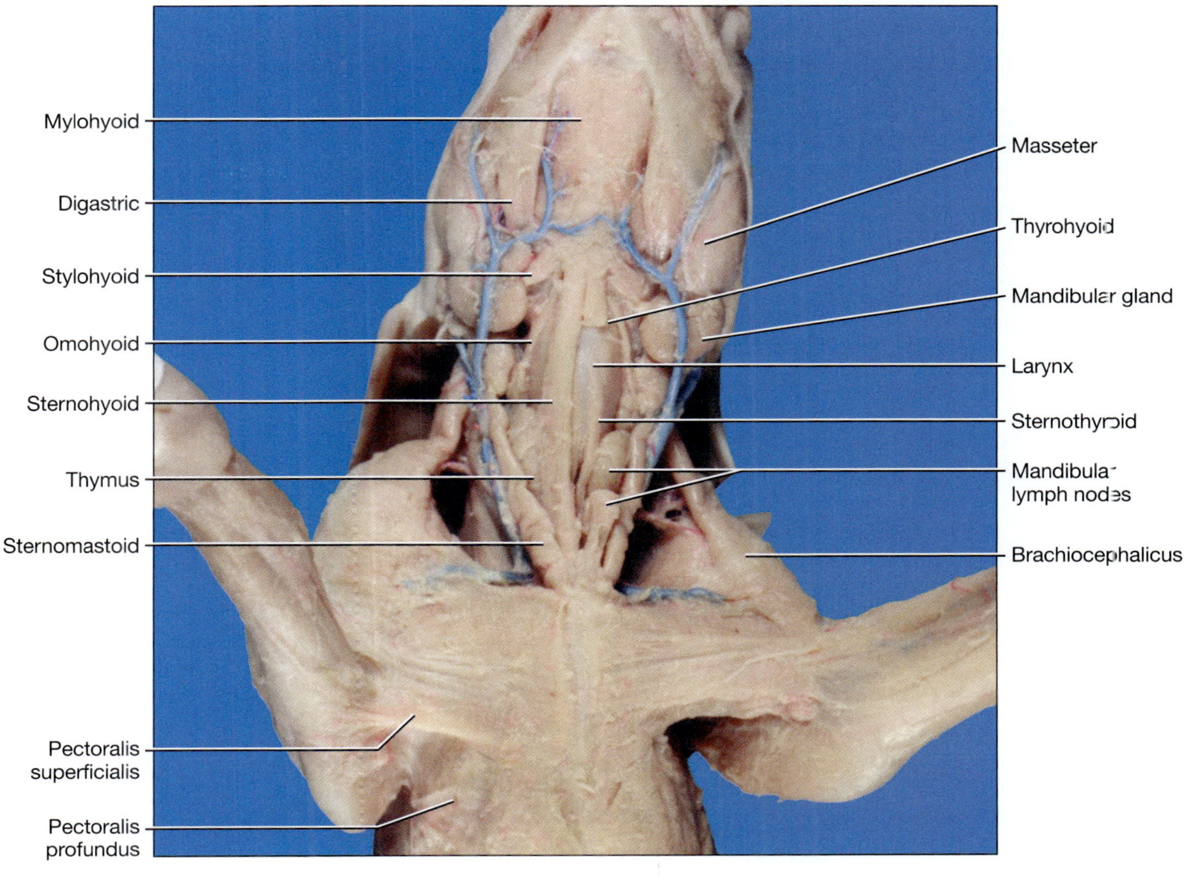

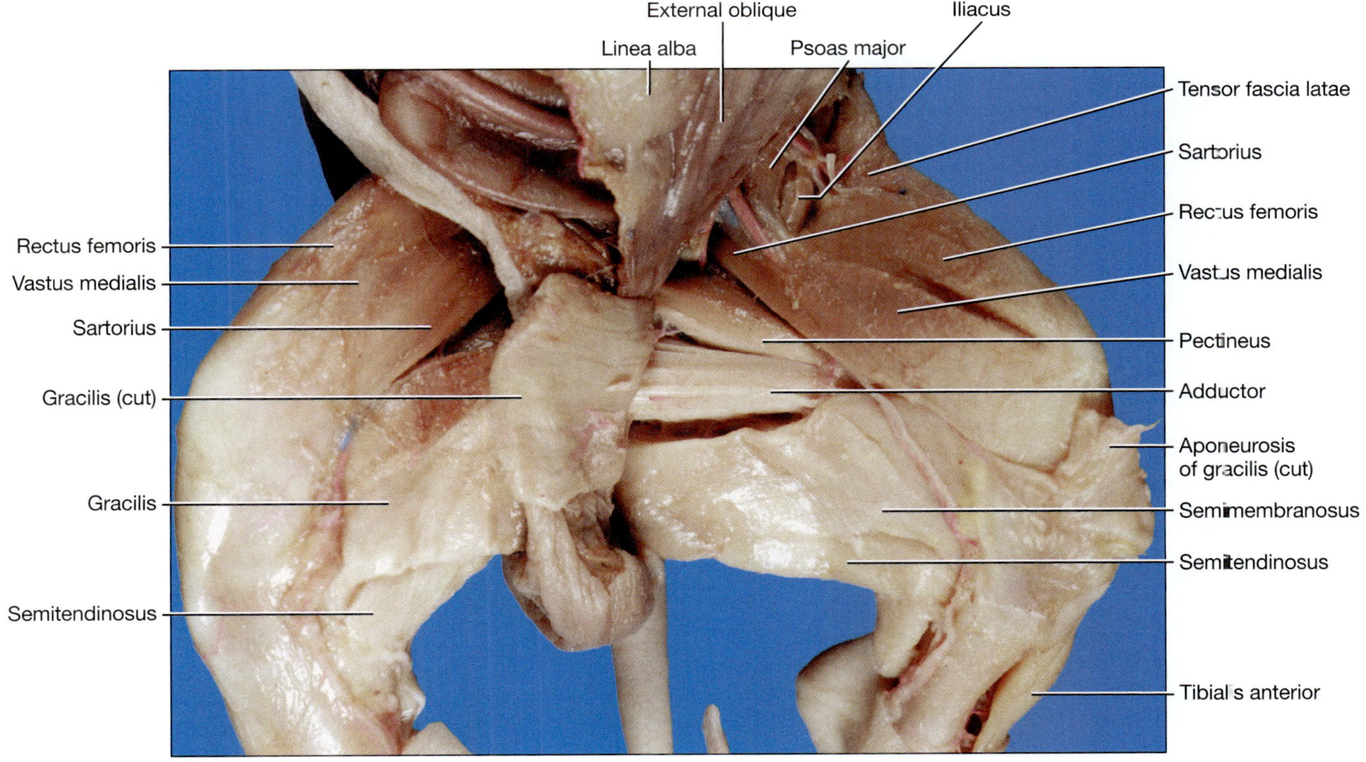

22.6 Superficial (pig's right) and selected deep muscles (pig's left) of the medial aspect of the neck, pectoral region, and hind limb.

Digestive System

The digestive system is responsible for breaking down food mechanically and chemically into smaller, usable compounds and absorbing and transporting these nutrients into the bloodstream for delivery to the individual cells of the body. This process provides the crucial raw materials and energy for all metabolic processes carried out by the organism. The extreme specialization of individual digestive organs and the efficiency of the digestive process permit mammals to sustain high metabolic rates and maintain an endothermic balance without the need for constant consumption of food.

Procedure

1 Place your pig on its side in your dissecting pan.

2 Using your scalpel, make a cut from the corner of the mouth toward the ear on each side of the pig's head. This will extend the opening of the mouth and allow you to view the deeper structures of the oral cavity.

3 Angle your incision along the lateral margins of the mouth, and depress the lower jaw with your fingers as you progress. Don't be afraid to cut too far; usually, if you cannot see the structures indicated, you have not cut far enough.

4 You may have to use a pair of heavy scissors to sever the mandible on each side of the oral cavity.

If your pig is sufficiently mature, teeth will be protruding from the roof of the mouth. In this region of the mouth, the roof is composed of a bony hard palate separating the oral cavity from the nasal cavity above (Fig. 22.7). The soft palate is a fleshy continuation caudally from the hard palate. The advent of the complete secondary palate allowed mammals to eat and breathe simultaneously—one characteristic that allows mammals to have such a high metabolic rate, making endothermy possible. Just caudal to the soft palate is the elliptical opening to the nasopharynx, leading rostrally to the external nares. The opening to the **esophagus** also should be visible.

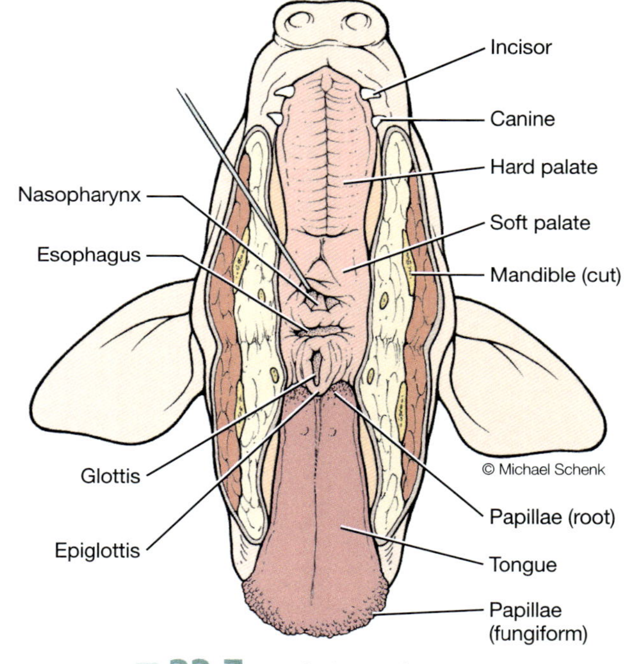

22.7 Fetal pig, oral cavity.

5 Next, locate the glottis, the opening into the larynx. When the pig swallows, this opening is protected by a thin flap of cartilage called the **epiglottis**.

6 Slowly close the oral cavity and notice how the glottis and epiglottis meet up with the opening to the nasopharynx.

7 On the lower jaw, locate the tongue. Notice the small bumps near the tip and base of the tongue. Called papillae, they help mammals manipulate food in their mouths.

8 To observe the internal anatomy, position your pig on its dorsal surface in the dissecting pan.

9 Using scissors, begin by making an incision from the base of the umbilical cord cranially along the ventral midline to the base of the chin.

10 Next make an incision around each side of the umbilical cord caudally toward the anus.

11 Then make lateral incisions in the abdomen just in front of the hips.

12 Next make lateral incisions along the base of the ribs toward each side.

13 Finally, make lateral incisions to each side of the neck.

14 Many preserved specimens contain large amounts of liquid preservatives in their body cavities. You may want to drain these out of your specimen, or use a paper towel or sponge to remove them before proceeding with identification of the digestive organs. You also may want to cut the ribs along the sides of the body and remove them to facilitate access to the thoracic cavity.

15 Use pins to secure the flaps of tissue to your dissecting pan to hold your pig open for identification of the underlying anatomy.

16 Once the internal organs of your specimen have been exposed completely, use Figure 22.8 and Table 22-3 to begin identifying the digestive anatomy.

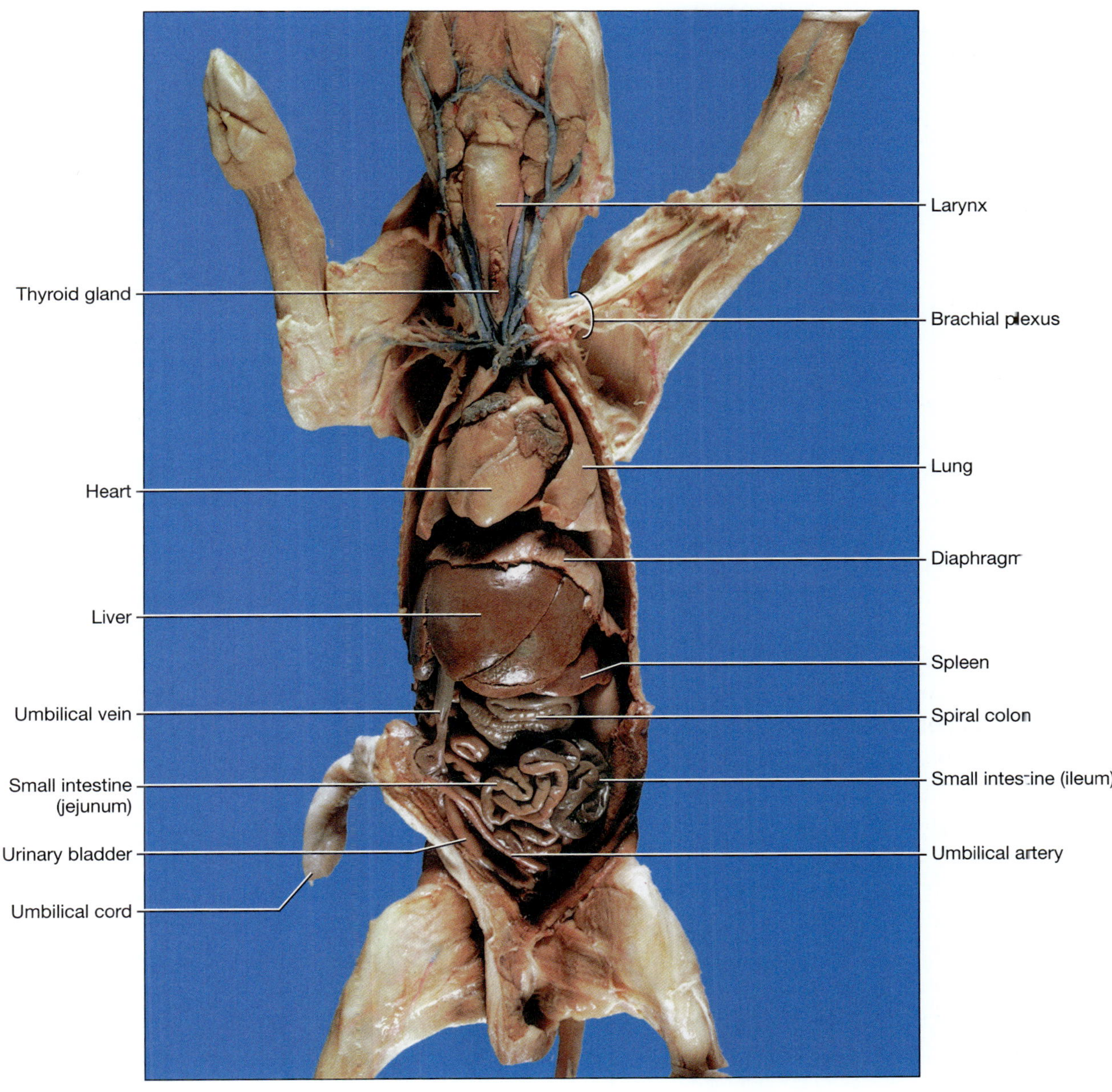

22.8 Fetal pig: major organs of thoracic and abdominal cavities, ventral view.

22

Table 22-3 Major Digestive Organs and Their Functions in the Fetal Pig

Structure	Function
Esophagus	Transports food to the stomach
Stomach	Produces hydrochloric acid and pepsinogen, which aid in the chemical breakdown of food
Pyloric sphincter (not pictured)	Muscular band that regulates the flow of chyme from the stomach into the duodenum
Liver	Produces bile, converts glucose to glycogen for storage, detoxifies many constituents of the absorbed digested compounds
Gallbladder	Stores bile produced by the liver
Bile duct	Transports bile from the gallbladder to the duodenum
Pancreas	Produces digestive enzymes and delivers them through the pancreatic duct to the duodenum
Duodenum	Receives chyme from the stomach along with bile and digestive enzymes from the gallbladder and pancreas
Jejunum	Responsible for the majority of nutrient absorption and reabsorption of water
Ileum	Continues the process of nutrient absorption and reabsorption of water
Cecum	Small, blind-ended outpocket demarcating the beginning of the large intestine that has a reduced appearance and function in carnivores and omnivores; in herbivores it contains anaerobic bacteria responsible for fermentation of cellulose and other plant materials
Spiral colon	Responsible for reabsorption of water and electrolytes; transports feces to the rectum via peristalsis
Rectum	Final site of water reabsorption and feces dehydration
Anus	Regulates egestion of undigested food (feces) from the body

The digestive system of the pig follows the basic mammalian blueprint. Digestion begins in the mouth where the teeth grind the food mechanically as it mixes with secretions produced by the **salivary glands**. This softened mixture passes down the esophagus to the **stomach**, where chemical secretions from the stomach lining facilitate the digestive process. The esophagus is a narrow tube containing smooth muscle that contracts to push food into the stomach. The stomach lies on the left side of the pig underneath the large, dark liver. It is a J-shaped sac responsible for storing large quantities of food, which relieves the pig of the need to eat constantly. A large stomach permits an animal to consume greater quantities of food in a short time span and then retire to a safe place to digest the meal over several hours. The stomach releases several chemical compounds that assist the digestive process, including hydrochloric acid and pepsinogen.

Bile, a compound containing bile salts produced in the **liver**, is routed to the **gallbladder** for storage (Fig. 22.9). From the gallbladder it is transported through the **bile duct** directly into the **duodenum**, the first portion of the small intestine. The **pancreas** produces digestive enzymes that also enter the duodenum in this region. The duodenum receives the partially digested foodstuffs and enzyme mix, known as **chyme**, and is primarily responsible for the final stages of enzymatic digestion.

Food leaves the duodenum and enters the second portion of the small intestine, the **jejunum**, a region of the small intestine that is highly convoluted and tightly bound together by mesentery. Absorption of nutrients and water occurs along the length of the jejunum, and the nutrients are delivered to the circulatory system through the hundreds of small blood vessels found throughout the intestinal mesentery.

The distal third of the small intestine, the **ileum**, is where further nutrient absorption and water reabsorption occur. Together, these three regions of the small intestine perform the majority of nutrient absorption in mammals.

22

At the juncture of the small intestine and the colon is a small, blind-ended outpocket of the intestine known as the **cecum** (Fig. 22.9). In carnivores and omnivores the cecum is very small and does not play a major role in digestion. In fact, in humans the cecum has been reduced to a vestigial remnant that we call the appendix. In herbivores, however, the cecum typically is quite large and serves as a fermentation chamber where symbiotic bacteria and protozoa reside.

These microorganisms produce an important enzyme, cellulase, naturally lacking in mammals, that breaks down the cellulose in plant cell walls and allows the mammal's own digestive enzymes access to the proteins and carbohydrates within the plant cells. Because pigs are omnivorous, the cecum is moderately reduced in this group of animals.

The mixture passes from the cecum through the **colon**, the organ primarily responsible for water reabsorption. In many mammals the colon is divided into three regions based on their relative positions in the body: the ascending colon, the transverse colon, and the descending colon. Functionally they are identical. Because of its unusually long length, the transverse portion of the colon in pigs is wound tightly into what is referred to more commonly as the spiral colon (Fig. 22.10). Locate the descending portion of the colon, which runs along the dorsal aspect of the abdominal cavity. Its distal portion is referred to as the rectum. The **rectum** is the final site of water reabsorption and feces dehydration, and the **anus** regulates the egestion of feces from the body.

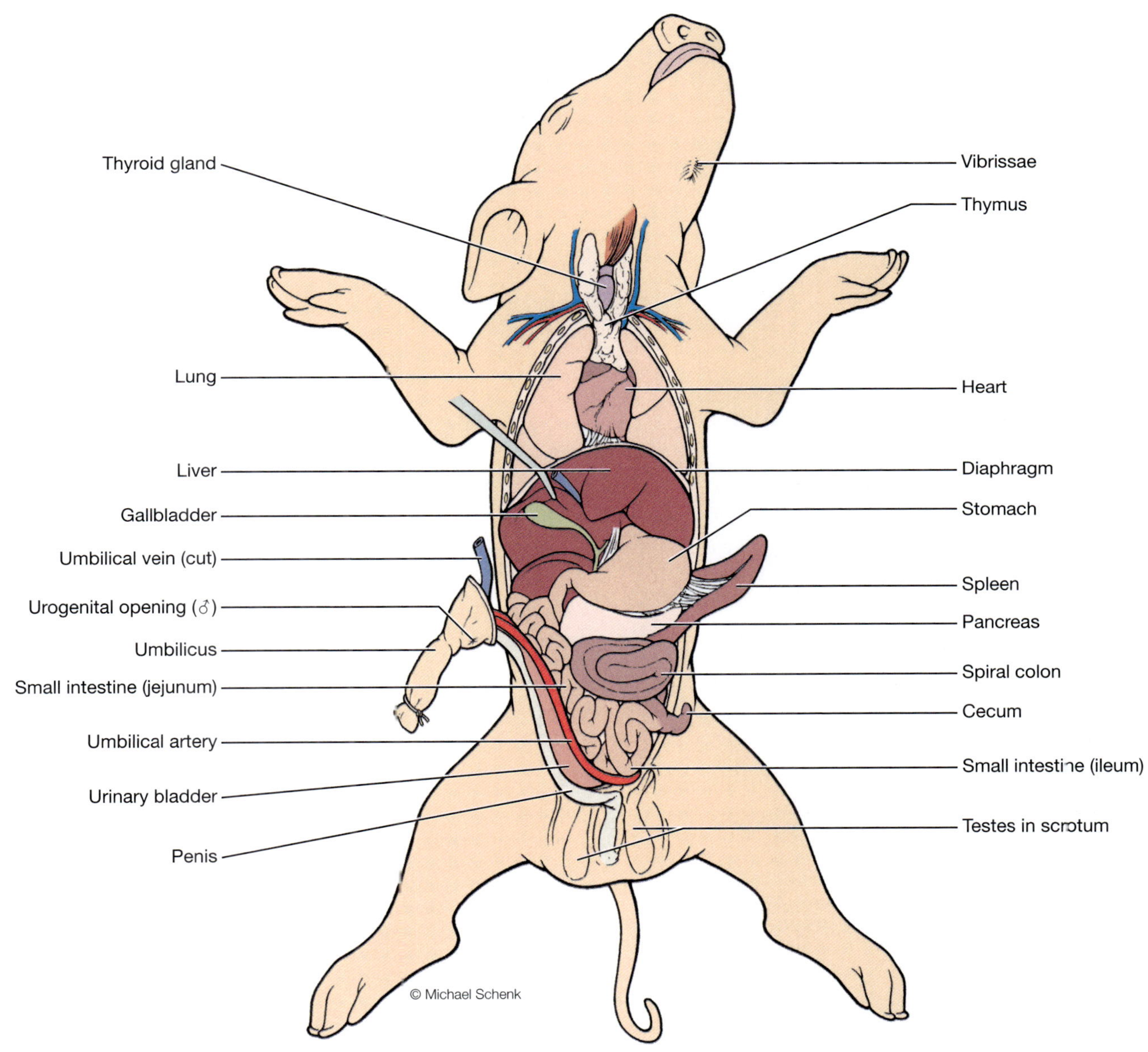

22.9 Fetal pig: organs of the thoracic and abdominal cavities, ventral view.

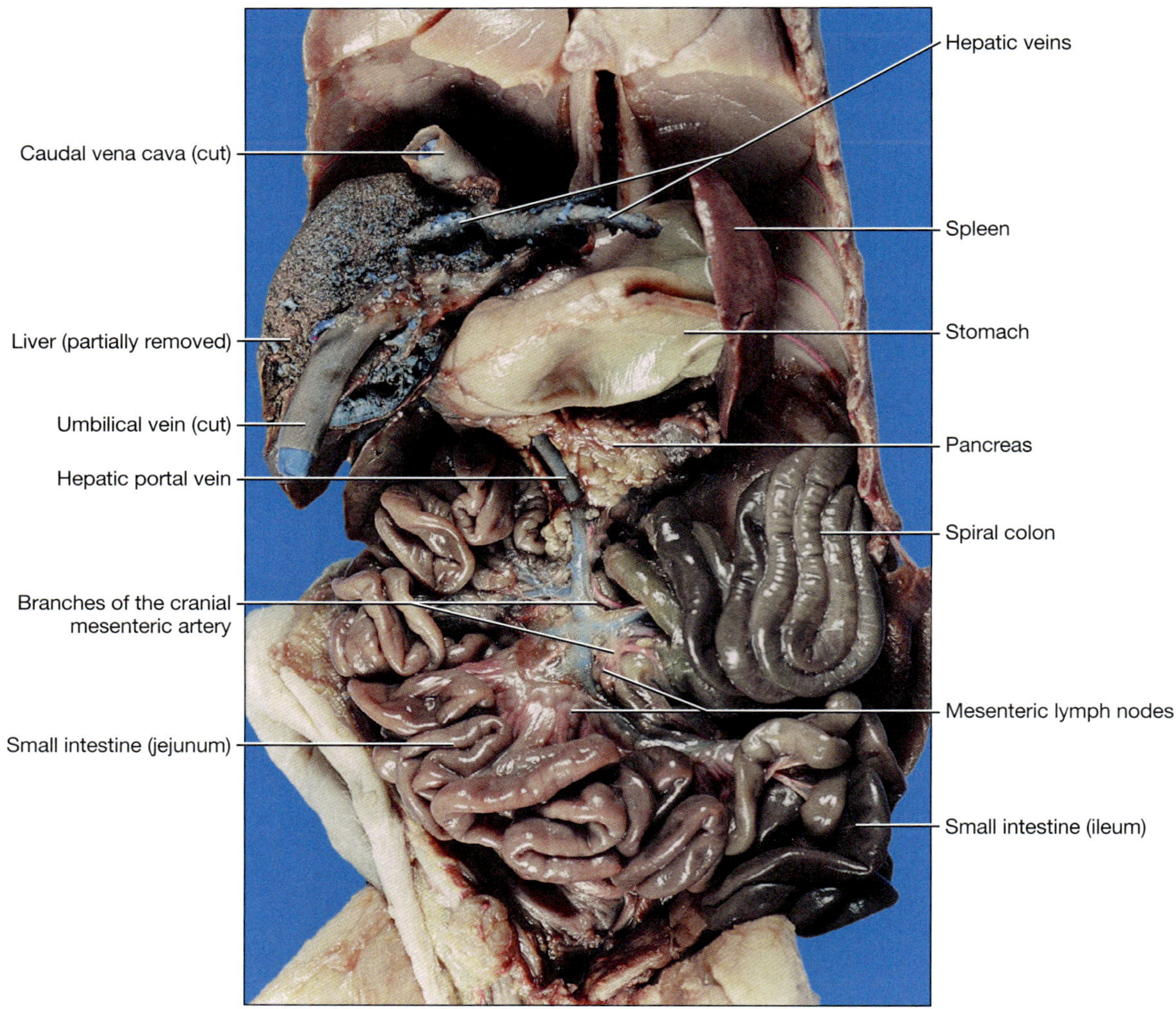

22.10 Fetal pig: abdominal cavity with organs displaced to expose underlying anatomy, ventral view.

Check Your Progress

3.1 Which structures/organs in the pig are responsible for the mechanical digestion of food?

3.2 Which structures/organs are responsible for chemical digestion?

3.3 Which structures/organs are responsible for nutrient or water absorption?

3.4 What role does the cecum play in the digestive process of herbivores? What structure in humans is homologous to the cecum?

Circulatory System

The circulatory system is responsible for transporting nutrients, gases, hormones, and metabolic wastes to and from the individual cells of an animal. Mammals are far too large for all of their individual cells to exchange nutrients, wastes, and gases with the external world by simple diffusion. Most cells are buried too deep inside the body to accomplish this task effectively. The circulatory system represents a series of vessels that diverge from the heart (arteries) to supply blood to the tissues and a confluence of vessels draining blood from the tissues (veins) and returning it to the heart.

Despite the extensive network of arteries and veins throughout the body, no actual exchange of water, nutrients, wastes, or gases occurs in arteries or veins; their walls are too thick to permit diffusion. Instead, extensive networks of capillary beds connecting branches of arteries and veins throughout the body transfer these dissolved substances between the bloodstream and the tissues.

To simplify identification of arteries and veins, two general principles you should remember are:

1. Arteries and veins tend to be paired, especially when the organs they supply or drain are paired.
2. A continuous vessel often undergoes several name changes along its length as it passes through different regions.

Therefore, to identify arteries and veins successfully, you will have to trace them along their entire length (typically from the heart outward).

Procedure

1 If you have not exposed the interior of the thoracic cavity by opening the rib cage, do so at this time.

2 The heart will be partially obscured from view by the thymus gland and will be encased in a thin, pericardial membrane (Fig. 22.11A). Gently move the thymus out of the way and carefully remove the pericardial membrane from the heart.

3 Using a teasing needle and forceps, dissect the muscle tissue and fatty tissue away from the major arteries and veins in the neck, pectoral, abdominal, and pelvic regions. This is a tedious process and will take some time.

4 If your specimen has been double-injected with latex, the arteries will appear red and the veins will appear blue. If your pig has not been injected with latex, the arteries will appear whiter and stiffer than the thin, collapsed veins. Remember—arteries are more heavily walled than veins (to accommodate higher blood pressures) and generally are thicker, and thus more evident during dissection.

The driving force behind the circulatory system of mammals is the heavily muscled, four-chambered heart that delivers blood through two major circulatory pathways—the pulmonary circuit (from the heart to the lungs and back) and the systemic circuit (from the heart through the rest of the body and back). A hallmark of the mammalian heart is that its internal design keeps the blood from these two circuits entirely separate, thus keeping oxygen-depleted blood from mixing with oxygen-rich blood and allowing for different pressures to be maintained on the two sides of the heart.

5 First, identify the four chambers of the heart.

Caudally there are two large, thick-walled ventricles, the **right ventricle** and the **left ventricle** (Fig. 22.11B). These chambers pump blood out of the heart to the lungs and to the rest of the body, respectively. Cranial to the ventricles and somewhat darker in color are the **right** and **left auricles**. Chambers within the right and left auricles (called atria) receive blood from the body and the lungs, respectively, and pass it to the ventricles. The term *auricle* is used to describe the small, outer, flap-like region that covers a portion of the atrial chamber. The term *atrium* refers to the actual chamber inside that collects the blood. Notice the large veins entering the heart. These are the cranial and caudal vena cavae, which bring deoxygenated blood to the right atrium from the cranial and caudal portions of the body.

6 On the ventral surface of the heart, locate the large pulmonary artery emanating from the right ventricle.

In the adult the pulmonary artery channels blood from the right ventricle through the right and left pulmonary arteries to the lungs. Notice that at the base of the pulmonary artery there is a connection to the aorta—the large artery leaving the cranial aspect of the left ventricle. This connection is called the ductus arteriosus, a short, temporary linkage found only in the fetus.

Lying adjacent to the pulmonary arteries are the pulmonary veins, the vessels that, in the adult, return oxygenated blood to the left atrium of the heart. Unlike most arteries that receive red latex during the injection process, the pulmonary arteries receive blue latex, because they carry deoxygenated blood. Likewise, the pulmonary veins are injected with red latex, because they carry oxygenated blood. Do not let the color difference confuse your identification of these vessels.

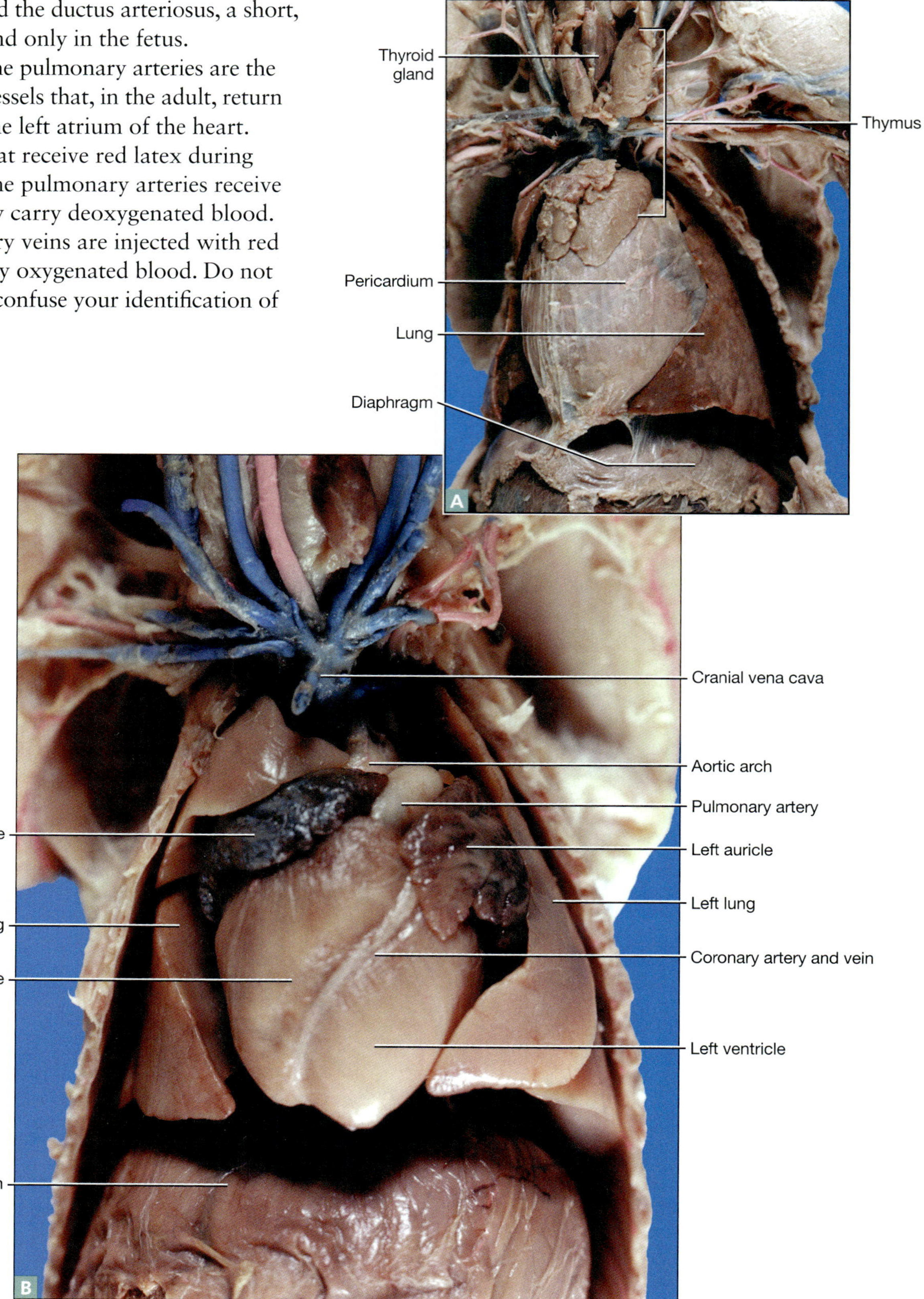

22.11 Fetal pig, heart: (A) surrounded by pericardial membrane, and (B) with pericardial membrane removed.

The circulatory systems of a fetus and adult mammal have several major differences (Fig. 22.12). The most obvious is the connection between the fetus and the mother through the umbilical cord. In placental mammals the unborn young are attached to the placenta of the mother by this connection, which serves as the lifeline for the fetus, transporting nutrients and oxygen to the growing fetus, and providing a channel for carbon dioxide and excess metabolic wastes to be eliminated from the fetus. The maternal blood is restricted to the maternal side of the placenta and never mixes with fetal blood under normal circumstances.

Exchange of gases, nutrients, and wastes is accomplished by diffusion across the placental barrier. The single umbilical vein carries oxygen- and nutrient-rich blood to the fetus from the fetal side of the placenta, and the two smaller umbilical arteries route deoxygenated blood from the fetus to the placenta. While the fetus remains in its mother's uterus, metabolic waste (urine) that collects in the bladder passes through the **allantoic stalk** and into the **allantois** for storage.

The fetus is not breathing with its own lungs, so fetal lungs do not oxygenate blood that passes through them. In fact, only a small fraction of the blood leaving the fetus' right ventricle travels through the pulmonary arteries to the lungs. The majority is redirected through the ductus arteriosus, a connection between the pulmonary artery and the aorta that channels blood into the aorta (Fig. 22.12A). Another structure inside the heart of the fetus, called the foramen ovale, also aids in rerouting blood to bypass the lungs. This opening in the septum between the right and left atria allows blood passing into the right atrium to be channeled into the left atrium and away from the lungs. Both of these adaptations ensure that the majority of oxygenated blood arriving via the umbilical vein is passed through the fetal circulatory system via the aorta, while permitting enough blood to reach the lungs to allow the lung tissue to develop.

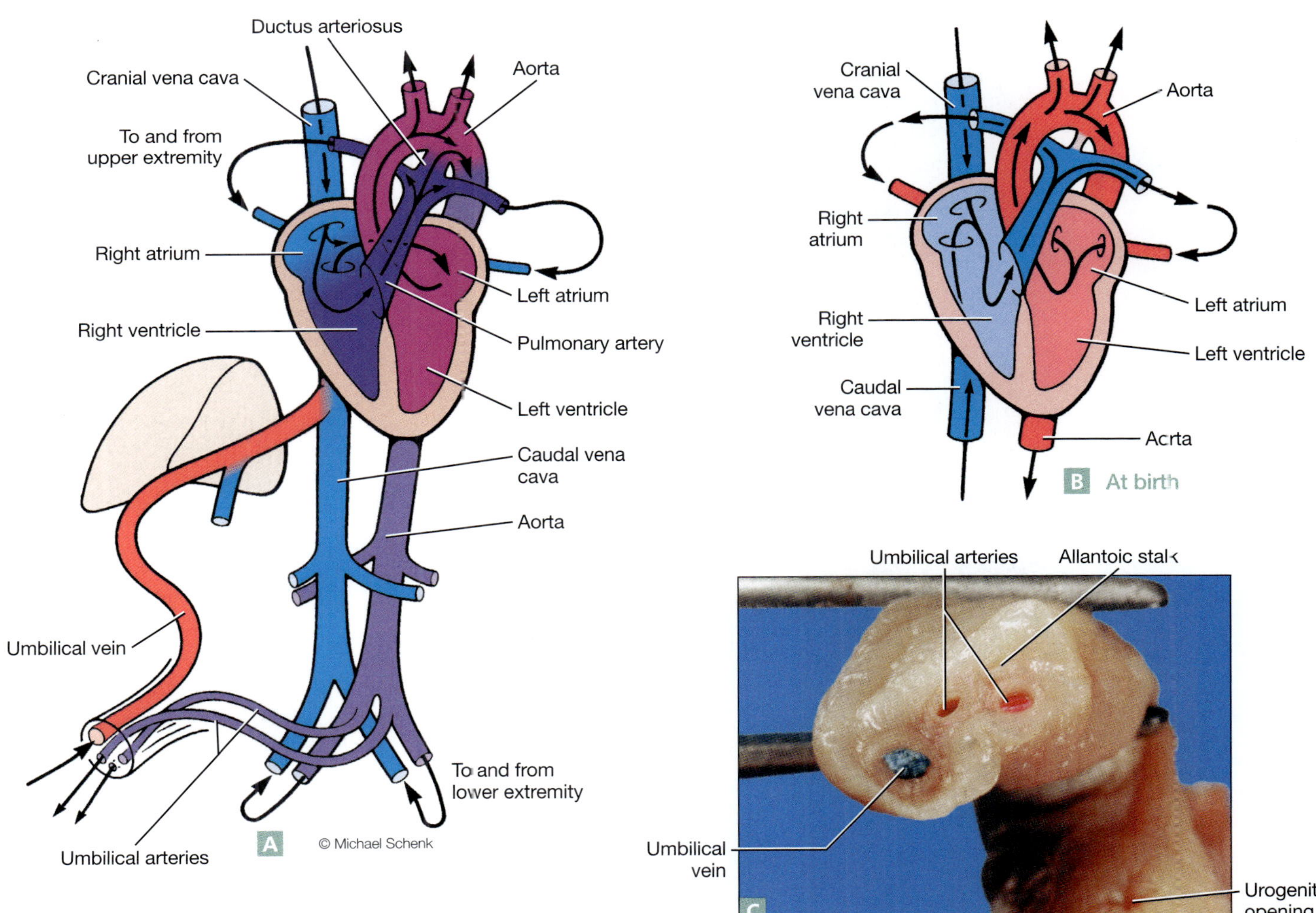

22.12 Fetal pig: schematic illustrations of (A) fetal circulatory pathway, and (B) pathway after birth, with inset photograph (C) of umbilical vessels.

22

7 Locate the spleen along the left side of the body adjacent to the stomach.

In addition to the heart, the **spleen** is a vascular, ductless organ that also plays a critical role in the circulatory system of vertebrates (Figs. 22.9–22.10). Because mammalian red blood cells do not contain nuclei, they cannot undergo cell division and, thus, have a finite life span. New red blood cells are produced continuously in the bone marrow and are delivered to the spleen for storage. The spleen stores these cells along with excess blood and releases these products into the bloodstream as needed. Through this mechanism the spleen regulates the body's total blood volume and the relative concentration of red blood cells. The spleen also manufactures white blood cells (lymphocytes) to fend off diseases and destroys and recycles worn-out blood cells.

8 Trace the paths of the major arteries emanating from the heart and the major veins returning to the heart as they diverge and converge throughout the body, using Tables 22-4–22-5 and Figures 22.13–22.15 to assist you in identifying these vessels in the pig.

Table 22-4 Major Arteries and Veins of the Thoracic Region and Their Functions in the Fetal Pig

Vessel Name	Function
External jugular veins	Drain deoxygenated blood away from the head region
Internal jugular veins	Drain deoxygenated blood away from the neck region
Subclavian veins	Drain deoxygenated blood away from the lateral portions of the thoracic cavity and forelimbs
Cephalic veins	Drain deoxygenated blood away from the lateral portions of the thoracic cavity and forelimbs
Subscapular veins	Drain deoxygenated blood away from the forelimbs
Axillary veins	Drain deoxygenated blood away from the forelimbs
Cranial vena cava	Returns deoxygenated blood from the cranial portion of the body to the heart
Caudal vena cava	Returns deoxygenated blood to the heart from all posterior parts of the body
Aortic arch	Supplies oxygenated blood from the heart to the entire body through numerous major branches
Brachiocephalic trunk	First major branch off aortic arch; supplies oxygenated blood to the right subclavian artery and carotid trunk
Left subclavian artery	Second major branch off aortic arch; supplies oxygenated blood to the left forelimb
Right suclavian artery	Branches off brachiocephalic trunk; supplies oxygenated blood to right forelimb
Common carotid arteries	Both branch off the brachiocephalic trunk; supply oxygenated blood to the head and neck regions
Axillary arteries	Supply oxygenated blood to the forelimbs

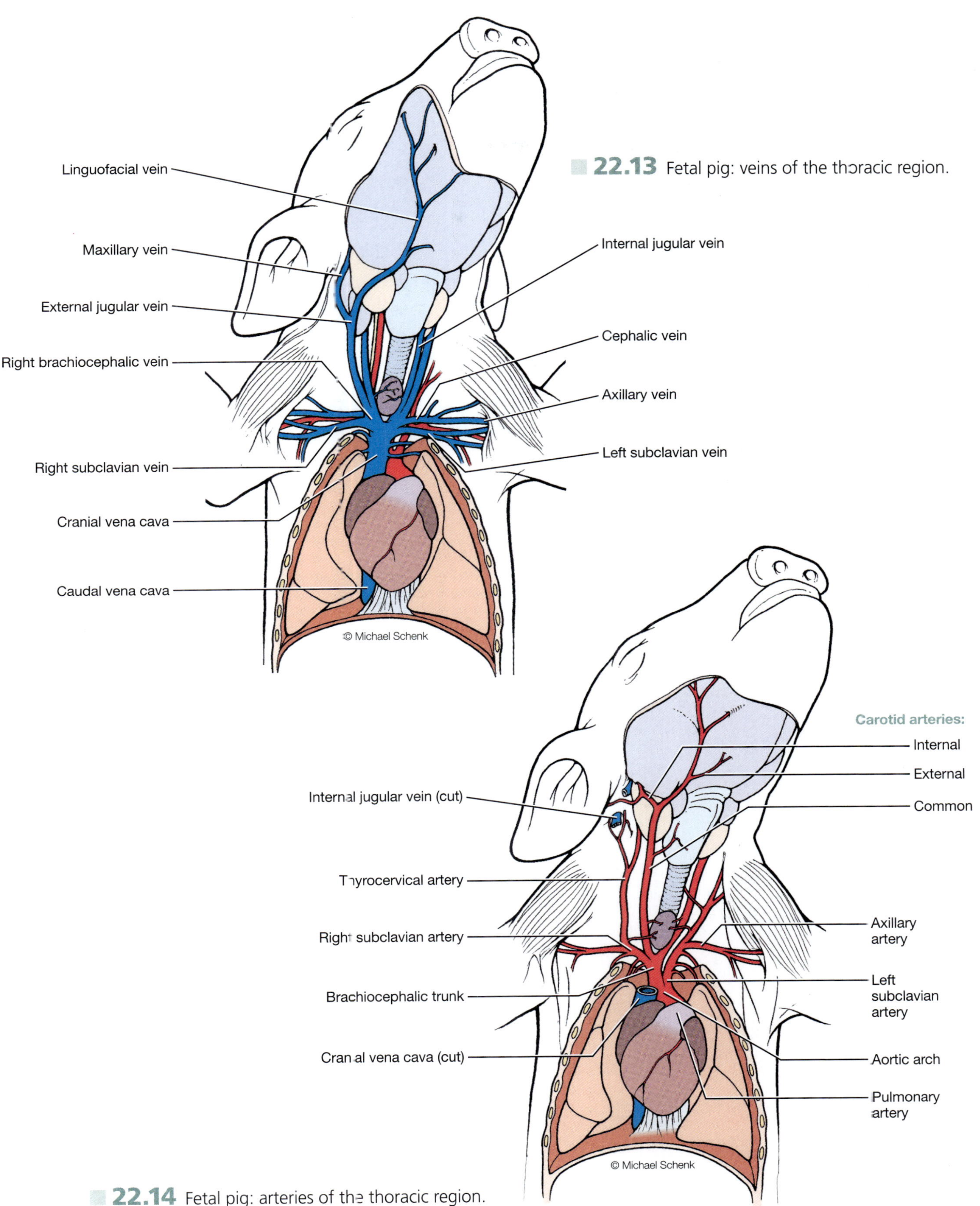

22.13 Fetal pig: veins of the thoracic region.

22.14 Fetal pig: arteries of the thoracic region.

22

Table 22-5 Major Arteries and Veins of the Abdominal Region and Their Functions in the Fetal Pig

Vessel Name	Function
Umbilical vein	In the fetus, returns oxygenated, nutrient-rich blood from the fetal side of the placenta to the caudal vena cava
Aorta	Supplies oxygenated blood to all posterior parts of the body
Celiac artery	Supplies oxygenated blood to the stomach, spleen, pancreas, duodenum, and liver
Cranial mesenteric artery	Supplies oxygenated blood to the jejunum, ileum, cecum, and colon
Caudal vena cava	Returns deoxygenated blood to the heart from all posterior parts of the body
Renal arteries and veins	Supply oxygenated blood to the kidneys (arteries) and return deoxygenated blood from the kidneys to caudal vena cava (veins)
Genital arteries and veins	Supply oxygenated blood to the testes (male) and ovaries (female) and return deoxygenated blood from the gonads to caudal vena cava
External iliac arteries and veins	Major divergences off aorta and vena cava that supply blood to the hind limbs (arteries) and drain blood from the hind limbs (veins)
Femoral arteries and veins	Divergences off external iliac arteries and veins that supply and drain blood from the lateral portions of the hind limbs
Deep femoral arteries and veins	Divergences off external iliac arteries and veins that supply and drain blood from the medial portions of the hind limbs
Umbilical arteries	In the fetus, transport blood to the fetal side of the placenta to receive nutrients and oxygen
Internal iliac arteries and veins	Supply oxygenated blood to the dorsal surface of the pelvic region (arteries) and return blood from the pelvic region to caudal vena cava (veins)

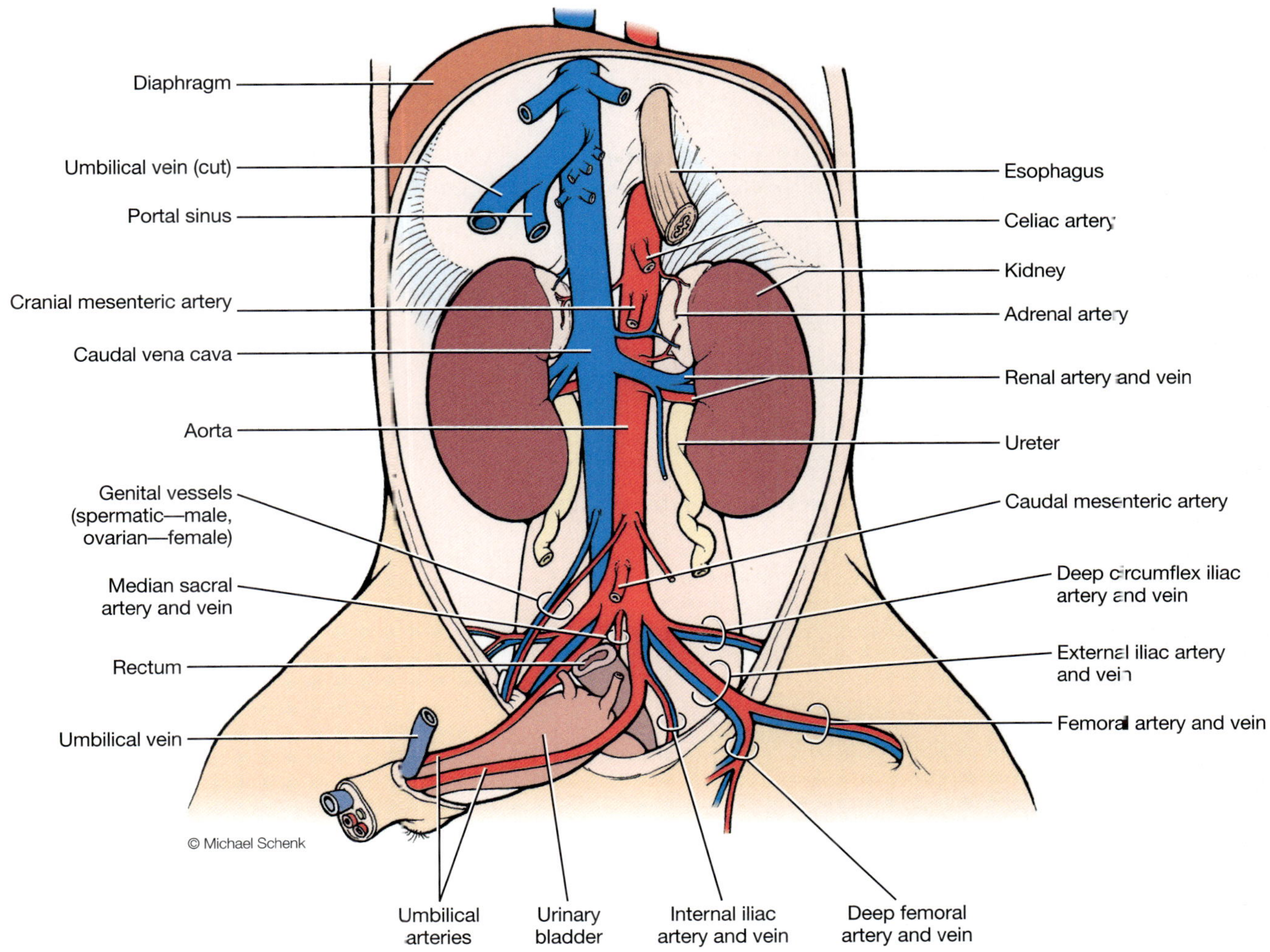

22.15 Fetal pig: arteries and veins of the abdominal region (ureters cut to expose vessels).

22

4.1 The mammalian heart has ______ chambers.

4.2 Arteries carry blood ______________ the heart, and veins carry blood ______________ the heart. Tiny blood vessels called ______________ connect arteries to veins and are the actual sites of gas and nutrient exchange with the body tissues.

4.3 True or false? In the fetus, the umbilical vein brings oxygenated blood from the placenta to the fetus.

4.4 What role does the ductus arteriosus play in fetal circulation?

4.5 What role does the spleen play in the circulatory system?

4.6 Describe at least two advantages that the four-chambered heart provides mammals.

Respiratory System

The respiratory system of mammals is responsible for bringing a fresh supply of oxygen to the bloodstream and carrying off excess carbon dioxide. The anatomy of the respiratory tract is designed to humidify and warm the air while filtering out dust particles and germs. The lining of the nasal epithelium is covered with fine hairs that capture these foreign particles and prevent them from passing into the lungs, where they may infect the body. Similarly, as air is exhaled, it is cooled and dried, thereby reducing the amount of heat and moisture that mammals lose through respiration.

Procedure

1 Locate the large larynx and trachea.

In mammals the **trachea** is a long tube reinforced with cartilaginous rings to prevent collapse as the animal inhales (Fig. 22.16). This tube leads from the nasopharynx through the larynx and into the lungs. The larynx should appear as an enlarged, oval-shaped protrusion toward the cranial end of the trachea. The **larynx** ("voice box") allows mammals to have a vast repertoire of vocalizations ranging from ultrasonic squeaks and chirps (in bats) and guttural barks or grunts (in dogs and pigs) to the highly complex sounds of human speech. The pitch of these vocalizations is controlled by muscles attached to the larynx that contract and relax, altering the shape of the voice box, thus changing the sounds that it produces.

2 Follow the trachea caudally toward the lungs.

Notice that the trachea first splits into two primary **bronchi**—the left and right bronchus (Fig. 22.16). These lead into the **left** and **right lung**, respectively. Notice that the right lung is divided into four lobes and the left lung has two divisions. In humans the right lung has three lobes and the left lung has two. These individual lobes are distinguished primarily by internal divisions of the bronchi that are not always apparent superficially.

Inside the lungs the bronchi are divided further into several branches called bronchioles. The bronchioles branch into smaller and smaller tubules, eventually terminating in microscopic, open sacs called **alveoli.** The alveoli are composed of squamous epithelial tissue and are surrounded by capillary networks. Here, in the adult, oxygen is picked up by the bloodstream and carbon dioxide is released into the lungs to be expelled from the body through exhalation.

3 Identify the thin, muscular diaphragm lying on the cranial margin of the liver (Fig. 22.16).

After the fetus is born and is breathing on its own, the **diaphragm** allows the thoracic cavity to expand and compress, drawing in fresh air with each expansion (as the diaphragm contracts) and expelling stale air with each compression (as the diaphragm relaxes). A muscular diaphragm is a predominately mammalian characteristic that improves the efficiency of the respiratory process and helps mammals maintain a high metabolic rate.

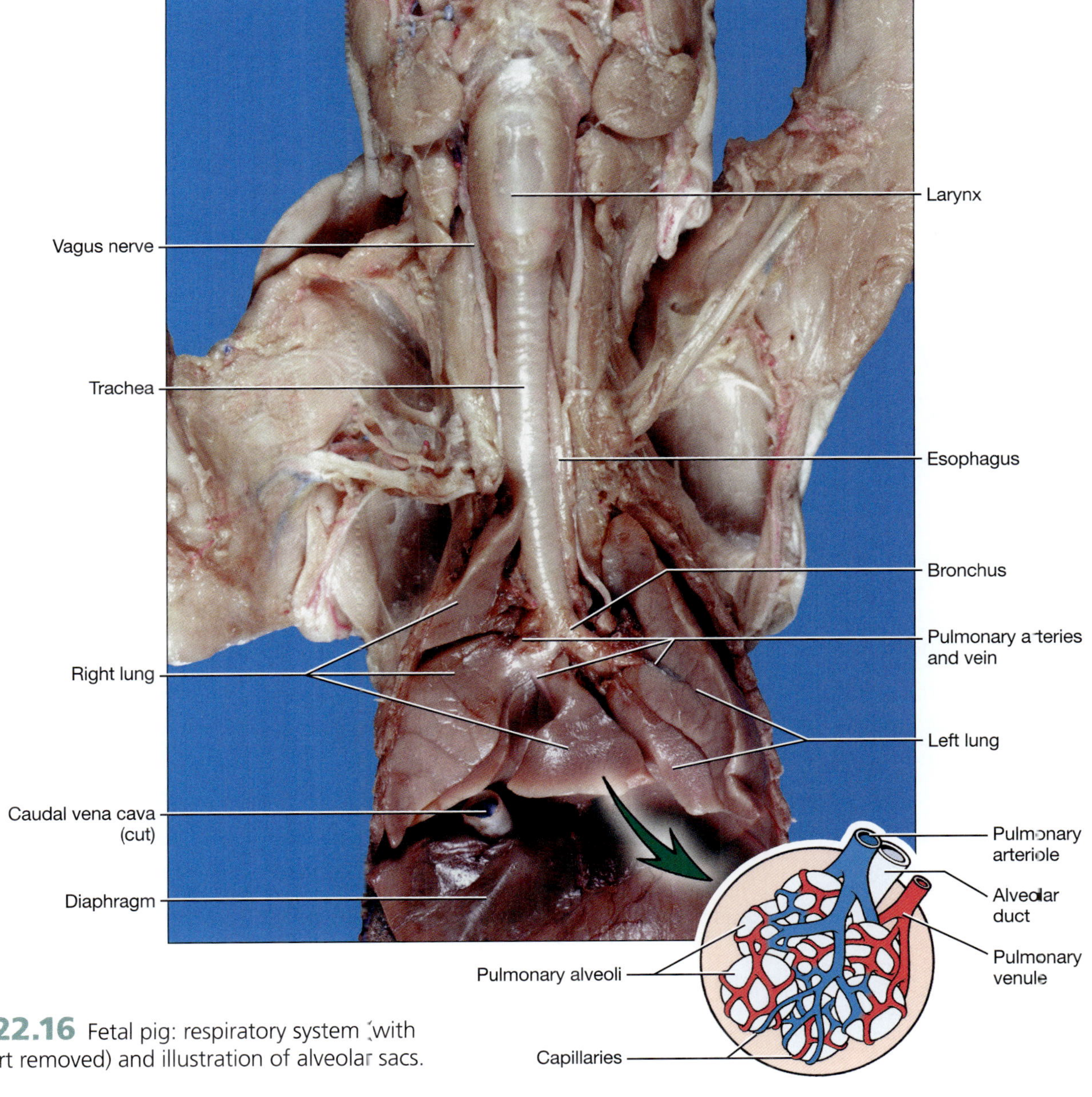

22.16 Fetal pig: respiratory system (with heart removed) and illustration of alveolar sacs.

22

5.1 How many distinct lobes are present on the right lung? The left lung?

5.2 Describe how the respiratory system of mammals differs from that of reptiles.

Excretory and Reproductive Systems

Excretory organs are responsible for eliminating metabolic wastes that the body produces from cellular respiration and for maintaining a homeostatic balance among the levels of fluids, electrolytes, sugars, hormones, and proteins in the body. Remember—excretion is an entirely different process from that which expels indigestible products through the anus. Excretion and egestion (or defecation) are different processes, handled by completely different systems in mammals. Reproductive organs are responsible for producing the gametes that ultimately will fuse with the corresponding gametes of the opposite sex. In addition to reproduction, the testes and ovaries produce many of the hormones that are associated with the development and maturation of primary and secondary sexual characteristics and which drive the complex repertoire of sexual behaviors indicative of most mammals.

Excretory System

Procedure

1 Using a teasing needle, carefully dissect away the membranous tissue surrounding one of the kidneys. Take care not to destroy the **adrenal gland**, which sits along the cranial margin of the kidney, or any of the ducts and blood vessels in the area.

2 If your specimen is a male, be careful not to damage the vas deferens, which "loops" around the ureter.

3 Clean the area around the kidney to expose the renal artery and vein and the ureter passing from the medial margin of the kidney caudally toward the urinary bladder.

4 To uncover all of the reproductive structures completely, you must cut longitudinally through the pubic symphysis with a scalpel. *Cut carefully* and start your incision slightly to one side of the median plane of the pelvis to avoid cutting through underlying structures. Preferably, cut through the symphysis only partially and then apply downward (lateral) pressure to each of the hind limbs to complete the separation.

5 Regardless of the sex of your specimen, you are expected to be familiar with the reproductive structures of both male and female pigs, so work closely with another group that has a specimen of the opposite sex.

The **kidneys** are large, bean-shaped organs that lie along the dorsal surface of the abdominal cavity on either side of the spine (Figs. 22.17–22.18). The kidneys filter blood from the circulatory system, removing the metabolic waste products produced in the tissues of the body during cellular respiration. The major function of the kidneys is to concentrate these toxins and eliminate them from the body while conserving water, salts, and other compounds that the body needs. In humans, the kidneys filter between 1,100 and 2,000 liters of blood each day! From this tremendous volume of blood only about 1.5 liters of urine are actually produced. The other 99.9% is reabsorbed into the bloodstream through a highly efficient system of semipermeable tubules that generates concentration gradients in the nephrons of the kidney. The urine is concentrated in the kidneys and passes down each **ureter**, tubes lined with smooth muscle that transport urine toward the **urinary bladder** for temporary storage.

The urinary bladder is a muscular reservoir that can expand to many times its "relaxed" size to accommodate large volumes of urine. When relaxed, the inner walls of the bladder appear folded, somewhat resembling the rugae of the stomach lining. In adults, the stored urine is eliminated through the **urethra**, but while the fetus remains in its mother's uterus, urine that collects in the bladder passes through the allantoic stalk and into the allantois for storage.

Male Reproductive System

In males, the **scrotum** houses the paired testes. Depending on the age of your pig and degree of testicular migration, the **testes** will be found along the dorsal side of the abdominal cavity somewhere between the kidneys and the scrotum, appearing as small, bean-shaped structures. If the testes have descended into the scrotum, they will be enclosed within thin, membranous cremasteric pouches (Fig. 22.17). At the cranial end of each cremasteric pouch, a narrow tube should be evident. This is a **spermatic cord** containing a vas deferens, a spermatic artery and vein, lymphatic vessels, and numerous nerves leading to the testis and the epididymis located within the cremasteric pouch.

Procedure

1 Carefully make a slit in the cremasteric pouch and peel it open, using scissors if necessary.

2 Leave the testis attached to the spermatic cord, but separate it from the tissue of the cremasteric pouch.

Cupped around the side of each testis is a highly coiled system of tubules known as the **epididymis** (Fig. 22.17). Sperm are produced within the **seminiferous tubules** of each testis and are stored along the length of the epididymis. Upon ejaculation, sperm leave each epididymis and travel through each **vas deferens** toward the urethra. Trace along the length of the spermatic cord to visualize the path sperm travel as they move out of the epididymis through the vas deferens (which loops around the ureter) toward the base of the urethra. If the testes on your male pig have descended into the scrotum, there will be a visible opening in the abdominal wall (the inguinal canal) through which each spermatic cord passes from the scrotum into the abdominal cavity.

3 Further dissection will reveal several accessory reproductive glands along this route.

At the juncture of the vas deferens and the urethra are the **seminal vesicles**. They lie on the dorsal side of this juncture and may be difficult to find unless you use a blunt probe to flip over the urethra and view its dorsal aspect (Fig. 22.17). The seminal vesicles secrete a viscous fluid that contains mucus (to prevent the sperm from drying out), large amounts of fructose (to provide energy for the sperm, thereby promoting sperm motility and viability), and hormones to stimulate uterine contractions. In addition, the secretions are highly alkaline to neutralize the acidic environment of the vagina and increase the chances of survival for the sperm. The **bulbourethral glands** lie more caudally on either side of the urethra. They produce alkaline secretions that assist in lubrication during intercourse and also aid in neutralizing vaginal acidity. Together, the seminal vesicles and bulbourethral glands contribute more than 60% of the total fluid volume of semen.

During ejaculation all of these secretions mix with the sperm as the mixture passes through the urethra along the length of the **penis**. Notice that the penis of the fetal pig does not protrude from the body yet. Instead, it is enclosed in an epithelial sheath and held along the ventral wall of the abdomen. To isolate it, you will have to tease away the penis carefully from the tissue along the ventral midline of the body.

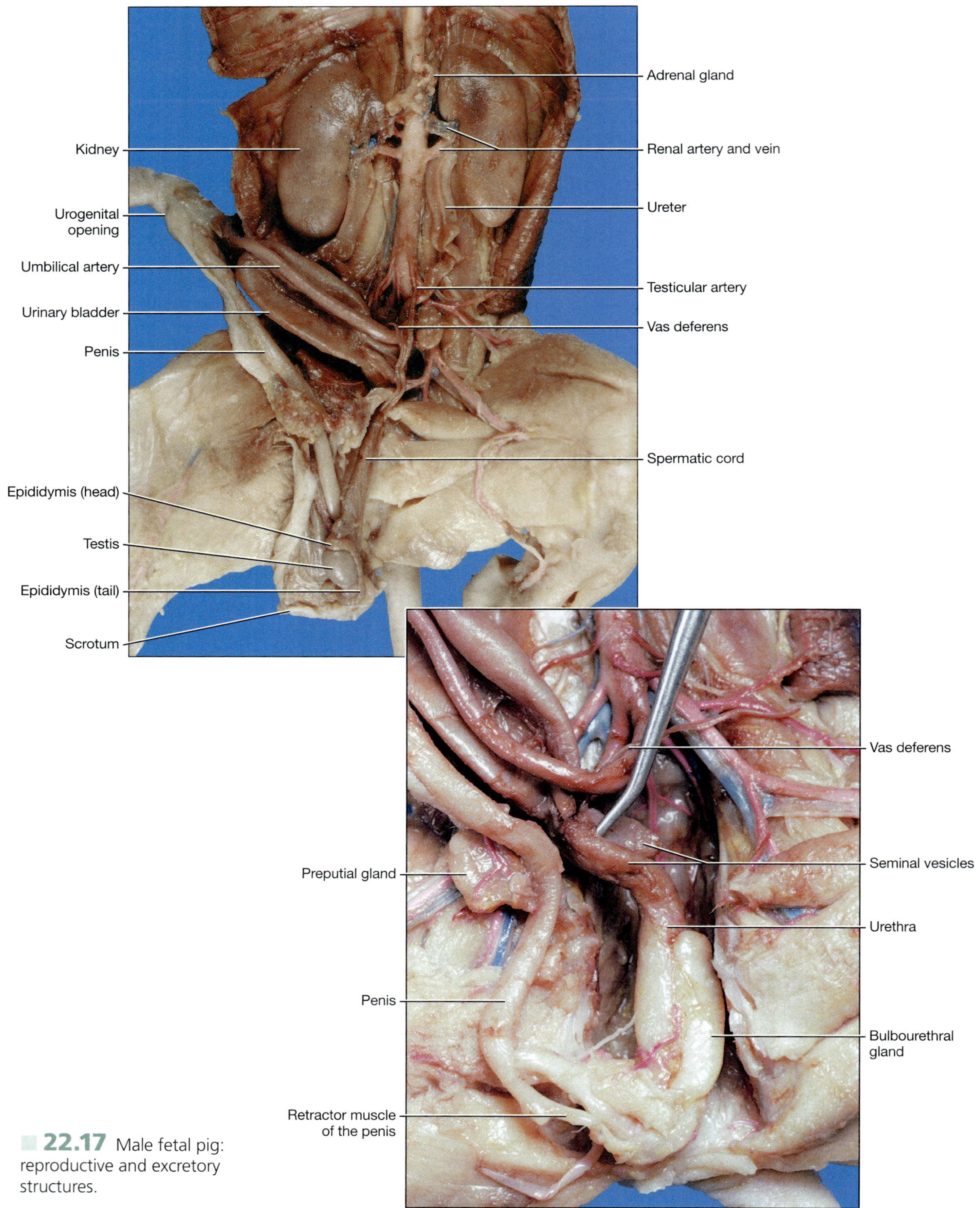

22

■ **22.17** Male fetal pig: reproductive and excretory structures.

Female Reproductive System

In the female, paired **ovaries** are located in the abdominal cavity caudal to the kidneys. They can be identified by their small, round appearance (Fig. 22.18). In direct communication with each ovary is a tiny, coiled **oviduct** that receives the mature oocyte (egg) when it is released from the ovary at the time of ovulation. No actual physical connection exists between the oviductal opening and the ovary. Instead, small, fingerlike projections of the oviduct generate movements that sweep the egg into the oviduct. The epithelial lining of the oviduct is ciliated and propels the egg along the length of the oviduct toward the uterine horn.

Fertilization typically takes place in the upper third of the oviduct, but implantation of the embryos occurs further along the uterus. In pigs, the uterus is divided into two conspicuous **uterine horns**, where embryonic development of the fetuses occurs, and a short uterine body, where the two uterine horns converge on the cervix (Fig. 22.18). In humans the uterine horns (known as fallopian tubes) are quite reduced, because the zygote implants and develops in the body of the uterus.

Locate the juncture of the cervix and the vagina. The **cervix** is a constriction of semi-cartilaginous tissue, and the **vagina** extends caudally from this constriction. The vagina is joined by the urethra, and the two open into a common chamber called the urogenital sinus—so named because it handles products of both the urinary and the reproductive systems. The urogenital sinus opens to the outside of the body through the **urogenital opening.** Unlike pigs, human females lack a single urogenital opening. Instead, the urethra and vagina have separate openings to the outside of the body in close proximity to one another. On the fetus the genital papilla will be visible as a small, fingerlike projection on the caudoventral surface of the abdominal cavity covering the urogenital opening (Fig. 22.18). As a homologue to the male penis, this structure plays a similar role in sexual sensation and sends information about sexual stimulation to the brain.

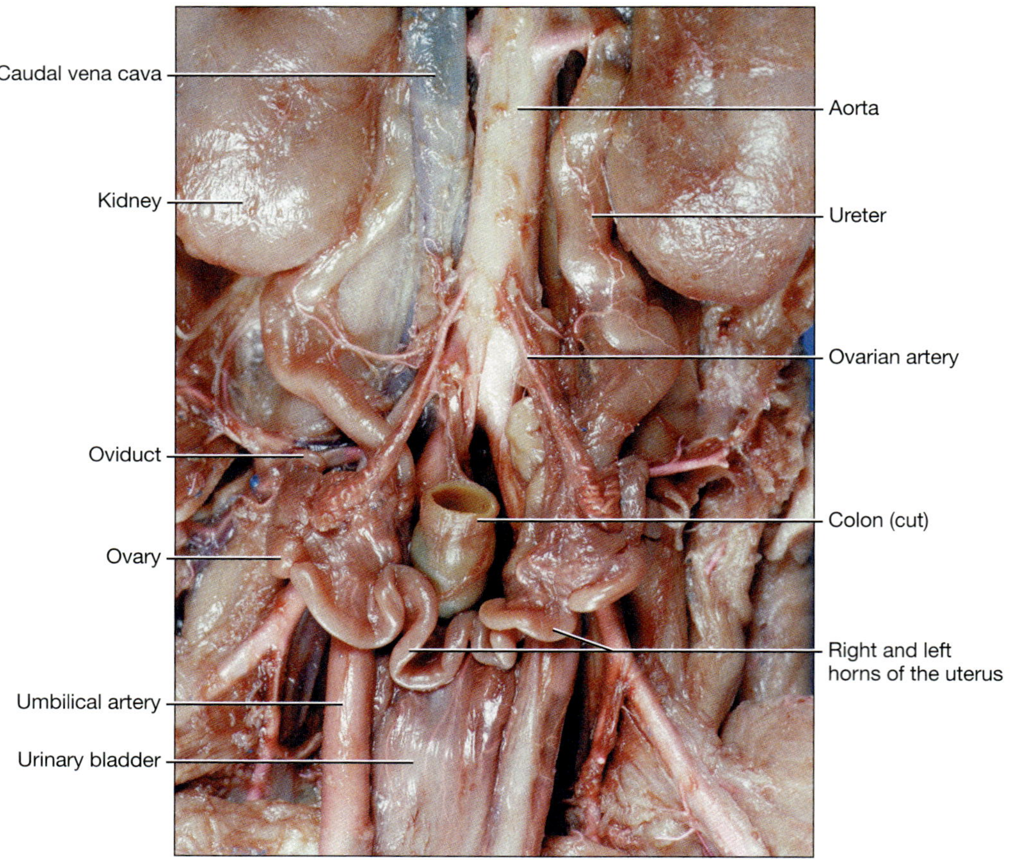

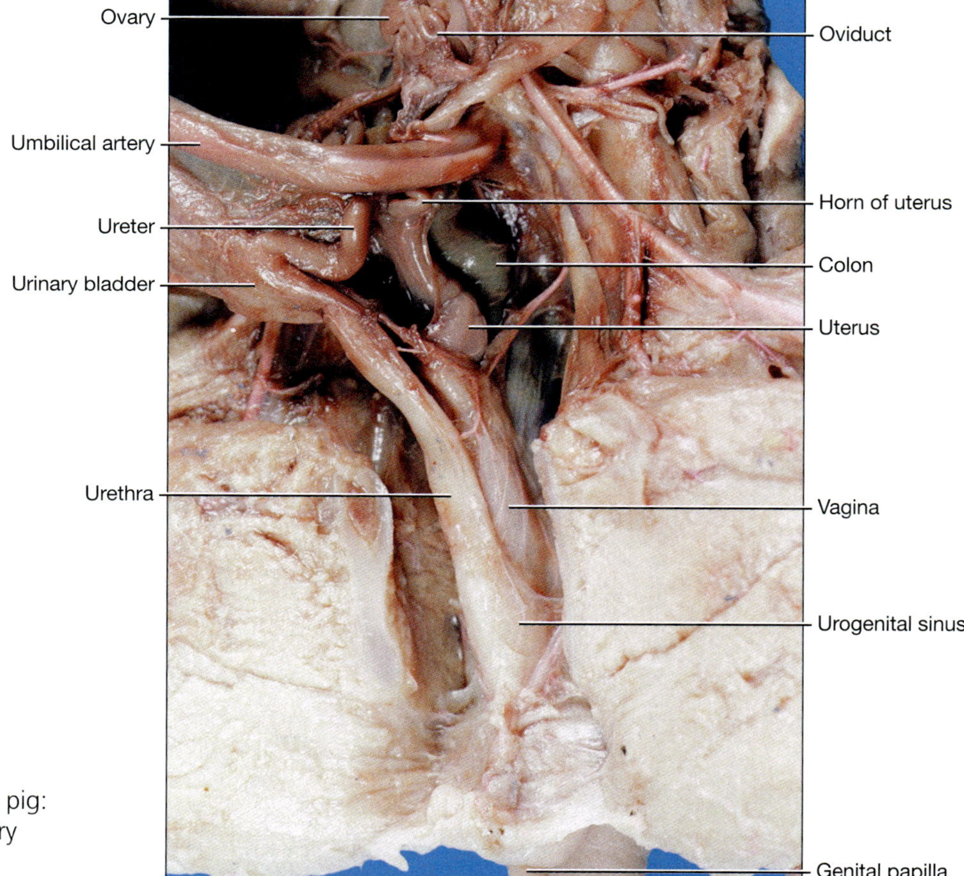

22.18 Female fetal pig: reproductive and excretory structures.

Endocrine System

The endocrine system is a major player in the body's attempt to coordinate the activities of its many organs and organ systems. In that respect, the endocrine and nervous systems are very much alike. The similarities between the two systems do not go very far beyond that, however. Unlike the nervous system, which has its own contained system for information transfer (the nerves), the endocrine system is ductless and, therefore, must rely on another neighboring system (the circulatory system) to send its messages throughout the body. The glands of the endocrine system produce and secrete their hormones directly into the bloodstream to be carried to their target organs.

Hormones are chemical compounds that interact with target cells in the body to produce myriad behavioral, neurological, and physiological responses. In this way they influence many of the same behaviors and processes that the nervous system regulates. Because of the nature of hormones, however, the effects produced by the endocrine system generally are not short-lived. Nervous responses are instantaneous and degrade immediately, but hormones circulating through the bloodstream may take some time to produce a response and anywhere from minutes to hours to break down. Thus, hormonal effects tend to be much longer in duration and the processes that are under hormonal control typically are processes that occur over hours, days, weeks, or even years (for example, sexual maturation, metabolic rate, growth rate, and ovulation).

In addition, the degree of response shown by the target organ is directly proportional to the amount of hormone released by the endocrine gland: The more hormone a gland releases, the more pronounced the effect. This is a fundamental distinction from the all-or-nothing response of nerve cells and illustrates why both systems—the nervous system and the endocrine system—are essential for complex organisms to coordinate different aspects of their lives.

The centralized control center of the endocrine system is the hypothalamus-pituitary complex of the brain, often referred to as the pituitary gland (Fig. 22.19). You will be able to see this organ only if you examine a commercially prepared sheep brain. The hypothalamus-pituitary complex produces many hormones that, in turn, stimulate the activity of many of the other endocrine glands in the body. Likewise, other endocrine organs produce hormones that stimulate or inhibit regions of the pituitary gland and hypothalamus. Through this feedback loop, the endocrine system is able to turn itself on and off in response to environmental or endogenous stimuli.

You already have identified some of the organs discussed in this section, because organs that function in the endocrine system often contain other tissue that functions in other body systems (for instance, digestion, reproduction, or nervous control). Return to your dissection, and identify the endocrine glands visible on your specimen. Figure 22.19 depicts the approximate location of these glands in the fetal pig, and Table 22-6 lists the hormones that each gland produces and their effects on the body.

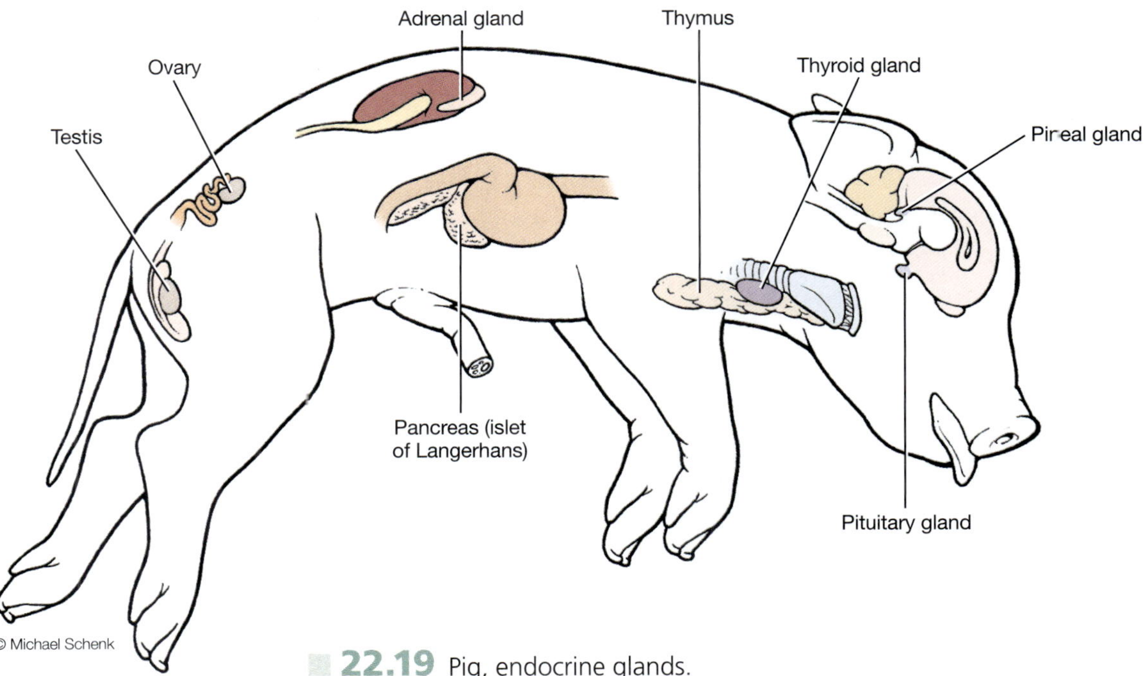

22.19 Pig, endocrine glands.

Table 22-6 Endocrine Glands, Hormone Products, and Their Functions in Mammals

Endocrine Gland	Hormone Product	Hormone Function
Hypothalamus		Regulates other endocrine glands
Anterior pituitary (adenohypophysis)	Growth hormone	Stimulates growth and metabolic functions
	Prolactin	Stimulates milk production and secretion
	Follicle-stimulating hormone	Stimulates sperm and ova production
	Luteinizing hormone	Stimulates testes and ovaries
	Thyroid-stimulating hormone	Stimulates thyroid gland
	Adrenocorticotropic hormone	Stimulates adrenal cortex to secrete steroid hormones and endorphins
Posterior pituitary (neurohypophysis)	Oxytocin	Stimulates contractions of uterus and mammary gland cells
	Antidiuretic hormone	Promotes water retention in kidneys
Thymus	Thymosin	Stimulates immune system
Thyroid	Thyroxine	Controls metabolism and growth rates
	Calcitonin	Lowers blood calcium levels
Parathyroid	Parathyroid hormone	Raises blood calcium levels
Pancreas	Insulin	Lowers blood glucose levels
	Glucagon	Raises blood glucose levels
	Somatostatin	Inhibits release of insulin and glucagon
Adrenal	Epinephrine and norepinephrine	Mediate responses to stressful situations
	Corticosteroids	Control carbohydrate and protein metabolism
	Aldosterone	Controls blood pressure
Ovaries (female)	Estrogen	Induces maturation of oocytes and ovulation; initiates thickening of uterine lining
	Progesterone	Increases thickening of uterine lining; causes negative feedback, which promotes disintegration of corpus luteum
Testes (male)	Testosterone	Maintains male sexual characteristics, sperm production, and sex drive

Check Your Progress

6.1 Discuss adaptations in the male and female mammalian reproductive systems that support internal development of offspring.

6.2 Compare and contrast the effects of the nervous system and the endocrine system.

Exercise 22-3 Rats (Order Rodentia)

Materials Needed

- ❑ preserved rat (double- or triple-injected recommended)
- ❑ dissecting tools
- ❑ dissecting pan
- ❑ dissecting pins
- ❑ latex gloves

Rats (*Rattus norvegicus*) are members of the order Rodentia, the largest order of mammals, both in population numbers and in species diversity (2,277 sp.). All rodents possess two pairs of chisel-like **incisors** that grow continuously throughout their lives and are adapted for gnawing. Rats are colonial in nature and typically occupy burrows along the foundations of buildings or beneath rubbish piles in close association to humans. Although primarily herbivorous, they will feed on anything edible. Commercially purchased rats usually are albino mutants of the Norway rat, a species that was introduced into North America unwittingly by early European travelers and since has grown dramatically in numbers. Today it generally is regarded as a pest and continues to cause millions of dollars of property damage each year.

External Anatomy

Procedure

1 Obtain a preserved specimen of a rat and use Figure 22.20 to identify the external features of your specimen.

2 Notice the sensory organs concentrated around the head—**eyes**, **ears**, **external nares** (used to sense chemicals dissolved in the air), and **vibrissae**.

These organs all play a collective role in the rat's ability to sense and respond to stimuli in its environment. The external coverings of the rat eye consist of upper and lower eyelids and a reduced nictitating membrane that moves laterally from the medial corner of the eye. The vibrissae (commonly called whiskers) are especially long and prominent in nocturnal mammals such as rats, which venture out under the cover of darkness in search of food. Used for tactile sensations, the base of each vibrissa is attached to a sensory nerve that is triggered by air movements or physical contact to the whisker. Other vibrissae are located above the eyes and on the cheeks and chin. Collectively, vibrissae provide important sensory cues for spatial orientation—especially critical for successful maneuvering in the dark.

3 Determine the sex of your specimen.

In addition to the **anus**, female rats have a separate urinary opening and genital opening (Fig. 22.20A). The urethral orifice lies just dorsal to the small, protruding clitoris, and the vaginal orifice lies in a slight depression referred to as the **vulva**. Males are identified most easily by the large **scrotal sacs** descending from the ventral portion of the abdomen near the tail (Fig. 22.20B). A prepuce covering the **penis** is located on the ventral surface of the abdomen between the thighs. **Mammary papillae** often are present in both sexes of mammals but are more prominent and functional only in females.

Nares
Incisors
Vibrissae
Ear
Forelimb
Mammary papillae:
Cranial
Caudal
Hind limb
Urethral orifice, clitoris, and vaginal orifice
Anus
Tail
A
Female

Hind limb
Prepuce covering penis
Urethral orifice
Scrotal sac
Anus
Tail
B
Male

22.20 Rat, external anatomy: ventral views of (**A**) pregnant female, and (**B**) male.

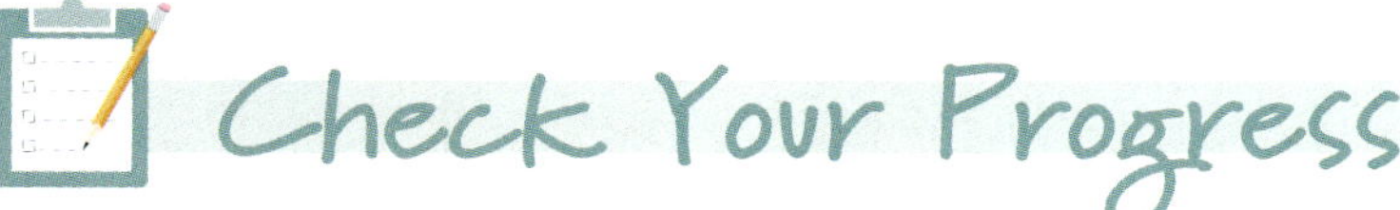

Check Your Progress

7.1 Is your rat male or female? How do you know?

7.2 What external features of the rat are characteristic of all mammals?

7.3 What sensory organs are present on the head?

22

Musculature

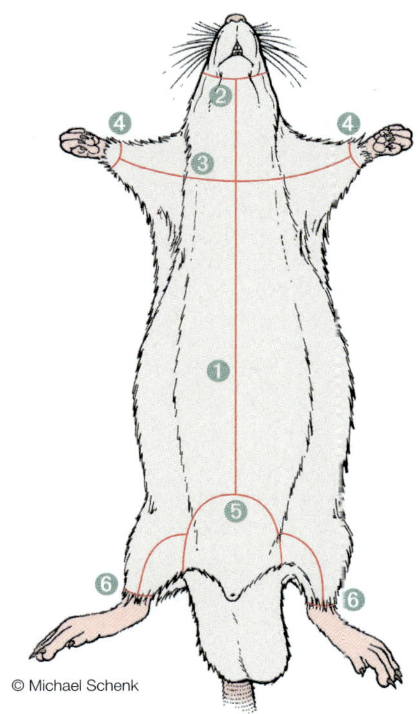

Dissection Diagram
Insert a finger (or blunt probe) in the opening created by the injection process and free the skin from the underlying fascia along the midline of the body. Use scissors to make the illustrated incisions through the skin and use a blunt probe to tease the skin away from the muscles. After skinning, clean your specimen with water or a wet paper towel to remove the bits of loose hair that accumulate during the skinning process.

Muscles are designed with one basic purpose in mind—movement. Muscles work to either move an animal through its environment or move substances through an animal. In vertebrates, there are three basic types of muscle tissue: (1) skeletal muscle and (2) cardiac muscle, both of which possess striated fibers, and (3) smooth muscle, sometimes called visceral muscle. Some of these muscles, such as skeletal muscle, can be controlled voluntarily by the animal, while others such as cardiac muscles and many smooth muscles produce involuntary actions that are regulated by the autonomic nervous system.

All skeletal muscles have a fixed end, the **origin**, and a movable end, the **insertion**. Most skeletal muscles taper from a thicker **muscle belly** in the middle to thinner, tough, white tendons on the ends, which anchor the muscle to bones or other muscles. The direction in which a muscle exerts force—its **action**—also plays a role in its shape and where it inserts and originates. A muscle that **adducts** moves a limb toward the midline of the body. Conversely, a muscle that **abducts** moves a limb away from the midline of the body.

Some muscles are **flexors**, bending one part of the body toward another part. Other muscles are **extensors**, straightening or extending a body part. Because muscles can contract in only one direction, and thus perform only unidirectional movements, they generally are arranged in antagonistic pairs or groups; that is, they work in opposition to one another. Because muscles usually are grouped closely together, it often is difficult to ascertain where one muscle ends and another begins. Thus, you will have to pay careful attention to the orientation of the muscle fibers. Often this will give you clues as to where two muscles cross or abut.

To observe the musculature of the rat, you first must remove all of the skin, fat, and fascia from the external body surface. Refer to the Dissection Diagram to assist you with this process.

Procedure

1. Place your rat on its dorsal surface in the dissecting pan and insert a finger or blunt probe in the opening created by the injection process to free the skin from the underlying fascia along the midline of the body.
2. Carefully separate the skin from the underlying muscles, using a blunt probe where necessary.
3. Use scissors to make incisions in the skin along the ventral midline, around each wrist, and longitudinal incisions along the ventral surface of each forelimb and hind limb toward the midventral incision.
4. Use a blunt probe to tease the skin carefully away from the muscles in the thoracic and abdominal areas. In most cases you should be able to "peel" the skin away with your fingers.
5. Continue until you have exposed the muscles of your specimen completely.
6. After the skin has been removed, you may have to clean away subcutaneous fat and membranous fascia covering the muscles in this region of the body.
7. Use Tables 22-7–22-9 and Figures 22.21–22.23 to identify the selected muscles and their actions in the rat.

Table 22-7 Superficial and Selected Deep Musculature of the Neck, Thoracic, and Abdominal Regions of the Rat

Body Region	Muscle Name	Action
Neck	**Masseter**	Elevates mandible
	Digastric	Depresses mandible
	Mylohyoid	Raises floor of mouth
	Sternomastoid	Extends forearm
	Omohyoid	Draws hyoid caudally and laterally
	Sternohyoid	Pulls tongue backward
	Cleidomastoid	Turns head
	Cleidocervicalis	Draws clavicle and scapula forward
Chest	**Cleidobrachialis** (clavodeltoid)	Pulls humerus forward
	Pectoralis superficialis	Adducts forelimb
	Pectoralis profundus	Adducts forelimb
Abdomen	**Rectus abdominis**	Compresses abdomen
	External abdominal oblique	Compresses abdomen and flexes trunk
	Internal abdominal oblique	Compresses abdomen and flexes trunk

Cleidobrachialis (clavodeltoid)
Pectoralis superficialis
Pectoralis profundus
Pectoralis profundus (abdominal part)

Masseter
Digastric (cranial part)
Digastric (caudal part)
Sternohyoid
Sternomastoid (superficial part)
Cleidocervicalis
Pectoralis superficialis
Pectoralis profundus

Transversus mandibularis
Mylohoid
Sternomastoid (superficial part) cut
Omohyoid
Sternomastoid (deep part)
Sternothyroid
Cleidomastoid
Pectoralis muscles (cut)

Rectus abdominis

External abdominal oblique
Internal abdominal oblique

22.21 Superficial (rat's right) and selected deep muscles (rat's left) of the medial aspect of the neck, thoracic region, and abdominal region.

22

Table 22-8 Superficial Musculature of the Lateral Aspect of the Neck, Pectoral Region, and Hind Limb of the Rat

Body Region	Muscle Name	Action
Neck	**Masseter**	Elevates mandible
	Temporalis	Flexes head, lifts mandible
	Omotransversaris	Draws scapula cranially
	Cervical trapezius	Adducts and stabilizes scapula
Pectoral region	**Cleidobrachialis** (clavodeltoid)	Pulls humerus forward
	Spinodeltoid	Abducts humerus and pulls humerus forward
	Triceps	Extends forearm
	Thoracic trapezius	Rotates scapula backward
	Latissimus dorsi	Pulls humerus backward
	Serratus ventralis	Depresses scapula
Hind limb	**External abdominal oblique**	Compresses abdomen and flexes trunk
	Gluteus superficialis	Abducts thigh
	Tibialis cranialis	Flexes hind foot
	Semitendinosus	Flexes hind limb
	Biceps femoris	Abducts thigh, flexes hind limb
	Tensor fascia latae	Extends hind limb

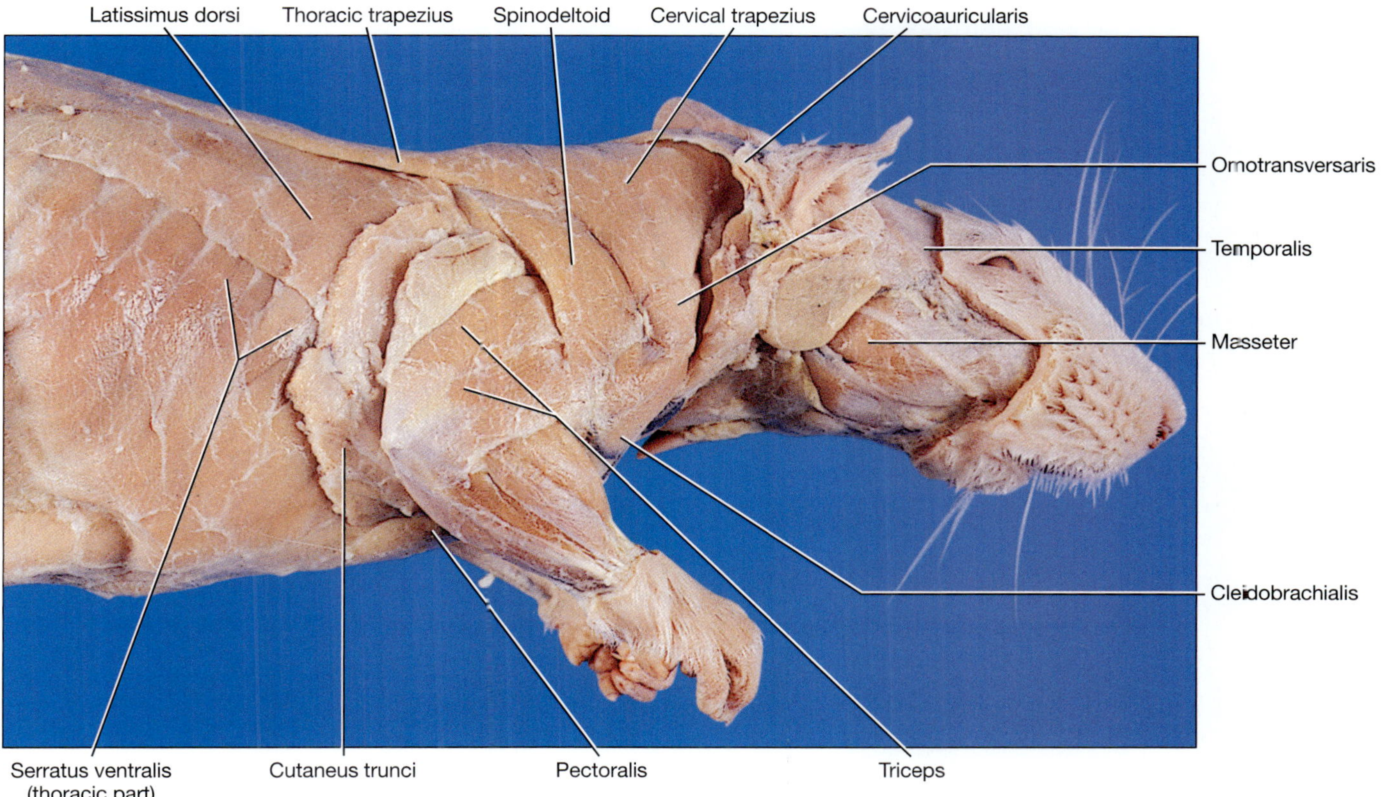

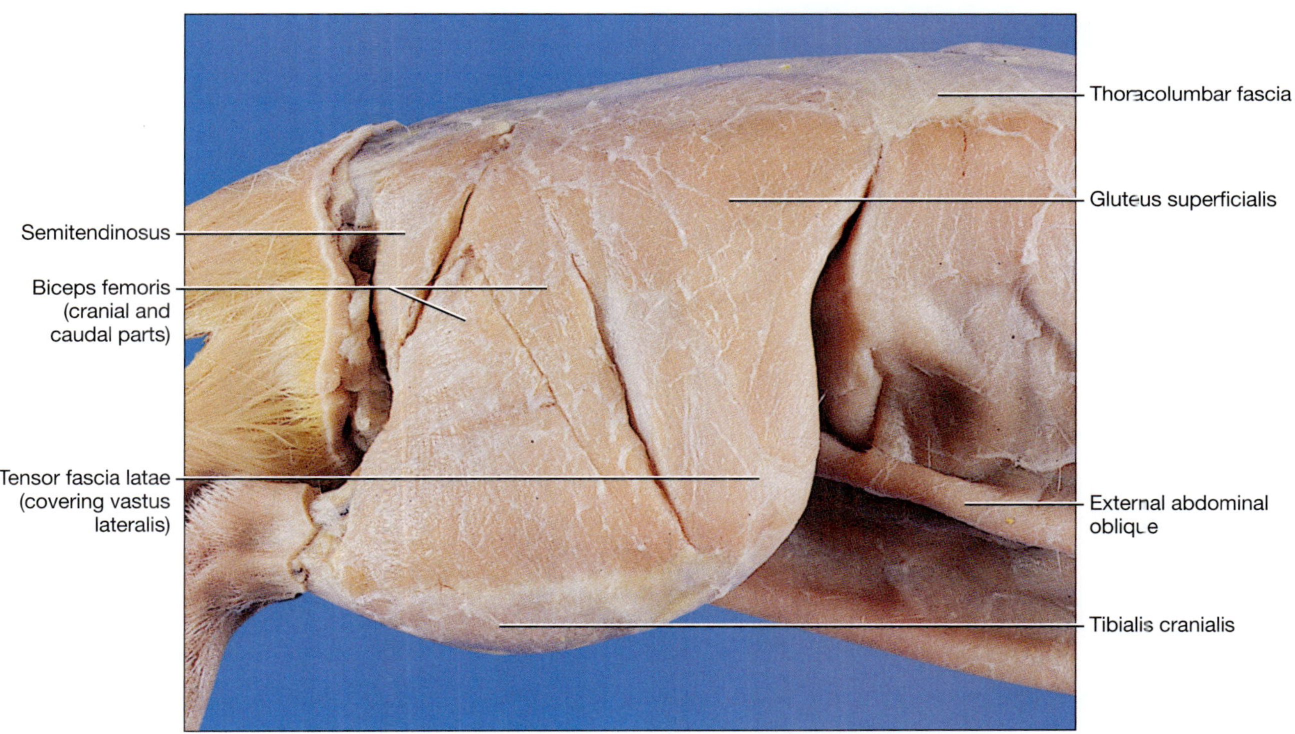

22.22 Rat: superficial musculature of the lateral aspect of the neck, pectoral region, and hind limb.

Table 22-9 Superficial and Selected Deep Musculature of the Medial Aspect of the Hind Limb of the Rat

Body Region	Muscle Name	Action
Hind limb	Tensor fascia latae	Extends hind limb
	Iliopsoas	Flexes thigh
	Rectus femoris	Extends hind limb
	Vastus medialis	Extends hind limb
	Adductor longus	Adducts thigh
	Pectineus	Flexes hip and adducts thigh
	Adductor magnus and brevis	Adducts thigh
	Gracilis cranialis	Adducts thigh
	Semimembranosus	Extends hind limb
	Gracilis caudalis	Adducts thigh
	Semitendinosus	Flexes hind limb

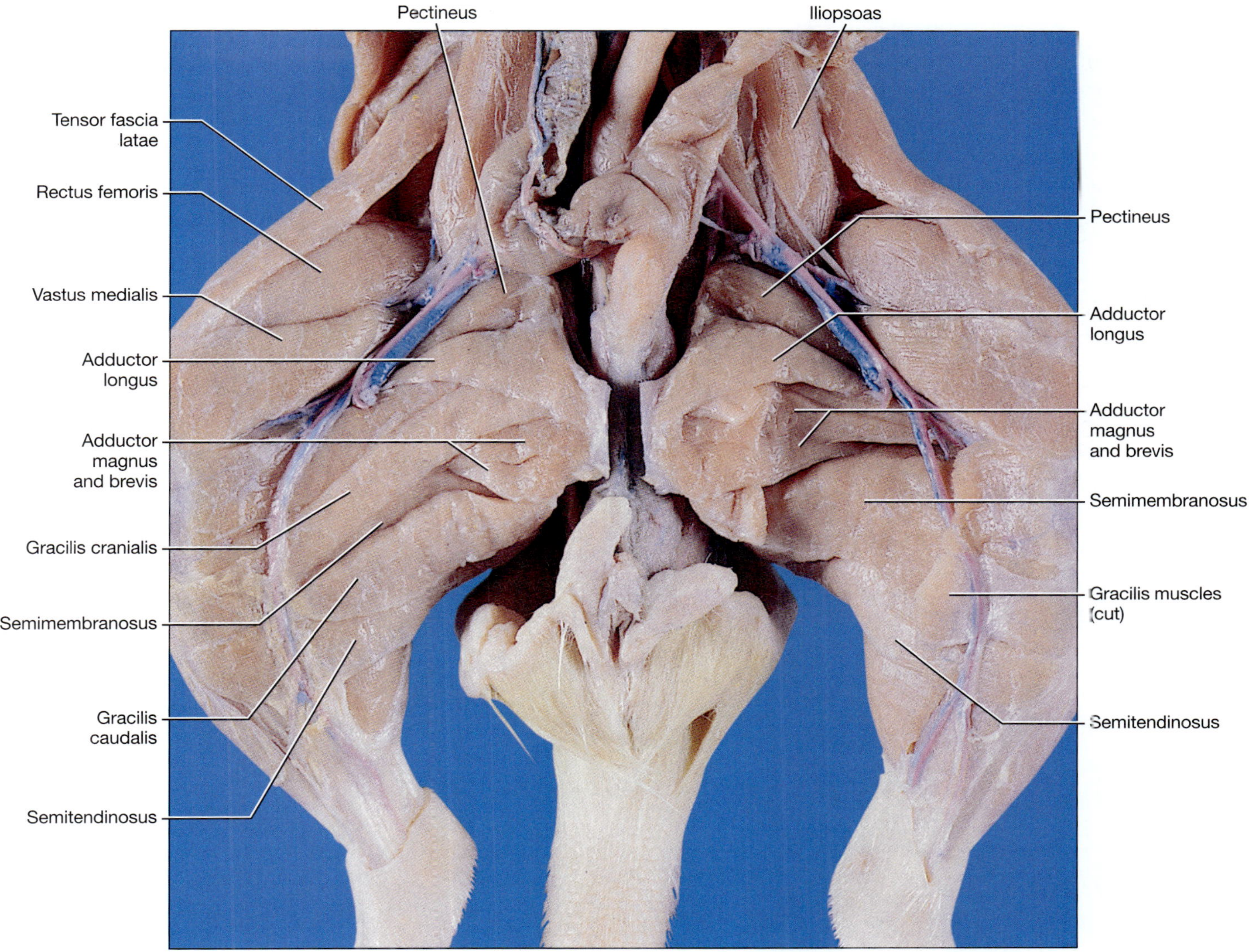

22.23 Superficial (rat's right) and selected deep muscles (rat's left) of the medial aspect of the hind limb.

22

Digestive System

The digestive system is responsible for breaking down food mechanically and chemically into smaller, usable compounds and absorbing and transporting these nutrients into the bloodstream for delivery to the individual cells of the body. This process provides the crucial raw materials and energy for all metabolic processes carried out by the organism. The extreme specialization of individual digestive organs and the efficiency of the digestive process permit mammals to sustain high metabolic rates and maintain an endothermic balance without the need for constant consumption of food.

Procedure

1 Place your rat on its dorsal surface in your dissecting pan.

2 Using scissors, make a shallow, longitudinal incision along the midventral line through the body wall (and rib cage) from the tip of the jaw caudally toward the anus. Orient your incision to one side of the reproductive organs to avoid damaging them.

3 Make two transverse incisions along the sides of the body just caudal to the ribs and just cranial to the hind limbs.

4 Spread apart the ribs and the flaps of tissue in the abdominal region and pin them to the dissecting tray to expose the internal organs.

5 To obtain a clear view of the trachea and esophagus, use a blunt probe to push aside the musculature and salivary glands in the neck region.

6 Once the internal organs of your specimen have been completely exposed, use Table 22-10 and Figure 22.24 to begin identifying the digestive anatomy.

Table 22-10 Major Digestive Organs and Their Functions in the Rat

Structure	Function
Esophagus	Transports food to the stomach
Stomach	Produces hydrochloric acid and pepsinogen that aid in the chemical breakdown of food
Pyloric sphincter (not pictured)	Muscular band that regulates the flow of chyme from the stomach into the duodenum
Liver	Produces bile, converts glucose to glycogen for storage, detoxifies many constituents of the absorbed digested compounds
Bile duct	Transports bile from the lobes of the liver directly to the duodenum
Pancreas	Produces digestive enzymes and delivers them through the pancreatic duct to the duodenum
Duodenum	Receives chyme from the stomach along with bile and digestive enzymes from the liver and the pancreas
Jejunum	Responsible for the majority of nutrient absorption and reabsorption of water
Ileum	Continues process of nutrient absorption and reabsorption of water
Cecum	Large, thin-walled pouch demarcating the beginning of the large intestine; contains anaerobic bacteria responsible for fermentation of cellulose and other plant materials; has a reduced appearance and function in carnivores and omnivores
Descending colon	Responsible for reabsorption of water and electrolytes; transports feces to the rectum via peristalsis
Rectum	Final site of water reabsorption and feces dehydration
Anus	Regulates egestion of undigested food (feces) from the body

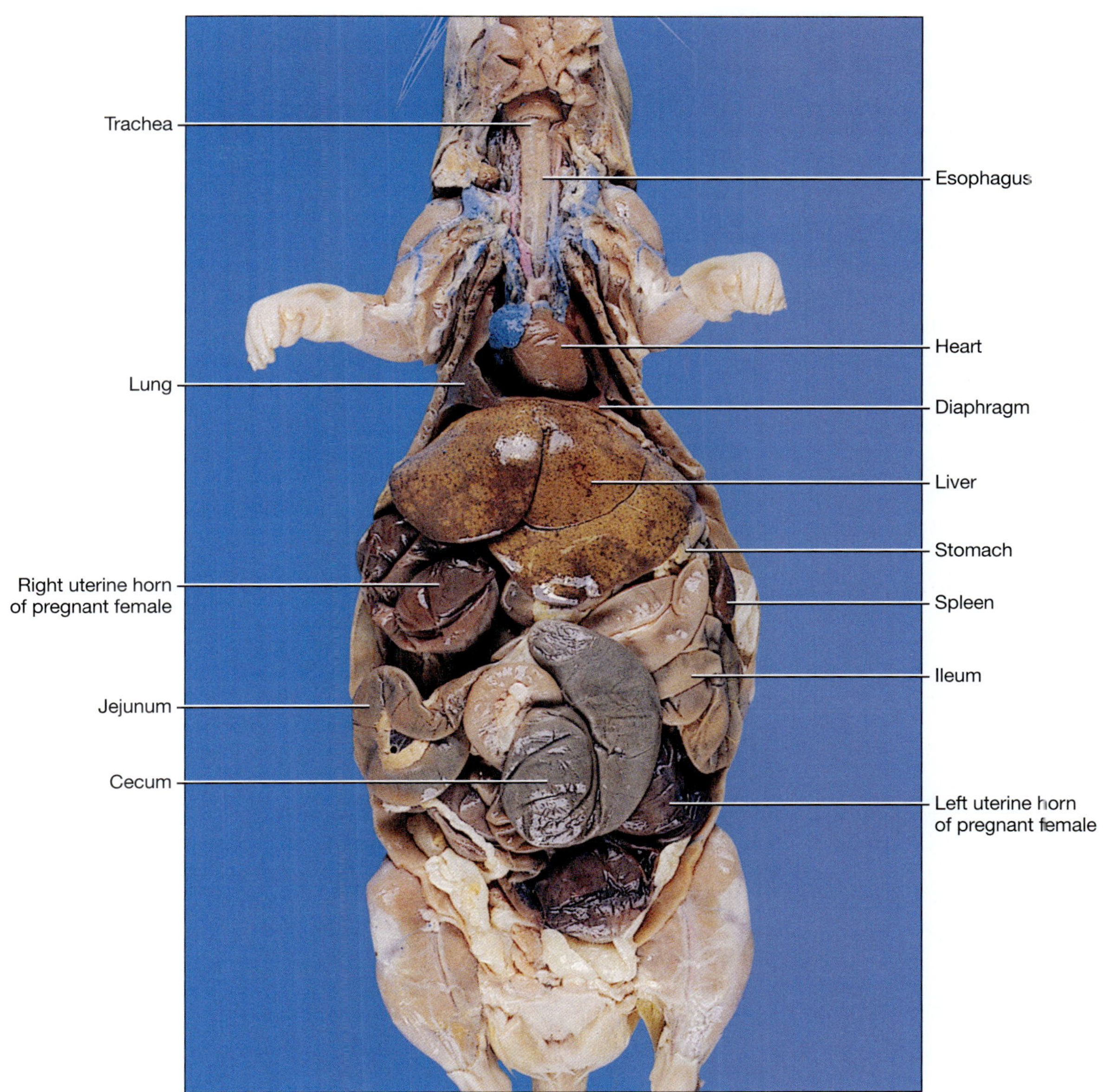

22.24 Rat: major organs of the thoracic and abdominal cavities, ventral view.

The digestive system of the rat follows the basic mammalian blueprint with only a few exceptions. Digestion begins in the **mouth** where the teeth mechanically grind the food as it mixes with secretions produced by the **salivary glands.** This softened mixture passes down the **esophagus** to the **stomach** where chemical secretions from the stomach lining facilitate the digestive process.

Rats lack a gallbladder; thus, **bile** from the **liver** is released directly into the **duodenum,** the first portion of the small intestine, through the **bile duct.** The **pancreas** produces digestive enzymes that also enter the duodenum in this region. The duodenum receives the partially digested foodstuffs and enzyme mix, known as **chyme,** from the stomach and is primarily responsible for the final stages of enzymatic digestion.

Food leaves the duodenum and enters the second portion of the small intestine, the **jejunum,** a region of the small intestine that is highly convoluted and tightly bound together by mesentery. Absorption of nutrients and water occurs along the length of the jejunum, and the nutrients are delivered to the circulatory system through the hundreds of small blood vessels throughout the intestinal mesentery. The distal third of the small intestine, the **ileum,** is where further nutrient absorption and water reabsorption occur. Together, these three regions of the small intestine perform the majority of nutrient absorption in mammals.

At the juncture of the ileum and the colon, a large, blind-ended pouch extends into the coelomic cavity. This thin-walled sac is the **cecum**, a fermentation chamber where symbiotic bacteria and protozoa reside. These microorganisms produce an important enzyme, cellulase, naturally lacking in mammals, that breaks down the cellulose in plant cell walls and allows the mammal's own digestive enzymes access to the proteins and carbohydrates within the plant cells (Fig. 22.25).

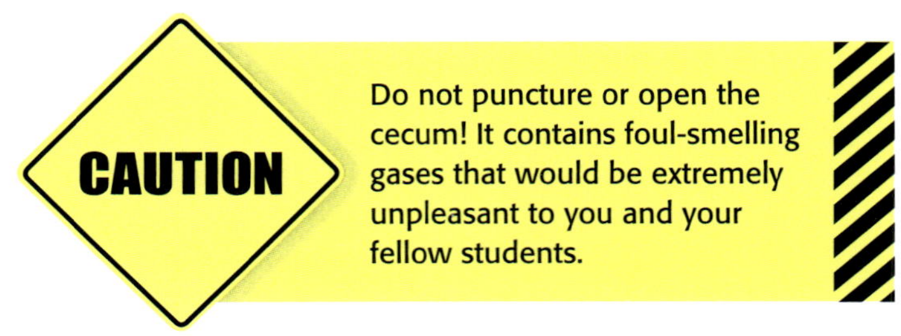

In rats and other herbivores the cecum is large and well developed to handle the considerable volume of cellulose in their diets. For example, the koala obtains virtually all of its nutrition from eucalyptus leaves and has an extraordinarily long cecum (2 meters!) for such a relatively small body size. In carnivorous and omnivorous mammals the cecum is greatly reduced in size and has a limited function in this capacity. Food leaving the cecum enters the **colon** (large intestine), where reabsorption of water and electrolytes occurs and relatively dry feces are produced and transported to the rectum via peristalsis. The **rectum** is the final site of water reabsorption and feces dehydration, and the anus regulates the egestion of feces from the body.

Ascending colon
Duodenum
Cecum
Ileum
Mesentery
Jejunum
Transverse colon
Rectum
Esophagus
Diaphragm
Stomach
Spleen
Pancreas
Greater omentum
Descending colon
Ureter
Urinary bladder
Penis
© Michael Schenk

22.25 Rat: abdominal cavity with organs displaced to show underlying anatomy, ventral view.

Check Your Progress

8.1 Which structures/organs in the mammal are responsible for the mechanical digestion of food?

8.2 Which structures/organs are responsible for chemical digestion?

8.3 Which structures/organs are responsible for nutrient or water absorption?

8.4 What role does the cecum play in the digestive process of herbivores? What structure in humans do you suppose is homologous to the cecum?

Circulatory System

The circulatory system is responsible for transporting nutrients, gases, hormones, and metabolic wastes to and from the individual cells of an animal. Mammals are far too large for all of their individual cells to exchange nutrients, wastes, and gases with the external world by simple diffusion. Most cells are buried too deep inside the body to accomplish this task effectively. The circulatory system represents a series of vessels that diverge from the heart (arteries) to supply blood to the tissues and a confluence of vessels draining blood from the tissues (veins) and returning it to the heart.

Despite the extensive network of arteries and veins throughout the body, no actual exchange of water, nutrients, wastes, or gases occurs in arteries or veins; their walls are too thick to permit diffusion. Instead, the body has extensive networks of capillary beds connecting branches of arteries and veins throughout the body to transfer these dissolved substances between the bloodstream and the tissues.

To simplify identification of arteries and veins, two general principles you should remember are:

1. Arteries and veins tend to be paired, especially when the organs they supply or drain are paired.
2. A continuous vessel often undergoes several name changes along its length as it passes through different regions.

Therefore, to identify arteries and veins successfully, you will have to trace them along their entire length (typically from the heart outward).

Procedure

1 If you have not exposed the interior of the thoracic cavity by opening the rib cage, do so at this time.

2 The heart will be partially obscured from view by the thymus gland and will be encased in a thin, pericardial membrane. Gently move the thymus out of the way, and carefully remove the pericardial membrane from the heart.

3 Using a teasing needle and forceps, dissect the muscle tissue and fatty tissue away from the major arteries and veins in the neck and the pectoral, abdominal, and pelvic regions. This is a tedious process and will take some time.

4 If your specimen has been double- or triple-injected with latex, the arteries will appear red and the veins will appear blue. If your rat has not been injected with latex, the arteries will appear whiter and stiffer than the thin, collapsed veins. Remember that arteries are more heavily walled than veins (to accommodate higher blood pressures) and generally are thicker and, thus, more evident during dissection.

The driving force behind the circulatory system of mammals is the heavily muscled, four-chambered heart that delivers blood through two major circulatory pathways—the pulmonary circuit (from the heart to the lungs and back) and the systemic circuit (from the heart through the rest of the body and back). A hallmark of the mammalian heart is that its internal design keeps the blood from these two circuits entirely separate, thereby keeping oxygen-depleted blood from mixing with oxygen-rich blood and allowing for different pressures to be maintained on the two sides of the heart.

5 First identify the four chambers of the heart.

Caudally there are two large, thick-walled ventricles, the **right ventricle** and the **left ventricle** (Fig. 22.26). These chambers pump blood out of the heart to the lungs and to the rest of the body, respectively. Cranial to the ventricles and somewhat darker in color are the **right** and **left auricles**. Chambers within the right and left auricles (called atria) receive blood from the body and the lungs, respectively, and pass it to the ventricles. The term *auricle* is used to describe the small, outer, flap-like region that covers a portion of the atrial chamber. The term *atrium* refers to the actual chamber inside that collects the blood.

6 Notice the large veins entering the heart. These are the cranial and caudal vena cavae, which bring deoxygenated blood to the right atrium from the cranial and caudal portions of the body.

7 On the ventral surface of the heart, locate the large pulmonary trunk emanating from the right ventricle. The pulmonary trunk is a large artery that channels blood from the right ventricle through the right and left pulmonary arteries to the lungs. Also notice the large, thick-walled aortic arch leaving the heart from the cranial aspect of the left ventricle.

8 Reflection of the heart to either side will reveal the pulmonary arteries and the pulmonary veins, the vessels that carry deoxygenated blood from the right ventricle to the lungs and return oxygenated blood to the left atrium of the heart. Unlike most arteries that receive red latex during the injection process, the pulmonary arteries receive blue latex because they carry deoxygenated blood. Likewise, the pulmonary veins are injected with red latex, because they carry oxygenated blood. Do not let the color difference confuse your identification of these vessels.

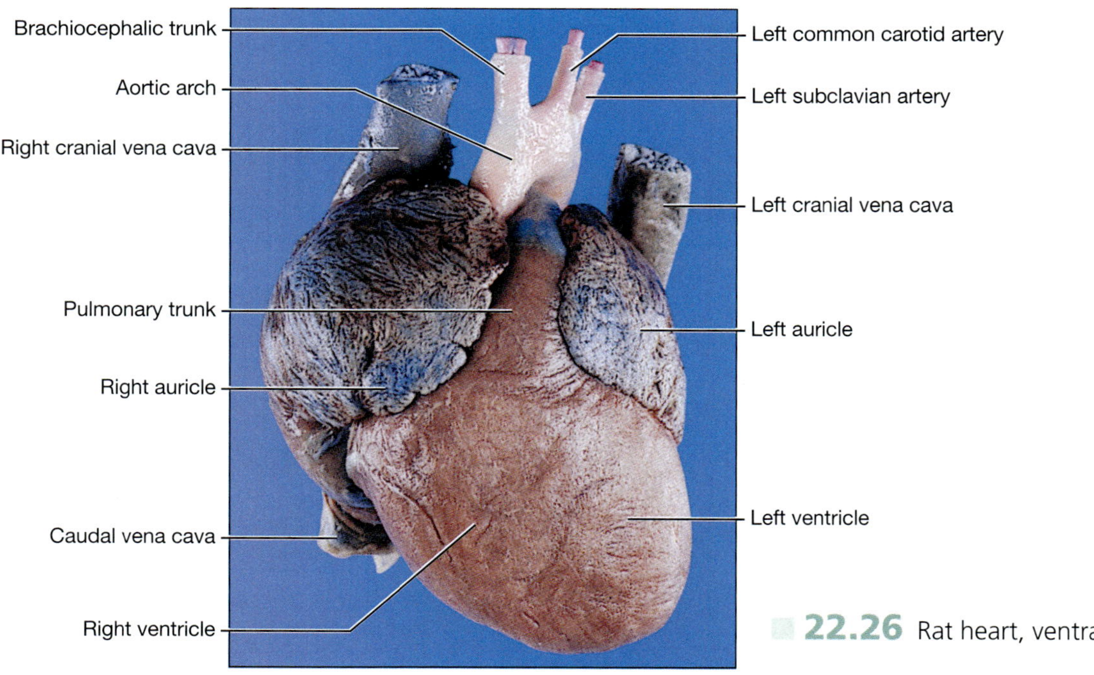

22.26 Rat heart, ventral aspect.

9 Locate the spleen along the left side of the body adjacent to the stomach.

In addition to the heart, the **spleen** is a vascular, ductless organ that also plays a critical role in the circulatory system of vertebrates (Figs. 22.24 and 22.25). Mammalian red blood cells do not contain nuclei, so they cannot undergo cell division and thus have a finite life span. New red blood cells are produced continuously in the bone marrow and are delivered to the spleen for storage. The spleen stores these cells along with excess blood and releases these products into the bloodstream as needed. Through this mechanism the spleen regulates the body's total blood volume and the relative concentration of red blood cells. The spleen also manufactures white blood cells (lymphocytes) to fend off diseases and destroys and recycles worn-out blood cells.

10 Trace the paths of the major arteries emanating from the heart and the major veins returning to the heart as they diverge and converge throughout the body using Table 22-11 and Figures 22.27 and 22.28 to assist you in identifying these vessels in the rat.

Table 22-11 Major Blood Vessels and Their Functions in the Rat

Vessel Name	Function
External jugular veins	Drain blood away from the head region
Internal jugular veins	Drain blood away from the neck region
Common carotid arteries	Supply oxygenated blood to the head and neck regions
Subclavian veins	Drain blood away from the lateral portions of the thoracic cavity and forelimbs
Left subclavian artery	Supplies oxygenated blood to the left forelimb
Aortic arch	Supplies oxygenated blood from the heart to the entire body
Aorta	Supplies oxygenated blood to all posterior parts of the body
Caudal vena cava	Returns blood to the heart from all posterior parts of the body
Renal arteries and veins	Supply oxygenated blood to the kidneys (arteries) and return deoxygenated blood from the kidneys to caudal vena cava (veins)
Gonadal arteries and veins	Supply oxygenated blood to the testes (male) and ovaries (female) and return deoxygenated blood from the gonads to caudal vena cava
Iliolumbar arteries and veins	Supply oxygenated blood to the dorsal surface of the pelvic region (arteries) and return blood from the pelvic region to caudal vena cava (veins)
Common iliac arteries and veins	Major divergences off aorta and vena cava that supply blood to the hind limbs (arteries) and drain blood from the hind limbs (veins)
Caudal epigastric arteries and veins	Branches of the common iliac arteries and veins that supply blood to the hind limbs (arteries) and drain blood from the hind limbs (veins)

22

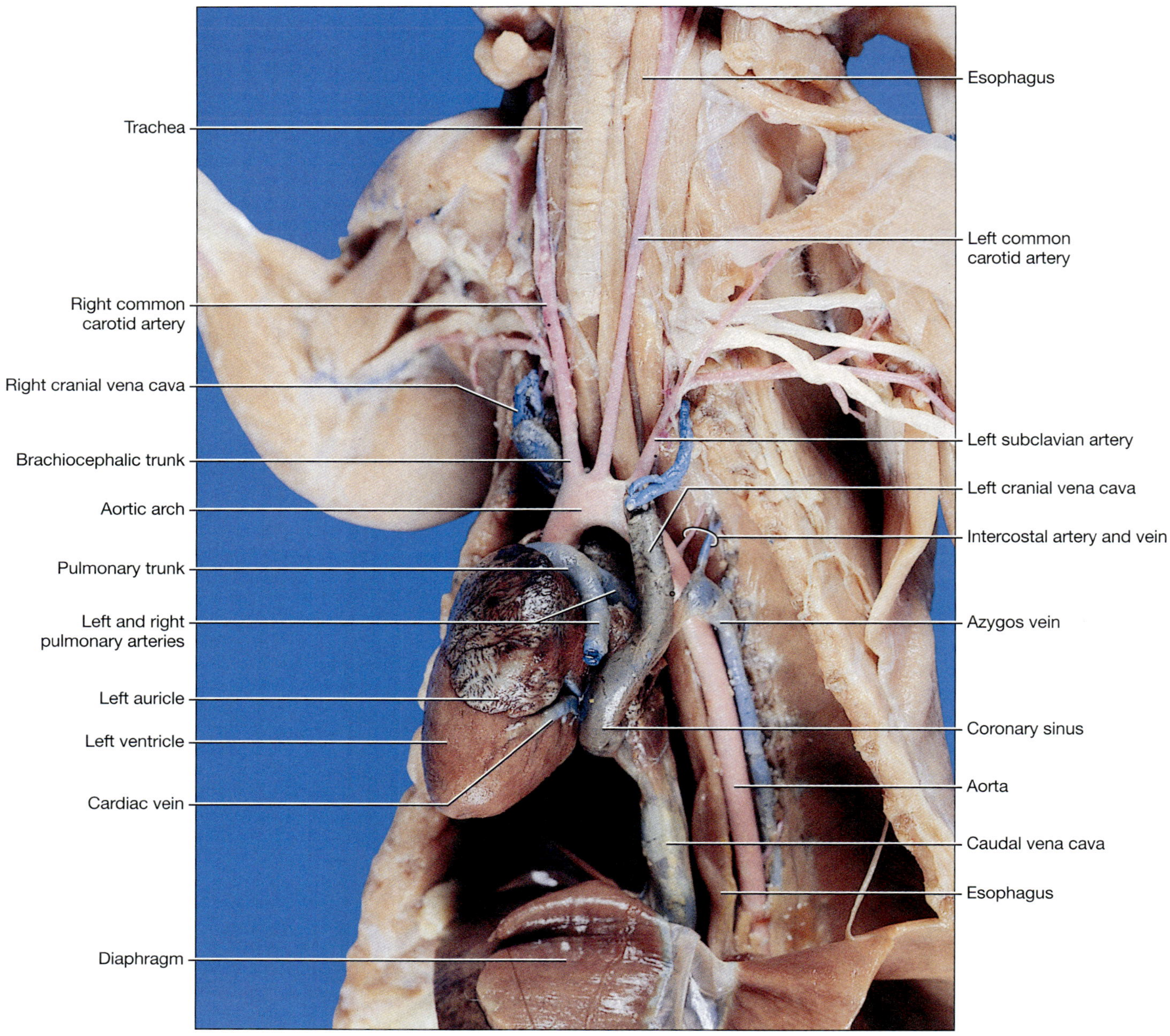

22.27 Rat: major arteries and veins of the thoracic region.

22

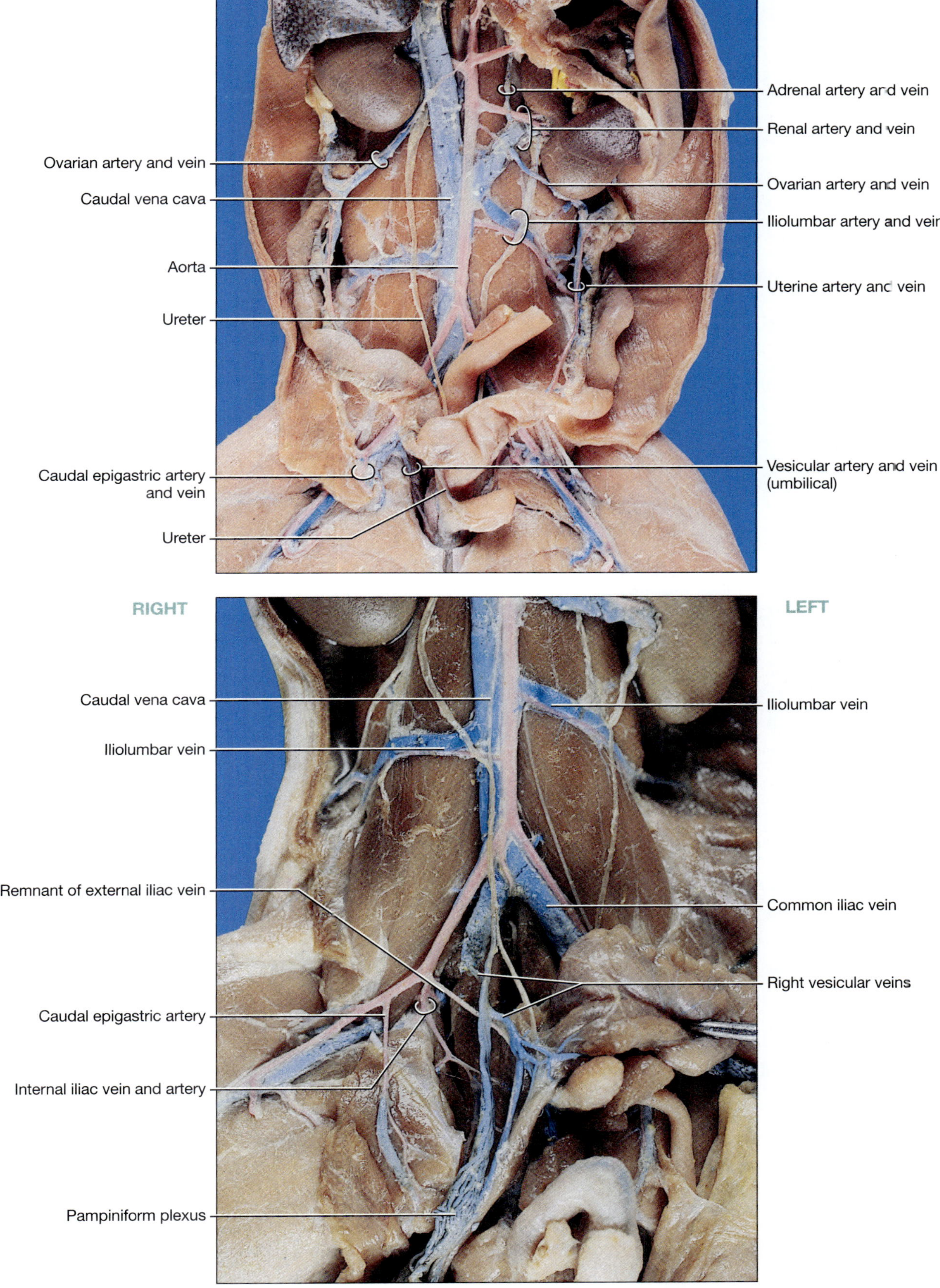

22.28 Rat: major arteries and veins of the abdominal region and hind limbs.

9.1 The mammalian heart has ________ chambers.

9.2 Arteries carry blood ________ the heart, and veins carry blood ________ the heart. Tiny blood vessels called ________ connect arteries to veins and are the actual sites of gas and nutrient exchange with the body tissues.

9.3 What role does the spleen play in the circulatory system?

____

9.4 Describe at least two advantages that the four-chambered heart provides mammals.

____

____

Respiratory System

The respiratory system of mammals is responsible for bringing a fresh supply of oxygen to the bloodstream and carrying off excess carbon dioxide. The anatomy of the respiratory tract is designed to humidify and warm the air while filtering out dust particles and germs. The lining of the nasal epithelium is covered with fine hairs that capture these foreign particles and prevent them from passing into the lungs, where they may infect the body. Similarly, as air is exhaled it is cooled and dried, reducing the amount of heat and moisture that mammals lose through respiration.

Procedure

1 Locate the long trachea descending from the oral region to the lungs.

Notice the cartilaginous rings that fully encircle the **trachea** to keep it from collapsing under the negative pressures generated during respiration (Fig. 22.29). The trachea branches into two primary **bronchi**, which enter the right and left lungs. Notice that the right lung is divided into four lobes and the left lung appears as one large lobe. In humans the right lung has three lobes and the left lung has two lobes.

2 Identify the cranial lobe, medial lobe, caudal lobe, and accessory lobe of the right lung. These individual lobes are distinguished primarily by internal divisions of the bronchi that are not always apparent superficially.

3 Identify the thin, muscular diaphragm lying on the cranial margin of the liver.

The **diaphragm** allows the thoracic cavity to expand and compress, drawing in fresh air with each expansion (as the diaphragm contracts) and expelling stale air with each compression (as the diaphragm relaxes). A muscular diaphragm is a predominately mammalian characteristic that improves the efficiency of the respiratory process and helps mammals maintain a high metabolic rate.

Inside the lungs, the bronchi are divided further into several branches called bronchioles. The bronchioles
22 branch into smaller and smaller tubules, eventually terminating in microscopic, open sacs called **alveoli.** The alveoli are composed of squamous epithelial tissue and are surrounded by capillary networks. Here, oxygen is picked up by the bloodstream and carbon dioxide is released into the lungs to be expelled from the body through exhalation.

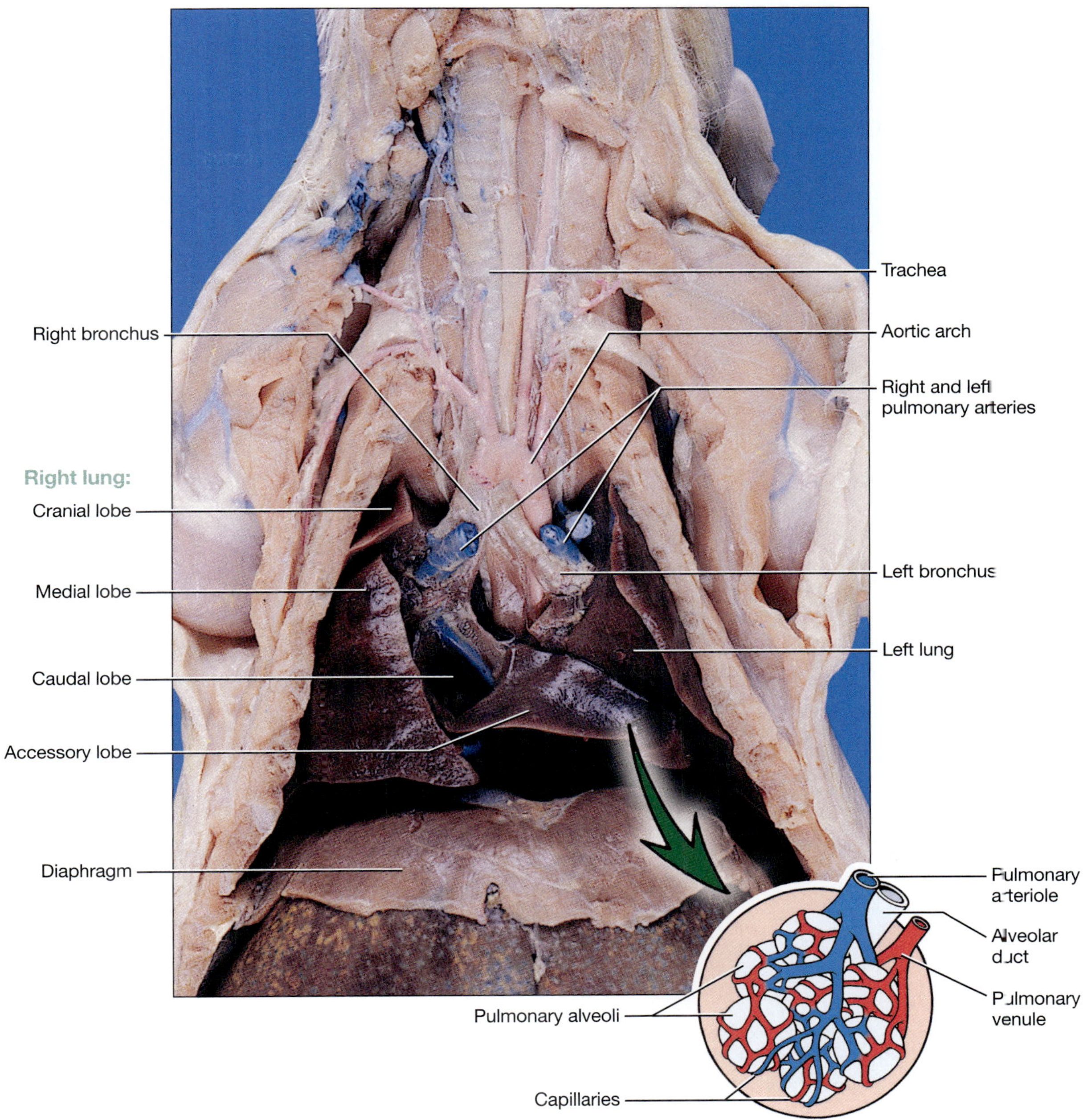

22.29 Rat: respiratory system (with heart removed) and illustration of alveolar sacs.

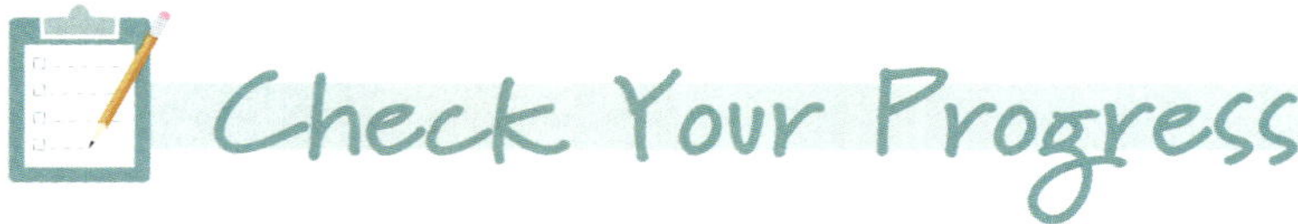

Check Your Progress

10.1 How many distinct lobes are present on the rat's right lung? The left lung?

10.2 Describe how the respiratory system of mammals differs from that of reptiles.

Excretory and Reproductive Systems

Excretory organs are responsible for eliminating metabolic wastes that the body produces from cellular respiration and for maintaining a homeostatic balance among the levels of fluids, electrolytes, sugars, hormones, and proteins in the body. Remember—excretion is an entirely different process from that which expels indigestible products through the anus. Excretion and egestion (or defecation) are different processes, handled by completely different systems in mammals.

Reproductive organs are responsible for producing the gametes that ultimately will fuse with the corresponding gametes of the opposite sex. In addition to reproduction, the testes and ovaries produce many of the hormones that are associated with the development and maturation of primary and secondary sexual characteristics and that drive the complex repertoire of sexual behaviors indicative of most mammals.

Excretory System

Procedure

1 Using a teasing needle, carefully dissect away the membranous tissue surrounding one of the kidneys. Take care not to destroy the **adrenal gland**, which sits along the cranial margin of the kidney (probably embedded in fat) or any of the ducts and blood vessels in the area.

2 If your specimen is a male, be careful not to damage the vas deferens, which "loops" around the ureter.

3 Clean the area around the kidney to expose the renal blood vessels and the ureter passing from the medial margin of the kidney caudally toward the urinary bladder.

4 To uncover all of the reproductive structures completely, you must cut longitudinally through the pubic symphysis with a scalpel. *Cutting carefully,* start your incision slightly to one side of the midline of the pelvis to avoid cutting through underlying structures. Preferably, cut through the symphysis only partially and then apply downward (lateral) pressure to each of the hind limbs to complete the separation.

5 Regardless of the sex of your specimen, you are expected to be familiar with the reproductive structures of both male and female rats, so work closely with another group that has a specimen of the opposite sex.

The **kidneys** are large, bean-shaped organs that lie along the dorsal surface of the abdominal cavity on either side of the spine (Figs. 22.30 and 22.31). Notice how they are offset in the rat, with the right kidney lying slightly more cranially than the left kidney. The kidneys filter blood from the circulatory system, removing the metabolic waste products produced in the tissues of the body during cellular respiration. The major function of the kidneys is to concentrate these toxins and eliminate them from the body while conserving water, salts, and other compounds that the body needs. In humans, the kidneys filter between 1,100 and 2,000 liters of blood each day! From this tremendous volume of blood only about 1.5 liters of urine are actually produced. The other 99.9% is reabsorbed into the bloodstream through a highly efficient system of semipermeable tubules that generates concentration gradients in the nephrons of the kidney.

Urine is concentrated in the kidneys and passes down each **ureter,** tubes lined with smooth muscle that transport urine toward the **urinary bladder** for temporary storage. The urinary bladder is a muscular reservoir that can expand to many times its "relaxed" size to accommodate large volumes of urine. When relaxed, the inner walls of the bladder appear folded, somewhat resembling the rugae of the stomach lining. Eventually the stored urine is eliminated through the **urethra.**

Male Reproductive System

The **scrotum** of the male houses the paired **testes**—small, bean-shaped structures that are the site of sperm production. Your first task in dissecting the male reproductive system is to locate the spermatic cords that leave the scrotum and enter the abdominal wall. Each testis is enclosed within a cremasteric pouch. At the cranial end of each cremasteric pouch, a narrow tube should be evident. This is the **spermatic cord,** containing the vas deferens, the spermatic artery and vein, lymphatic vessels, and numerous nerves leading to the testis and epididymis located within the cremasteric pouch.

Procedure

1 Using scissors, carefully make a slit in one of the cremasteric pouches and peel it open.

2 Leave the testis attached to the spermatic cord, but separate it from the tissue of the cremasteric pouch.

Cupped around the side of each testis is a highly coiled system of tubules known as the **epididymis** (Fig. 22.30). Sperm are produced within the **seminiferous tubules** of each testis and are stored along the length of the epididymis. Upon ejaculation, sperm leave each epididymis and travel through each **vas deferens** toward the urethra. Trace along the length of the spermatic cord to visualize the path that sperm travel as they move out of the epididymis through the vas deferens (which loops around the ureter) toward the prostate region. Notice the opening in the abdominal wall, the **inguinal canal,** through which the spermatic cord passes from the scrotum into the abdominal cavity.

The penis is enclosed in a sheath and held along the ventral wall of the abdomen. Careful dissection of this area should reveal several accessory reproductive glands. Together, the **prostate gland, seminal vesicles,** and **coagulating gland** contribute fluid to the sperm, accounting for more than 60% of the total volume of the semen. This fluid is thick and contains mucus (to prevent the sperm from drying out) and large amounts of fructose (to provide energy for the sperm). In addition, the semen is highly alkaline to neutralize the acidic environment of the vagina and prolong survival of the sperm.

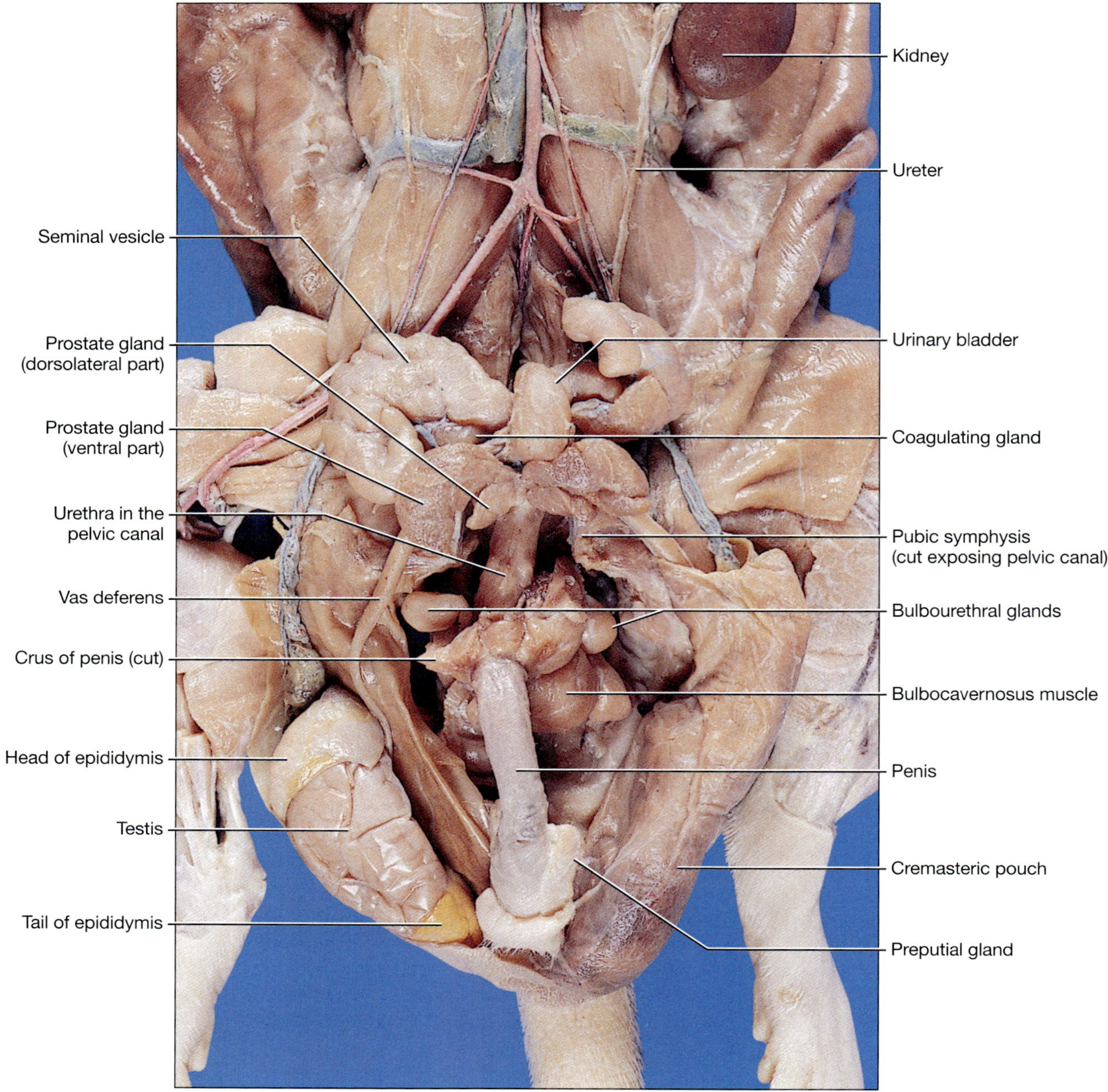

22.30 Male rat: reproductive and excretory structures.

Female Reproductive System

The paired female **ovaries**, located in the abdominal region caudal to the kidneys, can be identified by their small, round appearance (Fig. 22.31). They are anchored to the dorsal wall of the abdominal cavity and to each uterine horn by small ligaments. Attached to each ovary is a coiled oviduct. The **oviduct** receives the mature oocyte (egg) when it is released from the ovary at the moment of ovulation. Despite the close association, there is no actual physical connection between the oviductal opening and the ovary. Small, fingerlike projections of the oviduct sweep the egg into the ostium of the oviduct. The epithelial lining of the oviduct is ciliated and creates a current that propels the egg along the length of the oviduct toward the horn of the uterus.

Fertilization typically occurs in the oviduct, but implantation of the embryos occurs farther along in the uterus. In rats, the uterus is divided into two **uterine horns** in which embryonic development of the fetuses occurs. The two uterine horns converge on the **cervix**, a cartilaginous constriction marking the juncture with the **vagina**. In humans the uterine horns (known as fallopian tubes) are reduced substantially, because implantation and embryonic development occur in the body of the uterus. Because of larger litter sizes, rats require a larger area for young to develop, and the extensive size of the two uterine horns accommodates this need.

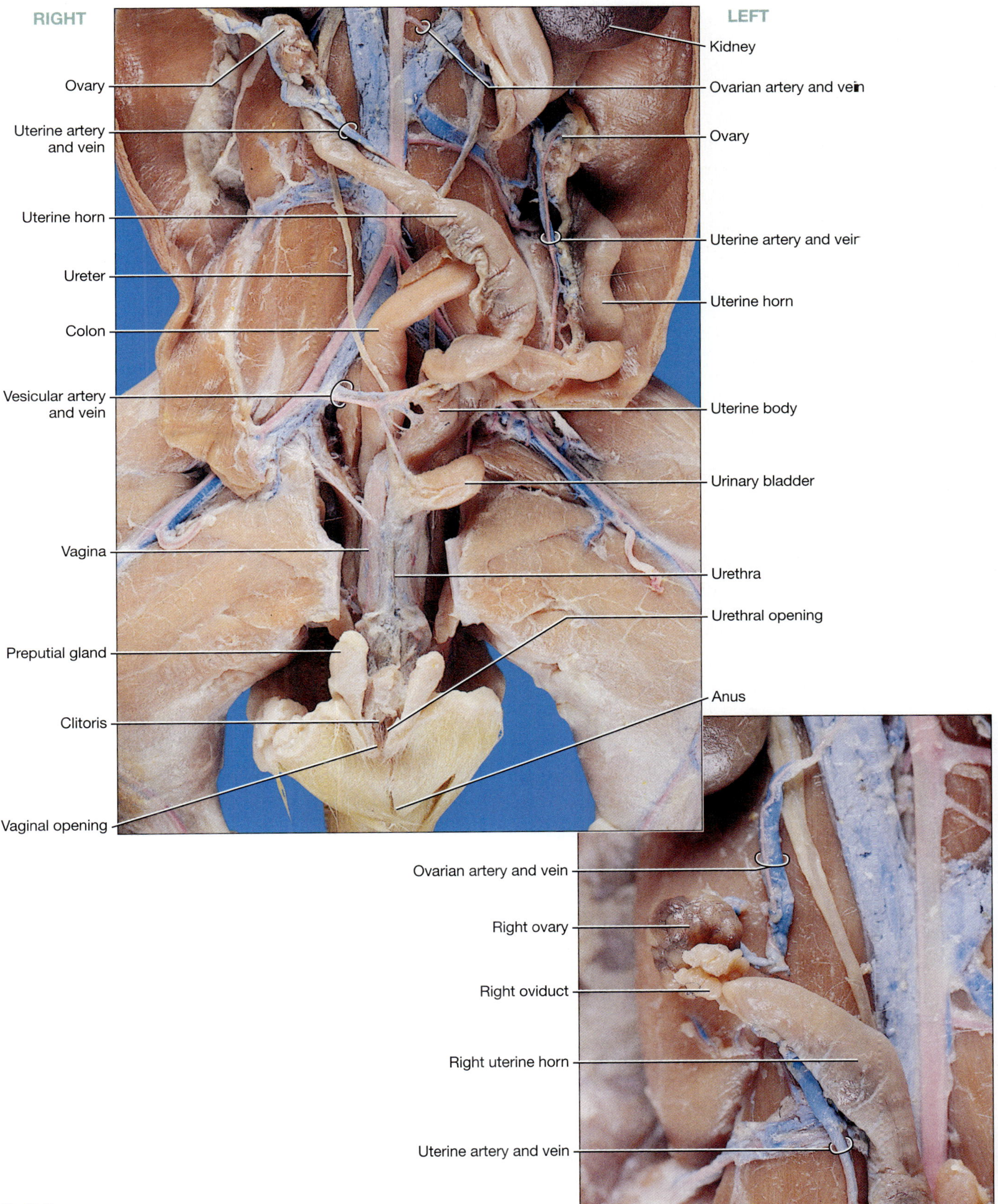

22.31 Female rat: reproductive and excretory structures.

Endocrine System

The endocrine system is a major player in the body's attempt to coordinate the activities of its many organs and organ systems. In that respect, the endocrine and nervous systems are very much alike. The similarities between the two systems do not go very far beyond that, however. Unlike the nervous system, which has its own contained system for information transfer (the nerves), the endocrine system is ductless and, therefore, must rely on another neighboring system (the circulatory system) to send its messages throughout the body. The glands of the endocrine system produce and secrete their hormones directly into the bloodstream to be carried to their target organs.

Hormones are chemical compounds that interact with target cells in the body to produce myriad behavioral, neurological, and physiological responses. In this way, they influence many of the same behaviors and processes that the nervous system regulates. Because of the nature of hormones, however, the effects produced by the endocrine system usually are not short-lived. Nervous responses are instantaneous and degrade immediately, but hormones circulating through the bloodstream may take some time to produce a response and anywhere from minutes to hours to break down. Thus, hormonal effects tend to be much longer in duration and the processes that are under hormonal control typically occur over hours, days, weeks, or even years (for example, sexual maturation, metabolic rate, growth rate, and ovulation).

In addition, the degree of response shown by the target organ is directly proportional to the amount of hormone released by the endocrine gland: The more hormone a gland releases, the more pronounced is the effect. This is a fundamental distinction from the all-or-nothing response of nerve cells and illustrates why both systems—the nervous system and the endocrine system—are essential for complex organisms to coordinate different aspects of their lives.

The centralized control center of the endocrine system is the hypothalamus–pituitary complex of the brain, often referred to as the pituitary gland (Fig. 22.32). You will be able to see this organ only if you examine a commercially prepared sheep brain. The hypothalamus–pituitary complex produces many hormones that, in turn, stimulate the activity of many of the other endocrine glands in the body. Likewise, other endocrine organs produce hormones that stimulate or inhibit regions of the pituitary gland and hypothalamus. Through this feedback loop the endocrine system is able to turn itself on and off in response to environmental or endogenous stimuli.

You already have identified some of the organs discussed in this section, because organs that function in the endocrine system often contain other tissue that functions in other body systems (for instance, digestion, reproduction, or nervous control). Return to your dissection, and identify the endocrine glands that are visible on your specimen. Figure 22.32 depicts the approximate location of these glands in the rat, and Table 22-12 lists the hormones that each gland produces and their effects on the body.

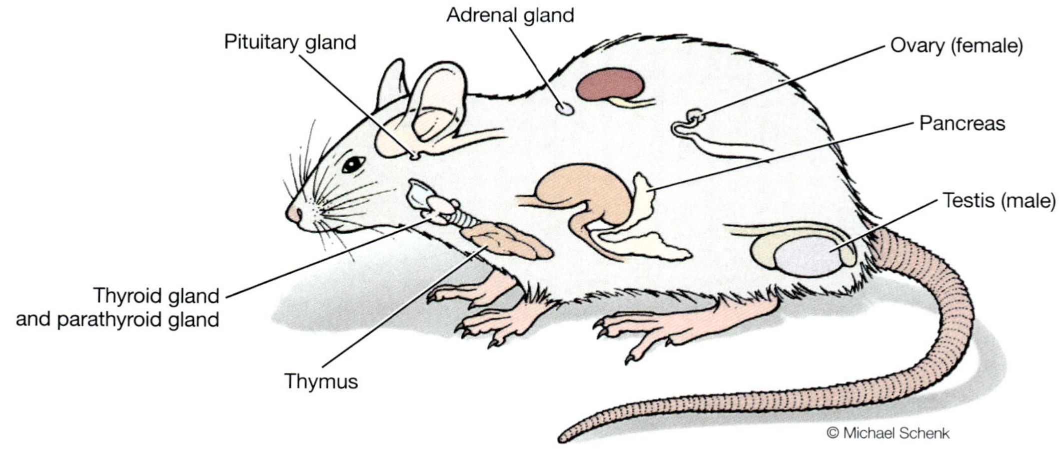

22.32 Rat, endocrine glands.

22

Table 22-12 Endocrine Glands, Hormone Products, and Their Functions in Mammals

Endocrine Gland	Hormone Product	Hormone Function
Hypothalamus		Regulates other endocrine glands
Anterior pituitary (adenohypophysis)	Growth hormone	Stimulates growth and metabolic functions
	Prolactin	Stimulates milk production and secretion
	Follicle-stimulating hormone	Stimulates sperm and ova production
	Luteinizing hormone	Stimulates testes and ovaries
	Thyroid-stimulating hormone	Stimulates thyroid gland
	Adrenocorticotropic hormone	Stimulates adrenal cortex to secrete steroid hormones and endorphins
Posterior pituitary (neurohypophysis)	Oxytocin	Stimulates contractions of uterus and mammary gland cells
	Antidiuretic hormone	Promotes water retention in kidneys
Thymus	Thymosin	Stimulates immune system
Thyroid	Thyroxine	Controls metabolism and growth rates
	Calcitonin	Lowers blood calcium levels
Parathyroid	Parathyroid hormone	Raises blood calcium levels
Pancreas	Insulin	Lowers blood glucose levels
	Glucagon	Raises blood glucose levels
	Somatostatin	Inhibits release of insulin and glucagon
Adrenal	Epinephrine and norepinephrine	Mediate responses to stressful situations
	Corticosteroids	Control carbohydrate and protein metabolism
	Aldosterone	Controls blood pressure
Ovaries (female)	Estrogen	Induces maturation of oocytes and ovulation; initiates thickening of uterine lining
	Progesterone	Increases thickening of uterine lining; causes negative feedback, which promotes disintegration of corpus luteum
Testes (male)	Testosterone	Maintains male sexual characteristics, sperm production, and sex drive

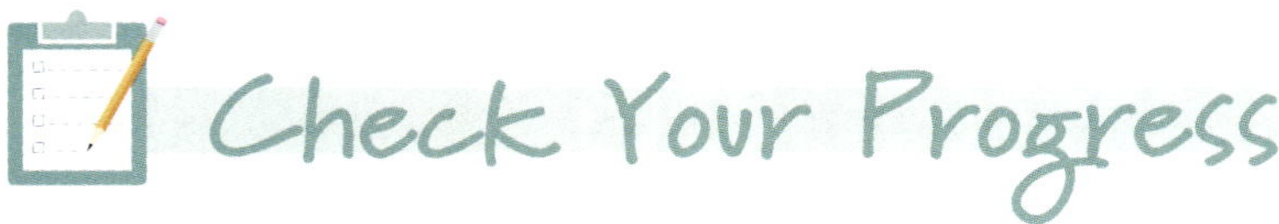

Check Your Progress

11.1 Discuss adaptations in the male and female mammalian reproductive systems that support internal development of offspring.

11.2 Compare and contrast the effects of the nervous system and the endocrine system.

Exercise 22-4 Nervous System

Materials Needed

- ❑ preserved sheep brain
- ❑ preserved sheep eye or cow eye
- ❑ dissecting tools
- ❑ dissecting pan
- ❑ latex gloves

The nervous system serves as the reconnaissance division of the body—receiving physical stimuli from the environment, converting it into electrical impulses, processing the information and affecting behavioral or physiological changes in response to the stimuli. The nervous system is divided into two main regions: the central nervous system, composed of the brain and spinal cord, and the peripheral nervous system, which includes the cranial nerves and spinal nerves emanating from the brain and spinal cord, respectively.

Peripheral nerves receive external stimuli (through sensory neurons) and produce motions in muscles and other organs (through motor neurons). The brain and the spinal cord are the sites of integration of the information picked up by the sensory neurons. These individual nerve cells are networked to produce a highly complex, intricately organized system for communication and information transfer.

Rather than dissecting the brain of a fetal pig or a rat (difficult and time consuming), we suggest that you use a commercially prepared sheep brain. This option is the preferred route of most laboratory courses. Therefore, photos and illustrations that accompany this section of the book will depict only the sheep brain.

The Mammalian Brain

Procedure

1 Obtain a preserved sheep brain and place it in your dissecting pan with the two large cerebral hemispheres facing upward, as depicted in Figure 22.33A.

22 The first apparent feature of the brain is its convoluted surface. The ridges you see are called gyri (singular = gyrus) and the grooves between the ridges are known as sulci (singular = sulcus). Because of this feature, the mammalian brain is referred to as gyrencephalic, as opposed to lissencephalic, which refers to a brain that has a smooth outer

surface. The advantage of a highly convoluted brain surface is the increase in total cortical surface area that can be accommodated in the same-sized cranial space; thus a larger brain capable of more complex behaviors and thought processes is possible, without the inherent problems of a larger skull.

The most prominent sulcus is the longitudinal fissure that divides the two hemispheres of the cerebrum into left and right cerebral hemispheres (Fig. 22.33). Internally, the two hemispheres are connected by the corpus callosum (not visible externally), which forms the floor of the longitudinal fissure on the exterior surface of the brain.

2 If a sagittal section through the brain is available for viewing, examine it to locate the corpus callosum.

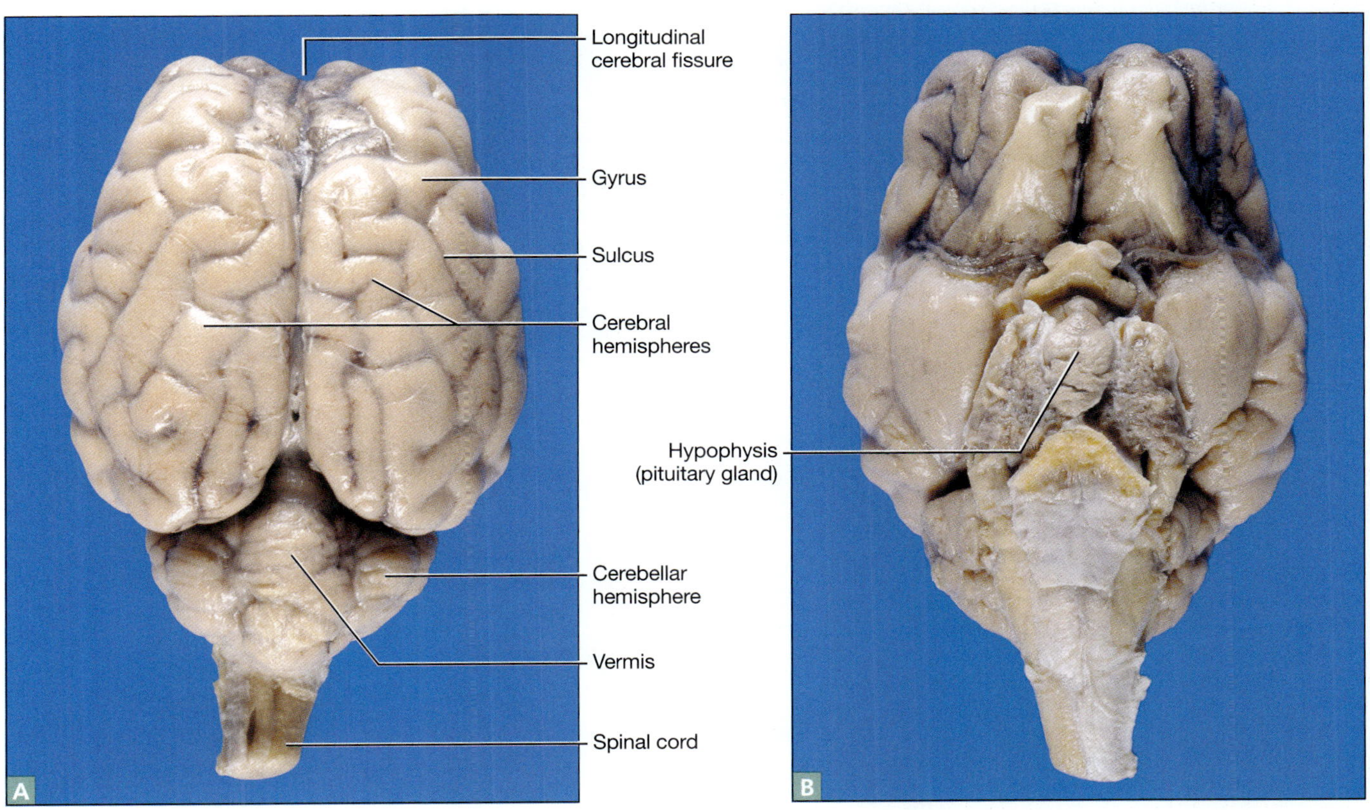

22.33 Sheep brain: (**A**) dorsal, and (**B**) ventral aspect.

The **cerebrum**, the largest portion of the brain, functions in the interpretation of sensory impulses and the coordination of voluntary movements. The parts of the brain responsible for more complex functions such as memory and learning also are located in the cerebrum. The cerebrum is composed of several regions (or lobes): the frontal lobe, the temporal lobe, the parietal lobe, and the occipital lobe (Fig. 22.34).

1. The frontal lobe primarily controls fine movements and is responsible for "higher" functions such as language, memory, emotional expression, and personality.
2. The temporal lobe processes auditory signals and some visual information.
3. The parietal lobe handles basic body information provided by touch receptors, muscle receptors, and joint receptors.
4. The occipital lobe processes visual information.

Caudal to the cerebrum is the smaller **cerebellum**, which also possesses gyri and sulci. The cerebellum consists of two lateral hemispheres that border a medial vermis. The cerebellum is primarily a reflex center for the integration of skeletal muscle movements. It is responsible for activities such as muscle coordination and balance.

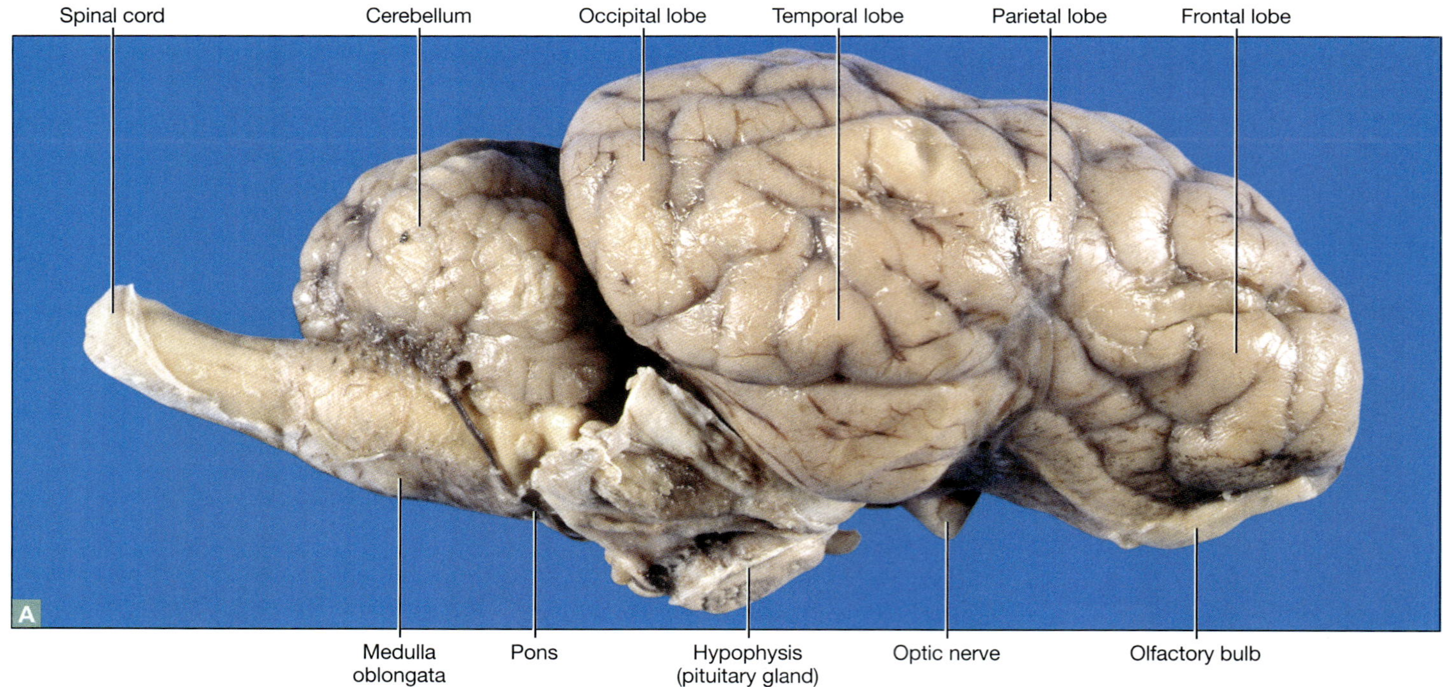

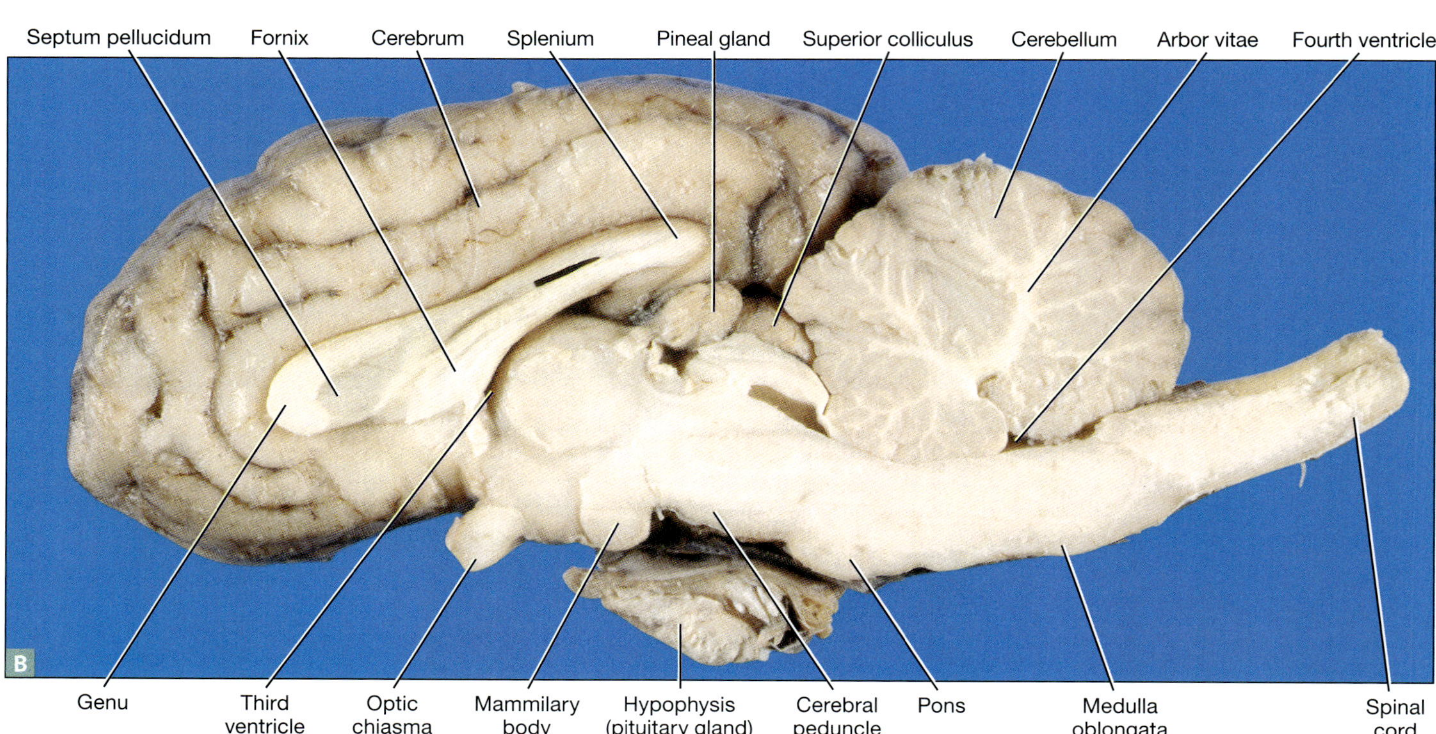

22.34 Sheep brain: (**A**) lateral aspect and (**B**) sagittal view.

22

3 At the base of the cerebellum, locate the brainstem, or **medulla oblongata**. This is the most caudal portion of the brain and leads directly into the spinal cord. The medulla oblongata regulates many autonomic functions such as breathing, heart rate, digestion, sweating, and vomiting.

4 On the ventral aspect of the brain, several other structures are visible. Moving from the medulla oblongata cranially, identify the pons (the enlarged portion of the medulla oblongata ventral to the cerebellum) and the pituitary gland, or hypophysis.

5 Moving rostrally from the pituitary gland, you will see the juncture where the optic nerves enter the brain, the optic chiasma. Although we don't often think of them as such, our eyes (at least the photoreceptors and intermediate ganglia of the retina) actually are extensions of our brain, much like the olfactory bulbs located in the nasal region. Basic processing of visual information actually starts in each retina, before these nerve impulses reach the optic lobes of the brain.

6 Find the olfactory tracts, located just dorsal to the optic nerves. If your specimen was carefully preserved, you may be fortunate enough to see the olfactory bulbs at the end of the olfactory tracts. They lie on the ventral portion of the cerebrum at the rostral end of the brain.

Check Your Progress

12.1 Explain how the mammalian brain benefits from a gyrencephalic organization.

12.2 True or false? In mammals, the cerebellum is the largest region of the brain.

12.3 What role does the medulla oblongata play in the brain? Why is it considered the most primitive region of the mammalian brain?

12.4 Describe the functions of the pons and pituitary regions of the brain.

The Mammalian Eye

The eye is a complex sensory organ specialized to receive external stimuli (in the form of light waves) and convert this light energy into chemical information that can be integrated (to some extent by the eye itself) and sent to the brain for interpretation. Because fetal pigs and rats have eyes that are quite small, we instead will dissect the eye of an adult sheep or cow to acquire an appreciation for the intricate structures that play a role in the vision of mammals. Most of the anatomical features of the sheep eye or cow eye are actually similar to those of the human eye.

Procedure

1. Obtain a preserved sheep eye or cow eye and place it in your dissecting pan.
2. Begin by removing the fatty tissue that covers the back of the eye. While preparing the external surface, be careful not to puncture the surface of the eye.
3. With forceps, locate the optic nerve (on the side of the eye opposite from the clear cornea) and use scissors to trim the fatty tissue and muscle remnants away from it, being careful not to damage the optic nerve.
4. Once you have removed all of the fatty tissue, glands, and muscle remnants from the surface of the eye and you are left with a smooth, spherical eyeball, you are ready to proceed.

Like many mammals, sheep possess a "third eyelid" that still may be present on your preserved eye. This eyelid, which is clear and remains invisible when closed, is known as the **nictitating membrane.** Humans lack a nictitating membrane; however, a vestigial remnant of this structure is present in the medial corner of each eye. Identify the **cornea**—a tough, transparent layer that allows light to enter the eye while protecting the underlying structures (Fig. 22.35). The cornea is composed of a special lamellar arrangement of cells that permits nearly perfect optical transparency. This property comes with a price, though. The cells of the cornea must pump out their interstitial fluid continuously to maintain the proper structural arrangement necessary for clear vision. The function of the cornea is to refract (bend) light rays striking its surface and direct them through the pupil.

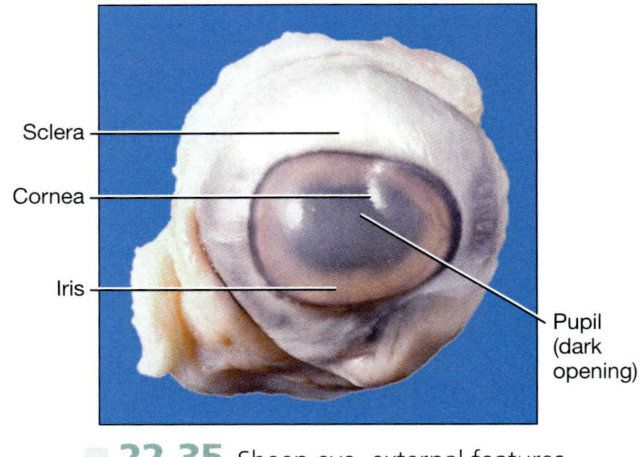

22.35 Sheep eye, external features.

The **optic nerve** in the back of the eye is the confluence at which the axons of all the photoreceptors contained in the retina converge and send their information from the eye to the brain. The optic nerve is continuous with the retina on the inner surface of the eye. Surrounding the remainder of the eye (exclusive of the cornea) and the optic nerve is a tough, white layer of tissue called the **sclera.** The tough sclera protects the eye from physical damage and deformation.

5. Using scissors, carefully cut the eye in half through its frontal plane (giving you one-half containing the cornea and iris and one-half with the sclera and the optic nerve.
6. Before attempting to separate the two halves, be sure you have cut completely through all the layers of the eye. A clear liquid usually will ooze out of the eye when you have cut deep enough.
7. Place the eye in your dissecting tray with the *cornea facing down,* and open it gently by lifting the back half of the eye away from the front half. This should result in the lens staying with the front half of the eye.

22

Inside the front half of the eye you should see the lens suspended in a fluid-filled chamber known as the **vitreous chamber** (Fig. 22.36A). The fluid contained in this chamber is a gelatinous mixture of water, called the vitreous humor, and fine transparent fibers suspended in the fluid. The **lens** is a fairly solid, biconvex structure composed of concentric sheets of clear cells, arranged much like the skin of an onion. While seemingly quite rigid, the lens is flexible in life and is capable of bending to focus images on the **retina** at the back of the eyeball. Small intrinsic muscles known as **ciliary bodies** attached to the lens accomplish this task (Fig. 22.36B).

The opening in front of the lens is known as the **pupil**. A thin sheet of pigmented tissue suspended between the cornea and the lens surrounds this opening. This is the **iris**, which contains two groups of smooth muscles (circular and radiating) that contract to change the size of the pupil opening and consequently regulate the amount of light entering the eye. When the circular fibers contract, the pupil becomes smaller; when the radiating fibers contract, the pupil enlarges. The retina diminishes in thickness from back to front, terminating at the margin of the ciliary bodies, identified by the scalloped junction known as the **ora serrata** (Fig. 22.36B). This region marks the division between the anterior portion of the retina and the ciliary bodies.

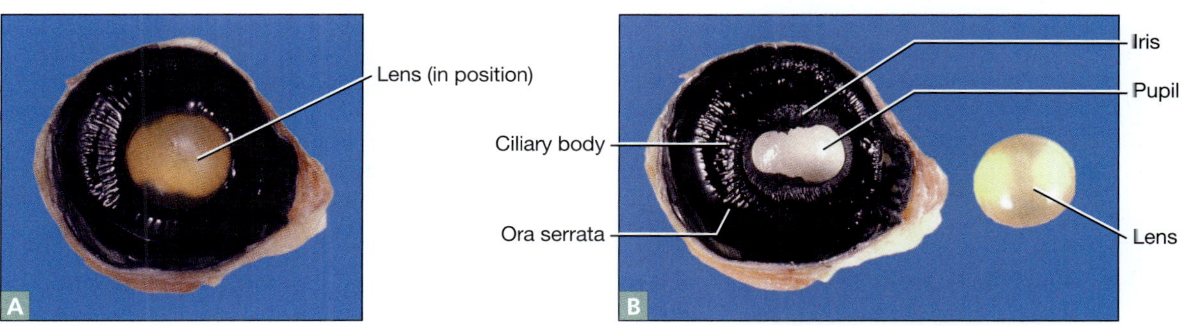

22.36 Front half of eye: (**A**) lens in place, and (**B**) lens removed.

The rods and cones (photoreceptors) are embedded in the retina. The back of the retina is covered by a jet black membrane called the **choroid layer** that absorbs light rays and reduces the amount of light that is scattered back onto the rods and cones embedded in the retinal layer (Fig. 22.37). You may remember, from our comparison of the squid eye to the human eye in Chapter 11, that because of evolutionary constraints, the photoreceptors of the mammalian eye are oriented in a bizarre manner. The rods and cones actually are along the back of the retina (farthest from the lens) and face away from the lens. Thus, light must pass through the bipolar and accessory nerve cells and stimulate the rods and cones on its way past them. For this to occur without any loss in visual acuity, rods, cones, and other associated nerve cells in the retina must be absolutely optically clear so distortion of the visible light rays entering the eye and passing through them is minimal.

Many mammals (including sheep) have a special coating on the choroid layer of the retina known as the **tapetum lucidum** that gives these mammals their traditional "eyeshine" when spotted at night by flashlight or in the headlights of a car (Fig. 22.37). This special layer increases the light-gathering ability of the eye and endows these mammals with enhanced night vision. Humans lack a tapetum lucidum and, therefore, do not demonstrate eyeshine at night, nor possess keen night vision. You should be able to distinguish the spot on the retina where the optic nerve exits the back of the eye. This spot is called the **optic disc** (Fig. 22.37). The surface of the retina has no photoreceptors at this point, and this confluence of nerves is responsible for the visual phenomenon known as the "blind spot."

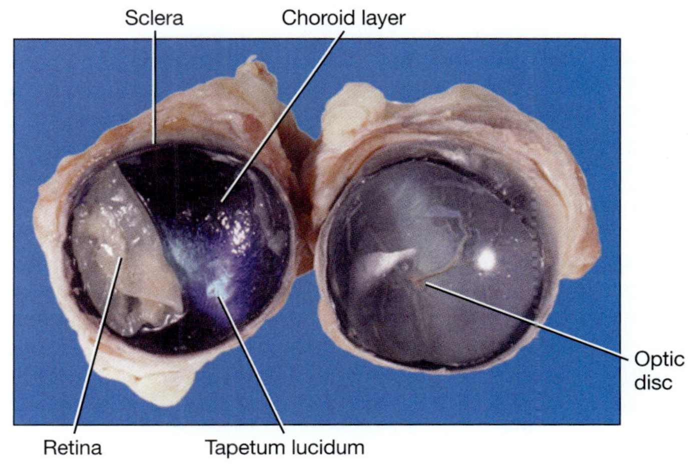

22.37 Rear halves of eye showing retina in place (right) and peeled back (left) to expose underlying layers.

22

Check Your Progress

13.1 True or false? The ciliary bodies of the mammalian eye focus images on the retina by moving the lens back and forth, like a camera lens.

13.2 True or false? In nocturnal mammals a special coating of the cornea known as the tapetum lucidum causes the eyes to shine at night when spotted by a bright light.

13.3 True or false? The area of the retina where the rods and cones are most numerous and visual acuity is the highest is known as the optic disc.

13.4 What role does the choroid layer play in the eye?

Exercise 22-5 Sheep Heart

Materials Needed

- ❑ preserved sheep heart
- ❑ dissecting tools
- ❑ dissecting pan
- ❑ latex gloves

The following section presents the anatomy of the sheep heart because of its popularity in laboratory courses as a model for studying the "typical" mammalian heart. Although much of this information was covered previously in the dissections of the fetal pig and rat, this optional section covers internal anatomy and more detailed information on external features than was presented earlier. Check with your instructor to see if you will be completing this section and, if so, to determine what level of depth you will be required to know.

Procedure

1 Obtain a preserved sheep heart and place it in your dissecting pan.

2 Begin by identifying the four chambers of the heart.

Caudally there are two large, thick-walled ventricles, the **right ventricle** and the **left ventricle** (Fig. 22.38). These chambers pump blood out of the heart to the lungs and to the rest of the body, respectively. The sheep heart has a superficial landmark that separates these two chambers, known as the interventricular groove, that runs obliquely down the ventral surface of the heart toward the apex, but runs more longitudinally on the dorsal surface of the heart.

Cranial to the ventricles and somewhat darker in color are the **right** and **left auricles.** Chambers within the right and left auricles receive blood from the body and the lungs, respectively, and route it to the ventricles. Running along the surface of the heart itself are the small coronary arteries and veins. The coronary arteries supply blood to the heart muscle, ensuring that it, too, receives nutrients and oxygen that maintain an energy supply to support its continuous, methodical beating throughout the entire life of the animal. In the sheep, the coronary vessels typically are buried under dense fat on the surface of the heart.

3 Locate the remnants of the two large veins entering the heart on the right side.

These are the remnants of the cranial and caudal vena cavae (Fig. 22.38B), which return deoxygenated blood to the right atrium from the cranial and caudal portions of the body. The most visible artery from the ventral surface leaving the heart is the large pulmonary artery (pulmonary trunk) emanating from the right ventricle. This artery channels blood from the right ventricle through the right and left pulmonary arteries to the lungs.

Notice the large, thick-walled aorta (aortic arch) leaving the heart from the cranial aspect of the left ventricle. The aorta and pulmonary artery are connected externally for a short distance by remnant tissue of the ductus arteriosus that diverted blood from the pulmonary artery to the aorta during fetal development. This band of solid connective tissue now superficially joining these two vessels is known as the ligamentum arteriosum.

In the sheep heart, the large brachiocephalic artery also will be visible as the first branch off the aorta. On the dorsal surface of the heart you should be able to identify the pulmonary veins leading back to the left auricle.

4 To cut the sheep heart in half, you will need a large, sharp knife with a blade that is several inches longer than the width of the heart.

5 Place the sheep heart in a dissecting pan and make a longitudinal cut along the frontal plane of the heart, dividing it into roughly equal dorsal and ventral halves. You will have to use a "sawing" motion to cut through the heart. Be precise in your motion so you do not "rip" through the heart but, rather, slice through it cleanly.

6 Use Figure 22.39 to assist you in identifying the internal features of the sheep heart.

Note

Throughout this section we use the terms *atrium* and *auricle* to refer to slightly different regions of the heart. The term *auricle* is used to describe the small, outer, flap-like region that covers a portion of the atrial chamber. The term *atrium* refers to the entire open space (or actual chamber) inside that collects the blood. The reason for this distinction is that part of each atrial chamber extends well beyond the boundaries of each flap-like auricle. This is evident when you view the interior of the heart.

Notice that the atrial chambers extend well beyond the boundaries of each auricle and that the walls of the atria are much thinner than the walls of the ventricles. Because the ventricles are responsible for pumping blood much longer distances, they have evolved into more muscular chambers capable of generating massive pressure to force blood out of the heart and throughout the body. Notice that inside the chambers of the heart are valves to prevent blood from flowing backward (Fig. 22.39). As blood enters the right atrium, it immediately flows into the right ventricle. Very little blood actually is *pumped* by the right atrium into the right ventricle. At the juncture of the right atrium and right ventricle is the tricuspid valve. As the right ventricle contracts and pushes blood out to the lungs, some blood is forced back up against the tricuspid valve, slamming its leaflets shut and preventing retrograde flow into the right atrium.

Upon entering the pulmonary trunk, the blood also passes through the pulmonary semilunar valve, which prevents backflow into the right ventricle as it relaxes and receives more blood from the right atrium. Fully oxygenated blood returns from the lungs into the left atrium via the pulmonary veins and then flows into the left ventricle through the bicuspid valve, or mitral valve. Blood leaving the left ventricle into the aorta passes through the aortic semilunar valve—another valve to prevent backflow as this ventricle relaxes. The bicuspid and tricuspid valves are prevented from being pushed too far backward (a condition known as "prolapse") by small string-like attachments of connective tissue called chordae tendineae, which have small muscular attachments to the inner wall of the heart called papillary muscles (Fig. 22.39).

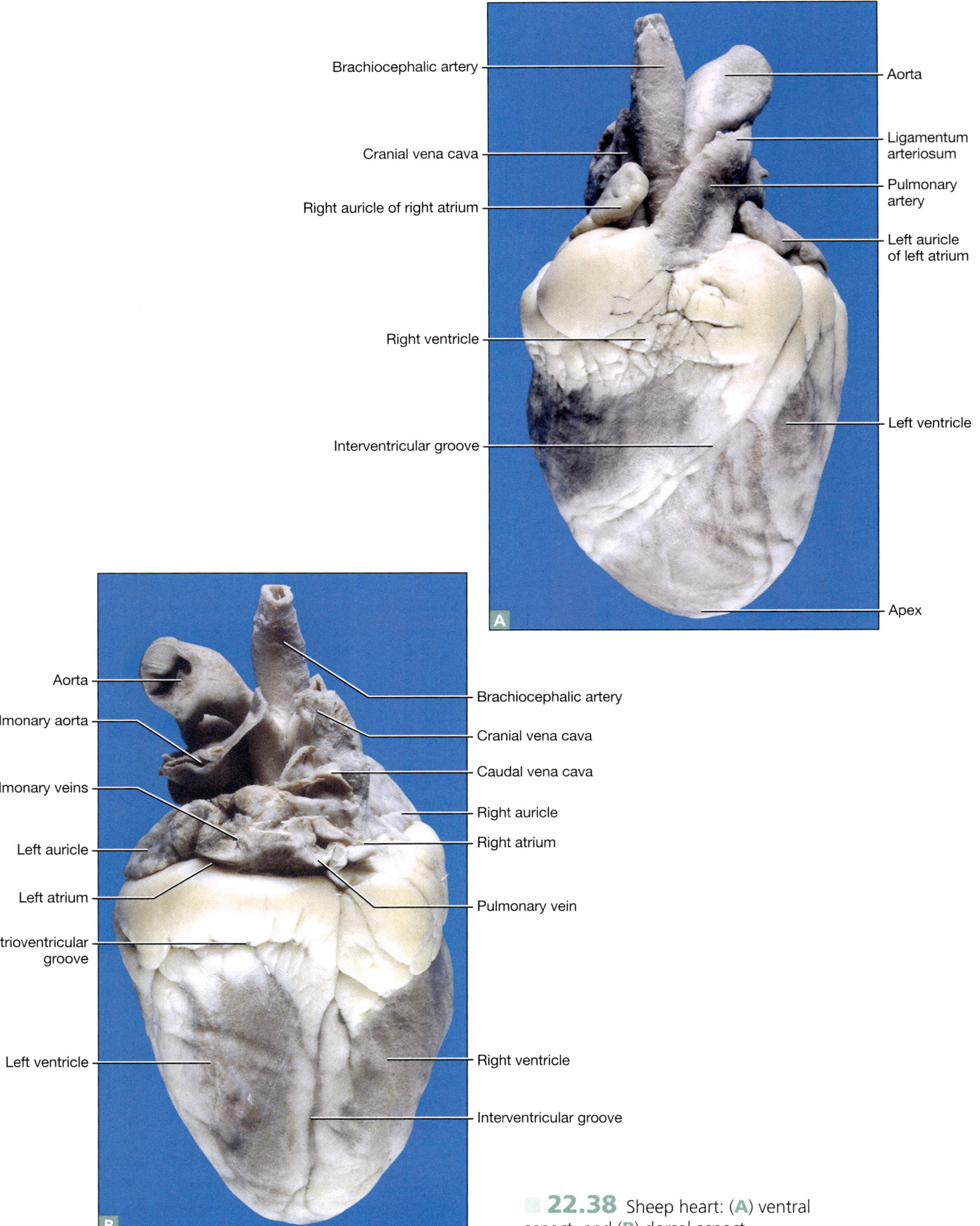

22.38 Sheep heart: (**A**) ventral aspect, and (**B**) dorsal aspect.

22

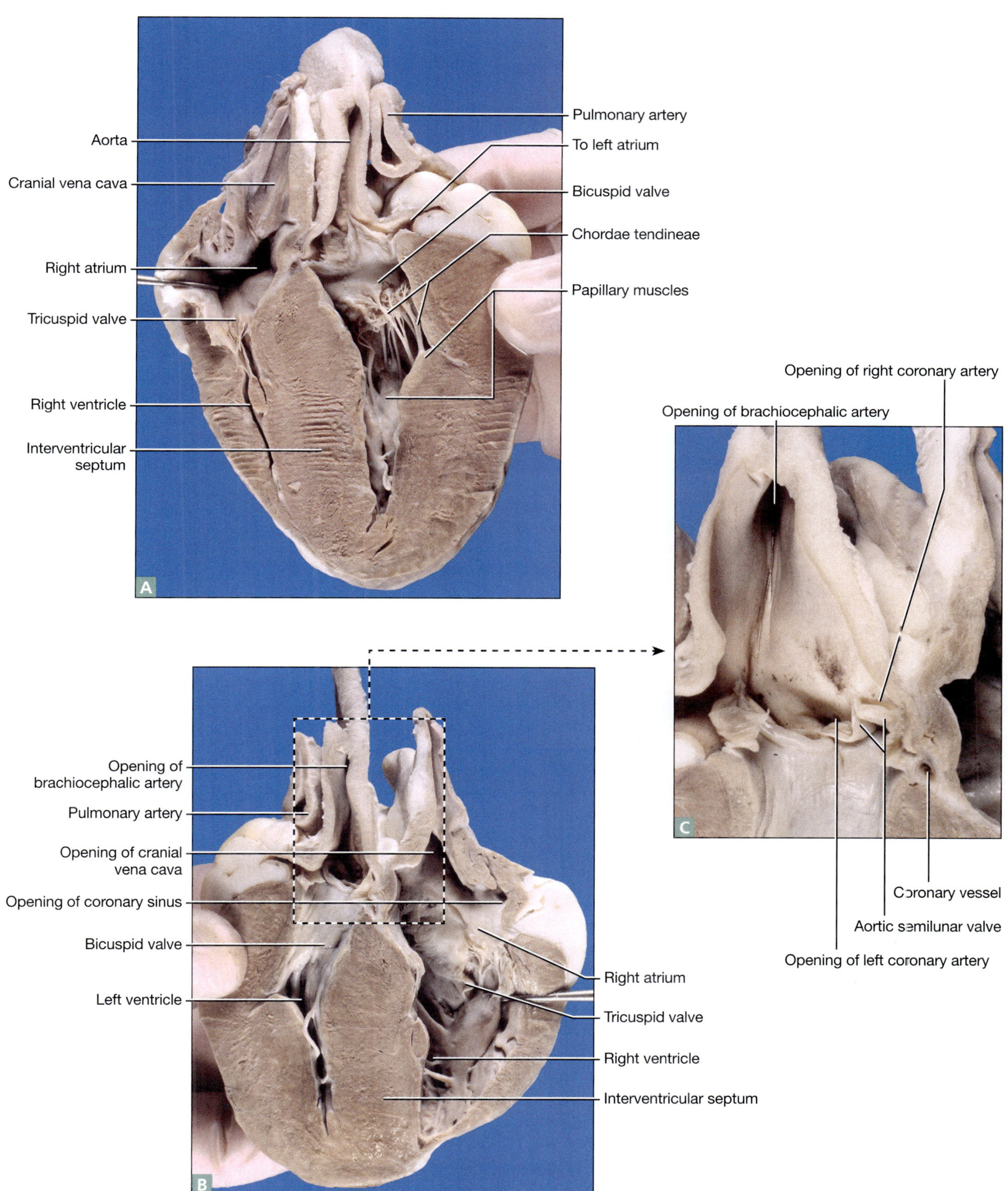

22.39 Dissected sheep heart: (**A**) dorsal half of frontal section, and (**B**) ventral half of frontal section, with inset (**C**) depicting openings to coronary arteries.

22

14.1 Trace the path of blood through the heart, beginning with blood entering the heart through the vena cavae.

14.2 Describe what problems might occur if the mammalian heart did not possess chordae tendineae.

22

Name ______________________ Section ______________

22 Questions for Review

1 Complete Summary Table A-1 in the Appendix, filling in the characteristics of chordates in the appropriate row.

2 List five evolutionary adaptations that have contributed to the success of mammals.

1. ______________________
2. ______________________
3. ______________________
4. ______________________
5. ______________________

3 Number the following organs in the order that represents the correct path of food through the digestive tract of a mammal:

_____ anus
_____ ileum
_____ stomach
_____ duodenum
_____ mouth
_____ jejunum
_____ esophagus
_____ rectum
_____ colon
_____ cecum

4 What is the adaptive value of an extremely long digestive tract?

____

5 What is the primary evolutionary reason for mammals to develop and house their lungs inside the body cavity?

____

6 Explain the functional relationship between the respiratory system and the circulatory system.

____

7 Distinguish between excretion and egestion.

____

22

8 In the mammalian excretory system, urine is produced in the kidneys and travels through the ______________ to the ______________, where it is temporarily stored.

9 Match the structure on the left with its one most appropriate function on the right. (Not all letters will be used.)

____ epididymis	a.	produces eggs
____ ovary	b.	releases urine to outside environment
____ oviduct	c.	captures recently "erupted" egg
____ testis	d.	site of embryo implantation/development
____ uterine horn	e.	produces sperm
____ vas deferens	f.	transports sperm to penis during ejaculation
	g.	stores sperm
	h.	contributes extra fluid to semen
	i.	cartilaginous constriction of cranial end of vagina

10 Match the region of the brain on the left with its most appropriate function on the right.

____ cerebellum	a.	controls fine movement, language, personality
____ frontal lobe	b.	controls breathing, heart rate, digestion, and sweating
____ parietal lobe	c.	processes auditory signals
____ medulla oblongata	d.	processes visual signals
____ occipital lobe	e.	controls muscle coordination and balance
____ temporal lobe	f.	processes information from touch receptors, muscles, and joints

11 Number the following regions of the eye in the correct order that light passes through them.

____ pupil
____ retina
____ cornea
____ vitreous chamber
____ lens
____ choroid layer

Appendix

Table A-1 Summary Table of Characteristics for Major Animals Covered in This Laboratory Guide

Phylum	Animal	Habitat	Symmetry	Tissue Organization	Body Cavity
Porifera	Sponge				
Cnidaria	*Hydra*, sea anemone				
Platyhelminthes	Planaria				
Nematoda	Roundworm				
Mollusca	Clam				
Mollusca	Squid				
Annelida	Earthworm				
Arthropoda	Crayfish				
Arthropoda	Grasshopper				
Echinodermata	Sea star				
Chordata Subphylum Cephalochordata	Lancelet				
Chordata Subphylum Urochordata	Tunicate				
Chordata Subphylum Vertebrata	Pig, rat				

	Locomotion	Support System	Digestive System	Respiratory System	Excretory System	Nervous System	Circulatory System

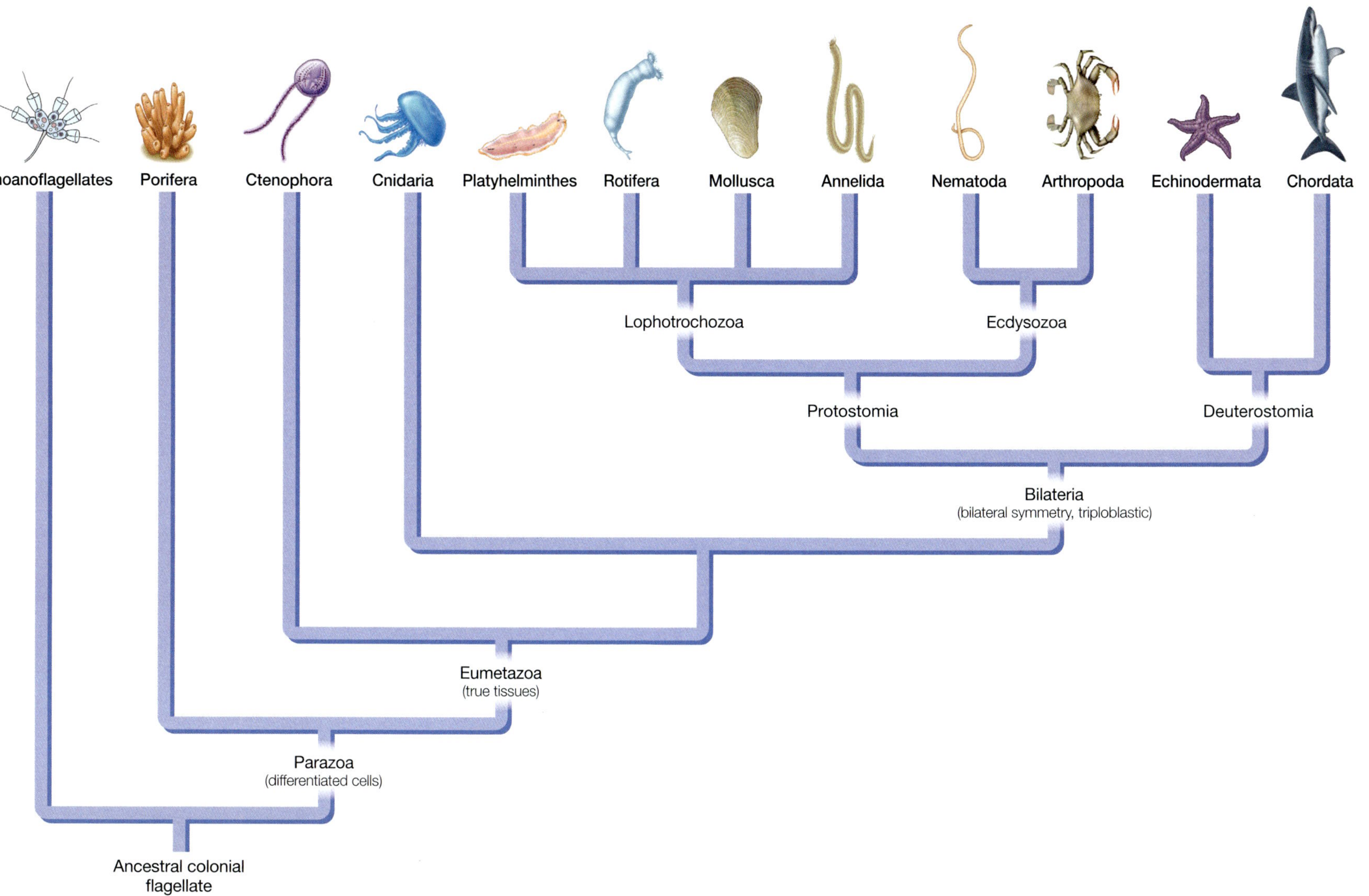

A.1 Phylogeny of the major animal phyla based on classical morphological evidence and molecular evidence that groups bilaterians into three major lineages: lophotrochozoans, ecdysozoans, and deuterostomes.

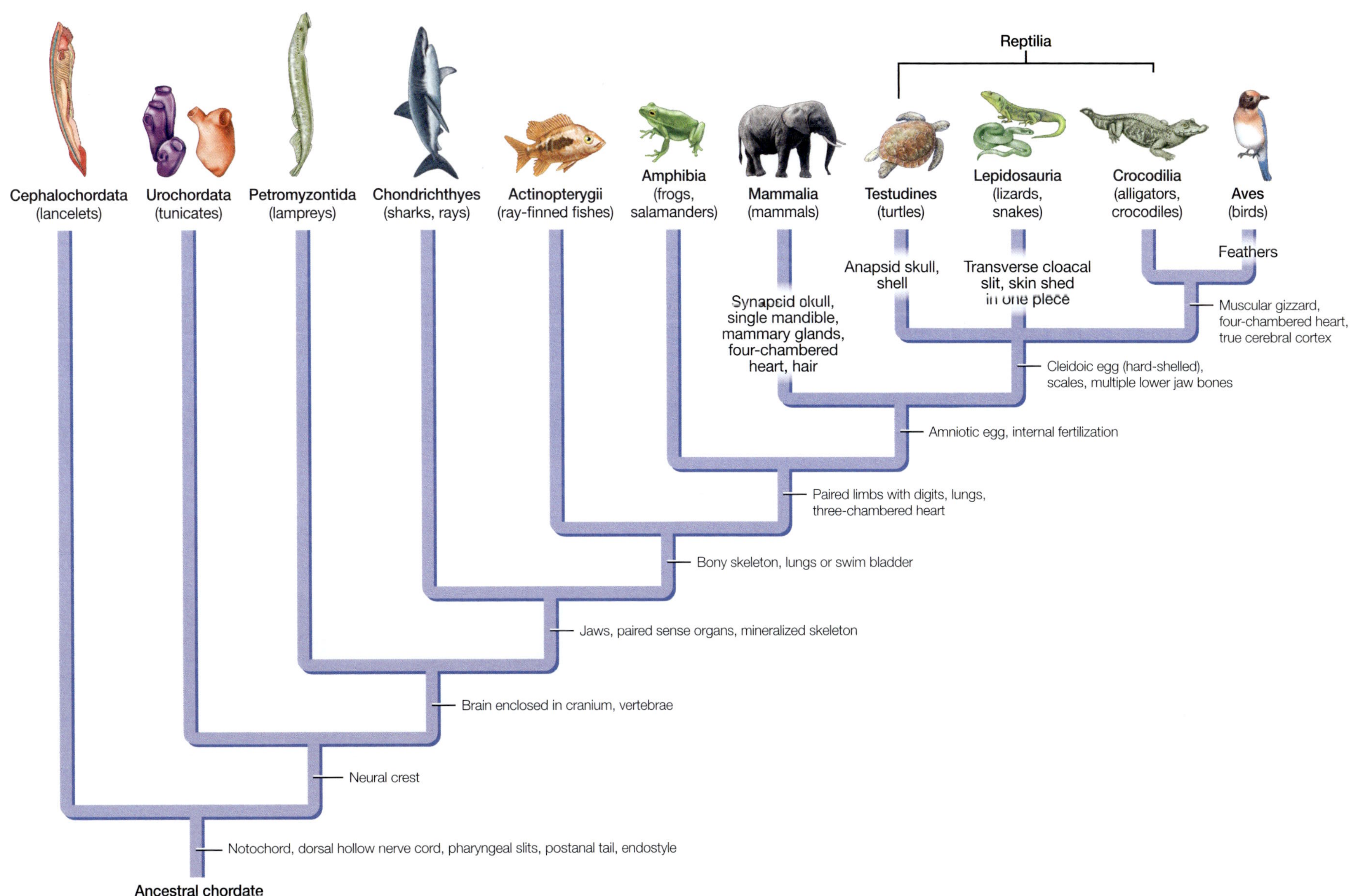

A.2 Phylogeny of the major chordate groups. Notice that reptiles are a paraphyletic group—meaning the class Reptilia, as we define it, does not include all the descendants of its ancestor. The separate class distinction for birds (Aves) emphasizes the dissimilarity between birds and crocodilians, rather than their true phylogenetic relationship. Many zoologists now consider Aves to be a subclass of Reptilia.

Table A-2 Geologic Time Scale

Era	Period	Epoch	Age (mya)	Important Events
Cenozoic	Quaternary	Holocene	0.01	Historical time
		Pleistocene	2.5	Ice ages; extinction of many large mammals; origin of genus *Homo*
	Tertiary	Pliocene	5.3	First bipedal hominids
		Miocene	23	Grazing mammals appear as grasses replace forests in drier areas; first apes
		Oligocene	35	Primates diversify; Europe separates from North America
		Eocene	56	Most present-day mammal orders evolved; flowering plants continue dominance
		Paleocene	65	Rapid diversification of mammals, birds, and pollinating insects; first primates appear
Mesozoic	Cretaceous		146	Flowering plants diversify and dominate; extinction of dinosaurs
	Jurassic		200	Dinosaurs at their peak; first bird fossils; first flowering plants
	Triassic		250	First dinosaurs and mammals, coniferous forests dominate landscape
Paleozoic	Permian		299	Diversification of reptiles; first mammal-like reptiles; origin of most insect groups
	Carboniferous		360	First reptiles; first seed plants; large coniferous forests
	Devonian		416	Diversification of bony fishes; first amphibians and insects appear; first forests
	Silurian		444	First jawed fishes; invertebrates colonize land
	Ordovician		500	Radiation of fishes; first land plants and fungi appear
	Cambrian		570	Rapid diversification of invertebrates into major phyla; first vertebrates (Cambrian explosion)
Precambrian			635	Soft-bodied invertebrates appear
			1,200	First evidence of multicellular life
			2,100	First eukaryotic cells appear
			2,500	Accumulation of atmospheric oxygen
			3,500	First prokaryotes appear
			3,800	Oldest known rocks on Earth's surface
			4,600	Formation of Earth

Representative Organisms
Extinction Events
Cretaceous Extinction
Triassic Extinction
Permian Extinction
Devonian Extinction
Ordovician Extinction

References

Adams, B. J. and J. L. Crawley. *Van De Graaff's Photographic Atlas for the Zoology Laboratory,* 7th ed. Englewood, CO: Morton Publishing, 2013.

Arnold, J. M., W. C. Summers, D. L. Gilbert, R. S. Manalis, N. W. Daw, and R. J. Lasek. *A Guide to Laboratory Use of the Squid* Loligo pealei. Woods Hole, MA: Marine Biological Laboratory, 1974.

Barnes, R. D. *Invertebrate Zoology*, 2nd ed. Philadelphia: W. B. Saunders, 1968.

Beck, D. E. and L. F. Braithwaite. *Invertebrate Zoology Laboratory Workbook*, 2nd ed. Minneapolis, MN: Burgess Publishing Company, 1962.

Berger, A. J. "The Musculature." In *Biology and Comparative Physiology of Birds*, vol. 1, edited by A. J. Marshall, 301–344. New York: Academic Press, 1960.

Brooke, M. and T. Birkhead. *The Cambridge Encyclopedia of Ornithology*. Cambridge: Cambridge University Press, 1991.

Brusca, R. C. and G. J. Brusca. *Invertebrates,* 2nd ed. Sunderland, MA: Sinauer Associates, 2003.

Chiasson, R. B. *Laboratory Anatomy of the Pigeon,* 3rd ed. Dubuque, IA: William C. Brown, 1984.

Dorit, R. L., W. F. Walker, and R. D. Barnes. *Zoology*. Philadelphia: W. B. Saunders, 1991.

Drewes, C. "Those Wonderful Worms." *Carolina Tips* 59, no. 3 (1996): 17–20.

Drewes, C. and K. Cain. "As the Worm Turns: Locomotion in a Freshwater Oligochaete Worm." *American Biology Teacher* 61, no. 3 (1999): 438–442.

Fishbeck, D. W. and A. Sebastiani. *Comparative Anatomy: Manual of Vertebrate Dissection,* 2nd ed. Englewood, CO: Morton Publishing, 2008.

Gilbert, S. G. *Pictorial Anatomy of the Dogfish*. Seattle: University of Washington Press, 1973.

Glase, J. C., M. C. Zimmerman, and J. A. Waldvogel. "Investigations in Orientation Behavior." In *Tested Studies for Laboratory Teaching*, vol. 6, edited by C. A. Goldman, S. E. Andrews, P. L. Hauta, and R. Ketcham, 1–16. Proceedings of the 6th Workshop/Conference of the Association for Biology Laboratory Education (ABLE), 1992.

Gunstream, S .E. *Explorations in Basic Biology,* 12th ed. Boston: Benjamin Cummings, 2011.

Guttman, B. S. *Biology*. Boston: WCB/McGraw-Hill, 1999.

Guyer, M. F. and C. E. Lane. *Animal Biology*, 5th ed. 1964. New York: Harper and Row, 1964.

Halliday, T. R. and K. Adler. *New Encyclopedia of Reptiles and Amphibians*. Oxford: Oxford University Press, 2002.

Hegner, R. W. and J. G. Engemann. *Invertebrate Zoology*, 2nd ed. New York: Macmillan, 1968.

Hickman Jr., C. P., L. B. Kats, and S. L. Keen. *Laboratory Studies in Integrated Principles of Zoology,* 14th ed. Boston: McGraw-Hill, 2008.

Hickman Jr., C .P., L. S. Roberts, and F. M. Hickman. *Biology of Animals*, 5th ed. St. Louis, MO: Times Mirror/Mosby, 1990.

Hickman Jr., C. P., L. S. Roberts, S. L. Keen, D. J. Eisenhour, A. Larson, and H. I'Anson. *Integrated Principles of Zoology*, 15th ed. Boston: McGraw-Hill, 2011.

Hildebrand, M. and G. E. Goslow, Jr. *Analysis of Vertebrate Structure*, 5th ed. New York: John Wiley and Sons, 2001.

Hopkins, P. M. and D. G. Smith. *Introduction to Zoology: A Laboratory Manual*, 3rd ed. Englewood, CO: Morton Publishing, 1997.

Kent, G. C. *Comparative Anatomy of the Vertebrates*, 2nd ed. St. Louis, MO: Mosby, 1969.

Lasiewski, R. C. "Respiratory Function in Birds." In *Avian Biology*, vol. 2, edited by D. S. Farner, J. R. King, and K. C. Parkes, 287–342. New York: Academic Press, 1972.

Leboffe, M. J. *A Photographic Atlas of Histology*. Englewood, CO: Morton Publishing, 2003.

Lytle, C. F. and J. R. Meyer. *General Zoology Laboratory Guide,* 15th ed. New York: McGraw-Hill, 2009.

Mader, D. R. "Reptilian Anatomy: Discover Exactly What's Inside Your Reptiles, and Where." *Reptiles* 3, no. 2 (1995): 84–93.

Miller, S. A., and J. P. Harley. *Zoology,* 8th ed. New York: McGraw-Hill, 2010.

Miller, S. A. *General Zoology Laboratory Manual,* 6th ed. New York: McGraw-Hill, 2004.

Mitchell, L. G., J. A. Mutchmor, and W. D. Dolphin. *Zoology*. Boston: Benjamin Cummings, 1988.

Nickel, R., A. Schummer, E. Seiferle, W. G. Siller, and P. A. L. Wight. *Anatomy of the Domestic Birds*. New York: Springer-Verlag, 1977.

Pechenik, J. A. *Biology of the Invertebrates,* 6th ed. New York: McGraw-Hill, 2009.

Petrunkevitch, A. *Morphology of Invertebrate Types*. New York: Macmillan, 1916.

Purves, W. K., D. Sadava, G. H. Orians, and H. C. Heller. *Life: The Science of Biology,* 7th ed. Sunderland, MA: Sinauer/W. H. Freeman, 2003.

Raven, P. H. and G. B. Johnson. *Biology*, 6th ed. Boston: McGraw-Hill, 2002.

Reece, J. B., L. A. Urry, M. L. Cain, S. A. Wasserman, P. V. Minorsky, and R. B. Jackson. *Campbell Biology,* 9th ed. Boston: Pearson, 2011.

Romer, A. S. *The Vertebrate Body*, 3rd ed. Philadelphia: W. B. Saunders, 1962.

Ruppert, E. E., R. S. Fox, and R. D. Barnes. *Invertebrate Zoology: A Functional Evolutionary Approach*, 7th ed. Belmont, CA: Brooks/Cole, 2003.

Sherman, I. W. and V. G. Sherman. *The Invertebrates: Function and Form, A Laboratory Guide*. New York: Macmillan, 1970.

Smith, D. G. *Exercises for the Zoology Laboratory*, 3rd ed. Englewood, CO: Morton Publishing, 2013.

Smith, D. G. and M. P. Schenk. *A Dissection Guide and Atlas to the Rat*. Englewood, CO: Morton Publishing, 2001.

Smith, D. G. and M. P. Schenk. 2011. *A Dissection Guide and Atlas to the Fetal Pig*, 3rd ed. Englewood, CO: Morton Publishing.

Solomon, E. P., L. R. Berg, and D. W. Martin. *Biology*, 6th ed. South Melbourne, Victoria: Brooks/Cole, 2002.

Stiles, K. A., R. W. Hegner and R. A. Boolootian. *College Zoology*, 8th ed. New York: Macmillan, 1969.

Summers, A. P. "Warm-hearted crocs." *Nature* 434 (2005): 833–834.

Tyler, M. S. *Developmental Biology: A Guide for Experimental Study*. Sunderland, MA: Sinauer, 1994.

Van De Graaff, K. M. and J. L. Crawley. *A Photographic Atlas for the Biology Laboratory,* 6th ed. Englewood, CO: Morton Publishing, 2009.

Vodopich, D. S. and R. Moore. *Biology Laboratory Manual,* 9th ed. New York: McGraw-Hill, 2009.

Wachtmeister, H. F. E. and L. J. Scott. *Encounters with Life: General Biology Laboratory Manual,* 7th ed. Englewood, CO: Morton Publishing, 2006.

Walker Jr., W. F. *Dissection of the Frog,* 2nd ed. New York: W. H. Freeman, 1998.

Walker Jr., W. F., and D. G. Homberger. *Anatomy and Dissection of the Rat,* 3rd ed. New York: W. H. Freeman, 1997.

Wischnitzer, S. and E. Wischnitzer. *Atlas and Dissection Guide for Comparative Anatomy,* 6th ed. New York: W. H. Freeman, 2006.

Wodsedalek, J.E. *General Zoology*. Dubuque, IA: Wm. C. Brown Company, 1963.

Wright, S. J. *A Photographic Atlas of Developmental Biology*. Englewood, CO: Morton Publishing, 2005.

Glossary

– A –

abdomen (1) posterior section of the body in arthropods; (2) body cavity caudal to or below the thoracic cavity containing the viscera

abduct movement away from the midline of the body

aboral referring to the side or surface of the body farthest from the mouth

acoelomate animal whose central space is filled with tissue (mesoderm) and in which no true body cavity exists

acontia "internal tentacles" that contain cnidocytes and aid in subduing live prey taken into the gastrovascular cavity of cnidarians

actin thin, proteinaceous filaments that overlap and alternate with myosin filaments to generate muscular contractions

action influence a muscle has on the motion of a joint or appendage

adaptive radiation evolution of numerous species from a common ancestor following migration into a new environment

adduct movement toward the midline of the body

adipose tissue type of connective tissue that stores or sequesters food for the body in the form of fat droplets

adrenal gland endocrine gland located on medial side of kidney that produces hormones that mediate responses to stressful situations and control blood pressure and carbohydrate and protein metabolism

air sacs (1) evolutionary precursors to lungs in early bony fishes; (2) thin, membranous pouches located throughout visceral cavity in birds; used in respiration

allantoic stalk tube passing through the umbilical cord of the fetus connecting it with the allantois in the uterus of the mother

allantois extra-embryonic sac that acts as a repository for metabolic wastes produced by the fetus during development

alveoli (1) small cavities found beneath the surface of protists in the candidate kingdom Alveolata; (2) small, air-containing compartments of the lungs in which the bronchioles terminate and from which respiratory gases are exchanged with the pulmonary capillaries

ambulacral grooves indentations along the arms of echinoderms that contain the tube feet

amniotic egg egg (of reptile, bird, or mammal) surrounded by a fluid-filled sac that encloses the embryo and protects it from desiccation

amoebocytes mobile cells specialized for distributing food throughout the sponge and for producing its skeleton

ampulla (*pl.* = ampullae) muscular, bulb-like reservoir at the internal end of each tube foot in echinoderms that controls the amount of fluid and flow of fluid into the tube foot

analogous referring to features or structures whose similarities are not due to common evolutionary origin

anal spot (*syn.* = cytoproct) opening through which indigestible matter is expelled from the cell

ancestral character state variant of a trait determined to have been present in the common ancestor of an entire group or lineage

animal pole portion of an egg where the embryo develops and that appears darker in coloration in comparison to the vegetal pole

antagonistic working in opposition to or limiting the movement of a paired structure

antennae movable, segmented organs of sensation located on the heads of arthropods

anus posterior opening of the digestive tract through which undigested food particles (feces) are egested from the body

aortae (*sing.* = aorta) major vessels of the arterial network that carry blood away from the heart

aortic arches muscular regions of the circulatory system in annelids that regulate blood pressure

apopyle opening through which water passes out of the radial canals into the spongocoel in sponges

appendicular skeleton portion of the skeleton consisting of the bones of the forelimbs, hind limbs, pectoral girdle, and pelvic girdle

archenteron (*syn.* = gastrocoel) hollow tube created by the arrangement of invaginated cells in the gastrula; embryonic precursor to the gut

arm supports microscope body, stage, and adjustment knobs

arteries vessels that conduct blood away from the heart

ascon simplest of the sponge body plans with canals leading directly from the outside to the interior with no folding of the body wall

asexual reproduction type of reproduction that occurs without genetic recombination and results in the production of offspring that are genetically identical to the parent; may occur by simple fission of a single cell, budding, or fragmentation

asters array of microtubule filaments by which the centriole pairs anchor themselves to the cell membrane at the conclusion of prophase

asymmetry lack of symmetry; irregular arrangement of body parts with no plane of symmetry to divide them into similar halves

atriopore cephalochordate structure that discharges water from the atrium to the external environment

atrium (*pl.* = atria) (1) internal chamber of the heart that receives blood from the body and channels it to a ventricle; (2) in tunicates and cephalochordates, the body chamber that receives water passed from the pharynx through gill slits

auricles (1) primitive chemoreceptors found in some invertebrate phyla (i.e., Platyhelminthes); (2) flap-like outer regions covering the cranial portions of the atria in many vertebrates' hearts

autotrophic characteristic of an organism that produces its own organic food molecules, usually through photosynthesis

axial skeleton portion of the skeleton consisting of the skull, vertebral column, and rib cage

axis (1) second cervical vertebra; (2) straight line that bisects the body into two equal halves; usually along the longer portion of the body

axon elongated portion of a neuron that transmits electrical impulses away from the cell body

– B –

basal disc terminal portion of stalk in cnidarian polyps that serves as the primary point of attachment to the substrate

base supports microscope unit

basophils least common type of white blood cell found in mammals; possess bilobed nuclei and are instrumental in the body's inflammatory response

behavior any overt response by an animal to a stimulus

bilateral symmetry body parts are divided into similar halves (mirror images) by a single plane of symmetry

bile digestive fluid secreted by liver and stored in gallbladder (when present); functions in the emulsification of fats in the intestine

bile duct canal that transports secretions from the liver (and from the pancreas in some animals) to the duodenum or stomach

binary fission type of cell division by which many unicellular protists and bacteria reproduce asexually; generally involves a simple mitotic division of the parent cell producing two genetically identical daughter cells

binomial nomenclature system of naming a species by the combination of the genus name and a species epithet

bipinnaria larva immature developmental stage of sea stars immediately following the gastrula stage; characterized by bilateral symmetry and the beginnings of organogenesis

biramous having two branches

blastocoel hollow, fluid-filled cavity within the blastula

blastomeres individual cells of a blastula

blastopore opening to the outside of the embryo marking the site of invagination of cells toward the center of the blastocoel

blastula embryonic stage of development containing ≥64 cells and characterized by a migration of cells toward the periphery of the embryonic sphere creating a hollow cavity deep within the spheroid of cells

blood fluid connective tissue that typically contains respiratory pigments and serves as a medium for nutrient and gas exchange between tissues and the external environment

blood vessel conduit of the circulatory network that routes blood throughout the body and enables exchange of water, chemicals, gases, nutrients, and toxins between the blood and body tissues

body microscope housing that keeps ocular and objective lenses in proper alignment

bone rigid connective tissue used to support the body; characterized by densely packed, hard, fibrous matrix composed of calcium salts surrounding osteocytes (bone-producing cells)

book gills respiratory structures found in aquatic chelicerates; consist of thin, overlapping, blood-filled flaps of tissue that exchange gases with the seawater flowing over them

book lungs respiratory structures found in arachnids; consist of thin-walled air flaps of tissue that extend into a blood-filled chamber in the abdomen

brain part of the central nervous system responsible for processing and integrating nerve impulses gathered from all sensory organs and receptors and for initiating motor impulses

branchial hearts smaller, muscular chambers in cephalopods that receive deoxygenated blood from all parts of the body and pump blood to the gills

bronchi (*sing.* = bronchus) major divisions of the trachea that supply oxygen (and remove carbon dioxide) from the lobes of the lungs

buccal cavity the first portion of the digestive canal in animals (for example, the mouth)

bud (budding) asexual outgrowth of the parent that pinches off when mature and lives independently; common in cnidarians

bulbourethral glands small glands in male mammals associated with the reproductive system that discharge a component of the seminal fluid that provides lubrication and pH neutralization of the vagina during intercourse

bulbus arteriosus large, tubelike structure that channels blood pumped from the ventricle of a fish's heart into the aorta

– C –

canaliculi tiny, fingerlike projections in the lamellae of bone tissue through which nutrients are transported to the osteocytes

carapace hard, bony, dorsal portion of the shell of turtles

cardiac chamber (*syn.* = cardiac stomach) first stomach region in some arthropods and echinoderms where mechanical digestion of food occurs

cardiac muscle type of muscle tissue that forms the walls of the heart; characterized by striated muscle fibers joined together with gap junctions called intercalated discs that relay each heartbeat

cartilage flexible connective tissue that is characterized by fibrous tissue surrounding individual chondrocytes (cartilage-producing cells)

caudal referring to a point of reference toward the tail

cecum (*pl.* = ceca) (1) cavity open at one end, usually associated with digestion; (2) in vertebrates, a blind projection located at the junction of the ileum and colon that serves as a sac for fermentation of cellulose; plays a prominent role in the digestive process of most herbivores, but is reduced in omnivores and carnivores

cell basic structural and functional unit of life consisting of a membrane-bound collection of genetic material and organelles

cell body portion of a neuron that contains the nucleus and other organelles

cell cycle events that encompass the entire life cycle of a cell from one division to the next

cell differentiation process by which newly formed embryonic cells take on specialized functions and undergo shifts in morphology to accommodate these new tasks

cell division formation of new cells through mitosis or meiosis

cell migration organization and movement of groups of cells to create an animal's body shape

cell theory scientific contention that the cell is the basic unit of life, all organisms are composed of cells, and all cells arise only by division of previously existing cells

cell wall protective, rigid, outer layer of the cells of plants, fungi, bacteria, and many protists; may consist of cellulose, calcium carbonate, silica, or other materials

cellulose structural polysaccharide of cell walls, consisting of glucose monomers joined together

cellulose plates structural element of dinoflagellates found beneath the outer plasma membrane providing support and shape to their cells

centriole one of a pair of small, cylindrical organelles lying at right angles to each other near the nucleus in the cytoplasm of animal cells; composed of nine triplets of microtubules

centrolecithal classification of eggs with a relatively large amount of centralized yolk, around which the developing embryo forms a thin layer

centromere centralized region joining adjacent sister chromatids

cephalization concentration of nervous tissue and sensory structures at the anterior end of the body

cephalothorax fused portion of the arthropod body consisting of the head and thorax regions

cercaria free-swimming, tadpole-shaped larva of a trematode that develops from a redia

cerebellum region of the vertebrate hindbrain that integrates the movements of skeletal muscles and controls coordination and balance

cerebral ganglia rudimentary "brain" in invertebrates responsible for processing and integrating nerve impulses gathered from all sensory organs and receptors and for initiating motor impulses

cerebrum portion of the brain devoted to the integration of sensory impulses, learning, memory, and voluntary movements; divided into two hemispheres and located in the upper portion of the cranial cavity

cervix cartilaginous constriction demarcating the cranial boundary of the vagina

character heritable trait or feature that can be used in phylogenetic analyses to determine the degree of relatedness among organisms

chelicera one of a pair of modified clawlike feeding appendages characteristic of chelicerates

cheliped pincerlike first pair of legs in crustaceans; specialized for capturing and crushing prey

chlorocruorin greenish, iron-containing respiratory pigment chemically related to hemoglobin and found in the blood of some marine polychaetes

chlorophyll C green pigment located within plastids in autotrophic protists and plants that functions in the light reactions of photosynthesis

chloroplast organelle found in autotrophic protists and plants that absorbs sunlight and drives photosynthetic reactions to produce sugars

choanocytes specialized cells lining the interior surface of the radial canals in sponges that trap and engulf small food particles

chondrocytes cartilage-producing cells

choroid layer vascular coating of the eye located between the sclera and the retina

chromatin the complex of DNA and protein that makes up eukaryotic chromosomes

chyme fluid produced by the action of digestive enzymes from the stomach mixing with and dissolving ingested food particles

cilia short, cellular appendages specialized for locomotion; found in many protists

ciliary bodies thickened, muscular portion of the tunic of the eye, connecting the choroid layer and iris

circular canal circular extension of the gastric pouches that distributes nutrients to outer rim of jellyfish

cleavage cell division

cleavage furrow shallow indentation formed along cell margin of the metaphase plate in a dividing cell

clitellum unsegmented band of tissue found in some groups of annelids; used in reproduction

cloaca common chamber for the elimination or release of digestive, excretory, and reproductive products; in reptiles and birds the cloaca is partitioned into three chambers: the urodeum, the coprodeum, and the proctodeum

closed circulatory system cardiovascular system in which blood is confined to and travels through a system of arteries, veins, and capillaries

cnidocytes stinging cells on cnidarian tentacles used for defense and for capturing food

coagulating gland accessory male reproductive glands that contribute fluid to the semen

coarse-focus adjustment knob moves microscope stage up or down to focus image

coelom body cavity lined on both sides by mesodermally derived tissue

coenosarc common gastrovascular cavity shared among colonial cnidarian polyps that transports nutrients throughout the colony

collagen structural protein in connective tissues; characterized by high content of the amino acids glycine, alanine, proline, and hydroxyproline

collar fleshy border separating head-foot from visceral mass (mantle) in some cephalopods

colon section of mammalian digestive tract extending from the cecum to the rectum, primarily responsible for reabsorption of water and electrolytes

complete digestive tract digestive tract that passes food through in one direction from the mouth to the anus

condenser lens mounted beneath microscope stage that focuses the light beam on the specimen

condenser adjustment knob moves microscope condenser lens up or down to focus light

conjugation process that occurs in some protists and bacteria in which genetic material from one cell is transferred to another cell, or genetic material from two cells is simultaneously transferred

connective tissues groups of cells that work together to bind, support, and protect body parts and systems

contractile stalk thin extension of the plasma membrane of certain protists that they use to extend and retract their cell bodies in relation to their substrate

contractile vacuole spherical organelle designed to pump excess water out of cells

conus arteriosus terminal constriction in the heart of fishes that receives blood from the ventricle and channels it into the afferent branchial arteries for delivery to the gills

convergent evolution independent development of similarity between species as a result of similar selection pressures typically generated by comparable ecological roles

coprodeum anterior-most chamber of the cloaca in reptiles and birds; receives fecal matter directly from the large intestine

copulatory organs structures used for sexual intercourse

cornea transparent convex anterior portion of the outer fibrous coating of the eyeball that covers the iris and the pupil and is a continuous extension of the sclera

corona ciliated disc that encircles the anterior end of rotifers

cranial referring to a point of reference toward the head

crop thin-walled, saclike portion of digestive system where ingested food is temporarily stored; found in some annelids, arthropods, and vertebrates

crossing over process in which small sections of chromosome are exchanged between neighboring homologous chromosomes during meiosis I

cuticle thin, epidermal secretion that covers the outer surface of the body and protects the organism from external chemical and mechanical dangers and/or desiccation

cyst dormant, resistant stage of cells that allows them to survive prolonged periods during which environmental conditions are unfavorable for growth or survival

cytokinesis period of cell division in which the cytoplasm divides to form two daughter cells

cytopharynx tubular opening in ciliates through which food passes and is incorporated into food vacuoles

cytoplasm living matter contained within the plasma membrane of an animal cell, excluding the nucleus

cytoplasmic streaming phenomenon in which cytoplasm of a cell flows from one region of the cell body to another, often as a means of distributing food, gases, or other substances

– D –

dendrites parts of a neuron that typically receive electrical impulses from neighboring neurons or sensory receptors and transmit them to the cell body

depth of field thickness of an image that is in focus at any point in time when looking through a microscope

derived character state variant of a trait that differs from the ancestral form and is thought to have arisen later in the evolutionary history of a lineage

dermal branchiae (*syn.* = skin gills) echinoderm organs of gas exchange and excretion

dermal exoskeleton in bony fishes, the fin rays and scales are composed of bony structures derived from intramembranous ossification, in contrast to dermal bone found in the bony endoskeleton that forms through the process of accretion by osteocytes

determinate development type of embryonic development in which the developmental "fate" of individual cells is established early in the embryonic process; characteristic of protostomes

deuterostome animal whose mouth develops from the second embryonic opening and whose embryonic cells divide by radial cleavage

diaphragm muscular sheet separating the thoracic and abdominal cavities; used to ventilate the lungs of mammals

digestive gland organ that secretes digestive enzymes into stomach and intestine in many invertebrate phyla to assist in the breakdown of food

digitigrade type of locomotion characterized by walking on the tips of the toes (digits); body weight is supported primarily by the phalanges

dioecious referring to an organism that contains only male or female reproductive structures

diploblastic animal whose body develops from only two embryonic tissue layers

direct development form of embryonic development in which no larval stage is present; young emerge resembling "miniature" adults

discoidal cleavage pattern of blastomere production in which the cleavage planes are restricted to a dome of cytoplasm at one end of the embryo

distal referring to a point of reference farther from the body's median plane or point of attachment than another structure

dorsal referring to a point of reference nearer the backbone

dorsal lip crescent-shaped groove along the surface of the blastula that marks the opening of the blastopore

ductus deferens (*syn.* = vas deferens) tubule in male animals that transports sperm

duodenum anterior portion of the digestive tract that receives secretions from the liver and pancreas through the common bile duct for further breakdown of food from the stomach

– E –

ecdysis (*syn.* = molting) periodic act of shedding the skin or exoskeleton

ectoderm outermost of the three primary embryonic germ layers that gives rise to the outer covering of animals and, in some phyla, the nervous system

ectoparasite parasite that attaches itself to the outside of the host organism to obtain its nourishment

ectotherm animal that derives the majority of its body heat from external sources

elastic cartilage gelatinous connective tissue that contains fine collagen fibers and many flexible fibers; found in the ear, nose, and voice box of humans

endoderm innermost of the three primary embryonic germ layers that gives rise to many major internal organs and linings

endoparasite parasite that lives within the body of its host

endoskeleton animal support system enclosed beneath the outer body surface (for example, bones of humans, ossicles of echinoderms)

endostyle evolutionary precursor to thyroid gland in higher chordates; in lower chordates it is a mucus-coated groove in the floor of the pharynx that accumulates food particles and moves them along the digestive tract

endosymbiosis mutually beneficial partnership between two dissimilar organisms in which one organism lives within the body of its partner

endotherm(ic) referring to an animal that derives the majority of its body heat from internal metabolic sources

eosinophils type of white blood cell found in mammals; possess bilobed nuclei and are instrumental in the body's inflammatory response

epidermis outer tissue covering in animals; derived from ectoderm

epididymis highly coiled tubule system that cups around the testis and serves as a storage unit and transportation canal for mature sperm

epiglottis cartilaginous flap that covers the glottis to prevent food from entering the larynx and trachea when swallowing

epithelial tissues cover external surfaces for protection or line the internal surfaces of body cavities and vessels

erythrocruorin oxygen-carrying respiratory pigment chemically similar to chlorocruorin and found in the blood plasma of many annelids

erythrocytes red blood cells; modified for transport of oxygen and carbon dioxide to and from body tissues

esophagus thin tube connecting the mouth to an adjacent portion of the digestive tract (for instance, stomach or crop)

eucoelomate animal with a central body cavity that lies between layers of mesodermally derived tissue

euglenoids category of protists characterized by an anterior pocket that bears one or two flagella extending from the anterior end of the organism

eukaryotic characteristic of cells that contain a membrane-enclosed nucleus and other membrane-enclosed organelles

excretion process of eliminating from the body metabolic waste products produced through cellular metabolism

excretory canals longitudinal channels running along the outer margins of the tapeworm body that expel metabolic waste products

excurrent siphon tube or opening through which water carrying bodily wastes is expelled from the body; present in many bivalve molluscs and tunicates

exoskeleton hard outer skeleton covering the body of an animal, such as the cuticle of arthropods or the shell of molluscs

extensor muscle that straightens or extends a body part

external nares (*syn.* = nostrils) paired openings in dorsum of head leading to olfactory receptors

eyes image-forming, photoreceptive organs of sight

eyespots rudimentary photoreceptors capable of detecting shadows and extremes in light; found in some invertebrate phyla (for example, Platyhelminthes)

– F –

fat bodies masses of fatty tissue that contain energy reserves for reproduction and metamorphosis; located near the gonads in amphibians

fertilization process of contact between sperm and egg, entry of the sperm into the egg, and fusion of the egg and sperm pronuclei to form a zygote

fertilization cone cytoplasmic protrusion of the egg at the site of initial sperm contact that draws the head of the sperm inside the egg's plasma membrane

fertilization membrane thin membrane encircling zygote; produced by the egg immediately following fertilization to prevent polyspermy (fertilization of the egg by multiple sperm)

fetal membranes embryonic tissues involved in pregnancy; chorion, amnion, and allantois

fibroblast specialized cell type that produces the fibers and proteinaceous matrix of connective tissues

field of view circular area seen when looking through a microscope

filopodia thin, protoplasmic extensions put out by clumps of aggregating sponge cells

fine-focus adjustment knob permits precise focusing of a microscope

flagellum (*pl.* = flagella) long, cellular appendage specialized for locomotion

flame cells excretory units common in many flatworms that collect metabolic wastes from nearby regions and channel wastes to excretory pores located on the body surface

flexor muscle that draws a limb or joint closer to the axis of the body

food vacuole small, spherical organelle containing enzymes to digest food

foot extensible, muscular organ used by molluscs for locomotion

foramina (*sing.* = foramen) any of the openings or orifices in bones through which blood vessels and nerves pass

fragmentation method of reproduction in which organism breaks into smaller pieces capable of surviving on their own

frontal plane longitudinal section dividing an animal into dorsal and ventral parts

fusiform tapering toward each end

– G –

gallbladder small, membranous, muscular sac that stores bile produced by the liver

gametogenesis production of egg or sperm cells

ganglion (*pl.* = ganglia) mass of neuron cell bodies

gastric ceca elongated, pouchlike projections of the upper end of the stomach in insects

gastric mill chitinous teeth that grind food; located within cardiac stomach of crustaceans

gastric pouch one of four divisions of the gastrovascular cavity of jellyfish for digestion of food

gastrocoel (*syn.* = archenteron) hollow tube created by the arrangement of invaginated cells in the gastrula; embryonic precursor of the gut

gastrodermis inner tissue lining of the digestive system in many animals; derived from endoderm

gastrovascular cavity (*syn.* = coelenteron, coenosarc) central digestive compartment in some invertebrates; characterized by a single opening that functions as both mouth and anus

gastrula/gastrulation embryonic stage of development characterized by the invagination of the cell surface and the migration of blastomeres toward the center of the gastrocoel

gemmule asexual cyst produced by freshwater sponges capable of overwintering through harsh environmental conditions

genital pore opening for the release and/or introduction of sperm or eggs

genus taxonomic category made up of related species

germ layers primary embryonic tissue layers that differentiate during gastrulation

gill bars semirigid structures in pharyngeal region of chordates; only functional in some groups (for example, tunicates and cephalochordates) and trap small food particles on their mucous coating

gill filaments highly vascular, feathery extensions of the respiratory structures in fishes that provide a large surface area for gas exchange underwater

gill rakers hard, serrated structures in fish gills that protect the gill apparatus from the passage of coarse material that could damage gill surfaces

gill slits spaces between gill bars that permit water to flow across gill bars

gills highly vascular, localized extensions of the body surface of aquatic animals modified for gas exchange

gizzard thick-walled, muscular digestive pouch in birds that pulverizes and churns food prior to passage into the intestine

glial cells assist in propagating nerve impulses and provide a nutritive role for neurons

glycogen converted form of glucose that is stored in the liver and muscles of animals

gonad reproductive organ that produces sperm or eggs

gonangium reproductive polyp in some colonial cnidarians

gravid pregnant; full of eggs or offspring

gray crescent delineating line between the animal pole and vegetal pole of amphibian eggs that forms on the opposite side of the egg from the entry point of the sperm

green gland excretory organ of crustaceans; responsible for filtering blood of metabolic waste and removing it from the body

– H –

haploid gametes cells that contain half the number of chromosomes as the somatic cells

Haversian canals tiny, narrow pathways around which lamellae form layered rings in bone tissue

heart muscular component of circulatory system responsible for pumping blood throughout body

hemipenes paired copulatory organs of male reptiles

hemocoel body space bathed in hemolymph (blood) in animals with an open circulatory system (arthropods)

hemocyanin colorless, copper-containing respiratory pigment found in the blood plasma of many arthropods and molluscs

hemoglobin iron-containing respiratory pigment of vertebrate red blood cells that functions in the transport of oxygen and carbon dioxide between the respiratory organs and body tissues

hemolymph fluid that bathes the tissues in animals with an open circulatory system (molluscs and arthropods)

hepatic cecum (*syn.* = cecum) lateral outpocket of intestine in cephalochordates that is responsible for intracellular digestion of small food particles and lipid and glycogen storage

hepatic portal system circulatory pathway in vertebrates that channels all blood from the capillary beds of the digestive organs to the liver before routing it through the posterior vena cava to the heart

hermaphroditic possessing both male and female sex organs, but not capable of self-fertilization (for example, earthworms)

heterotrophic characteristic of organisms that obtain nutrients by ingesting or absorbing organic material from external sources

hinges regions of flexion in the plastron of some turtle species

holdfast specialized, terminal region of algae that anchors them to the substrate

holoblastic cleavage cell division that completely separates the parent cell into roughly equal-sized, but distinct daughter cells

homeothermic maintaining a nearly constant internal body temperature

homologous referring to features or structures in different species that are similar due to common ancestry

hooks sharp, pointy attachments characteristic of many cestodes; modified for attachment of the parasite to its host

hormone chemical compound produced by endocrine tissue and distributed through the body via the circulatory system that communicates with target organs and tissues to produce a wide array of behavioral and physiological responses, depending on the specific hormone released

hyaline cartilage gelatinous connective tissue composed primarily of chondrin with thin collagen fibers; situated between bones where it cushions the surfaces of joints

hydranth feeding polyp in some colonial cnidarians

hydrostatic skeleton support system found in certain invertebrates in which contracting muscles push against a fluid-filled body cavity to generate body rigidity

hypophyseal sac terminal chamber of the nostril in lampreys that circulates water over the olfactory epithelium to facilitate olfaction

hypostome enlarged mound of tissue that surrounds the mouth in cnidarians

– I –

ileum distal portion of the small intestine in vertebrates extending from the jejunum to the cecum; primarily responsible for absorption of nutrients

illuminator source of light in a microscope

incisor front tooth adapted for cutting

incurrent canals small openings in the body surface of sponges through which water carrying oxygen and nutrients enters the sponge body

incurrent siphon tube or opening through which water carrying nutrients and oxygen is drawn into the body; present in many bivalve molluscs and tunicates

indeterminate development type of embryonic development in which each cell produced by early cell divisions retains the ability to develop into a complete embryo; characteristic of deuterostomes

ingestion process of taking in food through the oral cavity

inguinal canal opening in the abdominal wall through which the spermatic cord passes

ink sac reservoir in cephalopods that opens into the anus and secretes a dark brown or black fluid when the animal is alarmed

insertion distal point of attachment of a muscle, usually to the bone moved by that muscle

intercalated discs gap junctions between adjacent cells of cardiac muscle that permit expedient and synchronous contractions of the heart

internal fertilization reproductive method in which sperm are deposited within the female reproductive tract and fusion of sperm and egg occurs inside the body

interphase the period in the cell cycle when the cell is not dividing and cellular metabolic activity is high

intestine long, sometimes coiled, digestive organ that primarily absorbs nutrients, ions, and water

iris region of the eye that regulates the amount of light that enters the eye and reaches the retina by contraction of the sphincter muscles of the iris

iris diaphragm mounted beneath microscope stage near condenser; regulates amount of light illuminating specimen

isolecithal classification of eggs with relatively little yolk that is evenly distributed throughout the cytoplasm

– J –

jejunum middle portion of the small intestine in vertebrates extending from the duodenum to the ileum; primarily responsible for nutrient absorption

jointed appendages peripheral extensions of the body that contain hinged or pivoting points for increased flexibility and range of motion

– K –

keel prominent ventral protrusion of the sternum common in birds adapted for flight

keratinized scales shingle-like, overlapping body covering characteristic of reptiles; composed of fibrous protein to produce a hard, dry outer layer

kidney excretory organ that filters blood to create a highly concentrated metabolic by-product (urine); also responsible for maintaining a homeostatic balance of salts, fluids, and ions within the body (osmoregulation)

kinesis random behavior that affects the general rate of movement or degree of turning in an animal

kinetochore site of attachment of a chromosome at its centromere to a spindle fiber

kinetoplast characteristic organelle of kinetoplastids consisting of a single, large mitochondrion that contains extra-nuclear DNA

kinetoplastids group of protists characterized by the presence of kinetoplasts (see previous definition)

– L –

lacunae hollow chambers that contain chondrocytes (in cartilage) or osteocytes (in bone)

lamellae thin, concentric layers of a hard, calcified matrix that give bone its characteristic appearance

large intestine (*syn.* = colon) portion of the digestive tract that completes nutrient absorption and extracts water and ions from the fecal waste

larva free-living, sexually immature stage in certain animal life cycles

larynx portion of the respiratory tract containing the vocal cords

lateral referring to a point of reference farther from the median plane

lateral canals short branches of the radial canals in the echinoderm water vascular system that terminate at the tube feet

lateral lines (1) rudimentary excretory system of nematodes; (2) sensory system of fishes and some amphibians that detects vibrations in the water

lens transparent, biconvex body separating the posterior chamber and vitreous chamber of the eye that refracts and focuses light rays onto the retina

leucon most complex body plan of sponges; intricately folded canals contain chambers in which choanocytes reside

leukocytes white blood cells; typically function in immune system

ligaments tough, pliable bands of connective tissue that join bones to other bones in the body's joints

light-intensity adjustment dial permits the observer to regulate the amount of light projected through the specimen being viewed using a compound microscope for optimum viewing

lipids family of compounds, including fats, phospholipids, and steroids, that are insoluble in water

liver organ that produces bile and other secretions to facilitate digestion of food; also plays a role in lipid and glycogen storage and regulates blood glucose levels

loose connective tissue type of connective tissue composed of loosely scattered cells surrounded by a clear, gelatinous matrix

lung internal organ of respiration

lymphocytes type of white blood cells with spherical nuclei; play a central role in the body's immune response

– M –

macronucleus (*pl.* = macronuclei) organelle of ciliates containing many copies of a few genes; primarily responsible for metabolic processes of the cell

madreporite porous entrance to the water vascular system that serves as both pressure regulator and simple filter in most echinoderms

Malpighian tubules blind-ended tubules of insects that filter metabolic waste from the surrounding hemolymph and deposit this filtrate into the hindgut for elimination from the body

mammary glands modified tissues on the ventral surface of mammals that secrete milk to nourish young

mammary papillae small protuberances on the ventral surface of mammals; in adult females the papillae will develop into teats through which the mammary glands secrete milk

mantle thin, fleshy membrane characteristic of all molluscs; secretes the shell

manubrium stalk of fleshy tissue present in hydrozoan medusae that supports the mouth

marginal tentacles cnidarian tentacles that provide sensory information and are used for defense and locomotion

mastax muscular pharynx in rotifers

matrix ground substance found in connective tissue produced by the living cells that it contains; may be liquid (blood), gelatinous (cartilage), or solid (bone)

medial referring to a point of reference closer to the median plane

median plane the sagittal plane running along the midline of the animal

medulla oblongata most caudal region of the vertebrate brain; controls autonomic functions, such as breathing, heart rate, digestion, and swallowing

medusa stage in the cnidarian life cycle represented by a circular, free-swimming form resembling the familiar jellyfish in its morphology

medusa buds young, immature medusae produced within the reproductive polyps of some cnidarians

meiosis two-stage process of cell division in sexually reproductive organisms that results in gametes with half the chromosome number of the original cell

meroblastic cleavage cell division observed in telolecithal eggs in which cleavage is restricted to a small disc of cytoplasm at the animal pole

mesoderm middle primary embryonic germ layer that lines a true coelom (when present) and gives rise to muscle tissue, skeletal tissue, reproductive tissue, and most circulatory tissue

mesoglea acellular, gelatinous substance that fills the space between the outer and inner tissue layers of cnidarians; responsible for the "jellylike" feel of these organisms

mesolecithal eggs containing a moderate amount of yolk

metacercaria tailless, encysted, late larvae of a trematode that is usually the infective form for the definitive host

metamorphosis a type of development from egg to adult in which there is a series of distinct changes in body form

micronucleus (*pl.* = micronuclei) organelle of ciliates that represents a typical eukaryotic nucleus that contains the organism's entire genome and is responsible for genetic recombination

miracidia (*sing.* = miracidium) free-swimming, ciliated, early larvae of trematodes that infect an intermediate host in which they develop into sporocysts

mitosis process of nuclear division in eukaryotes whereby two genetically identical daughter cells are produced from one event of cell division

molting (*syn.* = ecdysis) periodic act of shedding the skin or exoskeleton

monocytes large white blood cells with relatively small, eccentric, oval or kidney-shaped nuclei; become phagocytic macrophages after moving into tissues

monoecious referring to an organism that contains both male and female reproductive structures

monophyletic pertaining to a taxon derived from a single ancestral species that gave rise to no species in any other taxa

morphogenesis embryonic development of body shape

morula embryonic stage of development characterized by 16–32 cells

motile having or pertaining to the ability to move

mouth external opening to the digestive tract

muscle belly thicker, middle section of muscle containing the bulk of muscle fibers

muscle tissues permit movement of the animal through its environment and/or movement of substances through the animal

myelin proteinaceous substance that coats the axon sheaths of nerve cells

myofibrils composites of many individual muscle cells, giving these fibers a multinucleated appearance

myomeres segmented, muscular bundles that provide movements for swimming in many aquatic chordates

myosin thick, proteinaceous filaments that overlap and alternate with actin filaments to produce muscular contractions

– N –

nematocyst stinging structure located within the cnidocyte in the epidermal layer of cnidarians that is discharged from the cell for defense or prey capture

nephridia excretory organs in many invertebrates and some chordates

nerve cord highly complex bundle of nerve fibers that, in many invertebrates, handles the majority of nervous coordination without intervention of the brain

nervous tissues initiate and transmit electrical nerve impulses to and from the body parts and store information in the form of biochemical compounds

neural folds thickening of ectodermal cells along the mid-dorsal region of the neurula to form two enlarged ridges on the surface

neural groove depression along the mid-dorsal region of the neurula bordered by the neural folds

neural tube enclosed cylinder that results from the meeting and fusing of the neural folds in the neurula; will develop into the brain and spinal cord

neuron nerve cell composed of a cell body, axons, and dendrites that transmits electrochemical information through the body

neurula early embryonic stage of development following the gastrula in which nervous tissue begins to differentiate and the neural tube begins to form

neurulation process of neural tube formation during embryonic development

neutrophils large, granular white blood cells with nuclei consisting of two to five lobes; principal phagocytic cells in mammalian blood

nictitating membrane thin fold of skin beneath the eyelid in birds, reptiles, and some mammals that can be drawn across the eye to protect and moisten the eye

nidamental gland large glandular organ found in the mantle cavity of female cephalopods that secretes egg cases or the gelatinous substance comprising egg masses

node branch point on a phylogeny that represents the divergence of a common ancestral species into multiple lineages of descendants

nosepiece revolving housing that supports microscope objective lenses

nostrils (*syn.* = external nares) paired openings in dorsum of head leading to olfactory receptors

notochord cylindrical section of differentiated mesodermal cells below the neural tube in the neurula; anchors myomeres to provide tension for muscular contractions generating body movements

nuclear membrane selectively permeable, double-membrane system that encloses the nucleus of eukaryotic cells

nucleolus organelle within the nucleus responsible for synthesis of ribosomal RNA

nucleus membrane-bound organelle of eukaryotic cell containing the organism's DNA

– O –

objective lenses microscope lenses of different magnification that work in conjunction with ocular lenses to magnify the image

ocellus (*pl.* = ocelli) minute simple eye or eyespot of an invertebrate

ocular lens microscope lens or lenses nearest the eye through which observer looks

olfactory sac enlarged, internal chamber of the nostril containing a sensory epithelium that permits detection of dissolved chemicals in the environment

oogenesis meiotic production of eggs (or ova)

oogonia female germ cells

ootid haploid cell receiving the bulk of the cytoplasm after the division (meiosis II) of a secondary oocyte

open circulatory system system in which the blood is not always confined within a network of vessels but rather flows through open sinuses in the tissues

operculum (*pl.* = opercula) paired bony plates that cover the gills of bony fish on either side of the head and allow for the release of water passing over the gills

optic disc location where axons from the ganglion cells exit the eye; the lack of photoreceptive cells in this area causes a break in the visual field known as the blind spot

optic nerve either of the second pair of cranial nerves that arise from the retina and carry visual information from the eye to the brain

ora serrata serrated junction between the retina and the ciliary body marking the transition from the non-photosensitive area of the retina to the complex, multilayered photosensitive region

oral referring to the mouth

oral arms cnidarian tentacles used for defense and prey capture

oral cirri cephalochordate feature that acts as a strainer to exclude large particles from entering the mouth during filter feeding

oral disc in anthozoans, the raised portion of the mouth (equivalent to the hypostome of hydrozoans)

oral groove depression in some unicellular protists into which food is swept by ciliary currents

oral hood flap-like protrusion of the anterior end of the lamprey larva that covers the oral lobes

oral lobes fingerlike projections that encircle the oral opening in lamprey larva and function as sensory organs and particle strainers

organogenesis differentiation of organ tissues

orientation behavior overt response that involves positioning or movements by an animal to a stimulus

origin fixed or less-movable end of a muscle attachment

osculum larger opening located at the top of sponges through which water that has collected in the spongocoel is pushed out

osmoregulation process of maintaining proper internal water and salt concentrations in a cell or body of a living organism

ossicles calcium carbonate structures embedded beneath the epidermis of echinoderms; used for skeletal support

osteocytes bone-producing cells

ostia (*sing.* = ostium) (1) openings in the cranial ends of vertebrate oviducts through which eggs released from the ovaries into the coelomic cavity enter the oviducts; (2) pores on the body surface of sponges through which water enters; (3) pores or openings that allow circulation of fluids between adjacent body or organ sections in sea anemones; (4) openings in the surface of the heart of invertebrates possessing an open circulatory system that allow blood to return to the heart

outgroup in phylogenetic analyses, a taxon that is closely related to but not included within the focal group being studied, and is used to "root" the phylogeny

ovary (*pl.* = ovaries) female gonad that produces eggs

oviduct tube that transports eggs from the ovaries to the uterus; in some vertebrates it secretes a gelatinous or calcified covering over the eggs for protection

ovipositor external reproductive opening found in females of some arthropod species; bordered by pointy, chitinous teeth that penetrate the soil and create burrows for egg deposition

– P –

pancreas granular organ usually located in close association to the stomach and intestine; produces digestive enzymes and a variety of hormones

paraphyletic pertaining to a taxon that does not include all the descendants of its ancestor

parapodia paired, feathery extensions along the lateral margins of polychaetes used primarily for respiration and, to some degree, for locomotion

parasite heterotrophic organism that obtains nourishment from the living tissue of another organism

parfocal condition in which the focal plane of a microscope does not change substantially when switching between different objective lenses

parsimony principle based on the experience that the simplest explanation is most probably the correct one and should be tested first; in phylogenetic analyses it refers to the methodological approach of assuming the smallest amount of evolutionary change to construct phylogenies

parthenogenesis form of unisexual reproduction in which the ovum develops into a new individual without fertilization

pedal disc tough, fleshy base that attaches cnidarian polyps to rocky substrates or sandy ocean floor

pedicellariae pincerlike structures of echinoderms believed to kill small organisms that might settle on body surfaces, thus keeping the epidermis free of parasites and algae

pellicle stiff, outer covering of some unicellular protists that maintains basic cellular shape

penis external male reproductive organ; deposits semen in the reproductive tract of the female and (in some animals) carries excretory wastes in the form of urine out of the body through the urethra

pentaradial symmetry type of radial symmetry based on five symmetrical divisions, or multiples of five

perisarc outer, epidermal covering forming a raised sheath surrounding the coenosarc in some cnidarians

peristome (1) opening at one end of *Vorticella* that serves as a mouth and is encircled by a ring of cilia; (2) circular oral membrane in sea stars that expands for eversion of the cardiac stomach during feeding

perpendicular grooves characteristic pattern of grooves in dinoflagellates in which their two flagella lie

phagocytosis a type of endocytosis in which large food particles are engulfed

pharyngeal of, or relating to, the pharynx

pharynx muscular portion of digestive tract usually responsible for pulling food into the digestive system

phylogenetic tree (*syn.* = phylogeny, cladogram) complete evolutionary history of a group of organisms

pinacocyte flattened cells making up the outer layer of the sponge body

pineal organ dorsal extension of the diencephalon region of the brain in many vertebrate groups that contains photosensory elements and regulates behaviors and physiological processes associated with diurnal and/or seasonal cycles

placenta vascular organ in mammals that unites fetus to the uterus during pregnancy and mediates exchange of nutrients and gases; absent in monotremes and marsupials

plankton mostly microscopic organisms that drift passively or swim weakly near the surface of ponds, lakes, and oceans

plantigrade type of locomotion characterized by walking on the soles of the feet; body weight is supported primarily by the tarsals

planula larva swimming, juvenile hydrozoan stage that settles to the bottom of the ocean floor where it develops into a new polyp

plasma fluid matrix of blood

plasma membrane membrane at the boundary of a cell that acts as a selective barrier regulating the cell's chemical composition

plastron hard, bony, ventral portion of the shell of turtles

platelets (*syn.* = thrombocytes) small, non-nucleated, colorless cell fragments present in mammalian blood that facilitate clotting

polar body haploid cell receiving virtually no cytoplasm after the division (meiosis I) of a primary oocyte

polyp stage in the cnidarian life cycle represented by a cylindrical organism that remains attached to the substrate by a short stalk

polyphyletic pertaining to a taxon whose members were derived from two or more ancestral forms not common to all members

polyspermy fertilization of an ovum by several sperm

porococyte elongated, doughnut-shaped cells distributed on outer body surface of the sponge containing pores that allow water to enter the sponge

postanal tail terminal portion of body containing an internal support structure that extends past the anus; one of the defining characteristics of chordates

power switch turns microscope light on or off and may contain a rheostat (dimmer switch) that permits further adjustment of light intensity

primary oocytes diploid cells resulting from the mitotic division of oogonia

primary spermatocytes diploid cells resulting from the mitotic division of spermatogonia

proctodeum posterior-most chamber of the cloaca in reptiles and birds; acts as a general collecting area for digestive and excretory wastes

proglottids serially repeated "segments" of a tapeworm's body containing primarily reproductive organs; each can live for a limited time after it detaches from the main body

prostate gland accessory male reproductive gland that contributes fluid to the semen

prostomium fleshy extension of the first body segment in oligochaetes that partially covers the opening to the mouth

protostome animal whose mouth develops from the first embryonic opening and whose embryonic cells divide by spiral cleavage

proventriculus digestive passageway in birds between the crop and gizzard that mixes peptic enzymes into the food

proximal referring to a point of reference nearer the median plane or point of attachment on the body than another structure

proximate cause series of immediate physiological events that lead to a behavior

pseudocoelom body cavity that lies between a layer of mesodermally derived tissue and a layer of gastrodermally derived tissue

pseudocoelomate animal with a central body cavity that lies between gastrodermis and mesoderm

pseudopodia cytoplasm-filled extensions of the plasma membrane employed by amoeboid species for locomotion and feeding

pupil opening in the center of the iris through which light enters the eye

pygostyle caudal portion of the vertebral column in birds formed by the fusion of several lumbar vertebrae

pyloric ceca tubular pouches that open from the pyloric end of the stomach into the digestive canal of most fishes

pyloric chamber (*syn.* = pyloric stomach) second stomach region in some arthropods and echinoderms where chemical digestion of food occurs

– Q –

quadrupedal animal that walks on four legs

– R –

radial canals (1) flagellated chambers into which water is channeled through the porocytes in sponges; (2) portions of the echinoderm water vascular system emanating from the ring canal

radial cleavage pattern of blastomere production in which the cells are located directly above or below one another

radial symmetry body parts arranged around a central axis such that any plane passing through the central axis divides the body into two similar halves

rectum distal end of the intestinal tract; primary function is to reabsorb water and produce dry, concentrated feces

red tides ecological phenomenon caused by massive population explosions of dinoflagellates

regeneration biological process of regrowing and reshaping tissues into exact replicas of missing body parts

renal portal system circulatory pathway in some vertebrates (amphibians and reptiles) that channels all blood from the lower half of the body to the kidneys before sending it through the rest of the systemic circulatory system

respiratory trees feathery, gill-like structures in sea cucumbers for gas exchange

retina innermost layer of the eyeball; contains the photoreceptive elements for reception of light rays and transmission of visual stimuli to the optic nerve

ring canal portion of the echinoderm water vascular system encircling the mouth

rostral referring to a point of reference closer to the tip of the nose

rugae ridges and folds of the inner wall of the stomach that increase the surface area of the stomach lining and provide texture for the manipulation of food as it is broken down

– S –

sagittal plane longitudinal section separating an animal into right and left sides

salivary glands special glands located within the oral cavity and neck that produce a variety of fluids and enzymes that facilitate swallowing and digestion

salt glands osmoregulatory organs embedded in the orbits of the eyes in birds; secrete excess sodium chloride

sarcomeres fundamental repeating units of striated muscle

scale epidermally derived, flattened plate forming part of the external body covering of an animal

sclera tough, white fibrous tissue that forms the outer covering of the eyeball

sclerocyte cell type found in sponges that secretes calcium- or silicon-based minerals that form into spicules

scolex anterior end of tapeworm; lacks sensory structures but possesses modifications for attachment to intestinal wall of host

scrotum (*syn.* = scrotal sacs) pouch extending from the caudal region of some male mammals that contains the testes (after they have descended from the abdominal cavity during embryonic development); its presence allows the temperature of the testes to be maintained at a slightly lower temperature than that of the abdominal cavity

scutes enlarged scales present in turtles

secondary oocyte haploid cell receiving the bulk of the cytoplasm after the division (meiosis I) of a primary oocyte

secondary septa incomplete partitions of the body wall that further subdivide the internal chambers of anthozoans

secondary spermatocyte haploid cell resulting from the division (meiosis I) of a primary spermatocyte

segmentation repetition of animal body regions containing similar organs

semen fluid composed of sperm and glandular secretions ejaculated by the male during orgasm

seminal receptacle organ that receives sperm during copulation and stores sperm until needed to fertilize eggs

seminal vesicles accessory male reproductive glands that contribute fluid to semen

seminiferous tubules tubule system located inside the testes where sperm are produced through meiosis; primary spermatocytes are formed along the outer margins of the seminiferous tubules and migrate inward as they mature

setae small, hairlike bristles on the surface of annelids used to gain traction on the substrate during locomotion

sexual reproduction type of reproduction in which offspring with unique genetic combinations are created by the union of sets of genetic material, usually, but not necessarily, from two separate parents

shell hard, exterior secretion produced by the mantle of molluscs for protection

shell glands specialized structures attached to the oviducts in certain animals that secrete the eggshell material

simple columnar epithelium single layer of columnar cells in a tissue covering or lining

simple cuboidal epithelium single layer of cuboidal cells in a tissue covering or lining

simple squamous epithelium single layer of flattened cells in a tissue covering or lining

sinus venosus thin-walled sac that directs blood collected from a fish's veins into the atrium of the heart

siphon (1) hollow tube found in cephalopods through which water is expelled from the mantle cavity at high velocity to propel the animal through the water; (2) aperture in bivalves and tunicates for water to enter (incurrent siphon) or exit (excurrent siphon) the body

sister chromatid any of the chromatids formed by replication of one chromosome during interphase of the cell cycle while they are still joined by a centromere

sister taxa group of taxa (usually a pair) that are each other's closest phylogenetic relatives

skeletal muscle striated muscle tissue that consists of long, unbranched myofibrils that have a multinucleated appearance

small intestine region of the digestive system responsible for nutrient absorption; often consists of the duodenum, the jejunum, and the ileum

smooth muscle non-striated muscle tissue that consists of long, spindle-shaped fibers containing single nuclei

species epithet second part of the name of a species; designates a specific species belonging to that genus

spectacle clear scale covering the eyes of snakes

sperm (*syn.* = spermatozoon) male gamete (sex cell), corresponding to the female ovum in organisms that reproduce sexually

spermatic cord long, narrow tube that leads from the testis through the abdominal wall and contains the vas deferens, the spermatic artery and vein, lymphatic vessels, and numerous nerves

spermatids haploid cells resulting from the division (meiosis II) of secondary spermatocytes

spermatogenesis meiotic production of sperm cells

spermatogonia male germ cells that divide mitotically to produce either primary spermatocytes or more spermatogonia

spermatophore capsule or mass containing spermatozoa (and sometimes nourishment) and transferred in entirety to the female during copulation

spermatophoric gland male reproductive organ that packages sperm into a capsule covered with membranes and accessory structures called a spermatophore

spermatozoa haploid, differentiated spermatids, or mature sperm cells

spicules (1) hard, crystalline calcium carbonate or silicon structures that form the skeleton in many sponges; (2) spiny projections found on the tail of male nematodes that are used during copulation

spindle fibers proteinaceous matrix of microtubules that form between the two pairs of centrioles during prophase

spines calcareous projections in echinoderms that afford protection and support and are used for locomotion in some groups

spiracles (1) external openings in abdomen of insects that allow air flow into and out of tracheae; (2) paired openings on the heads of some fishes that function as modified gill slits, allowing water to be drawn into the pharynx and across the gills

spiral cleavage pattern of blastomere production in which the cleavage planes occur at oblique angles to one another

spiral valve septum within the ileum of some fishes that increases the absorptive surface area of the intestinal tract and slows the passage of food through this length of small intestine; (2) septum within the conus arteriosus of the amphibian heart that directs the flow of blood to minimize the mixing of oxygenated and deoxygenated blood

spleen ductless, vascular organ in the abdominal cavity that is a component of the circulatory system; stores blood, recycles worn-out red blood cells, and produces lymphocytes

spongin proteinaceous, flexible material secreted by some sponges to form the skeleton

spongocoel large, central cavity that passes through the center of sponges

stage supports slides on microscope for viewing

stage adjustment knobs move microscope stage to center slide under objective lens

stage clips hold slide in steady, stationary position on microscope stage

starches insoluble polysaccharides composed of glucose subunits; serve as the primary food storage substance of plants

stomach digestive chamber that typically stores food and assists in mechanical and chemical breakdown of the food

stone canal portion of echinoderm water vascular system leading from the madreporite to the ring canal

stratified squamous epithelium several layers of flattened cells in a tissue covering or lining that collectively serve as a barrier against foreign substances and injury

subpharyngeal gland homologue to the endostyle; found in lamprey larva and represents the evolutionary precursor to the thyroid gland of vertebrates

suckers specialized structures common in trematodes and cestodes for attachment of the parasite to its host

superficial cleavage pattern of blastomere production in which cell division occurs only along the cell's periphery; characteristic of centrolecithal eggs

swim bladder gas-filled sac of fishes used to regulate buoyancy and, in some cases, for gas exchange

sycon intermediately complex body plan of sponges with simple infolding of the body wall and canals

synapsis process of physical association of homologous chromosomes during prophase of meiosis I

synsacrum skeletal structure formed by the fusion of the 13 caudal vertebrae in birds

systematics subdiscipline of biology that studies the diversity of organisms and their evolutionary relationships

systemic heart large, muscular chamber in cephalopods that receives oxygenated blood from the gills and pumps it throughout the body

– T –

tail projection from posterior end of body; in chordates the tail always contains some portion of the skeletal support system (for example, notochord or vertebrae)

tapetum lucidum reflective layer of tissue in the eye of many vertebrate animals that lies immediately behind the retina; reflects light back through the retina increasing the light available to the photoreceptors thereby improving vision in low-light conditions

taxis directed behavior in which an animal orients or moves toward or away from a stimulus

taxon formal taxonomic groups at any level (for example, phylum or genus)

taxonomy science of naming, describing, and classifying organisms

telson terminal body segment of an arthropod or segmented worm

tendon fibrous cord of connective tissue that typically serves as an attachment between a muscle and bone

tentacles long, extensible, prehensile arms found in cnidarians and some molluscs that aid in sensation, defense, prey capture, and locomotion

test support structure in sea urchins composed of numerous calcareous plates located beneath the epidermis that form the endoskeleton

testis (*pl.* = testes) male reproductive organ that produces sperm

tetrads homologous pairs of replicated double-stranded chromosomes joined together along the equator of the cell during meiosis I

tetrapods four-limbed vertebrates (amphibians, reptiles, birds, and mammals)

thallus (*pl.* = thalli) name for the leaflike body of many algae and lower, nonvascular plants

thrombocytes (*syn.* = platelets) small, non-nucleated, colorless cell fragments present in mammalian blood that facilitate clotting

tissue group of closely associated, similar cells that work together to carry out specific functions

trachea (*pl.* = tracheae) (1) cartilaginous tube extending from oral cavity to lungs through which air is transported during respiration; (2) one of many respiratory tubules of insects that conducts air flow from the outside environment directly to the tissues within the body

transverse plane section perpendicular to the long axis of the body separating the animal into anterior (cranial) and posterior (caudal) portions

triploblastic animal whose body develops from three embryonic tissue layers

tube feet small, extensible protrusions connected to the water vascular system that permit locomotion and prey capture in echinoderms

typhlosole inward fold of tissue projecting into the lumen of the intestine in certain annelid groups that serves to increase the absorptive surface area of the intestine and improve digestive efficiency

– U –

ultimate cause resultant evolutionary advantages that have promoted a particular behavior to remain in an animal's repertoire of possible responses

umbilical cord flexible, cordlike structure connecting a fetus at the abdomen with the placenta and containing vessels that transport nutrients and oxygen to the fetus and remove its wastes

undulating membrane region of trypanosome consisting of a fold in the plasma membrane and a single flagellum that together propel the animal through the host's bloodstream

unicellular consisting of or having one cell

unicellular heterotroph a single-celled organism that obtains nutrition by ingesting organic material from outside sources

uniramous unbranched

ureter tube that transports urine from the kidney to the urinary bladder for storage

urethra tube that leads from the urinary bladder to the outside of the body; transports urine and (in males) semen

uric acid primary insoluble nitrogenous waste product of insects, land snails, and many reptiles

urinary bladder membranous sac that serves as a receptacle for filtrate from the kidneys

urodeum middle chamber of the cloaca in reptiles and birds; receives reproductive and urinary products

urogenital opening (pore) external opening for the release of urine and reproductive products

uterine horns branched extensions of the body of the uterus; often the site of implantation and embryonic development

uterus (*pl.* = uteri) region of female reproductive tract where embryonic development of eggs or fetuses occurs

– V –

vacuole fluid-filled, membrane-bound sac found within the cytoplasm of cells; may function in storage, digestion, or water elimination

vagina female reproductive chamber that receives penis during copulation; also serves as part of the birth canal

vasa efferentia small ducts in male amphibians that transport sperm from the testes to the kidneys

vas deferens (*pl.* = vasa deferentia; *syn.* = ductus deferens) tubule in male animals that transports sperm

vector organism that acts as a secondary host to a parasite and transmits the parasite to its primary host, usually without suffering any ill effects itself

vegetal pole portion of an egg toward which the yolk is segregated and that appears lighter in coloration in comparison to the animal pole

veins vessels that conduct blood toward the heart

velum paired muscular flaps in the oral cavity of the lamprey larva that move water into the pharynx

ventral referring to a point of reference nearer the underside of the body

ventricle muscular chamber of the heart that forcefully contracts and expels blood through arterial system

vertebrae bony segments that constitute the spinal column in many animals

vertebrate subphylum of chordates characterized by the presence of a vertebral column

vibrissa (*pl.* = vibrissae) any of the long stiff hairs that project from the snout or brow of most mammals and perform a sensory function for the animal; synonymous with whiskers

visceral mass soft, fleshy pouch containing the major digestive and reproductive organs of molluscs

vitreous chamber fluid-filled space within the eye between the lens and the retina

viviparous producing live young instead of eggs

vomeronasal organ (*syn.* = Jacobson's organ) chemosensory organ located in the roof of a snake's mouth into which the forks of the tongue are placed, thereby depositing odor molecules for detection and interpretation by the brain

vulva most caudal region of the female urogenital tract consisting of the vestibule, clitoris, and labia

– W –

water vascular system hydraulic system composed of a network of canals and small pumps that move water through the body of echinoderms to affect locomotor movements

wet mount microscope slide preparation in which a small volume of water helps support the sample and fills the space between the coverslip and the slide, allowing light to pass through the slide, sample, and coverslip with less scatter and distortion

wheel organ cephalochordate structure in oral cavity; lined with cilia to produce water current that brings food into the mouth

wings modified appendages used for flight

working distance space between the objective lens and the slide when using a microscope

– Y –

yolk mass nutrient source for the developing embryo in egg-laying animals

yolk plug small remnant of nutritive source present at the opening of the blastopore during embryonic development in certain groups

– Z –

zygote diploid product of the fusion of two haploid gametes during fertilization

Photo Credits

Cover photo and pages 19, 37, 71, 107, 151, 197, 223, 249, 337, 359, 377, 407, and 441 by John Crawley

Chapter 1 Fundamental Laboratory Skills

Chapter opener: Photo courtesy of Nikon Instruments, Inc.

Figure 1.1: Photo courtesy of Nikon Instruments, Inc.

Figure 1.3: Photo courtesy of Nikon Instruments, Inc.

Chapter 2 Cells and Tissues

Figure 2.1: Shirley Wright, *Photographic Atlas of Developmental Biology* © Morton Publishing

Figure 2.3: Michael Leboffe, *Photographic Atlas of Histology*, 2e © Morton Publishing

Figures 2.8, 2.10, 2.12: Van De Graaff & Crawley, *Photographic Atlas for the Zoology Laboratory*, 5e © Morton Publishing

Figure 2.11: Van De Graaff & Crawley, *Photographic Atlas for the Anatomy & Physiology Laboratory*, 5e © Morton Publishing

Figure 2.13C: Michael Leboffe, *Photographic Atlas of Histology*, 2e © Morton Publishing

Figure 2.15: Van De Graaff & Crawley, *Photographic Atlas for the Zoology Laboratory*, 3e © Morton Publishing

Chapter 3 Reproduction, Regeneration, and Cell Division

Figures 3.1, 3.4: Van De Graaff & Crawley, *Photographic Atlas for the Anatomy & Physiology Laboratory*, 5e © Morton Publishing

Figures 3.3, 3.7: Shirley Wright, *Photographic Atlas of Developmental Biology* © Morton Publishing

Chapter 4 Animal Development

Chapter opener: Photo courtesy Will Kelley

Figures 4.2, 4.3, 4.6, 4.7, 4.9, 4.11: Shirley Wright, *Photographic Atlas of Developmental Biology* © Morton Publishing

Figure 4.3H: Van De Graaff & Crawley, *Photographic Atlas for the Zoology Laboratory*, 3e © Morton Publishing

Chapter 5 Taxonomy and Systematics

Figure 5.1: Van De Graaff & Crawley, *Photographic Atlas for the Zoology Laboratory*, 5e © Morton Publishing

Figure 5.2: Adams & Crawley, *Van De Graaff's Photographic Atlas for the Zoology Laboratory*, 7e © Morton Publishing

Figure 5.3: Adams & Crawley, *Van De Graaff's Photographic Atlas for the Zoology Laboratory*, 7e © Morton Publishing; Van De Graaff & Crawley, *Photographic Atlas for the Zoology Laboratory*, 6e © Morton Publishing

Table 5-3: Adams & Crawley, *Van De Graaff's Photographic Atlas for the Zoology Laboratory*, 7e © Morton Publishing

Chapter 6 Animal-like Protists

Chapter opener: Van De Graaff & Crawley, *Photographic Atlas for the Biology Laboratory*, 5e © Morton Publishing

Figure 6.2: Van De Graaff & Crawley, *Photographic Atlas for the Botany Laboratory*, 3e © Morton Publishing

Figure 6.3: Leboffe & Pierce, *Microbiology Laboratory Theory & Application, Brief*, 2e © Morton Publishing

Figures 6.4, 6.5, 6.9, 6.10, 6.11, 6.13: Van De Graaff & Crawley, *Photographic Atlas for the Biology Laboratory*, 5e © Morton Publishing

Chapter 7 Sponges

Figures 7.1, 7.8, 7.9: Van De Graaff & Crawley, *Photographic Atlas for the Zoology Laboratory*, 5e © Morton Publishing

Figure 7.6: Van De Graaff & Crawley, *Photographic Atlas for the Biology Laboratory*, 4e © Morton Publishing

Chapter 8 Ctenophores and Cnidarians

Chapter opener: Van De Graaff & Crawley, *Photographic Atlas for the Zoology Laboratory*, 5e © Morton Publishing

Figures 8.1, 8.12, 8.13: Van De Graaff & Crawley, *Photographic Atlas for the Zoology Laboratory*, 5e © Morton Publishing

Figures 8.5, 8.7, 8.10, 8.11: Van De Graaff & Crawley, *Photographic Atlas for the Biology Laboratory*, 5e © Morton Publishing

Figure 8.6: Van De Graaff & Crawley, *Photographic Atlas for the Zoology Laboratory*, 3e © Morton Publishing

Figure 8.14: Van De Graaff & Crawley, *Photographic Atlas for the Zoology Laboratory*, 6e © Morton Publishing

Figure 8.18: Adams & Crawley, *Van De Graaff's Photographic Atlas for the Zoology Laboratory*, 7e © Morton Publishing

Chapter 9 Flatworms

Figures 9.5, 9.7B–C, 9.8, 9.11: Van De Graaff & Crawley, *Photographic Atlas for the Biology Laboratory*, 5e © Morton Publishing

Figures 9.11, 9.13: Leboffe & Pierce, *Photographic Atlas for the Microbiology Laboratory*, 3e © Morton Publishing

Figure 9.12: Van De Graaff & Crawley, *Photographic Atlas for the Zoology Laboratory*, 3e © Morton Publishing

Chapter 10 Roundworms and Rotifers

Figure 10.1: Van De Graaff & Crawley, *Photographic Atlas for the Zoology Laboratory*, 5e © Morton Publishing

Figure 10.2: Leboffe & Pierce, *Photographic Atlas for the Microbiology Laboratory*, 3e © Morton Publishing

Figure 10.5: Van De Graaff & Crawley, *Photographic Atlas for the Zoology Laboratory*, 3e © Morton Publishing

Figure 10.7: Adams & Crawley, *Van De Graaff's Photographic Atlas for the Zoology Laboratory*, 7e © Morton Publishing

Chapter 11 Molluscs

Figures 11.1, 11.3, 11.4: Van De Graaff & Crawley, *Photographic Atlas for the Biology Laboratory*, 5e © Morton Publishing

Chapter 12 Annelids

Figures 12.3, 12.11: Van De Graaff & Crawley, *Photographic Atlas for the Zoology Laboratory*, 5e © Morton Publishing

Chapter 13 Arthropods

Figures 13.4, 13.8, 13.19: Adams & Crawley, *Van De Graaff's Photographic Atlas for the Zoology Laboratory*, 7e © Morton Publishing

Figures 13.5, 13.6, 13.7, 13.9: Van De Graaff & Crawley, *Photographic Atlas for the Zoology Laboratory*, 5e © Morton Publishing

Figures 13.19, 13.25: Van De Graaff & Crawley, *Photographic Atlas for the Zoology Laboratory*, 5e © Morton Publishing

Chapter 14 Echinoderms

Chapter opener: Adams & Crawley, *Van De Graaff's Photographic Atlas for the Zoology Laboratory*, 7e © Morton Publishing

Figure 14.2: Adams & Crawley, *Van De Graaff's Photographic Atlas for the Zoology Laboratory*, 7e © Morton Publishing

Figure 14.3: Van De Graaff & Crawley, *Photographic Atlas for the Zoology Laboratory*, 6e © Morton Publishing

Figure 14.4: Van De Graaff & Crawley, *Photographic Atlas for the Zoology Laboratory*, 5e © Morton Publishing; Adams & Crawley, *Van De Graaff's Photographic Atlas for the Zoology Laboratory*, 7e © Morton Publishing

Figures 14.5, 14.11, 14.12: Van De Graaff & Crawley, *Photographic Atlas for the Zoology Laboratory*, 5e © Morton Publishing

Figure 14.8C: Adams & Crawley, *Van De Graaff's Photographic Atlas for the Zoology Laboratory*, 7e © Morton Publishing

Chapter 15 Tunicates and Cephalochordates

Figure 15.1: Photo courtesy of the Carnegie Institution of Washington, D.C.

Figure 15.3: Van De Graaff & Crawley, *Photographic Atlas for the Zoology Laboratory*, 5e © Morton Publishing

Figures 15.4, 15.5, 15.6: Fishbeck & Sebastiani, *Comparative Anatomy: Manual of Vertebrate Dissection*, 2e © Morton Publishing

Chapter 16 Jawless Fishes

Chapter opener: Fishbeck & Sebastiani, *Comparative Anatomy: Manual of Vertebrate Dissection*, 2e © Morton Publishing

Figures 16.2, 16.3, 16.5, 16.6, 16.7, 16.8: Fishbeck & Sebastiani, *Comparative Anatomy: Manual of Vertebrate Dissection*, 2e © Morton Publishing

Chapter 17 Cartilaginous Fishes

Figures 17.1, 17.3: Van De Graaff & Crawley, *Photographic Atlas for the Zoology Laboratory*, 5e © Morton Publishing

Figures 17.3, 17.5, 17.6, 17.7, 17.8, 17.10, 17.11, 17.12: Fishbeck & Sebastiani, *Comparative Anatomy: Manual of Vertebrate Dissection*, 2e © Morton Publishing

Chapter 18 Ray-finned Fishes

Figure 18.1: Van De Graaff & Crawley, *Photographic Atlas for the Zoology Laboratory*, 5e © Morton Publishing; Van De Graaff & Crawley, *Photographic Atlas for the Biology Laboratory*, 5e © Morton Publishing

Chapter 19 Amphibians

Figures 19.1, 19.5: Van De Graaff & Crawley, *Photographic Atlas for the Biology Laboratory*, 5e © Morton Publishing

Figure 19.2: Adams & Crawley, *Van De Graaff's Photographic Atlas for the Zoology Laboratory*, 7e © Morton Publishing

Chapter 20 Reptiles

Figures 20.1, 20.12: Van De Graaff & Crawley, *Photographic Atlas for the Zoology Laboratory*, 5e © Morton Publishing

Figures 20.5, 20.13: Van De Graaff & Crawley, *Photographic Atlas for the Zoology Laboratory*, 3e © Morton Publishing

Figure 20.20: Adams & Crawley, *Van De Graaff's Photographic Atlas for the Zoology Laboratory*, 7e © Morton Publishing

Chapter 21 Birds

Figures 21.1, 21.4: Van De Graaff & Crawley, *Photographic Atlas for the Zoology Laboratory*, 5e © Morton Publishing

Figure 21.12: Adams & Crawley, *Van De Graaff's Photographic Atlas for the Zoology Laboratory*, 7e © Morton Publishing

Chapter 22 Mammals

Chapter opener: Van De Graaff & Crawley, *Photographic Atlas for the Biology Laboratory*, 5e © Morton Publishing

Figure 22.1: Van De Graaff & Crawley, *Photographic Atlas for the Biology Laboratory*, 5e © Morton Publishing

Figures 22.3–20.19: Smith & Schenk, *A Dissection Guide and Atlas to the Fetal Pig*, 2e © Morton Publishing

Figures 22.20–20.39: Smith & Schenk, *A Dissection Guide and Atlas to the Rat* © Morton Publishing

Index